CRM Series in Mathematical Physics

For other titles published in this series, go to
http://www.springer.com/series/3872

John Harnad
Editor

Random Matrices, Random Processes and Integrable Systems

Springer

Editor
John Harnad
Department of Mathematics and Statistics
Concordia University
1455 de Maisonneuve Blvd. West
Montréal, Québec, H3G 1M8
Canada
and
Centre de Recherches Mathématiques
Université de Montréal
C.P. 6128, Succ. Centre ville
Montréal, Québec H3C 3J7
Canada
harnad@crm.umontreal.ca

ISBN 978-1-4419-9513-1 e-ISBN 978-1-4419-9514-8
DOI 10.1007/978-1-4419-9514-8
Springer New York Dordrecht Heidelberg London

Library of Congress Control Number: 2011925846

Printed on acid-free paper

Springer is part of Springer Science+Business Media (www.springer.com)

Preface

John Harnad

Department of Mathematics and Statistics, Concordia University,
1455 de Maisonneuve Blvd. West, Montréal, Québec, H3G 1M8, Canada
Centre de Recherches Mathématiques, Université de Montréal, C.P. 6128, Succ.
Centre ville, Montréal, Québec H3C 3J7, Canada, harnad@crm.umontreal.ca

This volume is intended as an introduction and overview of the three domains featured in the title, with emphasis on the remarkable links between them. It has its origins in an intensive series of advanced courses given by the authors at the Centre de Recherches Mathématiques in Montréal in the summer of 2005. Since then, it has grown considerably beyond the material originally presented, and been extended in content and scope. The original courses were interlaced in schedule with an extended research workshop, dealing with related topics, whose proceedings, in enhanced form, have since been published as a special issue of Journal of Physics A [1].

The participants at the two events included a large number of graduate students and postdoctoral fellows, as well as many of the researchers who had made pioneering contributions to these domains, The original content of the lecture series, given by several of the subject's leading contributors, has since been considerably developed and polished, to the point where it may be viewed as both a research survey and pedagogical monograph providing a panoramic view of this very rich and still rapidly developing domain.

In Part I, we have combined the introductory chapters by Pierre van Moerbeke, covering nearly all the topics occurring in the rest of the volume, together with the further, more detailed chapters linking random matrices and integrable systems, concerning mainly their joint work, written by Mark Adler.

Van Moerbeke's part consists of nine chapters. The first of these concerns random permutations, random words and percolation, linked with random partitions through the Robinson–Schensted–Knuth (RSK) correspondence. This includes an introduction to the Ulam problem (concerning the longest increasing subsequence of a random permutation), the Plancherel measure on partitions, the relation to non-intersecting random walkers, as well as queueing problems and polynuclear growth. He then discusses Toeplitz determinants, infinite wedge product representations, and the non-positive generalized probability measure of Borodin–Okounkov–Olshanski, expressed as a matrix integral over $\mathrm{U}(n)$. Several examples follow, including the Poissonized Plancherel measure, bringing in the use of Fredholm determinants of integral operators with a variety of kernels (Bessel, Charlier and Meixner polynomial type kernels), with applications to the distributions arising in the above random processes.

There follows a discussion of limit theorems, such as the Vershik–Kerov limiting shape of a random partition and the Tracy-Widom distribution for the longest increasing subsequences, as well as geometrically distributed percolation problems. Next, it is shown that the multiple (N-fold) integrals obtained upon reducing $\mathrm{U}(n)$ invariant Hermitian matrix models with arbitrary exponential trace invariant series deformations are tau functions of the KP (Kadomtsev–Petviashvili) integrable hierarchy, as well as satisfying the usual Toda lattice equations for varying N's, and the Hirota bilinear relations. Next, Virasoro algebra constraints are deduced for the multiple integrals defining the generalized β-type integrals. There is also a review of the basic finite N Hermitian matrix model results, including the form of the reduced integrals over the eigenvalues, computation of the determinantal form of the correlation functions in terms of suitable (Christoffel–Darboux) correlation kernels, and Fredholm integral expressions for the gap probabilities. The PDEs satisfied by these Fredholm integrals when the endpoints of the support intervals are varied are derived, for a variety of limiting kernels.

In the subsequent chapters, by Mark Adler, further links between random matrix theory and integrable models are developed, using vertex operator constructions. A soliton-like tau function is constructed using a Fredholm determinant and shown to satisfy Virasoro constraints. For gap probabilities, these are used as a vehicle to deduce differential equations that they must satisfy. There are also a number of lattice systems that are constructed using as phase space variables that are defined as N-fold matrix-like integrals. Exponential trace series deformations of 2-matrix integrals are shown to satisfy the equations of the 2-Toda hierarchy, and bilinear identities. Using the Virasoro constraints, PDEs for the gap probabilities are also deduced.

There follows a discussion of the Dyson diffusion process and its relation to random matrices, and chains of random matrices, as well as the bulk and edge scaling limits (sine and Airy processes). Equations are derived for these processes similar to those for the gap probabilities, with respect to the edges of the windows where the nonintersecting random paths are excluded, as well as asymptotic expansions. The GUE with external source and its relation to conditioned non-intersecting Brownian motion, as developed by Aptekarev, Bleher and Kuijlaars is developed, together with its relation to the Riemann–Hilbert problem for multiple orthogonal polynomials. (See Bleher's chapters for further details.) Finally there is a derivation of PDEs for the Pearcey process again through the introduction of integrable deformations of the measure.

The second part of this monograph is mainly concerned with the spectral theory of random matrices, but ideas and methods from the theory of integrable systems plays a prominent role. The introductory chapter, by Harold Widom, begins with a review of basic operator theory definitions and results that are required for applications to random matrices. Then derivations are given for spacing distributions between consecutive eigenvalues, in term of gap probabilities. Using operatorial methods, these are expressed as Fredholm determinants, in suitable scaling limits, of integral operators with integrable kernels of sine type (for the bulk) and Airy type (leading to theTracy-Widom distributions) for the edge. All three cases, orthogonal ($\beta = 1$), unitary ($\beta = 2$) and symplectic ($\beta = 4$) ensembles are treated. Finally, differential equations for distribution functions are derived, in particular, equations of Painlevé type.

In his series of chapters, Pavel Bleher gives a detailed survey of the use of Riemann–Hilbert methods for the study of the asymptotics of spectral distributions of random matrices. First, unitary ensembles with polynomial potentials are treated, and their relation to orthogonal polynomials and the associated Christoffel–Darboux kernels determining the correlation functions at finite N, as well as the string equations determining the recurrence coefficients in the asymptotic $1/N$ series. The Riemann–Hilbert characterization of the orthogonal polynomials is then introduced, and it is shown that the equilibrium measure is supported on a finite union of intervals coinciding with the cuts defining a hyperelliptic algebraic curve. In particular, the Wigner semicircle law is derived for the Gaussian case, and the case of quartic potentials is treated in detail. There follows the treatment of scaled large N asymptotics of orthogonal polynomials, using the Riemann–Hilbert approach and the method of nonlinear steepest descent of Deift, Kricherbauer, Mclaughlin, Venakides Zhou (DKMVZ). A solution of the model Riemann–Hilbert problem is given in terms of Riemann theta functions. The Airy parametrix at the end points is constructed to complete the study of uniform asymptotics, and an indication of the proof of sine kernel universality in the bulk and Airy kernel universality at the edge.

Next, the double scaling limit for the critical point of the even quartic potential case is treated, and its relation to the Hastings–Mcleod solution of the $\mathrm{P_{II}}$ Painlevé equation derived. More generally, the asymptotics of the free energy in the one cut case is studied (assuming a special regularity property of the potential). Analyticity in a parameter defining the potential is examined for the q-cut case for the density function and the free energy. The quartic deviation in the free energy from the Gaussian case is expressed in terms of the first two terms of the large N asymptotic expansion, and related to the Tracy–Widom distribution with an error estimate.

Random matrix models with exponential external coupling are then analyzed in terms of multiple orthogonal polynomials, with emphasis on the case of two distinct eigenvalues in the externally coupled matrix. Correlation functions, at finite N, are expressed in determinantal form in terms of an analog of the Christoffel–Darboux kernel. The higher rank Riemann–Hilbert characterization of such multiple orthogonal polynomials is given, and the differential equations and recursion relations for these expressed in terms of these matrices. The relation to Brownian bridges is explained, and, finally, the Pearcey kernel is derived in the double scaling limit using the nonlinear steepest descent method.

Alexander Its, in his series, focuses upon the large N asymptotics of the spectra of random matrices. The reduced N-fold integral representation of the partition function of Hermitian matrix models is recalled, and the expression of eigenvalue correlation functions in terms of the Christoffel–Darboux kernel of the associated orthogonal polynomials. An introduction to the Its–Kitaev–Fokas Riemann–Hilbert approach to orthogonal polynomials is given, and a proof is given of its unique solvability under certain assumptions.

The asymptotic analysis, along the lines of the DKMVZ method, is then recalled, based on the introduction of the g-function (essentially, the log-Coulomb energy of the equilibrium distribution) to transform the exact RH problem into one which, in leading order, has only jump discontinuities, and hence may be solved exactly. A detailed analysis is then given for the case of even quartic potentials. This suffices to deduce the sine kernel asymptotic form for the correlation kernel in the bulk. The construction of the Airy parametrix at the the end points of the cuts is then discussed and an asymptotic solution given with uniform estimates in each region.

Bertrand Eynard reviews the relationship between convergent matrix integrals and formal matrix integrals, serving as generating functions for the combinatorics of maps. Essentially, the formal model is obtained by treating the integrand as a perturbation series about the Gaussian measure and interchanging the orders of integration and summation, without regard to convergence. He indicates the derivation of the loop equations relating the expectation values of invariant polynomials of various degrees as Dyson–Schwinger equations. This is illustrated by various examples, including 1 and 2 matrix models, as well as chains of matrices, the $\mathrm{O}(N)$ chain model and certain statistical models such as the Potts model. He ends with a number of interesting conjectures about the solution of the loop equations. In the current literature, this has led to a very remarkable program in which many results on random matrices, solvable statistical models, combinatorial, topological and representation theoretical generating functions may be included as part of a general scheme, based on the properties of Riemann surfaces, and their deformations.

The contribution of Momar Dieng and Craig Tracy deals in part with the earliest appearance of random matrices, due to Wishart, in the theory of mutivariate statistics, the so-called Wishart distribution. They present Johnstone's result relating the largest component variance to the F_1 Tracy–Widom distribution, as well as Soshnikov's generalization to lower components. The expression of the F_1, F_2 and F_4 distributions for the edge distributions in GOE, GUE and GSE respectively in terms of the Hastings–McLeod solution of the $\mathrm{P_{II}}$ Painlevé equation is recalled. There follow a discussion of the recurrence relations of Dieng which enter in the computation of the mth largest eigenvalues in GOE and GSE.

A derivation of the Airy kernel for the edge scaling limit of GUE from Plancherel–Rotach asymptotics of the Hermite polynomials is given, as well as the P_I equation that determines the Fredholm determinant of the Airy kernel integral operator supported on a semi-infinite interval. The computation of the mth largest eigenvalue distribution in the GSE and GOE is indicated, together with an interlacing property identifying the first sequence with the even terms of the second. Finally, numerical results are given comparing these distributions with empirical data.

This volume is a masterly combined effort by several of the leading contributors to this remarkable domain, covering a range of topics and applications that no individual author could hope to encompass. For the reader wishing to have a representative view of the fascinating ongoing developments in this domain, as well as a reliable account of the, many results that are by now classically established, it should provide an excellent reference and entry point to the subject.

References

1. J. Harnad and M. Bertola (eds.) *Special Issue on Random Matrices and Integrable Systems* J. Phys. A **39** (2006).

Contents

List of Contributors

Mark Adler
Department of Mathematics
Brandeis University
Waltham, MA 02454
USA
adler@brandeis.edu

Pavel M. Bleher
Department of Mathematical Sciences
Indiana University–Purdue University Indianapolis
402 N. Blackford St.
Indianapolis, IN 46202-3216
USA
bleher@math.iupui.edu

Momar Dieng
Department of Mathematics
The University of Arizona
617 N. Santa Rita Ave.
Tucson, AZ 85721-0089
USA
momar@math.arizona.edu

B. Eynard
Service de Physique Théorique
CEA/DSM/SPhT, CEA/Saclay
91191 Gif-sur-Yvette Cedex
France
eynard@spht.saclay.cea.fr

John Harnad
Department of Mathematics and Statistics
Concordia University
1455 de Maisonneuve Blvd. West
Montréal, Québec, H3G 1M8
Canada
and
Centre de Recherches Mathématiques
Université de Montréal
C.P. 6128, Succ. Centre ville
Montréal, Québec, H3C 3J7
Canada
harnad@crm.umontreal.ca

Alexander R. Its
Department of Mathematics Sciences
Indiana University–Purdue University Indianapolis
402 N. Blackford St.
Indianapolis, IN 46202-3216
USA
itsa@math.iupui.edu

Craig A. Tracy
Department of Mathematics
University of California
One Shields Ave.
Davis, CA 95616
USA
tracy@math.ucdavis.edu

Pierre van Moerbeke
Département de Mathématiques
Université Catholique de Louvain
1348 Louvain-la-Neuve
Belgium
vanmoerbeke@math.ucl.ac.be

Harold Widom
Mathematics Department
194 Baskin Engineering
University of California
Santa Cruz, CA 95064
USA
widom@math.ucsc.edu

Part I

Random Matrices, Random Processes and Integrable Models

1

Random and Integrable Models in Mathematics and Physics

Pierre van Moerbeke

Département de Mathématiques, Université Catholique de Louvain, 1348 Louvain-la-Neuve, Belgium vanmoerbeke@uclouvain.be
Department of Mathematics, Brandeis University, Waltham, MA 02454, USA vanmoerbeke@brandeis.edu

During the last 15 years or so, and since the pioneering work of E. Wigner, F. Dyson and M.L. Mehta, random matrix theory, combinatorial and percolation questions have merged into a very lively area of research, producing an outburst of ideas, techniques and connections; in particular, this area contains a number of strikingly beautiful gems. The purpose of these five Montréal lectures is to present some of these gems in an elementary way, to develop some of the basic tools and to show the interplay between these topics. These lectures were written to be elementary, informal and reasonably self-contained

J. Harnad (ed.), *Random Matrices, Random Processes and Integrable Systems*, CRM Series in Mathematical Physics, DOI 10.1007/978-1-4419-9514-8_1,

and are aimed at researchers wishing to learn this vast and beautiful subject. My purpose was to explain these topics at an early stage, rather than give the most general formulation. Throughout, my attitude has been to give what is strictly necessary to understand the subject. I have tried to provide the reader with plenty of references, although I may and probably will have omitted some of them; if so, my apologies!

As we now know, random matrix theory has reached maturity and occupies a prominent place in mathematics, being at the crossroads of many subjects: number theory (zeroes of the Riemann zeta functions), integrable systems, asymptotics of orthogonal polynomials, infinite-dimensional diffusions, communication technology, financial mathematics, just to name a few. Almost 200 years ago A. Quetelet tried to establish universality of the normal distribution (mostly by empirical means). Here we are, trying to prove universality of the many beautiful statistical distributions which come up in random matrix theory and which slowly will find their way in everyday life.

This set of five lectures were given during the first week of a random matrix 2005-summer school at the "Centre de recherches mathématiques" in Montréal; about half of them are devoted to combinatorial models, whereas the remaining ones deal with related random matrix subjects. They have grown from another set of ten lectures I gave at Leeds (2002 London Mathematical Society Annual Lectures), and semester courses or lecture series at Brandeis University, at the University of California (Miller Institute, Berkeley), at the Universiteit Leuven (Francqui chair, KULeuven) and at the Université de Louvain (UCLouvain).

I would like to thank many friends, colleagues and graduate students in the audience(s), who have contributed to these lectures, through their comments, remarks, questions, etc., especially Mark Adler, Ira Gessel, Alberto Grünbaum, Luc Haine, Arno Kuijlaars, Vadim Kuznetsov, Walter Van Assche, Pol Vanhaecke, and also Jonathan Delépine, Didier Vanderstichelen, Tom Claeys, Maurice Duits, Maarten Vanlessen, Aminul Huq, Dong Wang and many others....

Last, but not least, I would like to thank John Harnad for creating such a stimulating (and friendly) environment during the 2005-event on "random matrices" at Montréal. Finally, I would label it a success if this set of lectures motivated a few young people to enter this exciting subject.

1.1 Permutations, Words, Generalized Permutations and Percolation

1.1.1 Longest Increasing Subsequences in Permutations, Words and Generalized Permutations

(i) **Permutations** $\pi := \pi_n$ of $1, \dots, n$ are given by

$$\mathrm{S}_n \ni \pi_n = \begin{pmatrix} 1 & \dots & n \\ j_1 & \dots & j_n \end{pmatrix}, \quad 1 \le j_1, \dots, j_n \le n \text{ all distinct integers}$$

with $\pi_n(k) = j_k$. Then

$$\#\,\mathrm{S}_n = n!\,.$$

An *increasing subsequence* of $\pi_n \in \mathrm{S}_n$ is a sequence $1 \le i_1 < \dots < i_k \le n$, such that $\pi_n(i_1) < \dots < \pi_n(i_k)$. Define

$$L(\pi_n) = \text{length of a longest (strictly) increasing subsequence of } \pi_n\,. \quad (1.1)$$

Notice that there may be many longest (strictly) increasing subsequences!

Question (Ulam's problem 1961). Given uniform probability on S_n, compute

$$\mathrm{P}^n(L(\pi_n) \le k, \pi_n \in \mathrm{S}_n) = \frac{\#\{L(\pi_n) \le k, \pi_n \in \mathrm{S}_n\}}{n!} = ?$$

Example. For $\pi_7 = \left(\begin{smallmatrix} 1&2&3&4&5&6&7 \\ \underline{3}&1&\underline{4}&2&\underline{6}&\underline{7}&5 \end{smallmatrix}\right)$, we have $L(\pi_7) = 4$. A longest increasing sequence is underlined; it is not necessary unique.

(ii) **Words** $\pi := \pi_n^q$ of length n from an alphabet $1, \dots, q$ are given by integers

$$S_n^q \ni \pi_n^q = \begin{pmatrix} 1 & \dots & n \\ j_1 & \dots & j_n \end{pmatrix}, \quad 1 \le j_1, \dots, j_n \le q$$

with $\pi_n^q(k) = j_k$. Then

$$\#\, S_n^q = q^n\,.$$

An *increasing subsequence* of $\pi_n^q \in S_n^q$ is given by a sequence $1 \le i_1 < \dots < i_k \le n$, such that $\pi_n^q(i_1) \le \dots \le \pi_n^q(i_k)$. As before, define

$$L(\pi_n^q) = \text{length of the longest weakly increasing subsequence of } \pi_n^q\,. \quad (1.2)$$

Question. Given uniform probability on S_n^q, compute

$$\mathrm{P}_n^q(L(\pi_n^q) \le k, \pi_n^q \in S_n^q) = \frac{\#\{L(\pi_n^q) \le k, \pi_n^q \in S_n^q\}}{q^n} = ?$$

Example. for $\pi = \left(\begin{smallmatrix} 1&2&3&4&5 \\ 2&\underline{1}&\underline{1}&3&\underline{2} \end{smallmatrix}\right) \in \mathrm{S}_5^3$, we have $L(\pi) = 3$. A longest increasing sequence is underlined.

(iii) **Generalized permutations** $\pi := \pi_n^{p,q}$ are defined by an array of integers

$$\mathrm{GP}_n^{p,q} \ni \pi_n^{p,q} = \begin{pmatrix} i_1 & \dots & i_n \\ j_1 & \dots & j_n \end{pmatrix},$$

subjected to

$$1 \le i_1 \le i_2 \le \dots \le i_n \le p \quad \text{and} \quad 1 \le j_1, \dots, j_n \le q$$
$$\text{with } i_k = i_{k+1} \text{ implying } j_k \le j_{k+1}.$$

Then

$$\#\,\mathrm{GP}_n^{p,q} = \binom{pq+n-1}{n} .$$

An *increasing subsequence* of a generalized permutation π is defined as

$$\begin{pmatrix} i_{r_1} \\ j_{r_1} \end{pmatrix}, \dots, \begin{pmatrix} i_{r_m} \\ j_{r_m} \end{pmatrix} \subset \pi$$

with $r_1 \le \cdots \le r_m$ and $j_{r_1} \le j_{r_2} \le \cdots \le j_{r_m}$. Define

$$L(\pi) := \text{length of the longest weakly increasing subsequence of } \pi.$$

Example. For $\left(\begin{smallmatrix} 1 & 1 & 1 & 2 & 2 & 2 & 3 & 3 & 4 & 4 \\ 2 & 3 & 3 & \underline{1} & \underline{2} & \underline{2} & 1 & \underline{2} & 1 & \underline{3} \end{smallmatrix}\right) \in \mathrm{GP}_{10}^{4,3}$, we have $L(\pi) = 5$.

For more information on these matters, see Stanley [80, 81].

1.1.2 Young Diagrams and Schur Polynomials

Standard references to this subject are MacDonald [68], Sagan [78], Stanley [80, 81], Stanton and White [82]. To set the notation, we remind the reader of a few basic facts.

• A *partition* λ of n (noted $\lambda \vdash n$) or a *Young diagram* λ of weight n is represented by a sequence of integers $\lambda_1 \ge \lambda_2 \ge \cdots \ge \lambda_l \ge 0$, such that $n = |\lambda| := \lambda_1 + \cdots + \lambda_l$; $n = |\lambda|$ is called the weight. A *dual Young diagram* $\lambda^\top = \big((\lambda^\top)_1 \ge (\lambda^\top)_2 \ge \cdots\big)$ is the diagram obtained by flipping the diagram λ about its diagonal; set

$$\lambda_i^\top := (\lambda^\top)_i = \text{length of } i\text{th column of } \lambda. \tag{1.3}$$

Clearly $|\lambda| = |\lambda^\top|$. For future use, introduce the following notation:

$$\begin{aligned} \mathbb{Y} &:= \{\text{all partitions } \lambda\} \\ \mathbb{Y}_n &:= \{\text{all partitions } \lambda \vdash n\} \\ \mathbb{Y}^p &:= \{\text{all partitions, with } \lambda_1^\top \le p\} \\ \mathbb{Y}_n^p &:= \{\text{all partitions } \lambda \vdash n, \text{ with } \lambda_1^\top \le p\} . \end{aligned} \tag{1.4}$$

• A *semi-standard Young tableau* of shape λ is an array of integers $a_{ij} > 0$ placed in box (i, j) in the Young diagram λ, which are weakly increasing from left to right *and* strictly increasing from top to bottom.

• A *standard Young tableau* of shape $\lambda \vdash n$ is an array of integers $1, \dots, n = |\lambda|$ placed in the Young diagram, which are strictly increasing from left to right *and* from top to bottom. For $\lambda \vdash n$, define

$$f^\lambda := \#\{\text{standard tableaux of shape } \lambda \text{ filled with integers } 1, \dots, |\lambda|\} .$$

• The *Schur function* $\tilde{\mathbf{s}}_\lambda$ associated with a Young diagram $\lambda \vdash n$ is a symmetric function in the variables $x_1, x_2, \ldots,$ (finite or infinite), defined as

$$\tilde{\mathbf{s}}_\lambda(x_1, x_2, \ldots) = \sum_{\substack{\text{semi-standard} \\ \text{tableaux } P \\ \text{of shape } \lambda}} \prod_i x_i^{\#\{\text{times } i \text{ appears in } P\}} . \tag{1.5}$$

It equals a polynomial $\mathbf{s}_\lambda(t_1, t_2, \ldots)$ (which will be denoted without the tilde) in the symmetric variables $kt_k = \sum_{i \geq 1} x_i^k$,

$$\tilde{\mathbf{s}}_\lambda(x_1, x_2, \ldots) = \mathbf{s}_\lambda(t_1, t_2, \ldots) = \det\left(\mathbf{s}_{\lambda_i - i + j}(t)\right)_{1 \leq i,j \leq m} , \tag{1.6}$$

for any $m \geq n$. In this formula $s_i(t) = 0$ for $i < 0$ and $s_i(t)$ for $i \geq 0$ is defined as

$$\exp\left(\sum_1^\infty t_i z^i\right) := \sum_{i \geq 0} \mathbf{s}_i(t_1, t_2, \ldots) z^i .$$

Note for $\lambda \vdash n$,

$$\tilde{\mathbf{s}}_\lambda(x_1, x_2, \ldots) = f^\lambda x_1 \cdots x_n + \cdots . = f^\lambda \frac{t_1^{|\lambda|}}{|\lambda|!} + \cdots$$

• Given two partitions $\lambda \supseteq \mu$, (i.e., $\lambda_i \geq \mu_i$), the diagram $\lambda \backslash \mu$ denotes the diagram obtained by removing μ form λ. The *skew-Schur polynomial* $\mathbf{s}_{\lambda \backslash \mu}$ associated with a Young diagram $\lambda \backslash \mu \vdash n$ is a symmetric function in the variables $x_1, x_2, \ldots,$ (finite or infinite), defined by

$$\begin{aligned} \tilde{\mathbf{s}}_{\lambda \backslash \mu}(x_1, x_2, \ldots) &= \sum_{\substack{\text{semi-standard} \\ \text{skew-tableaux } P \\ \text{of shape } \lambda \backslash \mu}} \prod_i x_i^{\#\{\text{times } i \text{ appears in } P\}} \\ &= \det\left(\mathbf{s}_{\lambda_i - \mu_j - i + j}(t)\right)_{1 \leq i,j \leq n} , \end{aligned}$$

Similarly, for $\lambda \backslash \mu \vdash n$,

$$\tilde{\mathbf{s}}_{\lambda \backslash \mu}(x_1, x_2, \ldots) = f^{\lambda \backslash \mu} x_1 \cdots x_n + \cdots .$$

• The *hook length* of the (i,j)th box is defined by[1] $h_{ij}^\lambda := \lambda_i + \lambda_j^\top - i - j + 1$. Also define

$$h^\lambda := \prod_{(i,j) \in \lambda} h_{ij}^\lambda = \frac{\prod_1^m (m + \lambda_i - i)!}{\Delta_m(m + \lambda_1 - 1, \ldots, m + \lambda_m - m)} , \quad \text{for } m \geq \lambda_1^\top . \tag{1.7}$$

[1] $h_{ij}^\lambda := \lambda_i + \lambda_j^\top - i - j + 1$ is the *hook length* of the (i,j)th box in the Young diagram; i.e., the number of boxes covered by the hook formed by drawing a horizontal line emanating from the center of the box to the right and a vertical line emanating from the center of the box to the bottom of the diagram.

• The *number of standard Young tableaux* of a given shape $\lambda = (\lambda_1 \geq \cdots \geq \lambda_m)$ is given by[2]

$$\begin{aligned} f^\lambda &= \#\{\text{standard tableaux of shape } \lambda \text{ filled with the integers } 1,\dots,|\lambda|\} \\ &= \text{coefficient of } x_1\cdots x_n \text{ in } \tilde{\mathbf{s}}_\lambda(x) \\ &= \frac{|\lambda|!}{u^{|\lambda|}}\tilde{\mathbf{s}}_\lambda(x)\Big|_{\sum_{i\geq 1} x_i^k=\delta_{k1}u} = \frac{|\lambda|!}{u^{|\lambda|}}\mathbf{s}_\lambda(u,0,0,\dots) \\ &= \frac{|\lambda|!}{h^\lambda} = |\lambda|!\det\left(\frac{1}{(\lambda_i-i+j)!}\right)_{1\leq i,j\leq m} \\ &= |\lambda|!\frac{\Delta_m(m+\lambda_1-1,\dots,m+\lambda_m-m)}{\prod_1^m(m+\lambda_i-i)!}, \quad \text{for any } m\geq\lambda_1^\top . \end{aligned} \tag{1.8}$$

In particular, for any $m \geq \lambda_1^\top$ and arbitrary $u \in \mathbb{R}$,

$$\mathbf{s}_\lambda(u,0,0,\dots) = u^{|\lambda|}\frac{f^\lambda}{|\lambda|!} = u^{|\lambda|}\frac{\Delta_m(m+\lambda_1-1,\dots,m+\lambda_m-m)}{\prod_1^m(m+\lambda_i-i)!} . \tag{1.9}$$

• The *number of semi-standard Young tableaux* of a given shape $\lambda \vdash n$, filled with numbers 1 to q for $q \geq 1$:

$$\begin{aligned} &\#\{\text{semi-standard tableaux of shape } \lambda \text{ filled with numbers from 1 to } q\} \\ &\quad = \tilde{\mathbf{s}}_\lambda(\overbrace{1,\dots,1}^{q},0,0,\dots) = \mathbf{s}_\lambda\left(q,\frac{q}{2},\frac{q}{3},\dots\right) = \prod_{(i,j)\in\lambda}\frac{j-i+q}{h^\lambda_{i,j}} \\ &\quad = \begin{cases} \Delta_q(q+\lambda_1-1,\dots,q+\lambda_q-q)/\prod_{i=1}^{q-1} i!, & \text{when } q\geq\lambda_1^\top, \\ 0, & \text{when } q<\lambda_1^\top, \end{cases} \end{aligned} \tag{1.10}$$

using the fact that

$$\prod_{(i,j)\in\lambda}(j-i+q) = \frac{\prod_{i=1}^q(q+\lambda_i-i)!}{\prod_1^{q-1} i!}. \tag{1.11}$$

• Pieri's formula: given an integer $r \geq 0$ and the Schur polynomial $\mathbf{s}_\mu$, the following holds

$$\mathbf{s}_\lambda\mathbf{s}_r = \sum_{\substack{\mu\backslash\lambda = \text{ horizontal strip} \\ |\mu\backslash\lambda|=r}} \mathbf{s}_\mu . \tag{1.12}$$

Note $\mu\backslash\lambda$ is an horizontal r-strip, when they interlace, $\mu_1 \geq \lambda_1 \geq \mu_2 \geq \cdots$, and $|\mu\backslash\lambda| = r$.

---

[2] using the Vandermonde determinant $\Delta_m(z_1,\dots,z_m) = \prod_{1\leq i<j\leq m}(z_i - z_j)$

1.1.3 Robinson–Schensted–Knuth Correspondence for Generalized Permutations

Define the set of $p \times q$ integer matrices (see [80, 81])

$$\mathrm{Mat}_n^{p,q} := \left\{ W = (w_{ij})_{\substack{1\le i\le p\\ 1\le j\le q}}\ ,\ w_{ij} \in \mathbb{Z}_{\ge 0} \text{ and } \sum_{i,j} w_{ij} = n \right\} .$$

Theorem 1.1.1. *There is a* 1-1 *correspondence between the following three sets:*

$$\begin{array}{ccccc} \mathrm{GP}_n^{p,q} & \Longleftrightarrow & \left\{ \begin{array}{l} \textit{two semi-standard Young tableaux} \\ (P,Q), \textit{ of same, but arbitrary} \\ \textit{shape } \lambda \vdash n, \textit{ filled resp. with in-} \\ \textit{tegers } (1,\dots,p) \textit{ and } (1,\dots,q) \end{array} \right\} & \Longleftrightarrow & \mathrm{Mat}_n^{p,q} \\ \pi & \longleftrightarrow & (P,Q) & \longleftrightarrow & W(\pi) = (w_{ij})_{\substack{1\le i\le p\\ 1\le j\le q}} \end{array}$$

where

$$w_{ij} = \# \left\{ \textit{times that } \binom{i}{j} \in \pi \right\} .$$

Therefore, we have[3]

$$\binom{pq+n-1}{n} = \#\mathrm{GP}_n^{p,q} = \sum_{\lambda \vdash n} \tilde{\mathbf{s}}_\lambda(1^p)\tilde{\mathbf{s}}_\lambda(1^q) = \#\,\mathrm{Mat}_n^{p,q} \ . \tag{1.13}$$

Also, we have equality between the length of the longest weakly increasing subsequence of the generalized permutation π*, the length of the first row of the associated Young diagram and the weight of the optimal path:*

$$\begin{aligned} L(\pi) = \lambda_1 &= L(W) \\ &:= \max_{\substack{\textit{all such}\\ \textit{paths}}} \left\{ \sum w_{ij}, \begin{array}{l} \textit{over right/down paths starting} \\ \textit{from entry } (1,1) \textit{ to } (p,q) \end{array} \right\} . \end{aligned} \tag{1.14}$$

Sketch of proof. Given a generalized permutation

$$\pi = \begin{pmatrix} i_1 \dots i_n \\ j_1 \dots j_n \end{pmatrix} ,$$

the correspondence constructs two semi-standard Young tableaux P, Q having the same shape λ. This construction is inductive. Namely, having obtained two equally shaped Young diagrams P_k, Q_k from

$$\begin{pmatrix} i_1 \cdots i_k \\ j_1 \cdots j_k \end{pmatrix} , \quad 1 \le k \le n$$

[3] Use the notation $\tilde{\mathbf{s}}_\lambda(1^p) = \tilde{\mathbf{s}}_\lambda(\overbrace{1,\dots,1}^{p},0,0,\dots)$.

with the numbers $(j_1,\dots,j_k)$ in the boxes of P_k and the numbers $(i_1,\dots,i_k)$ in the boxes of Q_k, one forms a new diagram P_{k+1}, by creating a *new box in the first row of* P, *containing the next number* j_{k+1} , according to the following rule:

(i) if $j_{k+1} \geq$ all numbers appearing in the first row of P_k, then one creates a new box with j_{k+1} in that box to the right of the first row,

(ii) if not, place j_{k+1} in the box (of the first row) containing the smallest integer $> j_{k+1}$. The integer, which was in that box, then gets pushed down to the second row of P_k according to the rule (i) or (ii), and the process starts afresh at the second row.

The diagram Q is a bookkeeping device; namely, add a box (with the number i_{k+1} in it) to Q_k exactly at the place, where the new box has been added to P_k. This produces a new diagram Q_{k+1} of same shape as P_{k+1}. The inverse of this map is constructed by reversing the steps above.

Formula (1.13) follows from (1.10). $\square$

Example. For $n = 10$, $p = 4$, $q = 3$,

$$\mathrm{GP}_n^{p,q} \ni \pi = \begin{pmatrix} 1 & 1 & 1 & 2 & 2 & 2 & 3 & 3 & 4 & 4 \\ 2 & 3 & 3 & \underline{1} & \underline{2} & \underline{2} & 1 & \underline{2} & 1 & \underline{3} \end{pmatrix}, \qquad \text{with } L(\pi) = 5$$

$$\Updownarrow$$

$$(P,Q) = \left(\overbrace{\begin{array}{|c|c|c|c|c|} \hline 1 & 1 & 1 & 2 & 3 \\ \hline 2 & 2 & 2 & \multicolumn{2}{c}{} \\ \cline{1-3} 3 & 3 & \multicolumn{3}{c}{} \\ \cline{1-2} \end{array}}^{5} , \begin{array}{|c|c|c|c|c|} \hline 1 & 1 & 1 & 3 & 4 \\ \hline 2 & 2 & 2 & \multicolumn{2}{c}{} \\ \cline{1-3} 3 & 4 & \multicolumn{3}{c}{} \\ \cline{1-2} \end{array} \right) \qquad L(\pi) = \lambda_1 = 5$$

$$\Updownarrow$$

$$W = \begin{pmatrix} \textcircled{0} & \to & \textcircled{1} & & 2 \\ & & \downarrow & & \\ 1 & & \textcircled{2} & & 0 \\ & & \downarrow & & \\ 1 & & \textcircled{1} & \to & \textcircled{0} \\ & & & & \downarrow \\ 1 & & 0 & & \textcircled{1} \end{pmatrix}, \qquad \text{with } L(\pi) = \sum_{(i,j)\in\{\text{path}\}} w_{ij} = 5 \,.$$

The RSK algorithm proceeds as follows:

$$
\begin{array}{lccccc}
\text{adding} & & \binom{2}{1} & \binom{2}{2} & \binom{2}{2} & \binom{3}{1} \\
P & \begin{array}{l} 2\,3\,3 \end{array} & \begin{array}{l} 1\,3\,3 \\ 2 \end{array} & \begin{array}{l} 1\,2\,3 \\ 2\,3 \end{array} & \begin{array}{l} 1\,2\,2 \\ 2\,3\,3 \end{array} & \begin{array}{l} 1\,1\,2 \\ 2\,2\,3 \\ 3 \end{array} \\
 & & \Longrightarrow & \Longrightarrow & \Longrightarrow & \Longrightarrow \\
Q & \begin{array}{l} 1\,1\,1 \end{array} & \begin{array}{l} 1\,1\,1 \\ 2 \end{array} & \begin{array}{l} 1\,1\,1 \\ 2\,2 \end{array} & \begin{array}{l} 1\,1\,1 \\ 2\,2\,2 \end{array} & \begin{array}{l} 1\,1\,1 \\ 2\,2\,2 \\ 3 \end{array}
\end{array}
$$

$$
\begin{array}{cccc}
\binom{3}{2} & \binom{4}{1} & \binom{4}{3} & \\
\begin{array}{l} 1\,1\,2\,2 \\ 1\,2\,3 \\ 3 \end{array} & \begin{array}{l} 1\,1\,1\,2 \\ 1\,2\,2 \\ 3\,3 \end{array} & \begin{array}{l} 1\,1\,1\,2\,3 \\ 1\,2\,2 \\ 3\,3 \end{array} & \\
\Longrightarrow & \Longrightarrow & \Longrightarrow & = \binom{P}{Q} \\
\begin{array}{l} 1\,1\,1\,3 \\ 2\,2\,2 \\ 3 \end{array} & \begin{array}{l} 1\,1\,1\,3 \\ 2\,2\,2 \\ 3\,4 \end{array} & \begin{array}{l} 1\,1\,1\,3\,4 \\ 2\,2\,2 \\ 3\,4 \end{array} &
\end{array}
$$

yielding the set (P, Q) of semi-standard Young tableaux above.

1.1.4 The Cauchy Identity

Theorem 1.1.2. *Using the customary change of variables* $\sum_{k\geq 1} x_k^i = it_i$, $\sum_{l\geq 1} y_l^i = is_i$, *we have*

$$\prod_{i,j\geq 1} \frac{1}{1 - x_i y_j} = \sum_{\lambda} \tilde{\mathbf{s}}_\lambda(x) \tilde{\mathbf{s}}_\lambda(y) = \sum_{\lambda} \mathbf{s}_\lambda(t) \mathbf{s}_\lambda(s) = \exp\left(\sum_{i\geq 1} it_i s_i \right) .$$

Proof. On the one hand, to every $\pi \in \mathrm{GP} = \bigcup_{n,p,q} \mathrm{GP}_n^{p,q}$, we associate a monomial, as follows

$$\pi \longrightarrow \prod_{\binom{i}{j}\in\pi} x_i y_j \quad \text{(with multiplicities)}. \tag{1.15}$$

Therefore, taking into account the multiplicity of $\binom{i}{j} \in \pi$,

$$\sum_{\pi\in\mathrm{GP}} \prod_{\binom{i}{j}\in\pi} x_i y_j = \prod_{i,j\geq 1} (1 + x_i y_j + x_i^2 y_j^2 + x_i^3 y_j^3 + \dots) = \prod_{i,j\geq 1} \frac{1}{1 - x_i y_j}. \tag{1.16}$$

One must think of the product on the right-hand side of (1.16) in a definite order, as follows:

$$\prod_{i,j\geq 1}\frac{1}{1-x_iy_j}=\frac{1}{1-x_1y_1}\frac{1}{1-x_1y_2}\frac{1}{1-x_1y_3}\cdots$$
$$\times\frac{1}{1-x_2y_1}\frac{1}{1-x_2y_2}\frac{1}{1-x_2y_3}\cdots\times\cdots,$$

and similarly for the expanded version. Expanding out all the products,

$$\prod_{j\geq 1}(1+x_1y_j+x_1^2y_j^2+x_1^3y_j^3+\cdots)\prod_{j\geq 1}(1+x_2y_j+x_2^2y_j^2+x_2^3y_j^3+\cdots)\cdots, \quad (1.17)$$

leads to a sum of monomials, each of which can be interpreted as a generalized permutation, upon respecting the prescribed order. Vice-versa each generalized permutation can be found among those monomials. As an example illustrating identity (1.16), the monomial $x_1y_2x_1^2y_3^2x_3^2y_2^2x_3y_3x_4y_1x_4y_2$, appearing in the expanded version of (1.17), maps into the generalized permutation

$$\begin{pmatrix}1&1&1&3&3&3&4&4\\2&3&3&2&2&3&1&2\end{pmatrix},$$

and vice-versa. On the other hand, to every couple of semi-standard Young tableaux (P,Q), we associate

$$(P,Q)\longrightarrow\prod_i x_i^{\#\text{ times } i \text{ appears in } Q}\prod_j y_j^{\#\text{ times } j \text{ appears in } P}.$$

Therefore, the Robinson–Schensted–Knuth construction, mapping the generalized permutation π into two semi-standard Young tableaux (P,Q) of same shape λ, implies

$$\prod_{\binom{i}{j}\in\pi}x_iy_j=\prod_i x_i^{\#\text{ times } i \text{ appears in } Q}\prod_j y_j^{\#\text{ times } j \text{ appears in } P}.$$

Then, summing up over all $\pi\in\mathrm{GP}$, using the fact that RSK is a bijection, and using the definition of Schur polynomials, one computes

$$\begin{aligned}
&\sum_{\pi\in\mathrm{GP}}\prod_{\binom{i}{j}\in\pi}x_iy_j\\
&=\sum_{\substack{\text{all }(P,Q)\text{ with}\\ \text{shape }P=\text{shape }Q}}\prod_i x_i^{\#\text{ times } i \text{ appears in } Q}\prod_j y_j^{\#\text{ times } j \text{ appears in } P}\\
&=\sum_\lambda\sum_{\substack{\text{all }(P,Q)\text{ with}\\ \text{shape }P=\text{shape }Q=\lambda}}\prod_i x_i^{\#\text{ times } i \text{ appears in } Q}\prod_j y_j^{\#\text{ times } j \text{ appears in } P}\\
&=\sum_\lambda\Bigg(\sum_{\substack{\text{all }Q\text{ with}\\ \text{shape }Q=\lambda}}\prod x_i^{\#\text{ times } i \text{ appears in } Q}\Bigg)\Bigg(\sum_{\substack{\text{all }P\text{ with}\\ \text{shape }P=\lambda}}\prod y_j^{\#\text{ times } j \text{ appears in } P}\Bigg)\\
&=\sum_\lambda\tilde{\mathbf{s}}_\lambda(x)\tilde{\mathbf{s}}_\lambda(y),
\end{aligned}$$

using the definition (1.5) of the Schur polynomial. The proof is finished by observing that

$$\sum_{i\geq 1} it_i s_i = \sum_{i\geq 1} i \frac{\sum_{k\geq 1} x_k^i}{i} \frac{\sum_{l\geq 1} y_l^i}{i} = \sum_{k,l\geq 1}\sum_{i\geq 1} \frac{(x_k y_l)^i}{i} = \log \prod_{k,l\geq 1} (1-x_k y_l)^{-1} ,$$

ending the proof of Thm. 1.1.2. □

1.1.5 Uniform Probability on Permutations, Plancherel Measure and Random Walks

1.1.5.1 Plancherel Measure

In this section, one needs

$$\mathrm{Mat}_n^{n,n}(0,1) := \left\{ W = (w_{ij})_{1\leq i,j\leq n} \;\middle|\; \begin{array}{l}\text{with exactly one 1 in each row and}\\ \text{column and otherwise all zeros}\end{array} \right\}$$

See [23–25, 80, 81] and references within.

Proposition 1.1.1. *For* permutations, *we have a* 1-1 *correspondence between*

$$\begin{array}{ccccc} \mathrm{S}_n & \Longleftrightarrow & \left\{\begin{array}{l} \textit{two standard Young tableaux} \\ (P,Q),\ \textit{of same shape}\ \lambda\ \textit{and} \\ \textit{size}\ n,\ \textit{each filled with numbers} \\ (1,\dots,n) \end{array}\right\} & \Longleftrightarrow & \mathrm{Mat}_n^{n,n}(0,1) \\ \pi_n & \longleftrightarrow & (P,Q) & \longleftrightarrow & W(\pi) = (w_{ij})_{i,j\geq 1}\,. \end{array} \tag{1.18}$$

Uniform probability P^n *on* S_n *induces a probability* $\widetilde{\mathrm{P}}^n$ *(Plancherel measure) on Young diagrams* $\mathbb{Y}_n$*, given by* ($m := \lambda_1^\top$)

$$\begin{aligned} \widetilde{\mathrm{P}}^n(\lambda) &= \frac{1}{n!}\#\{\textit{permutations leading to shape}\ \lambda\} \\ &= \frac{(f^\lambda)^2}{n!} = n!\mathbf{s}_\lambda(1,0,\dots)^2 \\ &= n!\frac{\Delta_m(m+\lambda_1-1,\dots,m+\lambda_m-m)^2}{(\prod_1^m(m+\lambda_i-i)!)^2} \end{aligned} \tag{1.19}$$

and so

$$\#\,\mathrm{S}_n = \sum_{\lambda\vdash n}(f^\lambda)^2 = n!\,.$$

Finally, the length of the longest increasing subsequence in permutation π_n*, the length of the first row of the partition* λ *and the weight of the optimal path* $L(W)$ *in the percolation matrix* $W(\pi)$ *are all equal:*

$$L(\pi_n) = \lambda_1 = L(W) := \max_{\substack{\textit{all such}\\ \textit{paths}}} \left\{ \sum w_{ij},\ \begin{array}{l}\textit{over right/down paths starting}\\ \textit{from entry}\ (1,1)\ \textit{to}\ (n,n)\end{array}\right\}.$$

Hence

$$\mathrm{P}^n(L(\pi) \le l) = \sum_{\substack{\lambda \in \mathbb{Y}_n \\ \lambda_1 \le l}} \frac{(f^\lambda)^2}{n!} = n! \sum_{\substack{\lambda \in \mathbb{Y}_n \\ \lambda_1 \le l}} \mathbf{s}_\lambda(1,0,\dots)^2 .$$

Proof. A *permutation* is a generalized permutation, but with integers $i_1, \dots, i_n$ and $j_1, \dots, j_n$ all distinct and thus both tableaux P and Q are standard.

Consider now the uniform probability P^n on *permutations* in S_n; from the RSK correspondence we have the one-to-one correspondence, given a fixed partition λ,

$$\{\text{permutations leading to the shape } \lambda\}$$
$$\Longleftrightarrow \left\{ \begin{matrix} \text{standard tableaux of shape } \lambda, \\ \text{filled with integers } 1,\dots,n \end{matrix} \right\} \times \left\{ \begin{matrix} \text{standard tableaux of shape } \lambda \\ \text{filled with integers } 1,\dots,n \end{matrix} \right\}$$

and thus, using (1.8) and (1.10) and noticing that $\tilde{\mathbf{s}}_\lambda(1^q) = 0$ for $\lambda_1^\top > q$,

$$\widetilde{\mathrm{P}}^n(\lambda) = \frac{1}{n!} \#\{\text{permutations leading to the shape } \lambda\} = \frac{(f^\lambda)^2}{n!}, \quad \lambda \in \mathbb{Y}_n,$$

with

$$\sum_{\lambda \in \mathbb{Y}_n} \widetilde{P}^n(\lambda) = 1.$$

Formula (1.19) follows immediately from the explicit values (1.8) and (1.10) for f^λ. □

Example. For permutation $\pi_5 = \left(\begin{smallmatrix} 1 & 2 & 3 & 4 & 5 \\ 5 & \underline{1} & 4 & \underline{3} & 2 \end{smallmatrix}\right) \in \mathrm{S}_5$, the RSK algorithm gives

$$P \Longrightarrow \quad 5 \quad \begin{matrix} 1 \\ 5 \end{matrix} \quad \begin{matrix} 1\,4 \\ 5 \end{matrix} \quad \begin{matrix} 1\,3 \\ 4 \\ 5 \end{matrix} \quad \begin{matrix} 1\,2 \\ 3 \\ 4 \\ 5 \end{matrix}$$

$$Q \Longrightarrow \quad 1 \quad \begin{matrix} 1 \\ 2 \end{matrix} \quad \begin{matrix} 1\,3 \\ 2 \end{matrix} \quad \begin{matrix} 1\,3 \\ 2 \\ 4 \end{matrix} \quad \begin{matrix} 1\,3 \\ 2 \\ 4 \\ 5 \end{matrix} .$$

Hence

$$\pi_5 \Longrightarrow (P,Q) = \left(\overbrace{\begin{matrix} 1\,2 \\ 3 \\ 4 \\ 5 \end{matrix}}^{2}, \begin{matrix} 1\,3 \\ 2 \\ 4 \\ 5 \end{matrix} \atop \text{standard standard} \right) \Longrightarrow \begin{pmatrix} 0 & 0 & 0 & 0 & 1 \\ 1 & 0 & 0 & 0 & 0 \\ 0 & 0 & 0 & 1 & 0 \\ 0 & 0 & 1 & 0 & 0 \\ 0 & 1 & 0 & 0 & 0 \end{pmatrix} .$$

Remark. The Robinson–Schensted–Knuth correspondence has the following properties

- $\pi \mapsto (P, Q)$, then $\pi^{-1} \mapsto (Q, P)$
- length(longest increasing subsequence of π) = #(columns in P)
- length(longest decreasing subsequence of π) = #(rows in P)
- $\pi^2 = I$, then $\pi \mapsto (P, P)$
- $\pi^2 = I$, with k fixed points, then P has exactly k columns of odd length.

1.1.5.2 Solitaire Game

With Aldous and Diaconis [15], consider a deck of cards $1, \ldots, n$ thoroughly shuffled and put those cards one at a time into piles, as follows:

(1) a low card may be placed on a higher card, or can be put into a new pile to the right of the existing pile.
(2) only the top card of the pile is seen. If the card which turns up is higher than the card showing on the table, then start with that card a new pile to the right of the others.

Question. What is the optimal strategy which minimizes the number of piles?

Answer. Put the next card always on the leftmost possible pile!

Example. Consider a deck of 7 cards, appearing in the order 3, 1, 4, 2, 6, 7, 5. The optimal strategy is as follows:

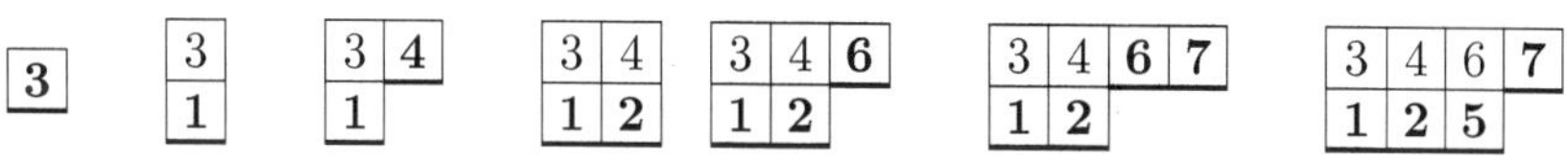

This optimal strategy leads to 4 piles! For a deck of 52 cards, you will find in the average between 10–13 piles and having 9 piles or less occurs approximately 5% of the times. It turns out that, for a given permutation,

$$\text{number of piles} = \text{length of longest increasing sequence.}$$

1.1.5.3 Anticipating Large n Asymptotics

Anticipating the results in Sects. 1.4.2 and 1.9.2, given the permutations of $(1, \ldots, n)$, given a percolation matrix of size n and given a card game of size n, the numbers fluctuate about $2\sqrt{n}$ like

$$\begin{aligned} L(\pi_n) &= \text{length of the longest increasing subsequence} \\ &= \text{weight of the optimal path in the } (0,1)\text{-percolation matrix} \\ &= \text{number of piles in the solitaire game} \\ &\simeq 2\sqrt{n} + n^{1/6}\mathcal{F} , \end{aligned}$$

where $\mathcal{F}$ is a probability distribution, the Tracy–Widom distribution, with

$$\mathrm{E}(\mathcal{F}) = -1.77109 \quad \text{and} \quad \sigma(\mathcal{F}) = 0.9018 .$$

In particular for $n = 52$ cards,

$$\mathrm{E}\big(L(\pi_{52})\big) \simeq 2\sqrt{52} + (52)^{1/6}(-1.77109) = 11.0005 .$$

The Tracy–Widom distribution will be discussed extensively in Sect. 1.9.2.

1.1.5.4 A Transition Probability and Plancherel Measure

Proposition 1.1.2 ([24, 92]). *P_n on $\mathbb{Y}_n$ can be constructed from P_{n-1} on $\mathbb{Y}_{n-1}$, by means of a transition probability, as follows*

$$P_n(\mu) = \sum_{\lambda\in\mathbb{Y}_{n-1}} P_{n-1}(\lambda)p(\lambda,\mu), \quad \mu \in \mathbb{Y}_n$$

where

$$p(\lambda,\mu) := \begin{cases} \dfrac{f^\mu}{f^\lambda}\dfrac{1}{|\mu|} & \text{if } \lambda\in\mathbb{Y}_{n-1} \text{ and } \mu\in\mathbb{Y}_n \text{ are such that } \mu=\lambda+\square \\ 0 & \text{otherwise} \end{cases}$$

is a transition probability, i.e.,

$$\sum_{\substack{\mu\in\mathbb{Y}_n\\ \mu=\lambda+\square}} p(\lambda,\mu) = 1 , \quad \text{for fixed } \lambda.$$

Proof. Indeed, for fixed μ, one computes

$$\begin{aligned}\sum_{\lambda\in\mathbb{Y}_{n-1}} P_{n-1}(\lambda)p(\lambda,\mu) &= \sum_{\substack{\lambda\in\mathbb{Y}_{n-1}\\ \mu=\lambda+\square\in\mathbb{Y}_n}} \frac{(f^\lambda)^2}{(n-1)!}\left(\frac{f^\mu}{f^\lambda}\frac{1}{|\mu|}\right) \\ &= \frac{f^\mu}{n!}\sum_{\substack{\lambda\in\mathbb{Y}_{n-1}\\ \mu=\lambda+\square\in\mathbb{Y}_n}} f^\lambda = \frac{(f^\mu)^2}{n!} = P_n(\mu) .\end{aligned}$$

Indeed, given a standard tableau of shape λ, filled with $1,\dots,n-1$, adjoining a box to λ such as to form a Young diagram μ and putting n in that box yield a new standard tableau (of shape μ).

That $p(\lambda,\mu)$ is a transition probability follows from Pieri's formula (1.12), applied to $r=1$, upon putting $t_i = \delta_{1i}$:

$$\sum_{\substack{\mu\in\mathbb{Y}_n\\ \mu=\lambda+\square}} p(\lambda,\mu) = \sum_{\substack{\lambda\in\mathbb{Y}_{n-1}\\ \lambda+\square\in\mathbb{Y}_n}} \frac{f^{\lambda+\square}}{|\lambda+\square|}\frac{1}{f^\lambda} = \sum_{\substack{\lambda\in\mathbb{Y}_{n-1}\\ \lambda+\square\in\mathbb{Y}_n}} \frac{f^{\lambda+\square}}{|\lambda+\square|!}\frac{|\lambda|!}{f^\lambda} = 1 . \qquad \square$$

Corollary 1.1.1. *The following probability*

$$P_n(\mu_1 \le x_1, \dots, \mu_k \le x_k)$$

decreases, when n increases.

Proof. Indeed

$$\begin{aligned}
P_n(\mu_1 \le x_1, \dots, \mu_k \le x_k) &= \sum_{\substack{\mu \in \mathbb{Y}_n \\ \text{all } \mu_i \le x_i}} \sum_{\lambda \in \mathbb{Y}_{n-1}} P_{n-1}(\lambda) p(\lambda, \mu) \\
&= \sum_{\substack{\lambda \in \mathbb{Y}_{n-1}, \mu \in \mathbb{Y}_n \\ \mu = \lambda + \square \\ \text{all } \mu_i \le x_i}} P_{n-1}(\lambda) p(\lambda, \mu) \le \sum_{\substack{\lambda \in \mathbb{Y}_{n-1} \\ \mu = \lambda + \square \\ \text{all } \mu_i \le x_i}} P_{n-1}(\lambda) \\
&= P_{n-1}(\lambda_1 \le x_1, \dots, \lambda_k \le x_k),
\end{aligned}$$

proving Cor. 1.1.1. □

1.1.5.5 Random Walks

Consider m random walkers in $\mathbb{Z}$, starting from distinct points $x := (x_1 < \cdots < x_m)$, such that, at each moment, only one walker moves either one step to the left or one step to the right. Notice that m walkers in $\mathbb{Z}$, obeying this rule, is tantamount to a random walk in $\mathbb{Z}^m$, where at each point the only moves are

$$\pm e_1, \dots, \pm e_m,$$

with all possible moves equally likely. That is to say the walk has at each point $2m$ possibilities and thus at time T there are $(2m)^T$ different paths. Denote by P_x the probability for such a walk, where x refers to the initial condition. *Requiring these walks not to intersect turns out to be closely related to the problem of longest increasing subsequences in random permutations*, as is shown in the proposition below. For skew-partitions, see Sect. 1.1.2. For references, see [11, 80, 81].

Proposition 1.1.3.

$$P_x \begin{pmatrix} \textit{that } m \textit{ walkers in } \mathbb{Z}, \textit{ reach } y_1 < \cdots < y_m \\ \textit{in } T \textit{ steps, without ever intersecting} \end{pmatrix} = \frac{1}{(2m)^T} \binom{T}{T_{\mathrm{L}} T_{\mathrm{R}}} \sum_{\substack{\lambda \textit{ with } \lambda \supset \mu, \nu \\ |\lambda \backslash \mu| = T_{\mathrm{L}} \\ |\lambda \backslash \nu| = T_{\mathrm{R}} \\ \lambda_1^\top \le m}} f^{\lambda \backslash \mu} f^{\lambda \backslash \nu} \tag{1.20}$$

where μ, ν are fixed partitions defined by the points x_i and y_i,

$$\mu_k = k-1-x_k\,, \quad \nu_k = k-1-y_k$$

$$T_{\mathrm{L}} = \frac{1}{2}\left(T+\sum_1^m (x_i-y_i)\right) = \tfrac{1}{2}(T-|\mu|+|\nu|)$$

$$T_{\mathrm{R}} = \frac{1}{2}\left(T-\sum_1^m (x_i-y_i)\right) = \tfrac{1}{2}(T+|\mu|-|\nu|)$$

$$T = T_{\mathrm{L}}+T_{\mathrm{R}}\,, \quad \sum_1^m (x_i-y_i) = T_{\mathrm{L}}-T_{\mathrm{R}}\,.$$

In particular, close packing of the walkers at times 0 *and* T *implies*

$$P_{1,\dots,m}\begin{pmatrix}\textit{that } m \textit{ walkers in } \mathbb{Z} \textit{ reach } 1,\dots,m \textit{ in } 2n\\ \textit{steps, without ever intersecting}\end{pmatrix}$$

$$= \frac{1}{(2m)^{2n}}\binom{2n}{n}\sum_{\substack{\lambda\vdash n\\ \lambda_1^\top\le m}}(f^\lambda)^2 = \frac{(2n)!}{n!}\frac{\#\{\pi_n\in \mathrm{S}_n : L(\pi_n)\le m\}}{(2m)^{2n}} \tag{1.21}$$

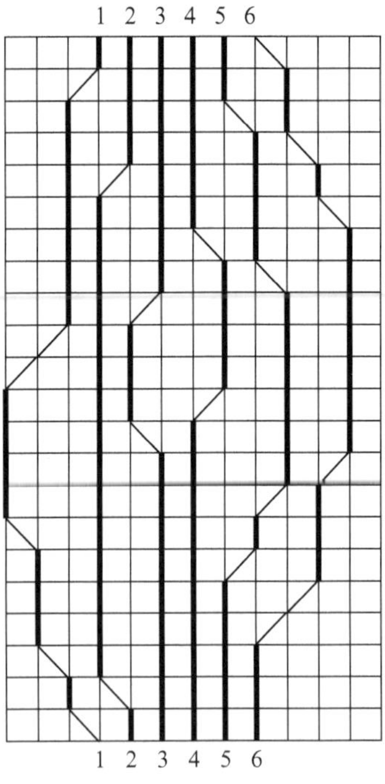

Fig. 1.1. Six nonintersecting walkers leaving from and returning to $1,\dots,6$.

Proof. *Step* 1. Associate to a given walk a sequence of T_{L} Ls and T_{R} Rs:

$$\mathrm{L}\quad \mathrm{R}\quad \mathrm{R}\quad \mathrm{R}\quad \mathrm{L}\quad \mathrm{R}\quad \mathrm{L}\quad \mathrm{L}\quad \mathrm{R}\ldots\mathrm{R}\,, \tag{1.22}$$

thus recording the nature of the move, left or right, at the first instant, at the second instant, etc.
If the kth walker is to go from x_k to y_k, then

$$y_k - x_k = \#\{\text{right moves for } k\text{th walker}\} - \#\{\text{left moves for } k\text{th walker}\}$$

and so, if

$$T_{\mathrm{L}} := \#\{\text{left moves for all } m \text{ walkers}\}$$

and

$$T_{\mathrm{R}} := \#\{\text{right moves for all } m \text{ walkers}\}\,,$$

we have, since at each instant exactly one walker moves,

$$T_{\mathrm{R}} + T_{\mathrm{L}} = T \quad \text{and} \quad T_{\mathrm{R}} - T_{\mathrm{L}} = \sum_1^m (y_k - x_k)\,,$$

from which

$$T_{\left\{ {\mathrm{L} \atop \mathrm{R}} \right\}} = \frac{1}{2}\left(T \pm \sum_1^m (x_k - y_k) \right).$$

Next, we show there is a *canonical* way to map a walk, corresponding to (1.22) into one with left moves only at instants $1,\ldots,T_{\mathrm{L}}$ and then right moves at instants $T_{\mathrm{L}}+1,\ldots,T_{\mathrm{L}}+T_{\mathrm{R}}=T$, thus corresponding to a sequence

$$\overbrace{\mathrm{L}\quad \mathrm{L}\quad \mathrm{L}\ldots\mathrm{L}}^{T_{\mathrm{L}}}\quad \overbrace{\mathrm{R}\quad \mathrm{R}\quad \mathrm{R}\ldots\mathrm{R}}^{T_{\mathrm{R}}}\,. \tag{1.23}$$

This map, originally found by Forrester [40] takes on here a slightly different (but canonical) form. Indeed, in a typical sequence, as (1.22),

$$\mathrm{L}\quad \mathrm{R}\quad \mathrm{R}\quad \underline{\mathrm{R}\quad \mathrm{L}}\quad \mathrm{R}\quad \mathrm{L}\quad \mathrm{L}\quad \mathrm{R}\ldots\mathrm{R}\,, \tag{1.24}$$

consider the first sequence R L (underlined) you encounter, in reading from left to right. It corresponds to one of the following three configurations (in the left column),

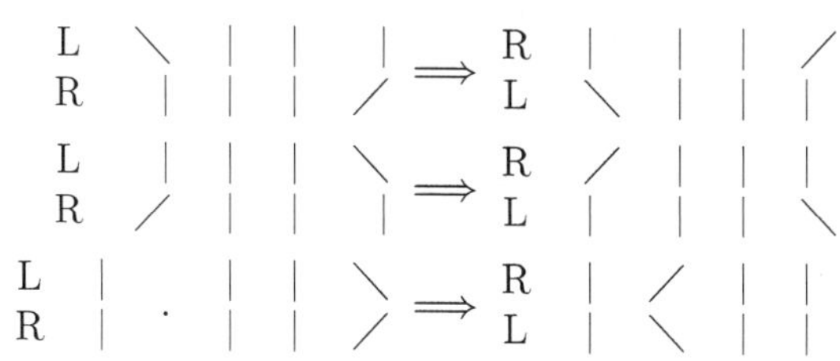

which then can be transformed into a new configuration LR, with same beginning and end, thus yielding a new sequence; in the third case the reflection occurs the first place it can. So, by the moves above, the original configuration (1.22) can be transformed in a new one. In the new sequence, pick again the first sequence RL, reading from left to right, and use again one of the moves. So, this leads again to a new sequence, etc.

$$
\begin{array}{cccccccccc}
\mathrm{L} & \mathrm{R} & \mathrm{R} & \underline{\mathrm{R} \;\; \mathrm{L}} & & \mathrm{R} & \mathrm{L} & \mathrm{L} & \mathrm{R}\ldots\mathrm{R} \\
\mathrm{L} & \mathrm{R} & \underline{\mathrm{R} \;\; \mathrm{L}} & & \mathrm{R} & \mathrm{R} & \mathrm{L} & \mathrm{L} & \mathrm{R}\ldots\mathrm{R} \\
\mathrm{L} & \underline{\mathrm{R} \;\; \mathrm{L}} & & \mathrm{R} & \mathrm{R} & \mathrm{R} & \mathrm{L} & \mathrm{L} & \mathrm{R}\ldots\mathrm{R} \\
\mathrm{L} & \mathrm{L} & \mathrm{R} & \mathrm{R} & \mathrm{R} & \underline{\mathrm{R} \;\; \mathrm{L}} & & \mathrm{L} & \mathrm{R}\ldots\mathrm{R} \\
\mathrm{L} & \mathrm{L} & \mathrm{R} & \mathrm{R} & \underline{\mathrm{R} \;\; \mathrm{L}} & & \mathrm{L} & \mathrm{L} & \mathrm{R}\ldots\mathrm{R} \\
 & & & & \vdots & & & & \\
\end{array}
\tag{1.25}
$$

$$
\underbrace{\mathrm{L} \;\; \mathrm{L} \;\; \mathrm{L} \;\; \mathrm{L}\ldots\mathrm{L}}_{T_{\mathrm{L}}} \;\; \underbrace{\mathrm{R} \;\; \mathrm{R} \;\; \mathrm{R} \;\; \mathrm{R}\ldots\mathrm{R}}_{T_{\mathrm{R}}} \; ;
$$

Since this procedure is invertible, it gives a *one-to-one* map between all the left-right walks corresponding to a given sequence, with T_{L} Ls and T_{R} RRs

$$
\mathrm{L} \;\; \mathrm{R} \;\; \mathrm{R} \;\; \mathrm{R} \;\; \mathrm{L} \;\; \mathrm{R} \;\; \mathrm{L} \;\; \mathrm{L} \;\; \mathrm{R}\ldots\mathrm{R}\,; \tag{1.26}
$$

and all the walks corresponding to

$$
\overbrace{\mathrm{L} \;\; \mathrm{L} \;\; \mathrm{L}\ldots\mathrm{L}}^{T_{\mathrm{L}}} \;\; \overbrace{\mathrm{R} \;\; \mathrm{R} \;\; \mathrm{R}\ldots\mathrm{R}}^{T_{\mathrm{R}}} \,. \tag{1.27}
$$

Thus, a walk corresponding to (1.27) will map into $\binom{T}{T_{\mathrm{L}} T_{\mathrm{R}}}$ different walks, corresponding to the $\binom{T}{T_{\mathrm{L}} T_{\mathrm{R}}}$ number of permutations of T_{L} Ls and T_{R} Rs.

Step 2. To two standard tableaux P, Q of shape $\lambda = (\lambda_1 \geq \cdots \geq \lambda_m > 0)$ we associate a random walk, going with (1.27), in the following way. Consider the situation where m walkers start at $0, 1, \ldots, m-1$.

instants of left move

1st walker	c_{11}	c_{12}	c_{13}	c_{14}	c_{15}	c_{16}
2nd walker	c_{21}	c_{22}	c_{23}			
3rd walker	c_{31}	c_{32}	c_{33}			
	c_{41}					

- The 1st walker starts at 0 and moves to the left, only at instants

$$
c_{1i} = \text{content of box } (1,i) \in P
$$

and thus has made, in the end, λ_1 steps to the left.

$$
\vdots
$$

• The kth walker ($1 \le k \le m$) starts at $k-1$ and moves to the left, only at instants

$$c_{ki} = \text{content of box } (k,i) \in P$$

and thus has made, in the end, λ_2 steps to the left.

$$\vdots$$

• Finally, walker $m = \lambda_1^\top$ walks according to the contents of the last row. Since the tableau is standard, filled with the numbers $1,\ldots,n$ the walkers never meet and at each moment exactly one walker moves, until instant $n = |\lambda|$, during which they have moved from position

$$0 < 1 < \cdots < k-1 < \cdots < m-1$$

to position

$$-\lambda_1 + 0 < -\lambda_2 + 1 < \cdots < -\lambda_k + k - 1 < \cdots < -\lambda_m + m - 1$$

That is to say the final position is given by unfolding the right hand boundary of λ, the horizontal (fat) segments refer to gaps between the final positions and vertical segments refer to contiguous final positions.
In the same fashion, one associates a similar walk to the other tableau Q, with the walkers also moving left. These walkers will have reached the same position as in the first case, since the final position only depends on the shape of P or Q. Therefore, reversing the time for the second set of walks, one puts the two walks together, thus yielding m nonintersecting walkers moving the first half of the time to the left and then the second half of the time to the right, as in the example below. Therefore the number of ways that m walkers start from and return to $0,\ldots,m-1$, without ever meeting each other, by first moving to the left and then to the right, is exactly $\sum_{\substack{\lambda\vdash n\\ \lambda_1^\top \le m}} (f^\lambda)^2$.

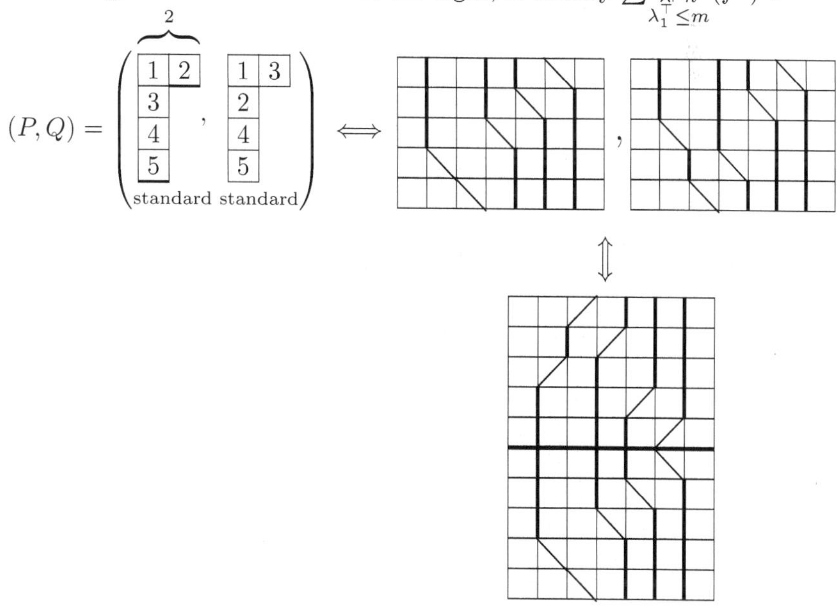

More generally, an analogous argument shows that the number of ways that walkers leave from $x_1 < \cdots < x_m$ and end up at $y_1 < \cdots < y_m$ at time T, without intersection and by first moving to the left and then to the right, is given by

$$\sum_{\substack{\lambda \vdash (T+|\mu|+|\nu|)/2 \\ \lambda_1^\top \le m}} f^{\lambda \backslash \mu} f^{\lambda \backslash \nu} \tag{1.28}$$

On the other hand, there are $\binom{T}{T_\mathrm{L} T_\mathrm{R}}$ sequences of T_L Ls and T_R Rs, which combined with (1.28) yields formula (1.20).
In the close packing situation, one has $\mu_k = \nu_k = 0$ for all k, and so $\mu = \nu = \varnothing$ and $T_\mathrm{L} = T_\mathrm{R} = T/2$. With these data, (1.21) is an immediate consequence of (1.20). □

1.1.6 Probability Measure on Words

Remember from Sect. 1.1.1 words $\pi := \pi_n^q$ of length n from an alphabet $1, \ldots, q$ and from Sect. 1.1.2, the set $\mathbb{Y}_n^q = \{$all of partitions $\lambda \vdash n$, with $\lambda_1^\top \le q\}$. Also define the set of $n \times q$ matrices,

$$\widetilde{\mathrm{Mat}}_n^{n,q}(0,1) := \left\{ W = (w_{ij})_{1 \le i,j \le n} \;\middle|\; \begin{array}{l} \text{with exactly one 1 in each row and oth-} \\ \text{erwise all zeros} \end{array} \right\}$$

For references, see [80, 81, 87].

Proposition 1.1.4. *In particular, for* words, *we have the* 1–1 *correspondence*

$$\begin{array}{ccccc} \mathrm{S}_n^q & \Longleftrightarrow & \left\{ \begin{array}{l} \textit{semi-standard and standard} \\ \textit{Young tableaux } (P,Q) \textit{ of same} \\ \textit{shape and of size } n, \textit{ filled resp.} \\ \textit{with integers } (1,\ldots,q) \textit{ and} \\ (1,\ldots,n) \end{array} \right\} & \Longleftrightarrow & \widetilde{\mathrm{Mat}}_n^{n,q}(0,1) \\ \pi & \longleftrightarrow & (P,Q) & \longleftrightarrow & W(\pi) = (w_{ij})_{i,j \ge 1} \,. \end{array} \tag{1.29}$$

Uniform probability $\mathrm{P}^{n,q}$ *on* S_n^q *induces a probability* $\widetilde{\mathrm{P}}^{n,q}$ *on Young diagrams* $\lambda \in \mathbb{Y}_n^{(q)}$, *given by*

$$\begin{aligned} \widetilde{\mathrm{P}}^{n,q}(\lambda) &= \frac{1}{q^n} \#\{\textit{words in } \mathrm{S}_n^q \textit{ leading to shape } \lambda\} \\ &= \frac{f^\lambda \tilde{s}_\lambda(1^q)}{q^n} = \frac{n!}{q^n} \mathbf{s}_\lambda(1,0,\ldots) \mathbf{s}_\lambda\left(q, \frac{q}{2}, \frac{q}{3}, \ldots\right) \\ &= \frac{n!}{q^n \prod_1^{q-1} i!} \frac{\Delta_q(q+\lambda_1-1, \ldots, q+\lambda_q-q)^2}{\prod_1^q (q+\lambda_i-i)!} \,. \end{aligned} \tag{1.30}$$

Also,

$$\# S_n^q = \sum_{\lambda \vdash n} f^\lambda \tilde{s}_\lambda(1^q) = q^n \,. \tag{1.31}$$

Finally, given the correspondence (1.29), *the length $L(\pi)$ of the longest weakly increasing subsequence of the word π equals*

$$L(\pi) = \lambda_1 = L(W) := \max_{\substack{\textit{all such}\\ \textit{paths}}} \left\{ \sum w_{ij}, \begin{array}{l}\textit{over right/down paths starting}\\ \textit{from entry } (1,1) \textit{ to } (n,q)\end{array} \right\} ,$$

and thus

$$\mathrm{P}^{n,q}(L(\pi) \le l) = \sum_{\substack{\lambda_1 \le l\\ \lambda \in \mathbb{Y}_n^{(q)}}} \frac{n!}{q^n} \mathbf{s}_\lambda(1,0,\dots) \mathbf{s}_\lambda(q, \frac{q}{2}, \frac{q}{3}, \dots) \,.$$

Proof. A word is a special instance of generalized permutation, where the numbers $i_1, \dots, i_n$ are all distinct. Therefore the RSK correspondence holds as before, except that Q becomes a standard tableau; thus, a word maps to a pair of arbitrary Young tableaux (P, Q), with P semi-standard and Q standard and converse. Also the correspondence with integer matrices is the same, with the extra requirement that the matrix contains 0 and 1's, with each row containing exactly one 1.

Consider now the uniform probability $\mathrm{P}^{n,q}$ on *words* in S_n^q; from the RSK correspondence we have the one-to-one correspondence, given a fixed partition λ,

$$\{\text{words in } \mathrm{S}_n^q \text{ leading to shape } \lambda\}$$
$$\Longleftrightarrow \left\{ \begin{array}{l}\text{semi-standard tableaux of}\\ \text{shape } \lambda, \text{ filled with integers}\\ 1, \dots, q\end{array} \right\} \times \left\{ \begin{array}{l}\text{standard tableaux of shape } \lambda\\ \text{filled with integers } 1, \dots, n\end{array} \right\}$$

and thus, using (1.8) and (1.10) and noticing that $\tilde{\mathbf{s}}_\lambda(1^q) = 0$ for $\lambda_1^\top > q$,

$$\widetilde{\mathrm{P}}^{n,q}(\lambda) = \frac{1}{q^n} \#\{\text{words leading to the shape } \lambda\} = \frac{\tilde{\mathbf{s}}_\lambda(1^q) f^\lambda}{q^n} , \quad \lambda \in \mathbb{Y}_n^q ,$$

with

$$\sum_{\lambda \in \mathbb{Y}_n^{(p)}} \widetilde{P}^{n,q}(\lambda) = 1 \,.$$

Formula (1.30) follows immediately from an explicit evaluation (1.8) and (1.10) for f^λ and $\mathbf{s}_\lambda(1^q)$. □

Example. For word

$$\pi = \begin{pmatrix} 1\ 2\ 3\ 4\ 5 \\ 2\ \underline{1}\ \underline{1}\ 3\ \underline{2} \end{pmatrix} \in \mathrm{S}_5^3$$

the RSK algorithm gives

2	1 2	1 1 2	1 1 3 2	1 1 2 2 3
1	1 2	1 3 2	1 3 4 2	1 3 4 2 5 .

Hence

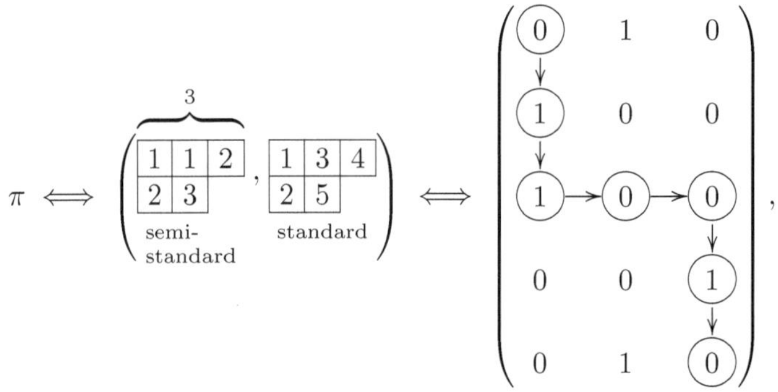

and $L(\pi) = \lambda_1 = L(W) = 3$.

1.1.7 Generalized Permutations, Percolation and Growth Models

The purpose of this section is to show that uniform probability, generalized permutations, percolation, queuing and growth models are intimately related. Their probabilities are ultimately given by the same formulae.

1.1.7.1 Probability on Young Diagrams Induced by Uniform Probability on Generalized Permutations

Proposition 1.1.5 (Johansson [57]). *Uniform probability* $\mathrm{P}_n^{p,q}$ *on* $\mathrm{GP}_n^{p,q}$ *induces a probability* $\widetilde{\mathrm{P}}_n^{p,q}$ *on Young diagrams* $\mathbb{Y}_n^{\min(p,q)}$*, given by*

$$\begin{aligned}\widetilde{P}_n^{p,q}(\lambda) &= \frac{1}{\#\,\mathrm{GP}_n^{p,q}}\,\#\{\textit{generalized permutations leading to shape } \lambda\}\\ &= \frac{\tilde{\mathrm{s}}_\lambda(1^p)\tilde{\mathrm{s}}_\lambda(1^q)}{\#\,\mathrm{GP}_n^{p,q}}\\ &= \frac{1}{\#\,\mathrm{GP}_n^{p,q}}\prod_{j=0}^{q-1}\frac{1}{j!(p-q+j)!}\Delta_q(q+\lambda_1-1,\dots,q+\lambda_q-q)^2\\ &\qquad\qquad\times\prod_{i=1}^{q}\frac{(p+\lambda_i-i)!}{(q+\lambda_i-i)!}\,, \quad \textit{for } q\le p,\end{aligned}$$

with[4]

[4] There is no loss of generality in assuming $q \le p$.

$$\#\,\mathrm{GP}_n^{p,q} = \sum_{\lambda\in\mathbb{Y}_n^{\min(p,q)}} \tilde{\mathbf{s}}_\lambda(1^p)\tilde{\mathbf{s}}_\lambda(1^q) = \binom{pq+n-1}{n} . \tag{1.32}$$

Proof. By the RSK correspondence, we have the one-to-one correspondence,

$$\{\pi\in\mathrm{GP}_n^{p,q}\text{ leading to the shape }\lambda\}$$
$$\longleftrightarrow \left\{\begin{array}{l}\text{semi-standard tableaux of}\\ \text{shape }\lambda\text{, filled with integers}\\ 1,\dots,q\end{array}\right\} \times \left\{\begin{array}{l}\text{semi-standard tableaux of}\\ \text{shape }\lambda\text{, filled with integers}\\ 1,\dots,p\end{array}\right\}$$

and thus, for $\lambda\in\mathbb{Y}_n^{\min(q,p)}$, we have[5]

$$\begin{aligned}\widetilde{\mathrm{P}}_n^{p,q}(\lambda) &= \frac{1}{\#\,\mathrm{GP}_n^{p,q}}\,\#\{\pi\in\mathrm{GP}_n^{p,q}\text{ leading to the shape }\lambda\}\\ &= \frac{1}{\#\,\mathrm{GP}_n^{p,q}}\tilde{\mathbf{s}}_\lambda(1^q)\tilde{\mathbf{s}}_\lambda(1^p)\ ;\end{aligned}$$

when $q<\lambda_1^\top$, we have automatically $\tilde{\mathbf{s}}_\lambda(1^q)=0$, and thus

$$\sum_{\lambda\in\mathbb{Y}_n^{\min(q,p)}} \widetilde{\mathrm{P}}_n^{p,q}(\lambda) = 1\,.$$

Notice that[6] for $m\ge\lambda_1^\top$,

$$\tilde{\mathbf{s}}_\lambda(1^m) = \frac{\Delta_m(m+\lambda_1-1,\dots,m+\lambda_m-m)}{\prod_1^{m-1} i!} = \prod_{\substack{i,j=1\\ i<j}}^{m}\frac{\lambda_i-\lambda_j+j-i}{j-i}\,. \tag{1.33}$$

Without loss of generality, assume $p\ge q$; then $\lambda_i=0$ if $q<i\le p$. Setting $h_j=\lambda_j+q-j$ for $j=1,\dots,q$, we have $h_1>\dots>h_q=\lambda_q\ge 0$. We now compute, using (1.10),

$$\begin{aligned}&\tilde{\mathbf{s}}_\lambda(1^p)\tilde{\mathbf{s}}_\lambda(1^q)\\ &= \prod_{\substack{i,j=1\\ i<j}}^{q}\left(\frac{\lambda_i-\lambda_j+j-i}{j-i}\right)^2\prod_{i=1}^{q}\prod_{j=q+1}^{p}\frac{\lambda_i+j-i}{j-i}\\ &= \prod_{\substack{i,j=1\\ i<j}}^{q}\frac{(h_i-h_j)^2}{(j-i)^2}\prod_{i=1}^{q}\prod_{j=q+1}^{p}\frac{h_i+j-q}{j-i}\\ &= \frac{1}{\prod_{i=1}^{q}\prod_{j=q+1}^{p}(j-i)\prod_{\substack{i,j=1\\ i<j}}^{q}(j-i)^2}\prod_{\substack{i,j=1\\ i<j}}^{q}(h_i-h_j)^2\prod_{i=1}^{q}\frac{(h_i+p-q)!}{h_i!}\end{aligned}$$

---

[5] Remember the notation from Sect. 1.1.2: $\lambda\in\mathbb{Y}_n^{\min(q,p)}$ means the partition $\lambda\vdash n$ satisfies $\lambda_1^\top\le p,q$.

[6] Note $\prod_{1\le i<j\le q}(j-i)=\prod_{j=0}^{q-1}j!$.

$$= \prod_{j=0}^{q-1} \frac{1}{j!(p-q+j)!} \prod_{\substack{i,j=1 \\ i<j}}^{q} (h_i - h_j)^2 \prod_{i=1}^{q} \frac{(h_i + p - q)!}{h_i!} ,$$

using

$$\prod_{\substack{i,j=1 \\ i<j}}^{q} (j-i) \prod_{i=1}^{q} \prod_{j=q+1}^{p} (j-i)$$
$$= \prod_{0}^{q-1} j! \prod_{i=1}^{q} (q+1-i)(q+2-i)(q+3-i) \cdots (p-i)$$
$$= q! \frac{(q+1)!}{1!} \frac{(q+2)!}{2!} \cdots \frac{(p-1)!}{(p-q-1)!} 1! \cdots (q-1)! = \prod_{j=0}^{q-1} (p-q+j)! .$$

This ends the proof of Prop. 1.1.5. □

1.1.7.2 Percolation Model with Geometrically Distributed Entries

Consider the ensemble

$$\mathrm{Mat}^{(p,q)} = \{p \times q \text{ matrices } M \text{ with entries } M_{ij} = 0, 1, 2, \ldots\}$$

with *independent and geometrically distributed entries*, for fixed $0 < \xi < 1$,

$$\mathrm{P}(M_{ij} = k) = (1-\xi)\xi^k, \qquad k = 0, 1, 2, \ldots .$$

Theorem 1.1.3 (Johansson [57]). *Then*

$$L(M) := \max_{\substack{\textit{all such} \\ \textit{paths}}} \left\{ \sum M_{ij}, \begin{array}{l} \textit{over right/down paths starting} \\ \textit{from entry } (1,1) \textit{ to } (p,q) \end{array} \right\}$$

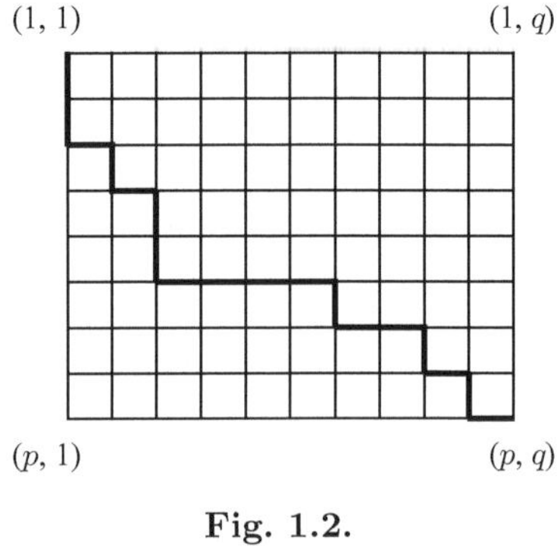

Fig. 1.2.

has the following distribution, assuming $q \le p$,

$$\begin{aligned} P(L(M) \le l) &= \sum_{\substack{\lambda \in \mathbb{Y}^{\min(q,p)} \\ \lambda_1 \le l}} (1-\xi)^{pq} \xi^{|\lambda|} \tilde{\mathbf{s}}_\lambda(1^q) \tilde{\mathbf{s}}_\lambda(1^p) \\ &= Z_{p,q}^{-1} \sum_{\substack{h \in \mathbb{N}^q \\ \max(h_i) \le l+q-1}} \Delta_q(h_1, \dots, h_q)^2 \prod_{i=1}^{q} \frac{(h_i + p - q)!}{h_i!} \xi^{h_i} \end{aligned}$$

where

$$Z_{p,q} = \xi^{q(q-1)/2}(1-\xi)^{-pq} q! \prod_{j=0}^{q-1} j!(p-q+j)! \,. \tag{1.34}$$

Proof. Then the probability that M be a given matrix $A = (a_{ij})_{\substack{1 \le i \le p \\ 1 \le j \le q}}$ equals

$$\begin{aligned} P\Big(M = (a_{ij})_{\substack{1 \le i \le p \\ 1 \le j \le q}}\Big) &= \prod_{\substack{1 \le i,j \le p \\ 1 \le j \le q}} P(M_{ij} = a_{ij}), \quad \text{using independence,} \\ &= \prod_{\substack{1 \le i \le p \\ 1 \le j \le q}} (1-\xi)\xi^{a_{ij}} \\ &= (1-\xi)^{pq} \xi^{\sum_{\substack{1 \le i \le p \\ 1 \le j \le q}} a_{ij}} = (1-\xi)^{pq} \xi^{|A|} \,. \end{aligned}$$

This probability only depends on the total weight $|A| = \sum_{i,j} a_{ij}$. Hence the matrices[7] $M \in \mathrm{Mat}_n^{pq}$ have all equal probability and, in particular, due to the fact that, according to Thm. 1.1.1, the matrices in Mat_n^{pq} are in one-to-one correspondence with generalized permutations of size n, with alphabets $1, \dots, p$ and $1, \dots, q$, one has

$$P(|M| = n) = \sum_{\substack{\text{all } a_{ij} \text{ with} \\ \sum a_{ij} = n}} P\Big(M = (a_{ij})_{\substack{1 \le i \le p \\ 1 \le j \le q}}\Big) = (\#\,\mathrm{GP}_n^{p,q})(1-\xi)^{qp}\xi^n \,.$$

We now compute

$$\begin{aligned} P(L(M) \le l \mid |M| = n) &= \frac{\#\{M \in \mathrm{Mat}_n^{p,q}, L(M) \le l\}}{\#\,\mathrm{Mat}_n^{p,q}} \\ &= \frac{\#\{\pi \in \mathrm{GP}_n^{p,q}, L(\pi) \le l\}}{\#\,\mathrm{GP}_n^{p,q}} \\ &= P_n^{p,q}(\lambda_1 \le l) = \frac{1}{\#\,\mathrm{GP}_n^{p,q}} \sum_{\substack{\lambda_1 \le l \\ |\lambda| = n}} \tilde{\mathbf{s}}_\lambda(1^q)\tilde{\mathbf{s}}_\lambda(1^p) \,. \end{aligned}$$

[7] $M \in \mathrm{Mat}_n^{pq} \subset \mathrm{Mat}^{pq}$, means that $\sum M_{ij} = n$.

Hence,

$$\begin{aligned}
\mathrm{P}(L(M) \le l) &= \sum_{n=0}^{\infty} \mathrm{P}(L(M) \le l \text{ with } |M| = n)\,\mathrm{P}(|M| = n) \\
&= \sum_{n=0}^{\infty} \sum_{\substack{\lambda_1 \le l \\ |\lambda| = n}} \frac{1}{\#\,\mathrm{GP}_n^{p,q}} \tilde{\mathrm{s}}_\lambda(1^q)\tilde{\mathrm{s}}_\lambda(1^p)(\#\,\mathrm{GP}_n^{p,q})(1-\xi)^{pq}\xi^n \\
&= \sum_{n=0}^{\infty} \sum_{\substack{\lambda_1 \le l \\ |\lambda| = n}} \tilde{\mathrm{s}}_\lambda(1^q)\tilde{\mathrm{s}}_\lambda(1^p)(1-\xi)^{pq}\xi^{|\lambda|} \\
&= \sum_{\substack{\lambda \in \mathbb{Y} \\ \lambda_1 \le l}} \tilde{\mathrm{s}}_\lambda(1^q)\tilde{\mathrm{s}}_\lambda(1^p)(1-\xi)^{pq}\xi^{|\lambda|} \,.
\end{aligned}$$

Now, using the expression for $\tilde{\mathrm{s}}_\lambda(1^p)\tilde{\mathrm{s}}_\lambda(1^q)$ in Prop. 1.1.5, one computes, upon setting $h_i = q + \lambda_i - i$, and noticing that $l \ge \lambda_1 \ge \cdots \ge \lambda_q \ge 0$ implies $l + q - 1 \ge h_1 > \cdots > h_q \ge 0$,

$$\begin{aligned}
\mathrm{P}(L(M) \le l) &= \sum_{\substack{\lambda \in \mathbb{Y}^q \\ \lambda_1 \le l}} (1-\xi)^{pq}\xi^{|\lambda|}\tilde{\mathrm{s}}_\lambda(1^p)\tilde{\mathrm{s}}_\lambda(1^q) \\
&= \sum_{l+q-1 \ge h_1 > \cdots > h_q \ge 0} \frac{\xi^{\sum_1^q h_i - q(q-1)/2}}{(1-\xi)^{-pq}\prod_{j=0}^{q-1} j!(p-q+j)!} \\
&\qquad\qquad \times \Delta_q(h_1, \dots, h_q)^2 \prod_{i=1}^{q} \frac{(h_i + p - q)!}{h_i!} \\
&= Z_{p,q}^{-1} \sum_{\substack{h \in \mathbb{N}^q \\ \max(h_i) \le l+q-1}} \Delta_q(h_1, \dots, h_q)^2 \prod_{i=1}^{q} \frac{(h_i + p - q)!}{h_i!}\xi^{h_i} \,,
\end{aligned}$$

since the expression in the sum is symmetric in $h_1, \dots, h_q$. The normalization $Z_{p,q}$ is as announced, ending the proof of Thm. 1.1.3. □

1.1.7.3 Percolation Model with Exponentially Distributed Entries

Theorem 1.1.4 ([57]). *Consider the ensemble*

$$\mathrm{Mat}^{p,q} = \{p \times q \text{ matrices } M \text{ with } \mathbb{R}^+\text{-entries}\}$$

with independent and exponentially distributed entries,

$$\mathrm{P}(M_{ij} \le t) = 1 - \mathrm{e}^{-t} \,, \quad t \ge 0 \,.$$

Then

$$L(M) = \max_{\substack{\textit{all such}\\\textit{paths}}} \left\{ \sum M_{ij}, \begin{array}{l}\textit{over right/lower paths starting}\\ \textit{from entry } (1,1) \textit{ to } (p,q)\end{array} \right\}$$

has the following distribution, (assuming $q \le p$, without loss of generality),

$$\begin{aligned} \mathrm{P}(L(M) \le t) &= \frac{\int_{(0,t)^q} \Delta_q(x_1,\dots,x_q)^2 \prod_{i=1}^q x_i^{p-q} \exp(-x_i)\,\mathrm{d}x_i}{\int_{(0,\infty)^q} \Delta_q(x_1,\dots,x_q)^2 \prod_{i=1}^q x_i^{p-q} \exp(-x_i)\,\mathrm{d}x_i} \\ &= \frac{1}{Z_n} \int_{\substack{M\in\mathcal{H}_q\\ M \textit{ positive definite}\\ \mathrm{Spectrum}(M)\le t}} (\det M)^{p-q} \mathrm{e}^{-\operatorname{Tr} M}\,\mathrm{d}M\,. \end{aligned}$$

Remark. It is remarkable that this percolation problem coincides with the probability that the spectrum of an Hermitian matrix does not exceed t, where the matrix is taken from a (positive definite) random Hermitian ensemble with the Laguerre distribution, as appears in the second formula; this ensemble will be discussed much later in Sect. 1.8.2.2.

Proof. For fixed $0 < \xi < 1$, let X_ξ have a geometric distribution

$$\mathrm{P}(X_\xi = k) = (1-\xi)\xi^k \quad 0 < \xi < 1,\ k = 0,1,2,\dots\ ;$$

then in distribution

$$(1-\xi)X_\xi \to Y, \quad \text{for } \xi \to 1$$

where Y is an exponential distributed random variable. Indeed, setting $\varepsilon := 1-\xi$,

$$\begin{aligned} P\big((1-\xi)X_\xi \le t\big) = \mathrm{P}(\varepsilon X_{1-\varepsilon} \le t) &= \sum_{0\le k\le t/\varepsilon} \varepsilon(1-\varepsilon)^k \\ &= \sum_{0\le k\le t/\varepsilon} \varepsilon(1-\varepsilon)^{(1/\varepsilon)k\varepsilon} \quad \text{(Riemann sum)} \\ &\to \int_0^t \mathrm{d}s\,\mathrm{e}^{-s} = \mathrm{P}(Y \le t)\,. \end{aligned}$$

Then, setting $\xi - 1 - \varepsilon$, $t = l\varepsilon$, $\varepsilon h_i = x_i$ in (1.34) of Thm. 1.1.3 and letting $\varepsilon \to 0$, one computes

$$\begin{aligned}
&\lim_{\varepsilon\to 0} P_\varepsilon(L(M)\le t)\\
&\qquad = \lim_{\varepsilon\to 0}\frac{Z^{-1}}{\varepsilon^{q(q-1)+q(p-q)+q}}\\
&\qquad\qquad \times \sum_{\max(h_i\varepsilon)\le l\varepsilon+(q-1)\varepsilon} \Delta_q(h_1\varepsilon,\dots,h_q\varepsilon)^2 \prod_{i=1}^{q}\left(\prod_{k=1}^{p-q}\varepsilon(h_i+k)\right)(1-\varepsilon)^{(1/\varepsilon)h_i\varepsilon}\varepsilon^q\\
&\qquad = \frac{1}{Z'}\int_{[0,t]^q}\Delta_q(x_1,\dots,x_q)^2\prod_{i=1}^{q}x_i^{p-q}\exp(-x_i)\,\mathrm{d}x_i\\
&\qquad = \frac{1}{Z_n}\int_{\substack{M\in\mathcal{H}_q\\ M\text{ positive definite}\\ \mathrm{Spectrum}(M)\le t}}(\det M)^{p-q}\mathrm{e}^{-\operatorname{Tr}M}\,\mathrm{d}M\ .
\end{aligned}$$

This last identity will be shown later in Prop. 1.7.2, ending the proof of Thm. 1.1.4. □

1.1.7.4 Queuing Problem

Consider servers $1,2,\dots,q$ waiting on customers $1,\dots,p$, with a *first-in first-out* service rule; see, e.g., [21] and references within. At first, the system is empty and then p customers are placed in the first queue. When a customer k is done with the ith server, he moves to the queue waiting to be served by the $(i+1)$st server; see Fig. 1.3. Let

$$V(k,l) = \text{service time of customer } k \text{ by server } l$$

be all geometrically independently distributed

$$\mathrm{P}(V(k,l)=t) = (1-\xi)\xi^t\ ,\quad t=0,1,2,\dots\ .$$

Theorem 1.1.5. *The distribution of*

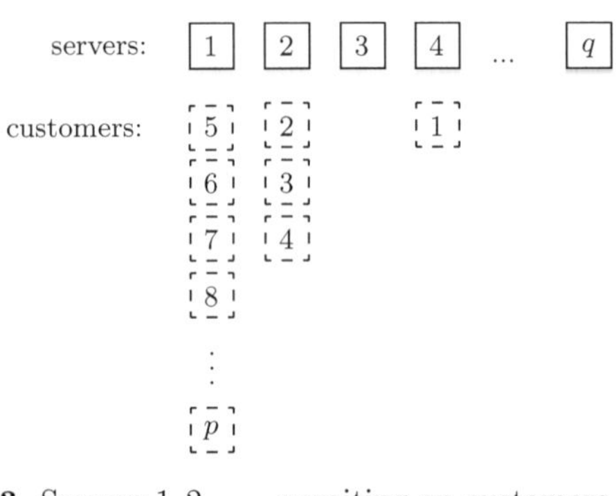

Fig. 1.3. Servers $1,2,\dots,q$ waiting on customers $1,\dots,p$

$$D(p,q) = \left\{ \begin{array}{l} \textit{departure time for the last customer } \fbox{p} \textit{ at the} \\ \textit{last server } \fbox{q} \end{array} \right\}$$

is given by (assuming $q \leq p$)

$$\begin{aligned} P(D(p,q) \leq l) &= \sum_{\substack{\lambda \in \mathbb{Y}^{\min(q,p)} \\ \lambda_1 \leq l}} (1-\xi)^{pq} \xi^{|\lambda|} \mathbf{s}_\lambda(1^q) \mathbf{s}_\lambda(1^p) \\ &= Z_{q,p}^{-1} \sum_{\substack{h \in \mathbb{N}^q \\ \max(h_i) \leq l+q-1}} \Delta_q(h_1, \ldots, h_q)^2 \prod_{i=1}^{q} \frac{(h_i+p-q)!}{h_i!} \xi^{h_i} . \end{aligned} \tag{1.35}$$

Proof. We show the problem is equivalent to the percolation problem discussed in Thm. 1.1.3. Indeed:

Step 1. *Setting $D(p,0) = D(0,q) = 0$ for all p, q, we have for all $p, q \geq 1$,*

$$D(p,q) = \max\big(D(p-1,q), D(p,q-1)\big) + V(p,q) . \tag{1.36}$$

Indeed, if $D(p-1,q) \leq D(p,q-1)$, then customer $p-1$ has left server q by the time customer p reaches q, so that customer p will not have to queue up and will be served immediately. Therefore

$$D(p,q) = D(p,q-1) + V(p,q) .$$

Now assume

$$D(p-1,q) \geq D(p,q-1) ;$$

then when customer p reaches server q, then customer $p-1$ is still being served at queue q. Therefore the departure time of customer p at queue q equals

$$D(p,q) = D(p-1,q) + V(p,q) .$$

In particular

$$D(p,1) = D(p-1,1) + V(p,1)$$

and

$$D(1,2) = D(1,1) + V(1,2),$$

establishing (1.36).

Step 2. We now prove

$$D(p,q) = \max_{\substack{\textit{all such} \\ \textit{paths}}} \left\{ \sum V(i,j) \middle| \begin{array}{l} \textit{over right/down paths from entry} \\ (1,1) \textit{ to } (p,q) \end{array} \right\},$$

where the paths are taken in the random matrix

$$V = \big(V(i,j)\big)_{\substack{1 \leq i \leq p \\ 1 \leq j \leq q}} .$$

By a straightforward calculation,

$$\begin{aligned} &D(p,q)\\ &= \max\bigl(D(p-1,q)+V(p,q), D(p,q-1)+V(p,q)\bigr)\\ &= \max\left(\max_{\substack{\text{all paths}\\ (0,0)\to(p-1,q)}} \sum_{\text{path}} V(i,j)+V(p,q),\ \max_{\substack{\text{all paths}\\ (0,0)\to(p,q-1)}} \sum_{\text{path}} V(i,j)+V(p,q)\right)\\ &= \max_{\substack{\text{all paths}\\ (0,0)\to(p,q)}} \left(\sum_{\text{path}} V(i,j)\right), \end{aligned}$$

ending the proof of Thm. 1.1.5. □

1.1.7.5 Discrete Polynuclear Growth Models

Consider geometric i.i.d. random variables $\omega(x,t)$, with $x \in \mathbb{Z}$, $t \in \mathbb{Z}_+$,

$$\mathrm{P}(\omega(x,t)=k) = (1-\xi)\xi^k\,, \quad k \in \mathbb{Z}_+\,,$$

except

$$\omega(x,t) = 0 \text{ if } t-x \text{ is even or } |x|>t\,.$$

Define inductively a growth process, with a height curve $h(x,t)$, with $x \in \mathbb{Z}$, $t \in \mathbb{Z}_+$, given by

$$\begin{aligned} h(x,0) &= 0\,,\\ h(x,t+1) &= \max\bigl(h(x-1,t), h(x,t), h(x+1,t)\bigr) + \omega(x,t+1)\,. \end{aligned}$$

For this model, see [59, 66, 77].

Theorem 1.1.6. *The height curve at even sites $2x$ at times $2t-1$ is given by*

$$h(2x,2t-1) = \max_{\substack{\textit{all such}\\ \textit{paths}}} \left\{\sum V(i,j), \begin{matrix}\textit{over right/down paths starting}\\ \textit{from entry } (1,1) \textit{ to } (t+x,t-x)\end{matrix}\right\}$$

where

$$V(i,j) := \omega(i-j, i+j-1)\,.$$

Thus $h(2x,2t-1)$ has again the same distribution as in (1.34).

Proof. It is based on the fact that, setting

$$G(q,p) := h(q-p, q+p-1)\,,$$

one computes

$$\begin{aligned} G(q,p) &= \max\bigl(G(q-1,p), G(q,p-1)\bigr) + V(q,p)\\ &= \max\left\{\sum V_{ij}, \begin{matrix}\text{over all right/down paths starting}\\ \text{from entry } (1,1) \text{ to } (q,p)\end{matrix}\right\}\,. \end{aligned}$$

So

$$h(2x, 2t-1) = G(t+x, t-x) ,$$

establishing Thm. 1.1.6. □

Figure 1.4 gives an example of such a growth process; the squares with stars refer to the contribution $\omega(x, t+1)$. It shows that $h(2x, 2t+1)$ is given by the maximum of right/down paths starting at the upper-left corner and going to site $(t+x, t-x)$, where x is the running variable along the anti-diagonal.

1.2 Probability on Partitions, Toeplitz and Fredholm Determinants

Consider variables $x = (x_1, x_2, \dots)$ and $y = (y_1, y_2, \dots)$, and the corresponding symmetric functions

$$t = (t_1, t_2, \dots) \quad \text{and} \quad s = (s_1, s_2, \dots)$$

with

$$kt_k = \sum_{i \geq 1} x_i^k \quad \text{and} \quad ks_k = \sum_{i \geq 1} y_i^k .$$

Following Borodin–Okounkov–Olshanski (see [24–26]), given arbitrary x, y, define the (not necessarily positive) *probability measure* on $\lambda \in \mathbb{Y}$,

$$\mathrm{P}_{x,y}(\lambda) := \frac{1}{Z}\tilde{\mathbf{s}}_\lambda(x)\tilde{\mathbf{s}}_\lambda(y) = \frac{1}{Z}\mathbf{s}_\lambda(t)\mathbf{s}_\lambda(s) \tag{1.37}$$

with

$$Z = \prod_{1 \leq i,j} (1 - x_i y_j)^{-1} = \exp\left(\sum_1^\infty k t_k s_k\right) .$$

Indeed, by Cauchy's identity,

$$\begin{aligned}\sum_{\lambda \in \mathbb{Y}} \mathrm{P}_{x,y}(\lambda) &= \sum_\lambda \mathbf{s}_\lambda(x)\mathbf{s}_\lambda(y) \prod_{1 \leq i,j} (1 - x_i y_j) \\ &= \prod_{i,j \geq 1} (1 - x_i y_j)^{-1} \prod_{i,j \geq 1} (1 - x_i y_j) = 1 .\end{aligned}$$

The main objective of this section is to compute

$$P_{x,y}(\lambda_1 \leq n), \tag{1.38}$$

which then will be specialized in the next section to specific x's and y's or, what is the same, to specific t's and s's. This probability (1.38) has three different expressions, one in terms of a Toeplitz determinant, another in terms

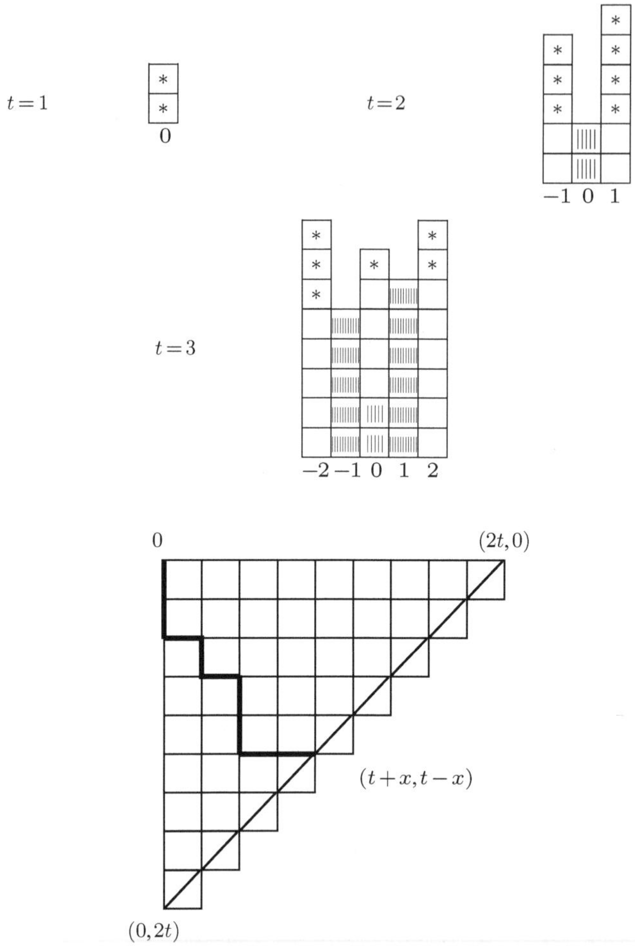

Fig. 1.4.

of an integral over the unitary group and still another in terms of a Fredholm determinant. The Toeplitz representation enables one to compute in an effective way the probability (1.38), whereas the Fredholm representation is useful, when taking limits for large permutations. In the statement below, we need the Fredholm determinant of a kernel $K(i,j)$, with $i,j \in \mathbb{Z}$ and restricted to $[n,\infty)$; it is defined as

$$\det\big(I - K(i,j)|_{[n,n+1,\dots]}\big) := \sum_{m=0}^{\infty}(-1)^m \sum_{\substack{n\le x_1<\dots<x_m\\ x_i\in\mathbb{Z}}} \det\big(K(x_i,x_j)\big)_{1\le i,j\le m}\,. \tag{1.39}$$

Now, one has the following statement [8, 23, 25, 45, 86]:

Theorem 1.2.1. *Given the "probability measure" (1.37), the following probability has three different expressions:*

$$\begin{aligned}
&\mathrm{P}(\lambda \textit{ with } \lambda_1 \le n)\\
&\quad = Z^{-1}\det\left(\oint_{S^1}\frac{\mathrm{d}z}{2\pi\mathrm{i}z}z^{k-l}\exp\Big(-\sum_1^{\infty}(t_iz^i+s_iz^{-i})\Big)\right)_{1\le k,l\le n}\\
&\quad = Z^{-1}\int_{U(n)}\exp\Big(-\operatorname{Tr}\sum_{i\ge 1}(t_iX^i+s_i\bar{X}^i)\Big)\,\mathrm{d}X\\
&\quad = \det\big(I-K(k,l)|_{[n,n+1,\dots]}\big)\,,
\end{aligned} \tag{1.40}$$

where $K(k,l)$ is a kernel

$$\begin{aligned}
&K(k,l)\\
&= \left(\frac{1}{2\pi\mathrm{i}}\right)^2\oint_{|w|=\rho<1}\oint_{|z|=\rho^{-1}>1}\frac{\mathrm{d}z\,\mathrm{d}w}{z^{k+1}w^{-l}}\frac{\exp\big(V(z)-V(w)\big)}{z-w}, \quad \textit{for } k,l\in\mathbb{Z}\\
&= \frac{1}{k-l}\left(\frac{1}{2\pi\mathrm{i}}\right)^2\oint\oint_{\substack{|w|=\rho<1\\|z|=\rho^{-1}>1}}\frac{\mathrm{d}z\,\mathrm{d}w}{z^{k+1}w^{-l}}\frac{z(\mathrm{d}/\mathrm{d}z)V(z)-w(\mathrm{d}/\mathrm{d}w)V(w)}{z-w}\\
&\qquad\qquad\times\exp\big(V(z)-V(w)\big)\\
&\qquad\qquad \textit{for } k,l\in\mathbb{Z},\ \textit{with } k\ne l,
\end{aligned} \tag{1.41}$$

with

$$V(z) := V_{t,s}(z) := -\sum_{i\ge 1}(t_iz^{-i}-s_iz^i).$$

1.2.1 Probability on Partitions Expressed as Toeplitz Determinants

In this subsection, the first part of Thm. 1.2.1 will be reformulated as Prop. 1.2.1 and also demonstrated: (see, e.g., Gessel [45], Tracy–Widom [86], Adler–van Moerbeke [8])

Proposition 1.2.1. *Given the "probability measure"*

$$\mathrm{P}(\lambda) = Z^{-1}\mathbf{s}_\lambda(t)\mathbf{s}_\lambda(s), \quad Z = \exp\Big(\sum_{i\ge 1} it_is_i\Big)\,,$$

the following holds

$$P(\lambda \text{ with } \lambda_1 \leq n) = Z^{-1} \det\left(\oint_{S^1} \frac{dz}{2\pi i z} z^{k-l} \exp\Big(-\sum_1^\infty (t_i z^i + s_i z^{-i})\Big)\right)_{1\leq k,l\leq n}$$

and

$$P(\lambda \text{ with } \lambda_1^\top \leq n) = Z^{-1} \det\left(\oint_{S^1} \frac{dz}{2\pi i z} z^{k-l} \exp\Big(\sum_1^\infty (t_i z^i + s_i z^{-i})\Big)\right)_{1\leq k,l\leq n}$$

Proof. Consider the semi-infinite Toeplitz matrix

$$m_\infty(t,s) = (\mu_{kl})_{k,l\geq 0}, \quad \text{with } \mu_{kl}(t,s) = \oint_{S^1} z^{k-l} \exp\Big(\sum_1^\infty (t_j z^j - s_j z^{-j})\Big)\frac{dz}{2\pi i z}.$$

Note that

$$\frac{\partial \mu_{kl}}{\partial t_m} = \oint_{S^1} z^{k-l+m} \exp\Big(\sum_1^\infty (t_j z^j - s_j z^{-j})\Big)\frac{dz}{2\pi i z} = \mu_{k+m,l}$$

$$\frac{\partial \mu_{kl}}{\partial s_m} = -\oint_{S^1} z^{k-l-m} \exp\Big(\sum (t_j z^j - s_j z^{-j})\Big)\frac{dz}{2\pi i z} = -\mu_{k,l+m},$$

with initial condition $\mu_{kl}(0,0) = \delta_{kl}$. In matrix notation, this amounts to the system of differential equations[8]

$$\frac{\partial m_\infty}{\partial t_i} = \Lambda^i m_\infty \quad \text{and} \quad \frac{\partial m_\infty}{\partial s_i} = -m_\infty (\Lambda^\top)^i \quad \text{with initial condition } m_\infty(0,0) = I_\infty.$$

The solution to this initial value problem is given by
(i)

$$m_\infty(t,s) = \big(\mu_{kl}(t,s)\big)_{k,l\geq 0} \tag{1.42}$$

and
(ii)

$$m_\infty(t,s) = \exp\Big(\sum_1^\infty t_i \Lambda^i\Big) m_\infty(0,0) \exp\Big(-\sum_1^\infty s_i \Lambda^{\top i}\Big), \tag{1.43}$$

where

[8] The operator Λ is the semi-infinite shift matrix, with zeroes everywhere, except for 1s just above the diagonal, i.e., $(\Lambda v)_n = v_{n+1}$. I_∞ is the semi-infinite identity matrix.

$$\exp\left(\sum_1^\infty t_i\Lambda^i\right) = \sum_0^\infty \mathbf{s}_i(t)\Lambda^i = \begin{pmatrix} 1 & \mathbf{s}_1(t) & \mathbf{s}_2(t) & \mathbf{s}_3(t) & \dots \\ 0 & 1 & \mathbf{s}_1(t) & \mathbf{s}_2(t) & \dots \\ 0 & 0 & 1 & \mathbf{s}_1(t) & \dots \\ 0 & 0 & 0 & 1 & \\ \vdots & \vdots & \vdots & \vdots & \end{pmatrix} = \left(\mathbf{s}_{j-i}(t)\right)_{\substack{1\le i<\infty \\ 1\le j<\infty}} .$$

Then, by the uniqueness of solutions of odes, the two solutions coincide, and in particular the $n\times n$ upper-left blocks of (1.42) and (1.43), namely

$$m_n(t,s) = E_n(t)m_\infty(0,0)E_n^\top(-s)\,, \tag{1.44}$$

where

$$E_n(t) = \begin{pmatrix} 1 & \mathbf{s}_1(t) & \mathbf{s}_2(t) & \mathbf{s}_3(t) & \dots & \mathbf{s}_{n-1}(t) & \dots \\ 0 & 1 & \mathbf{s}_1(t) & \mathbf{s}_2(t) & \dots & \mathbf{s}_{n-2}(t) & \dots \\ \vdots & & & & & & \\ & & & & & \mathbf{s}_1(t) & \dots \\ 0 & & \dots & & 0 & 1 & \dots \end{pmatrix} = \left(\mathbf{s}_{j-i}(t)\right)_{\substack{1\le i<n \\ 1\le j<\infty}} .$$

Therefore the determinants coincide:

$$\det m_n(t,s) = \det\left(E_n(t)m_\infty(0,0)E_n^\top(-s)\right) . \tag{1.45}$$

We shall need to expand the right-hand side of (1.45) in "Fourier series," which is based on the following lemma:

Lemma 1.2.1. *Given the semi-infinite initial condition $m_\infty(0,0)$, the expression below admits an expansion in Schur polynomials,*

$$\det\left(E_n(t)m_\infty(0,0)E_n^\top(-s)\right) = \sum_{\substack{\lambda,\nu \\ \lambda_1^\top,\nu_1^\top\le n}} \det(m^{\lambda,\nu})\mathbf{s}_\lambda(t)\mathbf{s}_\nu(-s)\,, \qquad \textit{for } n>0, \tag{1.46}$$

where the sum is taken over all Young diagrams λ and ν, with first columns $\lambda_1^\top$ and $\nu_1^\top \le n$ and where

$$m^{\lambda,\nu} := (\mu_{\lambda_i-i+n,\nu_j-j+n})_{1\le i,j\le n}. \tag{1.47}$$

Proof. The proof of this lemma is based on the Cauchy–Binet formula, which affirms that given two matrices $\underset{(m.n)}{A}$, $\underset{(n,m)}{B}$, for n large $\ge m$

$$\begin{aligned} \det(AB) &= \det\left(\sum_i a_{li}b_{ik}\right)_{1\le k,l\le m} \\ &= \sum_{1\le i_1<\dots<i_m\le n} \det(a_{k,i_l})_{1\le k,l\le m} \det(b_{i_k,l})_{1\le k,l\le m} . \end{aligned} \tag{1.48}$$

Note that every decreasing sequence $\infty > k_n > \dots > k_1 \ge 1$ can be mapped into a Young diagram $\lambda_1\ge\lambda_2\ge\dots\ge\lambda_n\ge 0$, by setting $k_j = j+\lambda_{n+1-j}$.

Relabeling the indices i, j with $1 \le i, j \le n$, by setting $j' := n - j + 1$, $i' := n-i+1$, we have $1 \le i', j' \le n$ and $k_j - i = \lambda_{j'} - j' + i'$ and $k_i - 1 = \lambda_{i'} - i' + n$. In other terms, the sequence of integers $\infty > k_n > \cdots > k_1 \ge 1$ leads to a partition $\lambda_1 = k_n - n > \lambda_2 = k_{n-1} - n + 1 > \cdots > \lambda_n = k_1 - 1 \ge 0$. The same can be done for the sequence $1 \le l_1 < \cdots < l_n < \infty$, leading to the Young diagram ν, using the same relabeling. Applying the Cauchy–Binet formula twice, expression (1.45)) leads to:

$$
\begin{aligned}
&\det\big(E_n(t) m_\infty(0,0) E_n^\top(-s)\big) \\
&\quad = \sum_{1 \le k_1 < \cdots < k_n < \infty} \det\big(\mathbf{s}_{k_j - i}(t)\big)_{1 \le i,j \le n} \det\Big(\big(m_\infty(0,0) E_n^\top(-s)\big)_{k_i, l}\Big)_{1 \le i,l \le n} \\
&\quad = \sum_{1 \le k_1 < \cdots < k_n < \infty} \det\big(\mathbf{s}_{k_j - i}(t)\big)_{1 \le i,j \le n} \det\Big(\big(\mu_{k_i - 1, j-1}\big)_{\substack{1 \le i \le n \\ 1 \le j < \infty}} \big(\mathbf{s}_{i-j}(-s)\big)_{\substack{1 \le i < \infty \\ 1 \le j \le n}}\Big) \\
&\quad = \sum_{1 \le k_1 < \cdots < k_n < \infty} \det\big(\mathbf{s}_{k_j - i}(t)\big)_{1 \le i,j \le n} \\
&\qquad\qquad \times \sum_{1 \le l_1 < \cdots < l_n < \infty} \det(\mu_{k_i - 1, l_j - 1})_{1 \le i,j \le n} \det\big(\mathbf{s}_{l_i - j}(-s)\big)_{1 \le i,j \le n} \\
&\quad = \sum_{\substack{\lambda \in \mathbb{Y} \\ \lambda_1^\top \le n}} \det\big(\mathbf{s}_{\lambda_{j'} - j' + i'}(t)\big)_{1 \le i', j' \le n} \\
&\qquad\qquad \times \sum_{\substack{\nu \in \mathbb{Y} \\ \nu_1^\top \le n}} \det(\mu_{\lambda_{i'} - i' + n, \nu_{j'} - j' + n})_{1 \le i', j' \le n} \det\big(\mathbf{s}_{\nu_{i'} - i' + j'}(-s)\big)_{1 \le i', j' \le n} \\
&\quad = \sum_{\substack{\lambda, \nu \in \mathbb{Y} \\ \lambda_1^\top, \nu_1^\top \le n}} \det(m^{\lambda,\nu}) \mathbf{s}_\lambda(t) \mathbf{s}_\nu(-s) \,,
\end{aligned}
$$

which establishes Lemma 1.2.1. □

Continuing the proof of Prop. 1.2.1, apply now Lemma 1.2.1 to $m_\infty(0,0) = I_\infty$, leading to

$$\det m^{\lambda,\nu} = \det(\mu_{\lambda_i - i + n, \nu_j - j + n})_{1 \le i,j \le n} \ne 0$$

if and only if

$$\lambda_i - i + n = \nu_i - i + n \quad \text{for all } 1 \le i \le n,$$

i.e., $\lambda = \nu$, in which case

$$\det m^{\lambda,\lambda} = 1 \,.$$

Therefore, from (1.45), it follows

$$\sum_{\substack{\lambda \in \mathbb{Y} \\ \lambda_1^\top \le n}} \mathbf{s}_\lambda(t) \mathbf{s}_\lambda(-s) = \det\left(\int_{S^1} \frac{\mathrm{d}z}{2\pi \mathrm{i} z} z^{k-l} \exp\Big(\sum_1^\infty (t_i z^i - s_i z^{-i})\Big)\right)_{1 \le k,l \le n} .$$

So, we have, changing $s \mapsto -s$,

$$\begin{aligned}
\mathrm{P}(\lambda \text{ with } \lambda_1^\top \le n) &= Z^{-1} \sum_{\substack{\lambda\in\mathbb{Y}\\ \lambda_1\le n}} \mathbf{s}_\lambda(t)\mathbf{s}_\lambda(s) \\
&= Z^{-1} \det\left(\oint_{S^1} \frac{\mathrm{d}z}{2\pi\mathrm{i}z} z^{k-l} \exp\left(\sum_1^\infty (t_i z^i + s_i z^{-i}) \right) \right)_{1\le k,l\le n},
\end{aligned}$$

and, using $\mathbf{s}_\lambda(-t) = (-1)^{|\lambda|}\mathbf{s}_{\lambda^\top}(t)$, we also have

$$\begin{aligned}
\mathrm{P}(\lambda \text{ with } \lambda_1 \le n) &= Z^{-1} \sum_{\substack{\lambda\in\mathbb{Y}\\ \lambda_1\le n}} \mathbf{s}_\lambda(t)\mathbf{s}_\lambda(s) \\
&= Z^{-1} \sum_{\substack{\lambda\in\mathbb{Y}\\ \lambda_1^\top\le n}} \mathbf{s}_{\lambda^\top}(t)\mathbf{s}_{\lambda^\top}(s) = Z^{-1} \sum_{\substack{\lambda\in\mathbb{Y}\\ \lambda_1^\top\le n}} \mathbf{s}_\lambda(-t)\mathbf{s}_\lambda(-s) \\
&= Z^{-1} \det\left(\oint_{S^1} \frac{\mathrm{d}z}{2\pi\mathrm{i}z} z^{k-l} \exp\left(-\sum_1^\infty (t_i z^i + s_i z^{-i}) \right) \right)_{1\le k,l\le n},
\end{aligned}$$

where

$$Z = \exp\left(\sum_1^\infty i t_i s_i \right),$$

ending the proof of Prop. 1.2.1. □

1.2.2 The Calculus of Infinite Wedge Spaces

The material in this section can be found in Kac [61] and Kac–Raina [62] and many specific results are due to Borodin–Okounkov–Olshanski (see [24–26]). Given a vector space $V = \bigoplus_{j\in\mathbb{Z}} \mathbb{C}v_j$ with inner-product $\langle v_i, v_j\rangle = \delta_{ij}$, the infinite wedge space $V^\infty = \bigwedge^\infty V$ is defined as

$$V^\infty = \operatorname{span}\{v_{s_1}\wedge v_{s_2}\wedge v_{s_3}\wedge\cdots : s_1 > s_2 > \cdots, s_k = -k \text{ for } k \gg 0\}$$

containing the "vacuum"

$$f_\varnothing = v_{-1}\wedge v_{-2}\wedge\cdots.$$

The vector space V^∞ comes equipped with an inner-product $\langle\ ,\ \rangle$, making the basis vectors $v_{s_1}\wedge v_{s_2}\wedge\cdots$ orthonormal. To each $k\in\mathbb{Z}$, we associate two operations, a *wedging* with v_k and a *contracting operator*, removing a v_k,

$$\begin{aligned}
&\psi_k\colon V^\infty \to V^\infty : f \mapsto \psi_k(f) = v_k\wedge f \\
&\psi_k^*\colon V^\infty \to V^\infty : \\
&\qquad v_{s_1}\wedge\cdots\wedge v_{s_i}\wedge\cdots \mapsto \sum_i (-1)^{i+1}\langle v_k, v_{s_i}\rangle v_{s_1}\wedge\cdots\wedge\hat{v}_{s_i}\wedge\cdots.
\end{aligned}$$

Note that

$$\begin{aligned} \psi_k(f) &= 0\,, && \text{if } v_k \text{ figures in } f \\ \psi_k^*(v_{s_1}\wedge\cdots) &= 0, && \text{if } k \notin (s_1, s_2, \dots). \end{aligned}$$

Define the shift

$$\Lambda^r := \sum_{k\in\mathbb{Z}} \psi_{k+r}\psi_k^*, \qquad r \in \mathbb{Z}$$

acting on V^∞ as follows

$$\begin{aligned} \Lambda^r v_{s_1}\wedge v_{s_2}\wedge\cdots = v_{s_1+r}\wedge v_{s_2}\wedge v_{s_3}\wedge\cdots + v_{s_1}\wedge v_{s_2+r}\wedge v_{s_3}\wedge\cdots \\ + v_{s_1}\wedge v_{s_2}\wedge v_{s_3+r}\wedge\cdots+\cdots\,. \end{aligned}$$

One checks that

$$[\Lambda^r, \psi_k] = \psi_{k+r}\,, \qquad [\Lambda^r, \psi_k^*] = -\psi_{k-r}^* \tag{1.49}$$

$$[\Lambda^k, \Lambda^l] = l\delta_{k,-l} \tag{1.50}$$

and hence

$$\left[\sum_{i\geq 1} t_i\Lambda^i, \sum_{j\geq 1} s_i\Lambda^{-i}\right] = -\sum_{i\geq 1} it_is_i\,. \tag{1.51}$$

Lemma 1.2.2 (Version of the Cauchy identity).

$$\begin{aligned} &\exp\Bigl(\sum_{i\geq 1} t_i\Lambda^i\Bigr)\exp\Bigl(-\sum_{j\geq 1} s_i\Lambda^{-j}\Bigr) \\ &\quad = \exp\Bigl(\sum_{i\geq 1} it_is_i\Bigr)\exp\Bigl(-\sum_{j\geq 1} s_j\Lambda^{-j}\Bigr)\exp\Bigl(\sum_{i\geq 1} t_i\Lambda^i\Bigr)\,. \end{aligned} \tag{1.52}$$

Proof. When two operators A and B commute with their commutator $[A,B]$, then (see Kac [61, p. 308])

$$\mathrm{e}^A\mathrm{e}^B = \mathrm{e}^B\mathrm{e}^A\mathrm{e}^{[A,B]}\,.$$

Setting $A = \sum_{i\geq i} t_i\Lambda^i$ and $B = -\sum_{j\geq 1} s_j\Lambda^{-j}$, we find

$$\begin{aligned} &\exp\Bigl(\sum_{i\geq 1} t_i\Lambda^i\Bigr)\exp\Bigl(-\sum_{j\geq 1} s_i\Lambda^{-j}\Bigr) \\ &= \exp\Bigl(-\sum_{j\geq 1} s_j\Lambda^{-j}\Bigr)\exp\Bigl(\sum_{i\geq 1} t_i\Lambda^i\Bigr)\exp\Bigl(-\Bigl[\sum_{i\geq 1} t_i\Lambda^i, \sum_{j\geq 1} s_j\Lambda^{-j}\Bigr]\Bigr) \\ &= \exp\Bigl(-\sum_{j\geq 1} s_j\Lambda^{-j}\Bigr)\exp\Bigl(\sum_{i\geq 1} t_i\Lambda^i\Bigr)\exp\Bigl(\sum_{i\geq 1} it_is_i\Bigr) \\ &= \exp\Bigl(\sum_{i\geq 1} it_is_i\Bigr)\exp\Bigl(-\sum -j\geq 1 s_j\Lambda^{-j}\Bigr)\exp\Bigl(\sum_{i\geq 1} t_i\Lambda^i\Bigr)\,. \end{aligned}$$

□

It is useful to consider the generators of ψ_i and ψ_i^*:

$$\psi(z) = \sum_{i\in\mathbb{Z}} z^i \psi_i , \quad \psi^*(w) = \sum_{j\in\mathbb{Z}} w^{-j}\psi_j^* . \tag{1.53}$$

From (1.49), it follows that

$$\begin{aligned}
[\Lambda^r, \psi(z)] &= \sum_k z^k [\Lambda^r, \psi_k] = \sum_k z^k \psi_{k+r} = \frac{1}{z^r}\psi(z) , \\
[\Lambda^r, \psi^*(w)] &= -\frac{1}{w^r}\psi^*(w) .
\end{aligned}$$

The two relations above lead to the following, by taking derivatives in t_i and s_i of the left hand side and setting all $t_i = s_i = 0$:

$$\begin{aligned}
&\exp\left(\pm\sum_1^\infty t_i\Lambda^i\right)\left\{\begin{matrix}\psi(z)\\ \psi^*(w)\end{matrix}\right\}\exp\left(\mp\sum_1^\infty t_i\Lambda^i\right) \\
&\qquad\qquad = \begin{cases}\exp(\pm\sum_1^\infty t_r/z^r)\psi(z)\\ \exp(\mp\sum t_r/w^r)\psi^*(w)\end{cases} \\
&\exp\left(\pm\sum_1^\infty s_i\Lambda^{-i}\right)\left\{\begin{matrix}\psi(z)\\ \psi^*(w)\end{matrix}\right\}\exp\left(\mp\sum_1^\infty s_i\Lambda^{-i}\right) \\
&\qquad\qquad = \begin{cases}\exp(\pm\sum_1^\infty s_r z^r)\psi(z)\\ \exp(\mp\sum_1^\infty s_r w^r)\psi^*(w)\end{cases}
\end{aligned} \tag{1.54}$$

To each partition $\lambda = (\lambda_1 \geq \lambda_2 \geq \cdots \geq \lambda_m \geq 0)$, associate a vector

$$f_\lambda := v_{\lambda_1 - 1}\wedge v_{\lambda_2-2}\wedge\cdots\wedge v_{\lambda_m - m}\wedge v_{-(m+1)}\wedge v_{-(m+2)}\wedge\cdots \in V^\infty .$$

The following lemma holds:

Lemma 1.2.3.

$$\begin{aligned}
\exp\left(\sum_{i\geq 1} t_i\Lambda^i\right) f_\lambda &= \sum_{\substack{\mu\in\mathbb{Y}\\ \mu\supset\lambda}} \mathbf{s}_{\mu\setminus\lambda}(t) f_\mu , \\
\exp\left(\sum_{i\geq 1} t_i\Lambda^{-i}\right) f_\lambda &= \sum_{\substack{\mu\in\mathbb{Y}\\ \mu\subset\lambda}} \mathbf{s}_{\lambda\setminus\mu}(t) f_\mu .
\end{aligned} \tag{1.55}$$

In particular,

$$\exp\left(\sum_{i\geq 1} t_i\Lambda^i\right) f_\varnothing = \sum_{\mu\in\mathbb{Y}} \mathbf{s}_\mu(t) f_\mu \quad \textit{and} \quad \exp\left(\sum_{i\geq 1} t_i\Lambda^{-i}\right) f_\varnothing = f_\varnothing . \tag{1.56}$$

Proof. First notice that a matrix $A \in \mathrm{GL}_\infty$ acts on V^∞ as follows

$$A(v_{s_1} \wedge v_{s_2} \wedge \cdots) = \sum_{s'_1 > s'_2 > \cdots} \det\left(A^{s_1, s_2, \ldots}_{s'_1, s'_2, \ldots}\right) v_{s'_1} \wedge v_{s'_2} \wedge \cdots ,$$

where

$$A^{s_1, s_2, \ldots}_{s'_1, s'_2, \ldots} = \left\{ \begin{matrix} \text{matrix located at the intersection of the rows } s'_1, s'_2, \ldots \\ \text{and columns } s_1, s_2, \ldots \text{ of } A \end{matrix} \right\} .$$

Here the rows and columns of the bi-infinite matrix are labeled by

$$\begin{array}{c} \\ 1 \\ 0 \\ -1 \\ -2 \\ -3 \end{array} \begin{array}{c} \begin{array}{ccccc} 1 & 0 & -1 & -2 & -3 \end{array} \\ \left(\begin{array}{ccccc} * & * & * & * & * \\ * & * & * & * & * \\ * & * & \bullet & * & * \\ * & * & * & * & * \\ * & * & * & * & * \end{array} \right) \end{array} .$$

Hence, for the bi-infinite matrix $\exp(\sum_1^\infty t_i \Lambda^i)$,

$$\det\left(\left(\exp\left(\sum_1^\infty t_i \Lambda^i\right)\right)^{s_1, s_2, \ldots}_{s'_1, s'_2, \ldots}\right) = \det\left(\mathbf{s}_{s'_i - s_j}\right)$$

Setting $s_i = \lambda_i - i$ and defining μ_i in the formula below by $s'_i = \mu_i - i$, one checks

$$\begin{aligned} \exp\left(\sum_1^\infty t_i \Lambda^i\right) f_\lambda &= \exp\left(\sum_1^\infty t_i \Lambda^i\right)(v_{s_1} \wedge v_{s_2} \wedge \cdots) \\ &\overset{*}{=} \sum_{s'_1 > s'_2 > \cdots} \det\left(\left(\exp\left(\sum_1^\infty t_i \Lambda^i\right)\right)^{s_1, s_2, \ldots}_{s'_1, s'_2, \ldots}\right) v_{s'_1} \wedge v_{s'_2} \wedge \cdots \\ &= \sum_{s'_1 > s'_2 > \cdots} \det\left(\mathbf{s}_{s'_i - s_j}(t)\right)_{1 \le i,j < \infty} v_{s'_1} \wedge v_{s'_2} \wedge \cdots \\ &= \sum_{\mu_1 - 1 > \mu_2 - 2 > \cdots} \det\left(\mathbf{s}_{(\mu_i - i) - (\lambda_j - j)}(t)\right)_{1 \le i,j \le \infty} v_{\mu_1 - 1} \wedge v_{\mu_2 - 2} \wedge \cdots \\ &= \sum_{\substack{\mu \in \mathbb{Y} \\ \mu \supset \lambda}} \mathbf{s}_{\mu \backslash \lambda}(t) f_\mu . \end{aligned}$$

The second identity in (1.55) is shown in the same way. Identities (1.56) follow immediately from (1.55), ending the proof of Lemma 1.2.3. □

We also need:

Lemma 1.2.4. *For*

$$\Psi^*_{b_k} = \sum_{i\geq 1} b_{ki}\psi^*_{-i} \quad \textit{and} \quad \Psi_{a_k} = \sum_{i\geq 1} a_{ki}\psi_{-i}\,, \tag{1.57}$$

the following identity holds:

$$\langle \Psi_{a_1}\cdots\Psi_{a_m}\Psi^*_{b_m}\cdots\Psi^*_{b_1} f_\varnothing, f_\varnothing\rangle = \det(\langle \Psi_{a_k}\Psi^*_{b_l} f_\varnothing, f_\varnothing\rangle)_{1\leq k,l\leq m}\,. \tag{1.58}$$

Proof. First one computes for the Ψ's as in (1.57),

$$\begin{aligned}
&\Psi^*_{b_m}\cdots\Psi^*_{b_1} f_\varnothing \\
&\quad = \sum_{i_1>\cdots>i_m\geq 1} (-1)^{\sum_1^m (i_k+1)} \det(b_{k,i_l})_{1\leq k,l\leq m} \\
&\qquad\qquad v_{-1}\wedge\cdots\wedge\hat{v}_{-i_m}\wedge\cdots\wedge\hat{v}_{-i_{m-1}}\wedge\cdots\wedge\hat{v}_{-i_2}\wedge\cdots\wedge\hat{v}_{-i_1}\wedge\cdots\,.
\end{aligned}$$

Then acting with the Ψ_{a_k} as in (1.57), it suffices to understand how it acts on the wedge products appearing in the expression above, namely:

$$\begin{aligned}
&\Psi_{a_1}\cdots\Psi_{a_m} v_{-1}\wedge\cdots\wedge\hat{v}_{-i_m}\wedge\cdots\wedge\hat{v}_{-i_{m-1}}\wedge\cdots\wedge\hat{v}_{-i_1}\wedge\cdots \\
&\qquad\qquad = (-1)^{\sum_1^m (i_k+1)} \det(a_{k,i_l})_{1\leq k,l\leq m} f_\varnothing\,.
\end{aligned}$$

Thus, combining the two, one finds, using the Cauchy–Binet formula (1.48) in the last equality,

$$\begin{aligned}
\Psi_{a_1}\cdots\Psi_{a_m}\Psi^*_{b_m}\cdots\Psi^*_{b_1} f_\varnothing &= \sum_{i_1>\cdots>i_m\geq 1} \det(a_{k,i_l})_{1\leq k,l\leq m} \det(b_{k,i_l})_{1\leq k,l\leq m} f_\varnothing \\
&= \det\left(\sum_i a_{li} b_{ki}\right)_{1\leq k,l\leq m} f_\varnothing\,.
\end{aligned}$$

In particular for $m = 1$,

$$\Psi_{a_l}\Psi^*_{b_k} f_\varnothing = \sum_i a_{li} b_{ki} f_\varnothing\,.$$

Hence

$$\begin{aligned}
\langle \Psi_{a_1}\cdots\Psi_{a_m}\Psi^*_{b_m}\cdots\Psi^*_{b_1} f_\varnothing, f_\varnothing\rangle &= \det\left(\sum_i a_{ki} b_{li}\right)_{1\leq k,l\leq m} \\
&= \det(\langle \Psi_{a_k}\Psi^*_{b_l} f_\varnothing, f_\varnothing\rangle)_{1\leq k,l\leq m}\,,
\end{aligned}$$

ending the proof of Lemma 1.2.4. □

1.2.3 Probability on Partitions Expressed as Fredholm Determinants

Remember the definition (1.39) of the Fredholm determinant of a kernel $K(i,j)$, with $i,j \in \mathbb{Z}$ and restricted to $[n,\infty)$. This statement has appeared in the literature in some form (see Geronimo–Case [44]) and a more analytic formulation has appeared in Basor–Widom [22]. The proof given here is an "integrable" one, due to Borodin–Okounkov [23].

Proposition 1.2.2.

$$\mathrm{P}(\lambda \text{ with } \lambda_1 \leq n) = \det\big(I - K(k,l)|_{[n,n+1,\dots]}\big),$$

where $K(k,l)$ is a kernel

$$\begin{aligned} &K(k,l)\\ &= \left(\frac{1}{2\pi \mathrm{i}}\right)^2 \oint_{|w|=\rho<1} \oint_{|z|=\rho^{-1}>1} \frac{\mathrm{d}z\,\mathrm{d}w}{z^{k+1}w^{-l}} \frac{\exp\big(V(z)-V(w)\big)}{z-w}\\ &= \frac{1}{k-l}\left(\frac{1}{2\pi \mathrm{i}}\right)^2 \oint\oint_{\substack{|w|=\rho<1\\ |z|=\rho^{-1}>1}} \frac{\mathrm{d}z\,\mathrm{d}w}{z^{k+1}w^{-l}} \frac{z(\mathrm{d}/\mathrm{d}z)V(z) - w(\mathrm{d}/\mathrm{d}w)V(w)}{z-w}\\ &\qquad\qquad \times \exp\big(V(z)-V(w)\big)\,, \quad \text{for } k \neq l, \end{aligned} \tag{1.59}$$

with

$$V(z) = -\sum_{i\geq 1}(t_i z^{-i} - s_i z^i).$$

The proof of Prop. 1.2.2 hinges on the following lemma, due to Borodin–Okounkov ([23, 74, 75]):

Lemma 1.2.5. *If $X = \{x_1,\dots,x_m\} \subset \mathbb{Z}$ and $S(\lambda) := \{\lambda_1 - 1, \lambda_2 - 2, \lambda_3 - 3, \dots\}$*

$$\mathrm{P}(\lambda \mid S(\lambda) \supset X) = \frac{1}{Z} \sum_{\substack{\lambda \text{ such that}\\ S(\lambda)\supset X}} s_\lambda(t)s_\lambda(s) = \det\big(K(x_i,x_j)\big)_{1\leq i,j\leq m}\,.$$

Proof of Prop. 1.2.2. Setting $A_k := \{\lambda \mid k \in S(\lambda)\}$ with $S(\lambda) := \{\lambda_1 - 1, \lambda_2 - 2, \lambda_3 - 3, \dots\}$, one computes (the x_i below are integers)

$$
\begin{aligned}
&\mathrm{P}(\lambda \text{ with } \lambda_1 \leq n)\\
&\quad = \mathrm{P}(\lambda \text{ with all } \lambda_i \leq n)\\
&\quad = \mathrm{P}(\lambda \text{ with } S(\lambda) \cap \{n, n+1, n+2, \dots\} = \varnothing)\\
&\quad = 1 - \mathrm{P}(\lambda \text{ with } S(\lambda) \text{ contains some } k \text{ for } k \geq n)\\
&\quad = 1 - \mathrm{P}(A_n \cup A_{n+1} \cup \cdots)\\
&\quad = 1 - \sum_{n \leq i} \mathrm{P}(A_i) + \sum_{n \leq i < j} \mathrm{P}(A_i \cap A_j) - \sum_{n \leq i < j < k} \mathrm{P}(A_i \cap A_j \cap A_k) + \cdots\\
&\qquad\qquad\qquad\qquad\qquad\qquad\qquad\qquad\qquad\qquad \text{using Poincaré's formula}\\
&\quad = \sum_{m=0}^{\infty} (-1)^m \sum_{n \leq x_1 < \cdots < x_m} \mathrm{P}(\lambda \text{ with } \{x_1, \dots, x_m\} \subset S(\lambda))\\
&\quad = \sum_{m=0}^{\infty} (-1)^m \sum_{n \leq x_1 < \cdots < x_m} \det\big(K(x_i, x_j)\big)_{1 \leq i,j \leq m}\\
&\quad = \det\big(I - K(i,j)\big|_{[n, n+1, \dots]}\big)\,,
\end{aligned}
$$

from which Prop. 1.2.2 follows. □

Proof of Lemma 1.2.5. Remembering the probability measure introduced in (1.37), we have that

$$
\mathrm{P}(\lambda \mid S(\lambda) \supset X) = \frac{1}{Z} \sum_{\substack{\lambda \text{ such that} \\ S(\lambda) \supset X}} s_\lambda(t) s_\lambda(s)\,.
$$

Next, from the wedging-contracting operation

$$
\psi_x \psi_x^* f_\lambda = \begin{cases} f_\lambda, & \text{if } x \in S_\lambda \\ 0, & \text{if } x \notin S_\lambda, \end{cases}
$$

and using both relations (1.55), one first computes:

$$\begin{aligned}
&\left\langle \exp\Big(\sum_1^\infty s_i\Lambda^{-i}\Big)\prod_{x\in X}\psi_x\psi_x^*\exp\Big(\sum_1^\infty t_i\Lambda^i\Big)f_\varnothing, f_\varnothing\right\rangle\\
&\qquad = \left\langle \exp\Big(\sum_1^\infty s_i\Lambda^{-i}\Big)\prod_{x\in X}\psi_x\psi_x^*\sum_\lambda \mathbf{s}_\lambda(t)f_\lambda, f_\varnothing\right\rangle\\
&\qquad = \sum_\lambda \mathbf{s}_\lambda(t)\left\langle \exp\Big(\sum_1^\infty s_i\Lambda^{-i}\Big)\prod_{x\in X}\psi_x\psi_x^* f_\lambda, f_\varnothing\right\rangle\\
&\qquad = \sum_{\substack{\lambda\\ S(\lambda)\supset X}} \mathbf{s}_\lambda(t)\left\langle \exp\Big(\sum_1^\infty s_i\Lambda^{-i}\Big) f_\lambda, f_\varnothing\right\rangle\\
&\qquad = \sum_{\substack{\lambda\\ S(\lambda)\supset X}} \mathbf{s}_\lambda(t)\sum_{\mu\subset\lambda}\langle \mathbf{s}_{\lambda\backslash\mu}(s)f_\mu, f_\varnothing\rangle\\
&\qquad = \sum_{\substack{\lambda\\ S(\lambda)\supset X}} \mathbf{s}_\lambda(t)\sum_{\mu\subset\lambda}\mathbf{s}_{\lambda\backslash\mu}(s)\langle f_\mu, f_\varnothing\rangle = \sum_{\substack{\lambda \text{ such that}\\ S(\lambda)\supset X}} \mathbf{s}_\lambda(t)\mathbf{s}_\lambda(s)\,.
\end{aligned}$$

Using this fact, one further computes (in the exponentials below the summation $\sum$ stands for $\sum_1^\infty$)

$$\begin{aligned}
&\frac{1}{Z}\sum_{\substack{\lambda \text{ such that}\\ S(\lambda)\supset X}} \mathbf{s}_\lambda(t)\mathbf{s}_\lambda(s)\\
&= \frac{1}{Z}\left\langle \exp\Big(\sum s_i\Lambda^{-i}\Big)\prod_{x\in X}\psi_x\psi_x^*\exp\Big(\sum t_i\Lambda^i\Big)f_\varnothing, f_\varnothing\right\rangle\\
&= \frac{1}{Z}\left\langle \exp\Big(\sum s_i\Lambda^{-i}\Big)\psi_{x_m}\cdots\psi_{x_1}\psi_{x_1}^*\cdots\psi_{x_m}^*\exp\Big(\sum t_i\Lambda^i\Big)f_\varnothing, f_\varnothing\right\rangle\\
&\hspace{20em}\text{using } \psi_{x_i}\psi_{x_j}^* = -\psi_{x_j}^*\psi_{x_i} \text{ for } i\neq j\\
&= \frac{1}{Z}\left\langle \exp\Big(\sum s_i\Lambda^{-i}\Big)\psi_{x_m}\cdots\psi_{x_1}\psi_{x_1}^*\cdots\psi_{x_m}^*\right.\\
&\qquad\left.\times\exp\Big(\sum t_i\Lambda^i\Big)\exp\Big(-\sum s_i\Lambda^{-i}\Big)f_\varnothing, \exp\Big(-\sum t_i\Lambda^{-i}\Big)f_\varnothing\right\rangle\\
&\hspace{24em}\text{using (1.56)}\\
&= \frac{1}{Z}\left\langle \exp\Big(-\sum t_i\Lambda^i\Big)\exp\Big(\sum s_i\Lambda^{-i}\Big)\psi_{x_m}\cdots\psi_{x_1}\psi_{x_1}^*\cdots\psi_{x_m}^*\right.\\
&\qquad\left.\times\exp\Big(\sum t_i\Lambda^i\Big)\exp\Big(-\sum s_i\Lambda^{-i}\Big)f_\varnothing, f_\varnothing\right\rangle\\
&= \left\langle \exp\Big(-\sum t_i\Lambda^i\Big)\exp\Big(\sum s_i\Lambda^{-i}\Big)\psi_{x_m}\cdots\psi_{x_1}\psi_{x_1}^*\cdots\psi_{x_m}^*\right.\\
&\qquad\left.\times\exp\Big(-\sum s_i\Lambda^{-i}\Big)\exp\Big(\sum t_i\Lambda^i\Big)f_\varnothing, f_\varnothing\right\rangle
\end{aligned}$$

using Cauchy's identity (1.52)

$$= \langle \Psi_{x_m} \cdots \Psi_{x_1} \Psi^*_{x_1} \cdots \Psi^*_{x_m} f_\varnothing, f_\varnothing \rangle ,$$

where

$$\Psi_k = \exp\Big(-\sum t_i \Lambda^i\Big) \exp\Big(\sum s_i \Lambda^{-i}\Big) \psi_k \exp\Big(-\sum s_i \Lambda^{-i}\Big) \exp\Big(\sum t_i \Lambda^i\Big)$$
$$\Psi^*_k = \exp\Big(-\sum t_i \Lambda^i\Big) \exp\Big(\sum s_i \Lambda^{-i}\Big) \psi^*_k \exp\Big(-\sum s_i \Lambda^{-i}\Big) \exp\Big(\sum t_i \Lambda^i\Big) .$$

Then, using Lemma 1.2.4, the expression above equals

$$\begin{aligned}\langle \Psi_{x_m} \cdots \Psi_{x_1} \Psi^*_{x_1} \cdots \Psi^*_{x_m} f_\varnothing, f_\varnothing \rangle &= \det(\langle \Psi_{x_k} \Psi^*_{x_l} f_\varnothing, f_\varnothing \rangle)_{1\le k,l\le m} \\ &= \det\big(K(x_k, x_l)\big)_{1\le k,l\le m} ,\end{aligned}$$

where

$$\begin{aligned}K(k,l) &= \langle \Psi_k \Psi^*_l f_\varnothing, f_\varnothing \rangle \\ &= \Big\langle \exp\Big(-\sum t_i \Lambda^i\Big) \exp\Big(\sum s_i \Lambda^{-i}\Big) \psi_k \psi^*_l \\ &\qquad \times \exp\Big(-\sum s_i \Lambda^{-i}\Big) \exp\Big(\sum t_i \Lambda^i\Big) f_\varnothing, f_\varnothing \Big\rangle .\end{aligned}$$

Using

$$\langle \psi_i \psi^*_j f_\varnothing, f_\varnothing \rangle = \begin{cases} 1 & \text{if } i = j < 0 \\ 0 & \text{otherwise} \end{cases}$$

and setting

$$V(z) = -\sum_{i\ge 1} (t_i z^{-i} - s_i z^i) , \tag{1.60}$$

the generating function of the $K(k,l)$ takes on the following simple form:

$$\begin{aligned}&\sum_{k,l\in\mathbb{Z}} z^k w^{-l} K(k,l) \\ &= \Big\langle \exp\Big(-\sum t_i \Lambda^i\Big) \exp\Big(\sum s_i \Lambda^{-i}\Big) \psi(z) \psi^*(w) \\ &\qquad \times \exp\Big(-\sum s_i \Lambda^{-i}\Big) \exp\Big(\sum t_i \Lambda^i\Big) f_\varnothing, f_\varnothing \Big\rangle \\ &= \Big\langle \exp\Big(-\sum t_i \Lambda^i\Big) \exp\Big(\sum s_i \Lambda^{-i}\Big) \psi(z) \exp\Big(-\sum s_i \Lambda^{-i}\Big) \exp\Big(\sum t_i \Lambda^i\Big) \\ &\qquad \times \exp\Big(-\sum t_i \Lambda^i\Big) \exp\Big(\sum s_i \Lambda^{-i}\Big) \psi^*(w) \\ &\qquad \times \exp\Big(-\sum s_i \Lambda^{-i}\Big) \exp\Big(\sum t_i \Lambda^i\Big) f_\varnothing, f_\varnothing \Big\rangle\end{aligned}$$

$$
\begin{aligned}
&= \exp\Big(\sum s_i z^i\Big)\exp\Big(-\sum t_i z^{-i}\Big)\exp\Big(-\sum s_i w^i\Big)\exp\Big(\sum t_i w^{-i}\Big) \\
&\qquad\qquad\qquad\qquad\qquad \times \langle \psi(z)\psi^*(w) f_\varnothing, f_\varnothing\rangle \quad \text{using (1.54)} \\
&= \exp\big(V(z)-V(w)\big)\langle \psi(z)\psi^*(w) f_\varnothing, f_\varnothing\rangle \\
&= \exp\big(V(z)-V(w)\big)\sum_{i,j\in\mathbb{Z}} z^i w^{-j}\langle \psi_i\psi_j^* f_\varnothing, f_\varnothing\rangle \\
&= \exp\big(V(z)-V(w)\big)\sum_{i\geq 1}\Big(\frac{w}{z}\Big)^i \\
&= \exp\big(V(z)-V(w)\big)\Big(\frac{w}{z}\Big)\Big(1+\frac{w}{z}+\cdots\Big) \\
&= \exp\big(V(z)-V(w)\big)\Big(\frac{w}{z}\Big)\frac{1}{1-w/z} \quad \text{for } |w|<|z| \\
&= \exp\big(V(z)-V(w)\big)\frac{w}{z-w},
\end{aligned}
$$

and so

$$
K(k,l) = \Big(\frac{1}{2\pi \mathrm{i}}\Big)^2 \oint_{|w|=\rho<1}\oint_{|z|=\rho^{-1}>1}\frac{\mathrm{d}z\,\mathrm{d}w}{z^{k+1}w^{-l}}\frac{\exp\big(V(z)-V(w)\big)}{z-w},
$$

ending the proof of Prop. 1.2.5. □

Formula (1.41) in the remark is obtained by setting $z\mapsto tz$, $w\mapsto tw$ in (1.59),

$$
K(k,l) = \Big(\frac{1}{2\pi \mathrm{i}}\Big)^2 \frac{1}{t^{k-l}}\oint\oint \frac{\mathrm{d}z\,\mathrm{d}w}{z^{k+1}w^{-l+1}}\frac{\exp\big(V(tz)-V(tw)\big)}{(z/w-1)}.
$$

and taking $(\partial/\partial t)|_{t=1}$ of both sides, using the t-independence of the left-hand side, yielding (1.41). □

1.2.4 Probability on Partitions Expressed as $\mathrm{U}(n)$ Integrals

This section deals with (1.40) in Thm. 1.2.1, stating that $\mathrm{P}(\lambda \mid \lambda_1 \leq n)$ can be expressed as a unitary matrix integral. First we need a lemma, whose proof can be found in [27]:

Lemma 1.2.6. *If f is a symmetric function of the eigenvalues $u_1,\dots,u_n$ of the elements in* $\mathrm{U}(n)$*, then*

$$
\int_{U(n)} f = \frac{1}{n!}\int_{(S^1)^n}|\Delta(u)|^2 f(u_1,\dots,u_n)\prod_1^n \frac{\mathrm{d}u_j}{2\pi \mathrm{i} u_j}.
$$

Proof. Set $G := \mathrm{U}(n)$ and $T := \{\mathrm{diag}(u_1, \ldots, u_n)$, with $u_k = \exp(\mathrm{i}\theta_k) \in S^1\}$, and let $\mathfrak{t}$ and $\mathfrak{g}$ denote the Lie algebras corresponding to T and G. An element in the quotient G/T can be identified with gT, because $g' \in gT$ implies $g^{-1}g' \in T$, and thus there is a natural map

$$T \times (G/T) \to G\colon (t, gT) \mapsto gtg^{-1}.$$

Note that the Jacobian J of this map (with respect to the invariant measures on T, G/T and G) only depends on $t \in \mathfrak{t}$, because of invariance of the measure under conjugation. For a function f as above

$$\int_G f = \frac{1}{n!}\int_{T\times(G/T)} f(gtg^{-1})J(t)\,\mathrm{d}t\,\mathrm{d}(gT) = \frac{\mathrm{vol}(G/T)}{n!}\int_T f(t)J(t)\,\mathrm{d}t\,.$$

Denote the tangent space to G/T at its base points by $\mathfrak{t}^\perp$, the orthogonal complement of $\mathfrak{t}$ in $\mathfrak{g} = \mathfrak{t} \oplus \mathfrak{t}^\perp$. Consider infinitesimal changes $t(1+\varepsilon\xi)$ of t with $\xi \in \mathfrak{t}$; also, an infinitesimal change of $1 \in G$, namely $1 \mapsto 1 + \varepsilon\eta$, with $\eta \in \mathfrak{t}^\perp$. Then

$$\begin{aligned} gtg^{-1} - t &= (1+\varepsilon\eta)t(1+\varepsilon\xi)\big(1 - \varepsilon\eta + \mathbf{O}(\varepsilon^2)\big) - t \\ &= \varepsilon(t\xi + \eta t - t\eta) + \mathbf{O}(\varepsilon^2) =: \varepsilon\rho + \mathbf{O}(\varepsilon^2) \end{aligned}$$

and so

$$t^{-1}\rho = \xi + (t^{-1}\eta t - \eta) \in \mathfrak{t} \oplus \mathfrak{t}^\perp = \mathfrak{g}\,.$$

Thus the Jacobian of this map is given by

$$J(t) = \det(A(t^{-1}) - I)\,,$$

where $A(t)$ denotes the adjoint action by means of the diagonal matrix t, denoted by $t = \mathrm{diag}(u_1, \ldots, u_n)$, with $|u_i| = 1$. Let E_{jk} be the matrix with 1 at the (j,k)th entry and 0 everywhere else. Then

$$(A(t^{-1}) - I)E_{jk} = (u_j^{-1}u_k - 1)E_{jk}$$

and thus the matrix $A(t^{-1}) - I$ has $n(n-1)$ eigenvalues $(u_j^{-1}u_k - 1)$, with $1 \le j,k \le n$ and $j \ne k$. Therefore

$$\det(A(t^{-1}) - I) = \prod_{j\ne k}(u_j^{-1}u_k - 1) = \prod_{j<k}|u_j - u_k|^2\,,$$

using the fact that $u_j^{-1} = \bar{u}_j$. □

Proposition 1.2.3. *Given the "probability measure"*

$$\mathrm{P}(\lambda) = Z^{-1}s_\lambda(t)s_\lambda(s), \quad Z = \exp\Big(\sum_{i\ge1} it_is_i\Big),$$

the following holds

$$\mathrm{P}(\lambda \mid \lambda_1 \le n) = Z^{-1}\int_{\mathrm{U}(n)} \exp\bigg(-\mathrm{Tr}\Big(\sum_1^\infty (t_iX^i + s_i\overline{X}^i)\Big)\bigg)\,\mathrm{d}X\,.$$

Proof. Using Lemma 1.2.6 and the following matrix identity,

$$\sum_{\sigma\in \mathrm{S}_n} \det\big(a_{i,\sigma(j)}\ b_{j,\sigma(j)}\big)_{1\le i,j\le n} = \det\big(a_{ik}\big)_{1\le i,k\le n} \det\big(b_{ik}\big)_{1\le i,k\le n},$$

one computes

$$\begin{aligned}
&n! \int_{\mathrm{U}(n)} \exp\bigg(-\operatorname{Tr}\Big(\sum_1^\infty (t_i X^i + s_i \bar{X}^i)\Big)\bigg)\,\mathrm{d}X \\
&= \int_{(S^1)^n} |\Delta_n(z)|^2 \prod_{k=1}^n \bigg(\exp\Big(-\sum_1^\infty (t_i z_k^i + s_i z_k^{-i})\Big)\frac{\mathrm{d}z_k}{2\pi \mathrm{i} z_k}\bigg) \\
&= \int_{(S^1)^n} \Delta_n(z)\Delta_n(\bar{z}) \prod_{k=1}^n \bigg(\exp\Big(-\sum_1^\infty (t_i z_k^i + s_i z_k^{-i})\Big)\frac{\mathrm{d}z_k}{2\pi \mathrm{i} z_k}\bigg) \\
&= \int_{(S^1)^n} \sum_{\sigma\in \mathrm{S}_n} \det\big(z_{\sigma(m)}^{l-1}\bar{z}_{\sigma(m)}^{m-1}\big)_{1\le l,m\le n} \prod_{k=1}^n \bigg(\exp\Big(-\sum_1^\infty (t_i z_k^i + s_i z_k^{-i})\Big)\frac{\mathrm{d}z_k}{2\pi \mathrm{i} z_k}\bigg) \\
&= \sum_{\sigma\in \mathrm{S}_n} \det\bigg(\oint_{S^1} z_k^{l-1}\bar{z}_k^{m-1} \exp\Big(-\sum_1^\infty (t_i z_k^i + s_i z_k^{-i})\Big)\frac{\mathrm{d}z_k}{2\pi \mathrm{i} z_k}\bigg)_{1\le l,m\le n} \\
&= n!\det\bigg(\oint_{S^1} z^{l-m} \exp\Big(-\sum_1^\infty (t_i z^i + s_i z^{-i})\Big)\frac{\mathrm{d}z}{2\pi \mathrm{i} z}\bigg)_{1\le l,m\le n} = n! \det m_n(t,s)
\end{aligned}$$

ending the proof of Prop. 1.2.3. □

1.3 Examples

1.3.1 Plancherel Measure and Gessel's Theorem

The point of this section will be to restrict the probability

$$\mathrm{P}_{x,y}(\lambda) := \frac{1}{Z}\tilde{\mathbf{s}}_\lambda(x)\tilde{\mathbf{s}}_\lambda(y) = \frac{1}{Z}\mathbf{s}_\lambda(t)\mathbf{s}_\lambda(s) \tag{1.61}$$

considered in Sect. 1.2 ((1.37)) to the locus $\mathcal{L}_1$, defined for real $\xi > 0$ (expressed in (x,y) and (s,t) coordinates),

$$\begin{aligned}
\mathcal{L}_1 &= \bigg\{(x,y) \text{ such that } \sum_{i\ge 1} x_i^k = \sum_{i\ge 1} y_i^k = \delta_{kl}\sqrt{\xi}\bigg\} \\
&= \{\text{all } s_k = t_k = 0, \text{ except } t_1 = s_1 = \sqrt{\xi}\}\,.
\end{aligned} \tag{1.62}$$

The reader is referred back to Sect. 1.1.5.1 for a number of basic formulae.

Theorem 1.3.1 ([8, 23, 45, 86]). *For the permutation group* S_k*, the generating function for distribution of the length of the longest increasing subsequence*

$$\widetilde{\mathrm{P}}^k(L(\pi_k) \le n)$$

is given by

$$\begin{aligned} \mathrm{e}^{-\xi} \sum_{k=0}^{\infty} \frac{\xi^k}{k!} \widetilde{\mathrm{P}}^k(L(\pi_k) \le n) &= \mathrm{e}^{-\xi} \det\left(\oint_{S^1} \frac{\mathrm{d}z}{2\pi \mathrm{i} z} z^{j-l} \exp\left(\sqrt{\xi}(z+z^{-1})\right)\right)_{1\le j,l\le n} \\ &= \mathrm{e}^{-\xi} \det\left(J_{j-l}(2\sqrt{-\xi})\right)_{1\le j,l \le n} \\ &= \mathrm{e}^{-\xi} \int_{U(n)} \exp\left(\sqrt{\xi}\,\mathrm{Tr}(X+X^{-1})\right) \mathrm{d}X \\ &= \det\left(I - K(j,l)|_{[n,n+1,\dots]}\right) \end{aligned}$$

where $J_i(z)$ *is the Bessel function and*

$$\begin{aligned} K(k,l) &= \frac{\sqrt{\xi}\left(J_k(2\sqrt{\xi})J_{l+1}(2\sqrt{\xi}) - J_{k+1}(2\sqrt{\xi})J_l(2\sqrt{\xi})\right)}{k-l} \qquad (k \ne l) \\ &= \sum_{n=1}^{\infty} J_{k+n}(2\sqrt{\xi}) J_{l+n}(2\sqrt{\xi}) \,. \end{aligned} \tag{1.63}$$

Proof. For an arbitrary partition $\lambda \in \mathbb{Y}$, and using (1.9), the restriction of $P_{x,y}$ to the locus $\mathcal{L}_1$, as in (1.62), reads as follows:

$$\begin{aligned} \mathrm{P}^{\xi}(\lambda) := \mathrm{P}_{x,y}(\lambda)|_{\mathcal{L}_1} &= \exp\left(-\sum_{k\ge 1} k t_k s_k\right) \mathrm{s}_\lambda(t)\mathrm{s}_\lambda(s)\Bigg|_{\substack{t_i=\sqrt{\xi}\delta_{i1}\\ s_i=\sqrt{\xi}\delta_{i1}}} \\ &= \mathrm{e}^{-\xi}\xi^{|\lambda|/2}\frac{f^\lambda}{|\lambda|!}\xi^{|\lambda|/2}\frac{f^\lambda}{|\lambda|!} = \mathrm{e}^{-\xi}\frac{\xi^{|\lambda|}}{|\lambda|!}\frac{(f^\lambda)^2}{|\lambda|!} \\ &= \mathrm{e}^{-\xi}\frac{\xi^{|\lambda|}}{|\lambda|!}\widetilde{\mathrm{P}}^{(n)}(\lambda)\,, \quad \text{for } n = |\lambda|, \end{aligned}$$

where $\widetilde{\mathrm{P}}^{(n)}(\lambda)$ can be recognized as *Plancherel measure* on partitions in $\mathbb{Y}_n$, as defined in Sect. 1.1.5.1,

$$\widetilde{\mathrm{P}}^n(\lambda) = \frac{(f^\lambda)^2}{n!}, \quad \lambda \in \mathbb{Y}_n \,.$$

It is clear that

$$\mathrm{P}^{\xi}(\lambda) = \mathrm{e}^{-\xi}\frac{\xi^{|\lambda|}}{|\lambda|!}\left(\frac{(f^\lambda)^2}{|\lambda|!}\right), \quad \lambda \in \mathbb{Y}\,,$$

is a genuine probability (≥ 0), called *Poissonized Plancherel measure.*

We now compute

$$\begin{aligned}
P^{\xi}(\lambda \text{ with } \lambda_1 \le n) &= e^{-\xi} \sum_{\substack{\lambda \in \mathbb{Y} \\ \lambda_1 \le n}} \frac{\xi^{|\lambda|}}{|\lambda|!} \frac{(f^{\lambda})^2}{|\lambda|!} \\
&= e^{-\xi} \sum_{k=0}^{\infty} \frac{\xi^k}{k!} \sum_{\substack{|\lambda|=k \\ \lambda_1 \le n}} \frac{(f^{\lambda})^2}{|\lambda|!} = e^{-\xi} \sum_{k=0}^{\infty} \frac{\xi^k}{k!} \widetilde{P}^{(k)}(L(\pi_k) \le n) ,
\end{aligned}$$

and thus using Thm. 1.2.1,[9]

$$\begin{aligned}
&P^{\xi}(\lambda \text{ with } \lambda_1 \le n) \\
&\quad = Z^{-1} \det \left(\oint_{S^1} \frac{dz}{2\pi i z} z^{k-l} \exp\left(- \sum_1^{\infty} (t_i z^i + s_i z^{-i}) \right) \right)_{1 \le k,l \le n} \Bigg|_{\mathcal{L}_1} \\
&\quad = e^{-\xi} \det \left(\oint_{S^1} \frac{dz}{2\pi i z} z^{k-l} \exp(-\sqrt{\xi}(z + z^{-1})) \right)_{1 \le k,l \le n} \\
&\quad = e^{-\xi} \int_{U(n)} \exp(\sqrt{\xi} \operatorname{Tr}(X + X^{-1})) \, dX \\
&\quad = e^{-\xi} \det(J_{k-l}(2\sqrt{-\xi}))_{1 \le k,l \le n} ,
\end{aligned}$$

where we used the fact that for a Toeplitz matrix

$$\det(a^{k-l} c_{k-l})_{1 \le k,l \le n} = \det(c_{k-l})_{1 \le k,l \le n} , \quad a \ne 0 .$$

It also equals

$$P^{\xi}(\lambda \text{ with } \lambda_1 \le n) = \det(I - K(i,j)|_{[n,n+1,\dots]})$$

where $K(i,j)$ is given by (1.41), where $V(z) = \sqrt{\xi}(z - z^{-1})$. Since

$$\frac{z(d/dz)V(z) - w(d/dw)V(w)}{z - w} = \sqrt{\xi}\left(1 - \frac{1}{wz}\right) ,$$

one checks

[9] The Bessel function $J_n(u)$ is defined by

$$\exp(u(t - t^{-1})) = \sum_{-\infty}^{\infty} t^n J_n(2u)$$

and thus

$$\exp(-\sqrt{\xi}(z + z^{-1})) = \exp\left(\sqrt{-\xi}((iz) - (iz)^{-1})\right) = \sum (iz)^n J_n(2\sqrt{-\xi}) .$$

$$
\begin{aligned}
&(k-l)K(k,l)\\
&= \frac{\sqrt{\xi}}{(2\pi\mathrm{i})^2}\oint\oint_{\substack{|z|=c_1\\|w|=c_2}} \mathrm{d}z\,\mathrm{d}w\Bigg(\frac{\exp\big(\sqrt{\xi}(z-z^{-1})\big)}{z^{k+1}}\frac{\exp\big(\sqrt{\xi}(w^{-1}-w)\big)}{w^{-l}}\\
&\qquad\qquad\qquad\qquad - \frac{\exp\big(\sqrt{\xi}(z-z^{-1})\big)}{z^{k+2}}\frac{\exp\big(\sqrt{\xi}(w^{-1}-w)\big)}{w^{-l+1}}\Bigg)\\
&= \sqrt{\xi}\big(J_k(2\sqrt{\xi})J_{l+1}(2\sqrt{\xi}) - J_{k+1}(2\sqrt{\xi})J_l(2\sqrt{\xi})\big)\\
&= (k-l)\sum_{n=1}^{\infty} J_{k+n}(2\sqrt{\xi})J_{l+n}(2\sqrt{\xi})\ . \qquad (1.64)
\end{aligned}
$$

The last equality follows from the recurrence relation of Bessel functions

$$
J_{k+1}(2z) = \frac{k}{z}J_k(2z) - J_{k-1}(2z)\ .
$$

Indeed, subtracting the two expressions

$$
\begin{aligned}
J_{l+1}(2z)J_k(2z) &= \frac{l}{z}J_k(2z)J_l(2z) - J_{l-1}(2z)J_k(2z)\\
J_{k+1}(2z)J_l(2z) &= \frac{k}{z}J_k(2z)J_l(2z) - J_{k-1}(2z)J_l(2z)\ ,
\end{aligned}
$$

one finds

$$
\begin{aligned}
(l-k)J_k(2z)J_l(2z) &= z\big(J_k(2z)J_{l+1}(2z) - J_{k+1}(2z)J_l(2z)\big)\\
&\quad - z\big(J_{k-1}(2z)J_l(2z) - J_k(2z)J_{l-1}(2z)\big)\ ,
\end{aligned}
$$

implying

$$
z\big(J_k(2z)J_{l+1}(2z) - J_{k+1}(2z)J_l(2z)\big) = (k-l)\sum_{n=1}^{\infty} J_{k+n}(2z)J_{l+n}(2z)\ ,
$$

thus proving (1.64). □

Remark. Incidentally, the fact that P^ξ is a probability shows that Plancherel measure itself is a probability; indeed, for all ξ,

$$
\begin{aligned}
\mathrm{e}^{-\xi}\sum_{n=0}^{\infty}\frac{\xi^n}{n!} = 1 = \sum_{\lambda\in\mathbb{Y}}\mathrm{P}^\xi(\lambda) &= \mathrm{e}^{-\xi}\sum_{n=0}^{\infty}\frac{\xi^n}{n!}\sum_{|\lambda|=n}\frac{(f^\lambda)^2}{|\lambda|!}\\
&= \mathrm{e}^{-\xi}\sum_{n=0}^{\infty}\frac{\xi^n}{n!}\sum_{\lambda\in\mathbb{Y}_n}\widetilde{\mathrm{P}}^n(\lambda)\ ;
\end{aligned}
$$

comparing the extremities leads to $\sum_{\lambda\in\mathbb{Y}_n}\widetilde{\mathrm{P}}^n(\lambda) = 1$.

1.3.2 Probability on Random Words

Here also, restrict the probability (1.61) to the locus below, for $\xi > 0$ and $p \in \mathbb{Z}_{\geq 1}$,

$$\begin{aligned}\mathcal{L}_2 &= \left\{\sum_{i\geq 1} x_i^k = \delta_{k1}\xi, y_1 = \cdots = y_p = \beta, \text{all other } y_j = 0\right\} \\ &= \{t_k = \delta_{k1}\xi, ks_k = p\beta^k\}\end{aligned}$$

Recall from Sect. 1.1.5.4 the probability $\widetilde{\mathrm{P}}^{k,p}$ on partitions, induced from uniform probability $\mathrm{P}^{k,p}$ on words. This was studied by Tracy–Widom [87] and Borodin–Okounkov [23]; see also Johansson [58] and Adler–van Moerbeke [8].

Theorem 1.3.2. *For the set of words S_n^p, the generating function for the distribution of the length of the longest increasing subsequence, is given by (setting $\beta = 1$)*

$$\begin{aligned}\mathrm{e}^{-p\xi}\sum_{k=0}^{\infty}\frac{(p\xi)^k}{k!}\widetilde{\mathrm{P}}^{k,p}(L(\pi_k)\leq n) &= \mathrm{e}^{-p\xi}\det\left(\oint_{S'}\frac{\mathrm{d}z}{2\pi\mathrm{i}z}z^{k-l}\exp(\xi z^{-1})(1+z)^p\right)_{1\leq k,l\leq n} \\ &= \mathrm{e}^{-p\xi}\int_{\mathrm{U}(n)}\exp(\xi\operatorname{Tr}\overline{M})\det(I+M)^p\,\mathrm{d}M \\ &= \det\big(I-K(j,k)\big)_{(n,n+1,\dots)},\end{aligned}$$

with $K(j,k)$ a Christoffel–Darboux kernel of Charlier polynomials:

$$\begin{aligned}&(j-k)K(j,k) \\ &= \frac{\xi}{(2\pi\mathrm{i})^2}\oint\oint_{\substack{|z|=c_1<1/|\xi| \\ |w|=c_2}}\mathrm{d}z\,\mathrm{d}w\left(p\frac{(1-\xi z)^{-p-1}\exp(-\xi z^{-1})}{z^{j+1}}\frac{(1-\xi w)^{p-1}\exp(\xi w^{-1})}{w^{-k}}\right. \\ &\qquad\left.-\frac{(1-\xi z)^{-p}\exp(-\xi z^{-1})}{z^{j+2}}\frac{(1-\xi w)^{p}\exp(\xi w^{-1})}{w^{-k+1}}\right) \\ &= (p)_{j+1}\exp(-\xi^2)\left(\frac{{}_1F_1(-p,j+1;\xi^2)}{j!}\frac{{}_1F_1(-p+1,k+2;\xi^2)}{(k+1)!}\right. \\ &\qquad\left.-\frac{{}_1F_1(-q+1,j+2;\xi^2)}{(j+1)!}\frac{{}_1F_1(-q,k+1;\xi^2)}{k!}\right)\end{aligned} \tag{1.65}$$

where

$$(a)_j := a(a+1)\cdots(a+j-1)\,, \quad (a)_0 = 1$$

and ${}_1F_1(a,c;x)$ is the confluent hypergeometric function:

$$\begin{aligned}\frac{1}{2\pi\mathrm{i}}\oint_{|z|=c_1<1/|\xi|}(1-\xi z)^{-p}\exp(\xi z^{-1})\frac{\mathrm{d}z}{z^{m+1}} &= \frac{(p)_m}{m!}\exp(-\xi^2){}_1F_1(1-p,m+1;\xi^2) \\ \frac{1}{2\pi\mathrm{i}}\oint_{|w|=c_2}(1-\xi w)^{q}\exp(\xi w^{-1})w^{m-1}\,\mathrm{d}w &= \frac{1}{m!}{}_1F_1(-q,m+1;\xi^2)\,.\end{aligned}$$

These functions are related to Charlier polynomials.[10]

Proof. This proof will be very sketchy, as more details will be given for the percolation case in the next section. One now computes

$$\begin{aligned}
\mathrm{P}^{\xi,p}(\lambda \text{ with } \lambda_1 \le n) &= \sum_{\substack{\lambda \\ \lambda_1 \le n}} \exp\Big(-\sum_{i\ge 1} it_i s_i\Big) \mathbf{s}_\lambda(t)\mathbf{s}_\lambda(s)\Big|_{\mathcal{L}_2} \\
&= \mathrm{e}^{-p\xi\beta} \sum_{\substack{\lambda \\ \lambda_1 \le n}} \mathbf{s}_\lambda(\xi, 0, \dots)\mathbf{s}_\lambda\Big(p\beta, \frac{p\beta^2}{2}, \frac{p\beta^3}{3}, \dots\Big) \\
&= \mathrm{e}^{-p\xi\beta} \sum_{\substack{\lambda \in \mathbb{Y}^p \\ \lambda_1 \le n}} \frac{(p\xi\beta)^{|\lambda|}}{|\lambda|!} \frac{\mathbf{s}_\lambda(1, 0, \dots)\mathbf{s}_\lambda(p, p/2, p/3, \dots)}{p^{|\lambda|}} \\
&= \mathrm{e}^{-p\xi\beta} \sum_{k=0}^{\infty} \frac{(p\xi\beta)^k}{k!} \sum_{\substack{\lambda \in \mathbb{Y}^p \\ |\lambda| = k \\ \lambda_1 \le n}} \frac{f^\lambda s_\lambda(1^p)}{p^{|\lambda|}} \\
&= \mathrm{e}^{-p\xi\beta} \sum_{k=0}^{\infty} \frac{(p\xi\beta)^k}{k!} \widetilde{\mathrm{P}}^{k,p}(L(\pi_k) \le n) \, .
\end{aligned}$$

In applying Thm. 1.2.1, one needs to compute

$$\exp\Big(-\sum_1^\infty (t_i z^i + s_i z^{-i})\Big)\Big|_{\mathcal{L}_2} = \mathrm{e}^{-\xi z} \exp\Big(-p\sum_1^\infty \frac{1}{i}\Big(\frac{\beta}{z}\Big)^i\Big) = \mathrm{e}^{-\xi z}\Big(1 - \frac{\beta}{z}\Big)^p$$

and

$$\mathrm{e}^V\big|_{\mathcal{L}_2} = \exp\Big(-\sum_1^\infty (t_i z^{-i} - s_i z^i)\Big)\Big|_{\mathcal{L}_2} = \exp(-\xi z^{-1})(1 - \beta z)^{-p} \, .$$

Therefore

[10] Charlier polynomials $P(k;\alpha)$, with $k \in \mathbb{Z}_{\ge 0}$, are discrete orthonormal polynomials defined by the orthonormality condition

$$\sum_{k=0}^\infty P_n(k;\alpha)P_m(k;\alpha)w_\alpha(k) = \delta_{nm} \, , \quad \text{for } w_\alpha(k) = \mathrm{e}^{-\alpha}\frac{\alpha^k}{k!} \, ,$$

with generating function

$$\sum_{n=0}^\infty \alpha^{n/2} \frac{1}{\sqrt{n!}} P_n(k;\alpha) w^n = \mathrm{e}^{-\alpha w}(1+w)^k \, .$$

$$\begin{aligned}
\mathrm{P}^{\xi,p}(\lambda_1 \le n) &= Z^{-1} \det\left(\oint_{S^1} \frac{\mathrm{d}z}{2\pi \mathrm{i} z} z^{k-l} \exp\left(-\sum_1^\infty (t_i z^i + s_i z^{-i})\right)\right)_{1\le k,l\le n}\Bigg|_{\mathcal{L}_2} \\
&= \mathrm{e}^{-\xi p\beta} \det\left(\oint_{S^1} \frac{\mathrm{d}z}{2\pi \mathrm{i} z} z^{k-l} \mathrm{e}^{-\xi z}(1-\beta z^{-1})^p\right)_{1\le k,l\le n} \\
&= \mathrm{e}^{-\xi p\beta} \det\left(\oint_{S^1} \frac{\mathrm{d}z}{2\pi \mathrm{i} z} z^{k-l} \exp(\xi z^{-1})(1+\beta z)^p\right)_{1\le k,l\le n} \\
&\qquad\qquad \text{using the change of variable } z \mapsto -z^{-1} \\
&= \mathrm{e}^{-\xi p\beta} \int_{\mathrm{U}(n)} \exp(\xi \operatorname{Tr} \overline{M}) \det(1+\beta M)^p \,\mathrm{d}M \,.
\end{aligned}$$

Then, one computes

$$\frac{z(\mathrm{d}/\mathrm{d}z)V(z) - w(\mathrm{d}/\mathrm{d}w)V(w)}{z-w} = \frac{\beta p}{(1-\beta z)(1-\beta w)} - \frac{\xi}{zw} \,.$$

and

$$\exp\big(V(z) - V(w)\big) = \exp(-\xi z^{-1})(1-\beta z)^{-p} \exp(\xi w^{-1})(1-\beta w)^p ,$$

leading to (1.65), upon using (1.59), combined an appropriate rescaling involving β. From footnote 10, the confluent hypergeometric functions turn out to be Charlier polynomials in this case. □

1.3.3 Percolation

Considering now the locus

$$\mathcal{L}_3 := \{kt_k = q\xi^{k/2}, ks_k = p\xi^{k/2}\} \,,$$

one is led to the probability appearing in the generalized permutations and percolations (Sects. 1.1.5.4 and 1.1.5.5), namely

$$\mathrm{P}(L(M) \le l) = \sum_{\substack{\lambda \\ \lambda_1 \le l}} (1-\xi)^{pq} \xi^{|\lambda|} \mathbf{s}_\lambda\left(q, \frac{q}{2}, \dots\right) \mathbf{s}_\lambda\left(p, \frac{p}{2}, \dots\right) .$$

We now state: (see [23, 57])

Theorem 1.3.3. *Assuming $q > p$, we have*

$$\begin{aligned}
&\mathrm{P}(L(M) \le l) \\
&\quad = \frac{(1-\xi)^{pq}}{l!} \int_{(S^1)^l} |\Delta_l(z)|^2 \prod_{j=1}^l (1+\sqrt{\xi} z_j)^q (1+\sqrt{\xi}\bar{z}_j)^p \frac{\mathrm{d}z_j}{2\pi \mathrm{i} z_j} \\
&\quad = (1-\xi)^{pq} \int_{\mathrm{U}(l)} \det(1+\sqrt{\xi} M)^q \det(1+\sqrt{\xi}\overline{M})^p \,\mathrm{d}M \\
&\quad = \det\big(I - K(i,j)\big)\big|_{[l,l+1,\dots]}
\end{aligned}$$

where (C is a constant depending on p, q, ξ*)*

$$K(i,j) = C\frac{M_p(p+i)M_{p-1}(p+j) - M_{p-1}(p+i)M_p(p+i)}{i-j} ,$$

where

$$M_p(k) := M_p(k; q-p+1, \xi) , \quad \text{with } k \in \mathbb{Z}_{\geq 0} ,$$

are Meixner polynomials.[11]

Proof. Using the restriction to locus $\mathcal{L}_3$, one finds

$$\exp\Big(-\sum_1^\infty (t_j z^j + s_j z^{-j})\Big)\Big|_{\mathcal{L}_3} = (1-\sqrt{\xi}z)^q (1-\sqrt{\xi}z^{-1})^p$$

and

$$\mathrm{e}^V\big|_{\mathcal{L}_3} = \exp\Big(-\sum_1^\infty (t_i z^{-i} - s_i z^i)\Big)\Big|_{\mathcal{L}_3} = (1-\sqrt{\xi}z^{-1})^q (1-\sqrt{\xi}z)^{-p} ,$$

and thus

$$V(z) = q\log(1-\sqrt{\xi}\, z^{-1}) - p\log(1-\sqrt{\xi}\, z) .$$

Hence, in view of (1.59)

$$\frac{z(\mathrm{d}/\mathrm{d}z)V(z) - w(\mathrm{d}/\mathrm{d}w)V(w)}{z-w} = \frac{p\sqrt{\xi}}{(1-\sqrt{\xi}z)(1-\sqrt{\xi}w)} - \frac{q\sqrt{\xi}}{zw(1-\sqrt{\xi}z^{-1})(1-\sqrt{\xi}w^{-1})}$$

and so

$$\begin{aligned}\frac{1}{z^{k+1}w^{-l}}&\frac{z(\mathrm{d}/\mathrm{d}z)V(z) - w(\mathrm{d}/\mathrm{d}w)V(w)}{z-w}\exp\big(V(z)-V(w)\big)\\ &= \sqrt{\xi}\Big(p\frac{(1-\sqrt{\xi}z)^{-p-1}(1-\sqrt{\xi}z^{-1})^q}{z^{k+1}}\frac{(1-\sqrt{\xi}w)^{p-1}(1-\sqrt{\xi}w^{-1})^{-q}}{w^{-l}}\\ &\qquad - q\frac{(1-\sqrt{\xi}z)^{-p}(1-\sqrt{\xi}z^{-1})^{q-1}}{z^{k+2}}\frac{(1-\sqrt{\xi}w)^{p}(1-\sqrt{\xi}w^{-1})^{-q-1}}{w^{-l+1}}\Big) .\end{aligned}$$

---

[11] Meixner polynomials are discrete orthogonal polynomials defined on $\mathbb{Z}_{\geq 0}$ by the orthogonality relations

$$\sum_{k=0}^\infty \binom{\beta+k-1}{k})\xi^k M_p(k;\beta,\xi)M_{p'}(k;\beta,\xi) = \frac{\xi^{-p}}{(1-\xi)^\beta}\binom{\beta+p-1}{p}^{-1}\delta_{pp'} ,$$
$$\beta > 0 ,$$

and with generating function

$$\sum_{p=0}^\infty \frac{(\beta)_p z^p}{p!}M_p(x;\beta,\xi) = \Big(1-\frac{z}{\xi}\Big)^x (1-z)^{-x-\beta} . \tag{1.66}$$

Expanding in Laurent series, one finds

$$
\begin{aligned}
&(1-\eta z)^{-\beta}(1-\eta z^{-1})^{\beta'}\\
&=\left(\sum_{i=0}^{\infty}\frac{(-\beta)(-\beta-1)\cdots(-\beta-i+1)}{i!}(-\eta)^i z^i\right)\\
&\qquad\qquad\times\left(\sum_{j=0}^{\infty}\frac{\beta'(\beta'-1)\cdots(\beta'-j+1)}{j!}(-\eta)^j\frac{1}{z^j}\right)\\
&=\sum_{m\in\mathbb{Z}} z^m \sum_{\substack{i-j=m\\ i,j\geq 0}}\frac{\beta(\beta+1)\cdots(\beta+i-1)}{i!}\frac{(-\beta')(-\beta'+1)\cdots(-\beta'+j-1)}{j!}\eta^{i+j}\\
&=\sum_{m\in\mathbb{Z}} z^m \sum_{j=0}^{\infty}\frac{\beta(\beta+1)\cdots(\beta+j+m-1)}{(j+m)!}\frac{(-\beta')\cdots(-\beta'+j-1)}{j!}\eta^{2j+m}\,.
\end{aligned}
$$

The coefficient of z^m for $m\geq 0$ reads

$$
\begin{aligned}
\sum_{j=0}^{\infty}&\frac{\beta(\beta+1)\cdots(\beta+j+m-1)(-\beta')\cdots(-\beta'+j-1)}{(j+m)!}\frac{\eta^{2j+m}}{j!}\\
&\qquad=\frac{(\beta)_m}{m!}\eta^m{}_2F_1(\beta+m,-\beta';m+1;\eta^2)\,,
\end{aligned}
$$

which is Gauss' hypergeometric function[12] and thus

$$
\oint_{z=0}(1-\eta z)^{-\beta}(1-\eta z^{-1})^{\beta'}\frac{\mathrm{d}z}{2\pi \mathrm{i} z^{k+1}}=\frac{(\beta)_k}{k!}\eta^k{}_2F_1(-\beta',\beta+k;k+1;\eta^2)\,.
$$

One does a similar computation for the other piece in the kernel, where one computes the coefficient of z^{-m}. Using this fact, using a standard linear transformation formula[13] for the hypergeometric function, and using a polynomial property for hypergeometric functions[14] and the formula for $0<p<q$

$$
\frac{\Gamma(1+p-q)}{\Gamma(1-q)}=(-1)^{p+1}\frac{(q-1)!}{(q-p-1)!}\,,
$$

one finds Meixner polynomials (assume $q>p>0$, $x\geq 0$ and $k\geq 0$) (see Nikiforov–Suslov–Uvarov [71] and Koekoek–Swartouw [65]):

---

12 ${}_2F_1(a,b;c;z):=\sum_0^\infty\big((a)_n(b)_n/(c)_n\big)z^n/n!$. Notice that when $a=-m<0$ is an integer, then ${}_2F_1(a,b;c;z)$ is a polynomial of degree m.

13 In general one has ${}_2F_1(a,b;c;z)=(1-z)^{-b}{}_2F_1\big(b,c-a;c;z/(z-1)\big)$.

14 When p is a positive integer ($<q$) and k an integer, then ${}_2F_1(-p,-q;k+1;z)$ is a polynomial of degree p in z, which satisfies ${}_2F_1(-p,-q;k+1;z)=\Gamma(k+1)\Gamma(1-q+p)/\big((k+p)!\Gamma(1-q)\big)(-z)^p{}_2F_1(-p,-k-p;q-p+1;1/z)$.

$$\begin{aligned}
&\frac{1}{2\pi \mathrm{i}}\oint_{|\sqrt{\xi}|<|z|=c_1}\frac{(1-\sqrt{\xi}z)^{-p-1}(1-\sqrt{\xi}z^{-1})^q}{z^{k+1}}\,\mathrm{d}z\\
&\quad=\frac{(p+1)_k}{k!}\xi^{k/2}{}_2F_1(p+1+k,-q;k+1;\xi)\\
&\quad=\frac{(p+1)_k}{k!}\xi^{k/2}(1-\xi)^q{}_2F_1\left(-p,-q;k+1;\frac{\xi}{\xi-1}\right)\\
&\quad=\frac{(p+1)_k\Gamma(1+p-q)}{(k+p)!\Gamma(1-q)}\frac{\xi^{p+k/2}}{(1-\xi)^{p-q}}{}_2F_1\left(-p,-k-p;q-p+1;\frac{\xi-1}{\xi}\right)\\
&\quad=\frac{\Gamma(1+p-q)}{p!\Gamma(1-q)}\frac{\xi^{p+k/2}}{(1-\xi)^{p-q}}M_p(p+k;q-p+1;\xi)\\
&\quad=\binom{q-1}{p}\frac{\xi^{p+k/2}}{(1-\xi)^{p-q}}M_p(p+k;q-p+1;\xi)\,,
\end{aligned}$$

where

$$\begin{aligned}
M_p(x;\beta,\xi)={}_2F_1\left(-p,-x;\beta;\frac{\xi-1}{\xi}\right)&=\left(\frac{1-\xi}{\xi}\right)^p\frac{1}{(\beta)_p}x^p+\cdots\,,\qquad x\in\mathbb{Z}_{\geq 0}\\
&=: a_px^p+\dots\,,\qquad \text{with } \beta=q-p+1
\end{aligned}$$

are Meixner polynomials in x, satisfying the following orthogonality properties:

$$\sum_{x=0}^{\infty} w(x)M_p(x;\beta,\xi)M_m(x;\beta,\xi)=\frac{p!}{\xi^p(1-\xi)^\beta(\beta)_p}\delta_{pm}=: h_p\delta_{pm}\,,$$

with weight

$$w(x):=\frac{(\beta)_x}{x!}\xi^x=\frac{(1+q-p)_x}{x!}\xi^x\,.$$

Using these facts, one computes

$$\begin{aligned}
&(k-l)K(k,l)\\
&=\frac{\sqrt{\xi}}{(2\pi\mathrm{i})^2}\iint_{\substack{|\sqrt{\xi}|<|z|=c_1\\ |w|=c_2<|1/\sqrt{\xi}|}}\mathrm{d}z\,\mathrm{d}w\left(p\frac{(1-\sqrt{\xi}z)^{-p-1}(1-\sqrt{\xi}z^{-1})^q}{z^{k+1}}\frac{(1-\sqrt{\xi}w)^{p-1}(1-\sqrt{\xi}w^{-1})^{-q}}{w^{-l}}\right.\\
&\qquad\left.-\,q\frac{(1-\sqrt{\xi}z)^{-p}(1-\sqrt{\xi}z^{-1})^{q-1}}{z^{k+2}}\frac{(1-\sqrt{\xi}w)^{p}(1-\sqrt{\xi}w^{-1})^{-q-1}}{w^{-l+1}}\right)\\
&=\frac{a_{p-1}}{a_ph_{p-1}}\sqrt{w(p+k)w(p+l)}\\
&\qquad\times\big(M_p(p+k)M_{p-1}(p+l)-M_{p-1}(p+k)M_p(p+l)\big)\,,
\end{aligned}$$

where

$$M_p(p+k):=M_p(p+k;q-p+1,\xi)\,.$$

We also have, using Props. 1.2.1 and 1.2.3 restricted to the locus $\mathcal{L}_3$,

$$
\begin{aligned}
&\mathrm{P}(L(M) \le l)\\
&= \sum_{\substack{\lambda\\ \lambda_1 \le l}} (1-\xi)^{pq} \xi^{|\lambda|} \mathbf{s}_\lambda\Big(q, \frac{q}{2}, \dots\Big) \mathbf{s}_\lambda\Big(p, \frac{p}{2}, \dots\Big)\\
&= \exp\Big(-\sum_1^\infty k t_k s_k\Big) \sum_{\substack{\lambda\\ \lambda_1 \le l}} \mathbf{s}_\lambda(t_1, t_2, \dots) \mathbf{s}_\lambda(s_1, s_2, \dots)\Big|_{\substack{kt_k = q\xi^{k/2}\\ ks_k = p\xi^{k/2}}}\\
&= (1-\xi)^{pq} \det\Big(\oint_{S^1} z^{\alpha-\alpha'} \exp\Big(-\sum_1^\infty (t_j z^j + s_j z^{-j})\Big) \frac{\mathrm{d}z}{2\pi \mathrm{i} z}\Big)_{1 \le \alpha, \alpha' \le l}\Big|_{\substack{kt_k = q\xi^{k/2}\\ ks_k = p\xi^{k/2}}}\\
&= (1-\xi)^{pq} \det\Big(\oint_{S^1} z^{\alpha-\alpha'} (1-\sqrt{\xi} z)^q (1-\sqrt{\xi} z^{-1})^p \frac{\mathrm{d}z}{2\pi \mathrm{i} z}\Big)_{1 \le \alpha, \alpha' \le l}\\
&= \frac{(1-\xi)^{pq}}{k!} \int_{(S^1)^l} |\Delta_l(z)|^2 \prod_{j=1}^k (1+\sqrt{\xi} z_j)^q (1+\sqrt{\xi} \bar{z}_j)^p \frac{\mathrm{d}z_j}{2\pi \mathrm{i} z_j}\\
&= (1-\xi)^{pq} \int_{\mathrm{U}(l)} \det(1+\sqrt{\xi} M)^q \det(1+\sqrt{\xi} \overline{M})^p \, \mathrm{d}M \ ,
\end{aligned}
$$

thus ending the proof of Thm. 1.3.3. □

1.4 Limit Theorems

1.4.1 Limit for Plancherel Measure

Stanislaw Ulam [89] raised in 1961 the question:

> How do you compute the probability $\mathrm{P}^{(n)}(L(\pi_n) \le k)$ that the length $L := L^{(n)}$ of the longest increasing sequence in a random permutation is smaller than k. What happens for very large permutations, i.e., when $n \to \infty$?

By Monte Carlo simulations, Ulam conjectured that

$$
c := \lim_{n \to \infty} \frac{\mathrm{E}^{(n)}(L)}{\sqrt{n}} \tag{1.67}
$$

exists ($\mathrm{E}^{(n)}$ denote the expectation with respect to $\mathrm{P}^{(n)}$). A much older argument of Erdős and Szekeres [35] implied that $\mathrm{E}^{(n)}(L) \ge \frac{1}{2}\sqrt{n-1}$ and so $c \ge \frac{1}{2}$. Numerical computation by Baer and Brock [17] suggested $c = 2$. Hammersley [50] showed the existence of the limit (1.67); in 1977, Logan and Shepp [67] proved $c \ge 2$ and, at the same time, Vershik and Kerov [93] showed $c = 2$. More recently other proofs have appeared by Aldous and Diaconis [14], Seppäläinen [79] and Johansson [56]. Meanwhile, Gessel [45] found a generating function for the probability (with respect to n) and connected

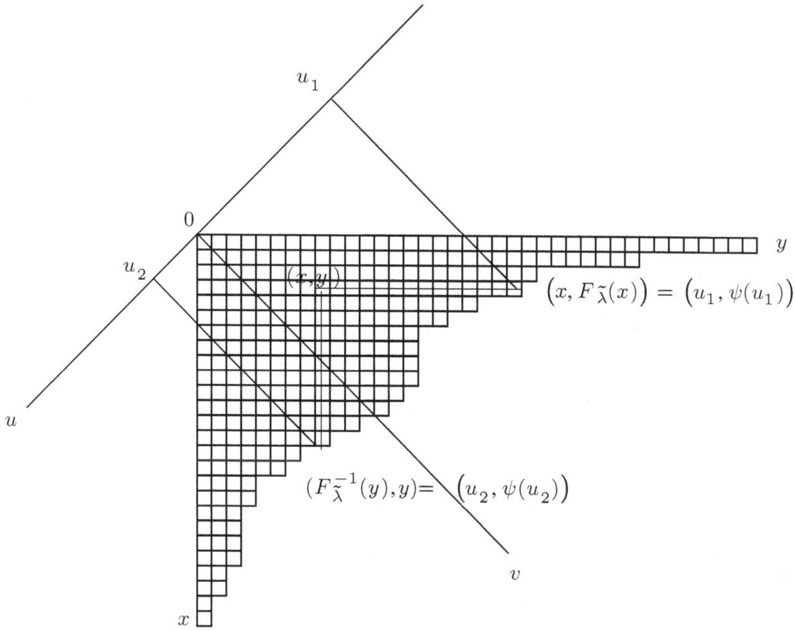

Fig. 1.5.

this problem with Toeplitz determinants. Monte Carlo simulations by Odlyzko and Rains [72] suggested

$$c_0 \sim \lim_{n\to\infty} \frac{\operatorname{Var} L^{(n)}}{n^{1/3}} \qquad c_1 = \lim \frac{\mathrm{E}^{(n)}(L^{(n)}) - 2\sqrt{n}}{n^{1/6}} \,,$$

with $c_0 \sim 0.819$ and $c_1 \sim -1.758$.

In this section, we explain some of the ideas underlying this problem. It is convenient to write a partition $\lambda \in \mathbb{Y}$ in (u, v)-coordinates, as shown in Fig. 1.5.

Remember from (1.7) the hook length h^λ, defined as

$$h^\lambda = \prod_{(i,j)\in\lambda} h_{ij}^\lambda, \quad \text{with } h_{ij}^\lambda = \text{hook length} = \lambda_i + \lambda_j^\top - i - j - 1,$$

where $\lambda_j^\top$ is the length of the jth column. Also remember from (1.8) the formula f^λ expressed in terms of h^λ. Then one has the following theorem, due to Vershik and Kerov [92]; see also Logan and Shepp [67]. A sketchy outline of the proof will be given here.

Consider Plancherel measure (see Sect. 1.3.1), which using (1.8), can be written in terms of the hook length,

$$\widetilde{\mathrm{P}}^{(n)}(\lambda) = \frac{(f^\lambda)^2}{n!} = \frac{n!}{(h^\lambda)^2} \quad \text{for } |\lambda| = n,$$

and define the function

$$\Omega(u) := \begin{cases} 2(u \arcsin(u/2) + \sqrt{4-u^2})/\pi, & \text{for } |u| \le 2 \\ |u|, & \text{for } |u| \ge 2. \end{cases} \tag{1.68}$$

Theorem 1.4.1 (Vershik–Kerov [92], Logan–Shepp [67]). *Upon expressing λ in (u,v)-coordinates, define the subset of partitions, for any $\varepsilon > 0$,*

$$\mathbb{Y}_n(\varepsilon) := \left\{ \lambda \in \mathbb{Y} \;\middle|\; \sup_u \left| \frac{1}{\sqrt{n}} \lambda(u\sqrt{n}) - \Omega(u) \right| < \varepsilon \right\}.$$

Then, for large n, Plancherel measure concentrates on a Young diagram whose boundary has shape $\Omega(u)$; i.e.,

$$\lim_{n\to\infty} \widetilde{\mathrm{P}}^{(n)}(\mathbb{Y}_n(\varepsilon)) = 1 .$$

Moreover, for the length λ_1 of the first row, one has

$$\lim_{n\to\infty} \widetilde{\mathrm{P}}^{(n)}\left(\left| \frac{\lambda_1}{2\sqrt{n}} - 1 \right| < \varepsilon \right) = 1 ,$$

and thus for uniform measure on permutations (remembering $L(\pi_n)$ = the length of the longest increasing sequence in π_n), one has

$$\lim_{n\to\infty} \mathrm{P}^{(n)}\left(\left| \frac{L(\pi_n)}{2\sqrt{n}} - 1 \right| < \varepsilon \right) = 1 .$$

Brief outline of the proof. In a first step, the following expression will be estimated, using Stirling's formula:[15]

$$\begin{aligned}
&-\frac{1}{n}\log \mathrm{P}_n(\lambda) \\
&= -\frac{1}{n}\log\frac{n!}{(h^\lambda)^2} \\
&= \frac{2}{n}\log\prod_{(i,j)\in\lambda} h^\lambda_{ij} - \frac{1}{n}\log n! \\
&= \frac{2}{n}\log\prod_{(i,j)\in\lambda} h^\lambda_{ij} - \log n - \frac{1}{n}\log\sqrt{2\pi n} + 1 + \mathbf{O}(1)\,, \quad \text{by Stirling} \\
&= 1 + \frac{2}{n}\left(\log\prod_{(i,j)\in\lambda} h^\lambda_{ij} - n\log n^{1/2}\right) - \frac{1}{n}\log\sqrt{2\pi n} + \mathbf{O}(1)\,, \text{using } |\lambda| = n, \\
&= 1 + 2\sum_{(i,j)\in\lambda}\frac{1}{n}\log\frac{h^\lambda_{ij}}{\sqrt{n}} + \mathbf{O}\left(\frac{\log\sqrt{n}}{n}\right) \quad \text{(Riemann sum)} \\
&\qquad\longrightarrow 1 + 2\iint_{\{(x,y),\,x,y\ge 0,\,y\le F_{\tilde\lambda}(x)\}} dx\,dy\,\log(F_{\tilde\lambda}(x) - y + F^{-1}_{\tilde\lambda}(y) - x)\,,
\end{aligned}$$

15 $\log n! = n\log n + \log\sqrt{2\pi n} - n + \cdots$ for $n \nearrow \infty$.

assuming the partition $\tilde{\lambda} = \lambda/\sqrt{n}$ tends to a continuous curve $y = F_{\tilde{\lambda}}(x)$ in (x, y) coordinates. Then the Riemann sum above tends to the expression above for $n \to \infty$; note $1/n$ is the area of a square in the Young diagram, after rescaling by $1/\sqrt{n}$ and thus turns into $\mathrm{d}x\,\mathrm{d}y$ in the limit. Clearly, the expression $F_{\tilde{\lambda}}(x) - y + F_{\tilde{\lambda}}^{-1}(y) - x$ is the hook length of the continuous curve, with respect to the point (x, y). In the (u, v)-coordinates, this hook length is particularly simple:

$$F_{\tilde{\lambda}}(x) - y + F_{\tilde{\lambda}}^{-1}(y) - x = \sqrt{2}(u_2 - u_1), \quad \text{with } u_1 \leq u_2,$$

where $\big(u_1, \psi(u_1)\big) = \big(x, F_{\tilde{\lambda}}(x)\big)$ and $\big(u_2, \psi(u_2)\big) = \big(F_{\tilde{\lambda}}^{-1}(y), y\big)$ and where $v = \psi(u)$ denotes the curve $y = F_{\tilde{\lambda}}(x)$ in (u, v)-coordinates.

Keeping in mind Fig. 1.5, consider two points $\big(u_1, \psi(u_1)\big)$ and $\big(u_2, \psi(u_2)\big)$ on the curve, with $u_1 < u_2$. The point (x, y) such that the hook (emanating from (x, y) parallel to the (x, y)-coordinates) intersects the curve at the two points $\big(u_1, \psi(u_1)\big)$ and $\big(u_2, \psi(u_2)\big)$, is given by a point on the line emanating from $\big(u_1, \psi(u_1)\big)$ in the $(1, -1)$-direction; to be precise, the point

$$(x, y) = \big(u_1, \psi(u_1)\big) + \tfrac{1}{2}\big(u_2 - \psi(u_2) - u_1 + \psi(u_1)\big)(1, -1)\,.$$

So the surface element $\mathrm{d}x\,\mathrm{d}y$ is transformed into the surface element $du_1 du_2$ by means of the Jacobian, $\mathrm{d}x\,\mathrm{d}y = \frac{1}{2}\big(1 + \psi'(u_1)\big)(1 - \psi'(u_2)\,\mathrm{d}u_1\,\mathrm{d}u_2$, and thus, further replacing $u_1 \mapsto \sqrt{2}u_1$ and $u_2 \mapsto \sqrt{2}u_2$, one is led to

$$\begin{aligned}
&-\frac{1}{n}\log \mathrm{P}_n(\lambda)\\
&\simeq 1 + 2\iint_{\{(x,y),x,y\geq 0, y\leq F(x)\}} \mathrm{d}x\,\mathrm{d}y\,\log\big(F(x) + F^{+}(y) - x - y\big)\\
&= 1 + \iint_{u_1<u_2} \mathrm{d}u_1\,\mathrm{d}u_2 \big(1 + \psi'(u_1)\big)\big(1 - \psi'(u_2)\big)\log\big(\sqrt{2}(u_2 - u_1)\big)\\
&= -\frac{1}{2}\iint_{\mathbb{R}^2} \log(\sqrt{2}|u_2 - u_1|) f'(u_1) f'(u_2)\,\mathrm{d}u_1\,\mathrm{d}u_2 + 2\int_{|u|>2} f(u)\operatorname{arc\,cosh}\left|\frac{u}{2}\right| du\\
&= \frac{1}{2}\iint_{\mathbb{R}^2}\left(\frac{f(u_1) - f(u_2)}{u_1 - u_2}\right)^2 \mathrm{d}u_1\,\mathrm{d}u_2 + 2\int_{|u|>2} f(u)\operatorname{arc\,cosh}\left|\frac{u}{2}\right| du
\end{aligned}$$

upon setting $\psi(u) = \Omega(u) + f(u)$, where the function $\Omega(u)$ is defined in (1.68). The last identity is obtained by using Plancherel's formula of Fourier analysis,

$$\int_{\mathbb{R}} g_1(v) g_2(v)\,\mathrm{d}v = \frac{1}{2\pi}\int_{\mathbb{R}} \widehat{g}_1(v)\widehat{g}_2(v)\,\mathrm{d}v$$

applied to the two functions

$$g_1(v) := \int_{\mathbb{R}} \mathrm{d}u_1\,\log(\sqrt{2}|v - u_1|) f'(u_1) = \int_{\mathbb{R}} \frac{f(u_1)}{\sqrt{2}|v - u_1|}\,\mathrm{d}u_1$$

and $g_2(v) := f'(v)$.

This shows the expression $-(1/n)\log \mathrm{P}_n(\lambda)$ above is minimal (and $= 0$), when $f(u) = 0$; i.e., when the curve $\psi(u) = \Omega(u)$ and otherwise the integrals above > 0. So, when the integral equals $\varepsilon > 0$, the expression $-(1/n)\log \mathrm{P}_n(\lambda) \simeq \varepsilon > 0$ and thus $\mathrm{P}_n(\lambda) \simeq \mathrm{e}^{-\varepsilon n}$, which tends to 0 for $n \nearrow \infty$. Only, when $\varepsilon = 0$, is there a chance that $\mathrm{P}_n(\lambda) = 1$; this happens only when $\psi = \Omega$. □

1.4.2 Limit Theorem for Longest Increasing Sequences

In Sect. 1.3.1, it was shown that a generating function for the probability of the length $L(\pi_k)$ of the longest increasing sequence in a random permutation is given in terms of a Bessel kernel:

$$\mathrm{e}^{-\xi}\sum_{k=0}^{\infty}\frac{\xi^k}{k!}\,\mathrm{P}^k(L(\pi_k)\le n) = \det\bigl(I - K(j,l)\bigr)\Big|_{[n,n+1,\dots]} ,$$

where

$$K(k,l) = \sum_{m=1}^{\infty} J_{k+m}(2\sqrt{\xi})J_{l+m}(2\sqrt{\xi}) . \tag{1.69}$$

In the statement below $\mathrm{A}(x)$ is the classical Airy function

$$\mathrm{A}(x) = \frac{1}{\pi}\int_0^{\infty}\cos\left(\frac{u^3}{3} + xu\right)\,du ;$$

this function is well known to satisfy the ordinary differential equation $\mathrm{A}''(x) = x\,\mathrm{A}(x)$ and to behave asymptotically as

$$\mathrm{A}(x) := \frac{\mathrm{e}^{-2x^{3/2}/3}}{2\sqrt{\pi}x^{1/4}}\bigl(1 + O(x^{3/2})\bigr) , \quad \text{as } x \to \infty .$$

Theorem 1.4.2 (J. Baik, P. Deift, and K. Johansson [19]). *The distribution of the length $L(\pi_n)$ of the longest increasing sequence in a random permutation behaves as*

$$\lim_{n\to\infty}\mathrm{P}^{(n)}\bigl(L(\pi_n) \le 2n^{1/2} + xn^{1/6}\bigr) = \det\bigl(I - \mathbf{A}\,\chi_{[x,\infty)}\bigr) ,$$

where[16]

$$\mathbf{A}(x,y) := \int_0^{\infty} du\ \mathrm{A}(x+u)\,\mathrm{A}(y+u) = \frac{\mathrm{A}(x)\,\mathrm{A}'(y) - \mathrm{A}'(x)\,\mathrm{A}(y)}{x-y} . \tag{1.70}$$

The proof of this theorem, presented here, is due to A. Borodin, A. Okounkov and G. Olshanski [24]; see also [20]. Before giving this proof, the following estimates on Bessel functions are needed:

[16] For more details on Fredholm determinants, see Sect. 1.7.4.

Lemma 1.4.1 ([24]).

(i) *The following holds for* $r \to \infty$,

$$\left| r^{1/3} J_{2r+xr^{1/3}}(2r) - \mathrm{A}(x) \right| = O(r^{-1/3}) \,,$$

uniformly in x, *when* $x \in$ *compact* $K \subset \mathbb{R}$.

(ii) *For any* $\delta > 0$, *there exists* $M > 0$ *such that for* $x, y > M$ *and large enough* r,

$$\left| \sum_{s=1}^{\infty} J_{2r+xr^{1/3}+s}(2r) J_{2r+yr^{1/3}+s}(2r) \right| < \delta r^{-1/3} \,.$$

Lemma 1.4.2 (de-Poissonization lemma; Johansson [56]). *Given* $1 \geq F_0 \geq F_1 \geq F_2 \geq \cdots \geq 0$, *and*

$$F(\xi) := \mathrm{e}^{-\xi} \sum_{0}^{\infty} \frac{\xi^k}{k!} F_k \,,$$

there exists $C > 0$ *and* k_0, *such that*

$$F(k + 4\sqrt{k \log k}) - \frac{C}{k^2} \leq F_k \leq F(k - 4\sqrt{k \log k}) + \frac{C}{k^2} \,, \quad \textit{for all } k > k_0 .$$

Sketch of proof of Thm. 1.4.2. Putting the following scaling in the Bessel kernel $K(k,l)$, as in (1.69), one obtains, setting $r := \sqrt{\xi}$,

$$\begin{aligned} &\xi^{1/6} K(2\xi^{1/2} + x\xi^{1/6}, 2\xi^{1/2} + y\xi^{1/6}) \\ &\quad = \left(\sum_{k=1}^{N} + \sum_{k=N+1}^{\infty} \right) \xi^{-1/6} \left[\xi^{1/6} J_{2\xi^{1/2}+(x+k\xi^{-1/6})\xi^{1/6}}(2\sqrt{\xi}) \right] \\ &\qquad\qquad \times \left[\xi^{1/6} J_{2\xi^{1/2}+(y+k\xi^{-1/6})\xi^{1/6}}(2\sqrt{\xi}) \right] \\ &\quad = \left(\sum_{k=1}^{N} + \sum_{k=N+1}^{\infty} \right) r^{-1/3} \left[r^{1/3} J_{2r+(x+kr^{-1/3})r^{1/3}}(2r) \right] \\ &\qquad\qquad \times \left[r^{1/3} J_{2r+(y+kr^{-1/3})r^{1/3}}(2r) \right] . \end{aligned} \tag{1.71}$$

Fix $\delta > 0$ and pick M as in Lemma 1.4.1(ii). Define $N := [(M - m + 1) r^{1/3}] = O(r^{1/3})$, where m is picked such that $x, y \geq m$ (which is possible, since x and y belong to a compact set). Then

$$\begin{aligned} &\left| \sum_{k=N+1}^{\infty} J_{2r+r^{1/3}(x+kr^{-1/3})}(2r) J_{2r+r^{1/3}(y+kr^{-1/3})}(2r) \right| \\ &\quad = \left| \sum_{s=1}^{\infty} J_{2r+r^{1/3}(x+M-m+1)+s}(2r) J_{2r+r^{1/3}(y+M-m+1)+s}(2r) \right| < \delta r^{-1/3} \,; \end{aligned} \tag{1.72}$$

the latter inequality holds, since

$$x + (M - m + 1) = M + (x - m) + 1 > M \ .$$

On the other hand,

$$\left| r^{1/3} J_{2r+r^{1/3}(x+kr^{-1/3})}(2r) - \mathrm{A}(x + kr^{-1/3}) \right| = O(r^{-1/3})$$

uniformly for $x \in$ compact $K \subset \mathbb{R}$, and all k such that $1 \leq k \leq N = [(M - m + 1)r^{1/3}]$. Indeed, for such k,

$$m \leq x \leq x + kr^{-1/3} \leq x + M - m + 1 = M + (x - m) + 1$$

and thus, for such k, the $x+kr^{-1/3}$'s belong to a compact set as well. Since the number of terms in the sum below is $N = [(M - m + 1)r^{1/3}]$, Lemma 1.4.1(i) implies

$$\left| \sum_{k=1}^{N} (r^{1/3} J_{2r+xr^{1/3}+k}(2r) r^{1/3} J_{2r+yr^{1/3}+k}(2r) - \mathrm{A}(x + kr^{-1/3}) \, \mathrm{A}(y + kr^{1/3})) \right| = O(1) \ . \quad (1.73)$$

But the Riemann sum tends to an integral

$$r^{-1/3} \sum_{k=1}^{[(M-m+1)r^{1/3}]} \mathrm{A}(x + kr^{-1/3}) \, \mathrm{A}(y + kr^{-1/3}) \to \int_0^{M-m+1} \mathrm{A}(x + t) \, \mathrm{A}(y + t) \, dt \ . \quad (1.74)$$

Hence, combining estimates (1.72), (1.73), (1.74) and multiplying with $r^{1/3}$ leads to

$$\left| r^{1/3} \sum_{k=1}^{\infty} J_{2r+xr^{1/3}+k}(2r) J_{2r+yr^{1/3}+k}(2r) - \int_0^{M-m+1} \mathrm{A}(x+t) \, \mathrm{A}(y+t) \, \mathrm{d}t \right| \leq \delta + o(1) \ .$$

for $r \to \infty$. Finally, letting $\delta \to 0$ and $M \to \infty$ leads to the result. Thus the expression (1.71) tends to the Airy kernel

$$\mathbf{A}(x, y) := \int_0^{\infty} du \ \mathrm{A}(x + u) \, \mathrm{A}(y + u) \ .$$

for $r = \sqrt{\xi} \to \infty$. Hence

$$\begin{aligned} \lim_{\xi \to \infty} \mathrm{e}^{-\xi} \sum_{n=0}^{\infty} \frac{\xi^n}{n!} \mathrm{P}(L(\pi_n) \leq 2\xi^{1/2} + x\xi^{1/6}) &= \lim_{\xi \to \infty} \det\left(I - K(l, l')\big|_{[k,k+1,\dots]}\right)\big|_{k=2\xi^{1/2}+x\xi^{1/6}} \\ &= 1 + \sum_{k=1}^{\infty} (-1)^k \int_{x \leq z_1 \leq \cdots \leq z_k} \det\left(\mathbf{A}(z_i, z_j)\right)_{1 \leq i,j \leq k} \prod_1^k \mathrm{d}z_i \\ &= \det\left(I - \mathbf{A} \, \chi_{[x,\infty)}\right) \ . \end{aligned}$$

Finally, one uses Johansson's de-Poissonization Lemma 1.4.2. From Cor. 1.1.1, Plancherel measure $\mathrm{P}_n(\lambda_1 \le x_1, \ldots, \lambda_k \le x_k)$ decreases, when n increases, which is required by Lemma 1.4.2. It thus follows that

$$\lim_{n\to\infty} \mathrm{P}(L(\pi_n) \le 2n^{1/2} + xn^{1/6}) = \det\big(I - \mathbf{A}\,\chi_{[x,\infty)}\big) ,$$

ending the proof of Thm. 1.4.2. □

1.4.3 Limit Theorem for the Geometrically Distributed Percolation Model, when One Side of the Matrix Tends to ∞

Consider Johansson's percolation model in Sect. 1.1.7.2, but with p and q interchanged. Consider the ensemble

$$\mathrm{Mat}^{(q,p)} = \{q \times p \text{ matrices } M \text{ with entries } M_{ij} = 0, 1, 2, \ldots\}$$

with *independent and geometrically distributed entries*, for fixed $0 < \xi < 1$,

$$\mathrm{P}(M_{ij} = k) = (1-\xi)\xi^k \ , \quad k = 0, 1, 2, \ldots \ .$$

$$L(M) := \max_{\substack{\text{all such}\\ \text{paths}}} \left\{ \sum M_{ij}, \begin{matrix}\text{over right/down paths starting}\\ \text{from entry } (1,1) \text{ to } (q,p)\end{matrix} \right\}$$

has the following distribution, assuming $q \le p$,

$$\mathrm{P}(L(M) \le l) = Z_{p,q}^{-1} \sum_{\substack{h \in \mathbb{N}^q \\ \max(h_i) \le l+q-1}} \Delta_q(h_1, \ldots, h_q)^2 \prod_{i=1}^{q} \frac{(h_i + p - q)!}{h_i!} \xi^{h_i} ,$$

where

$$Z_{p,q} = \xi^{q(q-1)/2}(1-\xi)^{-pq} q! \prod_{j=0}^{q-1} j!(p-q+j)! \ . \tag{1.75}$$

Assuming that the number of columns p of the $q \times p$ random M matrix above gets very large, as above, the maximal right/lower path starting from $(1,1)$ to (q,p) consists, roughly speaking, of many horizontal stretches and q small downward jumps. The M_{ij} have the geometric distribution, with mean and standard deviation

$$\mathrm{E}(M_{ij}) = \sum_{k=0}^{\infty} k\,\mathrm{P}(M_{ij} = k) = (1-\xi)\xi \sum_{1}^{\infty} k\xi^{k-1} = \frac{\xi}{1-\xi}$$

$$\begin{aligned}\sigma^2(M_{ij}) &= \mathrm{E}(M_{ij}^2) - \big(\mathrm{E}(M_{ij})\big)^2 = \sum k^2\,\mathrm{P}(M_{ij} = k) - \left(\frac{\xi}{1-\xi}\right)^2 \\ &= \frac{\xi(\xi+1)}{(1-\xi)^2} - \left(\frac{\xi}{1-\xi}\right)^2 = \frac{\xi}{(1-\xi)^2}\end{aligned}$$

So, in the average,

$$L(M) \simeq p\,\mathrm{E}(M_{ij}) = \frac{p\xi}{1-\xi}\,, \quad \text{for } p \to \infty$$

with

$$\sigma^2\left(L(M) - \frac{p\xi}{1-\xi}\right) \simeq p\sigma^2(M_{ij}) = \frac{p\xi}{(1-\xi)^2}\,.$$

Therefore, it seems natural to consider the variables

$$x_1 = \frac{L(M) - p\xi/(1-\xi)}{\sqrt{\xi p}/(1-\xi)} = \frac{\lambda_1 - p\xi/(1-\xi)}{\sqrt{\xi p}/(1-\xi)}$$

and, since $\lambda = (\lambda_1, \dots, \lambda_q)$, with q finite, all λ_i should be on the same footing. So, remembering from the proof of Thm. 1.1.3 that h_i in (1.75) and λ_i are related by

$$h_i = q + \lambda_i - i,$$

we set

$$x_i = \frac{\lambda_i - p\xi/(1-\xi)}{\sqrt{\xi p}/(1-\xi)}\,.$$

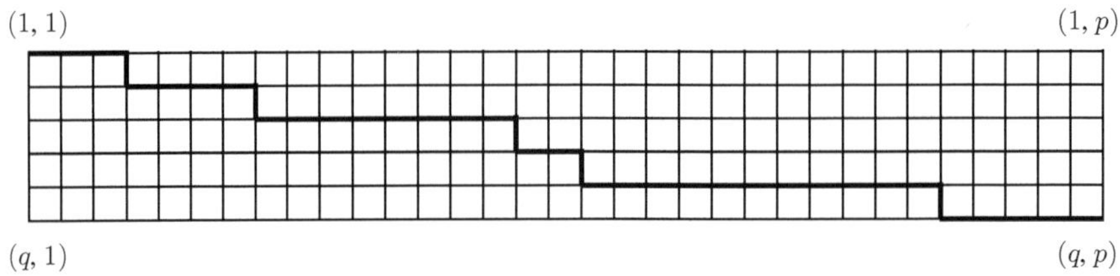

Fig. 1.6.

Theorem 1.4.3 (Johansson [58]). *The following limit holds:*

$$\lim_{p\to\infty} \mathrm{P}\left(\frac{L(M) - p\xi/(1-\xi)}{\sqrt{\xi p}/(1-\xi)} \le y\right) = \frac{\int_{(-\infty,y)^q} \Delta_q(x)^2 \prod_1^q \exp(-x_i^2/2)\,\mathrm{d}x_i}{\int_{\mathbb{R}^q} \Delta_q(x)^2 \prod_1^q \exp(-x_i^2/2)\,\mathrm{d}x_i}\,,$$

which coincides with the probability that a $q \times q$ matrix from the Gaussian Hermitian ensemble (GUE) has its spectrum less than y; see Sect. 1.8.2.1.

Proof. The main tool here is Stirling's formula.[17] Taking into account $h_i = q + \lambda_i - i$, we substitute

[17] Stirling's formula:

$$n! = \sqrt{2\pi n}\exp(n(\log n - 1))\left(1 + O\left(\frac{1}{n}\right)\right), \quad \text{for } n \to \infty\,.$$

$$\lambda_i = \frac{p\xi}{1-\xi} + \frac{\sqrt{\xi p}}{1-\xi} x_i$$

in (1.75). The different pieces will now be computed for large p and fixed q,

$$\begin{aligned}
&\prod_{i=1}^{q} (h_i + p - q)! \\
&= \prod_{i=1}^{q} (p + \lambda_i - i)! \\
&= \prod_{i=1}^{q} \left(\frac{1}{1-\xi}(p + x_i\sqrt{\xi p}) - i \right)! \\
&= \left(\frac{2\pi p}{1-\xi} \right)^{q/2} \prod_{i=1}^{q} \left(1 + x_i \sqrt{\frac{\xi}{p}} - \frac{i(1-\xi)}{p} \right)^{1/2} \left(1 + \mathbf{O}\left(\frac{1}{p}\right) \right) \\
&\times \exp\left(\sum_{i=1}^{q} \left(\frac{p}{1-\xi} + x_i \frac{\sqrt{\xi p}}{1-\xi} - i \right) \left(\log \frac{p}{1-\xi} - 1 + \log\left(1 + x_i \sqrt{\frac{\xi}{p}} - \frac{i(1-\xi)}{p} \right) \right) \right) \\
&= \left(\frac{2\pi p}{1-\xi} \right)^{q/2} \exp\left(\frac{\xi}{2(1-\xi)} \sum_1^q x_i^2 - \frac{qp}{1-\xi} + \mathbf{O}\left(\frac{1}{\sqrt{p}}\right) \right) \\
&\qquad \times \exp\left(\left(\frac{qp}{1-\xi} + \frac{\sqrt{\xi p}}{1-\xi} \sum_1^q x_i - \frac{q(q+1)}{2} \right) \log \frac{p}{1-\xi} \right) \left(1 + \mathbf{O}\left(\frac{1}{p}\right) \right),
\end{aligned}$$

upon expanding the logarithm in powers of $1/\sqrt{p}$. Similarly

$$\begin{aligned}
&\prod_{i=1}^{q} h_i \\
&= \prod_{i=1}^{q} (q + \lambda_i - i)! \\
&= \prod_{i=1}^{q} \left(\frac{1}{1-\xi}(p\xi + x_i\sqrt{\xi p}) + q - i \right)! \\
&= \left(\frac{2\pi p\xi}{1-\xi} \right)^{q/2} \prod_{i=1}^{q} \left(1 + \frac{x_i}{\sqrt{\xi p}} + \frac{(q-i)(1-\xi)}{p\xi} \right)^{1/2} \left(1 + \mathbf{O}\left(\frac{1}{p}\right) \right) \\
&\quad \times \exp\left(\sum_{i=1}^{q} \left(\frac{1}{1-\xi}(p\xi + x_i\sqrt{\xi p}) + q - i \right) \left(\log \frac{p\xi}{1-\xi} - 1 \right) \right) \\
&\quad \times \exp\left(\sum_{i=1}^{q} \left(\frac{1}{1-\xi}(p\xi + x_i\sqrt{\xi p}) + q - i \right) \log\left(1 + \frac{x_i}{\sqrt{\xi p}} + \frac{(q-i)(1-\xi)}{p\xi} \right) \right) \\
&= \left(\frac{2\pi p\xi}{1-\xi} \right)^{q/2} \exp\left(\frac{1}{2(1-\xi)} \sum_1^q x_i^2 - \frac{qp\xi}{1-\xi} \right) \left(1 + \mathbf{O}\left(\frac{1}{\sqrt{p}}\right) \right)
\end{aligned}$$

$$\times \exp\left(\left(\frac{q\xi p}{1-\xi}+\frac{\sqrt{\xi p}}{1-\xi}\sum_1^q x_i+\frac{q(q-1)}{2}\right)\log\frac{p\xi}{1-\xi}\right).$$

Also

$$\begin{aligned}\prod_1^q \xi^{h_i}=\prod_1^q \xi^{q+\lambda_i-i} &= \exp\left(\sum_1^q\left(\frac{1}{1-\xi}(p\xi+x_i\sqrt{\xi p})+q-i\right)\log\xi\right)\\ &= \exp\left(\left(\frac{q\xi p}{1-\xi}+\frac{\sqrt{\xi p}}{1-\xi}\sum_1^q x_i+\frac{q(q-1)}{2}\right)\log\xi\right).\end{aligned}$$

Besides

$$\begin{aligned}\prod_{j=0}^{q-1}(p-q+j)! &= (2\pi p)^{q/2}\prod_{j=0}^{q-1}\left(1-\frac{q-j}{p}\right)^{1/2}\left(1+\mathbf{O}\left(\frac{1}{p}\right)\right)\\ &\qquad\times\exp\left(\sum_{j=0}^{q-1}(p-q+j)\left(\log p-1+\log\left(1-\frac{q-j}{p}\right)\right)\right)\\ &= (2\pi p)^{q/2}\exp\left(-pq+qp\log p-\frac{q(q+1)}{2}\log p\right)\left(1+\mathbf{O}\left(\frac{1}{p}\right)\right)\\ &= (2\pi)^{q/2}p^{qp-q^2/2}\mathrm{e}^{-qp}\left(1+\mathbf{O}\left(\frac{1}{p}\right)\right)\end{aligned}$$

and

$$\begin{aligned}\Delta_q(h_1,\dots,h_q)=\prod_{1\le i<j\le q}(h_i-h_j) &= \prod_{1\le i<j\le q}\left((x_i-x_j)\frac{\sqrt{\xi p}}{1-\xi}-i+j\right)\\ &= \left(\frac{\sqrt{\xi p}}{1-\xi}\right)^{q(q-1)/2}\left(\prod_{1\le i<j\le q}(x_i-x_j)+\mathbf{O}\left(\frac{1}{\sqrt{p}}\right)\right).\end{aligned}$$

Remembering the relation $h_i=q+\lambda_i-i$, and using the estimates before in equality $\stackrel{*}{=}$ below, one computes

$$
\begin{aligned}
&Z_{p,q}^{-1} \sum_{\substack{h\in\mathbb{N}^q\\ \max h_i\le l+q-1}} \Delta_q(h_1,\dots,h_q)^2 \prod_1^q \frac{(h_i+p-q)!}{h_i!}\xi^{h_i}\\
&= \sum_{\substack{h\in\mathbb{N}^q\\ \max h_i\le l+q-1}} \frac{\xi^{-q(q-1)/2}}{\prod_{j=0}^q j!}(1-\xi)^{pq}\Delta_q(h_1,\dots,h_q)^2 \prod_{i=1}^q \frac{(p+\lambda_i-i)!\ \xi^{h_i}}{(q+\lambda_i-i)!(p-q+i-1)!}\\
&\overset{*}{=} \left(\frac{1}{(2\pi)^{q/2}\prod_1^q i!}\right)\sum_{x_i\le y}\left(\frac{1-\xi}{\sqrt{\xi p}}\right)^q \Delta(x)^2 \exp\left(-\frac{1}{2}\sum_1^q x_i^2\right)\left(1+\mathbf{O}\left(\frac{1}{\sqrt{p}}\right)\right)\\
&\qquad\longrightarrow \frac{\int_{x_i\le y}\Delta(x)^2\prod_1^q \exp(-x_i^2/2)\,\mathrm{d}x_i}{\int_{\mathbb{R}^q}\Delta_q(x)^2\prod_1^q \exp(-x_i^2/2)\,\mathrm{d}x_i}\,,
\end{aligned}
$$

using Selberg's formula

$$
\int_{\mathbb{R}^q} \Delta_q(x)^2 \prod_1^q \exp(-x_i^2/2)\,\mathrm{d}x_i = (2\pi)^{q/2}\prod_1^q i!
$$

and noticing that in

$$
h_i = q+\lambda_i - i = \frac{\xi p}{1-\xi} + \frac{\sqrt{\xi p}}{1-\xi}x_i + q - i
$$

an increment of one unit in λ_i implies an increment of x_i by

$$
\mathrm{d}x_i = \frac{1-\xi}{\sqrt{\xi p}}\,.
$$

Therefore, one has

$$
\left(\frac{1-\xi}{\sqrt{\xi p}}\right)^q \simeq \prod_1^q \mathrm{d}x_i\,.
$$

Finally, for p large, one has

$$
h_i \le l+q-1 = \frac{\xi p}{1-\xi} + \frac{\sqrt{\xi p}}{1-\xi}y + q - 1 \iff x_i \le y\,.
$$

The connection with the spectrum of Gaussian Hermitian matrices will be discussed in Sect. 1.8.2.1. This ends the proof of Thm. 1.4.3. □

1.4.4 Limit Theorem for the Geometrically Distributed Percolation Model, when Both Sides of the Matrix Tend to ∞

The model considered here is the percolation model as in Sect. 1.4.3, whose probability can, by Sect. 1.3.3, be written as a Fredholm determinant of a Meixner kernel; see Thm. 1.3.3.

Theorem 1.4.4 (Johansson [57]). *Given*

$$a = \frac{(1+\sqrt{\xi\gamma})^2}{1-\xi} - 1 \,, \quad \rho = \left(\frac{\xi}{\gamma}\right)^{1/6} \frac{(\sqrt{\gamma}+\sqrt{\xi})^{2/3}(1+\sqrt{\gamma\xi})^{2/3}}{1-\xi} \,,$$

the following holds:

$$\lim_{\substack{p,q\to\infty \\ q/p=\gamma\geq 1 \text{ fixed}}} \mathrm{P}\left(\frac{L(M_{q,p})-ap}{\rho p^{1/3}} \leq y\right) = \mathcal{F}(y) \,,$$

which is the Tracy–Widom distribution (see Sect. 1.9.2). In other terms, the random variable $L(M_{q,p})$ behaves in distribution, like

$$L(M_{q,p}) \sim ap + \rho p^{1/3} \mathcal{F} \,.$$

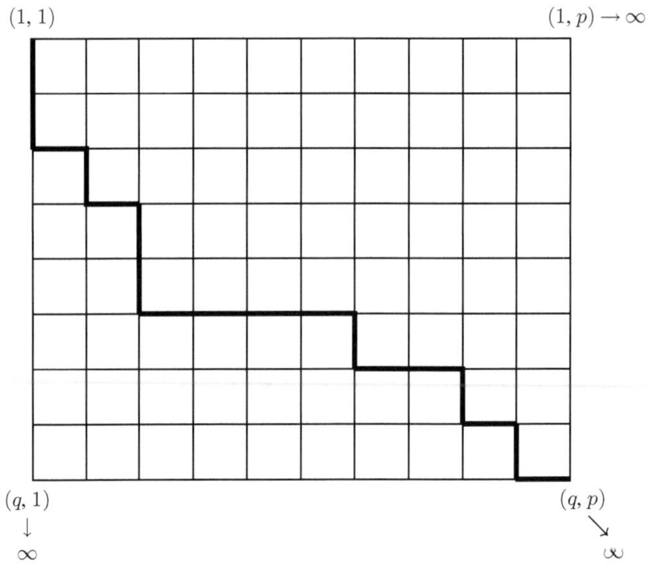

Fig. 1.7.

Sketch of proof. Remember from Thm. 1.3.3, the probability $\mathrm{P}(L(M_{q,p}) \leq z)$ is given by a Fredholm determinant of a Christoffel–Darboux kernel composed of Meixner polynomials. Proving the statement of Thm. 1.4.4 amounts to proving the limit of this Meixner kernel with the scaling mentioned in the theorem tends to the Airy kernel, i.e.,

$$\lim_{p\to\infty} \rho p^{1/3} K_p(ap+\rho p^{1/3}\eta, ap+\rho p^{1/3}\eta') = \mathbf{A}(\eta,\eta') \\ := \frac{\mathrm{A}(\eta)\,\mathrm{A}'(\eta') - \mathrm{A}'(\eta)\,\mathrm{A}(\eta')}{\eta-\eta'} ,$$

where[18]

$$K_p(x,y) = -\frac{\xi}{(1-\xi)d_{p-1}^2} \sqrt{\binom{x+\beta'-1}{x}\xi^x \binom{y+\beta'-1}{y}\xi^y} \\ \times \frac{m_p(x,\beta',\xi)m_{p-1}(y,\beta',\xi) - m_p(y,\beta',\xi)m_{p-1}(x,\beta',\xi)}{x-y}$$

for the Meixner polynomials, which have the following integral representation, a consequence of the the generating function (1.66), (slightly rescaled)

$$m_p(x,\beta',\xi) = p!\xi^{-x} \oint_{|z|=r<1} \frac{(\xi-z)^x}{z^p(1-z)^{x+\beta'}} \frac{\mathrm{d}z}{2\pi \mathrm{i} z}$$

Thus, it will suffice to prove the following limit

$$\lim_{p\to\infty} p^{1/3} \left(\frac{y^p(y-1)^{x+[\gamma p]-p+1}}{(y-\xi)^x} \right)\Bigg|_{y=-\sqrt{\xi/\gamma}} \\ \oint_{|z|=r<1} \frac{(z-\xi)^x}{z^p(z-1)^{x+[\gamma p]-p+1}} \frac{\mathrm{d}z}{2\pi \mathrm{i} z} \Bigg|_{x=\beta p+\rho p^{1/3}\eta} = C\,\mathrm{A}(\eta) , \quad (1.76)$$

where

$$\alpha := \frac{(\sqrt{\xi}+\sqrt{\gamma})^2}{1-\xi} , \quad \beta = \alpha - \gamma + 1 = \frac{(\sqrt{\gamma\xi}+1)^2}{1-\xi} , \\ \rho = \left(\frac{\xi}{\gamma}\right)^{1/6} \frac{(\sqrt{\gamma}+\sqrt{\xi})^{2/3}(1+\sqrt{\gamma\xi})^{2/3}}{1-\xi} \\ C := \gamma^{-1/3}\xi^{-1/6}(\sqrt{\xi}+\sqrt{\gamma})^{1/3}(1+\sqrt{\xi\gamma})^{1/3} .$$

In view of the saddle point method, define the function $F_p(z)$ such that

$$\exp\big(pF_p(z)\big) := \frac{(z-\xi)^x}{z^p(z-1)^{x+[\gamma p]-p+1}} \Bigg|_{x=\beta p+\rho p^{1/3}\eta} .$$

Then one easily sees that

$$F_p(z) = F(z) + \rho p^{-2/3}\eta \log\frac{z-\xi}{z-1} + \left(\gamma - \frac{[\gamma p]}{p}\right)\log(z-1) , \qquad (1.77)$$

where the p-independent function $F(z)$ equals,

[18] with $d_p = p!(p+\beta'-1)!/\big((1-\xi)^{\beta'}\xi^p(\beta'-1)!\big)$ and $\beta' = q-p+1 = p\gamma - p + 1$.

$$F(z) = \beta \log(z-\xi) - \alpha \log(z-1) - \log z \,. \tag{1.78}$$

For the specific values above of α and β, one checks the function $F(z)$ has a critical point at $z_c := -\sqrt{\xi/\gamma}$, i.e.,

$$F'(z_c) = F''(z_c) = 0\,, \quad F'''(z_c) = \frac{2\gamma^{5/2}}{\xi(\sqrt{\gamma}+\sqrt{\xi})(1+\sqrt{\xi\gamma})}$$

and thus

$$F(z) - F(z_c) = \tfrac{1}{6}(z-z_c)^3 F'''(z_c) + O\big((z-z_c)^4\big)\,. \tag{1.79}$$

Setting

$$z = z_c(1 - p^{-1/3} sC)\exp(itCp^{-1/3}) \tag{1.80}$$

in $F_p(z)$ as in (1.77), one first computes this substitution in $F(z)$, taking into account (1.79),

$$\begin{aligned} F(z) &= F(z_c) + \frac{\mathrm{i}(-z_c C)^3}{6p}(t+\mathrm{i}s)^3 F'''(z_c) + O(p^{-4/3}) \\ &= F(z_c) + \frac{\mathrm{i}}{3p}(t+\mathrm{i}s)^3 + O(p^{-4/3})\,, \end{aligned} \tag{1.81}$$

upon picking – in the last equality – the constant C such that

$$(-z_c C)^3 F'''(z_c) = 2\,.$$

Also, substituting the z of (1.80) in the part of $F(z)$ (see (1.78)) containing η,

$$\begin{aligned} &\rho p^{-2/3}\eta \log \frac{z-\xi}{z-1} \\ &\quad = \frac{\rho\eta}{p^{2/3}} \log\frac{z_c-\xi}{z_c-1} - \frac{\rho C z_c(\xi-1)}{(z_c-1)(z_c-\xi)} \frac{\mathrm{i}\eta(t+\mathrm{i}s)}{p} + O(p^{-4/3})\,. \end{aligned} \tag{1.82}$$

Thus, adding the two contributions (1.81) and (1.82), one finds

$$\begin{aligned} &pF_p\big(z_c(1-p^{-1/3}sC)\exp(itCp^{-1/3})\big) \\ &\quad = pF_p(z_c) + \frac{\mathrm{i}}{3}\big((t+\mathrm{i}s)^3 + 3\eta(t+\mathrm{i}s)\big) + O(p^{-1/3})\,. \end{aligned} \tag{1.83}$$

One then considers two contributions of the contour integral about the circle $|z| = r$ appearing in (1.76), a first one along the arc $(\pi-\delta_p, \pi+\delta_p)$, for δ_p tending to 0 with $p \to \infty$ and a second one about the complement of $(\pi - \delta_p, \pi+\delta_p)$. The latter tends to 0, whereas the former is the main contribution and tends to the Airy function (keeping s fixed, and in particular $= 1$)

$$C\,\mathrm{A}(\eta) = \frac{C}{2\pi}\int_{\infty}^{\infty} \exp\left(\mathrm{i}\left(\frac{(t+\mathrm{i}s)^3}{3} + \eta(t+\mathrm{i}s)\right)\right) \mathrm{d}t\,,$$

upon noticing that $\mathrm{d}z/z = \mathrm{i}Cp^{-1/3}\,\mathrm{d}t$ under the change of variable $z \mapsto t$, given in (1.80), establishing limit (1.76) and finally the limit of the Meixner kernel and its Fredholm determinant. Further details of this proof can be found in Johansson [57]. The Fredholm determinant of the Airy kernel is precisely the Tracy–Widom distribution, as will be shown in Sect. 1.9.2. □

1.4.5 Limit Theorem for the Exponentially Distributed Percolation Model, when Both Sides of the Matrix tend to ∞

Referring to the exponentially distributed percolation model, discussed in Sect. 1.1.4, we now state

Theorem 1.4.5 (Johansson [57]). *Given*

$$a = (1+\sqrt{\gamma})^2) \,, \quad \rho = \frac{(1+\sqrt{\gamma})^{4/3}}{\gamma^{1/6}} \,,$$

the following limit holds:

$$\begin{aligned} &\lim_{\substack{p,q\to\infty \\ q/p=\gamma\geq 1 \text{ fixed}}} \mathrm{P}\left(\frac{L(M_{q,p}) - (p^{1/2}+q^{1/2})}{(p^{1/2}+q^{1/2})(p^{-1/2}+q^{-1/2})^{1/3}} \leq y \right) \\ &\qquad = \lim_{\substack{p,q\to\infty \\ q/p=\gamma\geq 1 \text{ fixed}}} \mathrm{P}\left(\frac{L(M_{q,p}) - ap}{\rho p^{1/3}} \leq y \right) = \mathcal{F}(u) \,, \end{aligned}$$

which is again the Tracy–Widom distribution. Here again $L(M_{q,p})$ behaves, after some rescaling and in distribution, like the Tracy–Widom distribution, for large p and q such that $q/p = \gamma \geq 1$:

$$L(M_{q,p}) \sim ap + \rho p^{1/3} \mathcal{F} \,.$$

Proof. In Thm. 1.1.4, one has shown that $\mathrm{P}(L(M) \leq t)$ equals the ratio of two integrals; this ratio will be shown in Sect. 1.7 on random matrices (see Props. 1.7.5 and 1.7.3) to equal a Fredholm determinant of a kernel corresponding to Laguerre polynomials:

$$\begin{aligned} \mathrm{P}(L(M) \leq t) &= \frac{\int_{[0,t]^q} \Delta_p(x)^2 \prod_{i=1}^p x_i^{q-p} \exp(-x_i)\,\mathrm{d}x_i}{\int_{[0,\infty]^q} \Delta_p(x)^2 \prod_{i=1}^p x_i^{q-p} \exp(-x_i)\,\mathrm{d}x_i} \\ &= \det(I - K_p^{(\alpha)}(x,y)\chi_{[t,\infty]}) \end{aligned}$$

where

$$\begin{aligned} K_p^{(\alpha)}(x,y) = \sqrt{\frac{h_p}{h_{p-1}}}(xy)^{\alpha/2} \exp\left(-\tfrac{1}{2}(x+y)\right) \\ \times \frac{\mathcal{L}_p^{(\alpha)}(x)\mathcal{L}_{p-1}^{(\alpha)}(y) - \mathcal{L}_p^{(\alpha)}(y)\mathcal{L}_{p-1}^{(\alpha)}(x)}{x-y} \,; \end{aligned}$$

in the formula above, the $\mathcal{L}_p^{(\alpha)}(x) = 1/\sqrt{h_p}x^n + \cdots = (-1)^p(p!/(p+\alpha)!)^{1/2} \times L_p^\alpha(x)$ are the normalized Laguerre polynomials[19]:

$$\int_0^\infty \mathcal{L}_n^{(\alpha)}(x)\mathcal{L}_m^{(\alpha)}(x)x^\alpha \mathrm{e}^{-x}\,\mathrm{d}x = \delta_{nm}\,.$$

Therefore, using the precise values of a and ρ above,

$$\mathrm{P}\left(\frac{L(M_{q,p}) - ap}{\rho p^{1/3}} \le y\right) = \det\bigl(I - \mathcal{K}(\xi,\eta)\chi_{[y,\infty)}\bigr)$$

with

$$\mathcal{K}(\xi,\eta) = bp^{1/3}K_p^{(\gamma-1)p}(ap + \rho p^{1/3}\xi, ap + \rho p^{1/3}\eta)$$

The result follows from an asymptotic formula for Laguerre polynomials

$$\lim_{p\to\infty} K_p^{(\gamma-1)p}(ap + \rho p^{1/3}\xi, ap + \rho p^{1/3}\eta) = \mathbf{A}(\xi,\eta)\,,$$

with $\mathbf{A}(x,y)$ the Airy kernel, as in (1.70), namely

$$\mathbf{A}(x,y) := \frac{\mathrm{A}(x)\,\mathrm{A}'(y) - \mathrm{A}'(x)\,\mathrm{A}(y)}{x-y}\,.$$

The Fredholm determinant of the Airy kernel is the Tracy–Widom distribution. This ends the proof of Thm. 1.4.5. □

1.5 Orthogonal Polynomials for a Time-Dependent Weight and the KP Equation

1.5.1 Orthogonal Polynomials

The inner product with regard to the weight $\rho(z)$ over $\mathbb{R}$, assuming $\rho(z)$ decays fast enough at the boundary of its support[20]

$$\langle f,g\rangle = \int_{\mathbb{R}} f(z)g(z)\rho(z)\mathrm{d}z\,, \tag{1.84}$$

leads to a moment matrix

$$m_n = (\mu_{ij})_{0\le i,j<n} = (\langle z^i, z^j\rangle)_{0\le i,j\le n-1}\,. \tag{1.85}$$

Since the μ_{ij} depends on $i+j$ only, this is a *Hänkel matrix*, and thus symmetric. This is tantamount to the relation

[19] $L_n^{(\alpha)}(y) = \sum_{m=0}^n (-1)^m \binom{n+\alpha}{n-m} x^m/m! = \mathrm{e}^x/(2\pi\mathrm{i})\int_C \mathrm{e}^{-xz}z^{n+\alpha}/(z-1)^{n+1}\,\mathrm{d}z$, where C is a circle about $z=1$.

[20] In this section, the support of the weight $\rho(z)$ can be the whole of $\mathbb{R}$ or any other interval.

$$\Lambda m_\infty = m_\infty \Lambda^\top ,$$

where Λ denotes the semi-infinite shift matrix

$$\begin{pmatrix} 0\,1\,0\,0\,0\,0\cdots \\ 0\,0\,1\,0\,0\,0\cdots \\ 0\,0\,0\,1\,0\,0\cdots \\ \vdots \end{pmatrix} .$$

Define

$$\tau_n := \det m_n .$$

Consider the factorization of m_∞ into a lower times an upper triangular matrix[21]

$$m_\infty = S^{-1} S^{\top -1} , \tag{1.86}$$

with

$$S = \text{lower triangular with nonzero diagonal elements.}$$

For any $z \in \mathbb{C}$, define the semi-infinite column

$$\chi(z) := (1, z, z^2, \dots)^\top , \tag{1.87}$$

and functions $p_n(z)$ and $q_n(z)$,

$$p_n(z) := \big(S\chi(z)\big)_n , \quad q_n(z) := \big(S^{\top -1}\chi(z^{-1})\big)_n . \tag{1.88}$$

(1) *The $p_n(z)$ are polynomials of degree n, orthonormal with regard to $\rho(z)$, and $q_n(z)$ is the Stieltjes transform of $p_n(z)$,*

$$q_n(z) = z \int_{\mathbb{R}} \frac{p_n(u)\rho(u)}{z-u}\, \mathrm{d}u .$$

Indeed,

$$(\langle p_k, p_l\rangle)_{0\le k,l<\infty} = \int_{\mathbb{R}} S\chi(z)\big(S\chi(z)\big)^\top \rho(z)\, \mathrm{d}z = S m_\infty S^\top = I .$$

Note that $S\chi(z)\big(S\chi(z)\big)^\top$ is a semi-infinite matrix obtained by multiplying the semi-infinite column $S\chi(z)$ and row $\big(S\chi(z)\big)^\top$. The definition of q_n, together with the decomposition $S^{\top -1} = S m_\infty$, leads to

$$\begin{aligned} q_n(z) &= \big(S^{\top -1}\chi(z^{-1})\big)_n = \sum_{j\ge 0} (S m_\infty)_{nj} z^{-j} \\ &= \sum_{j\ge 0} z^{-j} \sum_{l=0}^{n} S_{nl}\mu_{lj} = \sum_{j\ge 0} z^{-j} \sum_{l=0}^{n} S_{nl} \int_{\mathbb{R}} u^{l+j}\rho(u)\, \mathrm{d}u \\ &= \int_{\mathbb{R}} \sum_{l=0}^{n} S_{nl} u^l \sum_{j\ge 0} \left(\frac{u}{z}\right)^j \rho(u)\, \mathrm{d}u = z \int_{\mathbb{R}} \frac{p_n(u)\rho(u)}{z-u}\, \mathrm{d}u . \end{aligned}$$

[21] This factorization is possible as long as $\tau_n := \det m_n \neq 0$ for all $n \ge 1$.

(2) *The orthonormal polynomials p_n have the following representation*

$$p_n(z) = \frac{1}{\sqrt{\tau_n \tau_{n+1}}} \det \left(\begin{array}{ccc|c} & & & 1 \\ & m_n & & z \\ & & & \vdots \\ \hline \mu_{n,0} & \cdots & \mu_{n,n-1} & z^n \end{array} \right) . \tag{1.89}$$

As a consequence, the monic orthogonal polynomials $\tilde{p}_n(z)$ are related to $p_n(z)$ as follows:

$$p_n(z) = \sqrt{\frac{\tau_n}{\tau_{n+1}}} \tilde{p}_n(z). \tag{1.90}$$

Defining $p_n'(z)$ to be the polynomial on the right-hand side of (1.89), it suffices to show that for $k < n$,

$$\langle p_n', p_k' \rangle = 0 \quad \text{and} \quad \langle p_n', p_n' \rangle = 1 ,$$

thus leading to $p_n = p_n'$. Indeed,

$$\begin{aligned} \langle p_n', z^k \rangle &= \frac{1}{\sqrt{\tau_n \tau_{n+1}}} \det \left(\begin{array}{ccc|c} & & & \langle 1, z^k \rangle \\ & m_n & & \langle z, z^k \rangle \\ & & & \vdots \\ \hline \mu_{n,0} & \cdots & \mu_{n,n-1} & \langle z^n, z^k \rangle \end{array} \right) \\ &= \begin{cases} 0 , & \text{for } k < n, \\ \sqrt{\tau_{n+1}/\tau_n} & \text{for } k = n, \end{cases} \end{aligned}$$

and thus for $k = n$,

$$\langle p_n', p_n' \rangle = \sqrt{\frac{\tau_n}{\tau_{n+1}}} \langle p_n', z^n + \cdots \rangle = \sqrt{\frac{\tau_n}{\tau_{n+1}}} \langle p_n', z^n \rangle = 1,$$

from which (1.89) follows. Formula (1.90) is a straightforward consequence.

(3) *The monic orthogonal polynomials $\tilde{p}_n$ and their Stieltjes transform have the following representation*

$$\begin{aligned} \tilde{p}_n(z) &= z^n \frac{\det(\mu_{ij} - \mu_{i,j+1}/z)_{0 \le i,j \le n-1}}{\det(\mu_{ij})_{0 \le i,j \le n-1}} \\ \int_{\mathbb{R}} \frac{\tilde{p}_n(u)\rho(u)}{z-u} \, du &= z^{-n-1} \frac{\det(\mu_{ij} + \mu_{i,j+1}/z + \mu_{i,j+2}/z^2 + \cdots)_{0 \le i,j \le n}}{\det(\mu_{ij})_{0 \le i,j \le n-1}} . \end{aligned} \tag{1.91}$$

Proof. Setting

$$\boldsymbol{\mu}_j = (\mu_{0j}, \ldots, \mu_{n-1,j}) \in \mathbb{R}^n ,$$

one computes

$$
\begin{aligned}
&z^n \det(\mu_{ij} - z^{-1}\mu_{i,j+1})_{0\le i,j\le n-1} \\
&\quad = \det(z\mu_{i,j} - \mu_{i,j+1})_{0\le i,j\le n-1} \\
&\quad = \det(z\boldsymbol{\mu}_0^\top - \boldsymbol{\mu}_1^\top, z\boldsymbol{\mu}_1^\top - \boldsymbol{\mu}_2^\top, \dots, z\boldsymbol{\mu}_{n-1}^\top - \boldsymbol{\mu}_n^\top) \\
&\quad = \det\left(\sum_0^{n-1} \frac{z\boldsymbol{\mu}_j^\top - \boldsymbol{\mu}_{j+1}^\top}{z^j}, \sum_0^{n-2} \frac{z\boldsymbol{\mu}_{j+1}^\top - \boldsymbol{\mu}_{j+2}^\top}{z^j}, \dots, z\boldsymbol{\mu}_{n-1}^\top - \boldsymbol{\mu}_n^\top\right) \\
&\qquad\qquad\qquad\qquad\qquad\qquad\qquad\qquad\text{by column operations} \\
&\quad = \det\left(z\boldsymbol{\mu}_0^\top - \frac{\boldsymbol{\mu}_n^\top}{z^{n-1}}, z\boldsymbol{\mu}_1^\top - \frac{\boldsymbol{\mu}_n^\top}{z^{n-2}}, \dots, z\boldsymbol{\mu}_{n-1}^\top - \boldsymbol{\mu}_n^\top\right) \\
&\quad = \frac{1}{z^n}\det\left(\begin{array}{c|c} z\boldsymbol{\mu}_0 - \boldsymbol{\mu}_n/z^{n-1} & 0 \\ z\boldsymbol{\mu}_1 - \boldsymbol{\mu}_n/z^{n-2} & 0 \\ \vdots & \vdots \\ z\boldsymbol{\mu}_{n-1} - \boldsymbol{\mu}_n & 0 \\ \boldsymbol{\mu}_n & z^n \end{array}\right) \\
&\qquad\qquad\qquad\qquad\qquad\text{enlarging the matrix by one row and column} \\
&\quad = \frac{1}{z^n}\det\left(\begin{array}{c|c} z\boldsymbol{\mu}_0 & z \\ z\boldsymbol{\mu}_1 & z^2 \\ \vdots & \vdots \\ z\boldsymbol{\mu}_{n-1} & z^n \\ \boldsymbol{\mu}_n & z^n \end{array}\right) \\
&\qquad\qquad\qquad\qquad\text{by adding a multiple of the last row to rows 1 to } n \\
&\quad = \tau_n \tilde{p}_n(z)\,.
\end{aligned}
$$

Setting this time

$$
\boldsymbol{\mu}_j := (\mu_{0j}, \dots, \mu_{nj}) \in \mathbb{R}^{n+1},
$$

one computes

$$
\begin{aligned}
\det&\left(\mu_{ij} + \frac{\mu_{i,j+1}}{z} + \frac{\mu_{i,j+2}}{z^2} + \cdots\right)_{0\le i,j\le n} \\
&= \det\left(\sum_0^\infty \frac{\boldsymbol{\mu}_j^\top}{z^j}, \sum_0^\infty \frac{\boldsymbol{\mu}_{j+1}^\top}{z^j}, \dots, \sum_0^\infty \frac{\boldsymbol{\mu}_{j+n}^\top}{z^j}\right) \\
&= \det\left(\boldsymbol{\mu}_0^\top, \boldsymbol{\mu}_1^\top, \dots, \boldsymbol{\mu}_{n-1}^\top, \sum_0^\infty \frac{\boldsymbol{\mu}_{j+n}^\top}{z^j}\right) \\
&= z^n \det\left(\boldsymbol{\mu}_0^\top, \boldsymbol{\mu}_1^\top, \dots, \boldsymbol{\mu}_{n-1}^\top, \sum_0^\infty \frac{\boldsymbol{\mu}_j^\top}{z^j}\right)
\end{aligned}
$$

$$= z^n \det \left(\begin{matrix} \boldsymbol{\mu}_0^\top & \boldsymbol{\mu}_1^\top & \cdots & \boldsymbol{\mu}_{n-1}^\top \end{matrix} \middle| \begin{matrix} \int_{\mathbb{R}} \sum_{j=0}^\infty (u/z)^j \rho(u)\,du \\ \int_{\mathbb{R}} \sum_{j=0}^\infty (u/z)^j u\rho(u)\,du \\ \vdots \\ \int_{\mathbb{R}} \sum_{j=0}^\infty (u/z)^j u^n \rho(u)\,du \end{matrix} \right)$$

$$= z^n \int_{\mathbb{R}} \sum_{j=0}^\infty \left(\frac{u}{z}\right)^j \rho(u)\,du \det \left(\begin{matrix} \boldsymbol{\mu}_0^\top & \boldsymbol{\mu}_1^\top & \cdots & \boldsymbol{\mu}_{n-1}^\top \end{matrix} \middle| \begin{matrix} 1 \\ u \\ \vdots \\ \vdots \\ u^n \end{matrix} \right)$$

$$= z^n \tau_n \int_{\mathbb{R}} \sum_{j=0}^\infty \left(\frac{u}{z}\right)^j \tilde{p}_n(u)\rho(u)\,du = z^{n+1}\tau_n \int_{\mathbb{R}} \frac{\tilde{p}_n(u)\rho_\mu(u)}{z-u}\,du\,. \quad \square$$

Remark. Representation (1.91) for orthogonal polynomials p_n can also be deduced from Heine's representation. However, representation (1.91) is a much simpler formula.

(4) *The vectors p and q are eigenvectors of the tridiagonal symmetric matrix*

$$L := S\Lambda S^{-1}. \tag{1.92}$$

Conjugating the shift matrix Λ by S yields a matrix

$$\begin{aligned} L &= S\Lambda S^{-1} \\ &= S\Lambda S^{-1} S^{\top -1} S^\top \\ &= S\Lambda m_\infty S^\top\,, && \text{using (1.86),} \\ &= S m_\infty \Lambda^\top S^\top\,, && \text{using } \Lambda m_\infty = m_\infty \Lambda^\top, \\ &= S(S^{-1}S^{\top -1})\Lambda^\top S^\top\,, && \text{using again (1.86),} \\ &= (S\Lambda S^{-1})^\top = L^\top\,, \end{aligned}$$

which is symmetric and thus tridiagonal. Remembering $\chi(z) = (1, z, z^2, \dots)^\top$, and the shift $(\Lambda v)_n = v_{n+1}$, we have

$$\Lambda\chi(z) = z\chi(z) \quad \text{and} \quad \Lambda^\top \chi(z^{-1}) = z\chi(z^{-1}) - ze_1\,, \text{ with } e_1 = (1, 0, 0, \dots)^\top\,.$$

Therefore, $p(z) = S\chi(z)$ and $q(z) = S^{\top -1}\chi(z^{-1})$ are eigenvectors, in the sense

$$\begin{aligned} Lp &= S\Lambda S^{-1} S\chi(z) = zS\chi(z) = zp \\ L^\top q &= S^{\top -1}\Lambda^\top S^\top S^{\top -1}\chi(z^{-1}) \\ &= zS^{\top -1}\chi(z^{-1}) - zS^{\top -1}e_1 = zq - zS^{\top -1}e_1\,. \end{aligned}$$

Then, using $L = L^\top$, one is led to

$$\big((L - zI)p\big)_n = 0\,, \text{ for } n \geq 0 \quad \text{and} \quad \big((L - zI)q\big)_n = 0\,, \text{ for } n \geq 1\,.$$

(5) *The off-diagonal elements of the symmetric tridiagonal matrix L are given by*

$$L_{n-1,n} = \sqrt{\frac{h_n}{h_{n-1}}}. \tag{1.93}$$

Since $\langle \tilde{p}_n, \tilde{p}_n \rangle = h_n$, one has $p_n(y) = (1/\sqrt{h_n})\tilde{p}_n(y)$. From the three step relation $Lp(y) = yp(y)$, it follows that

$$\begin{aligned}\left(\frac{1}{\sqrt{h_{n-1}}}y^n + \cdots\right) = yp_{n-1}(y) &= L_{n-1,n}p_n(y) + (\text{terms of degree} \leq n-1) \\ &= L_{n-1,n}\left(\frac{1}{\sqrt{h_n}}y^n + \cdots\right),\end{aligned}$$

leading to statement (1.93).

1.5.2 Time-Dependent Orthogonal Polynomials and the KP Equation

Introduce now into the weight $\rho(z)$ a dependence on parameters $t=(t_1, t_2, \dots)$, as follows

$$\rho_t(z) := \rho(z) \exp\left(\sum_1^\infty t_i z^i\right). \tag{1.94}$$

Consider the moment matrix $m_n(t)$, as in (1.85), but now dependent on t, and the factorization of m_∞ into lower- times upper-triangular t-dependent matrices, as in (1.86)

$$m_\infty(t) = \big(\mu_{ij}(t)\big)_{0\leq i,j<\infty} = S^{-1}(t)S^{\top -1}(t). \tag{1.95}$$

The Toda lattice mentioned in the theorem below will require the following Lie algebra splitting

$$\mathfrak{gl}(n) = \mathfrak{s} \oplus \mathfrak{b}, \tag{1.96}$$

into skew-symmetric matrices and (lower) Borel matrices.

Also, one needs in this section the Hirota symbol: given a polynomial $p(t_1, t_2, \dots)$ of a finite or infinite number of variables and functions $f(t_1, t_2, \dots)$ and $g(t_1, t_2, \dots)$, also depending on a finite or infinite number of variables t_i, define the symbol

$$p\left(\frac{\partial}{\partial t_1}, \frac{\partial}{\partial t_2}, \dots\right) f \circ g := p\left(\frac{\partial}{\partial y_1}, \frac{\partial}{\partial y_2}, \dots\right) f(t+y)g(t-y)\Big|_{y=0}. \tag{1.97}$$

The reader is reminded of the elementary Schur polynomials $\exp(\sum_1^\infty t_i z^i) := \sum_{i\geq 0} \mathrm{s}_i(t) z^i$ and for later use, set for $l = 0, 1, 2, \dots$,

$$\mathrm{s}_l(\tilde{\partial}) := \mathrm{s}_l\left(\frac{\partial}{\partial t_1}, \frac{1}{2}\frac{\partial}{\partial t_2}, \dots\right). \tag{1.98}$$

One also needs *Taylor's formula* for a $\mathcal{C}^{\infty}$-function f:

$$f(z+y) = \exp\left(y\frac{\partial}{\partial z}\right) f(z) , \tag{1.99}$$

which is seen by expanding the exponential. The following lemma will also be used later in the proof of the bilinear relations:

Lemma 1.5.1 ([5, 6]). *If $\oint_{\infty}$ denotes the integral along a small circle about ∞, the following identity holds (formal identity in terms of power series):*

$$\int_{\mathbb{R}} f(u)g(u)\,\mathrm{d}u = \frac{1}{2\pi\mathrm{i}} \oint_{\infty} \mathrm{d}z\, f(z) \int_{\mathbb{R}} \frac{g(u)}{z-u}\,\mathrm{d}u , \tag{1.100}$$

for holomorphic $f(z) = \sum_{i\geq 0} a_i z^i$ and $g(z)$, the latter assumed to have all its moments.

Proof. For holomorphic functions f in $\mathbb{C}$,

$$\begin{aligned}
\frac{1}{2\pi\mathrm{i}} \oint_{\infty} \mathrm{d}z f(z) \int_{\mathbb{R}} \frac{g(u)}{z-u}\,\mathrm{d}u &= \mathrm{Res}_{z=\infty}\left(\sum_{i\geq 0} a_i z^i\right)\left(\frac{1}{z}\sum_{j\geq 0} z^{-j}\int_{\mathbb{R}} g(u)u^j\,\mathrm{d}u\right) \\
&= \sum_{i\geq 0} a_i \int_{\mathbb{R}} g(u)u^i\,\mathrm{d}u \\
&= \int_{\mathbb{R}} g(u)\sum_{i\geq 0} a_i u^i\,\mathrm{d}u = \langle f, g\rangle ,
\end{aligned}$$

ending the proof of Lemma 1.5.1. □

The next theorem shows that the determinant of the *time*-dependent moment matrices satisfies the *KP hierarchy*, a nonlinear hierarchy, whereas in the next section, it will be shown that these same determinants satisfy *Virasoro equations*. These two features will play an important role in random matrix theory. Notice that this result is very robust: it can be generalized from orthogonal polynomials to multiple orthogonal polynomials, from the KP hierarchy to multi-component KP hierarchies; see [13].

Theorem 1.5.1 ([3,5]). *The determinants of the moment matrices, also representable as a multiple integral,*[22]

$$\tau_n(t) := \det m_n(t) = \frac{1}{n!}\int_{E^n} \Delta_n^2(z) \prod_{k=1}^{n} \rho(z_k) \exp\left(\sum_{i=1}^{\infty} t_i z_k^i\right) \mathrm{d}z_k \tag{1.101}$$

satisfy

[22] $\Delta_n(z) = \prod_{1\leq i<j\leq n}(z_i - z_j)$

(i) Eigenvectors of L: *The tridiagonal matrix* $L(t)$ *admits two independent eigenvectors*:

$$\begin{aligned} \big(L(t)p(t;z)\big)_n &= zp_n(t;z)\,, & n\geq 0\\ \big(L(t)q(t;z)\big)_n &= zq_n(t;z)\,, & n\geq 1\,. \end{aligned}$$

• $p_n(t;z)$ *are* n*th degree polynomials in* z, *depending on* $t\in\mathbb{C}^\infty$, *orthonormal with respect to* $\rho_t(z)$ (*defined in* (1.94)), *and enjoying the following representations*: (*define* $\chi(z) := (1,z,z^2,\dots)^\top$)

$$p_n(t;z) := \big(S(t)\chi(z)\big)_n = z^n h_n^{-1/2}\frac{\tau_n(t-[z^{-1}])}{\tau_n(t)}\,,\qquad h_n := \frac{\tau_{n+1}(t)}{\tau_n(t)} \tag{1.102}$$

• $q_n(t,z)$, $n\geq 0$, *are Stieltjes transforms of the polynomials* $p_n(t;z)$ *and have the following* τ*-function representations*:

$$\begin{aligned} q_n(t;z) := z\int_{\mathbb{R}^n}\frac{p_n(t;u)}{z-u}\rho_t(u)\,\mathrm{d}u &= \big(S^{\top-1}(t)\chi(z^{-1})\big)_n\\ &= z^{-n}h_n^{-1/2}\frac{\tau_{n+1}(t+[z^{-1}])}{\tau_n(t)}\,. \end{aligned}\tag{1.103}$$

(ii) *The* standard Toda lattice, *i.e., the symmetric tridiagonal matrix*

$$\begin{aligned} L(t) &:= S(t)\Lambda S(t)^{-1}\\ &= \begin{pmatrix} (\partial/\partial t_1)\log(\tau_1/\tau_0) & (\tau_0\tau_2/\tau_1^2)^{1/2} & 0 & \\ (\tau_0\tau_2/\tau_1^2)^{1/2} & (\partial/\partial t_1)\log(\tau_2/\tau_1) & (\tau_1\tau_3/\tau_2^2)^{1/2} & \\ 0 & (\tau_1\tau_3/\tau_2^2)^{1/2} & (\partial/\partial t_1)\log(\tau_3/\tau_2) & \\ & & & \ddots \end{pmatrix} \end{aligned}\tag{1.104}$$

satisfies the commuting equations[23]

$$\frac{\partial L}{\partial t_k} = \big[\tfrac12(L^k)_{\mathfrak{s}},L\big] = -\big[\tfrac12(L^k)_{\mathfrak{b}},L\big]\,. \tag{1.105}$$

(iii) *The functions* $\tau_n(t)$ *satisfy the following* bilinear *identity, for* $n\geq m+1$, *and all* $t,t'\in\mathbb{C}^\infty$, *where one integrates along a small circle about* ∞,

$$\oint_{z=\infty}\tau_n(t-[z^{-1}])\tau_{m+1}(t'+[z^{-1}])\exp\Big(\sum_1^\infty(t_i-t_i')z^i\Big)z^{n-m-1}\,\mathrm{d}z = 0\,. \tag{1.106}$$

(iv) *The* KP-hierarchy[24] *for* $k=0,1,2,\dots$ *and for all* $n=1,2\dots$,

$$\left(\mathbf{s}_{k+4}\Big(\frac{\partial}{\partial t_1},\frac12\frac{\partial}{\partial t_2},\frac13\frac{\partial}{\partial t_3},\dots\Big) - \frac12\frac{\partial^2}{\partial t_1\partial t_{k+3}}\right)\tau_n\circ\tau_n = 0\,,$$

[23] in terms of the Lie algebra splitting (1.96).

[24] Remember the Hirota symbol (1.97) and the Schur polynomial notation (1.98).

of which the first equation reads:

$$\left(\left(\frac{\partial}{\partial t_1}\right)^4+3\left(\frac{\partial}{\partial t_2}\right)^2-4\frac{\partial^2}{\partial t_1\partial t_3}\right)\log\tau_n+6\left(\frac{\partial^2}{\partial t_1^2}\log\tau_n\right)^2=0\,. \tag{1.107}$$

Remark. In order to connect with classical integrable theory, notice that, when τ satisfies the equation above, the function

$$q(t_1,t_2,\dots):=2\frac{\partial^2\log\tau_n}{\partial t_1^2}$$

satisfies the classical Kadomtsev–Petviashvili (KP) equation:

$$3\frac{\partial^2 q}{\partial t_2^2}-\frac{\partial}{\partial t_1}\left(4\frac{\partial q}{\partial t_3}-\frac{\partial^3 q}{\partial t_1^3}-6q\frac{\partial q}{\partial t_1}\right)=0\,. \tag{1.108}$$

If q happens to be independent of t_2, then q satisfies the Korteweg–de Vries equation

$$4\frac{\partial q}{\partial t_3}=\frac{\partial^3 q}{\partial t_1^3}+6q\frac{\partial q}{\partial t_1}\,. \tag{1.109}$$

Proof. Identity (1.101) follows from the general fact that the product of two matrices can be expressed as a symmetric sum of determinants,[25] in particular the square of a Vandermonde can be expressed as a sum of determinants:

$$\Delta^2(u_1,\dots,u_n)=\sum_{\sigma\in S_n}\det\left(u_{\sigma(k)}^{l+k-2}\right)_{1\le k,l\le n}\,.$$

Indeed,

$$\begin{aligned}
n!\tau_n(t)&=n!\det m_n(t)\\
&=\sum_{\sigma\in S_n}\det\left(\int_E z_{\sigma(k)}^{l+k-2}\rho_t(z_{\sigma(k)})\,\mathrm{d}z_{\sigma(k)}\right)_{1\le k,l\le n}\\
&=\sum_{\sigma\in S_n}\int_{E^n}\det\left(z_{\sigma(k)}^{l+k-2}\right)_{1\le k,l\le n}\rho_t(z_{\sigma(k)})\,\mathrm{d}z_{\sigma(k)}\\
&=\int_{E^n}\Delta_n^2(z)\prod_{k=1}^n\rho_t(z_k)\,\mathrm{d}z_k\,.
\end{aligned}$$

(i) At first, note

$$\begin{aligned}
\mu_{ij}(t\mp[z^{-1}])&=\int_{\mathbb{R}}u^{i+j}\exp\left(\sum_1^\infty\left(t_i\mp\frac{z^{-i}}{i}\right)u^i\right)\rho(u)\,\mathrm{d}u\\
&=\int_{\mathbb{R}}u^{i+j}\left(1-\frac{u}{z}\right)^{\pm1}\rho(u)\exp\left(\sum_0^\infty t_iu^i\right)\mathrm{d}u\\
&=\begin{cases}\mu_{i,j}(t)-\mu_{i,j+1}(t)/z\\ \mu_{i,j}(t)+\mu_{i,j+1}(t)/z+\mu_{i,j+2}(t)/z^2+\cdots\,,\end{cases}
\end{aligned}$$

[25] Indeed, $\sum_{\sigma\in S_n}\det\left(a_{i,\sigma(j)}b_{j,\sigma(j)}\right)_{1\le i,j\le n}=\det(a_{ik})_{1\le i,k\le n}\det(b_{ik})_{1\le i,k\le n}$.

which by formula (1.91) of Sect. 1.5.1 leads at once to the following representation for the monic orthogonal polynomials $\tilde{p}_n(t;z)$ and their Stieltjes transforms,

$$\begin{aligned} \tilde{p}_n(t;z) &= z^n \frac{\tau_n(t-[z^{-1}])}{\tau_n(t)} \\ z\int_{\mathbb{R}} \frac{\tilde{p}_n(t;u)\rho_t(u)}{z-u}\,du &= z^{-n}\frac{\tau_{n+1}(t+[z^{-1}])}{\tau_n(t)}\,. \end{aligned} \tag{1.110}$$

(ii) The matrix $L := S\Lambda S^{-1}$ satisfies the standard Toda lattice. One computes

$$\frac{\partial \mu_{ij}}{\partial t_k} = \mu_{i+k,j} \quad \text{implying} \quad \frac{\partial m_\infty}{\partial t_k} = \Lambda^k m_\infty\,.$$

Then, using the factorization (1.95) and the definition (1.92) of $L = S\Lambda S^{-1}$ of Sect. 1.5.1, one computes

$$\begin{aligned} 0 = S\left(\Lambda^k m_\infty - \frac{\partial m_\infty}{\partial t_k}\right)S^\top &= S\Lambda^k S^{-1} - S\frac{\partial}{\partial t_k}(S^{-1}S^{\top -1})S^\top \\ &= L^k + \frac{\partial S}{\partial t_k}S^{-1} + S^{\top -1}\frac{\partial S^\top}{\partial t_k}\,. \end{aligned}$$

Upon taking the $(\ )_-$ and $(\ )_0$ parts of this equation (A_- means the lower-triangular part of the matrix A, including the diagonal and A_0 the diagonal part) leads to

$$(L^k)_- + \frac{\partial S}{\partial t_k}S^{-1} + \left(S^{\top -1}\frac{\partial S^\top}{\partial t_k}\right)_0 = 0 \quad \text{and} \quad \left(\frac{\partial S}{\partial t_k}S^{-1}\right)_0 = -\tfrac{1}{2}(L^k)_0\,.$$

Upon observing that for any symmetric matrix the following holds,

$$\begin{pmatrix} a & c \\ c & b \end{pmatrix}_{\mathfrak{b}} = \begin{pmatrix} a & 0 \\ 2c & b \end{pmatrix} = 2\begin{pmatrix} a & c \\ c & b \end{pmatrix}_- - \begin{pmatrix} a & c \\ c & b \end{pmatrix}_0\,,$$

it follows that the matrices $L(t)$, $S(t)$ and the vector $p(t;z) = \big(p_n(t;z)\big)_{n\geq 0} = S(t)\chi(z)$ satisfy the (commuting) differential equations and the eigenvalue problem

$$\frac{\partial S}{\partial t_k} = -\tfrac{1}{2}(L^k)_{\mathfrak{b}}S\,, \quad L(t)p(t;z) = zp(t;z)\,, \tag{1.111}$$

and thus the tridiagonal matrix L satisfies the standard Toda lattice equations

$$\frac{\partial L}{\partial t_k} = \frac{\partial}{\partial t_k}S\Lambda S^{-1} = \frac{\partial S}{\partial t_k}S^{-1}S\Lambda S^{-1} - S\Lambda S^{-1}\frac{\partial S}{\partial t_k}S^{-1} = -[\tfrac{1}{2}(L^k)_{\mathfrak{b}}, L]$$

with $p(t;z)$ satisfying

$$\frac{\partial p}{\partial t_k} = \frac{\partial S}{\partial t_k}\chi(z) = -\tfrac{1}{2}(L^k)_{\mathfrak{b}}S\chi(z) = -\tfrac{1}{2}(L^k)_{\mathfrak{b}}p\,.$$

The two leading terms of $p_n(t;z)$ look as follows, upon using (1.89) and (1.110):

$$p_n(t;z) = \sqrt{\frac{\tau_n}{\tau_{n+1}}}\tilde{p}_n(t;z) = z^n \frac{\tau_n(t-[z^{-1}])}{\sqrt{\tau_n\tau_{n+1}}} = \sqrt{\frac{\tau_n}{\tau_{n+1}}} z^n \left(1 - z^{-1}\frac{\partial\tau_n/\partial t_1}{\tau_n} + \cdots\right). \tag{1.112}$$

Thus, z^n admits the following representation in terms of the orthonormal polynomials p_i:

$$z^n = \sqrt{\frac{\tau_{n+1}}{\tau_n}}\left(p_n + \frac{\partial\tau_n/\partial t_1}{\sqrt{\tau_n\tau_{n+1}}} z^{n-1} + O(z^{n-2})\right) = \sqrt{\frac{\tau_{n+1}}{\tau_n}} p_n + \frac{\partial\tau_n/\partial t_1}{\sqrt{\tau_{n-1}\tau_n}} p_{n-1} + O(z^{n-2})\,. \tag{1.113}$$

Then, using (1.112) in zp_n and then using the representation (1.113) for z^n and z^{n+1}, one checks that the diagonal entries b_n and nondiagonal entries a_n of L are given by

$$b_n = \langle zp_n, p_n\rangle = \sqrt{\frac{\tau_n}{\tau_{n+1}}}\left(\langle z^{n+1}, p_n\rangle - \langle z^n, p_n\rangle\frac{\partial\tau_n/\partial t_1}{\tau_n}\right) = \frac{\partial\tau_{n+1}/\partial t_1}{\tau_n} - \frac{\partial\tau_n/\partial t_1}{\tau_n} = \frac{\partial}{\partial t_1}\log\frac{\tau_{n+1}}{\tau_n}$$

and

$$a_n = \langle zp_n, p_{n+1}\rangle = \sqrt{\frac{\tau_n}{\tau_{n+1}}}\langle z^{n+1} + \cdots, p_{n+1}\rangle = \sqrt{\frac{\tau_n\tau_{n+2}}{\tau_{n+1}^2}}\,,$$

establishing (1.104).

(iii) *The bilinear identity*: The functions $\tau_n(t)$ satisfy the following identity, for $n \geq m+1$, $t, t' \in \mathbb{C}^\infty$, where one integrates along a small circle about ∞,

$$\oint_{z=\infty} \tau_n(t-[z^{-1}])\tau_{m+1}(t'+[z^{-1}])\exp\left(\sum_1^\infty (t_i - t'_i)z^i\right) z^{n-m-1}\,dz = 0\,. \tag{1.114}$$

Indeed, using the τ-function representation for the monic orthogonal polynomials and their Stieltjes transform (1.110), one checks:

$$\frac{1}{\tau_n(t)\tau_m(t')}\oint_{z=\infty}\tau_n(t-[z^{-1}])\tau_{m+1}(t'+[z^{-1}])\exp\Big(\sum_1^\infty(t_i-t_i')z^i\Big)z^{n-m-1}\,\mathrm{d}z$$
$$=\oint_{z=\infty}z^n\frac{\tau_n(t-[z^{-1}])}{\tau_n(t)}z^{-m}\frac{\tau_{m+1}(t'+[z^{-1}])}{\tau_m(t')}\exp\Big(\sum_1^\infty(t_i-t_i')z^i\Big)\frac{\mathrm{d}z}{z}$$
$$=\oint_{z=\infty}\mathrm{d}z\,\exp\Big(\sum_1^\infty(t_i-t_i')z^i\Big)\tilde p_n(t;z)\int_{\mathbb{R}}\frac{\tilde p_m(t';u)}{z-u}\exp\Big(\sum_1^\infty t_i'u^i\Big)\rho(u)\,\mathrm{d}u$$
$$=2\pi\mathrm{i}\int_{\mathbb{R}}\exp\Big(\sum_1^\infty(t_i-t_i')z^i\Big)\tilde p_n(t;z)\tilde p_m(t';z)\exp\Big(\sum_1^\infty t_i'z^i\Big)\rho(z)\,\mathrm{d}z\,,$$

using Lemma 1.5.1,

$$=2\pi\mathrm{i}\int_{\mathbb{R}}\tilde p_n(t;z)\tilde p_m(t';z)\exp\Big(\sum_1^\infty t_iz^i\Big)\rho(z)\,\mathrm{d}z=0\,,\quad\text{when } m\le n-1,$$

by orthogonality, establishing (1.114).

(iv) *The KP hierarchy* Setting $n=m+1$ in (1.106), shifting $t\mapsto t-y, t'\mapsto t+y$, evaluating the residue, Taylor expanding in y_k (see (1.99)) and using the notation

$$\tilde\partial=\Big(\frac{\partial}{\partial t_1},\frac{1}{2}\frac{\partial}{\partial t_2},\frac{1}{3}\frac{\partial}{\partial t_3},\dots\Big)\,,$$

one computes the following residue about $z=\infty$,

$$0=\frac{1}{2\pi\mathrm{i}}\oint_{z=\infty}\mathrm{d}z\,\exp\Big(-\sum_1^\infty 2y_iz^i\Big)\tau_n(t-y-[z^{-1}])\tau_n(t+y+[z^{-1}])$$
$$=\frac{1}{2\pi\mathrm{i}}\oint\mathrm{d}z\,\exp\Big(-\sum_1^\infty 2y_iz^i\Big)\exp\Big(\sum_1^\infty\frac{z^{-i}}{i}\frac{\partial}{\partial u_i}\Big)\exp\Big(\sum_1^\infty y_k\frac{\partial}{\partial u_k}\Big)\times\tau_n(t-u)\tau_n(t+u)\Big|_{u=0}$$
$$=\frac{1}{2\pi\mathrm{i}}\oint\mathrm{d}z\,\exp\Big(-\sum_1^\infty 2y_iz^i\Big)\exp\Big(\sum_1^\infty\frac{z^{-i}}{i}\frac{\partial}{\partial t_i}\Big)\exp\Big(\sum_1^\infty y_k\frac{\partial}{\partial t_k}\Big)\times\tau_n(t)\circ\tau_n(t)$$
$$=\frac{1}{2\pi\mathrm{i}}\oint\mathrm{d}z\,\Big(\sum_0^\infty z^i\mathbf{s}_i(-2y)\Big)\Big(\sum_0^\infty z^{-j}\mathbf{s}_j(\tilde\partial)\Big)\exp\Big(\sum_1^\infty y_k\frac{\partial}{\partial t_k}\Big)\tau_n\circ\tau_n$$
$$=\exp\Big(\sum_1^\infty y_k\frac{\partial}{\partial t_k}\Big)\sum_0^\infty\mathbf{s}_i(-2y)\mathbf{s}_{i+1}(\tilde\partial)\tau_n\circ\tau_n$$
$$=\Big(1+\sum_1^\infty y_j\frac{\partial}{\partial t_j}+O(y^2)\Big)\Big(\frac{\partial}{\partial t_1}+\sum_1^\infty\mathbf{s}_{i+1}(\tilde\partial)(-2y_i+O(y^2))\Big)\tau_n\circ\tau_n$$
$$=\Big(\frac{\partial}{\partial t_1}+\sum_1^\infty y_k\Big(\frac{\partial}{\partial t_k}\frac{\partial}{\partial t_1}-2\mathbf{s}_{k+1}(\tilde\partial)\Big)\Big)\tau_n\circ\tau_n+O(y^2)\,,$$

for arbitrary y_k, implying

$$\frac{\partial}{\partial t_1}\tau \circ \tau = 0 \quad \text{and} \quad \left(\frac{\partial^2}{\partial t_k \partial t_1} - 2\mathbf{s}_{k+1}(\tilde{\partial})\right)\tau_n \circ \tau_n = 0 \text{ for } k = 1, 2, \dots .$$

Taking into account the fact that trivially $(\partial/\partial t_1)\tau \circ \tau = 0$ and that the equation above is trivial for $k = 1$ and $k = 2$, one is led to the KP hierarchy:

$$\left(\mathbf{s}_{k+4}\left(\frac{\partial}{\partial t_1}, \frac{1}{2}\frac{\partial}{\partial t_2}, \frac{1}{3}\frac{\partial}{\partial t_3}, \dots\right) - \frac{1}{2}\frac{\partial^2}{\partial t_1 \partial t_{k+3}}\right)\tau_n \circ \tau_n = 0\,, \quad \text{for } k = 0, 1, 2, \dots .$$

In particular, for $k = 0$, one computes

$$\mathbf{s}_4(t) := \frac{t_1^4}{4!} + \frac{1}{2}t_2 t_1^2 + t_3 t_1 + \frac{1}{2}t_2^2 + t_4\,,$$

leading to the first equation in the hierarchy

$$\left(\left(\frac{\partial}{\partial t_1}\right)^4 + 3\left(\frac{\partial}{\partial t_2}\right)^2 - 4\frac{\partial^2}{\partial t_1 \partial t_3}\right)\log \tau_n + 6\left(\frac{\partial^2}{\partial t_1^2}\log \tau_n\right)^2 = 0\,. \qquad (1.115)$$

This ends the proof of Thm. 1.5.1. □

Remark. As mentioned earlier this method is very robust and can be generalized to other integrals, besides (1.101), upon using multiple orthogonal polynomials. Such integrals with appropriate multiple time-deformations lead to τ-functions for multi-component KP hierachies; see Adler–van Moerbeke–Vanhaecke [13].

1.6 Virasoro Constraints

1.6.1 Virasoro Constraints for β-Integrals

Consider weights $\rho(z)\,\mathrm{d}z = \mathrm{e}^{-V(z)}\,dz$ with rational logarithmic derivative and E a disjoint union of intervals:

$$-\frac{\rho'}{\rho} = V'(z) = \frac{g}{f} = \frac{\sum_0^\infty b_i z^i}{\sum_0^\infty a_i z^i} \quad \text{and} \quad E = \bigcup_1^r [c_{2i-1}, c_{2i}] \subset F \subseteq \mathbb{R}\,,$$

where $F = [A, B]$ is an interval such that

$$\lim_{z \to A,B} f(z)\rho(z)z^k = 0 \quad \text{for all } k \geq 0\,.$$

Consider an integral $I_n(t, c; \beta)$, generalizing (1.101), where $t := (t_1, t_2, \dots)$ and $c = (c_1, c_2, \dots, c_{2r})$; namely with a Vandermonde[26] to the power $2\beta > 0$ instead of a square, and omitting the $n!$ appearing in (1.101),

[26] $\Delta_n(z) = \prod_{1 \leq i < j \leq n}(z_i - z_j)$.

$$I_n(t,c;\beta) := \int_{E^n} |\Delta_n(z)|^{2\beta} \prod_{k=1}^{n} \left(\exp\left(\sum_1^\infty t_i z_k^i \right) \rho(z_k)\, \mathrm{d}z_k \right) \quad \text{for } n > 0\,. \tag{1.116}$$

Then the following theorem holds:

Theorem 1.6.1 (Adler–van Moerbeke [8]). *The multiple integrals* $I_n(t) := I_n(t,c;\beta)$ *with* $I_0 = 1$*, satisfy the following Virasoro constraints*[27] *for all* $k \geq -1$:

$$\left(-\sum_1^{2r} c_i^{k+1} f(c_i) \frac{\partial}{\partial c_i} + \sum_{i\geq 0} \left(a_i \mathbb{J}^{(2)}_{k+i,n}(t,n) - b_i \mathbb{J}^{(1)}_{k+i+1,n}(t,n) \right) \right) I_n(t) = 0\,, \tag{1.117}$$

where $\mathbb{J}^{(2)}_{k,n}(t,n)$ *and* $\mathbb{J}^{(1)}_{k,n}(t,n)$ *are combined differential and multiplication (linear) operators. For all* $n \in \mathbb{Z}$*, the operators* $\mathbb{J}^{(2)}_{k,n}(t,n)$ *and* $\mathbb{J}^{(1)}_{k,n}(t,n)$ *form a Virasoro and a Heisenberg algebra respectively, interacting as follows*

$$\begin{aligned} [\mathbb{J}^{(2)}_{k,n}, \mathbb{J}^{(2)}_{l,n}] &= (k-l)\mathbb{J}^{(2)}_{k+l,n} + c\left(\frac{k^3-k}{12} \right) \delta_{k,-l} \\ [\mathbb{J}^{(2)}_{k,n}, \mathbb{J}^{(1)}_{l,n}] &= -l\mathbb{J}^{(1)}_{k+l,n} + c' k(k+1)\delta_{k,-l} \\ [\mathbb{J}^{(1)}_{k,n}, \mathbb{J}^{(1)}_{l,n}] &= \frac{k}{2\beta}\delta_{k,-l}, \end{aligned} \tag{1.118}$$

with "central charge"

$$c = 1 - 6(\beta^{1/2} - \beta^{-1/2})^2 \quad \text{and} \quad c' = \frac{1}{2}\left(\frac{1}{\beta} - 1 \right). \tag{1.119}$$

Remark 1.6.1. The operators $\mathbb{J}^{(2)}_{k,n} = \mathbb{J}^{(2)}_{k,n}(t,n)$s are defined as follows: (the normal ordering symbol ": :" means: always pull differentiation to the right, ignoring commutation rules)

$$\mathbb{J}^{(2)}_{k,n} = \beta \sum_{i+j=k} {:}\mathbb{J}^{(1)}_{i,n}\mathbb{J}^{(1)}_{j,n}{:} + (1-\beta)\left((k+1)\mathbb{J}^{(1)}_{k,n} - k\mathbb{J}^{(0)}_{k,n} \right)\,, \tag{1.120}$$

in terms of the $\mathbb{J}^{(1)}_{k,n} = \mathbb{J}^{(1)}_{k,n}(t,n)$s. Componentwise, we have

$$\mathbb{J}^{(1)}_{k,n}(t,n) = J^{(1)}_k + nJ^{(0)}_k \quad \text{and} \quad \mathbb{J}^{(0)}_{k,n} = nJ^{(0)}_k = n\delta_{0k}$$

and

$$\mathbb{J}^{(2)}_{k,n}(t,n) = \beta J^{(2)}_k + \left(2n\beta + (k+1)(1-\beta) \right) J^{(1)}_k + n(n\beta + 1 - \beta) J^{(0)}_k\,,$$

[27] When E equals the whole range F, then the first term, containing the partials $\partial/\partial c_i$, are absent in (1.117).

where

$$J_k^{(0)} = \delta_{k0} \ , \quad J_k^{(1)} = \frac{\partial}{\partial t_k} + \frac{1}{2\beta}(-k)t_{-k} \ ,$$
$$J_k^{(2)} = \sum_{i+j=k} \frac{\partial^2}{\partial t_i \partial t_j} + \frac{1}{\beta} \sum_{-i+j=k} it_i \frac{\partial}{\partial t_j} + \frac{1}{4\beta^2} \sum_{-i-j=k} it_i jt_j \ . \tag{1.121}$$

The integer n appears explicitly in $\mathbb{J}_{k,n}^{(2)}(t,n)$ to indicate the explicit n-dependence of the nth component, besides t.

Remark 1.6.2. In the case $\beta = 1$, the Virasoro generators (1.121) take on a particularly elegant form, namely for $n \geq 0$,

$$\mathbb{J}_{k,n}^{(2)}(t) = \sum_{i+j=k} :\mathbb{J}_{i,n}^{(1)}(t)\mathbb{J}_{j,n}^{(1)}(t): = J_k^{(2)}(t) + 2nJ_k^{(1)}(t) + n^2\delta_{0k}$$
$$\mathbb{J}_{k,n}^{(1)}(t) = J_k^{(1)}(t) + n\delta_{0k} \ ,$$

with[28]

$$J_k^{(1)} = \frac{\partial}{\partial t_k} + \frac{1}{2}(-k)t_{-k} \ ,$$
$$J_k^{(2)} = \sum_{i+j=k} \frac{\partial^2}{\partial t_i \partial t_j} + \sum_{-i+j=k} it_i \frac{\partial}{\partial t_j} + \frac{1}{4} \sum_{-i-j=k} it_i jt_j \ . \tag{1.122}$$

One now establishes the following lemma:

Lemma 1.6.1. *Setting*

$$\mathrm{d}I_n(x) := |\Delta_n(x)|^{2\beta} \prod_{k=1}^{n} \left(\exp\left(\sum_1^{\infty} t_i x_k^i \right) \rho(x_k)\, \mathrm{d}x_k \right) \ ,$$

the following variational formula holds:

$$\frac{\mathrm{d}}{\mathrm{d}\varepsilon} \mathrm{d}I_n(x_i \mapsto x_i + \varepsilon f(x_i) x_i^{k+1}) \Big|_{\varepsilon=0} = \sum_{l=0}^{\infty} \left(a_l \mathbb{J}_{k+l,n}^{(2)} - b_l \mathbb{J}_{k+l+1,n}^{(1)} \right) \mathrm{d}I_n \ . \tag{1.123}$$

Proof. Upon setting

$$\mathcal{E}(x,t) := \prod_{k=1}^{n} \exp\left(\sum_{i=1}^{\infty} t_i x_k^i \right) \rho(x_k) \ , \tag{1.124}$$

the following two relations hold:

[28] The expression $J_k^{(1)} = 0$ for $k = 0$.

$$\left(\frac{1}{2}\sum_{\substack{i+j=k\\ i,j>0}}\frac{\partial^2}{\partial t_i\partial t_j}-\frac{n}{2}\delta_{k,0}\right)\mathcal{E}=\left(\sum_{\substack{1\le a<\beta\le n\\ i,j>0\\ i+j=k}}x_\alpha^i x_\beta^j+\frac{k-1}{2}\sum_{1\le\alpha\le n}x_\alpha^k\right)\mathcal{E}\,,$$
$$\left(\frac{\partial}{\partial t_k}+n\delta_{k,0}\right)\mathcal{E}=\left(\sum_{1\le\alpha\le n}x_\alpha^k\right)\mathcal{E}\,,\quad \text{all } k\ge 0\,. \tag{1.125}$$

So, the point now is to compute the ε-derivative

$$\frac{\mathrm{d}}{\mathrm{d}\varepsilon}\left(|\Delta_n(x)|^{2\beta}\times\exp\left(\sum_{k=1}^{n}\left(-V(x_k)+\sum_{i=1}^{\infty}t_i x_k^i\right)\right)\mathrm{d}x_1\cdots\mathrm{d}x_n\right)\Bigg|_{x_i\mapsto x_i+\varepsilon f(x_i)x_i^{k+1}}\Bigg|_{\varepsilon=0}\,, \tag{1.126}$$

which consists of three contributions:

Contribution 1.

$$\begin{aligned}
&\frac{1}{2\beta|\Delta(x)|^{2\beta}}\frac{\partial}{\partial\varepsilon}|\Delta(x+\varepsilon f(x)x^{k+1})|^{2\beta}\Big|_{\varepsilon=0}\\
&=\sum_{1\le\alpha<\gamma\le n}\frac{\partial}{\partial\varepsilon}\log(|x_\alpha-x_\gamma+\varepsilon(f(x_\alpha)x_\alpha^{k+1}-f(x_\gamma)x_\gamma^{k+1})|)\Big|_{\varepsilon=0}\\
&=\sum_{1\le\alpha<\gamma\le n}\frac{f(x_\alpha)x_\alpha^{k+1}-f(x_\gamma)x_\gamma^{k+1}}{x_\alpha-x_\gamma}\\
&=\sum_{l=0}^{\infty}a_l\sum_{1\le\alpha<\gamma\le n}\frac{x_\alpha^{k+l+1}-x_\gamma^{k+l+1}}{x_\alpha-x_\gamma}\\
&=\sum_{l=0}^{\infty}a_l\left(\sum_{\substack{i+j=l+k\\ i,j>0\\ 1\le\alpha<\gamma\le n}}x_\alpha^i x_\gamma^j+(n-1)\sum_{1\le\alpha\le n}x_\alpha^{l+k}-\frac{n(n-1)}{2}\delta_{l+k,0}\right)\\
&=\mathcal{E}^{-1}\sum_{l=0}^{\infty}a_l\left(\frac{1}{2}\sum_{\substack{i+j=k+l\\ i,j>0}}\frac{\partial^2}{\partial t_i\partial t_j}-\frac{n}{2}\delta_{k+l,0}\right.\\
&\qquad\left.+\left(n-\frac{k+l+1}{2}\right)\left(\frac{\partial}{\partial t_{k+l}}+n\delta_{k+l,0}\right)-\frac{n(n-1)}{2}\delta_{k+l,0}\right)\mathcal{E}\,,\\
&\qquad\qquad\text{using (1.125)},\\
&=\mathcal{E}^{-1}\sum_{l=0}^{\infty}a_l\left(\frac{1}{2}\sum_{\substack{i+j=k+l\\ i,j>0}}\frac{\partial^2}{\partial t_i\partial t_j}+\left(n-\frac{k+l+1}{2}\right)\frac{\partial}{\partial t_{k+l}}+\frac{n(n-1)}{2}\delta_{k+l,0}\right)\mathcal{E}\,.
\end{aligned} \tag{1.127}$$

Contribution 2. Using $f(x)=\sum_0^\infty a_i x^i$,

$$
\begin{aligned}
\frac{\partial}{\partial\varepsilon}\prod_1^n \mathrm{d}(x_\alpha+\varepsilon f(x_\alpha)x_\alpha^{k+1})\Big|_{\varepsilon=0}
&=\sum_1^n (f'(x_\alpha)x_\alpha^{k+1}+(k+1)f(x_\alpha)x_\alpha^k)\prod_1^n \mathrm{d}x_i\\
&=\sum_{l=0}^\infty (l+k+1)a_l\sum_{\alpha=1}^n x_\alpha^{k+l}\prod_1^n \mathrm{d}x_i\\
&=\mathcal{E}^{-1}\sum_{l=0}^\infty (l+k+1)a_l\left(\frac{\partial}{\partial t_{k+l}}+n\delta_{k+l,0}\right)\mathcal{E}\prod_1^n \mathrm{d}x_i\,. \qquad (1.128)
\end{aligned}
$$

Contribution 3. Again using $f(x)=\sum_0^\infty a_i x^i$,

$$
\begin{aligned}
&\frac{\partial}{\partial\varepsilon}\prod_{\alpha=1}^n \exp\Big(-V(x_\alpha+\varepsilon f(x_\alpha)x^{k+1})+\sum_{i=1}^\infty t_i\sum_{\alpha=1}^n (x_\alpha+\varepsilon f(x_\alpha)x_\alpha^{k+1})^i\Big)\Big|_{\varepsilon=0}\\
&\quad=\Big(-\sum_{\alpha=1}^n V'(x_\alpha)f(x_\alpha)x_\alpha^{k+1}+\sum_{i=1}^\infty it_i\sum_{\alpha=1}^n f(x_\alpha)x_\alpha^{i+k}\Big)\mathcal{E}\\
&\quad=\Big(-\sum_{l=0}^\infty b_l\sum_{\alpha=1}^n x_\alpha^{k+l+1}+\sum_{\substack{l\geq 0\\ i\geq 1}} a_l it_i\sum_{\alpha=1}^n x_\alpha^{i+k+l}\Big)\mathcal{E}\\
&\quad=\Big(-\sum_{l=0}^\infty b_l\Big(\frac{\partial}{\partial t_{k+l+1}}+n\delta_{k+l+1,0}\Big)\\
&\qquad\qquad+\sum_{l=0}^\infty a_l\sum_{i=1}^\infty it_i\Big(\frac{\partial}{\partial t_{i+k+l}}+n\delta_{i+k+l,0}\Big)\Big)\mathcal{E}\,. \qquad (1.129)
\end{aligned}
$$

As mentioned, for knowing (1.123), we must add up the three contributions 1, 2 and 3, resulting in:

$$
\begin{aligned}
&\frac{\partial}{\partial\varepsilon}\mathrm{d}I_n(x_i\mapsto x_i+\varepsilon f(x_i)x_i^{k+1})\Big|_{\varepsilon=0}\\
&=\Big(\sum_{l=0}^\infty a_l\big(\beta J_{k+l}^{(2)}+\big(2n\beta+(l+k+1)(1-\beta)\big)J_{k+l}^{(1)}+n\big((n-1)\beta+1\big)\delta_{k+l,0}\big)\\
&\qquad\qquad-\sum_{l=0}^\infty b_l(J_{k+l+1}^{(1)}+n\delta_{k+l+1,0})\Big)\mathrm{d}I_n(x)\,.
\end{aligned}
$$

Finally, the use of (1.121) ends the proof of Lemma 1.6.1. □

Proof of Thm. 1.6.1. The change of integration variable $x_i\mapsto x_i+\varepsilon f(x_i)x_i^{k+1}$ in the integral (1.116) leaves the integral invariant, but it induces a change of limits of integration, given by the inverse of the map above; namely the c_is in $E = \bigcup_1^r [c_{2i-1}, c_{2i}]$, get mapped as follows

$$c_i \mapsto c_i - \varepsilon f(c_i) c_i^{k+1} + O(\varepsilon^2).$$

Therefore, setting

$$E^\varepsilon = \bigcup_1^r [c_{2i-1} - \varepsilon f(c_{2i-1}) c_{2i-1}^{k+1} + O(\varepsilon^2), c_{2i} - \varepsilon f(c_{2i}) c_{2i}^{k+1} + O(\varepsilon^2)]$$

and $V(x,t) = V(x) + \sum_i^\infty t_i x^i$, we find, using Lemma 1.6.1 and the fundamental theorem of calculus,

$$\begin{aligned} 0 &= \frac{\partial}{\partial \varepsilon} \int_{(E^\varepsilon)^{2n}} |\Delta_{2n}(x + \varepsilon f(x) x^{k+1})| \\ &\qquad \times \prod_{i=1}^{2n} \exp\bigl(-V(x_i + \varepsilon f(x_i) x_i^{k+1}, t)\bigr) \, \mathrm{d}(x_i + \varepsilon f(x_i) x_i^{k+1}) \\ &= \left(- \sum_{i=1}^{2r} c_i^{k+1} f(c_i) \frac{\partial}{\partial c_i} + \sum_{l=0}^{\infty} (a_l \mathbb{J}_{k+l,2n}^{(2)} - b_l \mathbb{J}_{k+l+1,2n}^{(1)}) \right) I_n(t, c, \beta) \, , \end{aligned}$$

ending the proof of Thm. 1.6.1. □

1.6.2 Examples

These examples are taken from [1,3,9,90]; for the Laguerre ensemble, see also [49] and for the Jacobi ensemble, see [48].

Example 1.6.1 (*GUE*). Here we pick

$$\begin{gathered} \rho(z) = \mathrm{e}^{-V(z)} = \exp(-z^2) \, , \quad V' = g/f = 2z \, , \\ a_0 = 1 \, , \, b_0 = 0 \, , \, b_1 = 2 \, , \text{ and all other } a_i, b_i = 0 \, . \end{gathered}$$

Define the differential operators

$$\mathcal{B}_k := \sum_1^{2r} c_i^{k+1} \frac{\partial}{\partial c_i} \tag{1.130}$$

in terms of the end points of the set $E = \bigcup_1^{2r} [c_{2i-1}, c_{2i}] \subset \mathbb{R}$. From Thm. 1.6.1, the integrals

$$I_n = \int_{E^n} \Delta_n(z)^2 \prod_{k=1}^n \exp\left(-z_k^2 + \sum_{i=1}^\infty t_i z_k^i \right) \mathrm{d}z_k \tag{1.131}$$

satisfy the Virasoro constraints

$$-\mathcal{B}_k I_n = (-\mathbb{J}^{(2)}_{k,n} + 2\mathbb{J}^{(1)}_{k+2,n}) I_n \,, \quad k = -1, 0, 1, \ldots \,. \tag{1.132}$$

The first three constraints have the following form, upon setting $F = \log I_n$ (this will turn out to be more convenient in the applications),

$$\begin{aligned}
-\mathcal{B}_{-1}F &= \left(2\frac{\partial}{\partial t_1} - \sum_{i\geq 2} it_i \frac{\partial}{\partial t_{i-1}}\right)F - nt_1 \\
-\mathcal{B}_0 F &= \left(2\frac{\partial}{\partial t_2} - \sum_{i\geq 1} it_i \frac{\partial}{\partial t_i}\right)F - n^2 \\
-\mathcal{B}_1 F &= \left(2\frac{\partial}{\partial t_3} - 2n\frac{\partial}{\partial t_1} - \sum_{i\geq 1} it_i \frac{\partial}{\partial t_{i+1}}\right)F
\end{aligned}$$

For later use, take linear combinations such that each expression contains the pure differentiation term $\partial F/\partial t_i$, yielding

$$\begin{aligned}
-\frac{1}{2}\mathcal{B}_{-1}F =: \mathcal{D}_{-1}F &= \left(\frac{\partial}{\partial t_1} - \frac{1}{2}\sum_{i\geq 2} it_i \frac{\partial}{\partial t_{i-1}}\right)F - \frac{nt_1}{2} \\
-\frac{1}{2}\mathcal{B}_0 F =: \mathcal{D}_0 F &= \left(\frac{\partial}{\partial t_2} - \frac{1}{2}\sum_{i\geq 1} it_i \frac{\partial}{\partial t_i}\right)F - \frac{n^2}{2} \\
-\frac{1}{2}(\mathcal{B}_1 + n\mathcal{B}_{-1})F =: \mathcal{D}_1 F &= \left(\frac{\partial}{\partial t_3} - \frac{1}{2}\sum_{i\geq 1} it_i \frac{\partial}{\partial t_{i+1}} - \frac{n}{2}\sum_{i\geq 2} it_i \frac{\partial}{\partial t_{i-1}}\right)F - \frac{n^2 t_1}{2} \,.
\end{aligned} \tag{1.133}$$

Example 1.6.2 (*Laguerre ensemble*). Here, the weight is

$$\mathrm{e}^{-V} = z^a \mathrm{e}^{-z}, \quad V' = \frac{g}{f} = \frac{z-a}{z},$$
$$a_0 = 0 \,,\ a_1 = 1 \,,\ b_0 = -a \,,\ b_1 = 1 \,, \text{ and all other } a_i, b_i = 0 \,.$$

Here define the (slightly different) differential operators

$$\mathcal{B}_k := \sum_1^{2r} c_i^{k+2} \frac{\partial}{\partial c_i} \,. \tag{1.134}$$

Thus from Thm. 1.6.1, the integrals

$$I_n = \int_{E^n} \Delta_n(z)^2 \prod_{k=1}^n z_k^a \exp\left(-z_k + \sum_{i=1}^\infty t_i z_k^i\right) \mathrm{d}z_k \tag{1.135}$$

satisfy the Virasoro constraints, for $k \geq -1$,

$$-\mathcal{B}_k I_n = (-\mathbb{J}^{(2)}_{k+1,n} - a\mathbb{J}^{(1)}_{k+1,n} + \mathbb{J}^{(1)}_{k+2,n}) I_n \,. \tag{1.136}$$

Written out, the first three have the form, again upon setting $F = \log I_n$,

$$-\mathcal{B}_{-1}F = \left(\frac{\partial}{\partial t_1} - \sum_{i\geq 1} it_i \frac{\partial}{\partial t_i}\right)F - n(n+a)$$

$$-\mathcal{B}_0 F = \left(\frac{\partial}{\partial t_2} - (2n+a)\frac{\partial}{\partial t_1} - \sum_{i\geq 1} it_i \frac{\partial}{\partial t_{i+1}}\right)F$$

$$-\mathcal{B}_1 F = \left(\frac{\partial}{\partial t_3} - (2n+a)\frac{\partial}{\partial t_2} - \sum_{i\geq 1} it_i \frac{\partial}{\partial t_{i+2}} - \frac{\partial^2}{\partial t_1^2}\right)F - \left(\frac{\partial F}{\partial t_1}\right)^2 ;$$

Replacing the operators $\mathcal{B}_i$ by linear combinations

$$\begin{aligned}
\mathcal{D}_1 &= -\mathcal{B}_{-1} \\
\mathcal{D}_2 &= -\mathcal{B}_0 - (2n+a)\mathcal{B}_{-1} \\
\mathcal{D}_3 &= -\mathcal{B}_1 - (2n+a)\mathcal{B}_0 - (2n+a)^2\mathcal{B}_{-1}
\end{aligned} \tag{1.137}$$

yields expressions, each containing a pure derivative $\partial F/\partial t_i$

$$\begin{aligned}
\mathcal{D}_1 F &= \frac{\partial F}{\partial t_1} - \sum_{i\geq 1} it_i \frac{\partial F}{\partial t_i} - n(n+a) \\
\mathcal{D}_2 F &= \frac{\partial F}{\partial t_2} + \sum_{i\geq 1} it_i \left(-(2n+a)\frac{\partial}{\partial t_i} - \frac{\partial}{\partial t_{i+1}}\right)F - n(n+a)(2n+a) \\
\mathcal{D}_3 F &= \frac{\partial F}{\partial t_3} - \sum_{i\geq 1} it_i \left((2n+a)^2\frac{\partial}{\partial t_i} + (2n+a)\frac{\partial}{\partial t_{i+1}} + \frac{\partial}{\partial t_{i+2}}\right)F \\
&\quad - \left(\frac{\partial^2 F}{\partial t_1^2} + \left(\frac{\partial F}{\partial t_1}\right)^2\right) - n(n+a)(2n+a)^2 .
\end{aligned} \tag{1.138}$$

Notice the nonlinearity in this expression is due to the fact that one uses $F = \log I_n$ rather than I_n.

Example 1.6.3 (*Jacobi ensemble*). The weight is given by

$$\rho_{ab}(z) := \mathrm{e}^{-V} = (1-z)^a(1+z)^b , \quad V' = \frac{g}{f} = \frac{a-b+(a+b)z}{1-z^2}$$

$a_0 = 1$, $a_1 = 0$, $a_2 = -1$, $b_0 = a-b$, $b_1 = a+b$, and all other $a_i, b_i = 0$.

Here define

$$\mathcal{B}_k := \sum_1^{2r} c_i^{k+1}(1-c_i^2)\frac{\partial}{\partial c_i} .$$

The integrals

$$\int_{E^n} \Delta_n(z)^2 \prod_{k=1}^n (1-z_k)^a(1+z_k)^b \exp\left(\sum_{i=1}^\infty t_i z_k^i\right) \mathrm{d}z_k \tag{1.139}$$

satisfy the Virasoro constraints ($k \geq -1$):

$$-\mathcal{B}_k I_n = (\mathbb{J}^{(2)}_{k+2,n} - \mathbb{J}^{(2)}_{k,n} + b_0 \mathbb{J}^{(1)}_{k+1,n} + b_1 \mathbb{J}^{(1)}_{k+2,n}) I_n \,. \tag{1.140}$$

Introducing the following notation

$$\sigma = 2n + b_1$$

the first four having the following form, upon setting $F = \log I_n$,

$$\begin{aligned}
-\mathcal{B}_{-1}F &= \left(\sigma \frac{\partial}{\partial t_1} + \sum_{i\geq 1} i t_i \frac{\partial}{\partial t_{i+1}} - \sum_{i\geq 2} i t_i \frac{\partial}{\partial t_{i-1}}\right) F + n(b_0 - t_1) \\
-\mathcal{B}_0 F &= \left(\sigma \frac{\partial}{\partial t_2} + b_0 \frac{\partial}{\partial t_1} + \sum_{i\geq 1} i t_i \left(\frac{\partial}{\partial t_{i+2}} - \frac{\partial}{\partial t_i}\right) + \frac{\partial^2}{\partial t_1^2}\right) F \\
&\qquad + \left(\frac{\partial F}{\partial t_1}\right)^2 - \frac{n}{2}(\sigma - b_1) \\
-\mathcal{B}_1 F &= \left(\sigma \frac{\partial}{\partial t_3} + b_0 \frac{\partial}{\partial t_2} - (\sigma - b_1) \frac{\partial}{\partial t_1} + \sum_{i\geq 1} i t_i \left(\frac{\partial}{\partial t_{i+3}} - \frac{\partial}{\partial t_{i+1}}\right)\right. \\
&\qquad \left. + 2 \frac{\partial^2}{\partial t_1 \partial t_2}\right) F + 2 \frac{\partial F}{\partial t_1} \frac{\partial F}{\partial t_2} \\
-\mathcal{B}_2 F &= \left(\sigma \frac{\partial}{\partial t_4} + b_0 \frac{\partial}{\partial t_3} - (\sigma - b_1) \frac{\partial}{\partial t_2} + \sum_{i\geq 1} i t_i \left(\frac{\partial}{\partial t_{i+4}} - \frac{\partial}{\partial t_{i+2}}\right)\right. \\
&\qquad \left. + \frac{\partial^2}{\partial t_2^2} - \frac{\partial^2}{\partial t_1^2} + 2 \frac{\partial^2}{\partial t_1 \partial t_3}\right) F \\
&\quad + \left(\left(\frac{\partial F}{\partial t_2}\right)^2 - \left(\frac{\partial F}{\partial t_1}\right)^2 + 2 \frac{\partial F}{\partial t_1} \frac{\partial F}{\partial t_3}\right) .
\end{aligned} \tag{1.141}$$

1.7 Random Matrices

This whole section is a very standard chapter of random matrix theory. Most of the results can be found, e.g., in Mehta [69], Deift [31] and others.

1.7.1 Haar Measure on the Space $\mathcal{H}_n$ of Hermitian Matrices

Consider the most naïve measure (Haar measure)

$$\mathrm{d}M := \prod_1^n \mathrm{d}M_{ii} \prod_{\substack{i,j=1 \\ i<j}}^n \mathrm{d}\,\mathrm{Re}\, M_{ij}\, \mathrm{d}\,\mathrm{Im}\, M_{ij} \tag{1.142}$$

on the space of Hermitian matrices

$$\mathcal{H}_n := \{n \times n \text{ matrices such that } M^\top = \overline{M}\} .$$

The parameters M_{ij} in (1.142) are precisely the free ones in M. This measure turns out to be invariant under conjugation by unitary matrices: (see Mehta [69] and more recently Deift [31])

Proposition 1.7.1. *For a fixed $U \in \mathrm{SU}(n)$, the map*

$$\mathcal{H}_n \to \mathcal{H}_n \colon M \mapsto M' = U^{-1}MU$$

has the property

$$\mathrm{d}M = \mathrm{d}M' , \quad \textit{i.e.,} \left|\det\left(\frac{\partial M'}{\partial M}\right)\right| = 1 .$$

Proof. Setting $M' = U^{-1}MU$, we have

$$\operatorname{Tr} M^2 = \operatorname{Tr} M'^2$$

and so

$$\sum_{i,j} M_{ij}M_{ji} = \sum_{i,j} M'_{ij}M'_{ji} .$$

Working out this identity, one finds

$$\sum_1^n M_{ii}^2 + 2\sum_{\substack{i,j=1\\i<j}}^n \left((\operatorname{Re} M_{ij})^2 + (\operatorname{Im} M_{ij})^2\right) = \sum_1^n M_{ii}'^2 + 2\sum_{\substack{i,j=1\\i<j}}^n \left((\operatorname{Re} M'_{ij})^2 + (\operatorname{Im} M'_{ij})^2\right) . \quad (1.143)$$

Setting

$$\boldsymbol{M} := (M_{11}, \dots, M_{nn}, \operatorname{Re} M_{12}, \dots, \operatorname{Re} M_{n-1,n}, \operatorname{Im} M_{12}, \dots, \operatorname{Im} M_{n-i,n}) ,$$

identity (1.143) can be written, in terms of the usual inner product $\langle\ ,\ \rangle$ in $\mathbb{R}^{n(n+1)/2}$,

$$\langle \boldsymbol{M}, D\boldsymbol{M}\rangle = \langle \boldsymbol{M}', D\boldsymbol{M}'\rangle$$

for the $n^2 \times n^2$ diagonal matrix with n 1s and $n(n-1)$ 2s,

$$D = \begin{pmatrix} 1 & & & & & \\ & \ddots & & & \mathbf{O} & \\ & & 1 & & & \\ & & & 2 & & \\ & \mathbf{O} & & & \ddots & \\ & & & & & 2 \end{pmatrix} .$$

Let V be the matrix transforming the vectors M into M'

$$\boldsymbol{M}' = V\boldsymbol{M}$$

and so, (1.143) reads

$$\langle \boldsymbol{M}', D\boldsymbol{M}' \rangle = \langle V\boldsymbol{M}, DV\boldsymbol{M} \rangle = \langle \boldsymbol{M}, V^\top DV\boldsymbol{M} \rangle ,$$

from which it follows that $D = V^\top DV$ and so

$$0 \neq \det D = \det(V^\top DV) = (\det V)^2 \det D ,$$

implying

$$|\det V| = 1 .$$

But for a linear transformation $\boldsymbol{M}' = V\boldsymbol{M}$ the Jacobian of the map is V itself and so

$$\left| \det \frac{\partial \boldsymbol{M}'}{\partial M} \right| = |\det V| = 1 ,$$

ending the proof of Prop. 1.7.1. □

Proposition 1.7.2. *The diagonalization* $M = UzU^{-1}$ *leads to "polar" or "spectral" coordinates* $M \mapsto (U, z)$*, where* $z = \operatorname{diag}(z_1, \dots, z_n)$*,* $z_i \in \mathbb{R}$*. In these new coordinates*

$$\mathrm{d}M = \Delta^2(z)\, \mathrm{d}z_1 \cdots \mathrm{d}z_n\, \mathrm{d}U; . \tag{1.144}$$

Proof. Every matrix $M \in \mathcal{H}_n$ can be diagonalized,

$$M = UzU^{-1} ,$$

with $U = \mathrm{e}^A z \mathrm{e}^{-A} \in \mathrm{SU}(n)$ and[29]

$$A = \sum_{\substack{k,l=1 \\ k \le l}}^{n} \big(a_{kl}(e_{kl} - e_{lk}) + \mathrm{i}b_{kl}(e_{kl} + e_{lk})\big) \in \mathrm{su}(n) , \quad \text{with } a_{k,l}, b_{k,l} \in \mathbb{R} .$$

Then, using the definition of the measure $\mathrm{d}M$, as in (1.142), and using the fact that $[A, z]$ is a Hermitian matrix, one computes

$$\begin{aligned} \mathrm{d}M\big|_{M=z} = \mathrm{d}(\mathrm{e}^A z \mathrm{e}^{-A})\big|_{A=0} &= \mathrm{d}\big(z + [A, z] + \mathbf{O}(A^2)\big)\big|_{A=0} \\ &= \prod_1^n \mathrm{d}z_i \prod_{\substack{j,k=1 \\ j<k}}^{n} \mathrm{d}\,\mathrm{Re}[A, z]_{jk}\, \mathrm{d}\,\mathrm{Im}[A, z]_{jk} \Big|_{A=0} \\ &= \prod_1^n \mathrm{d}z_i \prod_{\substack{j,k=1 \\ j<k}}^{n} (z_j - z_k)^2 \prod_{\substack{j,k=1 \\ j<k}}^{n} \mathrm{d}a_{jk}\, \mathrm{d}b_{jk} \Big|_{A=0} ; \end{aligned}$$

---

[29] e_{kl} denotes the matrix having a 1 at the entry (k, l) and 0 everywhere else.

here one has used

$$[A, z] = \sum_{\substack{k,l=1 \\ k<l}}^{n} (a_{kl}[e_{kl} - e_{lk}, z] + \mathrm{i}b_{kl}[e_{kl} + e_{lk}, z])$$

$$= \sum_{\substack{k,l=1 \\ k<l}}^{n} \big(a_{kl}(z_l - z_k)(e_{kl} + e_{lk}) + \mathrm{i}b_{kl}(z_l - z_k)(e_{kl} - e_{lk})\big) ,$$

with

$$\mathrm{Re}[A, z]_{k,l} = a_{kl}(z_l - z_k) , \quad \mathrm{Im}[A, z]_{kl} = b_{kl}(z_l - z_k) .$$

This establish (1.144) near $M = z$. By Lemma 1.7.1, $\mathrm{d}M = \mathrm{d}(U'^{-1}MU')$ for any unitary matrix U', implying the result (1.144) holds everywhere, establishing Prop. 1.7.2. □

Remark. The set of Hermitian matrices is the tangent space to a symmetric space $G/K = \mathrm{SL}(n, \mathbb{C})/\mathrm{SU}(n)$. The argument in Prop. 1.7.2 can be generalized to many other symmetric spaces, as worked out in van Moerbeke [90, pp. 324–329].

1.7.2 Random Hermitian Ensemble

Consider the probability distribution on the space $\mathcal{H}_n$ of Hermitian matrices, in terms of Haar measure $\mathrm{d}M$, given by

$$\mathrm{P}(M \in \mathrm{d}M) = \frac{1}{Z_n} \exp\big(-\mathrm{Tr}\, V(M)\big)\, \mathrm{d}M .$$

Let $z_1 \le z_2 \le \cdots \le z_n$ be the real eigenvalues of M. Then

$$\mathrm{P}(z_1 \in \mathrm{d}z_1, \ldots, z_n \in \mathrm{d}z_n) = \mathrm{P}(z_1, \ldots, z_n)\, \mathrm{d}z_1 \cdots \mathrm{d}z_n$$
$$:= \frac{1}{Z_n} \Delta^2(z) \prod_{i=1}^{n} \exp\big(-V(z_i)\big)\, \mathrm{d}z_i ,$$

with

$$Z_n = \int_{z_1 \le \cdots \le z_n} \mathrm{P}(z_1, \ldots, z_n) \prod_{1}^{n} \mathrm{d}z_i .$$

Lemma 1.7.1.

$$\mathrm{P}(M \in \mathcal{H}_n, \text{spectrum}\, M \subseteq E) = \frac{\int_{E^n} \Delta^2(z) \prod_1^n \exp\big(-V(z_i)\big)\, \mathrm{d}z_i}{\int_{\mathbb{R}^n} \Delta^2(z) \prod_1^n \exp\big(-V(z_i)\big)\, \mathrm{d}z_i} . \tag{1.145}$$

Proof. Indeed

$$
\begin{aligned}
&P(M \in \mathcal{H}_n, \text{spectrum}\, M \subseteq E)\\
&\quad = \frac{1}{Z_n} \int_{z_1 < \cdots < z_n} \Delta_n^2(z_1, \dots, z_n) \prod_{i=1}^n \chi_E(z_i) \exp\bigl(-V(z_i)\bigr)\,\mathrm{d}z_i\\
&\quad = \frac{1}{Z_n n!} \sum_{\pi \in S_n} \int_{z_{\pi(1)} < \cdots < z_{\pi(n)}} \Delta_n^2\bigl(z_{\pi(1)}, \dots, z_{\pi(n)}\bigr) \prod_{i=1}^n \chi_E\bigl(z_{\pi(i)}\bigr) \exp\Bigl(-V\bigl(z_{\pi(i)}\bigr)\Bigr)\,\mathrm{d}z_{\pi(i)}\\
&\quad = \frac{1}{Z_n n!} \sum_{\pi \in S_n} \int_{z_{\pi(1)} < \cdots < z_{\pi(n)}} \Delta_n^2(z_1, \dots, z_n) \prod_{i=1}^n \chi_E(z_i) \exp\bigl(-V(z_i)\bigr)\,\mathrm{d}z_i\\
&\quad = \frac{\int_{E^n} \Delta_n^2(z) \prod_1^n \exp\bigl(-V(z_i)\bigr)\,\mathrm{d}z_i}{\int_{\mathbb{R}^n} \Delta_n^2(z) \prod_1^n \exp\bigl(-V(z_i)\bigr)\,\mathrm{d}z_i}\,,
\end{aligned}
$$

showing Lemma 1.7.1. □

Let $p_0(z), p_1(z), p_2(z), \dots$ be orthonormal polynomials with regard to the weight $\rho(z)$ defined on $\mathbb{R}$, as discussed in Sect. 1.5.1,

$$
\int_{\mathbb{R}} p_i(z) p_j(z) \rho(z)\,\mathrm{d}z = \delta_{ij}\,,
$$

and $\tilde{p}_n(z)$ be the monic orthogonal polynomials,

$$
\int_{\mathbb{R}} \tilde{p}_i(z) \tilde{p}_j(z) \rho(z)\,\mathrm{d}z = h_i \delta_{ij}\,.
$$

Then,

$$
p_n(z) = \frac{1}{\sqrt{h_n}}(z^n + \cdots) = \frac{1}{\sqrt{h_n}} \tilde{p}_n(z)\,. \tag{1.146}
$$

Proposition 1.7.3. *Setting*

$$
Z_n = \int_{\mathbb{R}^n} \Delta_n^2(z) \prod_1^n \rho(z_k)\,\mathrm{d}z_k\,,
$$

we have the identity

$$
Z_n^{-1} \Delta_n^2(z) \prod_1^n \rho(z_k) = \frac{1}{n!} \det\bigl(K_n(z_k, z_l)\bigr)_{1 \le k < l \le n}\,, \tag{1.147}
$$

where the symmetric kernel K_n is given by

$$
\begin{aligned}
K_n(y, z) &= \sqrt{\rho(y)\rho(z)} \sum_{j=0}^{n-1} p_j(y) p_j(z) \qquad (\textit{Christoffel–Darboux})\\
&= \sqrt{\frac{h_n}{h_{n-1}}} \sqrt{\rho(y)\rho(z)} \frac{p_n(y) p_{n-1}(z) - p_{n-1}(y) p_n(z)}{y - z}\,.
\end{aligned} \tag{1.148}
$$

The kernel $K_n(y,z)$ has the following "reproducing" property

$$\int_{\mathbb{R}} K_n(y,u)K_n(u,z)\,\mathrm{d}u = K_n(y,z) \quad \text{and} \quad \int_{\mathbb{R}} K_n(z,z)\,\mathrm{d}z = n\,.$$

Proof. Notice that the Vandermonde $\Delta_n(z)$ can also be expressed as $\det\big(\tilde{p}_{i-1}(z_j)\big)_{1\le i,j\le n}$ by row operations, where $\tilde{p}_i(z)$ can be chosen to be any monic polynomial of degree i, and in particular the monic orthogonal polynomial of degree i. Thus, one computes for the normalization Z_n,

$$\begin{aligned}
Z_n &= \int_{\mathbb{R}^n} \Delta^2(z) \prod_1^n \rho(z_i)\,\mathrm{d}z_i \\
&= \int_{\mathbb{R}^n} \det\big(\tilde{p}_{i-1}(z_j)\big)_{1\le i,j\le n} \det\big(\tilde{p}_{k-1}(z_l)\big)_{1\le k,l\le n} \prod_{i=1}^n \rho(z_i)\,\mathrm{d}z_i \\
&= \sum_{\pi,\pi'\in S_n} (-1)^{\pi+\pi'} \prod_{k=1}^n \int_{\mathbb{R}} \tilde{p}_{\pi(k)-1}(z_k)\tilde{p}_{\pi'(k)-1}(z_k)\rho(z_k)\,\mathrm{d}z_k \\
&= \sum_{\pi\in S_n} \prod_{k=1}^n \int_{\mathbb{R}} \tilde{p}^2_{\pi(k)-1}(z_k)\rho(z_k)\,\mathrm{d}z_k\,, \quad \text{using the orthogonality of the } \tilde{p}_i\text{s}, \\
&= n! \prod_0^{n-1} \int_{\mathbb{R}} \tilde{p}_k^2(z)\rho(z)\,\mathrm{d}z = n! \prod_0^{n-1} h_k\,.
\end{aligned}$$

Then using the expression obtained for Z_n and $(\det A)^2 = \det(AA^\top)$ in the third equality, one further computes

$$\begin{aligned}
&Z_n^{-1}\Delta_n^2(z)\prod_1^n \rho(z_k) \\
&\quad = \frac{1}{n!\prod_0^{n-1} h_k} \det\big(\tilde{p}_{i-1}(z_j)\big)_{1\le i,j\le n} \det\big(\tilde{p}_{k-1}(z_l)\big)_{1\le k,l\le n} \prod_{k=1}^n \rho(z_k) \\
&\quad = \frac{1}{n!} \det\left(\frac{\tilde{p}_{i-1}(z_j)}{\sqrt{h_{i-1}}}\sqrt{\rho(z_j)}\right)_{1\le i,j\le n} \det\left(\frac{\tilde{p}_{k-1}(z_l)}{\sqrt{h_{k-1}}}\sqrt{\rho(z_l)}\right)_{1\le k<l\le n} \\
&\quad = \frac{1}{n!} \det\left(\sum_{j=1}^n \frac{\tilde{p}_{j-1}(z_k)}{\sqrt{h_{j-1}}}\frac{\tilde{p}_{j-1}(z_l)}{\sqrt{h_{j-1}}}\sqrt{\rho(z_k)\rho(z_l)}\right)_{1\le k,l\le n} \\
&\quad = \frac{1}{n!} \det\big(K_n(z_k,z_l)\big)_{1\le k,l\le n}\,.
\end{aligned}$$

One finally needs the classical Christoffel–Darboux identity: setting $p_j(y) = 0$ for $j < 0$, one checks

$$
\begin{aligned}
(y-z)\sum_0^{n-1} p_j(y)p_j(z) \\
&= \sum_{j=0}^{n-1}\big(a_{j,j-1}p_{j-1}(y)+a_{jj}p_j(y)+a_{j,j+1}p_{j+1}(y)\big)p_j(z) \\
&\qquad - \sum_{j=0}^{n-1} p_j(y)\big(a_{j,j-1}p_{j-1}(z)+a_{jj}p_j(z)+a_{j,j+1}p_{j+1}(z)\big) \\
&= a_{n-1,n}\big(p_n(y)p_{n-1}(z)-p_{n-1}(y)p_n(z)\big) ,
\end{aligned}
$$

and one uses (1.146). The reproducing property follows immediately from the Christoffel–Darboux representation of the kernel in terms of orthogonal polynomials. This proves Prop. 1.7.3. □

1.7.3 Reproducing Kernels

Lemma 1.7.2. *Let $K(x,y)$ be a symmetric kernel satisfying the reproducing property*

$$\int_{\mathbb{R}} K(x,y)K(y,z)\,\mathrm{d}y = K(x,z) .$$

Then

$$
\begin{aligned}
\int \det\big(K(z_i,z_j)\big)_{1\le i,j\le n}\,\mathrm{d}z_n \\
&= \Big(\int_{\mathbb{R}} K(z,z)\,\mathrm{d}z - n + 1\Big)\det\big(K(z_i,z_j)\big)_{1\le i,j\le n-1} , \qquad (1.149)
\end{aligned}
$$

where dy *stands for any measure on* $\mathbb{R}$.

Proof. The proof of (1.149) is due to J. Verbaarschot [91], which proceeds in two steps:

Step 1. Let

$$M_n = \begin{pmatrix} M_{n-1} & m \\ \bar{m}^\top & \gamma \end{pmatrix}$$

be a $n\times n$ Hermitian matrix, with M_{n-1} a $n-1\times n-1$ Hermitian matrix. Then m is a $(n-1)\times 1$ column and $\gamma\in\mathbb{R}$. Then

$$\det M_n = \gamma \det M_{n-1} - \bar{m}^\top \tilde{M}_{n-1} m .$$

Indeed given the column $u\in\mathbb{C}^{n-1}$, one checks

$$
\begin{aligned}
\begin{pmatrix} I & 0 \\ \bar{u}^\top & 1 \end{pmatrix}\begin{pmatrix} M_{n-1} & m \\ \bar{m}^\top & \gamma \end{pmatrix}\begin{pmatrix} I & u \\ 0 & 1 \end{pmatrix} \\
&= \begin{pmatrix} M_{n-1} & M_{n-1}u + m \\ \bar{u}^\top M_{n-1} + \bar{m}^\top & \bar{u}^\top M_{n-1}u + \bar{m}^\top u + \bar{u}^\top m + \gamma \end{pmatrix} \\
&= \begin{pmatrix} M_{n-1} & 0 \\ 0 & \gamma - \bar{m}^\top M_{n-1}^{-1} m \end{pmatrix} , \quad \text{upon setting } u = -M_{n-1}^{-1} m .
\end{aligned}
$$

The latter assumes that M_{n-1} is invertible. Furthermore, using $M_{n-1}^{-1} = (\det M_{n-1})^{-1}\tilde{M}_{n-1}$,

$$\begin{aligned}\det M_n &= (\gamma - \bar{m}^\top M_{n-1}^{-1} m) \det M_{n-1}\\ &= (\gamma - \bar{m}^\top \tilde{M}_{n-1} m (\det M_{n-1})^{-1}) \det M_{n-1}\\ &= \gamma \det M_{n-1} - \bar{m}^\top \tilde{M}_{n-1} m\ ,\end{aligned}$$

finishing Step 1.

Step 2. Proof of identity (1.149). Define the matrix

$$M_k := \big(K(z_i, z_j)\big)_{1\le i,j\le k} \quad \gamma := K(z_k, z_k)$$

and

$$m = \begin{pmatrix} K(z_1, z_k)\\ \vdots \\ K(z_{k-1}, z_k)\end{pmatrix} ;$$

M_k is a symmetric matrix, since K is a symmetric kernel. One finds, upon integration over $\mathbb{R}$,

$$\begin{aligned}&\int_{\mathbb{R}} \det(M_n)\, \mathrm{d}z_n\\ &= \det(M_{n-1}) \int_{\mathbb{R}} K(z,z)\, \mathrm{d}z - \int_{\mathbb{R}} \mathrm{d}z_n \sum_{i,j=1}^{n-1} K(z_n, z_i)(\widetilde{M}_{n-1})_{ij} K(z_j, z_n)\\ &= \det(M_{n-1}) \int_{\mathbb{R}} K(z,z)\mathrm{d}z - \sum_{i,j=1}^{n-1} (\widetilde{M}_{n-1})_{ij} \int_{\mathbb{R}} K(z_n, z_i) K(z_j, z_n)\, \mathrm{d}z_n\\ &= \det(M_{n-1}) \int_{\mathbb{R}} K(z,z)\, \mathrm{d}z - \sum_{i,j=1}^{n-1} (\tilde{M}_{n-1})_{i,j} (M_{n-1})_{j,i}\\ &= \det(M_{n-1}) \int_{\mathbb{R}} K(z,z)\, \mathrm{d}z - \det(M_{n-1}) \sum_{i,j=1}^{n-1} (M_{n-1}^{-1})_{ij} (M_{n-1})_{ji}\\ &= \det(M_{n-1}) \bigg(\int_{\mathbb{R}} K(z,z)\, \mathrm{d}z - (n-1)\bigg),\end{aligned}$$

establishing Lemma 1.7.2. □

Lemma 1.7.3. *If*

(i) $\int_{\mathbb{R}} K(x,y)K(y,z)\, \mathrm{d}y = K(x,z)$
(ii) $\int_{\mathbb{R}} K(z,z)\, \mathrm{d}z = n$,

then

$$\begin{aligned}\int \cdots \int_{\mathbb{R}^{n-m}} \det\big(K(z_i, z_j)\big)_{1\le i,j\le n} \mathrm{d}z_{m+1} \cdots \mathrm{d}z_n\\ = (n-m)! \det\big(K(z_i,z_j)\big)_{1\le i,j\le m} \quad (1.150)\end{aligned}$$

Proof. The proof proceeds by induction on m. On the one hand, assuming (1.150) to be true, integrate with regard to z_m and use identity (1.149):

$$\begin{aligned}\int\cdots\int_{\mathbb{R}^{n-m+1}} &\det\big(K(z_i,z_j)\big)_{1\le i,j\le n}\,\mathrm{d}z_m\,\mathrm{d}z_{m+1}\cdots\mathrm{d}z_n\\ &=(n-m)!\int_{\mathbb{R}}\mathrm{d}z_m\,\det\big(K(z_i,z_j)\big)_{1\le i,j\le m}\\ &=(n-m)!\Big(\int_{\mathbb{R}}K(z,z)\,\mathrm{d}z-m+1\Big)\det\big(K(z_i,z_j)\big)_{1\le i,j\le m-1}\\ &=(n-m)!(n-m+1)\det\big(K(z_i,z_j)\big)_{1\le i,j\le m-1}\\ &=(n-m+1)!\det\big(K(z_i,z_j)\big)_{1\le i,j\le m-1}\,.\end{aligned}$$

On the other hand for $m+1=n$, the statement follows at once from Lemma 1.7.2. This ends the proof of Lemma 1.7.3.

1.7.4 Correlations and Fredholm Determinants

For this section, see M.L. Mehta [69], P. Deift [31], Tracy–Widom [84] and others. Returning now to the probability distribution on the space $\mathcal{H}_n$ of Hermitian matrices (setting $\rho(z):=\mathrm{e}^{-V(z)}$)

$$P(M\in\mathrm{d}M)=\frac{1}{Z_n}\exp\big(-\operatorname{Tr}V(M)\big)\,\mathrm{d}M\,,$$

remember from Lemma 1.7.1 and Prop. 1.7.3,

$$P(M\in\mathcal{H}_n,\operatorname{spectrum}M\subseteq E)=\int_{E^n}P_n(z)\,\mathrm{d}z_1\cdots\mathrm{d}z_n$$

with

$$\begin{aligned}P_n(z)=\frac{\Delta_n^2(z)\prod_1^n\rho(z_i)\,\mathrm{d}z_i}{\int_{\mathbb{R}^n}\Delta_n^2(z)\prod_1^n\rho(z_i)\,\mathrm{d}z_i}&=Z_n^{-1}\Delta_n^2(z)\prod_1^n\rho(z_k)\\ &=\frac{1}{n!}\det\big(K_n(z_k,z_l)\big)_{1\le k<l\le n}\,,\end{aligned}\tag{1.151}$$

with the kernel $K_n(y,z)$ defined in Prop. 1.7.3,

$$K_n(x,y)=\sqrt{\rho(x)\rho(y)}\sum_{j=0}^{n-1}p_j(x)p_j(y)\,.\tag{1.152}$$

Let $\mathbb{E}$ be the expectation associated with the probability P above. Then one has the following "classical" proposition for any subset $E\subset\mathbb{R}$ (for which a precise statement and proof was given by P. Deift [31]):

Proposition 1.7.4. *The* 1- *and* 2*-point correlations have the following meaning*[30]:

$$\int_E K_n(z,z)\,\mathrm{d}z = \mathbb{E}(\#\text{ of eigenvalues in } E)$$

$$\int_{E\times E} \det\bigl(K_n(z_i,z_j)\bigr)_{1\le i,j\le 2}\,\mathrm{d}z_1\,\mathrm{d}z_2 = \mathbb{E}(\#\text{ of pairs of eigenvalues in } E)\,,$$

and thus

$$K_n(z,z) = \frac{1}{\mathrm{d}z}\mathbb{E}(\#\text{ of eigenvalues in } \mathrm{d}z)\,. \tag{1.153}$$

Proof. Using (1.150) for $m=1$ and (1.147), one computes

$$\begin{aligned}
&\int_E K_n(z,z)\,\mathrm{d}z\\
&= \int_{\mathbb{R}} \chi_E(z_1)K_n(z_1,z_1)\,\mathrm{d}z_1\\
&= \frac{1}{(n-1)!}\int_{\mathbb{R}} \mathrm{d}z_1\chi_E(z_1)\int\cdots\int_{\mathbb{R}^{n-1}} \det\bigl(K_n(z_i,z_j)\bigr)_{1\le i,j\le n}\,\mathrm{d}z_2\cdots\mathrm{d}z_n\\
&= n\int_{\mathbb{R}} \mathrm{d}z_1\,\chi_E(z_1)\int\cdots\int_{\mathbb{R}^{n-1}} \frac{1}{Z_n}\Delta_n^2(z)\prod_1^n \rho(z_k)\,\mathrm{d}z_2\cdots\mathrm{d}z_n\\
&= \frac{n}{Z_n}\int_{\mathbb{R}^n} \chi_E(z_1)\Delta_n^2(z)\prod_1^n \rho(z_k)\,\mathrm{d}z_k\\
&= \frac{1}{Z_n}\int_{\mathbb{R}^n} \Bigl(\sum_{i=1}^n \chi_E(z_i)\Bigr)\Delta_n^2(z)\prod_1^n \rho(z_k)\,\mathrm{d}z_k\\
&= \frac{1}{Z_n}\int_{\mathbb{R}^n} \#\{i \text{ such that } z_i\in E\}\Delta_n^2(z)\prod_1^n \rho(z_k)\,\mathrm{d}z_k\\
&= \mathbb{E}(\#\text{ of eigenvalues in } E)\,.
\end{aligned}$$

A similar argument holds for the second identity of Prop. 1.7.4. □

Consider disjoint intervals $E_1,\dots,E_m$ and integers $1\le n_1,\dots,n_m\le n$ and set $n_{m+1} := n-\sum_1^m n_i$. Then for the $n\times n$ Hermitian ensemble with P_n as in (1.151), one has:

$$\begin{aligned}
&\mathrm{P}(\text{exactly } n_i \text{ eigenvalues} \in E_i,\ 1\le i\le m)\\
&= \binom{n}{n_1,\dots,n_m,n_{m+1}}\int_{\mathbb{R}^n} \prod_{i=1}^{n_1}\chi_{E_1}(x_i)\prod_{i=n_1+1}^{n_1+n_2}\chi_{E_2}(x_i)\cdots\prod_{i=n_1+\cdots+n_{m-1}+1}^{n_1+\cdots+n_m}\chi_{E_m}(x_i)\\
&\qquad\times\prod_{i=\sum_1^m n_k+1}^{n}\chi_{(\bigcup_{i=1}^m E_i)^c}(x_i)P_n(x)\,\mathrm{d}x_1\cdots\mathrm{d}x_n\,.
\end{aligned} \tag{1.154}$$

[30] If $x_1,x_2\in E$, then it counts for 2 in the second formula.

This follows from the symmetry of $P_n(x)$ under the permutation group; the multinomial coefficient takes into account the number of ways the event occurs.

Lemma 1.7.4. *The following identity holds*

$$\int_{\mathbb{R}^n} \prod_{k=1}^{n} \Big(1 + \sum_{i=1}^{m} \lambda_i \chi_{E_i}(x_k)\Big) P_n(x)\, \mathrm{d}x_1 \cdots \mathrm{d}x_n \\ = \det\Big[I + K_n(x,y) \sum_{i=1}^{m} \lambda_i \chi_{E_i}(y)\Big] .$$

Proof. Upon setting

$$\widetilde{K}(x,y) := K_n(x,y) \sum_{1}^{m} \lambda_i \chi_{E_i}(y) ,$$

and upon using the fact that $\widetilde{K}(x,y)$ has rank n in view of its special form (1.152), the Fredholm determinant can be computed as follows:

$$\begin{aligned}
&\det\big(I + \widetilde{K}(x,y)\big)\\
&= \sum_{l=0}^{\infty} \frac{1}{l!} \int_{\mathbb{R}^l} \det\big[\widetilde{K}(x_i,x_j)\big]_{1\le i,j\le l}\, \mathrm{d}x_1 \cdots \mathrm{d}x_l\\
&= \sum_{l=0}^{n} \frac{1}{l!} \int_{\mathbb{R}^l} \det\big[\widetilde{K}(x_i,x_j)\big]_{1\le i,j\le l}\, \mathrm{d}x_1 \cdots \mathrm{d}x_l\\
&= \sum_{l=0}^{n} \frac{1}{l!} \int_{\mathbb{R}^l} \det\Big[K_n(x_i,x_j) \sum_{k=1}^{m} \lambda_k \chi_{E_k}(x_j)\Big]_{1\le i,j\le l} \mathrm{d}x_1 \cdots \mathrm{d}x_l\\
&= \sum_{l=0}^{n} \frac{1}{l!} \sum_{1\le s_1,\dots,s_l\le m} \lambda_{s_1} \cdots \lambda_{s_l} \int_{\mathbb{R}^l} \mathrm{d}x_1 \cdots \mathrm{d}x_l \prod_{r=1}^{l} \chi_{E_{s_r}}(x_r) \det[K_n(x_i,x_j)]_{1\le i,j\le l}\\
&= \int_{\mathbb{R}^n} \sum_{l=0}^{n} \sum_{1\le s_1,\dots,s_l\le m} \lambda_{s_1} \cdots \lambda_{s_l} \chi_{E_{s_1}}(x_1) \cdots \chi_{E_{s_l}}(x_l)\\
&\qquad \times \frac{1}{l!(n-l)!} \det[K_n(x_i,x_j)]_{1\le i,j\le n}\, \mathrm{d}x_1 \cdots \mathrm{d}x_n ,
\end{aligned}$$

using Lemma 1.7.3,

$$\begin{aligned}
&= \int_{\mathbb{R}^n} \sum_{l=0}^{n} \binom{n}{l} \sum_{1\le s_1,\dots,s_l\le m} \lambda_{s_1} \cdots \lambda_{s_l} \chi_{E_{s_1}}(x_1) \cdots \chi_{E_{s_l}}(x_l) P_n(x)\, \mathrm{d}x_1 \dots \mathrm{d}x_n\\
&= \int_{\mathbb{R}^n} \sum_{l=0}^{n} \sum_{1\le s_1,\dots,s_l\le m} \sum_{1\le i_1<\dots<i_l\le n} \lambda_{s_1} \cdots \lambda_{s_l} \chi_{E_{s_1}}(x_{i_1}) \cdots \chi_{E_{s_l}}(x_{i_l})\\
&\qquad \times P_n(x)\, \mathrm{d}x_1 \cdots \mathrm{d}x_n\\
&= \int_{\mathbb{R}^n} \prod_{k=1}^{n} \Big(1 + \sum_{i=1}^{m} \lambda_i \chi_{E_i}(x_k)\Big) P_n(x)\, \mathrm{d}x_1 \cdots \mathrm{d}x_n ,
\end{aligned}$$

establishing Lemma 1.7.4. □

Proposition 1.7.5. *The Fredholm determinant is a generating function for the probabilities:*

$$P(\textit{exactly } n_i \textit{ eigenvalues} \in E_i,\ 1 \le i \le m)$$
$$= \prod_1^m \frac{1}{n_i!}\left(\frac{\partial}{\partial \lambda_i}\right)^{n_i} \det\left[I + \sum_1^m \lambda_i K_n(x,y)\chi_{E_i}(y)\right]\Bigg|_{\textit{all } \lambda_i = -1} . \quad (1.155)$$

In particular

$$\mathrm{P}(\textit{no eigenvalues} \in E_i,\ 1 \le i \le m) = \det\left[I - K_n(x,y)\chi_{\bigcup_1^m E_i}(y)\right] . \quad (1.156)$$

Proof. The first equality below follows from Lemma 1.7.4. Concerning the second equality, in order to carry out the differentiation $\prod_1^m (1/n_i!)(\partial/\partial\lambda_i)^{n_i}$, one chooses (keeping in mind the usual product rule of differentiation) a first group of n_1 factors, a (distinct) second group of n_2 factors,..., a mth group of n_m factors and finally the last group of $n - n_1 - \cdots - n_m$ remaining factors among the product $\prod_{k=1}^n \bigl(1 + \sum_{i=1}^m \lambda_i \chi_{E_i}(x_k)\bigr)$. Then one differentiates the first m groups, leaving untouched the last group, where one sets $\lambda_i = -1$. This explains the second equality below. Let $C^n_{n_1,\ldots,n_m}$ be the set of distinct committees of size $n_1, n_2, \ldots, n_m, n_{m+1}$, with $n_{m+1} := n - n_1 - \cdots - n_m$ formed with people $1, 2, \ldots, n$:

$$\prod_{i=1}^m \frac{1}{n_i!}\left(\frac{\partial}{\partial \lambda_i}\right)^{n_i} \det\left[I + \sum_1^m \lambda_i K_n(x,y)\chi_{E_i}(y)\right]\Bigg|_{\text{all } \lambda_i = -1}$$
$$= \prod_{i=1}^m \frac{1}{n_i!}\left(\frac{\partial}{\partial \lambda_i}\right)^{n_i} \int_{\mathbb{R}^n} \prod_{k=1}^n \left(1 + \sum_{i=1}^m \lambda_i \chi_{E_i}(x_k)\right) P_n(x)\,\mathrm{d}x_1 \cdots \mathrm{d}x_n \Bigg|_{\text{all } \lambda_i = -1}$$
$$= \sum_{\sigma \in C^n_{n_1,\ldots,n_m}} \int_{\mathbb{R}^n} \prod_{i=1}^{n_1} \chi_{E_1}(x_{\sigma(i)}) \cdots \prod_{i=n_1+\cdots+n_{m-1}}^{n_1+\cdots+n_m} \chi_{E_m}(x_{\sigma(i)})$$
$$\times \prod_{i=n_1+\cdots+n_m+1}^{n} \left(1 - \sum_{l=1}^m \chi_{E_l}(x_{\sigma(i)})\right) P_n(x)\,\mathrm{d}x_1 \cdots \mathrm{d}x_n$$
$$= \binom{n}{n_1, \ldots, n_m, n - \sum_1^m n_i} \int_{\mathbb{R}^n} \prod_{i=1}^{n_1} \chi_{E_1}(x_i) \cdots \prod_{i=n_1+\cdots+n_{m-1}}^{n_1+\cdots+n_m} \chi_{E_m}(x_i)$$
$$\times \prod_{i=n_1+\cdots+n_m+1}^{n} \left(1 - \sum_{l=1}^m \chi_{E_l}(x_i)\right) P_n(x)\,\mathrm{d}x_1 \cdots \mathrm{d}x_n$$
$$= \mathrm{P}(\text{exactly } n_i \text{ eigenvalues} \in E_i,\ 1 \le i \le m) ,$$

as follows from (1.154), thus establishing identity (1.155), whereas (1.156) follows from setting $n_1 = \cdots = n_m = 0$, completing the proof of Prop. 1.7.5. □

1.8 The Distribution of Hermitian Matrix Ensembles

1.8.1 Classical Hermitian Matrix Ensembles

1.8.1.1 The Gaussian Hermitian Matrix Ensemble (GUE)

Let $\mathcal{H}_n$ be the Hermitian ensemble

$$\mathcal{H}_n = \{n \times n \text{ matrices } M \text{ satisfying } M^\top = \overline{M}\} .$$

The real and imaginary parts of the entries M_{ij} of the $n \times n$ Hermitian matrix ($\overline{M} = M^\top$) are all independent and Gaussian; the variables M_{ii}, $1 \leq i \leq n$, $\operatorname{Re} M_{ij}$ and $\operatorname{Im} M_{ij}$, $1 \leq i < j \leq n$, which parametrize the full matrix have the following distribution (set $\mathcal{H}_n \ni H = (u_{ij})$, with real u_{ii} and off-diagonal elements $u_{ij} := v_{ij} + \mathrm{i} w_{ij}$)

$$\begin{aligned}
\mathrm{P}(M_{ii} \in \mathrm{d}u_{ii}) &= \frac{1}{\sqrt{\pi}} \exp(-u_{ii}^2)\, \mathrm{d}u_{ii} && 1 \leq i \leq n \\
\mathrm{P}(\operatorname{Re} M_{jk} \in \mathrm{d}v_{jk}) &= \frac{2}{\sqrt{\pi}} \exp(-2v_{jk}^2)\, \mathrm{d}v_{jk} && 1 \leq j < k \leq n \\
\mathrm{P}(\operatorname{Im} M_{jk} \in \mathrm{d}w_{jk}) &= \frac{2}{\sqrt{\pi}} \exp(-2w_{jk}^2)\, \mathrm{d}w_{jk} && 1 \leq j < k \leq n .
\end{aligned}$$

Hence, using Haar measure (1.142),

$$\begin{aligned}
&\mathrm{P}(M \in \mathrm{d}H) \\
&= \prod_1^n \mathrm{P}(M_{ii} \in \mathrm{d}u_{ii}) \prod_{\substack{j,k=1 \\ j<k}}^n \mathrm{P}(\operatorname{Re} M_{jk} \in \mathrm{d}v_{jk})\, \mathrm{P}(\operatorname{Im} M_{jk} \in \mathrm{d}w_{jk}) \\
&= c_n \prod_1^n \exp(-u_{ii}^2) \prod_{\substack{j,k=1 \\ j<k}}^n \exp\bigl(-2(v_{jk}^2 + w_{jk}^2)\bigr) \prod_1^n du_{ii} \prod_{\substack{j,k=1 \\ j<k}}^n \mathrm{d}v_{jk}\, \mathrm{d}w_{jk} \\
&= c_n \exp\biggl(-\sum_{i,j=1}^n |u_{ij}|^2\biggr) \prod_1^n \mathrm{d}u_{ii} \prod_{\substack{j,k=1 \\ j<k}}^n \mathrm{d}\operatorname{Re} u_{jk}\, \mathrm{d}\operatorname{Im} u_{jk} \\
&= c_n \exp(-\operatorname{Tr} H^2)\, DH = c_n \Delta_n^2(z) \prod_1^n \exp(-z_i^2)\, \mathrm{d}z_i , \quad H \in \mathcal{H}_n , \qquad (1.157)
\end{aligned}$$

using the representation of Haar measure in terms of spectral variables z_i (see Prop. 1.7.2) and where

$$c_n = \left(\frac{2}{\pi}\right)^{n^2/2} \frac{1}{2^{n/2}} .$$

This constant can be computed by representing the integral of (1.157) over the full range $\mathbb{R}^n$ as a determinant of a moment matrix (as in (1.101)) of Gaussian integrals.

1.8.1.2 Estimating Covariances of Complex Gaussian Populations and the Laguerre Hermitian Ensemble

Consider the complex Gaussian population $\boldsymbol{x} = (x_1, \ldots, x_p)^\top$, with mean and covariance matrix given by

$$\boldsymbol{\mu} = \mathbb{E}(\boldsymbol{x}) = (\mu_1, \ldots, \mu_p)^\top \,, \quad \Sigma = \big(\mathbb{E}(x_i - \mu_i)(\bar{x}_j - \bar{\mu}_j)\big)_{1 \le i,j \le p}$$

and density (for the complex inner-product $\langle \;,\; \rangle$),

$$\frac{1}{(2\pi)^{p/2}(\det \Sigma)^{1/2}} \exp(-\tfrac{1}{2}\langle \boldsymbol{x} - \boldsymbol{\mu}, \Sigma^{-1}(\boldsymbol{x} - \boldsymbol{\mu})\rangle) \,.$$

Let $\lambda_1 \ge \lambda_2 \ge \cdots \ge \lambda_p > 0$ be the eigenvalues of Σ. Taking n samples of $(x_1, \ldots, x_p)^\top$, consider the normalized $p \times n$ sample matrix:

$$x = \begin{pmatrix} x_{11} - (1/n)\big(\sum_1^n x_{1i}\big) & x_{12} - (1/n)\big(\sum_1^n x_{1i}\big) & \cdots & x_{1n} - (1/n)\big(\sum_1^n x_{1i}\big) \\ x_{21} - (1/n)\big(\sum_1^n x_{2i}\big) & x_{22} - (1/n)\big(\sum_1^n x_{2i}\big) & \cdots & x_{2n} - (1/n)\big(\sum_1^N x_{2i}\big) \\ \vdots & \vdots & & \vdots \\ x_{p1} - (1/n)\big(\sum_1^n x_{pi}\big) & x_{p2} - (1/n)\big(\sum_1^n x_{pi}\big) & \cdots & x_{pn} - (1/n)\big(\sum_1^n x_{pi}\big) \end{pmatrix}$$

and the $p \times p$ sample covariance matrix,

$$S := \frac{1}{N-1} x\bar{x}^\top \,, \quad \text{with eigenvalues } z_1, \ldots, z_p > 0 \,,$$

which is an unbiased estimator of Σ. It is a classical result that when $\Sigma = I$, the eigenvalues $z_1, \ldots, z_p > 0$ of S have the Wishart distribution, a special case of the Laguerre Hermitian ensemble (see Hotelling [53] and also Muirhead [70])

$$\begin{aligned} \mathbb{P}_{n,p}(S \in \mathrm{d}M) &= c_{np}\Delta_p^2(z) \prod_1^p \exp(-z_i) z_i^{n-p-1} \, \mathrm{d}z_i \, \mathrm{d}U \\ &= \mathrm{e}^{-\operatorname{Tr} M} (\det M)^{n-p-1} \, \mathrm{d}M \,. \end{aligned}$$

1.8.1.3 Estimating the Canonical Correlations Between two Gaussian Populations and the Jacobi Hermitian Ensemble

In testing the statistical independence of two complex Gaussian populations, one needs to know the distribution of *canonical correlation coefficients.* I present here the case of real Gaussian populations, not knowing whether the complex case has been worked out, although it should proceed in the same way. To set up the problem, consider $p+q$ normally distributed random variables $(X_1, \ldots, X_p)^\top$ and $(Y_1, \ldots, Y_q)^\top$ $(p \le q)$ with mean zero and covariance matrix

$$\operatorname{cov}\begin{pmatrix} X \\ Y \end{pmatrix} := \Sigma = \Bigg(\begin{matrix} \overbrace{\Sigma_{11}}^{p} & \overbrace{\Sigma_{12}}^{q} \\ \Sigma_{12}^\top & \Sigma_{22} \end{matrix} \Bigg) \begin{matrix} \}p \\ \}q \end{matrix} \,.$$

The method proposed by Hotelling [53] is to find linear transformations $U = L_1X$ and $V = L_2Y$ of X and Y having the property that the correlation between the first components U_1 and V_1 of the vectors U and V is maximal subject to the condition that $\operatorname{Var} U_1 = \operatorname{Var} V_1 = 1$; moreover, one requires the second components U_2 and V_2 to have maximal correlation subjected to

$$\begin{cases} \text{(i)} \;\; \operatorname{Var} U_2 = \operatorname{Var} V_2 = 1 \\ \text{(ii)} \;\; U_2 \text{ and } V_2 \text{ are uncorrelated with both } U_1 \text{ and } V_1 \,, \end{cases}$$

etc.

Then there exist orthogonal matrices $O_p \in \mathrm{O}(p)$, $O_q \in \mathrm{O}(q)$ such that

$$\Sigma_{11}^{-1/2}\Sigma_{12}\Sigma_{22}^{-1/2} = O_p^\top P O_q \,,$$

where P has the following form:

$$\begin{cases} P = \left(\underbrace{\overbrace{\begin{matrix} \rho_1 & & & & \\ & \ddots & & O & \\ & & \rho_k & & \\ & & & \rho_{k+1} & \\ & O & & & \ddots \\ & & & & & \rho_p \end{matrix}\Bigg|\; O}^{q}}_{p}\right)\Bigg\}\, p \,, \quad k = \operatorname{rank}\Sigma_{12}\,, \\ 1 \ge \rho_1 \ge \rho_2 \ge \cdots \ge \rho_k > 0, \rho_{k+1} = \cdots = \rho_p = 0 \\ \qquad\qquad\qquad\qquad \text{(canonical correlation coefficients)}, \\ \rho_i \text{ are solutions } (\ge 0) \text{ of } \det(\Sigma_{11}^{-1}\Sigma_{12}\Sigma_{22}^{-1}\Sigma_{12}^\top - \rho^2 I) = 0 \,. \end{cases}$$

Then the covariance matrix of the vectors

$$U = L_1X := O_p\Sigma_{11}^{-1/2}X \quad \text{and} \quad V = L_2Y := O_q\Sigma_{22}^{-1/2}Y$$

has the canonical form ($\det \Sigma_{\text{can}} = \prod_1^p(1-\rho_i^2)$)

$$\operatorname{cov}\begin{pmatrix} U \\ V \end{pmatrix} = \Sigma_{\text{can}} = \begin{pmatrix} I_p & P \\ P^\top & I_q \end{pmatrix} ,$$

with

$$\text{spectrum}\, \Sigma_{can} = \underbrace{1, \dots, 1}_{q-p}, 1-\rho_1, 1+\rho_1, \dots, 1-\rho_p, 1+\rho_p$$

and inverse

$$\Sigma_{\text{can}}^{-1} = \frac{1}{\prod_1^p(1-\rho_i^2)^2}\begin{pmatrix} I_p & -P \\ -P^\top & I_q \end{pmatrix} .$$

From here on, we may take $\Sigma = \Sigma_{\mathrm{can}}$. The n $(n \geq p+q)$ independent samples $(x_{11},\ldots,x_{1p},y_{11},\ldots,y_{1q})^\top,\ldots,(x_{n1},\ldots,x_{np},y_{n1},\ldots,y_{nq})^\top$, arising from observing $\binom{X}{Y}$ lead to a matrix $\binom{x}{y}$ of size $(p+q,n)$, having the normal distribution [70, pp. 79 and 539]

$$
\begin{aligned}
&(2\pi)^{-n(p+q)/2}(\det\Sigma)^{-n/2}\exp\left(-\frac{1}{2}\operatorname{Tr}(x^\top y^\top)\begin{pmatrix}\Sigma_{11} & \Sigma_{12}\\ \Sigma_{12}^\top & \Sigma_{22}\end{pmatrix}^{-1}\begin{pmatrix}x\\ y\end{pmatrix}\right)\\
&\quad=(2\pi)^{-n(p+q)/2}(\det\Sigma)^{-n/2}\\
&\qquad\qquad\times\exp\left(-\tfrac{1}{2}\operatorname{Tr}(x^\top(\Sigma^{-1})_{11}x+y^\top(\Sigma^{-1})_{22}y+2y^\top(\Sigma^{-1})_{12}^\top x)\right).
\end{aligned}
$$

The conditional distribution of $p\times n$ matrix x given the $q\times n$ matrix y is also normal:

$$
(\det 2\pi\Omega)^{-n/2}\exp(-\tfrac{1}{2}\operatorname{Tr}\Omega^{-1}(x-Py)(x-Py)^\top) \tag{1.158}
$$

with

$$
\begin{aligned}
\Omega &= \Sigma_{11}-\Sigma_{12}\Sigma_{22}^{-1}\Sigma_{21}=\operatorname{diag}(1-\rho_1^2,\ldots,1-\rho_p^2)\\
P &= \Sigma_{12}\Sigma_{22}^{-1}\,.
\end{aligned}
$$

Then the maximum likelihood estimates r_i of the ρ_i satisfy the determinantal equation

$$
\det(S_{11}^{-1}S_{12}S_{22}^{-1}S_{12}^\top - r^2 I)=0\,, \tag{1.159}
$$

corresponding to

$$
S=\begin{pmatrix}S_{11} & S_{12}\\ S_{12}^\top & S_{22}\end{pmatrix}:=\begin{pmatrix}xx^\top & xy^\top\\ yx^\top & yy^\top\end{pmatrix},
$$

where S_{ij} are the associated submatrices of the *sample* covariance matrix S.

Remark. T he r_i can also be viewed as $r_i=\cos\theta_i$, where the $\theta_1,\ldots,\theta_p$ are the *critical* angles between two planes in $\mathbb{R}^n$:

(i) a p-dimensional plane $=\operatorname{span}\{(x_{11},\ldots,x_{n1}),\ldots,(x_{1p},\ldots,x_{np})\}$
(ii) a q-dimensional plane $=\operatorname{span}\{(y_{11},\ldots,y_{n1})^\top,\ldots,(y_{1q},\ldots,y_{nq})\}$.

Since the (q,n)-matrix y has $\operatorname{rank}(y)=q$, there exists a matrix $H_n\in \mathrm{O}(n)$ such that $yH_n=(y_1\ O)$; therefore acting on x with H_n leads to

$$
yH_n=(\overbrace{\ y_1\ }^{q}\ \overbrace{\ O\ }^{n-q})\}_q\,,\quad xH_n=(\overbrace{\ u\ }^{q}\ \overbrace{\ v\ }^{n-q})\}_{p};. \tag{1.160}
$$

With this in mind,

$$\begin{aligned}
&S_{12}S_{22}^{-1}S_{12}^{\top} - r^2 S_{11} \\
&\quad = xy^{\top}(yy^{\top})^{-1}yx^{\top} - r^2 xx^{\top} \\
&\quad = xH(yH)^{\top}(yH(yH)^{\top})^{-1}yH(xH)^{\top} - r^2(xH)(xH)^{\top} \\
&\quad = (u\ v)\begin{pmatrix} y_1^{\top} \\ O \end{pmatrix}\left((y_1\ O)\begin{pmatrix} y_1^{\top} \\ O \end{pmatrix}\right)^{-1}(y_1\ O)\begin{pmatrix} u^{\top} \\ v^{\top} \end{pmatrix} - r^2\,(u\ v)\begin{pmatrix} u^{\top} \\ v^{\top} \end{pmatrix} \\
&\quad = (u\ v)\begin{pmatrix} I_q & O \\ O & 0_{n-q} \end{pmatrix}\begin{pmatrix} u^{\top} \\ v^{\top} \end{pmatrix} - r^2\,(u\ v)\begin{pmatrix} u^{\top} \\ v^{\top} \end{pmatrix} \\
&\quad = uu^{\top} - r^2(uu^{\top} + vv^{\top})\,,
\end{aligned}$$

and so (1.159) for the r_i can be rewritten

$$\det\bigl(uu^{\top} - r^2(uu^{\top} + vv^{\top})\bigr) = 0\,. \tag{1.161}$$

Then setting the forms (1.160) of x and y in the conditional distribution (1.158) of x given y, one computes the following, setting $H := H_n$,

$$\begin{aligned}
&\operatorname{Tr}\Omega^{-1}(x - Py)(x - Py)^{\top} \\
&\qquad = \operatorname{Tr}\Omega^{-1}(xH - PyH)(xH - PyH)^{\top} \\
&\qquad = \operatorname{Tr}\Omega^{-1}\bigl((u\ v) - P\,(y_1\ O)\bigr)\bigl((u\ v) - P\,(y_1\ O)\bigr)^{\top} \\
&\qquad = \operatorname{Tr}\Omega^{-1}(u - Py_1)(u - Py_1)^{\top} + \operatorname{Tr}\Omega^{-1}vv^{\top}\,; \\
&\qquad\qquad\qquad\qquad\qquad \Omega = \operatorname{diag}(1 - \rho_1^2, \dots, 1 - \rho_p^2)\,;
\end{aligned}$$

this establishes the independence of the normal distributions u and v, given the matrix y, with

$$u \equiv N(Py_1, \Omega)\,, \quad v \equiv N(O, \Omega)\,, \quad P = \operatorname{diag}(\rho_1, \dots, \rho_p)\,.$$

Hence $uu^{\top}$ and $vv^{\top}$ are conditionally independent and both Wishart distributed; to be precise:

• The $p \times p$ matrices $vv^{\top}$ are Wishart distributed, given y, with $n - q$ degrees of freedom and covariance Ω;

• The $p \times p$ matrices $uu^{\top}$ are noncentrally Wishart distributed, given y, with q degrees of freedom, with covariance Ω and with noncentrality matrix

$$\tfrac{1}{2}Py_1y_1^{\top}P^{\top}\Omega^{-1}\,.$$

• The marginal distribution of the $q \times q$ matrices $yy^{\top}$ are Wishart distributed, with n degrees of freedom and covariance I_q, because the marginal distribution of y is normal with covariance I_q.

To summarize, given the matrix y, the sample canonical correlation coefficients $r_1^2 > \cdots > r_p^2$ are the roots of

$$\begin{aligned}
(r_1^2 > \cdots > r_p^2) &= \text{roots of } \det(xy^{\top}(yy^{\top})^{-1}yx^{\top} - r^2 xx^{\top}) = 0 \\
&= \text{roots of } \det\bigl(uu^{\top} - r^2(uu^{\top} + vv^{\top})\bigr) = 0 \\
&= \text{roots of } \det(uu^{\top}(uu^{\top} + vv^{\top})^{-1} - r^2 I) = 0\,.
\end{aligned}$$

Then one shows that, knowing $uu^\top$ and $vv^\top$ are Wishart and conditiionally independent, the conditional distribution of $r_1^2 > \cdots > r_p^2$, given the matrix y is given by

$$\pi^{p^2/2} c_{n,p,q} \exp(-\tfrac{1}{2} \operatorname{Tr} P y y^\top P^\top \Omega^{-1}) \Delta_p(r^2) \prod_1^p (r_i^2)^{(q-p-1)/2} (1-r_i^2)^{(n-q-p-1)/2}$$
$$\times \sum_{\lambda \in \mathbb{Y}} \frac{(n/2)_\lambda C_\lambda(\frac{1}{2} P y y^\top P^\top \Omega^{-1})}{(q/2)_\lambda C_\lambda(I_p)\, |\lambda|!} C_\lambda(R^2) ,$$

where[31]

$$R^2 = \operatorname{diag}(r_1^2, \ldots, r_p^2) , \quad c_{n,p,q} = \frac{\Gamma_p(n/2)}{\Gamma_p(q/2) \Gamma_p\big((n-q)/2\big) \Gamma_p(p/2)} ,$$

and where the C_λ are proportional to Jack polynomials corresponding to the partition λ; for details see Muirhead [70] and Adler–van Moerbeke [10]. By taking the expectation with regard to y or, what is the same, by integrating over the matrix $yy^\top$, which is Wishart distributed (see Sect. 1.8.1.2), one obtains:

Theorem 1.8.1. *Let $X_1, \ldots, X_p, Y_1, \ldots, Y_q$ $(p \le q)$ be normally distributed random variables with zero means and covariance matrix $\Sigma = \left(\begin{smallmatrix} \Sigma_{11} & \Sigma_{12} \\ \Sigma_{21} & \Sigma_{22} \end{smallmatrix}\right)$. If $\rho_1^2, \ldots, \rho_p^2$ are the roots of $\det(\Sigma_{11}^{-1} \Sigma_{12} \Sigma_{22}^{-1} \Sigma_{12}^\top - \rho^2 I) = 0$, then the maximum likelihood estimates $r_1^2, \ldots, r_p^2$ from a sample of size n $(n \ge p+q)$ are given by the roots of*

$$\det(x y^\top (y y^\top)^{-1} y x^\top - r^2 x x^\top) = 0 .$$

Then, assuming $\rho_1^2 = \cdots = \rho_p^2 = 0$, the joint density of the $z_i = r_i^2$ is given by the following density:

$$\pi^{p^2/2} c_{n,p,q} \Delta_p(z) \prod_{i=1}^p z_i^{(q-p-1)/2} (1-z_i)^{(n-q-p-1)/2} \, \mathrm{d}z_i . \tag{1.162}$$

Remark. Taking complex Gaussian populations should introduce in the formula above $\Delta_p^2(z)$ instead of $\Delta_p(z)$ and should remove the $\frac{1}{2}$ in the exponent.

1.8.2 The Probability for the Classical Hermitian Random Ensembles and PDEs Generalizing Painlevé

1.8.2.1 The Gaussian Ensemble (GUE)

This section deals with the Gaussian Hermitian matrix ensemble, discussed in previous section. Given the disjoint union of intervals

[31] Using the standard notation, defined for a partition λ,

$$(a)_\lambda := \prod_i \big(a + (1-i)\big)_{\lambda_i} , \text{with } (x)_n := x(x+1)\cdots(x+n-1) , \ x_0 = 1 .$$

$$E := \bigcup_1^r [c_{2i-1}, c_{2i}] \subseteq \mathbb{R} ,$$

define the algebra of differential operators

$$\mathcal{B}_k = \sum_{i=1}^{2r} c_i^{k+1} \frac{\partial}{\partial c_i} . \tag{1.163}$$

The PDE (1.164) appearing below was obtained by Adler–Shiota–van Moerbeke [1, 2, 9], whereas the ODE (1.165) was first obtained by Tracy–Widom [85]. The method used here is different from the one of Tracy–Widom, who use the method proposed by Jimbo–Miwa–Mori–Sato [54]. John Harnad then shows in [51] the relationship between the PDEs obtained by Tracy–Widom and by Adler–van Moerbeke.

Theorem 1.8.2. *The* log *of the probability*

$$\mathbb{P}_n := \mathbb{P}_n(all\ z_i \in E) = \frac{\int_{E^n} \Delta_n^2(z) \prod_1^n \exp(-z_i^2)\, \mathrm{d}z_i}{\int_{\mathbb{R}^n} \Delta_n^2(z) \prod_1^n \exp(-z_i^2)\, \mathrm{d}z_i}$$

satisfies the PDE

$$(\mathcal{B}_{-1}^4 + 8n\mathcal{B}_{-1}^2 + 12\mathcal{B}_0^2 + 24\mathcal{B}_0 - 16\mathcal{B}_{-1}\mathcal{B}_1) \log \mathbb{P}_n + 6(\mathcal{B}_{-1}^2 \log \mathbb{P}_n)^2 = 0 . \tag{1.164}$$

In particular

$$f(x) := \frac{\mathrm{d}}{\mathrm{d}x} \log \mathbb{P}_n \Big(\max_i z_i \leq x \Big)$$

satisfies the 3rd-order ODE

$$f''' + 6f'^2 + 4(2n - x^2)f' + 4xf = 0 , \tag{1.165}$$

which can be transformed into the Painlevé IV *equation.*

Proof. In Thm. 1.5.1, it was shown that integral (here we indicate the t- and E-dependence)

$$\tau_n(t; E) = \frac{1}{n!} \int_{E^n} \Delta_n^2(z) \prod_1^n \exp\Big(-z_i^2 + \sum_1^\infty t_k z_i^k \Big)\, \mathrm{d}z_i \tag{1.166}$$

satisfies the KP equation, regardless of E,

$$\left(\left(\frac{\partial}{\partial t_1} \right)^4 + 3 \left(\frac{\partial}{\partial t_2} \right)^2 - 4 \frac{\partial^2}{\partial t_1 \partial t_3} \right) \log \tau_n + 6 \left(\left(\frac{\partial}{\partial t_1} \right)^2 \log \tau_n \right)^2 = 0 , \tag{1.167}$$

and in Thm. 1.6.1, $\tau_n(t; E)$ and $\tau_n(t; \mathbb{R})$ were shown to satisfy the same Virasoro constraints, and, in particular for the Gaussian case, the three equations (1.133), with the boundary term missing in the case $\tau_n(t; \mathbb{R})$.

Let T_i denote the pure t-differentiation appearing on the right-hand side of (1.133), with "principal symbol" $\partial/\partial t_{i+2}$,

$$\begin{aligned}T_{-1} &:= \frac{\partial}{\partial t_1} - \frac{1}{2}\sum_{i\geq 2} it_i \frac{\partial}{\partial t_{i-1}} \\ T_0 &:= \frac{\partial}{\partial t_2} - \frac{1}{2}\sum_{i\geq 1} it_i \frac{\partial}{\partial t_i} \\ T_1 &:= \frac{\partial}{\partial t_3} - \frac{1}{2}\sum_{i\geq 1} it_i \frac{\partial}{\partial t_{i+1}} - \frac{n}{2}\sum_{i\geq 2} it_i \frac{\partial}{\partial t_{i-1}} \,.\end{aligned} \tag{1.168}$$

Recall the differential operators $\mathcal{D}_i$ in terms of the boundary operators (1.163), appearing in the Virasoro constraints (1.133),

$$\begin{aligned}\mathcal{D}_{-1} &= -\tfrac{1}{2}\mathcal{B}_{-1} \\ \mathcal{D}_0 &= -\tfrac{1}{2}\mathcal{B}_0 \\ \mathcal{D}_1 &= -\tfrac{1}{2}(\mathcal{B}_1 + n\mathcal{B}_{-1}) \,.\end{aligned} \tag{1.169}$$

With this notation, the Virasoro constraints (1.133) can be summarized as ($F := \log \tau_n(t;E)$)

$$\mathcal{D}_{-1}F = T_{-1}F - \frac{nt_1}{2} \,, \quad \mathcal{D}_0 F = T_0 F - \frac{n^2}{2} \,, \quad \mathcal{D}_1 F = T_1 F - \frac{n^2 t_1}{2} \,.$$

Expressing the action of T_i on t_1, one finds

$$\begin{aligned}T_{-1}t_1 &= 1 - t_2 & T_{-1}^2 t_1 &= T_{-1}(1-t_2) = \tfrac{3}{2}t_3 \\ T_1 t_1 &= -nt_2 & T_{-1}^3 t_1 &= T_{-1}T_{-1}^2 t_1 = T_{-1}\left(\tfrac{3}{2}t_3\right) = -3t_4 \,,\end{aligned}$$

one computes consecutive powers of $\mathcal{D}_i$ and their products, and one notices that $\mathcal{D}_i$ involves differentiation with regard to the boundary terms only, implying in particular that *$\mathcal{D}_i$ and T_i commute.* In view of the form of the KP equation, containing only certain partials, and in view of the fact that the "principal symbol" of T_i equals $\partial/\partial t_{i+2}$, it suffices to compute

$$\begin{aligned}\mathcal{D}_{-1}F &= T_{-1}F - \frac{nt_1}{2} \\ \mathcal{D}_{-1}^2 F &= \mathcal{D}_{-1}T_{-1}F = T_{-1}\mathcal{D}_{-1}F = T_{-1}\left(T_{-1}F - \frac{nt_1}{2}\right) \\ &= T_{-1}^2 F - \frac{n}{2}(1-t_2) \\ \mathcal{D}_{-1}^3 F &= \mathcal{D}_{-1}T_{-1}^2 F = T_{-1}^2\mathcal{D}_{-1}F = T_{-1}^2\left(T_{-1}F - \frac{nt_1}{2}\right) \\ &= T_{-1}^3 F - \frac{3n}{4}t_3\end{aligned}$$

$$
\begin{aligned}
\mathcal{D}_{-1}^4 F = \mathcal{D}_{-1} T_{-1}^3 F = T_{-1}^3 \mathcal{D}_{-1} F &= T_{-1}^3 \left(T_{-1} F - \frac{nt_1}{2} \right) \\
&= T_{-1}^4 F + \frac{3n}{2} t_4
\end{aligned}
\tag{1.170}
$$

$$
\mathcal{D}_1 F = T_1 F - \frac{n^2 t_1}{2}
$$

$$
\begin{aligned}
\mathcal{D}_{-1}\mathcal{D}_1 F = \mathcal{D}_{-1} T_1 F = T_1 \mathcal{D}_{-1} F &= T_1 \left(T_{-1} F - \frac{n^2 t_1}{2} \right) \\
&= T_1 T_{-1} F + \frac{n^3}{2} t_2
\end{aligned}
$$

$$
\mathcal{D}_0 F = T_0 F - \frac{n^2}{2}
$$

$$
\mathcal{D}_0^2 F = \mathcal{D}_0 T_0 F = T_0 \mathcal{D}_0 F = T_0 \left(T_0 F - \frac{n^2}{2} \right) = T_0^2 F \,.
$$

Since one is actually interested in the integral (1.166) along the locus $\mathcal{L} := \{\text{all } t_i = 0\}$, and since readily from (1.168) one has $T_i|_{\mathcal{L}} = \partial/\partial t_{i+2}$, one deduces from the equations above (1.170)[32]

$$
\begin{aligned}
\mathcal{D}_{-1}^2 F\big|_{\mathcal{L}} &= T_{-1}^2 F\big|_{\mathcal{L}} - \frac{n}{2} = \left.\frac{\partial^2 F}{\partial t_1^2}\right|_{\mathcal{L}} - \frac{n}{2} \\
\mathcal{D}_{-1}^4 F\Big|_{\mathcal{L}} &= T_{-1}^4 F\big|_{\mathcal{L}} = \left.\frac{\partial}{\partial t_1} T_{-1}^3 F\right|_{\mathcal{L}} = T_{-1}^3 \left.\frac{\partial F}{\partial t_1}\right|_{\mathcal{L}} = \left.\frac{\partial^4 F}{\partial t_1^4}\right|_{\mathcal{L}} \\
\mathcal{D}_0^2 F\big|_{\mathcal{L}} &= T_0^2 F\big|_{\mathcal{L}} = \left.\frac{\partial}{\partial t_2} T_0 F\right|_{\mathcal{L}} = \left.\left(\frac{\partial^2}{\partial t_2^2} - \frac{\partial}{\partial t_2} \right) F\right|_{\mathcal{L}} \\
\mathcal{D}_{-1}\mathcal{D}_1 F\Big|_{\mathcal{L}} &= T_1 T_{-1} F\big|_{\mathcal{L}} = \left.\frac{\partial}{\partial t_3} T_{-1} F\right|_{\mathcal{L}} = \left.\left(\frac{\partial^2}{\partial t_3 \partial t_1} - \frac{3}{2}\frac{\partial}{\partial t_2} \right) F\right|_{\mathcal{L}} \\
\mathcal{D}_0 F\big|_{\mathcal{L}} &= T_0 F\big|_{\mathcal{L}} - \frac{n^2}{2} = \left.\frac{\partial F}{\partial t_2}\right|_{\mathcal{L}} - \frac{n^2}{2} \,.
\end{aligned}
$$

By solving the five expressions above linearly in terms of the left-hand side, one deduces

$$
\begin{aligned}
\left.\frac{\partial^2 F}{\partial t_1^2}\right|_{\mathcal{L}} &= \mathcal{D}_{-1}^2 F\big|_{\mathcal{L}} + \frac{n}{2} \,, &
\frac{\partial^4 F}{\partial t_1^4}\Big|_{\mathcal{L}} &= \mathcal{D}_{-1}^4 F\big|_{\mathcal{L}} \\
\left.\frac{\partial F}{\partial t_2}\right|_{\mathcal{L}} &= \mathcal{D}_0 F\big|_{\mathcal{L}} + \frac{n^2}{2} &
\left.\frac{\partial^2 F}{\partial t_2^2}\right|_{\mathcal{L}} &= \mathcal{D}_0^2 F\big|_{\mathcal{L}} + \mathcal{D}_0 F\big|_{\mathcal{L}} + \frac{n^2}{2} \\
\left.\frac{\partial^2 F}{\partial t_1 \partial t_3}\right|_{\mathcal{L}} &= \mathcal{D}_{-1}\mathcal{D}_1 F\big|_{\mathcal{L}} + \frac{3}{2}\mathcal{D}_0 F\big|_{\mathcal{L}} + \frac{3n^2}{4} \,.
\end{aligned}
$$

---

[32] Notice one also needs $\mathcal{D}_0 F$, because $\partial F/\partial t_2$ appears in the expressions $\mathcal{D}_0^2 F$ and $\mathcal{D}_{-1}\mathcal{D}_1 F$ below.

So, substituting into the KP equation and expressing the $\mathcal{D}_i$ in terms of the $\mathcal{B}_i$ as in (1.169), one finds

$$
\begin{aligned}
0 &= \left(\left(\frac{\partial}{\partial t_1}\right)^4 + 3\left(\frac{\partial}{\partial t_2}\right)^2 - 4\frac{\partial^2}{\partial t_1 \partial t_3}\right) F + 6\left(\left(\frac{\partial}{\partial t_1}\right)^2 F\right)^2 \Bigg|_{\mathcal{L}} \\
&= \mathcal{D}_{-1}^4 + 3\left(\mathcal{D}_0^2 F + \mathcal{D}_0 F + \frac{n^2}{2}\right) - 4\left(\mathcal{D}_{-1}\mathcal{D}_1 F + \frac{3}{2}\mathcal{D}_0 F + \frac{3n^2}{4}\right) \\
&\qquad\qquad + 6\left(\mathcal{D}_{-1}^2 F + \frac{n}{2}\right)^2 \Bigg|_{\mathcal{L}} \\
&= \mathcal{D}_{-1}^4 F + 6n\mathcal{D}_{-1}^2 F + 3\mathcal{D}_0^2 F - 3\mathcal{D}_0 F - 4\mathcal{D}_{-1}\mathcal{D}_1 F + 6(\mathcal{D}_{-1}^2 F)^2 \big|_{\mathcal{L}} \\
&= \frac{1}{16}(\mathcal{B}_{-1}^4 + 8n\mathcal{B}_{-1}^2 + 12\mathcal{B}_0^2 + 24\mathcal{B}_0 - 16\mathcal{B}_{-1}\mathcal{B}_1)F + \frac{3}{8}(\mathcal{B}_{-1}^2 F)^2 \big|_{\mathcal{L}} ,
\end{aligned}
$$

which establishes (1.164) for (remember notation (1.166))

$$
F = \log \tau_n(0, E) = \log \mathbb{P}_n(\text{all } z_i \in E) + \log \tau_n(0, \mathbb{R}) .
$$

Since the $\mathcal{B}_k$ are derivations with regard to the boundary points of the set E and since $\log \tau_n(0, \mathbb{R})$ is independent of those points, (1.164) is also valid for $\log \mathbb{P}_n$; it is an equation of order 4.

When E is a semi-infinite interval $(-\infty, x)$, then one has $\mathcal{B}_k = x^{k+1}\partial/\partial x$ and then, of course, PDE (1.164) turns into an ODE (1.165), of an order one less, involving $f(x) := (\mathrm{d}/Dx) \log \mathbb{P}_n(\max_i z_i \leq x)$. For the connection with Painlevé IV, see Sect. 1.8.3, thus ending the proof of Thm. 1.8.2. □

1.8.2.2 The Laguerre Ensemble

Given $E \subset \mathbb{R}^+$ and the boundary operators

$$
\mathcal{B}_k := \sum_1^{2r} c_i^{k+2} \frac{\partial}{\partial c_i} , \quad \text{for } k = -1, 0, 1, \dots ,
$$

the following statement holds: (see [1, 2, 9] for the PDE obtained below; the ODE was first obtained by Tracy–Widom [85])

Theorem 1.8.3. *The* log *of the probability*

$$
\mathbb{P}_n := \mathbb{P}_n(\textit{all } z_i \in E) = \frac{\int_{E^n} \Delta_n^2(z) \prod_1^n z_i^a \exp(-z_i)\,\mathrm{d}z_i}{\int_{(\mathbb{R}^+)^n} \Delta_n^2(z) \prod_1^n z_i^a \exp(-z_i)\,\mathrm{d}z_i}
$$

satisfies the PDE

$$
\begin{aligned}
(\mathcal{B}_{-1}^4 - 2\mathcal{B}_{-1}^3 + (1-a^2)\mathcal{B}_{-1}^2 - 4\mathcal{B}_1\mathcal{B}_{-1} + 3\mathcal{B}_0^2 \\
+ 2(2n+a)\mathcal{B}_0\mathcal{B}_{-1} - 2\mathcal{B}_1 - (2n+a)\mathcal{B}_0) \log \mathbb{P}_n \\
+ 6(\mathcal{B}_{-1}^2 \log \mathbb{P}_n)^2 - 4(\mathcal{B}_{-1}^2 \log \mathbb{P}_n)(\mathcal{B}_{-1} \log \mathbb{P}_n) = 0 . \quad (1.171)
\end{aligned}
$$

In particular, $f(x) := x(\mathrm{d}/\mathrm{d}x)\log P_n(\max_i z_i \le x)$ *satisfies*

$$x^2 f''' + xf'' + 6xf'^2 - 4ff' - \big((a-x)^2 - 4nx\big)f' - (2n+a-x)f = 0\,, \quad (1.172)$$

which can be transformed into the Painlevé V *equation.*

1.8.2.3 The Jacobi Ensemble

The Jacobi weight is given by $(1-z)^a(1+z)^b$. For $E \subset [-1,+1]$, the boundary differential operators $\mathcal{B}_k$ are now defined by

$$\mathcal{B}_k = \sum_1^{2r} c_i^{k+1}(1-c_i^2)\frac{\partial}{\partial c_i}\,.$$

Introduce the following parameters

$$\begin{aligned} r = 4(a^2+b^2)\,, \quad s = 2(a^2-b^2)\,, \quad q = 2(2n+a+b)^2 \\ r = a^2+b^2\,, \quad s = a^2-b^2\,, \quad q = (2n+a+b)^2\,. \end{aligned}$$

Theorem 1.8.4 (Haine–Semengue [49] and Adler–van Moerbeke [9]). *The following probability*

$$\mathbb{P}_n := \mathbb{P}_n(\mathit{all}\ z_i \in E) = \frac{\int_{E^n} \Delta_n(z)^2 \prod_{i=1}^n (1-z_i)^a(1+z_i)^b\,\mathrm{d}z_i}{\int_{[-1,1]^n} \Delta_n(z)^2 \prod_{i=1}^n (1-z_i)^a(1+z_i)^b\,\mathrm{d}z_i} \quad (1.173)$$

satisfies the PDE:

$$\begin{aligned} \big(\mathcal{B}_{-1}^4 &+ (q-2r+2)\mathcal{B}_{-1}^2 + q(3\mathcal{B}_0^2 - 2\mathcal{B}_0 + 2\mathcal{B}_2) + 4\mathcal{B}_0\mathcal{B}_{-1}^2 \\ &- 2(2q-1)\mathcal{B}_1\mathcal{B}_{-1} + (2\mathcal{B}_{-1}\log\mathbb{P}_n - s)(\mathcal{B}_1 - \mathcal{B}_{-1} + 2\mathcal{B}_0\mathcal{B}_{-1})\big)\log\mathbb{P}_n \\ &+ 2\mathcal{B}_{-1}^2\log\mathbb{P}_n(2\mathcal{B}_0\log\mathbb{P}_n + 3\mathcal{B}_{-1}^2\log\mathbb{P}_n) = 0\,. \quad (1.174) \end{aligned}$$

In particular, $f(x) := (1-x^2)(\mathrm{d}/\mathrm{d}x)\log\mathbb{P}_n(\max_i \lambda_i \le x)$ *for* $0 < x < 1$ *satisfies:*

$$\begin{aligned} (x^2-1)^2 f''' &+ 2(x^2-1)(xf'' - 3f'^2) \\ &+ (8xf - q(x^2-1) - 2sx - 2r)f' - f(2f - qx - s) = 0\,, \quad (1.175) \end{aligned}$$

which is a version of Painlevé VI.

Proof of Thms. 1.8.3 *and* 1.8.4. It goes along the same lines as Thm. 1.8.2 for GUE, namely using the Virasoro constraints (1.137) and (1.138), together with the KP equation (1.167). This then leads to the PDEs (1.171) and (1.174). The ODEs (1.172) and (1.175) are found by simple computation. For connections with the Painlevé equations see Sect. 1.8.3. □

1.8.3 Chazy and Painlevé Equations

Each of these three equations (1.165), (1.172), (1.175) is of the Chazy form

$$f''' + \frac{P'}{P} f'' + \frac{6}{P} f'^2 - \frac{4P'}{P^2} f f' + \frac{P''}{P^2} f^2 + \frac{4Q}{P^2} f' - \frac{2Q'}{P^2} f + \frac{2R}{P^2} = 0 \,, \quad (1.176)$$

with P, Q, R having the form:

Gauss	$P(x) = 1$	$4Q(x) = -4x^2 + 8n$	$R = 0$
Laguerre	$P(x) = x$	$4Q(x) = -(x-a)^2 + 4nx$	$R = 0$
Jacobi	$P(x) = 1 - x^2$	$4Q(x) = -(q(x^2-1) + 2sx + 2r)$	$R = 0$.

The differential equation (1.176) belongs to the general Chazy class

$$f''' = F(z, f, f', f'') \,,$$
$$\text{where } F \text{ is rational in } f, f', f'' \text{ and locally analytic in } z \,,$$

subjected to the requirement that the general solution be free of movable branch points; the latter is a branch point whose location depends on the integration constants. In his classification Chazy found thirteen cases, the first of which is given by (1.176), with arbitrary polynomials $P(z), Q(z), R(z)$ of degree 3, 2, 1 respectively. Cosgrove and Scoufis [28, 29], show that this third-order equation has a first integral, which is second order in f and quadratic in f'',

$$f''^2 + \frac{4}{P^2}\big((Pf'^2 + Qf' + R)f' - (P'f'^2 + Q'f' + R')f + \tfrac{1}{2}(P''f' + Q'')f^2 - \tfrac{1}{6}P'''f^3 + c\big) = 0 \,; \quad (1.177)$$

c is the integration constant. Equations of the general form

$$f''^2 = G(x, f, f'')$$

are invariant under the map

$$x \mapsto \frac{a_1 z + a_2}{a_3 z + a_4} \quad \text{and} \quad f \mapsto \frac{a_5 f + a_6 z + a_7}{a_3 z + a_4} \,.$$

Using this map, the polynomial $P(z)$ can be normalized to

$$P(z) = z(z-1),\ z,\ \text{or } 1 \,.$$

Equation (1.177) is a master Painlevé equation, containing the 6 Painlevé equations, replacing $f(z)$ by some new variable $g(z)$, e.g,

- $g''^2 = -4g'^3 - 2g'(zg' - g) + A_1$ (Painlevé II)
- $g''^2 = -4g'^3 + 4(zg' - g)^2 + A_1 g' + A_2$ (Painlevé IV)

- $(zg'')^2 = (zg' - g)\big(-4g'^2 + A_1(zg' - g) + A_2\big) + A_3 g' + A_4$ (Painlevé V)
- $(z(z-1)g'')^2 = (zg' - g)\big(4g'^2 - 4g'(zg' - g) + A_2\big) + A_1 g'^2 + A_3 g' + A_4$ (Painlevé VI)

Now, each of these Painlevé II, IV, V, VI equations can be transformed into the standard Painlevé equations, which are all differential equations of the form

$$f'' = F(z, f, f') \,, \quad \text{rational in } f, f', \text{ analytic in } z \,,$$

whose general solution has no movable critical points. Painlevé showed that this requirement leads to 50 types of equations, six of which cannot be reduced to known equations.

1.9 Large Hermitian Matrix Ensembles

1.9.1 Equilibrium Measure for GUE and Wigner's Semi-Circle

Remember according to (1.153), the average density of eigenvalues is given by $K_n(z,z)\,\mathrm{d}z$. Pastur and Marcenko [76] have proposed a method to compute the average density of eigenvalues (equilibrium distribution), when n gets very large. For a rigorous and very general approach, see Johansson [55], who also studies the fluctuations of the linear statistics of the eigenvalues about the equilibrium distribution.

Consider the case of a random Hermitian ensemble with probability defined by

$$\frac{1}{Z_n} \int_{\mathcal{H}_n(E)} \mathrm{d}M \exp\Big(-\frac{n}{2v^2} \operatorname{Tr}(M - A)^2\Big) \,\mathrm{d}M \,,$$

for a diagonal matrix $A = (a_1, \ldots, a_n)$. Consider then the spectral function of A, namely $\mathrm{d}\sigma(\lambda) := \frac{1}{n} \sum_i \delta(\lambda - a_i)$. The Pastur–Marcenko method tells us that the Stieltjes transform of the equilibrium measure of $\mathrm{d}\nu(\lambda)$, when $n \to \infty$, namely

$$f(z) = \int_{-\infty}^{\infty} \frac{\mathrm{d}\nu(\lambda)}{\lambda - z} \,, \quad \operatorname{Im} z \neq 0 \,,$$

satisfies the integral equation

$$f(z) = \int_{-\infty}^{\infty} \frac{\mathrm{d}\sigma(\lambda)}{\lambda - z - v^2 f(z)} \,. \tag{1.178}$$

The density of the equilibrium distribution is then given by

$$\frac{\mathrm{d}\nu(z)}{\mathrm{d}z} = \frac{1}{\pi} \operatorname{Im} f(z) \,.$$

When $A = 0$, the integral equation (1.178) becomes

$$f(z) = \frac{1}{-z - v^2 f(z)}$$

with solution $f(z) = (-z \pm \sqrt{z^2 - 4v^2})/(2v^2)$, and thus one finds the classical semi-circle law,

$$\frac{\mathrm{d}\nu(z)}{\mathrm{d}z} = \frac{1}{\pi} \operatorname{Im} f(z) = \begin{cases} \sqrt{4v^2 - z^2}/(2\pi v^2) & \text{for } -2v \le z \le 2v \\ 0 & \text{for } |z| \ge 2v, \end{cases}$$

concentrated on the interval $[-2v, 2v]$.

As an exercise, consider now the case, where $v = 1$ and where the diagonal matrix A has two distinct eigenvalues, namely

$$A = \operatorname{diag}(\underbrace{\alpha, \dots, \alpha}_{pn}, \underbrace{\beta, \dots, \beta}_{(1-p)n}) .$$

See, e.g., Adler–van Moerbeke [12]. The integral equation (1.178) becomes

$$f - \frac{1-p}{\beta - z - f} - \frac{p}{\alpha - z - f} = 0 , \quad \text{for } 0 < p < 1 ,$$

which, upon clearing, leads to a cubic equation for $g := f + z$,

$$g^3 - (z + \alpha + \beta)g^2 + \big(z(\alpha+\beta) + \alpha\beta + 1\big)g - \alpha\beta z - (1-p)\alpha - p\beta = 0 ,$$

having, as one checks, a quartic discriminant $D_1(z)$ in z. Since the roots of a cubic polynomial involve, in particular, the square root of the discriminant, the solution $g(z)$ of the cubic will have a non-zero imaginary part, if and only if $D_1(z) < 0$. Thus one finds the following equilibrium density,

$$\frac{\mathrm{d}\nu(z)}{\mathrm{d}z} = \frac{1}{\pi} \operatorname{Im} f(z) = \begin{cases} (1/\pi) \operatorname{Im} g(z) & \text{for } z \text{ such that } D_1(z) < 0 \\ 0 & \text{for } z \text{ such that } D_1(z) \ge 0. \end{cases}$$

Therefore the boundary of the support of the equilibrium measure will be given by the real roots of $D_1(z) = 0$. Depending on the values of the parameters α, β and p, there will be four real roots or two real roots, with a critical situation where there are three real roots, i.e., when two of the four real ones collide. The critical situation occurs exactly when the discriminant $D_2(\alpha, \beta, p)$ (with regard to z) of $D_1(z)$ vanishes, namely when

$$D_2(\alpha, \beta, p) = 4p(1-p)\gamma(\gamma^3 - 3\gamma^2 + 3\gamma(9p^2 - 9p + 1) - 1)^3\big|_{\gamma=(\alpha-\beta)^2} = 0 .$$

This polynomial has a positive root γ, which can be given explicitly, the others being imaginary, and one checks that, when one has the relationship

$$\alpha - \beta = \frac{q+1}{\sqrt{q^2 - q + 1}} , \quad \text{upon using the parametrization } p = \frac{1}{q^3 + 1} ,$$

two of the four roots of $D_1(z)$ collide. This is to say, this is the precise point at which the support of the equilibrium measure goes from two to one interval. This then occurs exactly at value

$$z = \beta + \frac{2q-1}{\sqrt{q^2-q+1}}$$

on the real line.

1.9.2 Soft Edge Scaling Limit for GUE and the Tracy–Widom Distribution

Consider the probability measure on eigenvalues z_i of the $n \times n$ Gaussian Hermitian ensemble (GUE)

$$\mathbb{P}_n(\text{all } z_i \in \widetilde{E}) = \frac{\int_{\widetilde{E}^n} \Delta_n^2(z) \prod_1^n \exp(-z_i^2)\, \mathrm{d}z_i}{\int_{\mathbb{R}^n} \Delta_n^2(z) \prod_1^n \exp(-z_i^2)\, \mathrm{d}z_i}\,.$$

Given the disjoint union $E := \bigcup_1^r [x_{2i-1}, x_{2i}] \subset \mathbb{R}$, define the gradient and the Euler operator with respect to the boundary points of the set E:

$$\nabla_x = \sum_1^{2r} \frac{\partial}{\partial x_i} \quad \text{and} \quad \mathcal{E}_x = \sum_1^{2r} x_i \frac{\partial}{\partial x_i}\,. \tag{1.179}$$

Remember the definition of the Fredholm determinant of a continuous kernel $K(x,y)$, the continuous analogue of the discrete kernel (1.39),

$$\det\bigl(I - K(x,y)\chi_E(y)\bigr) = 1 + \sum_{n=1}^{\infty} \frac{(-1)^n}{n!} \int_E \mathrm{d}z_1 \cdots \mathrm{d}z_n \det\bigl(K(z_i,z_j)\bigr)_{1\le i,j\le n}\,.$$

We now state: [1, 2, 85]

Theorem 1.9.1. *The gradient*

$$f(x_1,\dots,x_{2r}) := \nabla_x \log \mathbb{P}(E^{\mathrm{c}})\,,$$

with

$$\mathbb{P}(E^{\mathrm{c}}) := \lim_{n\to\infty} \mathbb{P}_n(\textit{all } \sqrt{2}n^{1/6}(z_i - \sqrt{2n}) \in E^{\mathrm{c}})\,,$$

satisfies the 3rd-order non-linear PDE:

$$\bigl(\nabla_x^3 - 4(\mathcal{E}_x - \tfrac{1}{2})\bigr)f + 6(\nabla_x f)^2 = 0\,. \tag{1.180}$$

In particular, for $E = (x,\infty)$,

$$\begin{aligned}
\mathcal{F}(x) &:= \lim_{n\nearrow\infty} \mathbb{P}_n(\sqrt{2}n^{1/6}(z_{\max} - \sqrt{2n}) \le x) \\
&= \det(I - \mathbf{A}\,\chi_{(x,\infty)}) \\
&= \exp\biggl(-\int_x^\infty (\alpha - x) g^2(\alpha)\,\mathrm{d}\alpha\biggr)\,,
\end{aligned} \tag{1.181}$$

is the Tracy–Widom distribution, *with*[33]

$$\mathbf{A}(x,y) := \frac{\mathrm{A}(x)\,\mathrm{A}'(y) - \mathrm{A}'(x)\,\mathrm{A}(y)}{x-y} = \int_0^\infty \mathrm{A}(u+x)\,\mathrm{A}(u+y)\,\mathrm{d}u$$

and $g(\alpha)$ *the Hastings–McLeod (unique) solution of*

$$\begin{cases} g'' = \alpha g + 2g^3 \\ g(\alpha) \sim \exp(-\frac{2}{3}\alpha^{3/2})/(2\sqrt{\pi}\alpha^{1/4}) \quad \textit{for } \alpha \nearrow \infty \end{cases} \quad \text{(Painlevé II)}.$$

Proof. *Step* 1. Applying Prop. 1.7.5, it follows that

$$\mathbb{P}_n(\text{all } z_i \in E^{\mathrm{c}}) = \det\bigl(I - K_n(y,z)\chi_E(z)\bigr) , \tag{1.182}$$

where the kernel $K_n(y,z)$ is given by Prop. 1.7.3,

$$K_n(y,z) = \left(\frac{n}{2}\right)^{1/2} \exp\bigl(-\tfrac{1}{2}(y^2+z^2)\bigr)\frac{p_n(y)p_{n-1}(z) - p_{n-1}(y)p_n(z)}{y-z} . \tag{1.183}$$

The p_ns are orthonormal polynomials with respect to $\exp(-z^2)$ and thus proportional to the classical Hermite polynomials:

$$p_n(y) := \frac{1}{2^{n/2}\sqrt{n!}\pi^{1/4}} H_n(y) = \frac{1}{\sqrt{h_n}} y^n + \cdots , \tag{1.184}$$

with

$$H_n(y) := \exp(y^2)\left(-\frac{\mathrm{d}}{\mathrm{d}y}\right)^n \exp(-y^2) , \quad h_n = \frac{\sqrt{\pi}n!}{2^n} .$$

Step 2. The Plancherel–Rotach asymptotic formula (see Szegő [83]) says that

$$\exp\left(-\frac{x^2}{2}\right)\frac{n^{1/12}H_n(x)}{2^{n/2+1/4}\sqrt{n!}\pi^{1/4}}\bigg|_{x=\sqrt{2n+1}+t/(\sqrt{2}n^{1/6})} = \mathrm{A}(t) + O(n^{-2/3}) ,$$

uniformly for $t \in$ compact $K \subset \mathbb{C}$ and thus, in view of (1.184),

$$\exp\left(-\frac{x^2}{2}\right)p_n(x)\bigg|_{x=\sqrt{2n+1}+t/(\sqrt{2}n^{1/6})} = 2^{1/4}n^{-1/12}\bigl(\mathrm{A}(t) + O(n^{-2/3})\bigr) . \tag{1.185}$$

Since the Hermite kernel (1.183) also involves $p_{n-1}(x)$, one needs an estimate like (1.185), with the same scaling but for p_n replaced by p_{n-1}. So, one needs the following:

---

[33] Remember the Airy function:

$$\mathrm{A}(x) = \frac{1}{\pi}\int_0^\infty \cos\left(\frac{u^3}{3} + xu\right)\,\mathrm{d}u , \quad \text{satisfying the ODE } \mathrm{A}''(x) = x\,\mathrm{A}(x) .$$

$$
\begin{aligned}
x &= \sqrt{2n+1} + \frac{t}{\sqrt{2}n^{1/6}} \\
&= \sqrt{(2n-1)\Big(1+\frac{2}{2n-1}\Big)} + \frac{t}{\sqrt{2}(n-1)^{1/6}\big(1+1/(n-1)\big)^{1/6}} \\
&= \sqrt{2n-1}\left(1+\frac{1}{2n-1}+O\Big(\frac{1}{n^2}\Big)\right) + \frac{t}{\sqrt{2}(n-1)^{1/6}}\left(1+O\Big(\frac{1}{n}\Big)\right) \\
&= \sqrt{2n-1} + \frac{t+1/n^{1/3}}{\sqrt{2}(n-1)^{1/6}} + O\Big(\frac{1}{n^{7/6}}\Big) .
\end{aligned}
$$

Hence, from (1.185) it follows that

$$
\begin{aligned}
&\exp\Big(-\frac{x^2}{2}\Big)p_{n-1}(x)\Big|_{x=\sqrt{2n+1}+t/(\sqrt{2}n^{1/6})} \\
&\qquad = \exp\Big(-\frac{x^2}{2}\Big)p_{n-1}(x)\Big|_{x=\sqrt{2n-1}+(t+n^{-1/3})/(\sqrt{2}(n-1)^{1/6})+\cdots} \\
&\qquad = 2^{1/4}n^{-1/12}\big(\mathrm{A}(t+n^{-1/3})+O(n^{-2/3})\big) . \qquad (1.186)
\end{aligned}
$$

From the definition of the Fredholm determinant, one needs to find the limit of $K_n(y;z)\,\mathrm{d}z$. Therefore, in view of (1.183) and using the estimates (1.185) and (1.186),

$$
\begin{aligned}
&\lim_{n\to\infty} K_n(y;z)\,\mathrm{d}z\Big|_{\substack{y=(2n+1)^{1/2}+t/(\sqrt{2}n^{1/6})\\ z=(2n+1)^{1/2}+s/(\sqrt{2}n^{1/6})}} \\
&= -\lim_{n\to\infty}\Big(\frac{n}{2}\Big)^{1/2} \exp\big(-\tfrac{1}{2}(y^2+z^2)\big) \\
&\quad \times \frac{p_n(y)(p_n(z)-p_{n-1}(z))-p_n(z)(p_n(y)-p_{n-1}(y))}{(y-z)\sqrt{2}n^{1/6}}\Big|_{\substack{y=(2n+1)^{1/2}+t/(\sqrt{2}\,n^{1/6})\\ z=(2n+1)^{1/2}+s/(\sqrt{2}n^{1/6})}} \\
&= \lim_{n\to\infty}\Big(\frac{n}{2}\Big)^{1/2}(2^{1/4}n^{-1/12})^2 n^{-1/3} \\
&\qquad \times \frac{\mathrm{A}(t)\big(\mathrm{A}(s+n^{-1/3})-\mathrm{A}(s)\big)-\mathrm{A}(s)\big(\mathrm{A}(t+n^{-1/3})-\mathrm{A}(t)\big)}{n^{-1/3}(t-s)}\,\mathrm{d}s \\
&= \frac{\mathrm{A}(t)\,\mathrm{A}'(s)-\mathrm{A}(s)\,\mathrm{A}'(t)}{t-s}\,\mathrm{d}s = \mathbf{A}(t,s)\,\mathrm{d}s
\end{aligned}
$$

and thus for $E := \bigcup_1^{2r}[x_{2i-1},x_{2i}]$ and $\widetilde{E} := \bigcup_1^{2r}[c_{2i-1},c_{2i}]$, related by

$$
\widetilde{E} = \sqrt{2n} + \frac{E}{\sqrt{2}n^{1/6}} , \qquad (1.187)
$$

one has shown (upon taking the limit term by term in the sum defining the Fredholm determinant)

$$\begin{aligned}
\lim_{n\to\infty} \mathbb{P}_n(\text{all } z_i \in \widetilde{E}^c) &= \lim_{n\to\infty} \mathbb{P}_n\left(\text{all } z_i \in \sqrt{2n} + \frac{E^c}{\sqrt{2}n^{1/6}}\right) \\
&= \lim_{n\to\infty} \mathbb{P}_n\left(\text{all } z_i \in (2n+1)^{1/2} + \frac{E^c}{\sqrt{2}n^{1/6}}\right) \\
&= \lim_{n\to\infty} \det(I - K_n \chi_{\widetilde{E}})\big|_{\widetilde{E}=(2n+1)^{1/2}+E/(\sqrt{2}n^{1/6})} \\
&= \det(I - \mathbf{A}\,\chi_E).
\end{aligned}$$

Step 3. From Thm. 1.8.2, $\mathbb{P}_n(\text{all } z_i \in \widetilde{E}^c)$ satisfies, with regard to the boundary points c_i of $\widetilde{E}$, the PDE (1.164); thus setting that scaling into this PDE yields (remember ∇_x and $\mathcal{E}_x$ are as in (1.179) and $\mathcal{B}_k = \sum_i c_i^{k+1}\partial/\partial c_i$)

$$\begin{aligned}
0 &= \left(\mathcal{B}_{-1}^4 + 8n\mathcal{B}_{-1}^2 + 12\mathcal{B}_0^2 + 24\mathcal{B}_0 - 16\mathcal{B}_{-1}\mathcal{B}_1\right)\log \mathbb{P}_n \\
&\qquad\qquad + 6\left(\mathcal{B}_{-1}^2 \log \mathbb{P}_n\right)^2\Big|_{c_i=\sqrt{2n}+x_i/(\sqrt{2}n^{1/6})} \\
&= 4n^{2/3}\left[\left(\nabla_x^3 - 4(\mathcal{E}_x - \tfrac{1}{2})\right)\nabla_x \log \mathbb{P} + 6(\nabla_x^2 \log \mathbb{P})^2\right] + o(n^{2/3})\ .
\end{aligned}$$

Note that in this computation, the terms of order $n^{4/3}$ cancel, because the leading term in

$$12\mathcal{B}_0^2 - 16\mathcal{B}_{-1}\mathcal{B}_1 = -4\sum_i c_i^2\left(\frac{\partial}{\partial c_i}\right)^2 + \cdots = -16n^{4/3}\nabla_x^2 + \cdots$$

cancels versus the leading term in $8n\mathcal{B}_{-1}^2 = 16n^{4/3}\nabla_x^2 + \cdots$; thus only the terms of order $n^{2/3}$ remain. Since in step 2 it was shown that the limit exists, the term in brackets vanishes, showing that $\log \mathbb{P}(E^c)$ satisfies the PDE (1.180).

Step 4. In particular, upon picking $E = (x, \infty)$, the PDE (1.180) for

$$f(x) = \frac{\partial}{\partial x}\log \mathcal{F}(x) = \frac{\partial}{\partial x}\log \lim_{n\nearrow\infty} \mathbb{P}_n\left(\sqrt{2}n^{1/6}\left(z_{\max} - \sqrt{2n}\right) \le x\right)$$

becomes an ODE:

$$f''' - 4xf' + 2f + 6f'^2 = 0\ .$$

Multiplying this equation with f'' and integrating from $-\infty$ to x lead to the differential equation (the nature of the solution shows the integration constant must vanish)

$$f''^2 + 4f'(f'^2 - xf' + f) = 0\ . \tag{1.188}$$

Then, setting

$$\begin{cases} f' = -g^2 \\ f = g'^2 - xg^2 - g^4 \end{cases} \tag{1.189}$$

and, since then $f'' = -2gg'$, an elementary computation shows that (1.189) is an obvious solution to (1.188). For (1.189) to be valid, the derivative of the right hand side of the second expression in (1.189) must equal the derivative of the right hand side of the first expression in (1.189), i.e., we must have:

$$0 = (f)' - f' = (g'^2 - xg^2 - g^4)' + g^2 = 2g'(g'' - 2g^3 - xg) ,$$

and so $g'' = 2g^3 + xg$. Hence

$$\frac{\partial^2}{\partial x^2} \log \mathcal{F}(x) = f' = -g^2 .$$

Integrating once yields (assuming that g^2 decays fast enough at ∞, which will be apparent later)

$$\frac{\partial}{\partial x} \log \mathcal{F}(x) = \int_x^\infty g^2(u)\, \mathrm{d}u ;$$

integrating once more and further integrating by parts yield

$$\begin{aligned} \log \mathcal{F}(x) &= -\int_x^\infty \mathrm{d}\alpha \int_\alpha^\infty g^2(u)\, \mathrm{d}u \\ &= -\int_x^\infty \mathrm{d}\alpha \left(\frac{d}{d\alpha}\alpha\right) \int_\alpha^\infty g^2(u)\, \mathrm{d}u \\ &= -\alpha \int_\alpha^\infty g^2(u)\, \mathrm{d}u \Big|_x^\infty - \int_x^\infty \alpha g^2(\alpha)\, \mathrm{d}\alpha \\ &= -\int_x^\infty (\alpha - x) g^2(\alpha)\, \mathrm{d}\alpha , \end{aligned} \tag{1.190}$$

confirming (1.181). For $x \to \infty$, one checks that, on the one hand, from the definition of the Fredholm determinant of $\mathbf{A}$, the two leading terms are given by

$$\mathcal{F}(x) = \det\bigl(I - \mathbf{A}\, \chi_{[x,\infty)}\bigr) = 1 - \int_x^\infty \mathbf{A}(z,z)\, \mathrm{d}z + \cdots \tag{1.191}$$

and, on the other hand, from (1.190), the two leading terms are

$$\mathcal{F}(x) = 1 - \int_x^\infty (\alpha - x) g^2(\alpha)\, \mathrm{d}\alpha + \cdots . \tag{1.192}$$

Therefore, comparing the two expressions above, (1.191) and (1.192), one has equality up to higher order terms

$$\int_x^\infty (\alpha - x) g^2(\alpha)\, \mathrm{d}\alpha = \int_x^\infty \mathrm{d}z\ \mathbf{A}(z,z) + \cdots$$

and upon taking two derivatives in x,

$$\begin{aligned} -g^2(x) &= \frac{\partial}{\partial x} \mathbf{A}(x,x) + \cdots \\ &= \frac{\partial}{\partial x} \int_0^\infty \mathrm{A}(u+x)^2\, \mathrm{d}u + \cdots \\ &= \frac{\partial}{\partial x} \int_x^\infty \mathrm{A}(u)^2\, \mathrm{d}u + \cdots \\ &= -\,\mathrm{A}(x)^2 + \cdots , \end{aligned}$$

showing that asymptotically $g(x) \sim A(x)$ for $x \to \infty$. It is classically known that asymptotically

$$\mathrm{A}(x) = \frac{\exp(-\frac{2}{3}x^{3/2})}{2\sqrt{\pi}x^{1/4}}\left(1 + \sum_1^\infty \frac{\alpha_i}{(x^{3/2})^i} + \dots\right), \quad \text{as } x \to \infty \, .$$

The solution $g(x)$ of $g'' = 2g^3 + xg$ behaving like the Airy function at ∞ is unique (Hastings–McLeod solution). It behaves like

$$\begin{aligned} g(x) &= \mathrm{Ai}(x) + \mathbf{O}\left(\frac{\exp(-\frac{4}{3}x^{3/2})}{x^{1/4}}\right) && \text{for } x \to \infty \\ &= \sqrt{\frac{-x}{2}}\left(1 + \frac{1}{8x^3} - \frac{73}{128x^6} + \mathbf{O}(|x|^{-9})\right) && \text{for } x \to -\infty \, . \end{aligned}$$

The Tracy–Widom distribution $\mathcal{F}$ of mean and standard deviation (see Tracy–Widom [88])

$$\mathrm{E}(\mathcal{F}) = -1.77109 \quad \text{and} \quad \sigma(\mathcal{F}) = 0.9018$$

has a density decaying for $x \to \infty$ as (since the probability distribution $\mathcal{F}(x)$ tends to 1 for $x \to \infty$)

$$\mathcal{F}'(x) = \mathcal{F}(x)\int_x^\infty g^2(\alpha)\,\mathrm{d}\alpha \sim \int_x^\infty \mathrm{A}^2(\alpha)\,\mathrm{d}\alpha \sim \frac{1}{8\pi x}\exp(-\tfrac{4}{3}x^{3/2}) \qquad \text{for } x \to \infty \, . \tag{1.193}$$

The last estimate is obtained by integration by parts:

$$\begin{aligned} \int_x^\infty \mathrm{A}^2(u)\,\mathrm{d}u &= \int_x^\infty \frac{-1}{8\pi u}\left(1 + \sum_1^\infty \frac{c_i}{(u^{3/2})^i}\right)\mathrm{d}\big(\exp(-\tfrac{4}{3}u^{3/2})\big) \\ &= \frac{\exp(-\frac{4}{3}x^{3/2})}{8\pi x}\left(1 + \sum_1^\infty \frac{c_i}{(x^{3/2})^i}\right) \\ &\qquad - \frac{1}{8\pi}\int_x^\infty \frac{\exp(-\frac{4}{3}u^{3/2})}{u^2}\left(1 + \sum_1^\infty \frac{c_i'}{(u^{3/2})^i}\right) . \end{aligned}$$

Then, following conjectures by Dyson [34] and Widom [94], Deift–Its–Krasovsky–Zhou [32] and Baik–Buckingham–DiFranco [18] give a representation of $\mathcal{F}(x)$ as an integral from $-\infty$ to x and thus this provides an estimate for $x \to -\infty$,

$$\begin{aligned} \mathcal{F}(x) &= 2^{1/24}\mathrm{e}^{\zeta'(-1)}\frac{exp(-|x|^3/12)}{|x|^{1/8}}\exp\left\{\int_{-\infty}^x \left(R(y) - \frac{1}{4}y^2 + \frac{1}{8y}\right)\mathrm{d}y\right\} \\ &= 2^{1/24}\mathrm{e}^{\zeta'(-1)}\frac{\exp(-|x|^3/12)}{|x|^{1/8}}\left(1 + \frac{3}{2^6|x|^3} + O(|x|^{-6})\right), \quad \text{for } x \to -\infty \, , \end{aligned}$$

where

$$R(y) = \int_y^\infty g(s)^2 ds = g'(y)^2 - yg(y)^2 - g(y)^4 \, . \qquad \square$$

Acknowledgements

The support of a National Science Foundation grant #DMS-07-06287, a European Science Foundation grant (MISGAM), a Marie Curie Grant (ENIGMA) and "Pôle d'attraction inter-universitaire" grants is gratefully acknowledged.

References

1. M. Adler, T. Shiota, P. van Moerbeke: Phys. Lett. A **208**, 67 (1995)
2. M. Adler, T. Shiota, P. van Moerbeke: Duke Math. J. **94**, 379 (1998)
3. M. Adler, P. van Moerbeke: Duke Math. J. **80**, 863 (1995)
4. M. Adler, P. van Moerbeke: Internat. Math. Res. Notices **1997**, 555 (1997)
5. M. Adler, P. van Moerbeke: Comm. Math. Phys. **203**, 185 (1999)
6. M. Adler, P. van Moerbeke: Comm. Math. Phys. **207**, 589 (1999)
7. M. Adler, P. van Moerbeke: Ann. of Math. **149**, 921 (1999)
8. M. Adler, P. van Moerbeke: Comm. Pure Appl. Math. **54**, 153 (2000)
9. M. Adler, P. van Moerbeke: Ann. of Math. **153**, 149 (2001)
10. M. Adler, P. van Moerbeke: Adv. Math **181**, 190 (2003)
11. M. Adler, P. van Moerbeke: Comm. Pure Appl. Math. **58**, 362 (2005)
12. M. Adler, P. van Moerbeke: Forthcoming
13. M. Adler, P. van Moerbeke, P. Vanhaecke: arXiv:math-ph/0612064
14. D. Aldous, P. Diaconis: Probab. Theory Related Fields **103**, 199 (1995)
15. D. Aldous, P. Diaconis: Bull. Amer. Math. Soc. (N.S.) **36**, 413 (1999)
16. K. Aomoto: SIAM J. Math. Anal. **18**, 545 (1987)
17. R.M. Baer, P. Brock: Math. Comp. **22**, 385 (1968)
18. J. Baik, R. Buckingham, J. DiFranco: arXiv:0704.3636
19. J. Baik, P. Deift, K. Johansson: J. Amer. Math. Soc., **36**, 1119 (1999)
20. J. Baik, P. Deift, K. Johansson: Geom. Funct. Anal. **10**, 702 (2000)
21. Yu. Baryshnikov: Probab. Theory Related Fields **119**, 256 (2001)
22. E. Basor, H. Widom: Integral Equations Operator Theory **37**, 397 (2000)
23. A. Borodin, A. Okounkov: Integral Equations Operator Theory **37**, 386 (2000)
24. A. Borodin, A. Okounkov, G. Olshanski: J. Amer. Math. Soc. **13**, 481 (2000)
25. A. Borodin, G. Olshanski: z-measures on partitions, Robinson–Schensted–Knuth correspondence, and $\beta = 2$ random matrix ensembles. In: *Random matrices and their applications*, ed by P. Bleher, A. Its (Cambridge Univ. Press, Cambridge 2001) pp 71–94
26. A. Borodin, G. Olshanski: Comm. Math. Phys. **211**, 335–358 (2001)
27. R. Carter, G. Segal, I. MacDonald: *Lectures on Lie Groups and Lie Algebras* (Cambridge Univ. Press, Cambridge 1995)
28. C.M. Cosgrove: Stud. Appl. Math. **104**, 171 (2000)
29. C.M. Cosgrove, G. Scoufis: Stud. Appl. Math. **88**, 25 (1993)
30. E. Date, M. Kashiwara, M. Jimbo, T. Miwa: Transformation groups for soliton equations, In: *Nonlinear Integrable Systems – Classical Theory and quantum Theory*, ed by M. Jimbo, T. Miwa (World Sci. Publishing, Singapore 1983) pp 39–119
31. P.A. Deift: *Orthogonal Polynomials and Random Matrices: A Riemann–Hilbert Approach* (Amer. Math. Soc., Providence RI 1999)
32. P. Deift, A. Its, I. Krasovsky, X. Zhou: arXiv:math/0601535v1

33. E. Dueñez: Random matrix ensembles associated to compact symmetric spaces. PhD Thesis, Princeton University, Princeton (2001)
34. F. Dyson: Comm. Math. Phys. **47**, 171 (1976)
35. P. Erdös, G. Szekeres: Compositio Math. **2**, 463 (1935)
36. J. Faraut: Introduction a l'analyse des matrices aléatoires. Cours de DEA, Université Pierre et Marie Curie (2001-2002)
37. M.E. Fisher: J. Stat. Phys. **34**, 669 (1984)
38. P.J. Forrester: Nucl. Phys. B **402**, 709 (1993)
39. P.J. Forrester: arXiv:nlin.SI/0005064
40. P.J. Forrester: J. Phys. A **34**, L417 (2001)
41. P.J. Forrester: *Log-Gases and Random Matrices.* Lecture notes
42. P.J. Forrester, N. Witte: Comm. Pure Appl. Math. **55**, 679 (2002)
43. P.J. Forrester, N. Witte: arXiv:math-ph/0204008
44. J.S. Geronimo, K.M. Case: J. Math. Phys. **20**, 299 (1979)
45. I.M. Gessel: J. Combin. Theory Ser. A **53**, 257 (1990)
46. D.J. Grabiner: J. Combin. Theory Ser. A **97**, 285 (2002)
47. C. Greene: Adv. in Math. **14**, 254 (1974)
48. L. Haine, E. Horozov: Bull. Sci. Math. **117**, 485 (1993)
49. L. Haine, J.-P. Semengue: J. Math. Phys. **40**, 2117 (1999)
50. J.M. Hammersley: A few seedlings of research. In: *Proceedings of the Sixth Berkeley Symposium on Mathematical Statistics and Probability*, vol 1, ed by L.M. Le Cam, J. Neyman, E.L. Scott (Univ. California Press, Berkeley CA 1972) pp 345–394
51. J. Harnad: J. Nonlinear Math Phys. **9**, 530 (2002)
52. S.P. Hastings, J.B. McLeod: Arch. Rational Mech. Anal. **73**, 31 (1980)
53. H. Hotelling: Biometrika, **28** 321 (1936)
54. M. Jimbo, T. Miwa, Y. Mori, M. Sato: Phys. D **1**, 80 (1980)
55. K. Johansson: Duke Math. J. **91**, 151 (1998)
56. K. Johansson: Math. Res. Lett. **5**, 63 (1998)
57. K. Johansson: Comm. Math. Phys., **209**, 437 (2000)
58. K. Johansson: Ann. of Math. **153**, 259 (2001)
59. K. Johansson: Comm. Math. Phys. **242**, 277 (2003)
60. K. Johansson: arXiv:math/0306216
61. V. Kac: *Infinite-Dimensional Lie Algebras* (Cambridge Univ. Press, Cammbridge 1983)
62. V. Kac, A. Raina: *Highest Weight Representation of Infinite-Dimensional Lie Algebras* (World Sci. Publishing, Singapore 1987)
63. J. Kaneko: SIAM J. Math. Anal. **24**, 1086 (1993)
64. S. Kerov: The boundary of Young tableaux and random Young tableaux. In: *Formal Power Series and Algebraic Combinatorics*, ed by L.J. Billera, C. Greene, R. Simion, R.P. Stanley (Amer. Math. Soc., Providence RI 1996) pp 133–158
65. R. Koekoek, R.F. Swarttouw: The Askey-scheme of hypergeometric orthogonal polynomials and its q-analogue. Report 98-17, Dept. of Technical Mathematics and Informatics, Delft Univ. of Technology (1998)
66. J. Krug, H. Spohn: Kinetic roughening of growing interfaces. In: *Solids far from Equilibrium: Growth, Morphology and Defects*, ed by C. Godrèche (Cambridge Univ. Press, Cambridge 1992) pp 479–582
67. B.F. Logan and L.A. Shepp: Adv. in Math. **26**, 206 (1977)
68. I.G. MacDonald: *Symmetric Functions and Hall Polynomials*, 2nd edn (Clarendon Press, New York 1995)

69. M.L. Mehta: *Random Matrices*, 2nd edn (Academic Press, Boston MA 1991)
70. R.J. Muirhead: *Aspects of Multivariate Statistical Theory* (John Wiley & Sons, New York 1982)
71. A.F. Nikiforov, S.K. Suslov, V.B. Uvarov: *Classical Orthogonal Polynomials of a Discrete Variable* (Springer, Berlin 1991)
72. A. Odlyzko, E.M. Rains: On longest increasing subsequences in random permutations. AT&T Labs, Florham Park NJ (1998)
73. A. Okounkov: *Random matrices and random permutations*, Internat. Math. Res. Notices **2000**, 1043 (2000)
74. A. Okounkov: SL(2) and z-measures. In: *Random Matrix Models and Their Applications*, ed by P. Bleher, A. Its (Cambridge Univ. Press, Cambridge, 2001) pp 407–420
75. A. Okounkov: Selecta Math. (N.S.) **7**, 57 (2001)
76. L.A. Pastur: Teoret. Mat. Fiz. **10**, 102 (1972); Theoret. and Math. Phys. **10**, 67 (1972)
77. M. Prähofer, H. Spohn: J. Stat. Phys. **108**, 1071 (2002)
78. B.E. Sagan: *The Symmetry Group* (Wadsworth & Brooks, Pacific Grove CA 1991)
79. T. Seppäläinen: Probab. Theory Related Fields, **112**, 221 (1998)
80. R.P. Stanley: *Enumerative Combinatorics*, vol 1 (Cambridge Univ. Press, Cambridge 1997).
81. R.P. Stanley: *Enumerative Combinatorics*, vol 2 (Cambridge Univ. Press, Cambridge 1999)
82. D. Stanton, D. White: *Constructive Combinatorics*, (Springer, New York 1986)
83. G. Szegő: *Orthogonal Polynomials*, 4th edn (Amer. Math. Soc., Providence RI 1975)
84. C. Tracy, H. Widom: Introduction to random matrices. In: *Geometric and Quantum Aspects of Integrable Systems*, ed by G.F. Helminck (Springer, Berlin, 1993) pp 103–130
85. C. Tracy, H. Widom: Comm. Math. Phys. **159**, 151 (1994)
86. C. Tracy, H. Widom: Comm. Math. Phys. **207**, 665 (1999)
87. C. Tracy, H. Widom: Probab. Theory Related Fields **119**, 350 (2001)
88. C. Tracy, H. Widom: Distribution functions for largest eigenvalues and their applications. In: *Proceedings of the International Congress of Mathematicians*, vol I, ed by T. Li (Higher Education Press, Beijing 2002) pp 587–596
89. S.M. Ulam, Monte-Carlo calculations in problems of mathematical physics. In: *Modern Mathematics for the Engineers*, ed by E.F. Beckenbach (McGraw-Hill, New York Toronto London 1961) pp 261 281
90. P. van Moerbeke: Integrable lattices: random matrices and random permutations. In: *Random Matrix Models and Their Applications*, ed by P. Bleher, A. Its (Cambridge Univ. Press, Cambridge, 2001) pp 321–406
91. J. Verbaarschot: Topics in random matrix theory. Lectures, Stony Brook University, Stony Brook NY
92. A.M. Vershik, S.V. Kerov: Soviet Math. Dokl. **18**, 527 (1977).
93. A.M. Vershik, S.V. Kerov: Funct. Anal. and Appl. **19**, 21 (1985)
94. H. Widom: Comm. Math. Phys. **171**, 159 (1995)

2

Integrable Systems, Random Matrices, and Random Processes

Mark Adler

Department of Mathematics, Brandeis University, Waltham, MA 02454, USA, adler@brandeis.edu

Introduction

Random matrix theory began in the 1950s, when E. Wigner [58] proposed that the local statistical behavior of scattering resonance levels for neutrons off heavy nuclei could be modeled by the statistical behavior of eigenvalues of a large random matrix, provided the ensemble had orthogonal, unitary or symplectic invariance. The approach was developed by many others, like Dyson [30, 31], Gaudin [34] and Mehta, as documented in Mehta's [44] famous treatise. The field experienced a revival in the 1980s due to the work of M. Jimbo, T. Miwa, Y. Mori, and M. Sato [36, 37], showing the Fredholm determinant involving the sine kernel, which had appeared in random matrix theory for large matrices, satisfied the fifth Painlevé transcendent; thus linking random matrix theory to integrable mathematics. Tracy and Widom soon applied their ideas, using more efficient function-theoretic methods, to the largest eigenvalues of unitary, orthogonal and symplectic matrices in the limit of large matrices, with suitable rescaling. This lead to the Tracy–Widom distributions for the 3 cases and the modern revival of random matrix theory (RMT) was off and running.

J. Harnad (ed.), *Random Matrices, Random Processes and Integrable Systems*, CRM Series in Mathematical Physics, DOI 10.1007/978-1-4419-9514-8_2,

This article will focus on integrable techniques in RMT, applying Virasoro gauge transformations and integrable equation (like the KP) techniques for finding Painlevé – like ODEs or PDEs for probabilities that are expressible as Fredholm determinants coming up in random matrix theory and random processes, both for finite and large n-limit cases. The basic idea is simple – just deform the probability of interest by some time parameters, so that, at least as a function of these new time parameters, it satisfies some integrable equations. Since in RMT you are usually dealing with matrix integrals, roughly speaking, it is usually fairly obvious which parameters to "turn on," although it always requires an argument to show you have produced "τ-functions" of an integrable system. Fortunately, to show a system is integrable, you ultimately only have to check bilinear identities and we shall present very general methods to accomplish this. Indeed, the bilinear identities are the actual source of a sequence of integrable PDEs for the τ-functions.

Secondly, because we are dealing with matrix integrals, we may change coordinates without changing the value of the integral (gauge invariance), leading to the matrix integrals being annihilated by partial differential operators in the artificially introduced time and the basic parameters of the problem – so-called Virasoro identities. Indeed, because the most useful coordinate changes are often "S^1-like" and because the tangent space of Diff(S^1) at the identity is the Virasoro Lie algebra (see [41]), the family of annihilating operators tends to be a subalgebra of the Virasoro Lie algebra. Integrable systems possess vertex algebras which infinitesimally deform them and the Virasoro algebras, as they explicitly appear, turn out to be generated by these vertex algebras. Thus while other gauge transformation are very useful in RMT, the Virasoro generating ones tend to mesh well with the integrable systems. Finally, the PDEs of integrable systems, upon feeding in the Virasoro relations, lead, upon setting the artificially introduced times to zero, to Painlevé-like ODE or PDE for the matrix integrables and hence for the probabilities, but in the original parameters of the problem! Sometimes we may have to introduce "extra parameters," so that the Virasoro relations close up, which we then have to eliminate by some simple means, like compatibility of mixed partial derivatives.

In RMT, one is particularly interested in large n (scaled) limits, i.e., central limit type results, usually called universality results; moreover, one is interested in getting Painlevé type relations for the probabilities in these limiting relations, which amounts to getting Painlevé type ODEs or PDEs for Fredholm determinants involving limiting kernels, analogous to the sine kernel previously mentioned. These relations are analogous to the heat equation for the Gaussian kernel in the central limit theorem (CLT), certainly a revealing relation. There are two obvious ways to derive such a relation; either, get a relation at each finite step for a particularly "computable or integrable" sequence of distributions (like the binomial) approaching the Gaussian, via the CLT, and then take a limit of the relation, or directly derive the heat equation for the actual limiting distribution. In RMT, the same can be said, and the integrable system step and Virasoro step mentioned previously are thus applied directly to the "integrable" finite approximations of the limiting case, which just involve matrix integrals. After deriving an equation at the finite n-step, we must ensure that estimates justify passage to the limit, which ultimately involves estimates of the convergence of the kernel of a Fredholm determinant. If instead we decide to directly work with the limiting case without passing through a limit, it is more subtle to add time parameters to get integrability and to derive Virasoro relations, as we do not have the crutch of finite matrix integrals. Nonetheless, working with the limiting case is usually quite insightful, and we include one such example in this article to illustrate how the limiting cases in RMT faithfully remember their finite $-$ n matrix integral ancestry.

In Sect. 2.1 we discuss random matrix ensembles and limiting distributions and how they directly link up with KP theory and its vertex operator, leading to PDEs for the limiting distribution. This is the only case where we work only with the limiting distribution. In Sect. 2.2 we derive recursion relations in n for n-unitary integrals which include many combinatoric generating functions. The methods involve bi-orthogonal polynomials on the circle and we need to introduce the integrable "Toeplitz Lattice," an invariant subsystem of the semi-infinite 2-Toda lattice of Ueno–Takasaki [55]. In Sect. 2.3 we study the coupled random matrix problem, involving bi-orthogonal polynomials for a nonsymmetric $\mathbb{R}^2$ measure, and this system involves the 2-Toda lattice, which leads to a beautiful PDE for the joint statistics of the ensemble. In Sect. 2.4 we study Dyson Brownian motion and 2 associated limiting processes – the Airy and Sine processes gotten by edge and bulk scaling. Using the PDE of Sect. 2.3, we derive PDEs for the Dyson process and then the 2 limiting processes, by taking a limit of the Dyson case, and then we derive for the Airy process asymptotic long time results. In Sect. 2.5 we study the Pearcey process, a limiting process gotten from the GUE with an external source or alternately described by the large n behavior of $2n$-non-intersecting Brownian motions starting at $x = 0$ at time $t = 0$, with n conditioned to go to $+\sqrt{n}$, and the other n conditioned to go to $-\sqrt{n}$ at $t = 1$, and we observe how the motions diverge at $t = \frac{1}{2}$ at $x = 0$, with a microscope of resolving power $n^{-1/4}$. The integrable system involved in the finite n problem is the 3-KP system and now instead of bi-orthogonal polynomials, multiple orthogonal polynomials (MOPS) are involved. We connect the 3-KP system and the associated Riemann–Hilbert (RH) problem and the MOPs.

The philosophy in writing this article, which is based on five lectures delivered at CRM, is to keep as much as possible the immediacy and flow of the lecture format through minimal editing and so in particular, the five sections can read in any order. It should be mentioned that although the first section introduces the basic structure of RMT and the KP equation, it in fact deals with the most sophisticated example. The next point was to pick five examples which maximized the diversity of techniques, both in applying Virasoro relations and using integrable systems. Indeed, this article provides a fair but sketchy introduction to integrable systems, although in point of fact, the only ones used in this particular article are invariant subsystems (reductions) and lattices generated from the n-KP, for $n = 1, 2, 3$. The point being that a lot of the skill is involved in picking precisely the right integrable system to deploy. Fortunately, if some sort of orthogonal polynomials are involved, this amounts to deforming the measure(s) intelligently. For further reading see [1–18].

Acknowledgement

These lectures present joint work, in the fullest sense, with Pierre van Moerbeke, Pol Vanhaecke and Taka Shiota, whom I must thank for all the fun we had and also for enduring my teasing. I also wish to thank Alexei Borodin, Persi Diaconis, Alexander Its, Kurt Johansson, Ken McLaughlin, Craig Tracy, Harold Widom, and Nick Witte for many helpful conversations and insightful remarks – the kind of remarks which make more sense as time goes on. Let me also thank Pol Vanhaecke for reading the manuscript and making helpful suggestions, in addition to producing the various figures that appear. Finally, I wish to thank Louise Letendre of CRM for her beautiful typing job of a messy and hard to read manuscript.

The support of a National Science Foundation grant #DMS-04-06287 is gratefully acknowledged.

2.1 Matrix Integrals and Solitons

2.1.1 Random Matrix Ensembles

Consider the probability ensemble over $n \times n$ Hermitian matrices

$$\begin{aligned} P\big(M \in \mathcal{H}(E)\big) &= \frac{\int_{\mathrm{sp}(M)\in E} \mathrm{e}^{-\operatorname{tr} V(M)}\,\mathrm{d}M}{\int_{\mathrm{sp}(M)\in \mathbb{R}} \mathrm{e}^{-\operatorname{tr} V(M)}\,\mathrm{d}M} \\ &= \frac{\int_{E^n} \Delta_n^2(z) \prod_1^n \exp\big(-V(z_i)\big)\,\mathrm{d}z_i}{\int_{\mathbb{R}^n} \Delta_n^2(z) \prod_1^n \exp\big(-V(z_i)\big)\,\mathrm{d}z_i}\,, \end{aligned} \tag{2.1}$$

with $\Delta(z) = \prod_{i,j=1}^n (z_i - z_j) = \det(z_j^{i-1})_{i,j=1}^n$ the Vandermonde and with $V(x)$ a "nice" function so the integral makes sense, $\mathrm{sp}(M)$ means the spectrum of M and $E \subset \mathbb{R}$,

$$\mathcal{H}_n(E) = \{M \text{ an Hermitian } n \times n \text{ matrix} \mid \mathrm{sp}(M) \subset E\}\,,$$

$$\mathrm{d}M = \prod_{i=1}^n \mathrm{d}M_{ii} \prod_{i<j} \mathrm{d}\,\mathrm{Real}\,M_{ij} \prod_{i<j} \mathrm{d}\,\mathrm{Imag}(M_{ij})\,,$$

the induced Haar measure on $n \times n$ Hermitian matrices, viewed as

$$T\big(\mathrm{SL}(n,\mathbb{C})/\,\mathrm{SU}(n)\big)\big|_I ,$$

and where we have used the Weyl integration formula [35] for

$$\begin{aligned} M &= U \operatorname{diag}(z_1,\dots,z_n)U^{-1} , \\ \mathrm{d}M &= \Delta_n^2(z_1,\dots,z_n)\prod_1^n \mathrm{d}z_i \, \mathrm{d}U . \end{aligned}$$

In order to recast $P(M)$ so that we may take a limit for large n and avoid ∞-fold integrals, we follow the reproducing method of Dyson [30, 31].

Let $p_k(z)$ be the monic orthogonal polynomials:

$$\int_{\mathbb{R}} p_i(z)p_j(z)\mathrm{e}^{-V(z)}\,\mathrm{d}z = h_i\delta_{ij} \tag{2.2}$$

and remember

$$(\det A)^2 = \det(AA^T) . \tag{2.3}$$

Compute

$$\begin{aligned} &\int_{\mathbb{R}^n} \Delta_n^2(z)\prod_1^n \exp\big(-V(z_i)\big)\,\mathrm{d}z_i \\ &= \int_{\mathbb{R}^n} \det\big(p_{i-1}(z_j)\big)_{i,j=1}^n \det\big(p_{k-1}(z_l)\big)_{k,l=1}^n \prod_1^n \exp\big(-V(z_i)\big)\,\mathrm{d}z_i \\ &= \sum_{\pi,\pi'\in S_n} (-1)^{\pi+\pi'} \prod_1^n \int_{\mathbb{R}} p_{\pi(k)-1}(z_k)p_{\pi'(k)-1}(z_k)\exp\big(-V(z_k)\big)\,\mathrm{d}z_k \\ &= n!\prod_0^{n-1}\int_{\mathbb{R}} p_k^2(z)\mathrm{e}^{-V(z)}\,\mathrm{d}z \qquad\qquad \text{(orthogonality)} \\ &= n!\prod_0^{n-1} h_k , \end{aligned} \tag{2.4}$$

and so using (2.3) and (2.4), conclude

$$\begin{aligned} &P\big(M \in \mathcal{H}_n(E)\big) \\ &\quad= \frac{1}{n!\prod_1^n h_{i-1}}\int_{E^n} \det\bigg(\sum_1^n p_{j-1}(z_k)p_{j-1}(z_l)\bigg)_{k,l=1}^n \prod_1^n \exp\big(-V(z_i)\big)\,\mathrm{d}z_i \\ &\quad= \frac{1}{n!}\int_{E^n} \det\big(K_n(z_k,z_l)\big)_{k,l=1}^n \prod_1^n \mathrm{d}z_i , \end{aligned}$$

with the Christoffel–Darboux kernel

$$K_n(y,z) := \sum_{j=1}^n \varphi_j(y)\varphi_j(z) = \frac{h_n}{h_{n-1}}\frac{\big(\varphi_n(y)\varphi_{n-1}(z) - \varphi_n(z)\varphi_{n-1}(y)\big)}{y-z} , \tag{2.5}$$

and

$$\varphi_j(x) = \mathrm{e}^{-V(x)/2} p_{j-1}(x)/\sqrt{h_{j-1}} \, .$$

Thus by (2.2)

$$\int_{\mathbb{R}} \varphi_i(x)\varphi_j(x)\,\mathrm{d}x = \delta_{ij} \, ,$$

and so we have

$$\left.\begin{aligned} &\int_{\mathbb{R}} K_n(y,z)K_n(z,u)\,\mathrm{d}z = K_n(y,u) \, , \\ &\int_{\mathbb{R}} K_n(z,z)\,\mathrm{d}z = n \, , \end{aligned}\right. \quad \text{– the reproducing property,} \tag{2.6}$$

which yields the crucial property:

$$\int_{\mathbb{R}^{n-m}} \det\big(K_n(z_i,z_j)\big)_{i,j=1}^{n} \mathrm{d}z_{m+1}\dots\mathrm{d}z_n = (n-m)!\det\big(K_n(z_i,z_j)\big)_{i,j=1}^{m} \, . \tag{2.7}$$

Replacing $E^n \to \prod_1^k \mathrm{d}z_i \, \mathbb{R}^{n-k}$ in (2.1) and integrating out $z_{k+1},\dots,z_n$ via the producing property yields:

$$\text{“}P(\text{one eigenvalue in each } [z_i, z_i + \mathrm{d}z_i],\ i=1,\dots,k) = \frac{1}{\binom{n}{k}} \det\big(K_n(z_i,z_j)\big)_{i,j=1}^{k} \prod_1^k \mathrm{d}z_i \, ,\text{”} \tag{2.8}$$

heuristically speaking. Setting

$$E = \bigcup_{\mathrm{d}z_i \in E} \mathrm{d}z_i = \bigcup E_i \, ,$$

and using Poincaré's formula

$$P\Big(\bigcup E_i\Big) = \sum_i P(E_i) - \sum_{i<j} P(E_i \cap E_j) + \sum_{i<j<k} P(E_i \cap E_j \cap E_k) + \cdots \, ,$$

yields the Fredholm determinant[1]

$$\begin{aligned} P\big(M \in \mathcal{H}_n(E^{\mathrm{c}})\big) &= 1 + \sum_{k=1}^{\infty} (-\lambda)^k \int_{z_1 \le \cdots \le z_k} \det\big(K_n^E(z_i,z_j)\big)_{i,j=1}^{k} \prod_1^k \mathrm{d}z_i \Big|_{\lambda=1} \\ &= \det(I - \lambda K_n^E)\big|_{\lambda=1} \end{aligned} \tag{2.9}$$

with kernel

$$K_n^E(y,z) = K_n(y,z) I_E(z)$$

and with $I_E(z)$ the indicator function of the set E, and more generally see [48],

$$P(\text{exactly } k\text{-eigenvalues} \in E) = \frac{(-1)^k}{k!} \left(\frac{\partial}{\partial\lambda}\right)^k \det(I - \lambda K_n^E)\Big|_{\lambda=1} \, . \tag{2.10}$$

---

[1] E^{c} is the complement of E in $\mathbb{R}$.

2.1.2 Large n-limits

The Fredholm determinant formulas (2.9) and (2.10), enable one to take large n-limits of the probabilities by taking large n-limits of the kernels $K_n(y,z)$. The first and most famous such law is for the Gaussian case $V(x) = x^2$, although it has been extended far beyond the Gaussian case [29, 38].

Wigner's semi-circle law: The density of eigenvalues converges (see Fig. 2.1) in the sense of measure:

$$K_n(z,z)\,\mathrm{d}z \to \begin{cases} \dfrac{1}{\pi}\sqrt{2n - z^2}\,\mathrm{d}z, & |z| \le \sqrt{2n}\,, \\ 0, & |z| > \sqrt{2n} \end{cases} \tag{2.11}$$

and so

$$\mathrm{Exp}(\#\text{eigenvalues in } E) = \int_E K_n(z,z)\,\mathrm{d}z.$$

This is a sort of Law of Large Numbers. Is there more refined universal behavior, a sort of Central Limit Theorem? The answer is as follows:

Bulk scaling limit: The density of eigenvalues near $z = 0$ is $\sqrt{2n}/\pi$ and so $\pi/\sqrt{2n}$ = average distance between eigenvalues. Magnifying at $z = 0$ with this rescaling

$$\begin{aligned} \lim_{n\to\infty} K_n\left(\frac{\pi x}{\sqrt{2n}}, \frac{\pi y}{\sqrt{2n}}\right)\mathrm{d}\left(\frac{y\pi}{\sqrt{2n}}\right) &= \frac{\sin \pi(x-y)}{\pi(x-y)}\,\mathrm{d}y \\ &=: K_{\sin}(x,y)\,\mathrm{d}y\,, \end{aligned} \tag{2.12}$$

with

$$\begin{aligned} \lim_{n\to\infty} P\left(\text{exactly } k \text{ eigenvalues} \in \frac{\pi}{\sqrt{2n}}[-2a,2a]\right) \\ = \frac{(-1)^k}{k!}\left(\frac{\partial}{\partial\lambda}\right)^k \det\left(I - \lambda K_{\sin}^{[-2a,2a]}\right)\Big|_{\lambda=1}. \end{aligned} \tag{2.13}$$

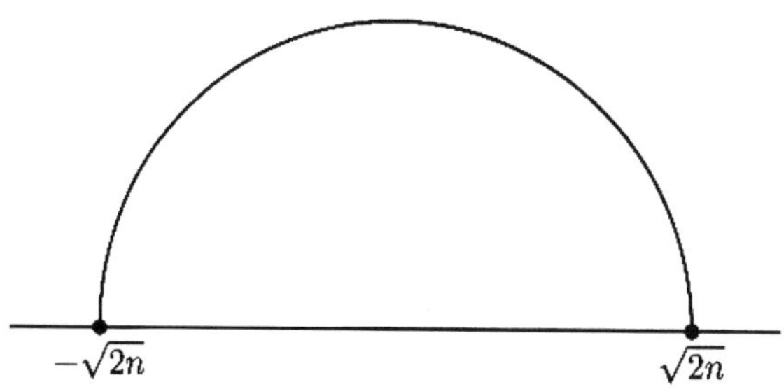

Fig. 2.1.

Moreover,

$$\det\Big(I - \lambda K_{\sin}^{[-2a,2a]}\Big) = \exp \int_0^{\pi a} \frac{f(x,\lambda)}{x}\, dx$$

with

$$(xf'')^2 + 4(xf' - f)(f'^2 + xf' - f) = 0 \quad \left(' = \frac{d}{dx}\right), \tag{2.14}$$

and boundary condition:

$$f(x,\lambda) \cong \frac{-\lambda x}{\pi} - \left(\frac{\lambda}{\pi}\right)^2 x^2 - \cdots, \qquad x \sim 0\,.$$

The ODE (2.14) is Painlevé V, and this is the celebrated result of Jimbo, Miwa, Mori, and Sato [36, 37].

Edge scaling limit: The density of eigenvalues near the edge, $z = \sqrt{2n}$ of the Wigner semi-circle is $\sqrt{2}n^{1/6}$. Magnifying at the edge with the rescaling $1/\sqrt{2}n^{1/6}$:

$$\lim_{n\to\infty} K_n\left(\sqrt{2n} + \frac{u}{\sqrt{2}n^{1/6}}, \sqrt{2n} + \frac{v}{\sqrt{2}n^{1/6}}\right) d\left(\sqrt{2n} + \frac{v}{\sqrt{2}n^{1/6}}\right) := K_{\text{Airy}}(u,v)\, dv\,, \tag{2.15}$$

with

$$\begin{aligned} K_{\text{Airy}}(u,v) &= \int_0^\infty \operatorname{Ai}(x+u)\operatorname{Ai}(x+v)\, dx\,, \\ \operatorname{Ai}(u) &= \frac{1}{\pi}\int_0^\infty \cos\left(\frac{u^3}{3} + xu\right) dv\,. \end{aligned} \tag{2.16}$$

Setting $\lambda_{\max} = \sqrt{2n} + u/\sqrt{2}n^{1/6}$

$$\begin{aligned} \lim_{n\to\infty} P(\sqrt{2}n^{1/6}(\lambda_{\max} - \sqrt{2n}) \le u) &= \det\Big(I - K_{\text{Airy}}^{[u,\infty]}\Big) \\ &= \exp\left(-\int_u^\infty (\alpha - u) g^2(\alpha)\, d\alpha\right) \\ &\quad - \text{the Tracy–Widom distribution}\,, \end{aligned} \tag{2.17}$$

with

$$g'' = xg + 2g^3 \tag{2.18}$$

and boundary condition:

$$g(x) \simeq \frac{\exp(-\frac{2}{3}x^{3/2})}{2\sqrt{\pi}x^{1/4}}, \qquad x \to \infty\,. \tag{2.19}$$

Equation (2.18) is Painlevé II and (2.17) is due to Tracy–Widom [49].

Hard edge scaling limit: Consider the Laguerre ensemble of $n \times n$ Hermitian matrices:

$$e^{-V(z)}\,dz = z^{\nu/2} e^{-z/2} I_{(0,\infty)}(z)\,dz\,. \tag{2.20}$$

Note $z = 0$ is called the hard edge, while $z = \infty$ is called the soft edge. The density of eigenvalues for large n has a limiting shape and the density of eigenvalues near $z = 0$ is $4n$. Rescaling the kernel with this magnification:

$$\begin{aligned}\lim_{n\to\infty} K_n^{(\nu)}\left(\frac{u}{4n},\frac{v}{4n}\right) d\left(\frac{v}{4n}\right) \\ &=: K_\nu(u,v)\,dv = \frac{1}{2}\int_0^1 sJ_\nu(s\sqrt{u})J_\nu(s\sqrt{v})\,ds\,dv \\ &= \frac{J_\nu(\sqrt{u})\sqrt{v}J'_\nu(\sqrt{v}) - J_\nu(\sqrt{v})\sqrt{u}J'_\nu(\sqrt{u})}{2(u-v)}\,dv\end{aligned} \tag{2.21}$$

yields the Bessel kernel, while one finds:

$$\begin{aligned}\lim_{n\to\infty} P\left(\text{no eigenvalue} \in \frac{1}{4n}[0,x]\right) \\ &= \det\left(I - K_\nu^{[0,x]}\right) = \exp\left(-\int_0^x \frac{f(v)}{u}\,du\right),\end{aligned} \tag{2.22}$$

with

$$(xf'')^2 - 4(xf'-f)f'^2 + \left((x-\nu^2)f' - f\right)f' = 0\,, \tag{2.23}$$

and boundary condition:

$$f(x) = c_\nu x^{1+\nu}\left(1 - \frac{1}{2(2+\nu)}x + \cdots\right), \quad c_\nu = \left[2^{2\nu+2}\Gamma(1+\nu)\Gamma(2+\nu)\right]^{-1}.$$

Equation (2.23) is Painlevé V and is due to Tracy–Widom [50].

2.1.3 KP Hierarchy

We give a quick discussion of the KP hierarchy. A more detailed discussion can be found in [28, 56]. Let $L = L(x,t)$ be a pseudo-differential operator and $\Psi^+(x,t,z)$ its eigenfunction (wave function). The KP hierarchy is an isospectral deformation of L:[2]

$$\begin{aligned}&\frac{\partial L}{\partial t_n} = [(L^n)_+, L], \quad L = D_x + a_{-1}(x,t)D_x^{-1} + \cdots\,, \qquad n = 1,2,\ldots\,, \\ &D_x = \frac{\partial}{\partial x}, \quad t = (t_1, t_2, \ldots)\end{aligned} \tag{2.24}$$

[2] Ψ^- is the eigenfunction of L adjoint $:= L^T$, $(A)_+ :=$ differential operator part of A.

with Ψ parametrized by Sato's τ-function and satisfying

$$\begin{aligned}\Psi^{\pm}(x,t,z) &= \exp\left(\pm\left(xz+\sum_1^{\infty} t_i z^i\right)\right)\frac{\tau(t\mp[z^{-1}])}{\tau(t)} \\ &= \exp\left(\pm\left(xz+\sum_1^{\infty} t_i z^i\right)\right)(1+O(1/z)), \quad z\to\infty \end{aligned} \tag{2.25}$$

and

$$\begin{aligned} z\Psi^+ &= L\Psi^+ , & \frac{\partial \Psi^+}{\partial t_n} &= (L^n)_+\Psi^+ , \\ z\Psi^- &= L^T\Psi^- , & \frac{\partial \Psi^-}{\partial t_n} &= -\left((L^T)^n\right)_+\Psi^- , \end{aligned} \tag{2.26}$$

with

$$[x] := \left(x, \frac{x^2}{2}, \frac{x^3}{3}, \dots\right).$$

Consequently the τ-function satisfies the crucial formal residue identity.

Bilinear identity for τ-function:

$$\oint_{\infty} \exp\left(\sum_1^{\infty}(t_i-t_i')z^i\right)\tau(t-[z^{-1}])\tau(t'+[z^{-1}])\,\mathrm{d}z = 0\,, \quad \forall t,t'\in\mathbb{C}^{\infty}\,, \tag{2.27}$$

which characterizes the KP τ-function. This is equivalent (see Appendix for proof) to the following generating function of Hirota relations ($a = (a_1, a_2, \dots)$ arbitrary):

$$\sum_{j=0}^{\infty} s_j(-2a)s_{j+1}(\tilde{\partial}_t)\exp\left(\sum_{l=1}^{\infty} a_l\frac{\partial}{\partial t_l}\right)\tau(t)\circ\tau(t) = 0\,, \tag{2.28}$$

where

$$\tilde{\partial}_t = \left(\frac{\partial}{\partial t_1}, \frac{1}{2}\frac{\partial}{\partial t_2}, \frac{1}{3}\frac{\partial}{\partial t_3}, \dots\right), \quad \partial_t = \left(\frac{\partial}{\partial t_1}, \frac{\partial}{\partial t_2}, \dots\right), \tag{2.29}$$

and

$$\exp\left(\sum_1^{\infty} t_i z_i\right) := \sum_{j=0}^{\infty} s_j(t)z^j \quad (s_j(t) \text{ elementary Schur polynomials}) \tag{2.30}$$

and

$$\begin{aligned} &p(\partial_n t)f(t)\circ g(t) \\ &\quad := p\left(\frac{\partial}{\partial y}, \frac{\partial}{\partial y_2}, \dots\right)f(t+y)g(t-y)\Big|_{y=0} \quad \text{(Hirota symbol)}. \end{aligned} \tag{2.31}$$

This yields the KP hierarchy

$$\left(\frac{\partial^4}{\partial t_1}+3\left(\frac{\partial}{\partial t_2}\right)^2-4\frac{\partial}{\partial t_1}\frac{\partial}{\partial t_3}\right)\tau(t)\circ\tau(t)=0 \tag{2.32}$$

$$\vdots$$

equivalent to

$$\left(\left(\frac{\partial}{\partial t_1}\right)^4+3\left(\frac{\partial}{\partial t_2}\right)^2-4\frac{\partial^2}{\partial t_1\partial t_3}\right)\log\tau+6\left(\frac{\partial^2}{\partial t_1^2}\log\tau\right)^2=0$$

$$\vdots \qquad \text{(KP equation)} \tag{2.33}$$

The p-reduced KP corresponds to the reduction:

$$L^p=D_x^p+\cdots=\text{diff. oper.}=(L^p)_+$$

and so

$$\frac{\partial L}{\partial t_{pn}}=[(L^{pn})_+,L]=[L^{pn},L]=0\,. \tag{2.34}$$

$p=2$: KdV

$$\tau=\tau(t_1,t_3,t_5,\dots) \qquad \text{(ignore } t_2,\, t_4,\dots) \tag{2.35}$$

$$\Psi^{\pm}(x,t,z)=\Psi(x,t,\pm z)\,. \tag{2.36}$$

2.1.4 Vertex Operators, Soliton Formulas and Fredholm Determinants

Vertex operators in integrable systems generate the tangent space of solutions and Darboux transformations; in other words, they yield the deformation theory. We now present
KP-*vertex operator*:[3]

$$X(t,y,z)=\frac{1}{z-y}\exp\left(\sum_1^\infty(z^i-y^i)t_i\right)\exp\left(\sum_1^\infty(y^{-i}-z^{-i})\frac{1}{i}\frac{\partial}{\partial t_i}\right), \tag{2.37}$$

and the "τ-space" near τ parametrized: $\tau+\varepsilon X(t,y,z)\tau$. Moreover

$$\tau+\varepsilon X(t,y,z)\tau$$

is a τ-function, *not just infinitesimally*, so it satisfies the bilinear identity. This vertex operator was used in [28] to generate solitons, but it also plays a role in generating Kac–Moody Lie algebras [40]. The identities that follow in Sects. 2.1.4, 2.1.5, 2.1.6 were derived by Adler–Shiota–van Moerbeke in [16], carefully and in great detail.

[3] $X(t,y,z)f(t)=\left(1/(z-y)\right)\exp(\sum_1^\infty(z^i-y^i)t_i)f(t+[y^{-1}]-[z^{-1}])$, $z\neq y$ and z, y are large complex parameters, and how we expand the operator X shall depend on the situation.

Link with kernels: We have the differential Fay identity (Adler–van Moerbeke [1])

$$\frac{1}{\tau(t)} X(t,y,z)\tau(t) = D_x^{-1}\left(\Psi^+(x,t,y)\Psi^-(x,t,z)\right) , \tag{2.38}$$

where since D_x^{-1} is integration, the r.h.s. of (2.38) has the same structure as the Airy and Bessel kernels of (2.16) and (2.21).

If $|y_i|, |z_i| < |y_{i+1}|, |z_{i+1}|$, $1 \le i \le n-1$, then we have the Fay identity:

$$\begin{aligned}\det\Big(D_x^{-1}\left(\Psi^+(x,t,y_i)\Psi^-(x,t,z_j)\right)\Big)_{i,j=1}^{n} &= \det\left(\frac{1}{\tau(t)} X(t,y_i,z_j)\tau(t)\right)_{i,j=1}^{n} \\ &= \frac{1}{\tau} X(t,y_1,z_1)\cdots X(t,y_n,z_n)\tau \,.\end{aligned} \tag{2.39}$$

We also have the following

Vertex identities:

$$X(y,z)X(y,z) = 0 \quad \text{and so} \quad \mathrm{e}^{aX(y,z)} = 1 + aX(y,z) , \tag{2.40}$$

and

$$[X(\alpha,\beta), X(\lambda,\mu)] = 0 \quad \text{if } \alpha \ne \mu,\ \beta \ne \lambda. \tag{2.41}$$

In addition we have the expansion of the vertex operator along the diagonal:

$$\begin{aligned} &X(t,y,z) = \frac{1}{z-y}\sum_{k=0}^{\infty}\frac{(z-y)^k}{k!}\sum_{l=-\infty}^{\infty} y^{-l-k} W_l^{(k)} , \\ &\text{Heisenberg:} \quad W_n^{(1)} := \frac{\partial}{\partial t_n} + (-n)t_{-n} \quad (W_n^{(0)} = \delta_{n0}) , \\ &\text{Virasoro:} \quad W_n^{(2)} := 2\sum_{i,i+n\ge 1} it_i \frac{\partial}{\partial t_{i+n}} + \sum_{i+j=n}\frac{\partial^2}{\partial t_i \partial t_j} + \sum_{i+j=-n}(it_i)(jt_j) \\ &\qquad\qquad - (n+1)\left(\frac{\partial}{\partial t_n} + (-n)t_{-n}\right) \end{aligned} \tag{2.42}$$

and from the commutation relations:

$$\begin{aligned} [X(\alpha,\beta), X(\lambda,\mu)] &= \left(-n^T(\alpha,\beta,\lambda) + n(\alpha,\beta,\mu)\right)X(\lambda,\mu) , \\ n(\lambda,\mu,z) &:= \delta(\lambda,z)\exp\left((\mu-\lambda)\frac{\partial}{\partial z}\right) , \\ \delta(\lambda,z) &:= \frac{1}{z}\sum_{-\infty}^{\infty}\left(\frac{z}{\lambda}\right)^l , \end{aligned} \tag{2.43}$$

conclude the vertex Lax equation:

$$\frac{\partial}{\partial z}\left(z^{l+1} Y(z)\right) = \left[\frac{1}{2} W_l^{(2)}, Y(z)\right] \tag{2.44}$$

with

$$Y(z) = X(t, \omega z, \omega' z) , \quad \omega, \omega' \in \xi_p,^4 \; l \geq -p \text{ and } p/l , \tag{2.45}$$

or a linear combination of such $X(t, \omega z, \omega' z)$.

A Fredholm determinant style soliton formula: A classical KP soliton formula [28] is as follows:

$$\begin{aligned}
\tau(t) &= \frac{1}{\tau_0} \Bigg(\prod_{\substack{1 \\ \text{ordered}}}^{n} \exp\big(a_i X(y_i, z_i)\big) \Bigg) \tau_0 \Bigg|_{\tau_0 = 1} \\
&= \det \Bigg(\delta_{ij} + \frac{a_j}{z_j - y_i} \exp\Bigg(\sum_{k=1}^{\infty} (z_j^k - y_i^k) t_k \Bigg) \Bigg)_{i,j=1}^{n} .
\end{aligned}$$

We now relax the condition that $\tau_0 = 1$ and setting $|y_i|, |z_i| < |y_{i+1}|, |z_{i+1}|$, $1 \leq i \leq n-1$, compute using the Fay identity (2.39) and differential Fay (2.38):

$$\begin{aligned}
\frac{\widetilde{\tau}(t)}{\tau(t)} &:= \frac{1}{\tau(t)} \exp\Bigg(\sum_{1}^{n} a_i X(y_i, z_i) \Bigg) \tau(t) \\
&= \frac{1}{\tau(t)} \Bigg(\prod_{\substack{1 \\ \text{ordered}}}^{n} \exp\big(a_i X(y_i, z_i)\big) \Bigg) \tau(t) \\
&= \frac{1}{\tau(t)} \Bigg(\prod_{\substack{1 \\ \text{ordered}}}^{n} \big(1 + a_i X(y_i, z_i)\big) \Bigg) \tau(t) \\
&= \frac{1}{\tau(t)} \Bigg(\tau + \sum_{i=1}^{n} a_i X(y_i, z_i) \tau + \sum_{\substack{i,j=1 \\ i<j}}^{n} a_i a_j X(y_i, z_i) X(y_j, z_j) \tau \\
&\qquad + \cdots + \prod_{1}^{n} a_i \Bigg(\prod_{\substack{1 \\ \text{ordered}}}^{n} X(y_i, z_i) \Bigg) \tau \Bigg) \\
&= 1 + \sum_{i=1}^{n} a_i \frac{X(y_i, z_i)\tau}{\tau} + \sum_{\substack{i,j=1 \\ i<j}}^{n} \det \begin{pmatrix} a_i \dfrac{X(y_i, z_i)\tau}{\tau}, & a_j \dfrac{X(y_i, z_j)\tau}{\tau} \\ a_i \dfrac{X(y_j, z_i)\tau}{\tau}, & a_j \dfrac{X(y_j, z_j)\tau}{\tau} \end{pmatrix} \\
&\qquad + \cdots + \det \Bigg(a_j \frac{X(y_i, z_j)\tau}{\tau} \Bigg)_{i,j=1}^{n} \\
&= \det \Bigg(\delta_{ij} + a_j \frac{X(y_i, z_j)\tau}{\tau} \Bigg)_{i,j=1}^{n} \quad \text{"Fredholm expansion" of determinant} \\
&= \det \Big(\delta_{ij} + a_j D_x^{-1} \big(\Psi^+(x, t, y_i) \Psi^-(x, t, y_j) \big) \Big)_{i,j=1}^{n} .
\end{aligned}$$

---

[4] ξ_p the pth roots of unity.

Replacing $y_i \to \omega z_i$, $z_i \to \omega' z_i$, $a_i = -\lambda\delta z$

$$z_i = a + (i-1)\delta z, \quad \delta z = \frac{b-a}{n-1}, \quad n \to \infty$$

yields the continuous (via the Riemann Integral).

Soliton Fredholm determinant:

$$\frac{\tau(t,E)}{\tau(t)} := \frac{1}{\tau(t)} \exp\left(-\lambda \int_E X(t,\omega z,\omega' z)\,\mathrm{d}z\right)\tau(t) = \det(I - \lambda k^E) \tag{2.46}$$

with kernel

$$k^E(y,z) = D_x^{-1}\big(\Psi^-(t,\omega y)\Psi^+(t,\omega' z)\big) I_E(z), \quad E = [a,b]\,.$$

More generally, for p-reduced KP, replace in (2.46)

$$X(t,y,z) \to Y(t,y,z) := \sum_{\omega,\omega'\in\xi_p} a_\omega b_{\omega'} X(t,\omega y,\omega' z)\,, \tag{2.47}$$

$$k^E(y,z) \to D_x^{-1} \sum_{\omega\in\xi_p} a_\omega \Psi^-(x,t,\omega y) \sum_{\omega'\in\xi_p} b_{\omega'}\Psi^+(x,t,\omega' z) I_E(z) \tag{2.48}$$

with

$$\sum_{\omega\in\xi_p} \frac{a_\omega b_\omega}{\omega} = 0 \quad (\text{so } Y(t,z,z) \text{ is pole free}) \tag{2.49}$$

and

$$X(t,\omega z,\omega' z) \to Y(t,z,z) := Y(z)\,, \tag{2.50}$$

$$E \to \bigcup_{i=1}^{k} [a_{2i-1}, a_{2i}]\,. \tag{2.51}$$

2.1.5 Virasoro Relations Satisfied by the Fredholm Determinant

It is a marvelous fact that the soliton Fredholm determinant satisfies a Virasoro relation as a consequence of the vertex Lax equation [16]; indeed, compute

$$\begin{aligned} 0 &= \int_a^b \left(\frac{\partial}{\partial z} z^{l+1} Y(z) - \left[\frac{1}{2} W_l^{(2)}, Y(z)\right]\right) \mathrm{d}z \\ &= b^{l+1} Y(b) - a^{l+1} Y(a) - \left[\frac{1}{2} W_l^{(2)}, \int_a^b Y(z)\,\mathrm{d}z\right] \\ &= \left(b^{l+1}\frac{\partial}{\partial b} + a^{l+1}\frac{\partial}{\partial a} - \left[\frac{1}{2}W_l^{(2)},\cdot\right]\right) \int_a^b Y(z)\,\mathrm{d}z := \delta(U)\,, \end{aligned}$$

with δ a derivation and hence

$$\delta \mathrm{e}^{-\lambda U} = 0 \, ,$$

or spelled out

$$\left(b^{l+1} \frac{\partial}{\partial b} + a^{l+1} \frac{\partial}{\partial a} - \left[\frac{1}{2} W_l^{(2)}, \cdot \right] \right) \exp\left(-\lambda \int_a^b Y(z) \, \mathrm{d}z \right) = 0 \, . \tag{2.52}$$

Let τ be a vacuum vector for p-KP:

$$W_l^{(2)} \tau = c_l \tau \, , \quad l = kp \, , \qquad k = -1, 0, 1, \ldots \, . \tag{2.53}$$

Remembering (2.46) with $Y(z)$ given by (2.50), and with $\tau(t)$ taken as a vacuum vector, yields

$$\frac{\tau(t, E)}{\tau(t)} = \frac{1}{\tau} \exp\left(-\lambda \int_E Y(z) \, \mathrm{d}z \right) \tau = \det(I - \lambda k^E) \, , \tag{2.54}$$

and setting $l = kp$, $k = -1, 0, 1, \ldots$, compute using, (2.52), (2.53) and (2.54):

$$\begin{aligned}
0 &= \left(b^{l+1} \frac{\partial}{\partial b} + a^{l+1} \frac{\partial}{\partial a} - \frac{1}{2} W_l^{(2)} \right) \exp\left(-\lambda \int_a^b Y(z) \, \mathrm{d}z \right) \tau \\
&\qquad\qquad + \exp\left(-\lambda \int_a^b Y(z) \, \mathrm{d}z \right) \left(\frac{1}{2} W_l^{(2)} \tau \right) \\
&= \left(b^{l+1} \frac{\partial}{\partial b} + a^{l+1} \frac{\partial}{\partial a} - \frac{1}{2} (W_l^{(2)} - c_l) \right) \left(\exp\left(-\lambda \int_a^b Y(z) \, \mathrm{d}z \right) \tau \right) \\
&= \left(b^{l+1} \frac{\partial}{\partial b} + a^{l+1} \frac{\partial}{\partial a} - \frac{1}{2} (W_l^{(2)} - c_l) \right) \left(\tau(t) \det(I - \lambda k^E) \right) \, .
\end{aligned} \tag{2.55}$$

More generally: setting

$$[a, b] \to E^{1/p} := \bigcup_{i=1}^{k} [a_{2i-1}, a_{2i}] \, ,$$

$$b^{l+1} \frac{\partial}{\partial b} + a^{l+1} \frac{\partial}{\partial a} \to \sum_{j=1}^{2k} a_j^{l+1} \frac{\partial}{\partial a_j} := B_l(a) \, ,$$

deduce

$$\left(B_{kp}(a) - \tfrac{1}{2} (W_{kp}^{(2)} - c_{kp}) \right) \left(\tau(t) \det\left(I - \lambda k^{E^{1/p}} \right) \right) = 0 \, .$$

with

$$W_{kp}^{(2)} \tau(t) = c_{kp} \tau(t), \quad k \geq -1 \, . \tag{2.56}$$

Since changing integration variables in a Fredholm determinant leaves the determinant invariant, change variables:

$$z \to z^p = \lambda, \quad a_i \to a_i^p = A_i \;, \quad \text{and}$$

$$E^{1/p} \to E = \bigcup_{i=1}^{k} [A_{2i-1}, A_{2i}] \;,$$

and

$$k^{E^{1/p}}(z,z') \to K^E(\lambda,\lambda') := \frac{k^{E^{1/p}}(\lambda^{1/p},\lambda'^{1/p})}{p\lambda^{(p-1)/2p}\lambda'^{(p-1)/2p}} I_E(\lambda') \;, \tag{2.57}$$

yielding

$$\begin{aligned} \det(I - \mu K^E) &= \det(I - \mu k^{E^{1/p}}) \\ &= \frac{1}{\tau}\exp\left(-\mu \int_{E^{1/p}} Y(z)\,\mathrm{d}z\right) = \frac{\tau(t,E)}{\tau(t)} \;, \end{aligned} \tag{2.58}$$

and so (2.56) yields the p *reduced Virasoro relation*:

$$\left(B_k(A) - \tfrac{1}{2}(W_{kp}^{(2)} - c_{kp})\right)\left(\tau(t)\det\left(I - \mu K^E\right)\right) = 0 \;, \tag{2.59}$$

with

$$W_{kp}^{(2)}\tau(t) = c_{kp}\tau(t), \quad k \geq -1 \;.$$

2.1.6 Differential Equations for the Probability in Scaling Limits

Now we shall derive differential equations for the limiting probabilities discussed in Sect. 2.1.2 using the integrable tools previously developed.

Airy edge limit: Remembering the edge limit for Hermitian eigenvalues of (2.15) and (2.16):

$$\lim_{n\to\infty} P(\sqrt{2}n^{1/16}(\lambda_{\max} - \sqrt{2n}) \in E^{\mathrm{c}}) = \det\left(I - K_{\mathrm{Airy}}^E\right) \tag{2.60}$$

with

$$\begin{aligned} K_{\mathrm{Airy}}(u,v) &= \int_0^\infty \mathrm{Ai}(x+u)\,\mathrm{Ai}(x+v)\,\mathrm{d}x \;, \\ \mathrm{Ai}(u) &= \frac{1}{\pi}\int_0^\infty \cos\left(\frac{u^3}{3} + xu\right)\mathrm{d}u \;. \end{aligned} \tag{2.61}$$

Consider the KdV reduction

$$\left(p = 2, \begin{matrix} t = (t_1, 0, t_3, 0, t_5, \dots) \\ t_0 = (0, 0, 2/3, 0, 0, \dots) \end{matrix}\right) \tag{2.62}$$

with initial conditions:

$$\begin{cases} \Psi(x,t_0,z) = 2\sqrt{\pi z}A(x + (-z)^2) \\ \qquad = \mathrm{e}^{xz + 2z^2/3}\left(1 + 0(1/z)\right) \;, \\ (D_x^2 - x)\Psi(x,t_0,z) = z^2\Psi(x,t_0,z) \;. \end{cases} \tag{2.63}$$

Under the KP (KdV) flow:

$$\begin{cases} x = q(x,t_0) \to q(x,t) = \dfrac{2\partial^2}{\partial t_1^2} \ln\bigl(\tau(t)\bigr) \,, \\ \Psi(x,t_0) \to \Psi(x,t) = \exp\Bigl(xz + \displaystyle\sum_1^\infty t_i z_i\Bigr) \dfrac{\tau(t-[z^{-1}])}{\tau(t)} \,, \\ \tau(t_0) \to \tau(t) \,. \end{cases} \tag{2.64}$$

where $\tau(t)$ turns out to be the well-known Konsevich integral [16, 18], satisfying a vacuum condition as a consequence of Grassmannian invariance, to wit:

Kontsevich integral:

$$\begin{cases} \tau(t) = \lim_{N\to\infty} \dfrac{\int_{\mathcal{H}_N} \mathrm{d}X \, \exp(-\operatorname{Tr}(X^3/3 + X^2 Z))}{\int_{\mathcal{H}_N} \mathrm{d}X \, \exp(-\operatorname{Tr}(X^2 Z))} \,, \\ Z \text{ diag}\colon t_n = -\dfrac{1}{n} \operatorname{Tr} Z^{-n} + \dfrac{2}{3}\delta_{n,3} \,. \end{cases} \tag{2.65}$$

Vacuum condition:

$$W_{2k}^{(2)}\tau = -\tfrac{1}{4}\delta_{k_0}\tau, \quad k \geq -1 \,. \tag{2.66}$$

Grassmannian invariance condition:

$$\begin{cases} W := \operatorname{span}_{\mathbb{C}}\biggl\{ \Bigl(\dfrac{\partial}{\partial x}\Bigr)^j \Psi(x,t_0,z)\Bigr|_{x=0}, j = 0,1,2,\dots \biggr\} \,, \\ z^2 W \subset W \ (KdV) \,, \quad AW \subset W \,, \\ A = \dfrac{1}{2z}\Bigl(\dfrac{\partial}{\partial z} + 2z^2\Bigr) - \dfrac{1}{4z^2} \,, \quad A^2\Psi(0,t_0,z) = z^2\Psi(0,t_0,z) \,. \end{cases} \tag{2.67}$$

We have the initial kernel:

$$\begin{aligned} K_{t_0}^E(\lambda,\lambda') &= \frac{I_E(\lambda')}{2\lambda^{1/4}\lambda'^{1/4}} \int_0^\infty \Psi(x,t_0,-\sqrt{\lambda})\Psi(x,t_0,\sqrt{\lambda'})\,\mathrm{d}x \\ &= 2\pi I_E(\lambda') \int_0^\infty \operatorname{Ai}(x+\lambda)\operatorname{Ai}(x+\lambda')\,\mathrm{d}x \xrightarrow{t_0\to t} K_t^E(\lambda,\lambda') \,. \end{aligned} \tag{2.68}$$

Conditions on $\tau(t,E)$:

$$\begin{aligned} \tau(t,E) &= \tau(t)\det\Bigl(I - \frac{1}{2\pi}K_t^E\Bigr) && \text{(by (2.58))} \\ &= \tau(t_0)\det(I - K_{\text{Airy}}^E) \quad \text{at } t = t_0 && \text{(by (2.68))} \,, \end{aligned} \tag{2.69}$$

which satisfies (2.33) and (2.35):

$$\biggl(\Bigl(\frac{\partial}{\partial t_1}\Bigr)^4 - 4\frac{\partial^2}{\partial t_1 \partial t_3}\biggr)\log\tau + 6\Bigl(\frac{\partial^2}{\partial t_1^2}\log\tau\Bigr)^2 = 0 \qquad \text{(KdV)} \,, \tag{2.70}$$

and by (2.59), (2.42) and (2.62) we find for $\tau(t,E)$

Virasoro constraints:

$$\begin{aligned} B_{-1}(A)\tau &= \left(\frac{\partial}{\partial t_1} + \frac{1}{2}\sum_{i\geq 3} it_i \frac{\partial}{\partial t_{i-2}} + \frac{t_1^2}{4}\right)\tau\,, \\ B_0(A)\tau &= \left(\frac{\partial}{\partial t_3} + \frac{1}{2}\sum_{i\geq 1} it_i \frac{\partial}{\partial t_i} + \frac{1}{16}\right)\tau\,. \end{aligned} \tag{2.71}$$

Replace t-derivatives of $\tau(t, E)$ at t_0 with A derivatives in KdV:

$$B_{-1}\tau = \frac{\partial\tau}{\partial t_1}, \quad B_{-1}^2\tau = \frac{\partial^2}{\partial t_1^2}\tau, \ldots, B_{-1}B_0\tau = \left(\frac{\partial^2}{\partial t_1 \partial t_3} + \frac{1}{2}\frac{\partial}{\partial t_1}\right)\tau, \ldots\,, \quad \text{at } t = t_0\,, \tag{2.72}$$

yielding

Theorem 2.1.1 (Adler–Shiota–van Moerbeke [16]).

$$\begin{aligned} R &:= B_{-1}\log \lim_{n\to\infty} P\big(\sqrt{2}n^{1/6}(\lambda_{\max} - \sqrt{2n}) \in E^{c}\big) \\ &= B_{-1}\log\det(I - K_{\text{Airy}}^{E}) = B_{-1}\log\frac{\tau(t_0, E)}{\tau(t_0)} = B_{-1}\log\tau(t_0, E) \end{aligned}$$

satisfies

$$\big(B_{-1}^3 - 4(B_0 - \tfrac{1}{2})\big)R + 6(B_{-1}R)^2 = 0\,. \tag{2.73}$$

Setting $E = (a, \infty)$ yields:

$$R''' - 4aR' + 2R + 6R'^2 = 0\,. \tag{2.74}$$

Setting

$$R = g'^2 - ag^2 - g^4, \quad R' = g^2$$

yields

$$g'' = 2g^3 + ag \qquad \text{(Painlevé II)}. \tag{2.75}$$

Hard edge limit: Remembering the hard edge limit (2.21) for the Hermitian Laguerre ensemble (2.20):

$$\begin{aligned} \lim_{n\to\infty} P\left(\text{no eigenvalues} \in \frac{1}{4n}E\right) &= \det(I - K_\nu^E)\,, \\ K_\nu(u, v) &= \frac{1}{2}\int_0^1 sJ_\nu(s\sqrt{u})J_\nu(s\sqrt{v})\,\mathrm{d}s \end{aligned} \tag{2.76}$$

defined in terms of Bessel functions, consider the KdV reduction with initial conditions:

$$\begin{cases} \Psi(x,0,z) = \mathrm{e}^{xz} B\big((1-x)z\big) = \mathrm{e}^{xz}\big(1 + 0(1/z)\big) , \\ \left(D_x^2 - \dfrac{(\nu^2 - \frac{1}{4})}{(x-1)^2} \right) \Psi(x,0,z) = z^2 \Psi(x,0,z) \end{cases} \tag{2.77}$$

with (see [19])

$$B(z) := \varepsilon \sqrt{z} H_\nu(\mathrm{i}z) = \frac{\mathrm{e}^z 2^{\nu+1/2}}{\Gamma(-\nu + \frac{1}{2})} \int_1^\infty \frac{z^{-\nu+1/2} \mathrm{e}^{-uz}}{(u^2-1)^{\nu+1/2}} \, \mathrm{d}u ,$$

where

$$\varepsilon = \mathrm{i}\sqrt{\frac{\pi}{2}} \mathrm{e}^{\mathrm{i}\pi\nu/2}, \quad -\frac{1}{2} < \nu < \frac{1}{2} .$$

Under the KdV flow:

$$\left(\frac{\nu^2 - \frac{1}{4}}{x^2 - 1}, \Psi(x,0,z), \tau(0) \right) \xrightarrow{t} \big(q(x,t), \Psi(x,t,z), \tau(t) \big) , \tag{2.78}$$

where $\tau(t)$ is both a Laplace matrix integral and a vacuum vector [16,18] due to Grassmannian invariance:

Laplace integral:

$$\tau(t) = \lim_{N\to\infty} S_1(t) \\ \times \frac{\int_{\mathcal{H}_N^+} \mathrm{d}X \, \det X^{\nu-1/2} \exp(-\operatorname{Tr} Z^2 X) \int_{\mathcal{H}_N^+} \mathrm{d}Y \, S_0(Y) \exp(-\operatorname{Tr} XY^2)}{\int_{\mathcal{H}_N^+} \mathrm{d}X \, \exp(-\operatorname{Tr} X^2 Z)}$$

with Z diag: $t_n = -1/n \operatorname{tr} Z^{-n}$, $\mathcal{H}_n^+ = \mathcal{H}_n \cap$ (matrices with non-negative spectrum) and $S_1(t)$, $S_0(Y)$ are symmetric functions,

Vacuum condition:

$$W_{2k}^{(2)} \tau = \big((2\nu)^2 - 1 \big) \tau \delta_{k0}, \quad k \geq -1 , \tag{2.79}$$

Grassmannian invariance:

$$\begin{cases} z^2 W \subset W, \quad AW \subset W, \\ A = \dfrac{1}{2} z \left(\dfrac{\partial}{\partial z} - 1 \right), \quad (4A^2 - 2A - \nu^2 + \frac{1}{4}) \Psi(0,0,z) = z^2 \Psi(0,0,z) . \end{cases}$$

We have the initial kernel:

$$\begin{aligned} &K_{(x,t)}^E(\lambda,\lambda') \\ &= \frac{I_E(\lambda')}{2\lambda^{1/4}\lambda'^{1/4}} D_x^{-1} Y(x,t,\sqrt{\lambda},\sqrt{\lambda'}) \qquad \text{(see (2.48), (2.49) and (2.57))} \\ &= \frac{I_E(\lambda')}{2\lambda^{1/4}\lambda'^{1/4}} D_x^{-1} \left(\sum_{\pm} a_\pm \Psi^-(x,t,\pm\sqrt{\lambda}) \sum_{\pm} b_\pm \Psi^+(x,t,\pm\sqrt{\lambda'}) \right) \end{aligned}$$

$$
\begin{aligned}
&:= \frac{I_E(\lambda')}{2\lambda^{1/4}\lambda'^{1/4}} \int_1^x \left(\frac{\mathrm{i}\mathrm{e}^{\mathrm{i}\pi\nu/2}}{\sqrt{2\pi}} \Psi(x,t,-\sqrt{\lambda}) + \frac{\mathrm{e}^{-\mathrm{i}\pi\nu/2}}{\sqrt{2\pi}} \Psi(x,t,\sqrt{\lambda}) \right) \\
&\qquad \times \left(-\frac{\mathrm{e}^{-\mathrm{i}\pi\nu/2}}{\sqrt{2\pi}} \Psi(x,t,\sqrt{\lambda'}) - \frac{\mathrm{i}\mathrm{e}^{\mathrm{i}\pi\nu/2}}{\sqrt{2\pi}} \Psi(x,t,-\sqrt{\lambda'}) \right) \mathrm{d}x \\
&= I_E(\lambda') \frac{1}{2} \int_0^1 s J_\nu(s\sqrt{\lambda}) J_\nu(s\sqrt{\lambda'})\, \mathrm{d}s \quad \text{at } (x,t_0) = (1+\mathrm{i}, -e_1), \qquad (2.80)
\end{aligned}
$$

and under the t-flow

$$
K^E_{(1+\mathrm{i},-e_1)}(\lambda,\lambda') \xrightarrow{t_0 \to t} K^E_{(x,t)}(\lambda,\lambda') .
$$

Conditions on $\tau(t,E)$:

$$
\tau(t,E) = \tau(t)\det(I - K^E_{(x,t)}) = \tau(t_0)(I - K^E_\nu) \quad \text{at } (x,t_0) = (1+\mathrm{i}, e_1) \\
\text{(by (2.58) and (2.80))} , \qquad (2.81)
$$

satisfies (2.33) and (2.35):

$$
\left(\left(\frac{\partial}{\partial t_1} \right)^4 - 4 \frac{\partial^2}{\partial t_1 \partial t_3} \right) \log \tau + 6 \left(\frac{\partial^2}{\partial t_1^2} \log \tau \right)^2 = 0 \qquad \text{(KdV)} , \quad (2.82)
$$

and by (2.59), (2.42), (2.62) and (2.81), the

Virasoro constraints:

$$
\begin{aligned}
B_0(A)\tau &= \frac{1}{2} \left(\sum_{i \ge 1} i t_i \frac{\partial}{\partial t_i} + \sqrt{-1} \frac{\partial}{\partial t_1} + 2 \left(\frac{1}{4} - \nu^2 \right) \right) \tau , \\
B_1(A)\tau &= \frac{1}{2} \left(\sum_{i \ge 1} i t_i \frac{\partial}{\partial t_{i+2}} + \frac{1}{2} \frac{\partial^2}{\partial t_1^2} + \sqrt{-1} \frac{\partial}{\partial t_3} \right) \tau .
\end{aligned}
$$

Replace t-derivatives of $\tau(t,E)$ at $(1+\mathrm{i},-e_1)$ with A derivatives in KdV:

$$
B_0(A)\tau = \frac{\mathrm{i}}{2} \frac{\partial \tau}{\partial t_1} + \left(\frac{1}{4} - \nu^2 \right) \tau , \quad B_1(A) = \frac{1}{4} \frac{\partial^2 \tau}{\partial t_1^2} + \frac{\mathrm{i}}{2} \frac{\partial \tau}{\partial t_3}, \ldots \quad \text{at } (1+\mathrm{i},-e_1)
$$

yielding

Theorem 2.1.2 (Adler–Shiota–van Moerbeke [16]).

$$
\begin{aligned}
R &:= \log \lim_{n \to \infty} P \left(\textit{no eigenvalues} \in \frac{E}{4n} \right) \\
&= \log\det(I - K^E_\nu) = \log \frac{\tau(\mathrm{i},0,0,\ldots;E)}{\tau(\mathrm{i},0,0,\ldots;R)}
\end{aligned}
$$

satisfies

$$
\left(B_0^4 - 2B_0^3 + (1-\nu^2)B_0^2 + B_1(B_0 - \tfrac{1}{2}) \right) R - 4(B_0 R)(B_0^2 R) + 6(B_0^2 R)^2 = 0 .
$$

Setting:

$$E = (0, x), \quad f = -x \frac{\partial R}{\partial x}$$

yields

$$f''' + \frac{f''}{x} - \frac{6f'^2}{x} + \frac{4ff'}{x^2} + \frac{(x-\nu^2)f'}{x^2} - \frac{f}{2x^2} = 0 \qquad \text{(Painlevé V)}.$$

2.2 Recursion Relations for Unitary Integrals

2.2.1 Results Concerning Unitary Integrals

Many generating functions in a parameter t for combinatorial problems are expressible in the form of unitary integrals $I_n(t)$ over $\mathrm{U}(n)$ (see [20,22,47,51]). Our methods can be used to either get a differential equation for $I_n(t)$ in t [2] or a recursion relation in n [3] and in the present case we concentrate on the latter. Borodin first got such results [26] using Riemann–Hilbert techniques. Consider the following basic objects ($\Delta_n(z)$ is the Vandermonde determinant, $\mathrm{i} = \sqrt{-1}$):

Unitary integral:

$$\begin{aligned} I_n^{(\varepsilon)} &= \int_{\mathrm{U}(n)} \det\big(M^\varepsilon \rho(M)\big)\,\mathrm{d}M \\ &= \frac{1}{n!}\int_{(S^1)^n} |\Delta_n(z)|^2 \prod_{k=1}^{n}\left(z_k^\varepsilon \rho(z_k)\frac{\mathrm{d}z_k}{2\pi \mathrm{i} z_k}\right) \\ &= \det\left(\int_{S^1} z^{\varepsilon+i'-j'}\rho(z)\frac{\mathrm{d}z}{2\pi\mathrm{i}z}\right)_{i',j'=1}^{n}, \end{aligned} \tag{2.83}$$

with weight $\rho(z)$:

$$\begin{aligned} \rho(z) = \exp\big(P_1(z) + P_2(z^{-1})\big) z^\gamma (1-d_1 z)^{\gamma_1'}(1-d_2 z)^{\gamma_2'} \\ \times (1-d_1^{-1}z^{-1})^{\gamma_1''}(1-d_2^{-1}z^{-1})^{\gamma_2''}, \end{aligned} \tag{2.84}$$

$$P_1(z) := \sum_1^{N_1} \frac{u_i z^i}{i}, \quad P_2(z) := \sum_1^{N_2} \frac{u_{-i} z^i}{i}, \tag{2.85}$$

and we introduce the

Basic recursion variables:

$$x_n := (-1)^n \frac{I_n^+}{I_n^{(0)}}, \quad y_n := (-1)^n \frac{I_n^-}{I_n^{(0)}}, \tag{2.86}$$

$$v_n := 1 - x_n y_n = \frac{I_{n-1}^{(0)} I_{n+1}^{(0)}}{I_n^{(0)2}}, \tag{2.87}$$

and so

$$I_n^{(0)} = (I_1^{(0)})^n \prod_1^{n-1} (1 - x_i y_i)^{n-i} \,, \tag{2.88}$$

thus

$$(x, y) \text{ recursively yields } \{I_n^{(0)}\} \,.$$

We also introduce the fundamental semi-infinite matrices:

Toeplitz matrices:

$$L_1(x,y) := \begin{pmatrix} -x_1y_0 & 1 - x_1y_1 & 0 & 0 & \cdots \\ -x_2y_0 & -x_2y_1 & 1 - x_2y_2 & 0 & \cdots \\ -x_3y_0 & -x_3y_1 & -x_3y_2 & 1 - x_3y_3 & \\ \vdots & \vdots & \vdots & \vdots & \cdots \end{pmatrix}, \tag{2.89}$$

$$L_2 := L_1^T(y,x) \,,$$

in terms of which we write the following:

Recursion matrices:[5]

$$\begin{aligned} \mathcal{L}_1^{(n)} &:= (aI + bL_1 + cL_1^2)P_1'(L_1) + c(n + \gamma_1' + \gamma_2' + \gamma)L_1 \,, \\ \mathcal{L}_2^{(n)} &:= (cI + bL_2 + aL_2^2)P_2'(L_2) + a(n + \gamma_1'' + \gamma_2'' - \gamma)L_2 \,. \end{aligned} \tag{2.90}$$

There exists the following basic involution $\sim$:

Basic involution:

$$\begin{gathered} \sim : z \to z^{-1}, \quad \rho(z) \to \rho(z^{-1}) \,, \\ I_n^{(0)} \leftrightarrow I_n^{(0)}, \quad I_n^+ \leftrightarrow I_n^- \,, \\ x_n \leftrightarrow y_n, \quad a \leftrightarrow c, \quad b \leftrightarrow b, \quad \gamma \to -\gamma \,, \\ L_1 \leftrightarrow L_2^T, \quad \mathcal{L}_1^{(n)} \leftrightarrow \mathcal{L}_2^{(n)T} \,. \end{gathered} \tag{2.91}$$

Self-dual case:

$$\rho(z) = \rho(z^{-1}) \,, \; x_n = y_n \implies L_1 = L_2^T, \; \mathcal{L}_1^{(n)} = \mathcal{L}_2^{(n)T} \,. \tag{2.92}$$

Let us define the "total discrete derivative":[6]

$$\partial_n f(n) = f(n+1) - f(n) \,. \tag{2.93}$$

We now state the main theorems of Adler–van Moerbeke [3]:

---

[5] I in (2.90) is the semi-infinite identity matrix and $P_i'(z) = \mathrm{d}P_i(z)/\mathrm{d}z$.

[6] The total derivative means you must take account, in writing $f(n)$, of *all* the places n appears either implicitly or explicitly, so $f(n) = g(n, n, n, \dots)$ in reality.

Rational relations for (x, y):

Theorem 2.2.1. *The* $(x_k, y_k)_{k\geq 1}$ *satisfy* 2 *finite step relations:*
Case 1. *In the generic situation, to wit:*

$$d_1, d_2, d_1 - d_2, |\gamma_1'| + |\gamma_1''|, |\gamma_2'| + |\gamma_2''| \neq 0 \quad \text{in } \rho(z) , \tag{2.94}$$

we find for $n \geq 1$ *that*

$$(+) \qquad \begin{cases} \partial_n(\mathcal{L}_1^{(n)} - \mathcal{L}_2^{(n)})_{n,n} + (cL_1 - aL_2)_{n,n} = 0 \\ \partial_n(v_n \mathcal{L}_1^{(n)} - \mathcal{L}_2^{(n)})_{n+1,n} + (cL_1^2 + bL_1)_{n+1,n+1} = (\mathit{same})_{n=0} , \end{cases}$$

with

$$(a, b, c) = \frac{1}{\sqrt{d_1 d_2}}(1, -d_1, -d_2, d_1 d_2) , \tag{2.95}$$

and the dual equations $(\tilde{+})$, *also hold.*
Case 2. *Upon rescaling,*

$$\rho(z) = z^\gamma (1 + z)^{\gamma_1'} \exp\bigl(P_1(z) + P_2(z^{-1})\bigr) , \tag{2.96}$$

and then $(+)$, $(\tilde{+})$ *are satisfied with* $(a, b, c) = (1, 1, 0)$ *or* $(0, 1, 1)$.
Case 3. *Upon rescaling,*

$$\rho(z) = z^\gamma \exp\bigl(P_1(z) + P_2(z^{-1})\bigr) , \tag{2.97}$$

and then $(+)$, $(\tilde{+})$ *are satisfied for* a, b, c *arbitrary and in addition finer relations hold:*

$$\Gamma_n(x, y) = 0 , \quad \widetilde{\Gamma}_n(x, y) = 0 ,$$

$$\Gamma_n(x, y) := \frac{v_n}{y_n} \begin{pmatrix} -\bigl(L_1 P_1'(L_1)\bigr)_{n+1,n+1} & -\bigl(L_2 P_2'(L_2)\bigr)_{n,n} \\ +\bigl(P_1'(L_1)\bigr)_{n+1,n} & + P_2'(L_2)_{n\ ,n+1} \end{pmatrix} + n x_n . \tag{2.98}$$

If $|N_1 - N_2| \leq 1$, where $N_i = \operatorname{degree} P_i$, the rational relations of Theorem 2.2.1 become, upon setting $z_n := (x_n, y_n)$:

Rational recursion relations:

Theorem 2.2.2. *For* $N_1 = N_2 \pm 1$ *or* $N_1 = N_2$, *the rational relations of Theorem* 2.2.1 *become recursion relations as follows:*
Case 1. *Yields inductive rational* $N_1 + N_2 + 4$ *step relations:*

$$z_n = F_n(z_{n-1}, z_{n-2}, \ldots, z_{n-N_1-N_2-3}) .$$

Case 2. *Yields inductive rational* $N_1 + N_2 + 3$ *step relations:*

$$z_n = F_n(z_{n-1}, z_{n-2}, \ldots, z_{n-N_1-N_2-2}) ,$$

with either

$$N_1 = N_2 \text{ or } N_2 + 1\colon \ (a, b, c) = (1, 1, 0) ,$$

or

$$N_1 = N_2 \text{ or } N_2 - 1\colon \ (a, b, c) = (0, 1, 1) .$$

Case 3. *Yields inductive rational N_1+N_2-1 step relations:*

$$z_n = F_n(z_{n-1}, z_{n-2}, \dots, z_{n-N_1-N_2}),$$

upon using Γ_n and $\widetilde{\Gamma}_n$ and in the case of the self dual weight:

$$\rho(z) = \exp\left(\sum_1^{N_1} \frac{u_i(z^i + z^{-i})}{i}\right).$$

One finds recursion relations:

$$x_n = F_n(x_{n-1}, x_{n-2}, \dots, x_{n-2N}).$$

2.2.2 Examples from Combinatorics

In this section, we give some well-known examples from combinatorics in the notation of the previous section. In the case of a permutation π_k, of k-numbers, $L(\pi_k)$ shall denote the length of the largest increasing subsequence of π_k. If π_k is only a word of size k from an alphabet of α numbers, $L^{(w)}(\pi_k)$ shall denote the length of the largest weakly increasing subsequence in the word π_k. We will also consider odd permutations π on $(-k,\dots,-1,1,\dots,k)$ or $(-k,\dots,-1,0,1,\dots,k)$, which means $\pi(-j) = -\pi(j)$, for all j.

Example 1. $\rho(z) = \exp\big(t(z+z^{-1})\big)$ (self-dual case)

$$I_n^{(0)}(t) := \sum_{k=0}^{\infty} \frac{t^{2k}}{k!} P(\pi_k \in S_k \mid L(\pi_k) \le n) = \int_{\mathrm{U}(n)} \exp\big(t\,\mathrm{Tr}(M+M^{-1})\big)\,\mathrm{d}M\,,$$

the latter equality due to I. Gessel, with

$$x_n(t) = (-1)^n \frac{\int_{\mathrm{U}(n)} \det(M) \exp\big(t\,\mathrm{Tr}(M+M^{-1})\big)\,\mathrm{d}M}{\int_{U(n)} \exp\big(t\,\mathrm{Tr}(M+M^{-1})\big)\,\mathrm{d}M}, \qquad \text{(as in (2.86))}$$

and

$$I_n^{(0)}(t) = \big(I_1^{(0)}(t)\big)^n \prod_{i=1}^{n-1} (1-x_i^2)^{n-i}.$$

One finds a

3-*step recursion relation for x_i:*

$$\begin{aligned} 0 &= nx_n - \frac{(1-x_n^2)}{x_n}\big(t(L_1)_{n+1,n+1} + t(L_1)_{nn}\big) && \text{(as in (2.98))} \\ &= nx_n + t(1-x_n^2)(x_{n+1}+x_{n-1}) && \text{(Borodin [26])} \end{aligned}$$

which possesses an

Invariant:
$$\Phi(x_{n+1},x_n)=\Phi(x_n,x_{n-1})\,,$$
$$\Phi(y,z)=(1-y^2)(1-z^2)-\frac{n}{t}yz \qquad \text{(McMillan [43])}\,.$$

The initial conditions in the recursion relation are as follows:

$$x_0=1,\quad x_1=-\frac{1}{2}\frac{\mathrm{d}}{\mathrm{d}t}\log I_0(2t),\quad I_1^{(0)}=I_0(2t)\,,$$

with I_0 the hyperbolic Bessel function of the first kind (see [19]).

Example 2. $\rho(z)=\exp\bigl(t(z+z^{-1})+s(z^2+z^{-2})\bigr)$ (self-dual case)

Set

$$S_{2k}^{\text{odd}}=\{\pi_{2k}\in S_{2k}\text{ acts on }(-k,\dots,-1,1,\dots,k)\text{ oddly}\}\,,$$
$$S_{2k+1}^{\text{odd}}=\{\pi_{2k+1}\in S_{2k+1}\text{ acts on }(-k,\dots,-1,0,1,\dots,k)\text{ oddly}\}$$

and then one finds:

$$\sum_{k=0}^{\infty}\frac{(\sqrt{2}s)^{2k}}{k!}P(\pi_{2k}\in S_{2k}^{\text{odd}}\mid L(\pi_{2k})\le n)=\int_{\mathrm{U}(n)}\exp\bigl(s\,\mathrm{Tr}(M^2+M^{-2})\bigr)\,\mathrm{d}M\,,$$
$$\sum_{k=0}^{\infty}\frac{(\sqrt{2}s)^{2k}}{k!}P(\pi_{2k+1}\in S_{2k+1}^{\text{odd}}\mid L(\pi_{2k+1})\le n)$$
$$=\frac{1}{4}\left(\frac{\partial}{\partial t}\right)^2\left(\int_{\mathrm{U}(n)}\exp\Bigl(\mathrm{Tr}\bigl(t(M+M^{-1})+s(M^2+M^{-2})\bigr)\Bigr)\,\mathrm{d}M\right.$$
$$\left.+\text{same }(t,-s)\right)\Bigg|_{t=0}\,,$$

as observed by M. Rains [47] and Tracy–Widom [51]. Moreover,

$$x_n(t,x)=(-1)^n\frac{\int_{U(n)}\det(M)\exp\Bigl(\mathrm{Tr}\bigl(t(M+M^{-1})+s(M^2+M^{-2})\bigr)\Bigr)\,\mathrm{d}M}{\int_{U(n)}\exp\Bigl(\mathrm{Tr}\bigl(t(M+M^{-1})+s(M^2+M^{-2})\bigr)\Bigr)\,\mathrm{d}M}\,,$$

and

$$I_n^{(0)}(t)=\left(I_1^{(0)}(t)\right)^n\prod_{i=1}^{n-1}(1-x_i^2)^{n-i}\,.$$

One finds a
5-step recursion relation for x_i:

$$nx_n+tv_n(x_{n-1}+x_{n+1})+2sv_n(x_{n+2}v_{n+1}+x_{n-2}v_{n-1}-x_n(x_{n+1}+x_{n-1})^2)=0$$
$$(v_n=1-x_n^2)$$

which possesses the

Invariant: $\Phi(x_{n-1},x_n,x_{n+1},x_{n+2})=\text{same}|_{n\to n+1}\,,$

$$\Phi(x,y,z,u):=nyz-(1-y^2)(1-z^2)\Bigl(t+2s\bigl(x(u-y)-z(u+y)\bigr)\Bigr)\,,$$

analogous to the McMillan invariant of the previous example.

Example 3. $\rho(z) = (1+z)\exp(sz^{-1})$ (Case 2 of Theorem 2.2.1)
Set

$$S_{k,\alpha} = \{\text{words } \pi_k \text{ of length } k \text{ from alphabet of size } \alpha\} ,$$

with

$$\begin{aligned} I_n^{(0)}(s) &= \sum_{k=0}^{\infty} \frac{(\alpha s)^k}{k!} P(\pi_k \in S_{k,\alpha} \mid L^{(w)}(\pi_k) \leq n) \\ &= \int_{U(n)} \det(I+M)^\alpha \exp(s \operatorname{tr} M^{-1})\, \mathrm{d}M , \end{aligned}$$

the latter identity observed by Tracy–Widom [52]. Then setting in Case 2

$$P_1(z) = 0 , \quad P_2(z) = sz , \quad N_1 = 0 , \quad N_2 = 1 , \quad (a,b,c) = (0,1,1) ,$$
$$\mathcal{L}_1^{(n)} = (n+\alpha)L_1 , \quad \mathcal{L}_2^{(n)} = s(I+L_2) ,$$

one finds the

Recursion relations:

$$\begin{gathered} \partial_n\big((n+\alpha)L_1 - sL_2\big)_{nn} + (L_1)_{nn} \\ \partial_n\big((n-1)+\alpha)v_{n-1}L_1 - sL_2\big)_{n,n-1} + (L_1^2 + L_1)_{n,n} = (\text{same})|_{n=1} \\ (v_n = 1 - x_n y_n) , \end{gathered}$$

with

$$\underset{+}{x_n}, \underset{-}{y_n} = \frac{(-1)^n \int_{U(n)} (\det M)^{\pm} \det(I+M)^\alpha \exp(s \operatorname{Tr} M^{-1})\, \mathrm{d}M}{\int_{U(n)} \det(I+M)^\alpha \exp(s \operatorname{tr} M^{-1})\, \mathrm{d}M}$$

and

$$I_n^{(0)}(s) = \big(I_1^{(0)}(s)\big)^n \prod_{i=1}^{n-1} (1 - x_i y_i)^{n-i} ,$$

leading to the

3 and 4 step relations for (x_i, y_i):

$$\begin{aligned} &-(n+\alpha+1)x_{n+1}y_n + sx_n y_{n+1} + (n+\alpha-1)x_n y_{n-1} - sx_{n-1}y_n = 0 , \\ &-v_n\big((n+\alpha+1)x_{n+1}y_{n-1} + s\big) + v_{n-1}\big((n+\alpha-2)x_n y_{n-2} + s\big) \\ &\qquad + x_n y_{n-1}(x_n y_{n-1} - 1) = -v_1\big((2+\alpha)x_2 + s\big) + x_1(x_1 - 1) \\ &\qquad\qquad (x_0 = y_0 = 1) . \end{aligned}$$

2.2.3 Bi-orthogonal Polynomials on the Circle and the Toeplitz Lattice

It turns out that the appropriate integrable system for our problem is the Toeplitz lattice, an invariant subsystem of the 2-Toda lattice. Indeed x_n and y_n of Sect. 2.2.1 turn out to be dual Hamiltonian variables for the integrable system; moreover x_n and y_n are nothing but the constant term of the nth bi-orthogonal polynomials on the circle, generated by a natural time deformation of our measure $\rho(z)$ of (2.84). These things are discussed by Adler–van Moerbeke in [2,3] in detail. Consider the bi-orthogonal polynomials and inner product generated by the following measure on S^1:

$$\rho(z,t,s)\frac{\mathrm{d}z}{2\pi\mathrm{i}z} = \exp\left(\sum_1^\infty (t_k z^k - s_k z^{-k})\right)\frac{\mathrm{d}z}{2\pi\mathrm{i}z}\,, \tag{2.99}$$

with

Inner product:

$$\langle f(z), g(z)\rangle := \oint_{S^1} \frac{\mathrm{d}z}{2\pi\mathrm{i}z} f(z) g(z^{-1}) \exp\left(\sum_1^\infty (t_i z^i - siz^{-i})\right). \tag{2.100}$$

Bi-orthogonal polynomials:

$$\langle p_n^{(1)}, p_m^{(2)}\rangle := \delta_{n,m} h_n, \quad h_n = \frac{\tau_{n+1}(t,s)}{\tau_n(t,s)}, \quad n,m = 0,1,\dots\,. \tag{2.101}$$

The polynomials are parametrized by

2-*Toda* τ *functions*:

$$\begin{aligned}\tau_n(t,s) &= \det(\langle z^k, z^l\rangle)_{k,l=0}^{n-1} \qquad \text{(Toeplitz determinant)}\\ &= \frac{1}{n!}\int_{(S^1)^n} |\Delta_n(z)|^2 \prod_{k=1}^n \exp\left(\sum_{j=1}^\infty (t_j z_k^j - s_j z_k^{-j})\right)\frac{\mathrm{d}z_k}{2\pi\mathrm{i}z_k}\\ &= \int_{\mathrm{U}(n)} \exp\left(\mathrm{Tr}\sum_1^\infty (t_i M^i - s_i M^{-i})\right)\mathrm{d}M, \quad n\ge 1, \tau_0\equiv 1\,,\end{aligned} \tag{2.102}$$

as follows:

$$\begin{aligned}&(p_n^{(1)}, p_n^{(2)})(u;t,s)\\ &\quad= \frac{u^n}{\tau_n(t,s)}\left(\tau_n(t-[u^{-1}],s), \tau_n(t,s+[u^{-1}])\right)\\ &\quad= \left(\frac{\int_{\mathrm{U}(n)}\det(uI-M)\exp\left(\mathrm{Tr}(\sum_1^\infty t_i M^i - s_i M^{-i})\right)\mathrm{d}M}{\int_{\mathrm{U}(n)}\exp\left(\mathrm{Tr}(\sum t_i M_i - s_i M^{-i})\right)\mathrm{d}M}\,,\right.\\ &\quad\left.\frac{\int_{U(n)}\det(uI-M^{-1})\exp\left(\mathrm{Tr}(\sum_1^\infty t_i M^i - s_i M^{-i})\right)\mathrm{d}M}{\int_{U(n)}\exp\left(\mathrm{Tr}(\sum t_i M_i - s_i M^{-i})\right)\mathrm{d}M}\right),\\ &\qquad [x] = (x, x^2/2, x^3/3, \dots)\,.\end{aligned} \tag{2.103}$$

The constant term of our polynomials yield the

Dynamical variables:

$$x_n(t,x) = p_n^{(1)}(0;t,s), \quad y_n(t,s) = p_n^{(2)}(0;t,s)\,, \tag{2.104}$$

$$v_n = 1 - x_n y_n = \frac{\tau_{n+1}\tau_{n-1}}{\tau_n^2}\,; \tag{2.105}$$

giving rise to an invariant subsystem of the

Toda equations:

$$L_1(x,y) := \begin{pmatrix} -x_1y_0 & 1-x_1y_1 & 0 & 0 & \dots \\ -x_2y_0 & -x_2y_1 & 1-x_2y_2 & 0 & \dots \\ -x_3y_0 & -x_3y_1 & -x_3y_2 & 1-x_3y_3 & \\ \vdots & \vdots & \vdots & \vdots & \ddots \end{pmatrix}, \tag{2.106}$$

$$L_2(x,y) := L_1^T(y,x)\,, \quad h_n = \tau_{n+1}/\tau_n\,, \quad h = \operatorname{diag}(h_0, h_1\,, \dots)\,, \tag{2.107}$$

$$\tilde{L}_1 = hL_1h^{-1}\,, \quad \tilde{L}_2 = L_2\,, \tag{2.108}$$

$$\text{(T)} \qquad \begin{cases} \dfrac{\partial \tilde{L}_i}{\partial t_n} = [(\tilde{L}_1^n)_+, \tilde{L}_i]^7 \\ \dfrac{\partial \tilde{L}_i}{\partial s_n} = [(\tilde{L}_2^n)_-, \tilde{L}_i] \end{cases}, \quad i = 1,2,\ n = 1,2,\dots \tag{2.109}$$

with

$$\begin{cases} (A)_+ = \text{upper tri}(A) + \operatorname{diag}(A)\,, \\ (A)_- = \text{lower tri}(A)\,. \end{cases}$$

We find, (T) $\Leftrightarrow$ *Toeplitz lattice*:[7]

$$\begin{cases} \dfrac{\partial x_n}{\partial t_i} = v_n \dfrac{\partial H_i^{(1)}}{\partial y_n}, & \dfrac{\partial y_n}{\partial t_i} = -v_n \dfrac{\partial H_i^{(1)}}{\partial x_n} \\ \dfrac{\partial x_n}{\partial s_i} = v_n \dfrac{\partial H_i^{(2)}}{\partial y_n}, & \dfrac{\partial y_n}{\partial s_i} = -v_n \dfrac{\partial H_i^{(2)}}{\partial x_n} \end{cases}, \quad i = 1,2,\dots\,, \tag{2.110}$$

$$H_i^{(j)} := -\frac{\operatorname{tr} L_j^i}{i}, \quad \omega = \sum_{k=1}^{\infty} \frac{dx_k \wedge dy_k}{v_k}, \quad v_n = 1 - x_n y_n\,, \tag{2.111}$$

with

$$\begin{cases} x_n(0,0) = y_n(0,0) = 0, & n \geq 1\,, \\ x_0(t,s) = y_0(t,s) = 1, & \forall t, s\,, \end{cases} \quad \text{(see [2,3]).} \tag{2.112}$$

[7] Equations (T) are the 2-Toda equations of Ueno–Takasaki [55], but equations (T) with precisely the initial conditions (2.106) and (2.107) are an invariant subsystem of the 2-Toda equations which are equivalent to the Toeplitz lattice, (2.110), (2.111). The latter equations are Hamiltonian, with ω the symplectic form.

Example

$$\begin{aligned}\frac{\partial x_n}{\partial t_1} &= (1 - x_n y_n)x_{n+1} \,, \quad \frac{\partial y_n}{\partial t_1} = -(1 - x_n y_n)y_{n-1} \,, \\ \frac{\partial x_n}{\partial s_1} &= (1 - x_n y_n)x_{n-1} \,, \quad \frac{\partial y_n}{\partial s_1} = -(1 - x_n y_n)y_{n+1} \,,\end{aligned} \tag{2.113}$$

yielding the

Ladik–Ablowitz lattice:

$$\frac{\partial}{\partial t} = \frac{\partial}{\partial t_1} - \frac{\partial}{\partial s_1} \,. \tag{2.114}$$

2.2.4 Virasoro Constraints and Difference Relations

In this section, using a 2-Toda vertex operator, which generates a subspace of the tangent space of the sequence of 2-Toda τ-functions $\{\tau_n(t,s)\}$, we derive Virasoro relations in our special case. Indeed, deriving a Lax-equation for the vertex operator leads to a fixed point theorem for our particular sequence of matrix integral τ-functions, fixed under an integrated version of the vertex operator, integrated over the unit circle. The Virasoro relations coupled with the Toeplitz lattice identities then lead to difference relations after some manipulation.

We present the following operators:

Toda vertex operator:

$$\begin{aligned}&X(u,v)\big(f_n(t,s)\big)_{n\geq 0} \\ &:= \left(\exp\left(\sum_1^\infty (t_i u^i - s_i v^{-i})\right)(uv)^{n-1} f_{n-1}(t - [u^{-1}], s + [v^{-1}])\right)_{n\geq 0} \,.\end{aligned} \tag{2.115}$$

Integrated version:

$$Y^\gamma := \int_{S^1} \frac{\mathrm{d}u}{2\pi \mathrm{i} u} u^\gamma X(u, u^{-1}) \,. \tag{2.116}$$

Virasoro operator:

$$\begin{aligned}\mathbb{V}_k^\gamma = (\mathbb{V}_{k,n}^\gamma)_{n\geq 0} &:= \mathbb{J}_k^{(2)}(t) - \mathbb{J}_{-k}^{(2)}(-s) - (k-\gamma)\big(\theta \mathbb{J}_k^{(1)}(t) + (1-\theta)\mathbb{J}_{-k}^{(1)}(-s)\big) \\ &\quad \text{(vector differential operator in } t,\ s, \text{ acting diagonally} \\ &\quad \text{as defined explicitly in (2.123), (2.124), and (2.125))}\,.\end{aligned} \tag{2.117}$$

Main facts:

$$\text{Lax-equation:} \qquad u^{-\gamma} u \frac{\mathrm{d}}{\mathrm{d}u} u^{k+\gamma} X(u, u^{-1})) = [\mathbb{V}_k^\gamma, X(u, u^{-1})] \,, \tag{2.118}$$

$$\text{Commutativity:} \qquad [\mathbb{V}_k^\gamma, Y^\gamma] = 0 \,, \tag{2.119}$$

$$\text{Fixed Point:} \qquad Y^\gamma I = I, \quad I := \left(n! \tau_n^{(\gamma)}(t,s)\right)_{n \geq 0} \,, \tag{2.120}$$

where

$$\tau_n^{(\gamma)}(t,s) = \frac{1}{n!} \int_{(S^1)^n} |\Delta_n(z)|^2 \prod_{k=1}^n \left(z_k^\gamma \exp\left(\sum_{j=1}^\infty (t_j z_k^j - s_j z_k^{-j}) \right) \frac{\mathrm{d}z_k}{2\pi \mathrm{i} z_k} \right) \,,$$
$$(\tau_0 = 1) \tag{2.121}$$

and $\{\tau_n^{(\gamma)}(t,s)\}$ *satisfy an* $\mathrm{sl}(2,\mathbb{Z})$ *algebra of Virasoro constraints*:

$$\mathbb{V}_{k,n}^\gamma \tau_n^{(\gamma)}(t,s) = 0 \quad \text{for } k = \begin{cases} -1, & \theta = 0 \,, \\ 0, & \theta \text{ arbitrary} \,, \\ 1, & \theta = 1, \end{cases} \quad (n \geq 0) \,. \tag{2.122}$$

Proof of main facts. The proof of the Lax-equation is a lengthy calculation (see [3]). Integrating the Lax-equation with regard to

$$\int_{S^1} \frac{\mathrm{d}u \, u^\gamma}{2\pi \mathrm{i} u}$$

immediately leads to the commutativity statement. To see the fixed point statement, compute (setting $u = z_n$)

$$\begin{aligned}
I_n(t,s) &= n! \tau_n^{(\gamma)}(t,s) \\
&= \int_{S^1} \frac{u^\gamma \, \mathrm{d}u}{2\pi \mathrm{i} u} \exp\left(\sum_1^\infty (t_j u^j - s_j u^{-j}) \right) u^{n-1} u^{-n+1} \\
&\quad \times \left(\int_{(S^1)^{n-1}} \Delta_{n-1}(z) \overline{\Delta_{n-1}(z)} \prod_{k=1}^{n-1} \left(1 - \frac{z_k}{u}\right)\left(1 - \frac{u}{z_k}\right) \right. \\
&\qquad\qquad \left. \times \exp\left(\sum_1^\infty (t_j z_k^j - s_j z_k^{-j}) \right) \frac{z_k^\gamma \mathrm{d}z_k}{2\pi \mathrm{i} z_k} \right) \\
&= \int_{S^1} \frac{u^\gamma \, \mathrm{d}u}{2\pi \mathrm{i} u} \exp\left(\sum_1^\infty (t_j u^j - s_j u^{-j}) \right) (u u^{-1})^{n-1} \\
&\qquad\qquad \times \exp\left(- \sum_1^\infty \left(\frac{u^{-j}}{j} \frac{\partial}{\partial t_j} - \frac{u^j}{j} \frac{\partial}{\partial s_j} \right) \right) \\
&\quad \times \left(\int_{(S^1)^{n-1}} \Delta_{n-1}(z) \overline{\Delta_{n-1}(z)} \prod_{k=1}^{n-1} \exp\left(\sum_1^\infty (t_j z_k^j - s_j z_k^{-j}) \right) \frac{z_k^\gamma \, \mathrm{d}z_k}{2\pi \mathrm{i} z_k} \right) \\
&= (Y^\gamma I)_n \,,
\end{aligned}$$

yielding the fixed point statement. Finally, to see the $\mathrm{sl}(2,\mathbb{Z})$ Virasoro statement, observe that from the other main facts, we have

$$\begin{aligned}
0 &= ([\mathbb{V}_k^\gamma, (Y^\gamma)^n]I)_n \\
&= (\mathbb{V}_k^\gamma (Y^\gamma)^n I - (Y^\gamma)^n \mathbb{V}_k^\gamma I)_n \\
&= (\mathbb{V}_k^\gamma I - (Y^\gamma)^n \mathbb{V}_k^\gamma I)_n \\
&= \mathbb{V}_{k,n}^\gamma I_n - \int_{S^1} u^\gamma \frac{\mathrm{d}u}{2\pi \mathrm{i} u} \exp\Big(\sum_1^\infty (t_j u^j - s_j u^{-j})\Big) \\
&\qquad\qquad \times \exp\bigg(-\sum_1^\infty \Big(\frac{u^{-j}}{j}\frac{\partial}{\partial t_j} - \frac{u^j}{j}\frac{\partial}{\partial s_j}\Big)\bigg) \\
&\qquad \cdots \int_{S^1} u^\gamma \frac{\mathrm{d}u}{2\pi \mathrm{i} u} \Big(\sum_1^\infty (t_j u^j - s_j u^{-j})\Big) \\
&\qquad\qquad \times \exp\bigg(-\sum_1^\infty \Big(\frac{u^{-j}}{j}\frac{\partial}{\partial t_j} - \frac{u^j}{j}\frac{\partial}{\partial s_j}\Big)\bigg) V_k^\gamma I_0 ,
\end{aligned}$$

upon using the backward shift (see (2.115)) present in Y^γ. Note $I_0 = 1$, and so it follows from the explicit Virasoro formulas given below, that $V_{k,0}^\gamma 1 = 0$ precisely if $k = -1, 0, 1$ and θ is as specified in (2.122), yielding the Virasoro constraints (2.122). We now make explicit the Virasoro relations.

Explicit Virasoro formulas:

$$\begin{cases} \mathbb{J}_k^{(1)} = (J_k^{(1)} + n\delta_{0k})_{n\geq 0} , \\ \mathbb{J}_k^{(2)} = \frac{1}{2}(J_k^{(2)} + (2n+k+1)J_k^{(1)} + n(n+1)\delta_{0k})_{n\geq 0} . \end{cases} \tag{2.123}$$

$$\begin{cases} \mathbb{J}_k^{(1)} = \dfrac{\partial}{\partial t_k} + (-k)t_{-k} \\ \mathbb{J}_k^{(2)} = 2\sum i t_i \dfrac{\partial}{\partial t_{i+k}} \end{cases} , \quad \text{if } k = 0, \pm 1 . \tag{2.124}$$

Virasoro relations:

$$\begin{cases} \mathbb{V}_{-1,n}^\gamma \tau_n^{(\gamma)} = \Big(\dfrac{1}{2}J_{-1}^{(2)}(t) - \dfrac{1}{2}J_1^{(2)}(s) + n\Big(t_1 + \dfrac{\partial}{\partial s_1}\Big) - \gamma\frac{\partial}{\partial s_1}\Big)\tau_n^{(\gamma)} = 0 , \\ \mathbb{V}_{0,n}^\gamma \tau_n^{(\gamma)} = \Big(\dfrac{1}{2}J_0^{(2)}(t) - \dfrac{1}{2}J_0^{(2)}(s) + n\gamma\Big)\tau_n^{(\gamma)} = 0 , \\ \mathbb{V}_{1,n}^\gamma \tau_n^{(\gamma)} = \Big(-\dfrac{1}{2}J_{-1}^{(2)}(s) + \dfrac{1}{2}J_1^{(2)}(t) + n\Big(s_1 + \dfrac{\partial}{\partial t_1}\Big) + \gamma\dfrac{\partial}{\partial t_1}\Big)\tau_n^{(\gamma)} = 0 . \end{cases} \tag{2.125}$$

We now derive the first difference relation (the second is similar) in Case 2 of Theorem 2.2.1. Setting for arbitrary a, b, c, t, and s,

$$\begin{aligned}\alpha_i(t) &:= a(i+1)t_{i+1} + bit_i + c(i-1)t_{i-1} + c(n+\gamma)\delta_{i1} \, , \\ \beta_i(s) &:= a(i-1)s_{i-1} + bis_i + c(i+1)s_{i+1} - a(n-\gamma)\delta_{i1} \, ,\end{aligned} \tag{2.126}$$

$$\mathcal{L}_1^{(n)} := \sum_{i\geq 1} \alpha_i(t) L_1^i \quad \text{and} \quad \mathcal{L}_2^{(n)} := -\sum_{i\geq 1} \beta_i(t) L_2^i \, , \tag{2.127}$$

and remembering

$$(x_n, y_n) = (-1)^n \left(\frac{\tau_n^+}{\tau_n^0}, \frac{\tau_n^-}{\tau_n^0} \right), \quad v_n = 1 - x_n y_n \, ,$$

compute, using the Virasoro relations and the Toeplitz flow:

$$\begin{aligned}
0 &= \frac{x_n y_n}{v_n} \left\{ \frac{1}{\tau_n^+} (a\mathbb{V}_{-1,n}^+ + b\mathbb{V}_{0,n}^+ + c\mathbb{V}_{1,n}^+)\tau_n^+ + \frac{1}{\tau_n^-} (a\mathbb{V}_{-1,n}^- + b\mathbb{V}_{0,n}^- + c\mathbb{V}_{1,n}^-)\tau_n^- \right. \\
&\qquad \left. - \frac{2}{\tau_n} (a\mathbb{V}_{-1,n} + b\mathbb{V}_{0,n} + c\mathbb{V}_{1,n})\tau_n \right\} \\
&= \frac{x_n y_n}{v_n} \left(\sum_{i\geq 1} \left(a_i(t) \frac{\partial}{\partial t_i} - \beta_i(s) \frac{\partial}{\partial s_i} \right) \log x_n y_n \right. \\
&\qquad \left. + \left(c\frac{\partial}{\partial t_1} - a\frac{\partial}{\partial s_1} \right) \log \frac{x_n}{y_n} \right) \\
&= \frac{x_n y_n}{v_n} \left(\sum_{i\geq 1} \left(a_i \frac{\partial}{\partial t_i} - \beta_i \frac{\partial}{\partial s_i} \right) (\log x_n + \log y_n) \right. \\
&\qquad \left. + \left(c\frac{\partial}{\partial t_1} - a\frac{\partial}{\partial s_1} \right) (\log x_n - \log y_n) \right) \\
&= (\mathcal{L}_1^{(n)} - \mathcal{L}_2^{(n)} + aL_2 - cL_1)_{nn} \\
&\qquad - (\mathcal{L}_1^{(n)} - \mathcal{L}_2^{(n)} - aL_2 + cL_1)_{n+1,n+1} \, , \\
&= -\partial_n (\mathcal{L}_1^{(n)} - \mathcal{L}_2^{(n)})_{n,n} + (aL_2 - cL_1)_{nn} \, .
\end{aligned} \tag{2.128}$$

We then find our result by specializing the above identity to the precise locus $\mathcal{L}$ in t, s space corresponding to the measure $\rho(z)$ of Case 2 of Theorem 2.2.1:

$$\mathcal{L} := \left\{ \begin{array}{l} it_i = it_i^{(0)} := \begin{cases} u_i - (\gamma_1' d_1^i + \gamma_2' d_2^i), & \text{for } 1 \leq i \leq N_1 \\ -(\gamma_1' d_1^i + \gamma_2' d_2^i), & \text{for } N_1 + 1 \leq i < \infty \end{cases} \\ \\ is_i = is_i^{(0)} := \begin{cases} -u_{-i} + (\gamma_1'' d_1^{-i} + \gamma_2'' d_2^{-i}), & \text{for } 1 \leq i \leq N_2 \\ (\gamma_1'' d_1^{-i} + \gamma_2'' d_2^{-i}), & \text{for } N_2 + 1 \leq i < \infty \end{cases} \end{array} \right\} \tag{2.129}$$

with

$$(a, b, c) = \frac{1}{\sqrt{d_1 d_2}} (1, -d_1 - d_2, d_1 d_2) \, , \tag{2.130}$$

and so in (2.129) all $\alpha_i(t) = 0$ for all $i \geq N_1 + 2$ and all $\beta_i(t) = 0$ for $i \geq N_2 + 2$.

2.2.5 Singularity Confinement of Recursion Relations

Since for the combinatorial examples the unitary integral $I_n^{(0)}(t)$ satisfies Painlevé differential equations in t, it is natural to expect they satisfy a discrete version of the Painlevé property regarding the development of poles. For instance, algebraically integrable systems (a.c.i.) $\dot{z} = f(z)$, $z \in \mathbb{C}^n$, admit Laurent solutions depending on the maximal number, $n-1$, of free parameters (see [14]). An analogous property for rational recursion relations

$$z_n = F(z_{n-1}, \dots, z_{n-\delta}), \quad z_n \in \mathbb{C}^k ,$$

would be that there exists solutions of the recursion relation $\{z_i(\lambda)\}$ which are "formal Laurent" solutions in λ developing a pole which disappears after a finite number of steps, and such that these "formal Laurent" solutions depend on the maximal number of free parameters $\delta \times k$ (counting λ) and moreover the coefficients of the expansions depend rationally on these free parameters. We shall give results for Case 2 of Theorem 2.2.1 and 2.2.2, where $N_1 = N_2 = N$. The results in this section, Theorem 2.2.3–2.2.6, are due to Adler–van Moerbeke–Vanhaecke and can be found with proofs in [13].

Self-dual case:

$$\rho(z) = \mathrm{e}^{\sum_1^N u_i(z^i+z^{-i})/i} . \tag{2.131}$$

Theorem 2.2.3 (Singularity confinement: self-dual case). *For any $n \in \mathbb{Z}$,*[8] *the difference equations $\Gamma_k(x) = 0$, $(k \in \mathbb{Z})$ admit two formal Laurent solution $x = \left(x_k(\lambda)\right)_{k\in\mathbb{Z}}$ in a parameter λ, having a (simple) pole at $k = n$ only and $\lambda = 0$. These solutions depend on $2N$ non-zero free parameters*

$$\alpha = (a_{n-2N}, \dots, a_{n-2}) \quad \textit{and} \quad \lambda .$$

Explicitly, these series with coefficients rational in α are given by ($\varepsilon = \pm 1$):

$$\begin{aligned}
x_k(\lambda) &= \sum_{i=0}^{\infty} x_k^{(i)}(\alpha)\lambda^i , && k < n - 2N , \\
x_k(\lambda) &= \alpha_k , && n - 2N \le k \le n - 2 , \\
x_{n-1}(\lambda) &= \varepsilon + \lambda , \\
x_n(\lambda) &= \frac{1}{\lambda}\sum_{i=0}^{\infty} x_n^{(i)}(\alpha)\lambda^i , \\
x_{n+1}(\lambda) &= -\varepsilon + \sum_{i=1}^{\infty} x_{n+1}^{(i)}(\alpha)\lambda^i , \\
x_k(\lambda) &= \sum_{i=0}^{\infty} x_k^{(i)}(\alpha)\lambda^i , && n + 1 < k .
\end{aligned}$$

[8] We consider the obvious *bi*-infinite extension of $L_i(x, y)$ (2.89) which through (2.98) defines a *bi*-infinite extension of $\Gamma_k(x, y)$, $\widetilde{\Gamma}_k(x, y)$.

General case:

$$\rho(z) = \exp\left(\sum_1^N \frac{(u_i z^i + u_{-i} z^{-i})}{i}\right) .$$

Theorem 2.2.4 (Singularity confinement: general case). *For any $n \in \mathbb{Z}$, the difference equations $\Gamma_k(x,y) = \tilde{\Gamma}_k(x,y) = 0$, $(k \in \mathbb{Z})$ admit a formal Laurent solution $x = \big(x_k(\lambda)\big)_{k\in\mathbb{Z}}$ and $y = \big(y_k(\lambda)\big)_{k\in\mathbb{Z}}$ in a parameter λ, having a (simple) pole at $k = n$ and $\lambda = 0$, and no other singularities. These solutions depend on $4N$ non-zero free parameters*

$$\alpha_{n-2N}, \ldots, \alpha_{n-2}, \alpha_{n-1}, \beta_{n-2N}, \ldots, \beta_{n-2} \quad \textit{and} \quad \lambda .$$

Setting $z_n := (x_n, y_n)$ and $\gamma_i := (\alpha_i, \beta_i)$, and $\gamma := (\gamma_{n-2N}, \ldots, \gamma_{n-2}, \alpha_{n-1})$, the explicit series with coefficients rational in γ read as follows:

$$\begin{aligned}
z_k(\lambda) &= \sum_{i=0}^{\infty} z_k^{(i)}(\gamma)\lambda^i , && k < n - 2N , \\
z_k(\lambda) &= \gamma_k , && n - 2N \le k \le n-2 , \\
x_{n-1}(\lambda) &= \alpha_{n-1} , && \\
y_{n-1}(\lambda) &= \frac{1}{\alpha_{n-1}} + \lambda , && \\
z_n(\lambda) &= \frac{1}{\lambda}\sum_{i=0}^{\infty} z_n^{(i)}(\gamma)\lambda^i , && \\
z_k(\lambda,\gamma) &= \sum_{i=0}^{\infty} z_k^{(i)}(\gamma)\lambda^i , && n < k .
\end{aligned}$$

Singularity confinement is consequence of:

(1) Painlevé property of a.c.i. Toeplitz lattice.
(2) Rational difference relations as a whole define an invariant manifold of the Toeplitz lattice.
(3) Formal Laurent solutions of Toeplitz lattice with maximal parameters restrict to the above invariant manifold, restricting the parameters.
(4) Reparametrizing the "restricted" Laurent solutions by $t \to \lambda$ and "restricted parameters" $\to \gamma$ yields the confinement result.

We discuss (1) and (2). Indeed, consider the Toeplitz lattice with the Hamiltonian $H = H_1^{(1)} - H_1^{(2)}$, yielding the flow of (2.114):

General case:

$$\begin{aligned}
\frac{\partial x_k}{\partial t} &= (1 - x_k y_k)(x_{k+1} - x_{k-1}) , \\
\frac{\partial y_k}{\partial t} &= (1 - x_k y_k)(y_{k+1} - x_{k-1}) ,
\end{aligned} \qquad k \in \mathbb{Z} .$$

Self-dual case:

$$\frac{\partial x_k}{\partial t} = (1 - x_k^2)(x_{k+1} - x_{k-1}) , \quad k \in \mathbb{Z} .$$

Then we have

Maximal formal Laurent solutions:

Theorem 2.2.5. *For arbitrary but fixed n, the first Toeplitz lattice vector field* (2.114) *admits the following formal Laurent solutions,*

$$\begin{aligned} x_n(t) &= \frac{1}{(a_{n-1} - a_{n+1})t}\big(a_{n-1}a_{n+1}(1 + at) + O(t^2)\big) , \\ y_n(t) &= \frac{1}{(a_{n-1} - a_{n+1})t}\left(-1 + \left(a + \frac{a_{n+1}a_+ - a_{n-1}a_-}{a_{n+1} - a_{n-1}}\right)t + O(t^2)\right) , \\ x_{n\pm 1}(t) &= a_{n\pm 1} + a_{n\pm 1}a_\pm t + O(t^2) , \\ y_{n\pm 1}(t) &= \frac{1}{a_{n\pm 1}} - \frac{a_\pm t}{a_{n\mp 1}} + O(t^2) \end{aligned}$$

whereas for all remaining k such that $|k - n| \geq 2$,

$$\begin{aligned} x_k(t) &= a_k + (1 - a_k b_k)(a_{k+1} - a_{k-1})t + O(t^2) , \\ y_k(t) &= b_k + (1 - a_k b_k)(b_{k+1} - b_{k-1})t + O(t^2) , \end{aligned}$$

where a, $a_\pm$, $a_{n\pm 1}$ *and all* a_i, b_i, *with* $i \in \mathbb{Z}\backslash\{n-1, n, n+1\}$ *and with* $b_{n\pm 1} = 1/a_{n\pm 1}$, *are arbitrary free parameters, and with* $(a_{n-1} - a_{n+1})a_{n-1}a_{n+1} \neq 0$. *In the self-dual case it admits the following two formal Laurent solutions, parametrized by* $\varepsilon = \pm 1$,

$$\begin{aligned} x_n(t) &= -\frac{\varepsilon}{2t}\big(1 + (a_+ - a_-)t + O(t^2)\big) , \\ x_{n\pm 1}(t) &= \varepsilon\big(\mp 1 + 4a_\pm t + O(t^2)\big) , \\ x_k(t) &= \varepsilon\big(a_k + (1 - a_k^2)(a_{k+1} - a_{k-1})t + O(t^2)\big) , \quad |k - n| \geq 2 , \end{aligned}$$

where a_+, a_- *and all* a_i, *with* $i \in \mathbb{Z}\backslash\{n-1, n, n+1\}$ *are arbitrary free parameter and* $a_{n-1} = -a_{n+1} = 1$.

Together with time t these parameters are in bijection with the phase space variables; we can put for the general Toeplitz lattice for example $z_k \leftrightarrow (a_k, b_k)$ for $|k - n| \geq 1$ and $x_{n\pm 1} \leftrightarrow a_{n\pm 1}$ and $y_{n\pm 1}$, x_n, $y_n \leftrightarrow a_\pm$, a, t. Consider the locus $\widehat{\mathcal{L}}$ defined by the difference relations (2.98) of Case 3 of Theorem 2.2.1, namely:

General case:

$$\widehat{\mathcal{L}} = \{(x, y) \mid \Gamma_n(x, y, u) = 0 , \widetilde{\Gamma}_n(x, y, u) = 0 , \forall n\} .$$

Self-dual case:

$$\widehat{\mathcal{L}} = \{x \mid \Gamma_n(x,y,u) = 0,\ \forall n\} ,$$

where we now explicitly exhibit the dependence of Γ_n, $\widetilde{\Gamma}_n$ on the coefficients of the measure, namely u. The point of the following theorem is that $\widehat{\mathcal{L}}$ is an invariant manifold for the flow generated by $H = H_1^{(1)} - H_1^{(2)}$, upon our imposing the following u dependence on t ($v_n = 1 - x_n y_n$).

Theorem 2.2.6. *Upon setting $\partial u_{\pm i}/\partial t = \delta_{1i}$, the recursion relations satisfy the following differential equations*

$$\dot{\Gamma}_k = v_k(\Gamma_{k+1} - \Gamma_{k-1}) + (x_{k+1} - x_{k-1})(x_k \widetilde{\Gamma}_k - y_k \Gamma_k) ,$$

$$\dot{\widetilde{\Gamma}}_k = v_k(\widetilde{\Gamma}_{k+1} - \widetilde{\Gamma}_{k-1}) - (y_{k+1} - y_{k-1})(x_k \widetilde{\Gamma}_k - y_k \Gamma_k) ,$$

which specialize in the self-dual case to

$$\dot{\Gamma}_k = v_k(\Gamma_{k+1} - \Gamma_{k-1}) .$$

Sketch of proof: The proof is based on the crucial identities:

$$\Gamma_n = \mathcal{V}_0 x_n + n x_n \quad \text{and} \quad \widetilde{\Gamma}_n = -\mathcal{V}_0 y_n + n y_n ,$$

where

$$\mathcal{V}_0 := \sum_{i \geq 1} \left(u_i \frac{\partial}{\partial t_i} + u_{-i} \frac{\partial}{\partial s_i} \right) ,$$

and

$$\frac{\partial x_n}{\partial \begin{Bmatrix} t_1 \\ s_1 \end{Bmatrix}} = v_n x_{n \pm 1} , \qquad \frac{\partial y_n}{\partial \begin{Bmatrix} t_1 \\ s_1 \end{Bmatrix}} = -v_n y_{\mp 1} , \qquad \frac{\partial u_k}{\partial \begin{Bmatrix} t_1 \\ s_1 \end{Bmatrix}} = \pm \delta_{k, \pm 1} ,$$

which implies

$$\frac{\partial \Gamma_n}{\partial \begin{Bmatrix} t_1 \\ s_1 \end{Bmatrix}} = v_n \Gamma_{n \pm 1} + x_{n \pm 1}(x_n \widetilde{\Gamma}_n - y_n \Gamma_n) ,$$

$$\frac{\partial \widetilde{\Gamma}_n}{\partial \begin{Bmatrix} t_1 \\ s_1 \end{Bmatrix}} = -v_n \widetilde{\Gamma}_{n \mp 1} + y_{n \mp 1}(x_n \widetilde{\Gamma}_n - y_n \Gamma_n) ,$$

which, upon using $\partial/\partial t = \partial/\partial t_1 - \partial/\partial s_1$, yields the theorem.

2.3 Coupled Random Matrices and the 2-Toda Lattice

2.3.1 Main Results for Coupled Random Matrices

The study of coupled random matrices will lead us to the 2-Toda lattice and bi-orthogonal polynomials, which are essentially 2 of the 4 wave functions for the 2-Toda lattice. This problem will lead to many techniques which will come up again, as well as a PDE for the basic probability in coupled random matrices.

Let $M_1, M_2 \in \mathcal{H}_n$, Hermitian $n \times n$ matrices and consider the probability ensemble of

Coupled random matrices:

$$P\big((M_1, M_2) \subset S\big) = \frac{\int_S \mathrm{d}M_1\,\mathrm{d}M_2 \exp\big(-\frac{1}{2}\operatorname{Tr}(M_1^2 + M_2^2 - 2cM_1M_2)\big)}{\int_{\mathcal{H}_n \times \mathcal{H}_n} \mathrm{d}M_1\,\mathrm{d}M_2 \exp\big(-\frac{1}{2}\operatorname{Tr}(M_1^2 + M_2^2 - 2cM_1M_2)\big)} \tag{2.132}$$

with

$$\mathrm{d}M_1 = \Delta_n^2(x)\prod_1^n \mathrm{d}x_i\,\mathrm{d}U_1, \quad = \mathrm{d}M_2 = \Delta_n^2(y)\prod_1^n \mathrm{d}y_i\,\mathrm{d}U_2\,.$$

Given $E = E_1 \times E_2 = \bigcup_1^r [a_{2i-1}, a_{2i}] \times \bigcup_1^s [b_{2i-1}, b_{2i}]$, define the boundary operators:

$$\begin{gathered}\mathcal{A}_1 = \frac{1}{1-c^2}\left(\sum_1^{2r}\frac{\partial}{\partial a_j} + c\sum_1^{2s}\frac{\partial}{\partial b_j}\right), \quad \mathcal{A}_2 = \sum_1^{2r} a_j\frac{\partial}{\partial a_j} - c\frac{\partial}{\partial c}\,, \\ \mathcal{B}_1 = \mathcal{A}_1\big|_{a\leftrightarrow b}\,, \quad \mathcal{B}_2 = \mathcal{A}_2\big|_{a\leftrightarrow b}\,,\end{gathered} \tag{2.133}$$

which form a Lie algebra:

$$\begin{aligned}&[\mathcal{A}_1, \mathcal{B}_1] = 0\,, \quad [\mathcal{A}_1, \mathcal{A}_2] = \frac{1+c^2}{1-c^2}\mathcal{A}_1\,, \quad [\mathcal{A}_2, \mathcal{B}_1] = -\frac{2c}{1-c^2}\mathcal{A}_1\,, \\ &[\mathcal{A}_2, \mathcal{B}_2] = 0\,, \quad [\mathcal{B}_1, \mathcal{B}_2] = \frac{1+c^2}{1-c^2}\mathcal{B}_1\,, \quad [\mathcal{B}_2, \mathcal{A}_1] = -\frac{2c}{1-c^2}\mathcal{B}_1\,,\end{aligned} \tag{2.134}$$

We can now state the main theorem of Sect. 2.3:

Theorem 2.3.1 (Adler–van Moerbeke [4]). *The statistics*

$$\begin{aligned}F_n &:= \frac{1}{n}\log P_n(E) \\ &:= \frac{1}{n}\log P(\textit{all } (M_1\textit{-eigenvalues}) \in E_1,\ \textit{all } (M_2\textit{-eigenvalues}) \in E_2)\end{aligned} \tag{2.135}$$

satisfies the third order nonlinear PDE:

$$\mathcal{A}_1\left(\frac{\mathcal{B}_2\mathcal{A}_1F_n}{\mathcal{B}_1\mathcal{A}_1F_n+c/(1-c^2)}\right)=\mathcal{B}_1\left(\frac{\mathcal{A}_2\mathcal{B}_1F_n}{\mathcal{A}_1\mathcal{B}_1F_n+c/(1-c^2)}\right)=0\,. \tag{2.136}$$

In particular for $E=(-\infty,a]\times(-\infty,b]$, *setting:* $x:=-a+cb$, $y:=-ac+b$, $\mathcal{A}_1\to-\partial/\partial x$, $\mathcal{B}_1\to\partial/\partial y$, (2.136) *becomes*

$$\frac{\partial}{\partial x}\left(\frac{(c^2-1)^2\partial^2F_n/\partial x\partial c+2cx-(1+c^2)y}{(c^2-1)\partial^2F_n/\partial x\partial y+c}\right)$$
$$=\frac{\partial}{\partial y}\left(\frac{(c^2-1)^2\partial^2F_n/\partial y\partial c+2cy-(1+c^2)x}{(c^2-1)\partial^2F_n/\partial y\partial x+c}\right)\,. \tag{2.137}$$

2.3.2 Link with the 2-Toda Hierarchy

In this section we deform the coupled random matrix problem in a natural way to introduce the 2-Toda hierarchy into the problem. We first need the celebrated Harish-Chandra–Itzkson–Zuber formula [23]:

HCIZ: $x=\operatorname{diag}(x_1,\dots,x_n)$, $y=\operatorname{diag}(y_1,\dots,y_n)$

$$\int_{U(n)}\mathrm{d}U\,\exp(c\operatorname{Tr}xUyU^{-1})=\left(\frac{2\pi}{c}\right)^{n(n-1)/2}\frac{\det\big(\exp(cx_iy_j)\big)_{i,j=1}^{n}}{n!\Delta_n(x)\Delta_n(y)}\,. \tag{2.138}$$

Compute, using HCIZ:

$$\iint_{\mathcal{H}_n(E_1)\times\mathcal{H}_n(E_2)} \mathrm{d}M_1\,\mathrm{d}M_2 \exp(c\operatorname{Tr} M_1M_2)\exp\left(\operatorname{Tr}\left(\sum_1^\infty (t_iM_1^i - s_iM_2^i)\right)\right)$$

$(t,\ s \text{ deformation})$

$$= \int_{E_1^nE_2^n}\left\{\Delta_n^2(x)\Delta_n^2(y)\prod_{k=1}^n \exp\left(\sum_1^\infty (t_ix_k^i - s_iy_k^i)\right)\mathrm{d}x_k\,\mathrm{d}y_k\right\}$$
$$\times\iint_{U(n)\times U(n)} \exp(c\operatorname{Tr} U_1 \times U_1^{-1}U_2yU_2^{-1})\,\mathrm{d}U_1\,\mathrm{d}U_2$$
$$= \int_{E_1^nE_2^n}\left\{\Delta_n^2(x)\Delta_n^2(y)\prod_{k=1}^n \exp\left(\sum_1^\infty (t_ix_k^i - s_iy_k^i)\right)\mathrm{d}x_k\,\mathrm{d}y_k\right\}$$
$$\times\int_{U(n)}\mathrm{d}U_1\int_{U(n)} \exp(c\operatorname{Tr} \times U_1^{-1}U_2yU_2^{-1}U_1)\,\mathrm{d}U_2$$
$$= \int_{E_1^n}\int_{E_2^n}\left\{\Delta_n^2(x)\Delta_n^2(y)\prod_{k=1}^n \exp\left(\sum_1^\infty (t_ix_k^i - s_iy_k^i)\right)\mathrm{d}x_k\,\mathrm{d}y_k\right\}$$
$$\times\int_{U(n)}\mathrm{d}U_1\int_{U(n)} \exp(c\operatorname{Tr} \times UyU^{-1})\mathrm{d}U \quad \begin{pmatrix} U = U_1^{-1}U_2 \\ \mathrm{d}U = \mathrm{d}U_2\end{pmatrix}$$
$$= c_n\int_{E_1^n}\int_{E_2^n}\Delta_n(x)\Delta_n(y)\det\bigl(\exp(cx_iy_j)\bigr)_{i,j=1}^n$$
$$\times\prod_1^n \exp\left(\sum_1^\infty (t_ix_k^i - s_iy_k^i)\right)\mathrm{d}x_k\,\mathrm{d}y_k$$
$$:= c_n\int_{E_1^n}\int_{E_2^n}\Delta_n(x)\Delta_n(y)\left(\sum_{\sigma\in S_n}(-1)^\sigma\prod_1^n \exp(cx_iy_{\sigma(i)})\right)\prod_1^n \mathrm{d}\mu(x_k)\,\mathrm{d}\psi(y_k)$$

$(\text{setting: } y_{\sigma(i)} \to y_i,\ E_2^n \to E_2^n)$

$$= n!c_n\int_{E_1^n}\int_{E_2^n}\Delta_n(x)\Delta_n(y)\prod_1^n \exp(cx_iy_i)\prod_1^n \mathrm{d}\mu(x_k)\,\mathrm{d}\psi(y_k)$$
$$= n!c_n\int_{E^n}\Delta_n(x)\Delta_n(y)\prod_{k=1}^n \exp\left(\sum_1^\infty (t_ix_k^i - s_iy_k^i) + cx_ky_k\right)\mathrm{d}x_k\,\mathrm{d}y_k\ ,$$
$$E = E_1 \times E_2\ ,$$

where we have used in the above that Haar measure is translation invariant.

We now make a further c-deformation of this matrix integral.

Define τ*-function*:

$$\tau_n(t,s,C,E)$$
$$:= \frac{1}{n!}\int_{E^n}\Delta_n(x)\Delta_n(y)\prod_{k=1}^n \exp\left(\sum_1^\infty (t_ix_k^i - s_iy_k^i) + \sum_{i,j\geq 1} c_{ij}x_k^iy_k^j\right)\mathrm{d}x_k\,\mathrm{d}y_k\ ,$$

$(\tau_0 = 1)$ which is *not* a matrix integral! (2.139)

Thus

$$\frac{\tau_n(t,s,C,E)}{\tau_n(t,s,C,\mathbb{R})} \text{ is } t,s,C \text{ deformation of } P_n(E)\,.$$

It is quite crucial to recast the τ-function using the de Bruijn trick:

Moment matrix form of τ-function:

$$\begin{aligned}
&\tau_n(t,s,C,E)\\
&\quad = \frac{1}{n!}\int_{E^n} \Delta_n(x)\Delta_n(y)\prod_{k=1}^{n}\exp\Bigl(\sum_{1}^{\infty}(t_i x_k^i - s_i y_k^i) + \sum_{i,j\geq 1} c_{ij}x_k^i y_k^i\Bigr)\mathrm{d}x_k\,\mathrm{d}y_k\\
&\quad = \frac{1}{n!}\int_{E^n} \det\bigl(f_i(x_j)\bigr)_{i,j=1}^{n}\det\bigl(g_i(y_j)\bigr)_{i,j=1}^{n}\prod_{1}^{n}\mathrm{d}\psi(x_k,y_k)\\
&\qquad\qquad\qquad\qquad\qquad\qquad\qquad\qquad (f_i(x)=g_i(x)=x^{i-1})\\
&\quad = \frac{1}{n!}\sum_{\sigma,\mu\in S_n}(-1)^{\sigma+\mu}\int_{E^n}\prod_{1}^{n} f_{\sigma(i)}(x_i)g_i(y_{\mu(i)})\mathrm{d}\psi(x_{\mu(i)},y_{\mu(i)})\\
&\quad = \frac{1}{n!}\sum_{\sigma,\mu\in S_n}(-1)^{\sigma+\mu}\int_{E^n}\prod_{1}^{n} f_{\sigma\circ\mu(i)}(x_{\mu(i)})g_i(y_{\mu(i)})d\psi(x_{\mu(i)},y_{\mu(i)})\\
&\quad = \frac{1}{n!}\sum_{\mu\in S_n}\sum_{\sigma'\in S_n}(-1)^{\sigma'}\prod_{1}^{n}\int_E f_{\sigma'(i)}(x)g_i(y)\mathrm{d}\psi(x,y)\\
&\quad = \frac{1}{n!}\sum_{\mu\in S_n}\det\Bigl(\int_E f_i(x)g_j(y)\mathrm{d}\psi(x,y)\Bigr)_{i,j=1}^{n}\\
&\quad = \det(\mu_{ij})_{i,j=0}^{n-1}\,,
\end{aligned}\tag{2.140}$$

with

$$\begin{aligned}
\mu_{ij}(t,s,C,E) &= \int_E x^i y^j \exp\Bigl(\sum_{k=1}^{\infty}(t_k x^k - s_k y_k^k) + \sum_{\alpha,\beta\geq 1} c_{\alpha\beta}x^\alpha y^\beta\Bigr)\mathrm{d}x\,\mathrm{d}y\\
&:= \langle x^i, y^j\rangle\,.
\end{aligned}\tag{2.141}$$

Thus we have shown:

$$\tau_n(t,s,C,E) = \det m_n\,,\quad m_n = (\mu_{ij})_{i,j=0}^{n-1}\,.\tag{2.142}$$

This immediately leads to the

2-Toda differential equations – Moment form:

$$\frac{\partial\mu_{ij}}{\partial t_k} = \mu_{i+k,j}\,,\quad \frac{\partial\mu_{ij}}{\partial s_k} = -\mu_{i,j+k}\,,\tag{2.143}$$

which we reformulate in terms of the moment matrix m_∞

$$\frac{\partial m_\infty}{\partial t_k} = \Lambda^k m_\infty\,,\quad \frac{\partial m_\infty}{\partial s_k} = -m_\infty(\Lambda^T)^k\,,\tag{2.144}$$

or equivalently

$$m_\infty(t,s) = \exp\Bigl(\sum_1^\infty t_k \Lambda^k\Bigr) m_\infty(0,0) \exp\Bigl(-\sum_1^\infty s_k (\Lambda^T)^k\Bigr) , \tag{2.145}$$

with the semi-infinite shift matrix

$$\Lambda := \begin{pmatrix} 0\,1\,0\,0\dots \\ 0\,0\,1\,0\dots \\ 0\,0\,0\,1\dots \\ 0\,0\,0\,0\dots \\ \vdots\,\vdots\,\vdots\,\vdots\;\vdots \end{pmatrix} . \tag{2.146}$$

Thus

$$\tau_n(t,s,C,E) = \det m_n(t,s) = \det\bigl(E_n(t) m_\infty(0,0) E_n^T(-s)\bigr) , \tag{2.147}$$

with

$$E_n(t) = (I + s_1(t)\Lambda + s_2(t)\Lambda^2 + \cdots)_{n\times\infty}$$

and

$$\exp\Bigl(\sum_1^\infty t_i z^i\Bigr) := \sum_0^\infty s_i(t) z^i , \tag{2.148}$$

with $s_i(t)$ the elementary Schur polynomials.

2.3.3 *L–U* Decomposition of the Moment Matrix, Bi-orthogonal Polynomials and 2-Toda Wave Operators

The L–U decomposition of m_∞ is equivalent to bi-orthogonal polynomials; indeed, consider the L–U decomposition of m_∞, as follows (see [9]):

$$m_\infty = LhU := S^{-1}(m_\infty)\bigl(h(m_\infty)\bigl(S^{-1}(m_\infty^T)\bigr)^T\bigr) := S_1^{-1} S_2 \tag{2.149}$$

where we define the string orthogonal polynomials

$$\bigl(p_n^{(1)}(y)\bigr)_{n\ge 0} := S(m_\infty)\begin{pmatrix} 1 \\ y \\ y^2 \\ \vdots \end{pmatrix} := \left(\frac{\det \left(\begin{array}{ccc|c} & m_n & & \begin{matrix} 1 \\ y \\ \vdots \end{matrix} \\ \hline \mu_{n0} & \cdots & \mu_{n,n-1} & y^n \end{array} \right)}{\det m_n} \right)_{n\ge 0} \tag{2.150}$$

and

$$\bigl(p_n^{(2)}(y)\bigr)_{n\ge 0} := S(m_\infty^T)\begin{pmatrix} 1 \\ y \\ y^2 \\ \vdots \end{pmatrix} . \tag{2.151}$$

Setting

$$\langle\ ,\ \rangle\colon \langle x^i, y^j\rangle := \mu_{ij}(t,s,C)\ , \tag{2.152}$$

conclude (as a tautology) the defining relations of the monic bi-orthogonal polynomials, namely

$$\langle p_i^{(1)}, p_j^{(2)}\rangle = h_i\delta_{ij} \Longleftrightarrow S(m_\infty)m_\infty\big(S(m_\infty^T)\big)^T = h(m_\infty)\ , \tag{2.153}$$

with the first identity a consequence of (2.150) and (2.151), which also implies

$$h(m_\infty) := \operatorname{diag}\left(h_0, h_1, \ldots, h_i = \frac{\det m_{i+1}}{\det m_i}, \ldots\right), \tag{2.154}$$

and by (2.149)

$$S_1 = S(m_\infty)\ , \quad S_2 = h(m_\infty)\big(S^{-1}(m_\infty^T)\big)^T\ . \tag{2.155}$$

We now define the 2-Toda operators:

$$L_1 = S_1 \Lambda S_1^{-1}\ , \quad L_2 = S_2\Lambda^T S_2^{-1}\ . \tag{2.156}$$

Since $m_\infty = S_1^{-1}S_2$, with

$$S_1 \in I + g_-\ , \quad S_2 \in g_+\ ,$$

then

$$\dot S_1 S_1^{-1} \in g_-\ , \quad \dot S_2 S_2^{-1} \in g_+\ ,$$

with g_- strictly lower triangular matrices and g_+ upper triangular matrices, including the diagonal, and $g_- + g_+ = g :=$ all semi-infinite matrices. Compute

$$S_1 \dot m_\infty S_2^{-1} = S_1(S_1^{-1}S_2)^{\cdot}\, S_2^{-1} = -\dot S_1 S_1^{-1} + \dot S_2 S_2^{-1} \in g_- + g_+\ , \tag{2.157}$$

where we have used $(S_1^{-1})^{\cdot} = -S_1^{-1}\dot S_1 S_1^{-1}$. On the other hand, one computes, using (2.144), for $\partial/\partial t_n$ or $\partial/\partial s_n$ separately, that:

$$\begin{aligned} S_1\frac{\partial m_\infty}{\partial t_n}S_2^{-1} &= S_1\Lambda^n m_\infty S_2^{-1} = S_1\Lambda^n S_1^{-1}S_2S_2^{-1} = S_1\Lambda^n S_1^{-1} \\ &= L_1^n := (L_1^n)_- + (L_1^n)_+ \in g_- + g_+\ , \end{aligned} \tag{2.158}$$

and

$$\begin{aligned} S_1\frac{\partial m_\infty}{\partial s_n}S_2^{-1} &= -S_1 m_\infty(\Lambda^T)^n S_2^{-1} = -S_1S_1^{-1}S_2(\Lambda^T)^n S_2^{-1} \\ &= -S_2(\Lambda^T)^n S_2^{-1} \\ &= -L_2^n := -(L_2^n)_- - (L_2^n)_+ \in g_- + g_+\ , \end{aligned} \tag{2.159}$$

and so (2.157), (2.158), and (2.159) yield the differential equations

$$\begin{aligned} \frac{\partial S_1}{\partial t_n}S_1^{-1} &= -(L_1^n)_-\ , & \frac{\partial S_2}{\partial t_n}S_2^{-1} &= \ \ (L_1^n)_+\ , \\ \frac{\partial S_1}{\partial s_n}S_1^{-1} &= \ \ (L_2^n)_-\ , & \frac{\partial S_2}{\partial s_n}S_2^{-1} &= -(L_2^n)_-\ . \end{aligned} \tag{2.160}$$

Setting $\chi(x) = (1, z, z^2, \ldots)^T$ we now connect the bi-orthogonal polynomials with the 2-Toda wave functions:

2-*Toda wave functions*:

$$\begin{cases} \Psi_1(z) := \exp\Big(\sum_1^\infty t_k z^k\Big) p^{(1)}(z) = \exp\Big(\sum_1^\infty t_k z^k\Big) S_1 \chi(z) , \\ \Psi_2^*(z) := \exp\Big(-\sum_1^\infty s_k z^{-k}\Big) h^{-1} p^{(2)}(z^{-1}) \\ \qquad = \exp\Big(-\sum_1^\infty s_k z^{-k}\Big) (S_2^{-1})^T \chi(z^{-1}) . \end{cases}$$

Eigenfunction identities:

$$L_1 \Psi_1(z) = z \Psi_1(z), \quad L_2^T \Psi_2^*(z) = z^{-1} \Psi_2^*(z) .$$

Formulas (2.160) and (2.156) yield,

$t - s$ flows for L_i and Ψ:

$$\begin{cases} \dfrac{\partial L_i}{\partial t_n} = [(L_1^n)_+, L_i] , & \dfrac{\partial L_i}{\partial s_n} = [(L_2^n)_-, L_i] , \qquad i = 1,2,\ n = 1,2,\dots \\ \dfrac{\partial \Psi_1}{\partial t_n} = (L_1^n)_+ \Psi_1 , & \dfrac{\partial \Psi_1}{\partial s_n} = (L_2^n)_- \Psi_1 , \\ \dfrac{\partial \Psi_2^*}{\partial t_n} = -(L_1^n)_+^T \Psi_1^* , & \dfrac{\partial \Psi_2^*}{\partial s_n} = -(L_2^n)_-^T \Psi_2^* . \end{cases}$$

Wave operators:

$$W_1 := S_1 \exp\Big(\sum_1^\infty t_k \Lambda^k\Big) , \quad W_2 := S_2 \exp\Big(\sum_1^\infty s_k (\Lambda^T)^k\Big) , \tag{2.161}$$

satisfy

$$W_1(t,s) W_1(t',s')^{-1} = W_2(t,s) W_2(t',s')^{-1} , \quad \forall t,s,t',s' . \tag{2.162}$$

All the data in 2-Toda is parametrized by τ-functions, to wit:

L_1, L_2, Ψ_1, Ψ_2^* parametrized by τ-functions:

$$\begin{cases} \Psi_1(z,t,s) = \left(\dfrac{\tau_n(t-[z^{-1}],s)}{\tau_n(t,s)} \exp\Big(\sum_1^\infty t_i z^i\Big) z^n \right)_{n\ge 0} \\ \Psi_2^*(z,t,s) = \left(\dfrac{\tau_n(t,s+[z])}{\tau_{n+1}(t,s)} \exp\Big(-\sum_1^\infty s_i z^{-i}\Big) z^{-n} \right)_{n\ge 0} , \\ \qquad [x] = (x, x^2/2, \dots) , \end{cases} \tag{2.163}$$

$$\begin{cases} L_1^k = \displaystyle\sum_{l=0}^\infty \operatorname{diag}\left(\frac{s_l(\tilde\partial_t)\tau_{n+k-l+1} \circ \tau_n}{\tau_{n+k-l+1}\tau_n} \right)_{n\ge 0} \Lambda^{k-l} , \\ (h(L_2^T)^k h^{-1}) = \displaystyle\sum_{l=0}^\infty \operatorname{diag}\left(\frac{s_l(\tilde\partial_t)\tau_{n+k-l+1} \circ \tau_n}{\tau_{n+k-l+1}\tau_n} \right)\Bigg|_{\tilde\partial_t \to -\tilde\partial_s} , \end{cases} \tag{2.164}$$

with

$$s_l(t) \text{ the elementary Schur polynomials, } \tilde{\partial}_t = \left(\frac{\partial}{\partial t_1}, \frac{1}{2}\frac{\partial}{\partial t_2}, \frac{1}{3}\frac{\partial}{\partial t_3}, \dots\right),$$

and the

Hirota symbol:

$$p\left(\frac{\partial}{\partial t}\right) f \circ g := p\left(\frac{\partial}{\partial y}\right) f(t+y)g(t-y)\Big|_{y=0} \,. \tag{2.165}$$

2.3.4 Bilinear Identities and τ-function PDEs

Just like in KP theory (see Sect. 2.1.3), where the bilinear identity generates the KP hierarchy of PDEs for the τ-function, the same situation holds for the 2-Toda Lattice. In general 2-Toda theory (see [17]) the bilinear identity is a consequence of (2.163) and (2.162), but in the special case of 2-Toda being generated from bi-orthogonal polynomials, we can and will, at the end of this section, sketch a quick direct alternate proof of Adler–van Moerbeke–Vanhaecke [15] based on the bi-orthogonal polynomials, which has been vastly generalized. Since all we ever need of integrability in any problem is the PDE hierarchy, it is clearly of great practical use to have in general a quick proof of just the bilinear identities, but without all the usual integrable baggage. We now give the

2-*Toda bilinear identities*:

$$\oint_{z=\infty} \tau_n(t-[z^{-1}],s)\tau_{m+1}(t'+[z^{-1}],s') \exp\left(\sum_1^\infty (t_i - t_i')z^i\right) z^{n-m-1}\, dz$$
$$= \oint_{z=0} \tau_{n+1}(t, s-[z^{-1}])\tau_m(t', s'+[z^{-1}]) \exp\left(\sum_1^\infty (s_i - s_i')z^i\right) z^{m-n-1}\, dz\,,$$
$$\forall s,t,s',t',m,n\,. \tag{2.166}$$

The identities are a consequence of (2.162) and (2.163) and they yield, as in Sect. 2.1.3 (see Appendix) a generating function involving elementary Schur polynomials $s_j(\cdot)$ and arbitrary parameters a, b, in the following[9]

Hirota form:

[9] Hopefully there will be no confusion in this section between the elementary Schur polynomials, $s_j(\cdot)$, which are *functions*, and the time variables s_j, which are parameters, but the situation is not ideal.

$$
\begin{aligned}
0 &= -\sum_{j=0}^{\infty} s_{m-n+j}(-2a)s_j(\tilde{\partial}_t)\exp\left(\sum_1^{\infty}\left(a_k\frac{\partial}{\partial t_k}+b_k\frac{\partial}{\partial s_k}\right)\right)\tau_{m+1}\circ\tau_n \\
&\quad +\sum_{j=0}^{\infty} s_{-m+n+j}(-2b)s_j(\tilde{\partial}_s)\exp\left(\sum_1^{\infty}\left(a_k\frac{\partial}{\partial t_k}+b_k\frac{\partial}{\partial s_k}\right)\right)\tau_m\circ\tau_{n+1}
\end{aligned}
\tag{2.167}
$$

$$
= a_{j+1}\left(2s_j(\tilde{\partial}_t)\tau_{n+2}\circ\tau_n+\frac{\partial^2}{\partial s_1\partial t_{j+1}}\tau_{n+1}\circ\tau_{n+1}\right)+0(a_{j+1}^2)\,, \tag{2.168}
$$

upon setting $m = n+1$, and all b_k, $a_k = 0$, except a_{j+1}. Note:

$$
s_0(t)=1\,,\quad s_1(t)=t_1\,,\quad s_1(\tilde{\partial}_t)f\circ g = g\frac{\partial f}{\partial t_1}-f\frac{\partial g}{\partial t_1}\,,
$$

and $s_k(t) = t_k + \text{poly.}\,(t_1,\ldots,t_{k-1})$. This immediately yields the

2-*Toda* τ-*function identities*:

$$
-\frac{\partial^2}{\partial s_1\partial t_k}\log\tau_{n+1} = \frac{s_{k-1}(\tilde{\partial}_t)\tau_{n+2}\circ\tau_n}{\tau_{n+1}^2} \tag{2.169}
$$

$$
= \begin{cases} \dfrac{\tau_{n+2}\tau_n}{\tau_{n+1}^2}\,, & k=1\,,\\[2ex] \dfrac{\tau_{n+2}\tau_n}{\tau_{n+1}^2}\dfrac{\partial}{\partial t_1}\log\dfrac{\tau_{n+2}}{\tau_n}\,, & k=2\,, \end{cases} \tag{2.170}
$$

from which we deduce, by forming the ratio of the $k = 1, 2$ identities, and using (2.164):

Fundamental identities:

$$
(L_1^2)_{n-1,n} = \frac{\partial}{\partial t_1}\log\frac{\tau_{n+1}}{\tau_{n-1}} = \frac{\partial^2/\partial s_1\partial t_2\log\tau_n}{\partial^2/\partial s_1\partial t_1\log\tau_n}\,, \tag{2.171}
$$

(and by duality $t \leftrightarrow -s$, $L_1 \leftrightarrow hL_2^T h^{-1}$)

$$
(hL_2^T h^{-1})_{n-1,n}^2 = -\frac{\partial}{\partial s_1}\log\frac{\tau_{n+1}}{\tau_{n-1}} = \frac{\partial^2/\partial t_1\partial s_2\log\tau_n}{\partial^2/\partial t_1\partial s_1\log\tau_n}\,. \tag{2.172}
$$

As promised we now give, following Adler–van Moerbeke–Vanhaecke [15]:

Sketch of alternate proof of bilinear identities: The proof is based on the following identities concerning the bi-orthogonal polynomials and their Cauchy transforms with regard to the measure $d\rho$ defining the moments of (2.141) and (2.152) μ_{ij}:

$$
d\rho(x,y,t,s,c) = \exp\left(\sum_1^{\infty}(t_i x^i - s_i y^i) + \sum_{\alpha,\beta\geq 1} c_{\alpha\beta}x^{\alpha}y^{\beta}\right)\mathrm{d}x\,\mathrm{d}y\,.
$$

Namely, the bi-orthogonal polynomials of (2.150) and (2.151) and their formal Cauchy transforms with regard to $d\rho$ can be expressed in terms of τ-functions as follows:[10]

$$\begin{cases} p_n^{(1)}(z,t,s) = z^n \dfrac{\tau_n(t-[z^{-1}],s)}{\tau_n(t,s)}\,, \quad \text{(suppressing } c \text{ and } E \text{ in } p_n^{(i)} \text{ and } \tau_n\text{)}\,, \\ p_m^{(2)}(z,t,s) = z^m \dfrac{\tau_m(t,s+[z^{-1}])}{\tau_m(t,s)}\,, \\ \left\langle p_n^{(1)}(x), \dfrac{1}{z-y} \right\rangle = z^{-n-1} \dfrac{\tau_{n+1}(t,s-[z^{-1}])}{\tau_n(t,s)}\,, \\ \left\langle \dfrac{1}{z-x}, p_m^{(2)}(y) \right\rangle = z^{-m-1} \dfrac{\tau_{m+1}(t+[z^{-1}],s)}{\tau_m(t,s)}\,. \end{cases} \tag{2.173}$$

These formulas are not hard to prove, they depend on substituting

$$\exp\left(\pm \sum_{i\geq 1} \frac{(x/z)^i}{i}\right) = \left(1-\frac{x}{z}\right)^{\mp 1} = \sum_{i\geq 0}\left(\frac{x}{z}\right)^i \quad \text{or} \quad 1-\frac{x}{z} \tag{2.174}$$

into formula (2.142), which express $\tau_n(t,s,C,E)$ in terms of the moments $\mu_{ij}(t,s,c,E)$ of (2.141). Then one must expand the rows or columns of the ensuing determinants using (2.174) and make the identification of (2.173), using (2.150) and (2.151), an amusing exercise. If one then substitutes (2.173) into the bilinear identity (2.166) divided by $\tau_n(t,s)\tau_m(t,s)$, thus eliminating τ-functions, and if we make use of the following self-evident *formal* residue identities:

$$\begin{aligned} \oint_{z=\infty} f(z)\left\langle \frac{h(x)}{z-x}, g(y)\right\rangle \frac{\mathrm{d}z}{2\pi\mathrm{i}} &= \langle f(x)h(x), g(y)\rangle\,, \\ \oint_{z=\infty} f(z)\left\langle g(x), \frac{h(y)}{z-y}\right\rangle \frac{\mathrm{d}z}{2\pi\mathrm{i}} &= \langle g(x), f(y)h(y)\rangle\,, \end{aligned} \tag{2.175}$$

with

$$f(z) = \sum_{i-0}^{\infty} a_i z^i\,, \tag{2.176}$$

the bilinear identity becomes a tautology.

2.3.5 Virasoro Constraints for the τ-functions

In this section we derive the Virasoro constraints for our τ-functions $\tau(t,s,C,E)$ using their integral form:

$$V_k^{(1)}\tau(t,s,C,E) = 0\,, \quad V_k^{(2)}\tau(t,s,C,E) = 0\,, \qquad k \geq -1\,. \tag{2.177}$$

[10] $1/(z-x) := (1/z)\sum_{i=0}^{\infty}(x/z)^i$, etc., $1/(z-y)$, and thus z is viewed as large.

Explicitly:

$$\begin{cases} \displaystyle\sum_1^r a_i^{k+1}\frac{\partial}{\partial a_i}\tau_n^E = \Big(\mathbb{J}_{k,n}^{(2)}(t) + \sum_{i,j\geq 1} ic_{ij}\frac{\partial}{\partial c_{i+k,j}}\Big)\tau_n^E := \mathbb{V}_k\tau_n^E \\ \displaystyle\sum_1^s b_i^{k+1}\frac{\partial}{\partial b_i}\tau_n^E = \Big(\mathbb{J}_{k,n}^{(2)}(-s) + \sum_{i,j\geq 1} jc_{ij}\frac{\partial}{\partial c_{i,j+k}}\Big)\tau_n^E := \widetilde{\mathbb{V}}_k\tau_n^E \,, \end{cases}$$

$$k \geq -1 \,,\ n \geq 0 \,, \quad (2.178)$$

with

$$E = \bigcup_1^r [a_{2i-1}, a_{2i}] \times \bigcup_1^s [b_{2i-1}, b_{2i}] \,, \quad (2.179)$$

$$\begin{aligned} \tau_n^E &= \tau_n(t,s,C,E) \\ &= \frac{1}{n!}\int_{E^n}\Big(\Delta_n(x)\prod_{k=1}^n \exp\Big(\sum_1^\infty t_i x_k^i\Big)\,\mathrm{d}x_k\Big) \\ &\quad\times\Big(\Delta_n(y)\prod_{k=1}^n \exp\Big(-\sum_1^\infty s_i y_k^i\Big)\,\mathrm{d}y_k\Big)\prod_{k=1}^n \exp\Big(\sum_{i,j\geq 1} c_{ij}x_k^i y_k^j\Big) \,, \end{aligned} \quad (2.180)$$

and

$$\begin{aligned} \mathbb{J}_k^{(1)}(t) &:= \big(\mathbb{J}_{k,n}^{(1)}(t)\big)_{n\geq 0} := (J_k^{(1)}(t) + nJ_k^{(0)})_{n\geq 0} \,, \\ \mathbb{J}_k^{(2)}(t) &:= \big(\mathbb{J}_{k,n}^{(2)}(t)\big)_{n\geq 0} \\ &\quad := \tfrac{1}{2}\big(J_k^{(2)}(t) + (2n+k+1)J_k^{(1)}(t) + n(n+1)J_k^{(0)}\big)_{n\geq 0} \,, \\ J_k^{(1)}(t) &:= \frac{\partial}{\partial t_k} + (-k)t_{-k}, \quad J_k^{(0)} = \delta_{0k} \,, \\ J_k^{(2)}(t) &:= \sum_{i+j=k}\frac{\partial^2}{\partial t_i\partial t_j} + 2\sum_{i\geq 1} it_i\frac{\partial}{\partial t_{i+k}} + \sum_{i+j=-k}(it_i)(jt_j) \,. \end{aligned} \quad (2.181)$$

Main fact: $\mathbb{J}_k^{(2)}$ forms a Virasoro algebra of charge $c=-2$,

$$\big[\mathbb{J}_k^{(2)}, \mathbb{J}_l^{(2)}\big] = (k-l)\mathbb{J}_{k+l}^{(2)} + (-2)\frac{(k^3-k)}{12}\delta_{k,-l} \,. \quad (2.182)$$

To prove (2.178) we need the following lemma:

Lemma 2.3.1 (Adler–van Moerbeke [5]). *Given*

$$\rho = \mathrm{e}^{-V} \quad \textit{with} \ -\frac{\rho'}{\rho} = V' = \frac{g}{f} = \frac{\sum_0^\infty \beta_i z^i}{\sum_0^\infty \alpha_i z^i} \,,$$

the integrand

$$\mathrm{d}I_n(x) := \Delta_n(x) \prod_{k=1}^{n} \left(\exp\left(\sum_1^{\infty} t_i x^i \right) \rho(x_k) \mathrm{d}x_k \right) , \tag{2.183}$$

satisfies the following variational formula:

$$\frac{\mathrm{d}}{\mathrm{d}\varepsilon} \mathrm{d}I_n(x_i \mapsto x_i + \varepsilon f(x_i) x_i^{m+1})\Big|_{\varepsilon=0} = \sum_{l=0}^{\infty} (\alpha_l \mathbb{J}^{(2)}_{m+l,n} - \beta_l \mathbb{J}^{(1)}_{m+l+1,n}) \, \mathrm{d}I_n \, . \tag{2.184}$$

The contribution coming from $\prod_1^n \mathrm{d}x_j$ *is given by*

$$\sum_{l=0}^{\infty} a_l (l+m+1) \mathbb{J}^{(1)}_{m+l,n} \mathrm{d}I_n \, . \tag{2.185}$$

Proof of (2.178). First make the change of coordinates $x_i \to x_i + \varepsilon x_i^{k+1}$, $1 \le i \le n$, in the integral (2.180), which remains unchanged, and then differentiate the results by ε, at $\varepsilon = 0$, which of course yields 0, i.e.,

$$\frac{\mathrm{d}}{\mathrm{d}\varepsilon} \tau_n^E \Big|_{x_i \to x_i + \varepsilon x_i^{k+1}} \Bigg|_{\varepsilon=0} = 0 \, . \tag{2.186}$$

Now, by (2.180), there are precisely 3 contributions to the l .h.s. of (2.186), namely one of the form (2.184), with $\rho(x) = 1$, yielding $\mathbb{J}^{(2)}_{k,n}(t)\tau_n^E$, one coming from

$$\begin{aligned} &\frac{\mathrm{d}}{\mathrm{d}\varepsilon} \left(\prod_{l=1}^{n} \exp\left(\sum_{i,j \ge 1} c_{ij} x_l^i y_l^j \right) \right) \Bigg|_{x_s \to x_s + \varepsilon x_s^{k+1}} \Bigg|_{\varepsilon=0} \\ &\qquad = \sum_{i,j\ge 1} i c_{ij} \sum_{l=1}^{n} x_l^{i+k} y_l^j \prod_{l=1}^{n} \exp\left(\sum_{i,j\ge 1} c_{ij} x_l^i y_l^j \right) \\ &\qquad = \sum_{i,j\ge 1} i c_{ij} \frac{\partial}{\partial c_{i+k,j}} \prod_{l-1}^{n} \exp\left(\sum_{i,j\ge 1} c_{ij} x_l^i y_l^j \right) , \end{aligned}$$

yielding $\sum_{i,j\ge 1} i c_{ij} \partial / \partial c_{i+k,j} \tau_n^E$.[11] Finally, we have a third contribution, since the limits of integration the integral (2.180) must change:

$$a + i \to a_i - \varepsilon a_i^{k+1} + 0(\varepsilon^2) \, , \quad 1 \le i \le 2r \, .$$

Upon differentiating τ_n^E with respect to the ε in these *new* limits of integration, we have by the chain rule, the contribution $-\sum_1^{2r} a_i^{k+1} \partial/\partial a_i \tau_n^E$. Thus altogether we have:

[11] It must be noted that: $\partial/\partial c_{0,n} = -\partial/\partial s_n$, $\partial/\partial c_{n,0} = \partial/\partial t_n$.

$$
\begin{aligned}
0 &= \left.\frac{\mathrm{d}}{\mathrm{d}\varepsilon}\tau_n^E\Big|_{x_i\to x_i+\varepsilon x_i^{k+1}}\right|_{\varepsilon=0} \\
&= -\sum_1^{2r} a_i^{k+1}\frac{\partial}{\partial a_i}\tau_n^E + \mathbb{J}_{k,n}^{(2)}(t)\tau_n^E + \sum_{i,j\geq 1} ic_{ij}\frac{\partial}{\partial c_{i+k,n}}\tau_n^E ,
\end{aligned}
$$

yielding the first expression (2.178). The second expression follows from the first by duality, $t \leftrightarrow -s$, $a \leftrightarrow b$, $c_{ij} \leftrightarrow c_{ji}$.

2.3.6 Consequences of the Virasoro Relations

Observe from (2.132), (2.139) that $\big(e_2 = (0,1,0,0,\dots)\big)$

$$
P_n(E) = \left.\frac{\tau_n^E(t-e_2/2, s+e_2/2, C)}{\tau_n^{\mathbb{R}}(t-e_2/2, s+e_2/2, C)}\right|_{\mathcal{L}} , \tag{2.187}
$$

$$
\mathcal{L} := \{t = s = 0, \text{all } c_{ij} = 0, \text{ but } c_{11} = c\} , \tag{2.188}
$$

and so computing (2.178) for $\tau_n^E(t-\frac{1}{2}e_2, s+\frac{1}{2}e_2, C)$ requires us to shift the t, s in $\mathbb{J}_{k,n}^{(2)}(t)$, $\mathbb{J}_{k,n}^{(2)}(-s)$ accordingly, and we find from (2.181) shifted, the following:

$$
\mathcal{A}_k\tau_n = \mathcal{V}_k\tau_n, \quad \mathcal{B}_k\tau_n = \mathcal{W}_k\tau_n, \quad k = 1,2 \tag{2.189}
$$

with

$$
\tau_n = \tau_n^E\big(t - \tfrac{1}{2}e_2, s + \tfrac{1}{2}e_2, C\big) ,
$$

and

$$
\begin{aligned}
\mathcal{A}_1 &= \frac{1}{1-c^2}\left(\sum_1^{2r}\frac{\partial}{\partial a_j} + c\sum_1^{2s}\frac{\partial}{\partial b_j}\right), &
\mathcal{B}_1 &= \frac{1}{1-c^2}\left(\sum_1^{2s}\frac{\partial}{\partial b_j} + c\sum_1^{2r}\frac{\partial}{\partial a_j}\right), \\
\mathcal{A}_2 &= \sum_1^{2r} a_j\frac{\partial}{\partial a_j} , &
\mathcal{B}_2 &= \sum_1^{2s} b_j\frac{\partial}{\partial b_j} ,
\end{aligned} \tag{2.190}
$$

with

$$\begin{aligned}
\mathcal{V}_1 &:= \frac{1}{1-c^2}(\mathbb{V}_{-1} + c\widetilde{\mathbb{V}}_{-1}) = \widehat{\mathcal{V}}_1 + v_1 \\
&:= -\frac{\partial}{\partial t_1} - \frac{n(t_1 - cs_1)}{c^2-1} \\
&\quad - \frac{1}{c^2-1}\left(\sum_{i\geq 2} i\left(t_i\frac{\partial}{\partial t_{i-1}} + cs_i\frac{\partial}{\partial s_{i-1}}\right) + \sum_{\substack{i,j\geq 1,\\ i,j\neq(1,1)}} c_{ij}\left(i\frac{\partial}{\partial c_{i-1,j}} + jc\frac{\partial}{\partial c_{i,j-1}}\right)\right),
\end{aligned}$$

$$\begin{aligned}
\mathcal{W}_1 &:= \frac{1}{1-c^2}(c\mathbb{V}_{-1} + \widetilde{\mathbb{V}}_{-1}) = \widehat{\mathcal{W}}_1 + w_1 \\
&:= \frac{\partial}{\partial s_1} - \frac{n(ct_1 - s_1)}{c^2-1} \\
&\quad - \frac{1}{c^2-1}\left(\sum_{i\geq 2} i\left(ct_i\frac{\partial}{\partial t_{i-1}} + s_i\frac{\partial}{\partial s_{i-1}}\right) + \sum_{\substack{i,j\geq 1,\\ i,j\neq(1,1)}} c_{ij}\left(ci\frac{\partial}{\partial c_{i-1,j}} + j\frac{\partial}{\partial c_{i,j-1}}\right)\right),
\end{aligned}$$

$$\begin{aligned}
\mathcal{V}_2 &:= \mathbb{V}_0 - c\frac{\partial}{\partial c} := \widehat{\mathcal{V}}_2 + v_2 \\
&:= -\frac{\partial}{\partial t_2} + \sum_{i\geq 1} it_i\frac{\partial}{\partial t_i} + \frac{n(n+1)}{2} + \sum_{\substack{i,j\geq 1,\\ (i,j)\neq(1,1)}} ic_{ij}\frac{\partial}{\partial c_{ij}},
\end{aligned}$$

$$\begin{aligned}
\mathcal{W}_2 &:= \widetilde{\mathbb{V}}_0 - c\frac{\partial}{\partial c} := \widehat{\mathcal{W}}_2 + w_2 \\
&:= \frac{\partial}{\partial s_2} + \sum_{i\geq 1} is_i\frac{\partial}{\partial s_i} + \frac{n(n+1)}{2} + \sum_{\substack{i,j\geq 1,\\ (i,j)\neq(1,1)}} jc_{ij}\frac{\partial}{\partial c_{ij}}.
\end{aligned}$$

Note $\widehat{\mathcal{V}}_1$, $\widehat{\mathcal{W}}_1$, $\widehat{\mathcal{V}}_2$, $\widehat{\mathcal{W}}_2$ are first order operators such that (and this is the point):

$$\widehat{\mathcal{V}}_1|_{\mathcal{L}} = -\frac{\partial}{\partial t_1}, \quad \widehat{\mathcal{W}}_1|_{\mathcal{L}} = \frac{\partial}{\partial s_1}, \quad \widehat{\mathcal{V}}_2|_{\mathcal{L}} = -\frac{\partial}{\partial t_2}, \quad \widehat{\mathcal{W}}_2|_{\mathcal{L}} = \frac{\partial}{\partial s_2}, \tag{2.191}$$

and

$$v_1 = \frac{n(t_1 - cs_1)}{1-c^2}, \quad w_1 = \frac{n(ct_1 - s_1)}{1-c^2}, \quad v_2 = w_2 = \frac{n(n+1)}{2}. \tag{2.192}$$

Because of (2.191) we call this a principal symbol analysis. Hence

$$\mathcal{A}_k \log\tau_n = \widehat{\mathcal{V}}_k \log\tau_n + v_k, \quad \mathcal{B}_k \log\tau_n = \widehat{\mathcal{W}}_k \log\tau_n + w_k, \qquad k = 1, 2, \tag{2.193}$$

and so on the locus $\mathcal{L}$ using (2.191)–(2.193) we find:

$$\begin{aligned}
\frac{\partial}{\partial t_1}\log\tau_n|_{\mathcal{L}} &= -\mathcal{A}_1\log\tau_n|_{\mathcal{L}}, & \frac{\partial}{\partial s_1}\log\tau_n|_{\mathcal{L}} &= \mathcal{B}_1\log\tau_n|_{\mathcal{L}}, \\
\frac{\partial}{\partial t_2}\log\tau_n|_{\mathcal{L}} &= -\mathcal{A}_2\log\tau_n|_{\mathcal{L}} & \frac{\partial}{\partial s_2}\log\tau_n|_{\mathcal{L}} &= \mathcal{B}_2\log\tau_n|_{\mathcal{L}} \\
&\quad + \frac{n(n+1)}{2}, & &\quad - \frac{n(n+1)}{2}.
\end{aligned} \tag{2.194}$$

Extending this analysis to second derivatives, compute:

$$\begin{aligned}\mathcal{B}_1\mathcal{A}_1 \log \tau_n|_{\mathcal{L}} &= \mathcal{B}_1(\widehat{\mathcal{V}}_1 \log \tau_n + v_1)|_{\mathcal{L}} = \mathcal{B}_1\widehat{\mathcal{V}}_1 \log \tau_n|_{\mathcal{L}} + \mathcal{B}_1(v_1)|_{\mathcal{L}} \\ &\overset{(x)}{=} \widehat{\mathcal{V}}_1\mathcal{B}_1 \log \tau_n|_{\mathcal{L}} + \mathcal{B}_1(v_1)|_{\mathcal{L}} \\ &\overset{\binom{x}{x}}{=} -\frac{\partial}{\partial t_1}(\widehat{\mathcal{W}}_1 \log \tau_n + w_1)|_{\mathcal{L}} + \mathcal{B}_1(v_1)|_{\mathcal{L}} \\ &= -\frac{\partial}{\partial t_1}\left(\frac{\partial}{\partial s_1} + \cdots\right) \log \tau_n|_{\mathcal{L}} - \frac{\partial}{\partial t_1} w_1|_{\mathcal{L}} + \mathcal{B}_1(v_1)|_{\mathcal{L}} \end{aligned} \tag{2.195}$$

where we have used in (x) that $[\mathcal{B}_1, \widehat{\mathcal{V}}_1]\big|_{\mathcal{L}} = 0$ and in $\binom{x}{x}$ that $\widehat{\mathcal{V}}_1\big|_{\mathcal{L}} = -\partial/\partial t_1$. So we must compute

$$-\frac{\partial}{\partial t_1}\widehat{\mathcal{W}}_1|_{\mathcal{L}} = -\frac{\partial^2}{\partial t_1 \partial s_1}\,, \quad -\frac{\partial}{\partial t_1} w_1|_{\mathcal{L}} = \frac{nc}{c^2-1}\,, \quad \mathcal{B}_1(v_1)|_{\mathcal{L}} = 0\,, \tag{2.196}$$

and so conclude that $\left(\tau_n = \tau_n^E(t - \frac{1}{2}e_2, s + \frac{1}{2}e_2, C)\right)$

$$\frac{\partial^2}{\partial t_1 \partial s_1} \log \tau_n|_{\mathcal{L}} = -\mathcal{B}_1\mathcal{A}_1 \log \tau_n + \frac{nc}{c^2-1}\,. \tag{2.197}$$

The crucial points in this calculation were:

$$[\mathcal{B}_1, \widehat{\mathcal{V}}_1]|_{\mathcal{L}} = 0\,, \quad \widehat{\mathcal{V}}_1|_{\mathcal{L}} = -\frac{\partial}{\partial t_1}, \quad \widehat{\mathcal{W}}_1 = \frac{\partial}{\partial s_1} + \cdots\,, \tag{2.198}$$

and indeed this is a model calculation, which shall be repeated over and over again. And so in the same fashion, conclude:

$$\frac{\partial^2}{\partial t_1 \partial s_2} \log \tau_n|_{\mathcal{L}} = -\mathcal{B}_2\mathcal{A}_1 \log \tau_n\,, \quad \frac{\partial^2}{\partial s_1 \partial t_2} \log \tau_n|_{\mathcal{L}} = -\mathcal{A}_2\mathcal{B}_1 \log \tau_n\,, \tag{2.199}$$

where we have used $\partial/\partial t_1(n(n+1)/2) = \partial/\partial s_1\ (n(n+1)/2) = 0$.

2.3.7 Final Equations

We have derived in Sect. 2.3.6, the following

Relations on Locus $\mathcal{L}$:

$$\begin{aligned} &\frac{\partial}{\partial t_1} \log \tau_n^E = -\mathcal{A}_1 \log \tau_n^E, && \frac{\partial}{\partial t_2} \log \tau_n^E = -\mathcal{A}_2 \log \tau_n^E + \frac{n(n+1)}{2}\,, \\ &\frac{\partial}{\partial s_1} \log \tau_n^E = \mathcal{B}_1 \log \tau_n^E, && \frac{\partial}{\partial s_2} \log \tau_n^E = \mathcal{B}_2 \log \tau_n^E - \frac{n(n+1)}{2}\,, \\ &\frac{\partial^2}{\partial t_1 \partial s_2} \log \tau_n^E = -\mathcal{B}_2\mathcal{A}_1 \log \tau_n^E\,, && \frac{\partial^2}{\partial s_1 \partial t_2} \log \tau_n^E = -\mathcal{A}_2\mathcal{B}_1 \log \tau_n^E\,, \\ &\frac{\partial^2}{\partial t_1 \partial s_1} \log \tau_n^E = -\mathcal{B}_1\mathcal{A}_1 \log \tau_n^E + \frac{nc}{c^2-1}\,. \end{aligned} \tag{2.200}$$

Remember from Sect. 2.3.4:

2-*Toda relations*

$$\begin{aligned} \frac{\partial}{\partial t_1} \log \frac{\tau_{n+1}^E}{\tau_{n-1}^E} &= \frac{\partial^2/\partial s_1 \partial t_2 \log \tau_n^E}{\partial^2/\partial t_1 \partial s_1 \log \tau_n^E}, \\ -\frac{\partial}{\partial s_1} \log \frac{\tau_{n+1}^E}{\tau_{n-1}^E} &= \frac{\partial^2/\partial t_1 \partial s_2 \log \tau_n^E}{\partial^2/\partial t_1 \partial s_1 \log \tau_n^E}, \end{aligned} \tag{2.201}$$

Substitute the relations on $\mathcal{L}$ into the 2-Toda relations, which yields:

Pure boundary relations on the locus $\mathcal{L}$

$$\begin{aligned} -\mathcal{A}_1 \log \frac{\tau_{n+1}^E}{\tau_{n-1}^E} &= \frac{\mathcal{A}_2 \mathcal{B}_1 \log \tau_n^E}{\mathcal{A}_1 \mathcal{B}_1 \log \tau_n^E + nc/(1-c^2)}, \\ -\mathcal{B}_1 \log \frac{\tau_{n+1}^E}{\tau_{n-1}^E} &= \frac{\mathcal{B}_2 \mathcal{A}_1 \log \tau_n^E}{\mathcal{B}_1 \mathcal{A}_1 \log \tau_n^E + nc/(1-c^2)}. \end{aligned} \tag{2.202}$$

Since $\mathcal{A}_1 \mathcal{B}_1 = \mathcal{B}_1 \mathcal{A}$, conclude that

$$\mathcal{A}_1 \left(\frac{\mathcal{B}_2 \mathcal{A}_1 \log \tau_n^E}{\mathcal{B}_1 \mathcal{A}_1 \log \tau_n^E + nc/(1-c^2)} \right) = \mathcal{B}_1 \left(\frac{\mathcal{A}_2 \mathcal{B}_1 \log \tau_n^E}{\mathcal{A}_1 \mathcal{B}_1 \log \tau_n^E + nc/(1-c^2)} \right). \tag{2.203}$$

Notice that since $\tau_n^{\mathbb{R}}$ is independent of a_i and b_i, we find that

$$\begin{aligned} \mathcal{A}_1 \log \tau_n^E &= \mathcal{A}_1 \log \tau_n^E - \mathcal{A}_1 \log \tau_n^{\mathbb{R}} \\ &= \mathcal{A}_1 \log \frac{\tau_n^E}{\tau_n^{\mathbb{R}}} = \mathcal{A}_1 \log P_n(E), \end{aligned} \tag{2.204}$$

and so (2.203) is true with $\log \tau_n^E \to \log P_n(E)$, yielding the final equation for $F_n(E) = 1/n \log P_n(E)$, and proving Theorem 2.3.1.

2.4 Dyson Brownian Motion and the Airy Process

2.4.1 Processes

The joint distribution for the Dyson process at 2-times deforms naturally to the 2-Toda integrable system, as it is described by a coupled Hermitian matrix integral, analyzed in the previous section. Taking limits of the Dyson process leads to the Airy and Sine processes. We describe the processes in this section in an elementary and intuitive fashion. A good reference for this discussion would be [33] and Dyson's celebrated papers [30, 31] on Dyson diffusion.

A random walk corresponds to a particle moving either left or right at time n with probability p. If the particle is totally drunk, one may take $p = \frac{1}{2}$. In that case, if X_n is its location after n steps,

$$E(X_n) = 0, \quad E(X_n^2) = n, \tag{2.205}$$

and in any case, this discrete process has no memory:

$$P\big(X_{n+1} = j \mid (X_n = i) \cap (\text{arbitrary past event})\big) = P(X_1 = j \mid X_0 = i) ,$$

i.e. it is Markovian. In the continuous version of this process (say with $p = \frac{1}{2}$), $[t/\delta]$ steps are taken in time t and each step is of magnitude $\sqrt{\delta}$, consistent with the scaling of (2.205). By the central limit theorem (CLT) for the binomial distribution, or in other words by Stirlings formula, it follows immediately that

$$\begin{aligned} \lim_{\delta\to 0} P(X_t \in (X, X+\mathrm{d}X) \mid X_0 = \overline{X}) &= \frac{\exp(-(X-\overline{X})^2/2t)}{\sqrt{2\pi t}}\,\mathrm{d}X \\ &=: P(t, \overline{X}, X)\,\mathrm{d}X . \end{aligned} \tag{2.206}$$

Note that $P(t, \overline{X}, X)$ satisfies the (heat) diffusion equation

$$\frac{\partial P}{\partial t} = \frac{1}{2}\frac{\partial^2 P}{\partial X^2} . \tag{2.207}$$

The limiting motion where the particle moves $\pm\sqrt{\delta}$ with equal probability $\frac{1}{2}$ in time δ, is in the limit, as $\delta \to 0$, Brownian motion. The process is scale invariant and so infinitesimally its fluctuations in t are no larger than $\sqrt{\Delta(t)}$ and hence while the paths are continuous, they are nowhere differentiable (for almost all initial conditions.) We may also consider Brownian motion in n directions, all independent, and indeed, it was first observed in $n = 2$ directions, under the microscope by Robert Brown, an English botanist, in 1828. In general, by independence,

$$\begin{aligned} P(t, \overline{X}, X) &= \prod_1^n P(t, \overline{X}_i, X_i) \\ &= \frac{1}{(2\pi t/\beta)^{n/2}} \exp\bigg(-\sum_1^n \frac{(X_i - \overline{X}_i)^2}{2t/\beta}\bigg) , \end{aligned} \tag{2.208}$$

hence

$$\frac{\partial P}{\partial t} = \frac{1}{2\beta}\sum_1^n \frac{\partial^2}{\partial X_i^2} P , \tag{2.209}$$

where we have changed the variance and hence the diffusion constant from $1 \to \beta$.

In addition (going back to $n = 1$), besides changing the rate of diffusion, we may also subject the diffusing particle located at X, to a harmonic force $-\rho X$, pointing toward the origin. Thus you have a drunken particle executing Brownian motion under the influence of a steady wind pushing him towards the origin – the Ornstein–Uhlenbeck (see [33]) process – where now the probability density $P(t, \overline{X}, X)$ is given by the diffusion equation:

$$\frac{\partial P}{\partial t} = \left(\frac{1}{2\beta} \frac{\partial^2}{\partial X^2} - \frac{\partial}{\partial X}(-\rho X) \right) P. \tag{2.210}$$

This can immediately be transformed to the case $\rho = 0$, yielding ($c = \mathrm{e}^{-\rho t}$)

$$P(t, \overline{X}, X) = \frac{1}{\sqrt{2\pi}((1-c^2)/2\rho\beta)^{1/2}} \exp\left(\frac{-(X - c\overline{X})^2}{(1-c^2)/\rho\beta} \right), \tag{2.211}$$

which as $\rho \to 0$, $(1-c^2)/2\rho \to t$, transforms to the old case. This process becomes stationary, i.e. the probabilities at a fixed time do not change in t, if and only if the initial distribution of $\overline{X}$ is given by the limiting $t \to \infty$ distribution of (2.211):

$$\frac{\exp(-\rho\beta\overline{X}^2)}{\sqrt{\pi/\rho\beta}} \mathrm{d}\overline{X}\,, \tag{2.212}$$

and this is the only "normal Markovian process" with this property.

Finally, consider a Hermitian matrix $B = (B_{ij})$ with n^2 real quantities B_{ij} undergoing n^2 *independent* Ornstein–Uhlenbeck processes with $\rho = 1$, but

$$\begin{cases} \beta = 1 & \text{for } B_{ii} \text{ (on diagonal)}\,, \\ \beta = 2 & \text{for } B_{ij} \text{ (off diagonal)}\,, \end{cases}$$

and so the respective probability distribution P_{ii}, P_{ij} satisfy by (2.210):

$$\begin{cases} \dfrac{\partial P_{ii}}{\partial t} = \left(\dfrac{1}{2} \dfrac{\partial^2}{\partial B_{ii}^2} - \dfrac{\partial}{\partial B_{ii}}(-B_{ii}) \right) P_{ii}\,, \\ \dfrac{\partial P_{ij}}{\partial t} = \left(\dfrac{1}{2 \times 2} \dfrac{\partial^2}{\partial B_{ij}^2} - \dfrac{\partial}{\partial B_{ij}}(-B_{ij}) \right) P_{ij}\,, \end{cases} \tag{2.213}$$

with solution, by (2.211) ($c = \mathrm{e}^{-t}$), given by:

$$\begin{cases} P_{ii}(t, \overline{B}_{ii}, B_{ii}) = \dfrac{1}{\sqrt{2\pi}\sqrt{(1-c^2)/2}} \exp\left(\dfrac{-(B_{ii} - c\overline{B}_{ii})^2}{1-c^2} \right), \\ P_{ij}(t, \overline{B}_{ij}, B_{ij}) = \dfrac{1}{\sqrt{2\pi}\sqrt{(1-c^2)/4}} \exp\left(\dfrac{-(B_{ij} - c\overline{B}_{ij})^2}{(1-c^2)/2} \right). \end{cases} \tag{2.214}$$

By the independence of the processes the joint probability distribution is given by:

$$P(t, \overline{B}, B) = \prod_{i=1}^{n} P_{ii} \prod_{\substack{1 \le i,j \le n, \\ i \ne j}} P_{ij} = \frac{Z^{-1}}{(1-c^2)^{n^2/2}} \exp\left(-\frac{\operatorname{Tr}(B - c\overline{B})^2}{1-c^2} \right), \tag{2.215}$$

with $Z = (2\pi)^{n^2/2} 2^{(-n^2+n/2)}$, which by (2.213) and (2.215) evolves by the Ornstein–Uhlenbeck process:

$$\begin{aligned}\frac{\partial P}{\partial t} &= \sum_{i,j=1}^{n}\left(\frac{1}{4}(1+\delta_{ij})\frac{\partial^2}{\partial B_{ij}^2}+\frac{\partial}{\partial B_{ij}}B_{ij}\right)P \\ &= \sum_{i,j=1}^{n}\left(\frac{1}{4}(1+\delta_{ij})\frac{\partial}{\partial B_{ij}}\Phi(B)\frac{\partial}{\partial B_{ij}}\frac{1}{\Phi(B)}\right)P(B)\,,\end{aligned} \tag{2.216}$$

with $\Phi(B) = \exp(-\operatorname{tr} B^2)$. Note that the *most general solution* (2.215) to (2.216) is invariant under the unitary transformation

$$(\overline{B}, B) \to (U\overline{B}U^{-1}, UBU^{-1})\,, \tag{2.217}$$

which forces the actual process (2.216) to possess this unitary invariance and in fact (2.216) induces a random motion purely on the spectrum of B. This motion, discovered by Dyson in [30,31], is called Dyson diffusion, and indeed the Ornstein–Uhlenbeck process

$$B(t) = \big(B_{ij}(t)\big)$$

given by (2.216) with solution (2.215), induces Dyson Brownian motion: $\big(\lambda_1(t), \dots, \lambda_n(t)\big) \in \mathbb{R}^n$ on the eigenvalues of $B(t)$.

The transition probability $P(t, \bar{\lambda}, \lambda)$ satisfies the following diffusion equation:

$$\begin{aligned}\frac{\partial P}{\partial t} &= \sum_{1}^{n}\left(\frac{1}{2}\frac{\partial^2}{\partial \lambda_i^2} - \frac{\partial}{\partial \lambda_i}\frac{\partial \log\sqrt{\Phi(\lambda)}}{\partial \lambda_i}\right)P \\ &= \frac{1}{2}\sum_{i=1}^{n}\frac{\partial}{\partial \lambda_i}\Phi(\lambda)\frac{\partial}{\partial \lambda_i}\frac{1}{\Phi(\lambda)}P\end{aligned} \tag{2.218}$$

with

$$\Phi(\lambda) = \Delta_n^2(\lambda)\prod_{1}^{n} \mathrm{e}^{-\lambda_i^2}\,,$$

which is a Brownian motion, where instead of the particle at λ_i feeling only the harmonic restoring force $-\lambda_i$, as in the Ornstein–Uhlenbeck process, it feels the full force

$$F_i(\lambda) := \frac{\partial \log\sqrt{\Phi(\lambda)}}{\partial \lambda_i} = \sum_{j\neq i}\frac{1}{\lambda_i - \lambda_j} - \lambda_i\,, \tag{2.219}$$

which acts to keep the particles apart. In short, the Vandermonde in $\Phi(\lambda)$ creates n-repelling Brownian motions, while the exponential term keeps them from flying out to infinity. The equation (2.218) was shown by Dyson in [30,31] by observing that Brownian motion with a force term $F = (F_i)$ is, in general, completely characterized infinitesimally by the dynamics:

$$E(\delta\lambda_i) = F_i(\lambda)\delta t, \quad E\big((\delta\lambda_i)^2\big) = \delta t\,, \tag{2.220}$$

and so in particular (2.216) yields

$$E(\delta B_{ij}) = -B_{ij}\delta t, \quad E\big((\delta B_{ij})^2\big) = \tfrac{1}{2}(1+\delta_{ij})\delta t\,. \tag{2.221}$$

Then by the unitary invariance (2.217) of the process (2.216), one may set at time t: $B(t) = \operatorname{diag}\big(\lambda_1(t),\dots,\lambda_n(t)\big)$, and then using the perturbation formula:

$$\delta\lambda_i|_t = \delta B_{ii} + \sum_{j\neq i} \frac{(\delta B_{ji})^2 + (\delta B_{ij})^2}{(\lambda_i - \lambda_j)} + \cdots\,,$$

compute $E(\delta\lambda_i)$ and $E\big((\delta\lambda_i)^2\big)$ by employing (2.221), immediately yielding (2.220) with $F_i(\lambda)$ given by (2.219). Thus by the characterization of (2.218) by (2.220), we have verified Dysons result (2.218).

Remember an Ornstein–Uhlenbeck process has a stationary measure precisely if we take for the initial measure the equilibrium measure at $t \to \infty$. So consider our Ornstein–Uhlenbeck transition density (2.215) with $t \to \infty$ stationary distribution:

$$Z^{-1}\exp(-\operatorname{tr} B^2)\,\mathrm{d}B\,,$$

and with this invariant measure as initial condition, one finds for the joint distribution ($c = \exp(-(t_2 - t_1))$)

$$\begin{aligned} &P(B(t_1)\in \mathrm{d}B_1, B(t_2)\in \mathrm{d}B_2) \\ &\quad = Z^{-1}\frac{\mathrm{d}B_1\,\mathrm{d}B_2}{(1-c^2)^{n^2/2}}\exp\left(\frac{-1}{1-c^2}\operatorname{Tr}(B_1^2 - 2cB_1B_2 + B_2^2)\right)\,, \end{aligned} \tag{2.222}$$

and similarly ($c_i = \exp(-(t_{i+1} - t_i))$) compute

$$\begin{aligned} &P(B(t_1)\in \mathrm{d}B_1,\dots,B(t_k)\in \mathrm{d}B_k) \\ &= Z_k^{-1}\exp(-\operatorname{tr} B_1^2)\prod_{i=2}^{k}\exp\left(\frac{-1}{1-c_{i-1}^2}\operatorname{Tr}(B_i - c_{i-1}B_{i-1})^2\right)\mathrm{d}B_1,\dots,\mathrm{d}B_k \\ &= Z_k^{-1}\prod_{i=1}^{k}\exp\left(-\left(\frac{1}{1-c_{i-1}^2} + \frac{c_i^2}{1-c_i^2}\right)\right)\left(\sum_{j=1}^{n}\lambda_{j,i}^2\right)\prod_{j=1}^{n}\mathrm{d}\lambda_{j,i} \\ &\quad \prod_{i=1}^{k-1}\det\left(\exp\left(\frac{2c_i}{1-c_i^2}\lambda_{l,i+1}\lambda_{m,i}\right)\right)_{l,m=1}^{n} \Delta_n(\lambda_1)\Delta_n(\lambda_k), \\ &\qquad\qquad (\lambda_i = (\lambda_{1,i},\lambda_{2,i},\dots,\lambda_{n,i}))\,, \end{aligned} \tag{2.223}$$

using the HCIZ formula (2.138).

The distribution of the eigenvalues for GUE is expressible as a Fredholm determinant (2.9) involving the famous *Hermite kernel* (2.5) and Eynard and Mehta [32] showed that you have for the Dyson process an analogous extended Hermite kernel, specifically the matrix kernel [39]:

$$K^{\mathrm{H},n}_{t_it_j} := \begin{cases} \displaystyle\sum_{k=1}^{\infty} \exp\bigl(-k(t_i-t_j)\bigr)\varphi_{n-k}(x)\varphi_{n-k}(y), & t_i \geq t_j\,, \\ \displaystyle -\sum_{k=-\infty}^{0} \exp\bigl(k(t_j-t_i)\bigr)\varphi_{n-k}(x)\varphi_{n-k}(y), & t_i < t_j\,, \end{cases} \tag{2.224}$$

with

$$\int_{\mathbb{R}} \varphi_i(x)\varphi_j(x)\,\mathrm{d}x = \delta_{ij}, \quad \varphi_i(x) = p_i(x)\exp\left(\frac{-x^2}{2}\right),$$

where

$$\varphi_k(x) = \begin{cases} \exp\left(\dfrac{-x^2}{2}\right)p_k(x), & \text{for } k \geq 0, \text{ with } p_k(x) = \dfrac{H_k(x)}{2^{k/2}\sqrt{k!}\pi^{1/4}}, \\ 0, & \text{for } k < 0\,; \end{cases}$$

so $p_k(x)$ are the normalized Hermite polynomials. Then we have

$$\begin{aligned} \operatorname{Prob}(\text{all } B(t_i) \text{ eigenvalues} \notin E_i,\ 1 \leq i \leq m) &= \det\bigl(I - K^{\mathrm{H},E}\bigr): \\ K^{\mathrm{H},E}_{ij}(x,y) &= I_{E_i}(x)K^{\mathrm{H},n}_{t_it_j}(x,y)I_{E_j}(y)\,, \end{aligned} \tag{2.225}$$

the above being a Fredholm determinant with a matrix kernel.

Remark. In general such a Fredholm determinant is given by:

$$\begin{aligned} &\det\bigl(I - z(K_{t_it_j})_{i,j=1}^m\bigr)\Big|_{z=1} \\ &= 1 + \sum_{N=1}^{\infty}(-z)^N \\ &\times \sum_{\substack{0\leq r_i\leq N,\\ \sum_1^m r_i = N}} \int_R \prod_1^{r_1} \mathrm{d}\alpha_i^{(1)} \cdots \prod_1^{r_m} \mathrm{d}\alpha_i^{(m)} \det\left(\bigl(K_{t_kt_l}(\alpha_i^{(k)},\alpha_j^{(l)})\bigr)_{\substack{1\leq i\leq r_k,\\ 1\leq j\leq r_l}} \right)_{k,l=1}^m \Bigg|_{z=1}, \end{aligned}$$

where the N-fold integral above is taken over the range

$$R = \left\{ \begin{matrix} -\infty < \alpha_1^{(1)} \leq \cdots \leq \alpha_{r_1}^{(1)} < \infty \\ \vdots \\ -\infty < \alpha_1^{(m)} \leq \cdots \leq \alpha_{r_m}^{(m)} < \infty \end{matrix} \right\},$$

with integrand equal to the determinant of an $N \times N$ matrix, with blocks given by the $r_k \times r_l$ matrices $\bigl(K_{t_kt_l}(\alpha_i^{(k)},\alpha_j^{(l)})\bigr)_{\substack{1\leq i\leq r_k,\\ 1\leq j\leq r_l}}$. In particular, for $m = 2$, we have

$$1+\sum_{N=1}^{\infty}(-z)^N \sum_{\substack{0\le r,s\le N,\\ r+s=N}} \int_{\left\{\substack{-\infty<\alpha_1\le\cdots\le\alpha_r<\infty\\ -\infty<\beta_1\le\cdots\le\beta_s<\infty}\right\}} \prod_1^r \mathrm{d}\alpha_i \prod_1^s \mathrm{d}\beta_i$$

$$\times \det\begin{pmatrix} \left(\widehat{K}_{t_1t_1}(\alpha_i,\alpha_j)\right)_{1\le i,j\le r} & \left(\widehat{K}_{t_1t_2}(\alpha_i,\beta_j)\right)_{\substack{1\le i\le r,\\ 1\le j\le s}} \\ \left(\widehat{K}_{t_2t_1}(\beta_i,\alpha_j)\right)_{\substack{1\le i\le s,\\ 1\le j\le r}} & \left(\widehat{K}_{t_2t_2}(\beta_i,\beta_j)\right)_{1\le i,j\le s} \end{pmatrix}\Bigg|_{z=1} . \tag{2.226}$$

These processes have scaling limits corresponding to the bulk and edge scaling limits in the GUE.

The *Airy process* is defined by rescaling in the extended Hermite kernel:

$$x=\sqrt{2n}+\frac{u}{\sqrt{2}n^{1/6}}\,,\quad y=\sqrt{2n}+\frac{v}{\sqrt{2}n^{1/6}}\,,\quad t=\frac{\tau}{n^{1/3}}\,, \tag{2.227}$$

and the *Sine process* by rescaling in the extended Hermite kernel:

$$x=\frac{u\pi}{\sqrt{2n}}\,,\quad y=\frac{v\pi}{\sqrt{2n}}\,,\quad t=\pi^2\frac{\tau}{2n}\,. \tag{2.228}$$

This amounts to following, in slow time, the eigenvalues at the edge and in the bulk, but with a microscope specified by the above rescalings. Then the extended kernels have well-defined limits as $n\to\infty$:

$$K^{\mathrm{A}}_{t_it_j}(x,y)=\begin{cases} \displaystyle\int_0^{\infty} \exp\bigl(-z(t_i-t_j)\bigr)\,\mathrm{Ai}(x+z)\,\mathrm{Ai}(y+z)\,\mathrm{d}z\,, & t_i\ge t_j\,,\\ \displaystyle -\int_{-\infty}^{0} \bigl(z(t_j-t_i)\bigr)\,\mathrm{Ai}(x+z)\,\mathrm{Ai}(y+z)\,\mathrm{d}z\,, & t_i<t_j\,, \end{cases} \tag{2.229}$$

$$K^{\mathrm{S}}_{t_it_j}=\begin{cases} \displaystyle\frac{1}{\pi}\int_0^{\pi} \exp\Bigl(\frac{z^2}{2}(t_i-t_j)\Bigr)\cos z(x-y)\,\mathrm{d}z\,, & t_i\ge t_j\,,\\ \displaystyle -\frac{1}{\pi}\int_{\pi}^{\infty} \exp\Bigl(-\frac{z^2}{2}(t_j-t_i)\Bigr)\cos z(x-y)\,\mathrm{d}z\,, & t_i<t_j\,, \end{cases} \tag{2.230}$$

with Ai the Airy function. Letting $A(t)$ and $S(t)$ denote the Airy and Sine processes, we define them below by

$$\begin{aligned} \mathrm{Prob}(A(t_i)\notin E_i, 1\le i\le k) &= \det\bigl(I-K^{\mathrm{A},E}\bigr)\,,\\ \mathrm{Prob}(S(t_i)\notin E_i, 1\le i\le k) &= \det\bigl(I-K^{\mathrm{S},E}\bigr)\,, \end{aligned} \tag{2.231}$$

where the determinants are matrix Fredholm determinants defined by the matrix kernels (2.229) and (2.230) in the same fashion as (2.225). The Airy process was first defined by Prähofer and Spohn in [46] and the Sine process was first defined by Tracy and Widom in [53].

2.4.2 PDEs and Asymptotics for the Processes

It turns out that the 2-time joint probabilities for all three processes, Dyson, Airy and Sine satisfy PDEs, which moreover lead to long time $t = t_2 - t_1$ asymptotics in, for example, the Airy case. In this section we state the results of Adler and van Moerbeke [6], sketching the proofs in the next section. The first result concerns the Dyson process:

Theorem 2.4.1 (Dyson process). *Given* $t_1 < t_2$ *and* $t = t_2 - t_1$*, the logarithm of the joint distribution for the Dyson Brownian motion* $(\lambda_1(t), \dots, \lambda_n(t))$*,*

$$G_n(t; a_1, \dots a_{2r}; b_1, \dots, b_{2s}) := \log P(\text{all } \lambda_i(t_1) \in E_1, \text{ all } \lambda_i(t_2) \in E_2) ,$$

satisfies a third-order nonlinear PDE *in the boundary points of* E_1 *and* E_2 *and* t*, which takes on the simple form, setting* $c = \mathrm{e}^{-t}$*,*

$$\mathcal{A}_1 \frac{\mathcal{B}_2 \mathcal{A}_1 G_n}{\mathcal{B}_1 \mathcal{A}_1 G_n + 2nc} = \mathcal{B}_1 \frac{\mathcal{A}_2 \mathcal{B}_1 G_n}{\mathcal{A}_1 \mathcal{B}_1 G_n + 2nc} . \tag{2.232}$$

The sets E_1 *and* E_2 *are the disjoint union of intervals*

$$E_1 := \bigcup_{i=1}^{r} [a_{2i-1}, a_{2i}] \text{ and } E_2 := \bigcup_{i=1}^{s} [b_{2i-1}, b_{2i}] \subseteq \mathbb{R} ,$$

which specify the linear operators

$$\begin{aligned}
\mathcal{A}_1 &= \sum_1^{2r} \frac{\partial}{\partial a_j} + c \sum_1^{2s} \frac{\partial}{\partial b_j} , \\
\mathcal{B}_1 &= c \sum_1^{2r} \frac{\partial}{\partial a_j} + \sum_1^{2s} \frac{\partial}{\partial b_j} , \\
\mathcal{A}_2 &= \sum_1^{2r} a_j \frac{\partial}{\partial a_j} + c^2 \sum_1^{2s} b_j \frac{\partial}{\partial b_j} + (1 - c^2) \frac{\partial}{\partial t} - c^2 , \\
\mathcal{B}_2 &= c^2 \sum_1^{2r} a_j \frac{\partial}{\partial a_j} + \sum_1^{2s} b_j \frac{\partial}{\partial b_j} + (1 - c^2) \frac{\partial}{\partial t} - c^2 .
\end{aligned}$$

The duality $a_i \leftrightarrow b_j$ *reflects itself in the duality* $\mathcal{A}_i \leftrightarrow \mathcal{B}_i$*.*

The next result concerns the Airy process:

Theorem 2.4.2 (Airy process). *Given* $t_1 < t_2$ *and* $t = t_2 - t_1$*, the joint distribution for the Airy process* $A(t)$*,*

$$G(t; u_1, \dots u_{2r}; v_1, \dots, v_{2s}) := \log P(A(t_1) \in E_1, A(t_2) \in E_2) ,$$

satisfies a third-order nonlinear PDE *in the* u_i*,* v_i *and* t*, in terms of the Wronskian* $\{f(y), g(y)\}_y := f'(y)g(y) - f(y)g'(y)$*,*

$$((L_u+L_v)(L_uE_v-L_vE_u)+t^2(L_u-L_v)L_uL_v)G$$
$$=\tfrac{1}{2}\{(L_u^2-L_v^2)G,(L_u+L_v)^2G\}_{L_u+L_v}\,. \tag{2.233}$$

The sets E_1 and E_2 are the disjoint union of intervals

$$E_1:=\bigcup_{i=1}^{r}[u_{2i-1},u_{2i}] \text{ and } E_2:=\bigcup_{i=1}^{s}[v_{2i-1},v_{2i}]\subseteq\mathbb{R}\,,$$

which specify the set of linear operators

$$L_u:=\sum_1^{2r}\frac{\partial}{\partial u_i}\,,\qquad L_v:=\sum_1^{2s}\frac{\partial}{\partial v_i}\,,$$
$$E_u:=\sum_1^{2r}u_i\frac{\partial}{\partial u_i}+t\frac{\partial}{\partial t}\,,\qquad E_v:=\sum_1^{2s}v_i\frac{\partial}{\partial v_i}+t\frac{\partial}{\partial t}\,.$$

The duality $v_i\leftrightarrow v_j$ reflects itself in the duality $L_u\leftrightarrow L_v$, $E_u\leftrightarrow E_v$.

Corollary 2.4.1. *In the case of semi-infinite intervals E_1 and E_2, the* PDE *for the Airy joint probability*

$$H(t;x,y):=\log P\left(A(t_1)\le\frac{y+x}{2},A(t_2)\le\frac{y-x}{2}\right)\,,$$

takes on the following simple form in x, y and t^2, with $t=t_2-t_1$, also in terms of the Wronskian,

$$2t\frac{\partial^3H}{\partial t\partial x\partial y}=\left(t^2\frac{\partial}{\partial x}-x\frac{\partial}{\partial y}\right)\left(\frac{\partial^2H}{\partial x^2}-\frac{\partial^2H}{\partial y^2}\right)+8\left\{\frac{\partial^2H}{\partial x\partial y},\frac{\partial^2H}{\partial y^2}\right\}_y\,. \tag{2.234}$$

Remark. Note for the solution $H(t;x,y)$,

$$\lim_{t\searrow0}H(t;x,y)=\log F_2\left(\min\left(\frac{y+x}{2},\frac{y-x}{2}\right)\right)\,.$$

The following theorem concerns the Sine process and uses the same sets and operators as Theorem 2.4.2:

Theorem 2.4.3 (Sine process). *For $t_1<t_2$, and compact E_1 and $E_2\subset\mathbb{R}$, the* log *of the joint probability for the Sine processes $S_i(t)$,*

$$G(t;u_1,\dots u_{2r};v_1,\dots,v_{2s}):=\log P(\textit{all } S_i(t_1)\in E_1^c,\ \textit{all } S_i(t_2)\in E_2^c)\,,$$

satisfies

$$L_u\frac{(2E_vL_u+(E_v-E_u-1)L_v)G}{(L_u+L_v)^2G+\pi^2}$$
$$=L_v\frac{(2E_uL_v+(E_u-E_v-1)L_u)G}{(L_u+L_v)^2G+\pi^2}\,. \tag{2.235}$$

Corollary 2.4.2. *In the case of a single interval, the logarithm of the joint probability for the Sine process,*

$$H(t;x,y) = \log P(S(t_1) \notin [x_1+x_2, x_1-x_2], S(t_2) \notin [y_1+y_2, y_1-y_2])$$

satisfies

$$\frac{\partial}{\partial x_1}\frac{(2E_y\partial/\partial x_1 + (E_y - E_x - 1)\partial/\partial y_1)H}{(\partial/\partial x_1 + \partial/\partial y_1)^2 H + \pi^2} = \frac{\partial}{\partial y_1}\frac{(2E_x\partial/\partial y_1 + (E_x - E_y - 1)\partial/\partial x_1)H}{(\partial/\partial x_1 + \partial/\partial y_1)^2 H + \pi^2}. \tag{2.236}$$

Asymptotic consequences: Prähofer and Spohn showed that the Airy process is a stationary process with continuous sample paths; thus the probability $P(A(t) \le u)$ is independent of t, and is given by the Tracy–Widom distribution

$$P(A(t) \le u) = F_2(u) := \exp\left(-\int_u^\infty (\alpha - u) q^2(\alpha)\, \mathrm{d}\alpha\right), \tag{2.237}$$

with $q(\alpha)$ the solution of the *Painlevé* II equation,

$$q'' = \alpha q + 2q^3 \quad \text{with} \quad q(\alpha) \simeq \begin{cases} -\dfrac{\exp(-\frac{2}{3}\alpha^{3/2})}{2\sqrt{\pi}\alpha^{1/4}}, & \text{for } \alpha \nearrow \infty, \\ \sqrt{-\alpha/2}, & \text{for } a \searrow -\infty. \end{cases} \tag{2.238}$$

The PDEs obtained above provide a very handy tool to compute large time asymptotics for these different processes, with the disadvantage that one usually needs, for justification, a nontrivial assumption concerning the interchange of sums and limits, which can be avoided upon directly using the Fredholm determinant formula for the joint probabilities (see Widom [57]) the latter method, however, tends to be quite tedious and quickly gets out of hand. We now state the following asymptotic result:

Theorem 2.4.4 (Large time asymptotics for the Airy process). *For large $t = t_2 - t_1$, the joint probability admits the asymptotic series*

$$P(A(t_1) \le u, A(t_2) \le v) = F_2(u)F_2(v) + \frac{F_2'(u)F_2'(v)}{t^2} + \frac{\Phi(u,v) + \Phi(v,u)}{t^4} + O\left(\frac{1}{t^6}\right), \tag{2.239}$$

with the function $q = q(\alpha)$ given by (2.238) *and*

$$\Phi(u,v) := F_2(u)F_2(v) + \left(\frac{1}{4}\left(\int_u^\infty q^2\,\mathrm{d}\alpha\right)^2\left(\int_v^\infty q^2\,\mathrm{d}\alpha\right)^2 + q^2(u)\left(\frac{1}{4}q^2(v) - \frac{1}{2}\left(\int_v^\infty q^2\,\mathrm{d}\alpha\right)^2\right) + \int_v^\infty \mathrm{d}\alpha\,(2(v-\alpha)q^2 + q'^2 - q^4)\int_u^\infty q^2\,\mathrm{d}\alpha\right). \tag{2.240}$$

Moreover, the covariance for large $t = t_2 - t_1$ *behaves as*

$$E\big(A(t_2)A(t_1)\big) - E\big(A(t_2)\big)E\big(A(t_1)\big) = \frac{1}{t^2} + \frac{c}{t^4} + \cdots , \tag{2.241}$$

where

$$c := 2 \iint_{\mathbb{R}^2} \Phi(u, v)\, \mathrm{d}u\, \mathrm{d}v .$$

Conjecture. The Airy process satisfies the nonexplosion condition for fixed x:

$$\lim_{z\to\infty} P(A(t) \geq x + z \mid A(0) \leq -z) = 0 . \tag{2.242}$$

2.4.3 Proof of the Results

In this section we sketch the proofs of the results of the prior section, but all of these proofs are ultimately based on a fundamental theorem that we have proven in Sect. 2.3, which we now restate.

Let $M_1, M_2 \in \mathcal{H}_n$, Hermitian $n \times n$ matrices and consider the ensemble:

$$P\big((M_1, M_2) \subset S\big) = \frac{\int_S \mathrm{d}M_1\, \mathrm{d}M_2 \exp\big(-\frac{1}{2}\operatorname{Tr}(M_1^2 + M_2^2 - 2cM_1M_2)\big)}{\int_{\mathcal{H}_n\times\mathcal{H}_n} \mathrm{d}M_1\, \mathrm{d}M_2 \exp\big(-\frac{1}{2}\operatorname{Tr}(M_1^2 + M_2^2 - 2cM_1M_2)\big)} , \tag{2.243}$$

with

$$\mathrm{d}M_1 = \Delta_n^2(x)\prod_1^n \mathrm{d}x_i\, \mathrm{d}U_1 , \quad \mathrm{d}M_2 = \Delta_n^2(y)\prod_1^n \mathrm{d}y_i\, \mathrm{d}U_2 .$$

Given

$$E = E_1 \times E_2 = \bigcup_1^r [a_{2i-1}, a_{2i}] \times \bigcup_1^s [b_{2i-1}, b_{2i}] ,$$

define the boundary operators:

$$\tilde{\mathcal{A}}_1 = -\frac{1}{c^2 - 1}\left(\sum_1^r \frac{\partial}{\partial a_j} + c\sum_1^s \frac{\partial}{\partial b_j}\right) , \qquad \tilde{\mathcal{A}}_2 = \sum_1^r a_j \frac{\partial}{\partial a_j} - \frac{\partial}{\partial c} ,$$

$$\widetilde{\mathcal{B}}_1 = \tilde{\mathcal{A}}_1\big|_{a\leftrightarrow b} , \qquad \widetilde{\mathcal{B}}_2 = \tilde{\mathcal{A}}_2\big|_{a\leftrightarrow b} .$$

Note $\tilde{\mathcal{A}}_1\widetilde{\mathcal{B}}_1 - \widetilde{\mathcal{B}}_1\tilde{\mathcal{A}}_1$

The following theorem was proven in Sect. 2.3:

Theorem 2.4.5. *The statistics*

$$\begin{aligned} F_n(c; a_1, \ldots, a_{2r}; b_1, \ldots, b_{2s}) &:= \log P_n(E) \\ &:= \log P(\textit{all } (M_1\textit{-eigenvalues}) \in E_1,\ \textit{all } (M_2\textit{-eigenvalues}) \in E_2) , \end{aligned}$$

satisfies the third order nonlinear PDE:

$$\tilde{\mathcal{A}}_1\left(\frac{\widetilde{\mathcal{B}}_2\tilde{\mathcal{A}}_1 F_n}{\widetilde{\mathcal{B}}_1\tilde{\mathcal{A}}_1 F_n + nc/(1-c^2)}\right) = \widetilde{\mathcal{B}}_1\left(\frac{\tilde{\mathcal{A}}_2\widetilde{\mathcal{B}}_1 F_n}{\tilde{\mathcal{A}}_1\widetilde{\mathcal{B}}_1 F_n + nc/(1-c^2)}\right) . \tag{2.244}$$

Proof of Theorem 2.4.1. Changing limits of integration in the integral F_n defined by the measure (2.243) to agree with the integral G_n defined by the measure (2.222), we find the function G_n of Theorem 2.4.1 is related to the function F_n of Theorem 2.4.5 by a trivial rescaling:

$$\begin{aligned} &G_n(t; a_1, \dots, a_{2r}; b_1, \dots, b_{2s}) \\ &\quad = F_n\left(c; \frac{a_1}{\sqrt{(1-c^2)/2}}, \dots, \frac{a_{2r}}{\sqrt{(1-c^2)/2}}; \frac{b_1}{\sqrt{(1-c^2)/2}}, \dots, \frac{b_{2s}}{\sqrt{(1-c^2)/2}}\right) \end{aligned} \tag{2.245}$$

and applying the chain rule to (2.244) using (2.245) leads to Theorem 2.4.1 immediately, upon clearing denominators.

In order to prove the theorems concerning the Airy and Sine processes, we need a rigorous statement concerning the asymptotics of our Dyson, Airy and Sine kernels. To that end, letting

$$\begin{aligned} \mathcal{S}_1 &:= \left\{ t \mapsto \frac{t}{n^{1/3}}, s \mapsto \frac{s}{n^{1/3}}, x \mapsto \sqrt{2n+1} + \frac{u}{\sqrt{2}n^{1/6}}\right. \\ &\qquad\qquad\qquad\qquad \left. y \mapsto \sqrt{2n+1} + \frac{v}{\sqrt{2}n^{1/6}} \right\} \\ \mathcal{S}_2 &:= \left\{ t \mapsto \frac{\pi^2 t}{2n}, s \mapsto \frac{\pi^2 s}{2n}, x \mapsto \frac{\pi u}{\sqrt{2n}}, y \mapsto \frac{\pi v}{\sqrt{2n}} \right\}, \end{aligned} \tag{2.246}$$

we have:

Proposition 2.4.1. *Under the substitutions $\mathcal{S}_1$ and $\mathcal{S}_2$, the extended Hermite kernel tends with derivative, respectively, to the extended Airy and Sine kernel, when $n \to \infty$, uniformly for $u, v \in$ compact subsets $\subset \mathbb{R}$:*

$$\begin{aligned} \lim_{n\to\infty} K^{\mathrm{H},n}_{t,s}(x,y)\,\mathrm{d}y\big|_{\mathcal{S}_1} &= K^{\mathrm{A}}_{t,s}(u,v)\,\mathrm{d}v\,, \\ \lim_{n\to\infty} K^{\mathrm{H},n}_{t,s}(x,y)\,\mathrm{d}y\big|_{\mathcal{S}_2} &= \exp\left(-\frac{\pi^2}{2}(t-s)\right) K^{\mathrm{S}}_{t,s}(u,v)\,\mathrm{d}v\,. \end{aligned} \tag{2.247}$$

Remark. The proof involves careful estimating and Riemann–Hilbert techniques and is found in [6].

Proof of Theorem 2.4.2. Rescale in Theorem 2.4.1

$$a_i = \sqrt{2n} + \frac{u_i}{\sqrt{2}n^{1/6}}\,, \quad b_i = \sqrt{2n} + \frac{v_i}{\sqrt{2}n^{1/6}}\,, \quad t = \frac{\tau}{n^{1/3}} \tag{2.248}$$

and then from Proposition 2.4.1 it follows that, with derivatives that,

$$G_n\left(\frac{\tau}{n^{1/3}}, a; b\right) = G(\tau, u, v) + O\left(\frac{1}{k}\right), \quad k = n^{1/6}\,. \tag{2.249}$$

We now do large n asymptotics on the operators $\mathcal{A}_i$, $\mathcal{B}_i$, setting $L = L_u + L_v$, $E = E_u + E_v$, with L_u, L_v, E_u, E_v defined in Theorem 2.4.2; we find:

$$\begin{aligned}
\mathcal{A}_1 &= \sqrt{2}k\left(L - \left(\frac{\tau}{k^2} - \frac{\tau^2}{2k^4} + \frac{\tau^3}{6k^6}\right)L_v + O\left(\frac{1}{k^8}\right)\right),\\
\mathcal{B}_1 &= \sqrt{2}k\left(L - \left(\frac{\tau}{k^2} - \frac{\tau^2}{2k^4} + \frac{\tau^3}{6k^6}\right)L_u + O\left(\frac{1}{k^8}\right)\right),\\
\mathcal{A}_2 &= 2k^4\left(L - \frac{2\tau}{k^2}L_v + \frac{1}{2k^4}(E - 1 + 4\tau^2 L_v)\right.\\
&\qquad\left. - \frac{\tau}{k^6}\left(E_v - 1 + \frac{4}{3}\tau^2 L_v\right) + O\left(\frac{1}{k^8}\right)\right),\\
\mathcal{B}_2 &= 2k^4\left(L - \frac{2\tau}{k^2}L_u + \frac{1}{2k^4}(E - 1 + 4\tau^2 L_u)\right.\\
&\qquad\left. - \frac{\tau}{k^6}\left(E_u - 1 + \frac{4}{3}\tau^2 L_u\right) + O\left(\frac{1}{k^8}\right)\right),
\end{aligned} \tag{2.250}$$

and consequently

$$\begin{aligned}
&\frac{1}{2\sqrt{2}k^5}\mathcal{B}_2\mathcal{A}_1\\
&= L^2 - \frac{\tau}{k^2}(L + L_u)L + \frac{1}{2k^4}\left(L(E-2) + \tau^2\left(4L_u(L + L_v) + LL_v\right)\right)\\
&\quad - \frac{\tau}{k^6}\left(L(E_u - 2) + \frac{1}{2}L_v(E+2) + \frac{\tau^2}{6}(8LL_u + 18L_uL_v + LL_v)\right) + O\left(\frac{1}{k^8}\right)
\end{aligned} \tag{2.251}$$

$$\frac{1}{2k^2}\mathcal{B}_1\mathcal{A}_1 = L^2 - \frac{\tau}{k^2}L^2 + \frac{\tau^2}{k^4}\left(\frac{1}{2}L^2 + L_uL_v\right) - \frac{\tau^3}{k^6}\left(\frac{1}{6}L^2 + L_uL_v\right) + O\left(\frac{1}{k^8}\right).$$

Feeding these estimates and (2.249) into the relation (2.232) of Theorem 2.4.1, multiplied by $(\mathcal{B}_1\mathcal{A}_1G_n + 2nc)^2$, which by the quotient rule becomes an identity involving Wronskians, we then find ($\{f, g\}_X = gXf - fXg$)

$$
\begin{aligned}
0 &= \left\{\frac{1}{2\sqrt{2}k^5}\mathcal{B}_2\mathcal{A}_1 G_n, \frac{1}{2k^2}\left(\mathcal{B}_1\mathcal{A}_1 G_n + 2k^6\exp\left(\frac{-\tau}{k^2}\right)\right)\right\}_{\mathcal{A}_1/(\sqrt{2}k)} \\
&\qquad - \left\{\frac{1}{2\sqrt{2}k^5}\mathcal{A}_2\mathcal{B}_1 G_n, \frac{1}{2k^2}\left(\mathcal{A}_1\mathcal{B}_1 G_n + 2k^6\exp\left(\frac{-\tau}{k^2}\right)\right)\right\}_{\mathcal{B}_1/(\sqrt{2}k)} \\
&= \frac{2\tau}{k^2}\Big[\big((L_u+L_v)(L_uE_v - L_vE_u) + \tau^2(L_u-L_v)L_uL_v\big)G_n \\
&\qquad\qquad - \frac{1}{2}\{(L_u^2-L_v^2)H_n, (L_u+L_v)^2 G_n\}_{L_u+L_v}\Big] + O\left(\frac{1}{k^3}\right) \\
&= \frac{2\tau}{k^2}\Big[\big((L_u+L_v)(L_uE_v - L_vE_u) + \tau^2(L_u-L_v)L_uL_v\big)G \\
&\qquad\qquad - \frac{1}{2}\{(L_u^2-L_v^2)G, (L_u+L_v)^2 G\}_{L_u+L_v}\Big] + O\left(\frac{1}{k^3}\right) .
\end{aligned}
$$

In this calculation, we used the linearity of the Wronskian $\{X,Y\}_Z$ in its three arguments and the following commutation relations:

$$
[L_u, E_u] = L_u \ , \quad [L_u, E_v] = [L_u, L_v] = [L_u, \tau] = 0 \ , \quad [E_u, \tau] = \tau \ ,
$$

including their dual relations by $u \leftrightarrow v$; also we have $\{L^2G, 1\}_{L_u-L_v} = \{L(L_u-L_v)G, 1\}_L$. It is also useful to note that the two Wronskians in the first expression are dual to each other by $u \leftrightarrow v$. The point of the computation is to preserve the Wronskian structure up to the end. This proves Theorem 2.4.2, upon replacing $\tau \to t$.

Proof of Corollary 2.4.1. Equation (2.233) for the probability

$$
G(\tau; u, v) := \log P(A(\tau_1) \le u, A(\tau_2) \le v) \ , \quad \tau = \tau_2 - \tau_1 \ ,
$$

takes on the explicit form

$$
\begin{aligned}
\tau\frac{\partial}{\partial\tau}\left(\frac{\partial^2}{\partial u^2} - \frac{\partial^2}{\partial v^2}\right)G &= \frac{\partial^3 G}{\partial u^2\partial v}\left(2\frac{\partial^2 G}{\partial v^2} + \frac{\partial^2 G}{\partial u\partial v} - \frac{\partial^2 G}{\partial u^2} + u - v - \tau^2\right) \\
&\quad - \frac{\partial^3 G}{\partial v^2\partial u}\left(2\frac{\partial^2 G}{\partial u^2} + \frac{\partial^2 G}{\partial u\partial v} - \frac{\partial^2 G}{\partial v^2} - u + v - \tau^2\right) \\
&\quad + \left(\frac{\partial^3 G}{\partial u^3}\frac{\partial}{\partial v} - \frac{\partial^3 G}{\partial v^3}\frac{\partial}{\partial u}\right)\left(\frac{\partial}{\partial u} + \frac{\partial}{\partial v}\right)G \ . \qquad (2.252)
\end{aligned}
$$

This equation enjoys an obvious $u \leftrightarrow v$ duality. Finally the change of variables in the statement of Corollary 2.4.1 leads to (2.234).

The proof of Theorem 2.4.3 is done in the same spirit as that of Theorem 2.4.2 and Corollary 2.4.2 follows immediately by substitution in Theorem 2.4.3. Next, we need some preliminaries to prove Theorem 2.4.4. The first being:

Proposition 2.4.2. *The following ratio of probabilities admits the asymptotic expansion for large $t > 0$ in terms of functions $f_i(u,v)$, symmetric in u and v*

$$\frac{P(A(0) \le u, A(t) \le v)}{P(A(0) \le u)P(A(t) \le v)} = 1 + \sum_{i \ge 1} \frac{f_i(u,v)}{t^i}, \tag{2.253}$$

from which it follows that

$$\lim_{t \to \infty} P(A(0) \le u, A(t) \le v) = P(A(0) \le u)P(A(t) \le v) = F_2(u)F_2(v),$$

this means that the Airy process decouples at ∞.

The proof necessitates using the extended Airy kernel. Note, since the probabilities in (2.253) are symmetric in u and v, the coefficients f_i are symmetric as well. The last equality in the formula above follows from stationarity and (2.237).

Conjecture. The coefficients $f_i(u,v)$ have the property

$$\lim_{u \to \infty} f_i(u,v) = 0 \quad \textit{for fixed } v \in \mathbb{R}, \tag{2.254}$$

and

$$\lim_{z \to \infty} f_i(-z, z+x) = 0 \quad \textit{for fixed } x \in \mathbb{R}. \tag{2.255}$$

The justification for this plausible conjecture will now follow. First, considering the following conditional probability:

$$\begin{aligned} P(A(t) \le v \mid A(0) \le u) &= \frac{P(A(0) \le u, A(t) \le v)}{P(A(0) \le u)} \\ &= F_2(v)\left(1 + \sum_{i \ge 1} \frac{f_i(u,v)}{t^i}\right), \end{aligned}$$

and letting $v \to \infty$, we have automatically

$$\begin{aligned} 1 = \lim_{v \to \infty} P(A(t) \le v \mid A(0) \le u) &= \lim_{v \to \infty} \left[F_2(v)\left(1 + \sum_{i \ge 1} \frac{f_i(u,v)}{t^i}\right)\right] \\ &= 1 + \lim_{v \to \infty} \sum_{i \ge 1} \frac{f_i(u,v)}{t^i}, \end{aligned}$$

which would imply, assuming the interchange of the limit and the summation is valid,

$$\lim_{v \to \infty} f_i(u,v) = 0, \tag{2.256}$$

and, by symmetry

$$\lim_{u \to \infty} f_i(u,v) = 0.$$

To deal with (2.255) we assume the following *nonexplosion* condition for any fixed $t > 0$, $x \in \mathbb{R}$, namely, that the conditional probability satisfies

$$\lim_{z\to\infty} P(A(t) \geq x + z \mid A(0) \leq -z) = 0 \,.$$

Hence, the conditional probability satisfies, upon setting

$$v = z + x \,, \quad u = -z \,,$$

and using $\lim_{z\to\infty} F_2(z + x) = 1$, the following:

$$1 = \lim_{z\to\infty} P(A(t) \leq z + x \mid A(0) \leq -z) = 1 + \lim_{z\to\infty} \sum_{i\geq 1} \frac{f_i(-z, z + x)}{t^i} \,,$$

which, assuming the validity of the same interchange, implies that

$$\lim_{z\to\infty} f_i(-z, z + x) = 0 \quad \text{for all } i \geq 1 \,.$$

Proof of Theorem 2.4.4. Putting the log of the expansion (2.253)

$$\begin{aligned} G(t; u, v) &= \log P(A(0) \leq u, A(t) < v) \\ &= \log F_2(u) + \log F_2(v) + \sum_{i\geq 1} \frac{h_i(u, v)}{t^i} \\ &= \log F_2(u) + \log F_2(v) + \frac{f_1(u, v)}{t} + \frac{f_2(u, v) - f_1^2(u, v)/2}{t^2} + \cdots \,, \end{aligned} \tag{2.257}$$

into (2.252) leads to:

(i) a leading term of order t, given by

$$\mathcal{L} h_1 = 0 \,, \tag{2.258}$$

where

$$\mathcal{L} := \left(\frac{\partial}{\partial u} - \frac{\partial}{\partial v}\right) \frac{\partial^2}{\partial u \partial v} \,. \tag{2.259}$$

The most general solution to (2.258) is given by

$$h_1(u, v) = r_1(u) + r_3(v) + r_2(u + v) \,,$$

with arbitrary functions r_1, r_2, r_3. Hence,

$$P(A(0) \leq u, A(t) \leq v) = F_2(u) F_2(v) \left(1 + \frac{h_1(u, v)}{t} + \cdots\right)$$

with $h_1(u, v) = f_1(u, v)$ as in (2.253). Applying (2.254)

$$r_1(u) + r_3(\infty) + r_2(\infty) = 0 \quad \text{for all } u \in \mathbb{R} \,,$$

implying

$$r_1(u) = \text{ constant } = r_1(\infty)\,,$$

and similarly

$$r_3(u) = \text{ constant } = r_3(\infty)\,.$$

Therefore, without loss of generality, we may absorb the constants $r_1(\infty)$ and $r_3(\infty)$ in the definition of $r_2(u+v)$. Hence, from (2.257),

$$f_1(u,v) = h_1(u,v) = r_2(u+v)$$

and using (2.255),

$$0 = \lim_{z\to\infty} f_1(-z, z+x) = r_2(x)\,,$$

implying that the $h_1(u,v)$-term in the series (2.257) vanishes.

(ii) One computes that the term $h_2(u,v)$ in the expansion (2.257) of $G(t;u,v)$ sastisfies

$$\mathcal{L}h_2 = \frac{\partial^3 g}{\partial u^3}\frac{\partial^2 g}{\partial v^2} - \frac{\partial^3 g}{\partial v^3}\frac{\partial^2 g}{\partial u^2} \quad \text{with } g(u) := \log F_2(u)\,. \tag{2.260}$$

This is the term of order t^0, by putting the series (2.257) in (2.252). The most general solution to (2.260) is

$$h_2(u,v) = g'(u)g'(v) + r_1(u) + r_3(v) + r_2(u+v)\,.$$

Then

$$\begin{aligned} P(A(0)\le u, A(t)\le v) &= \mathrm{e}^{G(t;u,v)} \\ &= F_2(u)F_2(v)\exp\Bigl(\sum_{i\ge 2}\frac{h_i(u,v)}{t^i}\Bigr) \\ &= F_2(u)F_2(v)\Bigl(1+\frac{h_2(u,v)}{t^2}+\cdots\Bigr)\,. \end{aligned} \tag{2.261}$$

In view of the explicit formula for the distribution F_2 (2.237) and the behavior of $q(\alpha)$ for $\alpha \nearrow \infty$, we have that

$$\begin{aligned} \lim_{u\to\infty} g'(u) &= \lim_{u\to\infty}\bigl(\log F_2(u)\bigr)' \\ &= \lim_{u\to\infty}\int_u^\infty q^2(\alpha)\,\mathrm{d}\alpha = 0\,. \end{aligned}$$

Hence

$$0 = \lim_{u\to\infty} f_2(u,v) = \lim_{u\to\infty} h_2(u,v) = r_1(\infty) + r_3(v) + r_2(\infty)\,,$$

showing r_3 and similarly r_1 are constants. Therefore, by absorbing $r_1(\infty)$ and $r_3(\infty)$ into $r_2(u+v)$, we have

$$f_2(u,v) = h_2(u,v) = g'(u)g'(v) + r_2(u+v) \,.$$

Again, by the behavior of $q(x)$ at $+\infty$ and $-\infty$, we have for large $z>0$,

$$g'(-z)g'(z+x) = \int_{-z}^{\infty} q^2(\alpha)\,\mathrm{d}\alpha \int_{z+x}^{\infty} q^2(\alpha)\,\mathrm{d}\alpha \le cz^{3/2}\mathrm{e}^{-2z/3} \,.$$

Hence

$$0 = \lim_{z\to\infty} f_2(-z,z+x) = r_2(x)$$

and so

$$f_2(u,v) = h_2(u,v) = g'(u)g'(v) \,,$$

yielding the $1/t^2$ term in the series (2.257), and so it goes.

Finally, to prove (2.241), we compute from (2.239), after integration by parts and taking into account the boundary terms using (2.238):

$$\begin{aligned}
&E\big(A(0)A(t)\big) \\
&= \iint_{\mathbb{R}^2} uv \frac{\partial^2}{\partial u \partial v} P(A(0)\le u, A(t)\le v)\,\mathrm{d}u\,\mathrm{d}v \\
&= \int_{-\infty}^{\infty} uF_2'(u)\,\mathrm{d}u \int_{-\infty}^{\infty} vF_2'(v)\,\mathrm{d}v + \frac{1}{t^2}\int_{-\infty}^{\infty} F_2'(u)\,\mathrm{d}u \int_{-\infty}^{\infty} F_2'(v)\,\mathrm{d}v \\
&\qquad + \frac{1}{t^4}\iint_{\mathbb{R}^2} \big(\varPhi(u,v)+\varPhi(v,u)\big)\,\mathrm{d}u\,\mathrm{d}v + O\Big(\frac{1}{t^6}\Big) \\
&= \Big(E\big(A(0)\big)\Big)^2 + \frac{1}{t^2} + \frac{c}{t^4} + O\Big(\frac{1}{t^6}\Big) \,,
\end{aligned}$$

where

$$c := \iint_{\mathbb{R}^2} \big(\varPhi(u,v)+\varPhi(v,u)\big)\,\mathrm{d}u\,\mathrm{d}v = 2\iint_{\mathbb{R}^2} \varPhi(u,v)\,\mathrm{d}u\,\mathrm{d}v \,,$$

thus ending the proof of Theorem 2.4.4.

2.5 The Pearcey Distribution

2.5.1 GUE with an External Source and Brownian Motion

In this section we discuss the equivalence of GUE with an external source, introduced by Brézin–Hikami [27] and a conditional Brownian motion, following Aptkarev, Bleher and Kuijlaars [21].

Non-intersecting Brownian paths: Consider n-non-intersecting Brownian paths with predetermined endpoints at $t = 0, 1$, as specified in Figure 2.2.

By the Karlin–McGregor formula [42], the above situation has probability density

$$
\begin{aligned}
&p_n(t, x_1, \dots, x_n) \\
&\quad = \frac{1}{Z_n} \det\big(p(\alpha_i, x_j, t)\big)_{i,j=1}^n \det\big(p(x_i, a_j, 1-t)\big)_{i,j=1}^n \\
&\quad = \frac{1}{Z_n'} \prod_1^n \exp\left(\frac{-x_i^2}{t(1-t)}\right) \det\left(\exp\left(\frac{2a_i x_j}{t}\right)\right)_{i,j=1}^n \det\left(\exp\left(\frac{2a_i x_j}{1-t}\right)\right)_{i,j=1}^n ,
\end{aligned}
\tag{2.262}
$$

with[12]

$$
p(x, y, t) = \frac{\exp(-(x-y)^2/t)}{\sqrt{\pi t}} . \tag{2.263}
$$

For example,[13] let all the particles start out at $x = 0$, at $t = 0$, with n_1 particles ending up at a, n_2 ending up at $-a$ at $t = 1$, with $n = n_1 + n_2$.

Here

$$
p_{n_1,n_2}(t, x_1, \dots, x_n) = \frac{1}{Z_{n_1,n_2,a}} \Delta_{n_1+n_2}(x) \det \begin{pmatrix} \big(\psi_{i_+}^+(x_j)\big)_{\substack{1 \le i_+ \le n_1, \\ 1 \le j \le n_1+n_2}} \\ \big(\psi_{i_-}^-(x_j)\big)_{\substack{1 \le i_- \le n_2, \\ 1 \le j \le n_1+n_2}} \end{pmatrix} ,
$$

with

$$
\psi_i^{\pm}(x) = x^{i-1} \exp\left(\frac{-x^2}{t(1-t)} \pm \frac{2ax}{1-t}\right) .
$$

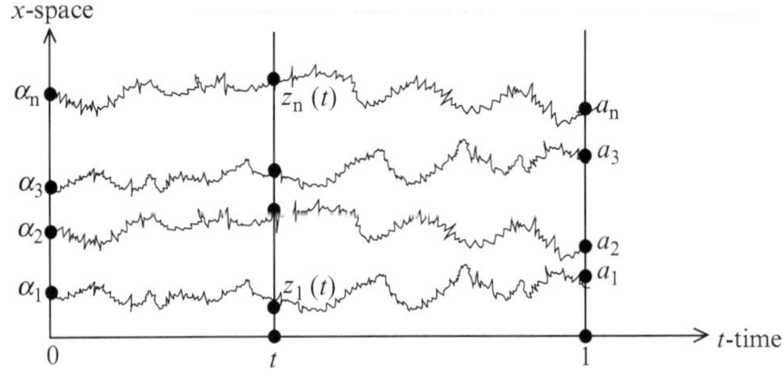

Fig. 2.2.

[12] Here $Z_n = Z_n(a, \alpha)$.

[13] Obviously, implicit in this example is a well-defined limit as the endpoints come together.

So setting $E = \bigcup_{1\le i\le r}[b_{2i-1}, b_{2i}]$, we find

$$\begin{aligned} P^a_{n_1,n_2}(t,b) &:= \mathrm{Prob}^a_{n_1,n_2}(\text{all } x_i(t) \subset E) \\ &:= \mathrm{Prob}\left(\text{all } x_i(t) \subset E \;\middle|\; \begin{array}{l} n_1 \text{ left-most paths end up at } a,\ n_2 \text{ right-} \\ \text{most paths end up at } -a, \text{ and all start at} \\ 0, \text{ with all paths non-intersecting.} \end{array}\right) \\ &= \frac{\int_{E^n} \prod_1^n \mathrm{d}x_i\, \Delta_n(x) \det \begin{pmatrix} \left(\psi^+_{i_+}(x_j)\right)_{\substack{1\le i_+\le n_1, \\ 1\le j\le n_1+n_2}} \\ \left(\psi^-_{i_-}(x_j)\right)_{\substack{1\le i_-\le n_2, \\ 1\le j\le n_1+n_2}} \end{pmatrix}}{\int_{\mathbb{R}^n} \prod_1^n \mathrm{d}x_i\, \Delta_n(x) \det \begin{pmatrix} \left(\psi^+_{i_+}(x_j)\right)_{\substack{1\le i_+\le n_1, \\ 1\le j\le n_1+n_2}} \\ \left(\psi^-_{i_-}(x_j)\right)_{\substack{1\le i_-\le n_2, \\ 1\le j\le n_1+n_2}} \end{pmatrix}} . \end{aligned} \tag{2.264}$$

Random matrix with external source: Consider the ensemble, introduced by Brézin–Hikami [27], on $n \times n$ Hermitian matrices $\mathcal{H}_n$

$$P\big(M \in (M, M + \mathrm{d}M)\big) = \frac{1}{Z_n} \exp\big(\mathrm{tr}(-V(M) + AM)\big)\, \mathrm{d}M , \tag{2.265}$$

with

$$A = \mathrm{diag}(a_1, \dots, a_n) .$$

By HCIZ (see (2.138)), we find

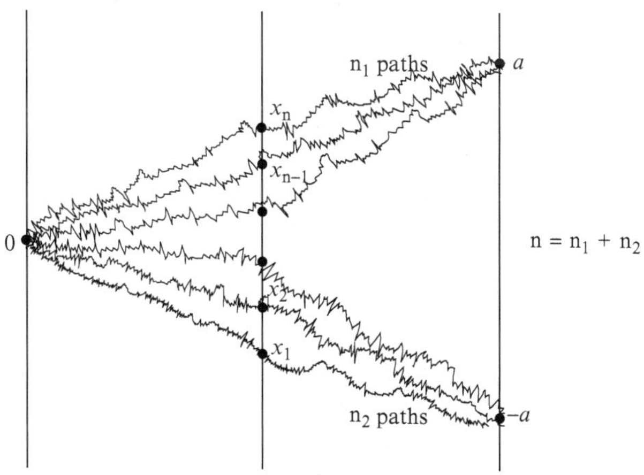

Fig. 2.3.

$$
\begin{aligned}
&P(\operatorname{spec}(M) \subset E)) \\
&= \frac{1}{Z_n} \int_{E^n} \Delta_n^2(z) \prod_1^n \exp(-V(z_i))\, \mathrm{d}z_i \int_{U(n)} \exp(\operatorname{tr} AU\Lambda U^{-1})\, \mathrm{d}U \\
&= \frac{1}{Z_n'} \int_{E^n} \Delta_n^2(z) \prod_1^n \exp(-V(z_i))\, \mathrm{d}z_i \frac{\det[\exp(a_i z_j)]_{i,j=1}^n}{\Delta_n(z)\Delta_n(a)} \\
&= \frac{1}{Z_n''} \int_{E^n} \Delta_n(z) \prod_1^n \exp(-V(z_i))\, \mathrm{d}z_i \det[\exp(a_i z_j)]_{i,j=1}^n . \qquad (2.266)
\end{aligned}
$$

For example: consider the limiting case

$$
A = \operatorname{diag}\,(\underbrace{-a, -a, \ldots, -a}_{n_2}, \underbrace{a, a, \ldots, a}_{n_1}), \quad E = \bigcup_{i=1}^r [b_{2i-1}, b_{2i}], \qquad (2.267)
$$

$$
P(\operatorname{spec}(M) \subset E) := P_{n_1,n_2}(a,b) = \frac{\int_{E^n} \prod_1^n \mathrm{d}x_i \Delta_n(z) \det \begin{pmatrix} (\rho_{i_+}^+(x_j))_{\substack{1\le i_+\le n_1, \\ 1\le j\le n_1+n_2}} \\ (\rho_{i_-}^-(x_j))_{\substack{1\le i_-\le n_2, \\ 1\le j\le n_1+n_2}} \end{pmatrix}}{\int_{\mathbb{R}^n} \prod_1^n \mathrm{d}x_i \Delta_n(z) \det \begin{pmatrix} (\rho_{i_+}^+(x_j))_{\substack{1\le i_+\le n_1, \\ 1\le j\le n_1+n_2}} \\ (\rho_{i_-}^-(x_j))_{\substack{1\le i_-\le n_2, \\ 1\le j\le n_1+n_2}} \end{pmatrix}} . \qquad (2.268)
$$

where

$$
\rho_i^{\pm}(z) = z^{i-1} \exp(-V(z) \pm az), \quad n = n_1 + n_2 . \qquad (2.269)
$$

Then we have by Aptkarev, Bleher, Kuijlaars [21]:

Non-intersecting Brownian motion ⇔ GUE *with external source*

$$
P_{n_1,n_2}^a(t,b) = P_{n_1,n_2}\left(\sqrt{\frac{2t}{1-t}}a, \sqrt{\frac{2}{t(1-t)}}b\right)\bigg|_{V(z)=z^2/2}, \qquad (2.270)
$$

so the two problems: non-intersecting Brownian motion and GUE with an external source, are equivalent!

2.5.2 MOPS and a Riemann–Hilbert Problem

In this section, we introduce multiple orthogonal polynomials (MOPS), following Bleher and Kuijlaars [24]. This lead them to a determinal m-point correlation function for the GUE with external source, in terms of a "Christoffel–Darboux" kernel for the MOPS, as in the pure GUE case. In addition, they formulated a Riemann–Hilbert (RH) problem for the MOPS, analogous to that for classical orthogonal polynomials, thus enabling them to understand universal behavior for the MOPS and hence universal behavior for the GUE with external source (see [21, 25]).

Let us first order the spectrum of A of (2.267) in some definite fashion, for example

Ordered spectrum of $(A) = (-a, a, a, -a, \dots, -a) := (\alpha_1, \alpha_2, \dots, \alpha_n)$.

For each $k = 0, 1, \dots, n$, let $k = k_1 + k_2$, k_1, k_2 defined as follows:

$$\begin{aligned} k_1 &:= \#\text{ times } a \text{ appears in } \alpha_1, \dots, \alpha_k \;, \\ k_2 &:= \#\text{ times } -a \text{ appears in } \alpha_1, \dots, \alpha_k \;. \end{aligned} \tag{2.271}$$

We now define the 2 kinds of MOPS.

MOP II: Define a unique monic kth degree polynomial $p_k = p_{k_1,k_2}$:

$$p_k(x) = p_{k_1,k_2}(x)\colon \int_{\mathbb{R}} p_{k_1,k_2}(x)\rho_{i_\pm}^{\pm}(x)\,\mathrm{d}x = 0$$
$$\rho_i^{\pm}(x) = x^{i-1}\exp(-V(x) \pm ax) \;, \quad 1 \le i_+ \le k_1 \;,\; 1 \le i_- \le k_2 \;, \tag{2.272}$$

MOP I: Define unique polynomials $q^+_{k_1-1,k_2}(x)$, $q^-_{k_1,k_2-1}(x)$ of respective degrees $k_1 - 1$, $k_2 - 1$:

$$q_{k-1}(x) := q_{k_1,k_2}(x) = q^+_{k_1-1,k_2}(x)\rho_1^+(x) + q^-_{k_1,k_2-1}(x)\rho_1^-(x)\colon$$
$$\int_{\mathbb{R}} x^j q_{k-1}(x)\,\mathrm{d}x = \delta_{j,k-1} \;, \quad 0 \le j \le k-1 \;, \tag{2.273}$$

which immediately yields:

Bi-orthonal polynomials:

$$\int_{\mathbb{R}} p_j(x)q_k(x)\,\mathrm{d}x = \delta_{j,k} \quad j,k = 0,1,\dots,n-1 \;. \tag{2.274}$$

This leads to a Christoffel–Darboux kernel, as in (2.5):

$$K^{(a)}_{n_1,n_2}(x,y) := K_n(x,y) := \exp\!\left(-\frac{1}{2}V(x) + \frac{1}{2}V(y)\right)\sum_0^{n-1} p_k(x)q_k(y) \;, \tag{2.275}$$

which is independent of the ad hoc ordering of the spectrum of A and which, due to bi-orthonality, has the usual reproducing properties:

$$\int_{-\infty}^{\infty} K_n(x,x)\,\mathrm{d}x = n \;, \quad \int_{-\infty}^{\infty} K_n(x,y)K_n(y,z)\,\mathrm{d}y = K_n(x,z) \;. \tag{2.276}$$

The joint probability density can be written in terms of K_n,

$$\frac{1}{Z_n}\exp\bigl(\mathrm{Tr}(-V(\Lambda) + A\Lambda)\bigr)\Delta_n(\lambda) = \frac{1}{n!}\det[K_n(\lambda_i,\lambda_j)]_{i,j=1}^n \;, \tag{2.277}$$

with $\Lambda = \mathrm{diag}(\lambda_1, \dots, \lambda_n)$, yielding the m-point correlation function:

$$R_m(\lambda_1, \dots, \lambda_m) = \det[K_n(\lambda_i,\lambda_j)]_{i,j=1}^m \;, \tag{2.278}$$

and we find the usual Fredholm determinant formula:

$$P(\mathrm{spec}(M) \subset E^c) = \det\bigl(I - K_n(x,y)I_E(y)\bigr) \;. \tag{2.279}$$

Finally, we have a Riemann–Hilbert (RH) problem for the MOPS.

Riemann–Hilbert problem for MOPS: MOP II:

$$Y(z) := \begin{bmatrix} p_{n_1,n_2}(z) & C_+ p_{n_1,n_2} & C_- p_{n_1,n_2} \\ c_1 p_{n_1-1,n_2}(z) & c_1 C_+ p_{n_1-1,n_2} & c_1 C_- p_{n_1-1,n_2} \\ c_2 p_{n_1,n_2-1}(z) & c_2 C_+ p_{n_1,n_2-1} & c_2 C_- p_{n_1,n_2-1} \end{bmatrix} \tag{2.280}$$

with $C_\pm$ Cauchy transforms:

$$C_\pm f(z) := \frac{1}{2\pi \mathrm{i}} \int_{\mathbb{R}} \frac{f(s)\rho_1^\pm(s)\,\mathrm{d}s}{s-z}\,, \qquad \rho_1^\pm(z) = \exp(-V(z) \pm az)\,. \tag{2.281}$$

Then $Y(z)$ satisfies the RH problem:

1. $Y(z)$ analytic on $\mathbb{C} \setminus \mathbb{R}$.
2. Jump condition for $x \in \mathbb{R}$:

$$Y_+(x) = Y_-(x) \begin{pmatrix} 1 & \rho_1^+(x) & \rho_1^-(x) \\ 0 & 1 & 0 \\ 0 & 0 & 1 \end{pmatrix}\,.$$

3. Behavior as $z \to \infty$

$$Y(z) = \left(I + O\left(\frac{1}{z}\right)\right) \begin{pmatrix} z^{n_1+n_2} & & \\ & z^{-n_1} & \\ & & z^{-n_2} \end{pmatrix}\,. \tag{2.282}$$

MOP I: A dual RH problem for $q^\pm_{k_1,k_2}$ and $(Y^{-1})^T$.

Finally we have a Christoffel–Darboux type formula (see (2.5)) for the kernel $K^{(a)}_{n,n}(x,y)$ of (2.275) expressed in terms of the RH matrix (2.280):

$$K^{(a)}_{n,n}(x,y) = \frac{\exp(-\frac{1}{4}(x^2+y^2))}{2\pi \mathrm{i}(x-y)}(0, \mathrm{e}^{ay}, \mathrm{e}^{-ay})Y^{-1}(y)Y(x) \begin{pmatrix} 1 \\ 0 \\ 0 \end{pmatrix}\,. \tag{2.283}$$

Thus to understand the large n asymptotics of the GUE with external source, from (2.277), (2.278), and (2.279), it suffices to understand the asymptotics of $K^{(a)}_{n,n}(x,y)$ given by (2.275). Thus by (2.283) it suffices to understand the asymptotics of the solution $Y(z)$ to the RH problem of (2.282), which is the subject of [21] and [25].

2.5.3 Results Concerning Universal Behavior

In this section we will first discuss universal behavior for the equivalent random matrix and Brownian motion models, leading to the Pearcey process. We will then give a PDE of Adler–van Moerbeke [7] governing this behavior, and finally a PDE for the n-time correlation function of this process, deriving the first PDE in the following sections. The following pictures illustrate the situation.

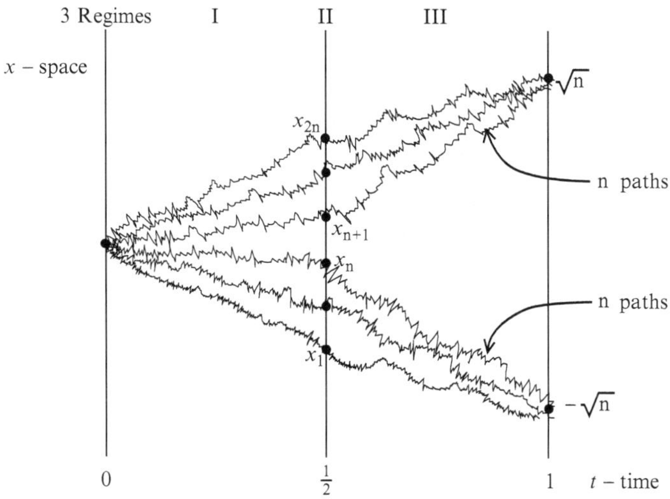

Fig. 2.4.

Universal behavior:

I. *Brownian motion:* $2n$ paths, $a = \sqrt{n}$
   At $t = \frac{1}{2}$ the Brownian paths start to separate into 2 distinct groups.
II. *Random matrices:* $n_1 = n_2 = n$, $V(z) = z^2/2$, $a := \hat{a}\sqrt{2n}$.

   Density of eigenvalues: $\rho(x) := \lim_{n\to\infty}(K_{n,n}^{\hat{a}\sqrt{2n}}(\sqrt{2n}x, \sqrt{2n}x))/2n$

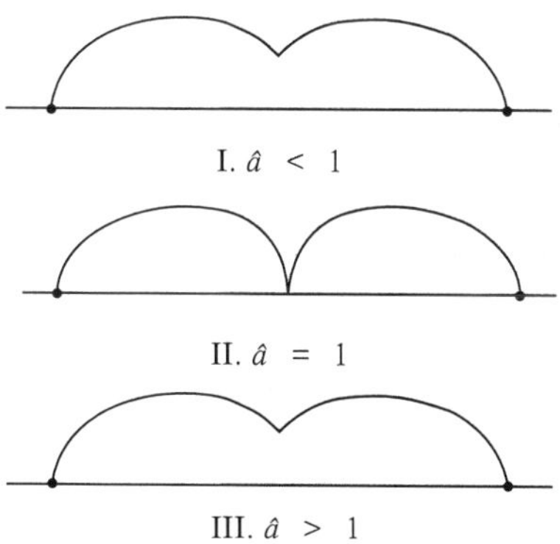

Fig. 2.5.

The 3 corresponding regimes, I, II and III, for the random matrix density of states $\rho(x)$ and thus the corresponding situation for the Brownian motion, are explained by the following theorem:

Theorem 2.5.1 (Aptkarev, Bleher, Kuijlaars [21]). *For the* GUE *with external source, the limiting mean density of states for* $\hat{a} > 0$ *is:*

$$\rho(x) := \lim_{n\to\infty} \frac{K_{n,n}^{\hat{a}\sqrt{2n}}(\sqrt{2n}x, \sqrt{2n}x)}{2n} = \frac{1}{\pi}|\operatorname{Im}\xi(x)| , \tag{2.284}$$

with

$$\xi(x)\colon \xi^3 - x\xi^2 - (\hat{a}^2 - 1)\xi + x\hat{a}^2 = 0 \quad (\textit{Pastur's equation } [45]) ,$$

yielding the density of eigenvalues pictures.

It is natural to look for universal behavior near $\hat{a} = 1$ by looking with a microscope about $x = 0$. Equivalently, thinking in terms of the $2n$-Brownian motions, one sets $a = \sqrt{n}$ and about $t = \frac{1}{2}$ one looks with a microscope near $x = 0$ to see the $2n$-Brownian motions slowly separating into two distinct groups. Rescale as follows:

$$t = \frac{1}{2} + \frac{\tau}{\sqrt{n}}, \quad x = \frac{u}{n^{1/4}}, \quad a = \sqrt{n} . \tag{2.285}$$

Remembering the equivalence between Brownian motion and the GUE with external source, namely (2.270), the Fredholm determinant formula (2.279), for the GUE with external source, yields:

$$\operatorname{Prob}_{n_1,n_2}^{a}(\text{all } x_i(t) \subset E^{c}) = \det(I - \widetilde{K}_n^E) ,$$

$$\widetilde{K}_n^E(x,y) = \sqrt{\frac{2}{t(1-t)}}K_{n_1,n_2}^{\sqrt{2ta/(1-t)}}\left(\sqrt{\frac{2}{t(1-t)}}\,x, \sqrt{\frac{2}{t(1-t)}}\,y\right)I_E(y) . \tag{2.286}$$

Universal behavior for $n \to \infty$ amounts to understanding $\widetilde{K}_n^E(x,y)$ for $n \to \infty$, under the rescaling (2.285) and indeed we have the results:

Universal Pearcey limiting behavior:

Theorem 2.5.2 (Tracy–Widom [54]). *Upon rescaling the Brownian kernel, we find the following limiting behavior, with derivates*:

$$\lim_{n\to\infty} n^{-1/4}\sqrt{\frac{2}{t(1-t)}}K_{n,n}^{\sqrt{2ta/(1-t)}}\left(\sqrt{\frac{2}{t(1-t)}}\,x, \sqrt{\frac{2}{t(1-t)}}\,y\right)\Bigg|_{\substack{a=\sqrt{n},\\ t=\frac{1}{2}+\tau/\sqrt{n},\\ (x,y)=(u,v)/n^{1/4}}} = K_\tau^P(u,v) ,$$

with the Pearcy kernel $K_\tau(u,v)$ *of Brézin–Hikami* [27] *defined as follows*:

$$\begin{aligned} K_\tau(x,y) &:= \frac{p(x)q''(y) - p'(x)q'(y) + p''(x)q(y) - \tau p(x)q(y)}{x-y} \\ &= \int_0^\infty p(x+z)q(y+z)\,\mathrm{d}z , \end{aligned} \tag{2.287}$$

where (note $\omega = \mathrm{e}^{\mathrm{i}\pi/4}$)

$$\begin{aligned} p(x) &:= \frac{1}{2\pi} \int_{-\infty}^{\infty} \exp\left(-\frac{u^4}{4} - \tau\frac{u^2}{2} - \mathrm{i}ux\right) \mathrm{d}u \,, \\ q(y) &:= \frac{1}{2\pi} \int_X \exp\left(\frac{u^4}{4} - \tau\frac{u^2}{2} + uy\right) \mathrm{d}u \\ &= \operatorname{Im}\left[\frac{\omega}{\pi} \int_0^{\infty} \mathrm{d}u \, \exp\left(-\frac{u^4}{4} - \mathrm{i}\tau\frac{u^2}{2}\right)(\mathrm{e}^{\omega uy} - \mathrm{e}^{-\omega uy})\right] \end{aligned} \tag{2.288}$$

satisfy the differential equations (adjoint to each other)

$$p''' - \tau p' - xp = 0 \quad \text{and} \quad q''' - \tau q' + yq = 0 \,.$$

The contour X is given by the ingoing rays from $\pm\infty \exp(\mathrm{i}\pi/4)$ to 0 and the outgoing rays from 0 to $\pm\infty \exp(-\mathrm{i}\pi/4)$.

Theorem 2.5.2 allows us to define the Pearcey process $\mathcal{P}(\tau)$ as the motion of an infinite number of non-intersecting Brownian paths, near $t = \frac{1}{2}$, upon taking a limit in (2.286), using the precise scaling of (2.285), to wit:

$$\begin{aligned} \lim_{n\to\infty} \operatorname{Prob}_{n,n}^{\sqrt{n}}\left(\text{all } n^{1/4}x_i\left(\frac{1}{2} + \frac{\tau}{\sqrt{n}}\right) \notin E\right) &= \det(I - K_\tau^P I_E) \\ &=: \operatorname{Prob}(\mathcal{P}(\tau) \notin E) \,, \end{aligned} \tag{2.289}$$

which defines for us the Pearcey process. Note the pathwise interpretation of $\mathcal{P}(\tau)$ certainly needs to be resolved. The Pearcey distribution with the parameter τ can also be interpreted as the transitional probability for the Pearcey process. We now give a PDE for the distribution, which shall be derived in the following section:

Theorem 2.5.3 (Adler–van Moerbeke [7]).
For compact $E = \bigcup_{i=1}^r [u_{2i-1}, u_{2i}]$,

$$F(\tau; u_1, \ldots, u_{2r}) := \log \operatorname{Prob}(\mathcal{P}(\tau) \notin E) \tag{2.290}$$

satisfies the following 4th order, 3rd degree PDE in τ and the u_i:

$$B_{-1}\left(\frac{\frac{1}{2}\partial^3 F/\partial\tau^3 + (B_0 - 2)B_{-1}^2 F + \{B_{-1}\partial F/\partial\tau, B_{-1}^2 F\}_{B_{-1}}/16}{B_{-1}^2 \partial F/\partial\tau}\right) = 0 \,, \tag{2.291}$$

where

$$B_{-1} = \sum_1^{2r} = \frac{\partial}{\partial u_i} \,, \quad B_0 = \sum_1^{2r} = u_i \frac{\partial}{\partial u_i} \,. \tag{2.292}$$

It is natural to ask about the joint Pearcey distribution involving k-times, namely:

$$\lim_{n\to\infty} \operatorname{Prob}_{n,n}^{\sqrt{n}}\left(\text{all } n^{1/4}x_i\left(\frac{1}{2}+\frac{\tau_j}{\sqrt{n}}\right)\notin E_j, 1\le j\le k\right)$$
$$= \operatorname{Prob}(\mathcal{P}(\tau_j)\notin E_j, 1\le i\le k)$$
$$= \det\left(I-(I_{E_i}K^P_{\tau_i\tau_j}I_{E_j})^k_{i,j=1}\right), \quad (2.293)$$

where the above is a Fredholm determinant involving a matrix kernel, and the extended Pearcey kernel of Tracy–Widom [54] $K^P_{\tau_i\tau_j}$, is given by

$$K^P_{\tau_i\tau_j}(x,y) = -\frac{1}{4\pi^2}\int_X\int_{-i\infty}^{i\infty}\exp\left(-\frac{s^4}{4}+\tau_j\frac{s^2}{2}-ys+\frac{t^4}{4}-\tau_i\frac{t^2}{2}+xt\right)\frac{\mathrm{d}s\,\mathrm{d}t}{s-t}, \quad (2.294)$$

with X the same contour of (2.288). We note $K^P_{\tau\tau}(x,y)=K^P_\tau(x,y)$ of (2.287), the Brézin–Hikami Pearcey kernel. We then have the analogous theorem to Theorem 2.5.3 (which we will not prove here) namely:

Theorem 2.5.4 (Adler–van Moerbeke [12]).
For compact $E_j=\bigcup_{i=1}^{r_j}[u^{(j)}_{2i-1}u^{(j)}_{2i}]$, $1\le j\le k$,

$$F(\tau_1,\tau_2,\ldots,\tau_k;u^{(1)},u^{(2)},\ldots,u^{(k)}) := \log\operatorname{Prob}(P(\tau_j)\notin E_j, 1\le j\le k) \quad (2.295)$$

satisfies the following 4*th order,* 3*rd degree* PDE *in* τ_i *and* $u^{(j)}$: $\mathcal{D}_{-1}X=0$, *with*

$$X:=\frac{4(E_{-1}^2-\widetilde{\mathcal{D}}_{-1}\mathcal{D}_{-1})E_{-1}F+(2E_0+\mathcal{D}_0-2)\mathcal{D}_{-1}^2F+\frac{1}{8}\{\mathcal{D}_{-1}E_{-1}F,\mathcal{D}_{-1}^2F\}_{\mathcal{D}_{-1}}}{\mathcal{D}_{-1}^2E_{-1}F} \quad (2.296)$$

where

$$\mathcal{D}_j:=\sum_{i=1}^k B_j(u^{(i)}), \qquad \widetilde{\mathcal{D}}_{-1}:=\sum_{i=1}^k \tau_i B_{-1}(u^{(i)}),$$
$$B_j(u^{(i)}):=\sum_{l=1}^{2r_i}(u_l^{(i)})^{j+1}\frac{\partial}{\partial u_l^{(i)}}, \qquad E_j:=\sum_{i=1}^k \tau_i^{j+1}\frac{\partial}{\partial\tau_i}.$$

2.5.4 3-KP Deformation of the Random Matrix Problem

In this section we shall deform the measures (2.269) in the random matrix problem, for $V(z)=z^2/2$, so as to introduce 3-KP τ-functions into the picture, and using the bilinear identities, we will derive some useful 3-KP PDE for these τ-functions. The probability distribution for the GUE with external source was given by (2.268), to wit:

$$
\begin{aligned}
P(\operatorname{spec}(M) \subset E) &:= P_{n_1,n_2}(E) \\
&= \frac{1}{Z_n} \int_{E^n} \prod_1^n \mathrm{d}z_i \, \Delta_n(z) \det \begin{pmatrix} \left(\rho^+_{i_+}(z_j)\right)_{\substack{1 \le i_+ \le n_1, \\ 1 \le j \le n_1+n_2}} \\ \left(\rho^-_{i_-}(z_j)\right)_{\substack{1 \le i_- \le n_2, \\ 1 \le j \le n_1+n_2}} \end{pmatrix}
\end{aligned} \tag{2.297}
$$

where $\rho_i^{\pm}(z) = z^{i-1}\exp(-z^2/2 \pm az)$. Let us deform $\rho_i^{\pm}(z)$ as follows:

$$
\rho_i^{\pm}(z) \to \hat{\rho}_i^{\pm}(z) := z^{i-1}\exp\left(-\frac{z^2}{2} \pm az \pm \beta z^2\right)\exp\left(\sum_{k=1}^{\infty}(t_k - s_k^{\pm})z^k\right), \tag{2.298}
$$

yielding a deformation of the probability:

$$
P_{n_1,n_2}(E) \to \frac{\tau_{n_1,n_2}(t,s^+,s^-,E)}{\tau_{n_1,n_2}(t,s^+,s^-,\mathbb{R})} . \tag{2.299}
$$

Where, by the same argument used to derive (2.140),

$$
\tau_{n_1,n_2}(t,s^+,s^-,E) := \det m_{n_1,n_2}(t,s^+,s^-,E) , \tag{2.300}
$$

with

$$
m_{n_1,n_2}(t,s^+,s^-,E) := \begin{Bmatrix} [\mu^+_{ij}]_{\substack{1 \le i_+ \le n_1, \\ 0 \le j \le n_1+n_2-1}} \\ [\mu^-_{ij}]_{\substack{1 \le i_- \le n_2, \\ 0 \le j \le n_1+n_2-1}} \end{Bmatrix} , \tag{2.301}
$$

and

$$
\mu^{\pm}_{ij}(t,s^+,s^-) := \int_E \hat{\rho}^{\pm}_{i+j}(z)\,\mathrm{d}z .
$$

We also need the identity $(n = n_1 + n_2)$

$$
\begin{aligned}
&\tau_{n_1,n_2}(t,s^+,s^-,E) \\
&:= \det m_{n_1,n_2}(t,s^+,s^-,E) \\
&= \frac{1}{n_1!n_2!} \int_{E^n} \Delta_n(x,y) \prod_{j=1}^{n_1} \exp\left(\sum_1^{\infty} t_i x_j^i\right) \prod_{j=1}^{n_2} \exp\left(\sum_1^{\infty} t_i y_j^i\right) \\
&\quad \times \left(\Delta_{n_1}(x) \prod_{j=1}^{n_1} \exp\left(-\frac{x_j^2}{2} + ax_j + \beta x_j^2\right) \exp\left(-\sum_1^{\infty} s_i^+ x_j^i\right) \mathrm{d}x_j\right) \\
&\quad \times \left(\Delta_{n_2}(y) \prod_{j=1}^{n_2} \exp\left(-\frac{y_j^2}{2} - ay_j - \beta y_j^2\right) \exp\left(-\sum_1^{\infty} s_i^- y_j^i\right) \mathrm{d}y_j\right) .
\end{aligned} \tag{2.302}
$$

That the above is a 3-KP deformation is the content of the following theorem.

Theorem 2.5.5 (Adler–van Moerbeke–Vanhaecke [15]). *Given the functions τ_{n_1,n_2} as in (2.300), the wave matrices*

$$
\begin{aligned}
\Psi^{+}_{n_1,n_2}(\lambda;t,s^+,s^-) &:= \frac{1}{\tau_{n_1,n_2(t,s^+,s^-)}} \\
&\times \begin{pmatrix} \psi^{(1)^+}_{n_1,n_2} & (-1)^{n_2}\psi^{(2)^+}_{n_1+1,n_2} & \psi^{(3)^+}_{n_1,n_2+1} \\ (-1)^{n_2}\psi^{(1)^+}_{n_1-1,n_2} & \psi^{(2)^+}_{n_1,n_2} & (-1)^{n_2}\psi^{(3)^+}_{n_1-1,n_2+1} \\ \psi^{(1)}_{n_1,n_2-1} & (-1)^{n_2+1}\psi^{(2)^+}_{n_1+1,n_2-1} & \psi^{(3)+}_{n_1,n_2} \end{pmatrix}, \\
\Psi^{-}_{n_1,n_2}(\lambda;t,s^+,s^-) &:= \frac{1}{\tau_{n_1,n_2(t,s^+,s^-)}} \\
&\times \begin{pmatrix} \psi^{(1)^-}_{n_1,n_2} & (-1)^{n_2+1}\psi^{(2)^-}_{n_1-1,n_2} & -\psi^{(3)^-}_{n_1,n_2-1} \\ (-1)^{n_2+1}\psi^{(1)^-}_{n_1+1,n_2} & \psi^{(2)^-}_{n_1,n_2} & (-1)^{n_2}\psi^{(3)^-}_{n_1+1,n_2-1} \\ -\psi^{(1)^-}_{n_1,n_2+1} & (-1)^{n_2+1}\psi^{(2)^-}_{n_1-1,n_2+1} & \psi^{(3)^-}_{n_1,n_2} \end{pmatrix},
\end{aligned} \tag{2.303}
$$

with wave functions

$$
\begin{aligned}
\psi^{(1)\pm}_{n_1,n_2}(\lambda;t,s^+,s^-) &:= \lambda^{\pm(n_1+n_2)} \exp\Big(\pm\sum_1^\infty t_i\lambda^i\Big)\tau_{n_1,n_2}(t\mp[\lambda^{-1}],s^+,s^-)\,, \\
\psi^{(2)\pm}_{n_1,n_2}(\lambda;t,s^+,s^-) &:= \lambda^{\mp n_1} \exp\Big(\pm\sum_1^\infty s_i^+\lambda^i\Big)\tau_{n_1,n_2}(t,s^+\mp[\lambda^{-1}],s^-)\,, \\
\psi^{(3)}_{n_1,n_2}(\lambda;t,s^+,s^-) &:= \lambda^{\mp n_2} \exp\Big(\pm\sum_1^\infty s_i^-\lambda^i\Big)\tau_{n_1,n_2}(t,s^+,s^-\mp[\lambda^{-1}])\,,
\end{aligned} \tag{2.304}
$$

satisfy the bilinear identity,

$$
\oint_\infty \Psi^+_{k_1,k_2}\Psi^{-T}_{l_1,l_2}\mathrm{d}\lambda = 0\,, \quad \forall k_1k_2,l_1,l_2,\ \forall t,s^\pm,\underline{t},\underline{s}^\pm\,, \tag{2.305}
$$

of which the $(1,1)$ *component spelled out is:*

$$
\begin{aligned}
&\oint_\infty \tau_{k_1,k_2}(t-[\lambda^{-1}],s^+,s^-)\tau_{l_1,l_2}(\underline{t}+[\lambda^{-1}],\underline{s}^+,\underline{s}^-) \\
&\qquad\qquad \times \lambda^{k_1+k_2-l_1-l_2}\exp\Big(\sum_1^\infty(t_i-\underline{t}_i)\lambda^i\Big)\mathrm{d}\lambda \\
&\quad -(-1)^{k_2+l_2}\oint_\infty \tau_{k_1+1,k_2}(t,s^+-[\lambda^{-1}],s^-)\tau_{l_1-1,l_2}(\underline{t},\underline{s}^++[\lambda^{-1}],\underline{s}^-) \\
&\qquad\qquad \times \lambda^{l_1-k_1-2}\exp\Big(\sum_1^\infty(s_i^+-\underline{s}_i^+)\lambda^i\Big)\mathrm{d}\lambda \\
&\quad -\oint_\infty \tau_{k_1,k_2+1}(t,s^+,s^--[\lambda^{-1}])\tau_{l_1,l_2-1}(\underline{t},\underline{s}^+,\underline{s}^-+[\lambda^{-1}]) \\
&\qquad\qquad \times \lambda^{l_2-k_2-2}\exp\Big(\sum_1^\infty(s_i^--\underline{s}_i^-)\lambda^i\Big)\mathrm{d}\lambda = 0\,.
\end{aligned} \tag{2.306}
$$

Sketch of Proof: The proof is via the MOPS of Sect. 2.5.2. We use the formal Cauchy transform, thinking of z as large:

$$C_{\pm}f(z) := \int_{\mathbb{R}} \frac{f(s)\hat{\rho}_1^{\pm}(s)}{z-s}\,\mathrm{d}s := \sum_{i\geq 1} \frac{1}{z^i} \int \hat{\rho}_1^{\pm}(s) s^{i-1}\,\mathrm{d}s\ , \tag{2.307}$$

which should be compared with the Cauchy transform of (2.281), which we used in the Riemann Hilbert problem involving MOPS, and let C_0 denote the Cauchy transform with 1 instead of $\hat{\rho}_1^{\pm}$. We now make the following identification between the MOPS of (2.272), (2.273) defined with $\rho_i^{\pm} \to \hat{\rho}_i^{\pm}$ (and so dependent on t, s^+, s^-) and their Cauchy transforms and shifted τ-functions, namely:

$$\begin{aligned}
p_{k_1,k_2}(\lambda) &= \lambda^{k_1+k_2} \frac{\tau_{k_1,k_2}(t-[\lambda^{-1}], s^+, s^-)}{\tau_{k_1,k_2}(t, s^+, s^-)}\ ,\\
C_+ p_{k_1,k_2}(\lambda) &= \lambda^{-k_1-1} \frac{\tau_{k_1+1,k_2}(t, s^+ - [\lambda^{-1}], s^-)(-1)^{k_2}}{\tau_{k_1,k_2}(t, s^+, s^-)}\ ,\\
C_- p_{k_1,k_2}(\lambda) &= \lambda^{-k_2-1} \frac{\tau_{k_1,k_2+1}(t, s^+, s^- - [\lambda^{-1}])}{\tau_{k_1,k_2}(t, s^+, s^-)}\ ,\\
C_0 q_{k_1,k_2}(\lambda) &= \lambda^{-k_1-k_2} \frac{\tau_{k_1,k_2}(t+[\lambda^{-1}], s^+, s^-)}{\tau_{k_1,k_2}(t, s^+, s^-)}\ ,\\
q^+_{k_1-1,k_2}(\lambda) &= \lambda^{k_1-1} \frac{\tau_{k_1-1,k_2}(t, s^+ + [\lambda^{-1}], s^-)(-1)^{k_2+1}}{\tau_{k_1,k_2}(t, s^+, s^-)}\ ,\\
q^-_{k_1,k_2-1}(\lambda) &= -\lambda^{k_2-1} \frac{\tau_{k_1,k_2-1}(t, s^+, s^- + [\lambda^{-1}])}{\tau_{k_1,k_2}(t, s^+, s^-)}\ ,
\end{aligned} \tag{2.308}$$

where $p_{k_1,k_2}(\lambda)$ was the MOP of the second kind and $q_{k_1,k_2}(\lambda) = q^+_{k_1-1,k_2}(\lambda)\hat{\rho}_1^+ + q^-_{k_1,k_2-1}(\lambda)\hat{\rho}_1^-$ was the MOP of the first kind. This in effect identifies all the elements in the RH matrix $Y(\lambda)$ given in (2.280) and $(Y^{-1})^T$, the latter which also satisfies a dual RH problem in terms of ratios of τ-functions; indeed, $\Psi^+_{k_1,k_2}(\lambda)$ without the exponenials is precisely $Y(\lambda)$, etc. for $\Psi^-_{k_1,k_2}$. Then using a self-evident formal residue identity, to wit:

$$\frac{1}{2\pi \mathrm{i}} \oint_{\infty} \left(f(z) \times \int_{\mathbb{R}} \frac{g(s)}{s-z}\,\mathrm{d}\mu(s) \right) \mathrm{d}z = \int_{\mathbb{R}} f(s)g(s)\,\mathrm{d}\mu(s)\ , \tag{2.309}$$

with $f(z) = \sum_0^\infty a_i z^i$, and designating $f(t, s^+, s^-)' := f(\underline{t}, \underline{s}^+, \underline{s}^-)$, we immediately conclude that

$$
\begin{aligned}
&\oint_{\infty} \exp\Big(\sum_1^{\infty}(t_i-\underline{t}_i)\lambda^i\Big)p_{k_1,k_2}\big(C_0 q_{l_1,l_2}(\lambda)\big)'\,\mathrm{d}\lambda\\
&\quad=\int_{\mathbb{R}} \exp\Big(\sum_1^{\infty}(t_i-\underline{t}_i)\lambda^i\Big)p_{k_1,k_2}(\lambda,t,s^+,s^-)q_{l_1,l_2}(\lambda,\underline{t},\underline{s}^+,\underline{s}^-)\,\mathrm{d}\lambda\\
&\quad=\int_{\mathbb{R}} \exp\Big(\sum_1^{\infty}(t_i-\underline{t}_i)\lambda^i\Big)p_{k_1,k_2}(\lambda,t,s^+,s^-)\\
&\qquad\qquad\times\big(q^+_{l_1-1,l_2}(\lambda,\underline{t},\underline{s}^+,\underline{s}^-)\hat{\rho}^+_1+q^-_{l_1,l_2-1}(\lambda,\underline{t},\underline{s}^+,\underline{s}^-)\hat{\rho}^-_1\big)\,\mathrm{d}\lambda\\
&\quad=\oint_{\infty}\big(C_+p_{k_1,k_2}(\lambda)\big)\big(q^+_{l_1-1,l_2}(\lambda)\big)'\exp\Big(\sum(s^+_i-\underline{s}^+_i)\lambda^i\Big)\,\mathrm{d}\lambda\\
&\qquad+\oint_{\infty}\big(C_-p_{k_1,k_2}(\lambda)\big)\big(q^-_{l_1,l_2-1}(\lambda)\big)'\exp\Big(\sum(s^-_i-\underline{s}^-_i)\lambda^i\Big)\,\mathrm{d}\lambda\,. \qquad (2.310)
\end{aligned}
$$

By (2.308), this is nothing but the bilinear identity (2.306). In fact, all the other entries of (2.305) are just (2.306) with its subscripts shifted. To say a quick word about (2.308), all that really goes into it is solving explicitly the linear systems defining p_{k_1,k_2}, $q^+_{k_1-1,k_2}$ and $q^-_{k_1,k_2-1}$, namely (2.272) and (2.273), and making use of the identity $\exp(\pm\sum_1^{\infty}x^i/i)=(1-x)^{\mp 1}$ for x small in the formula (2.300) for $\tau_{k_1,k_2}(t,s^+,s^-)$.

An immediate consequence of Theorem 2.5.5 is the following:

Corollary 2.5.1. *Given the above τ-functions $\tau_{k_1,k_2}(t,s^+,s^-)$, they satisfy the following bilinear identities*[14]

$$
\begin{aligned}
&\sum_{j=0}^{\infty}\mathbf{s}_{l_1+l_2-k_1-k_2+j-1}(-2a)\mathbf{s}_j(\tilde{\partial}_t)\exp\Big(\sum_1^{\infty}\Big(a_k\frac{\partial}{\partial t_k}+b_k\frac{\partial}{\partial s^+_k}+c_k\frac{\partial}{\partial s^-_k}\Big)\Big)\\
&\qquad\qquad\times\tau_{l_1,l_2}\circ\tau_{k_1,k_2}\\
&\quad-\sum_{j=0}^{\infty}\mathbf{s}_{k_1-l_1+1+j}(-2b)\mathbf{s}_j(\tilde{\partial}_{s^+})\exp\Big(\sum_1^{\infty}\Big(a_k\frac{\partial}{\partial t_k}+b_k\frac{\partial}{\partial s^+_k}+c_k\frac{\partial}{\partial s^-_k}\Big)\Big)\\
&\qquad\qquad\times\tau_{l_1-1,l_2}\circ\tau_{k_1+1,k_2}(-1)^{k_2+l_2}\\
&\quad-\sum_{j=0}^{\infty}\mathbf{s}_{k_2-l_2+1+j}(-2c)\mathbf{s}_j(\tilde{\partial}_{s^-})\exp\Big(\sum_1^{\infty}\Big(a_k\frac{\partial}{\partial t_k}+b_k\frac{\partial}{\partial s^+_k}+c_k\frac{\partial}{\partial s^-_k}\Big)\Big)\\
&\qquad\qquad\times\tau_{l_1,l_2-1}\circ\tau_{k_1,k_2+1}=0\,, \qquad (2.311)
\end{aligned}
$$

with $a,b,c\in\mathbb{C}^{\infty}$ arbitrary.

Upon specializing, these identities imply PDEs *expressed in terms of Hirota's symbol for $j=1,2\ldots$:*

[14] With $\exp(\sum_1^{\infty}t_iz^i)=:\sum_0^{\infty}\mathbf{s}_i(t)z^i$ defining the elementary Schur polynomials.

$$\mathbf{s}_j(\tilde{\partial}_t)\tau_{k_1+1,k_2} \circ \tau_{k_1-1,k_2} = -\tau^2_{k_1,k_2} \frac{\partial^2}{\partial s_1^+ \partial t_{j+1}} \log \tau_{k_1,k_2} , \tag{2.312}$$

$$\mathbf{s}_j(\tilde{\partial} s^+)\tau_{k_1-1,k_2} \circ \tau_{k_1+1,k_2} = -\tau^2_{k_1,k_2} \frac{\partial^2}{\partial t_1 \partial s^+_{j+1}} \log \tau_{k_1,k_2} , \tag{2.313}$$

yielding

$$\frac{\partial^2 \log \tau_{k_1,k_2}}{\partial t_1 \partial s_1^+} = -\frac{\tau_{k_1+1,k_2}\tau_{k_1-1,k_2}}{\tau^2_{k_1,k_2}} , \tag{2.314}$$

$$\frac{\partial}{\partial t_1} \log \frac{\tau_{k_1+1,k_2}}{\tau_{k_1-1,k_2}} = \frac{\partial^2/\partial t_2 \partial s_1^+ \log \tau_{k_1,k_2}}{\partial^2/\partial t_1 \partial s_1^+ \log \tau_{k_1,k_2}} , \tag{2.315}$$

$$-\frac{\partial}{\partial s_1^+} \log \frac{\tau_{k_1+1,k_2}}{\tau_{k_1-1,k_2}} = \frac{\partial^2/\partial t_1 \partial s_2^+ \log \tau_{k_1,k_2}}{\partial^2/\partial t_1 \partial s_1^+ \log \tau_{k_1,k_2}} , \tag{2.316}$$

Proof. Applying Lemma A.1 to the bilinear identity (2.306) immediately yields (2.311). Then Taylor expanding in a, b, c and setting in equation (2.311) all $a_i, b_i, c_i = 0$, except a_{j+1}, and also setting $l_1 = k_1 + 2$, $l_2 = k_2$, equation (2.311) becomes

$$a_{j+1}\left(-2\mathbf{s}_j(\tilde{\partial}_t)\tau_{k_1+2,k_2} \circ \tau_{k_1,k_2} - \frac{\partial^2}{\partial s_1^+ \partial t_{j+1}} \tau_{k_1+1,k_2} \circ \tau_{k_1+1,k_2}\right) + \mathbf{O}(a^2_{j+1}) = 0,$$

and the coefficient of a_{j+1} must vanish identically, yielding equation (2.312) upon setting $k_1 \to k_1 - 1$. Setting in equation (2.311) all $a_i, b_i, c_i = 0$, except b_{j+1}, and $l_1 = k_1$, $l_2 = k_2$, the vanishing of the coefficient of b_{j+1} in equation (2.311) yields equation (2.313). Specializing equation (2.312) to $j = 0$ and 1 respectively yields (since $\mathbf{s}_1(t) = t_1$ implies $\mathbf{s}_1(\tilde{\partial}_t) = \partial/\partial t_1$; also $\mathbf{s}_0 = 1$):

$$\frac{\partial^2 \log \tau_{k_1,k_2}}{\partial t_1 \partial s_1^+} = -\frac{\tau_{k_1+1,k_2}\tau_{k_1-1,k_2}}{\tau^2_{k_1,k_2}}$$

and

$$\frac{\partial^2}{\partial s_1^+ \partial t_2} \log \tau_{k_1,k_2} = -\frac{1}{\tau^2_{k_1,k_2}}\left[\left(\frac{\partial}{\partial t_1}\tau_{k_{1+1},k_2}\right)\tau_{k_{1-1},k_2} - \tau_{k_{1+1},k_2}\left(\frac{\partial}{\partial t_1}\tau_{k_1-1,k_2}\right)\right] .$$

Upon dividing the second equation by the first, we find (2.315) and similarly (2.316) follows from (2.313).

2.5.5 Virasoro Constraints for the Integrable Deformations

Given the Heisenberg and Virasoro operators, for $m \geq -1$, $k \geq 0$:

$$\begin{aligned}\mathbb{J}^{(1)}_{m,k} &= \frac{\partial}{\partial t_m} + (-m)t_{-m} + k\delta_{0,m}\;,\\ \mathbb{J}^{(2)}_{m,k}(t) &= \frac{1}{2}\Big(\sum_{i+j=m}\frac{\partial^2}{\partial t_i\partial t_j} + 2\sum_{i\geq 1} it_i\frac{\partial}{\partial t_{i+m}} + \sum_{i+j=-m} it_ijt_j\Big)\\ &\quad + \Big(k+\frac{m+1}{2}\Big)\Big(\frac{\partial}{\partial t_m} + (-m)t_{-m}\Big) + \frac{k(k+1)}{2}\delta_{m,0}\;,\end{aligned} \tag{2.317}$$

we now state (explicitly exhibiting the dependence of τ_{k_1,k_2} on β):

Theorem 2.5.6. *The integral $\tau_{k_1,k_2}(t,s^+,s^-;\beta;E)$, given by (2.302) satisfies*

$$\mathcal{B}_m\tau_{k_1,k_2} = \mathbb{V}^{k_1,k_2}_m\tau_{k_1,k_2} \quad \textit{for } m\geq -1\;, \tag{2.318}$$

where $\mathcal{B}_m$ and $\mathbb{V}_m$ are differential operators:

$$\mathcal{B}_m = \sum_1^{2r} b_i^{m+1}\frac{\partial}{\partial b_i}, \quad \textit{for } E=\bigcup_1^{2r}[b_{2i-1},b_{2i}]\subset\mathbb{R} \tag{2.319}$$

and

$$\begin{aligned}\mathbb{V}^{k_1,k_2}_m := \{&\mathbb{J}^{(2)}_{m,k_1+k_2}(t) - (m+1)\mathbb{J}^{(1)}_{m,k_1+k_2}(t)\\ &+\mathbb{J}^{(2)}_{m,k_1}(-s^+) + a\mathbb{J}^{(1)}_{m+1,k_1}(-s^+) - (1-2\beta)\mathbb{J}^{(1)}_{m+2,k_1}(-s^+)\\ &+\mathbb{J}^{(2)}_{m,k_2}(-s^-) - a\mathbb{J}^{(1)}_{m+1,k_2}(-s) - (1+2\beta)\mathbb{J}^{(1)}_{m+2,k_2}(-s^-)\}\;.\end{aligned} \tag{2.320}$$

Lemma 2.5.1. *Setting*

$$\begin{aligned}\mathrm{d}I_n &= \Delta_n(x,y)\prod_{j=1}^{k_1}\exp\Big(\sum_1^\infty t_ix_j^i\Big)\prod_{j=1}^{k_2}\exp\Big(\sum_1^\infty t_iy_j^i\Big)\\ &\times\Big(\Delta_{k_1}(x)\prod_{j=1}^{k_1}\exp\Big(-\frac{x_j^2}{2}+ax_j+\beta x_j^2\Big)\exp\Big(-\sum_1^\infty s_i^+x_j^i\Big)\,\mathrm{d}x_j\Big)\\ &\times\Big(\Delta_{k_2}(y)\prod_{j=1}^{k_2}\exp\Big(-\frac{y_j^2}{2}-ay_j-\beta y_j^2\Big)\exp\Big(-\sum_1^\infty s_i^-y_j^i\Big)\,\mathrm{d}y_j\Big)\;,\end{aligned} \tag{2.321}$$

the following variational formula holds for $m\geq -1$:

$$\frac{\mathrm{d}}{\mathrm{d}\varepsilon}\mathrm{d}I_n\begin{pmatrix}x_i\mapsto x_i+\varepsilon x_i^{m+1}\\ y_i\mapsto y_i+\varepsilon y_i^{m+1}\end{pmatrix}\Bigg|_{\varepsilon=0} = \mathbb{V}^{k_1,k_2}_m(\mathrm{d}I_n)\;. \tag{2.322}$$

Proof. The variational formula (2.322) is an immediate consequence of applying the variational formula (2.184) separately to the three factors of $\mathrm{d}I_n$, and in addition applying formula (2.185) to the first factor, to account for the fact that $\prod_{j=1}^{k_1}\mathrm{d}x_j\prod_{j=1}^{k_2}\mathrm{d}y_j$ is missing from the first factor.

Proof of Theorem 2.5.6: Formula (2.318) follows immediately from formula (2.322) by taking into account the variation of ∂E under the change of coordinates.

From (2.317) and (2.320) and from the identity when acting on on τ_{k_1,k_2},

$$\frac{\partial}{\partial t_n} = -\frac{\partial}{\partial s_n^+} - \frac{\partial}{\partial s_n^-} , \tag{2.323}$$

compute that

$$\begin{aligned}
\mathbb{V}_m^{k_1,k_2} &= \frac{1}{2}\sum_{i+j=m}\left(\frac{\partial^2}{\partial t_i \partial t_j} + \frac{\partial^2}{\partial s_i^+ \partial s_j^+} + \frac{\partial^2}{\partial s_i^- \partial s_j^-}\right) \\
&+ \sum_{i\geq 1}\left(it_i \frac{\partial}{\partial t_{i+m}} + is_i^+ \frac{\partial}{\partial s_{i+m}^+} + is_i^- \frac{\partial}{\partial s_{i+m}^-}\right) \\
&+ (k_1+k_2)\left(\frac{\partial}{\partial t_m} + (-m)t_{-m}\right) - k_1\left(\frac{\partial}{\partial s_m^+} + (-m)s_{-m}^+\right) \\
&- k_2\left(\frac{\partial}{\partial s_m^-} + (-m)s_{-m}^-\right) + (k_1^2 + k_1k_2 + k_2^2)\delta_{m,0} \\
&+ a(k_1-k_2)\delta_{m+1,0} + \frac{m(m+1)}{2}(-t_{-m} + s_{-m}^+ + s_{-m}^-) - \frac{\partial}{\partial t_{m+2}} \\
&+ a\left(-\frac{\partial}{\partial s_{m+1}^+} + \frac{\partial}{\partial s_{m+1}^-} + (m+1)(s_{-m-1}^+ - s_{-m-1}^-)\right) \\
&+ 2\beta\left(\frac{\partial}{\partial s_{m+2}^-} - \frac{\partial}{\partial s_{m+2}^+}\right), \quad m \geq -1 .
\end{aligned} \tag{2.324}$$

The following identities, valid when acting on $\tau_{k_1,k_2}(t, s^+, s^-; \beta, E)$, will also be used:

$$\begin{aligned}
\frac{\partial}{\partial s_1^+} &= -\frac{1}{2}\left(\frac{\partial}{\partial t_1} + \frac{\partial}{\partial a}\right), & \frac{\partial}{\partial s_2^+} &= -\frac{1}{2}\left(\frac{\partial}{\partial t_2} + \frac{\partial}{\partial \beta}\right), \\
\frac{\partial}{\partial s_1^-} &= -\frac{1}{2}\left(\frac{\partial}{\partial t_1} - \frac{\partial}{\partial a}\right), & \frac{\partial}{\partial s_2^-} &= -\frac{1}{2}\left(\frac{\partial}{\partial t_2} - \frac{\partial}{\partial \beta}\right).
\end{aligned} \tag{2.325}$$

Corollary 2.5.2. *The* τ*-function* $\tau = \tau_{k_1,k_2}(t, s^+, s^-; \beta, E)$ *satisfies the following differential identities, with* $\mathcal{B}_m = \sum_1^{2r} b_i^{m+1}\partial/\partial b_i$.

$$\begin{aligned}
-B_{-1}\tau &= \mathcal{V}_1\tau + v_1\tau \\
&=: \left(\frac{\partial}{\partial t_1} - 2\beta\frac{\partial}{\partial a}\right)\tau \\
&\quad - \sum_{i\geq 2}\left(it_i\frac{\partial}{\partial t_{i-1}} + is_i^+\frac{\partial}{\partial s_{i-1}^+} + is_i^-\frac{\partial}{\partial s_{i-1}^-}\right)\tau \\
&\quad + (a(k_2-k_1) + k_1s_1^+ + k_2s_1^- - (k_1+k_2)t_1)\tau ,
\end{aligned}$$

$$
\begin{aligned}
\frac{1}{2}\left(B_{-1}-\frac{\partial}{\partial a}\right)\tau &= \mathcal{W}_1\tau + w_1\tau \\
&=: \left(\frac{\partial}{\partial s_1^+}+\beta\frac{\partial}{\partial a}\right)\tau \\
&\quad + \frac{1}{2}\sum_{i\geq 2}\left(it_i\frac{\partial}{\partial t_{i-1}}+is_i^+\frac{\partial}{\partial s_{i-1}^+}+is_i^-\frac{\partial}{\partial s_{i-1}^-}\right)\tau \\
&\quad + \left(\frac{a}{2}(k_1-k_2)+\frac{1}{2}(k_1+k_2)t_1-k_1s_1^+-k_2s_1^-\right)\tau \\
-\left(B_0-a\frac{\partial}{\partial a}\right)\tau &= \mathcal{V}_2\tau+v_2\tau \qquad (2.326)\\
&=: \frac{\partial\tau}{\partial t_2}-2\beta\frac{\partial\tau}{\partial\beta} \\
&\quad - \sum_{i\geq 1}\left(it_i\frac{\partial}{\partial t_i}+is_i^+\frac{\partial}{\partial s_i^+}+is_i^-\frac{\partial}{\partial s_i^-}\right)\tau \\
&\quad - (k_1^2+k_2^2+k_1k_2)\tau \\
\frac{1}{2}\left(B_0-a\frac{\partial}{\partial a}-\frac{\partial}{\partial\beta}\right)\tau &= \mathcal{W}_2\tau+w_2\tau \\
&=: \frac{\partial\tau}{\partial s_2^+}+\beta\frac{\partial\tau}{\partial\beta} \\
&\quad + \frac{1}{2}\sum_{i\geq 1}\left(it_i\frac{\partial}{\partial t_i}+is_i^+\frac{\partial}{\partial s_i^+}+is_i^-\frac{\partial}{\partial s_i^-}\right)\tau \\
&\quad + \frac{1}{2}(k_1^2+k_2^2+k_1k_2)\tau
\end{aligned}
$$

where $\mathcal{V}_1$, $\mathcal{W}_1$, $\mathcal{V}_2$, $\mathcal{W}_2$ are first order operators and v_1, w_1, v_2, w_2 are functions, acting as multiplicative operators.

Corollary 2.5.3. *On the locus $\mathcal{L} := \{t = s^+ = s^- = 0, \beta = 0\}$, the function $f = \log\tau_{k_1,k_2}(t, s^+, s^-; \beta, E)$ satisfies the following differential identities:*

$$
\begin{aligned}
\frac{\partial f}{\partial t_1} &= -\mathcal{B}_{-1}f + a(k_1-k_2)\,, \\
\frac{\partial f}{\partial s_1^+} &= \frac{1}{2}\left(\mathcal{B}_{-1}-\frac{\partial}{\partial a}\right)f+\frac{a}{2}(k_2-k_1)\,, \\
\frac{\partial f}{\partial t_2} &= \left(-\mathcal{B}_0+a\frac{\partial}{\partial a}\right)f+k_1^2+k_1k_2+k_2^2\,, \\
\frac{\partial f}{\partial s_2^+} &= \frac{1}{2}\left(\mathcal{B}_0-a\frac{\partial}{\partial a}-\frac{\partial}{\partial\beta}\right)f-\frac{1}{2}(k_1^2+k_2^2+k_1k_2)\,, \qquad (2.327)\\
2\frac{\partial^2 f}{\partial t_1\partial s_1^+} &= \mathcal{B}_{-1}\left(\frac{\partial}{\partial a}-\mathcal{B}_{-1}\right)f-2k_1\,,
\end{aligned}
$$

$$
\begin{aligned}
2\frac{\partial^2 f}{\partial t_1 \partial s_2^+} &= \left(a\frac{\partial}{\partial a} + \frac{\partial}{\partial \beta} - \mathcal{B}_0 + 1\right)\mathcal{B}_{-1} f - 2\frac{\partial f}{\partial a} - 2a(k_1 - k_2)\,, \qquad (2.328)\\
2\frac{\partial^2 f}{\partial t_2 \partial s_1^+} &= \frac{\partial}{\partial a}\left(\mathcal{B}_0 - a\frac{\partial}{\partial a} + a\mathcal{B}_{-1}\right) f - \mathcal{B}_{-1}(\mathcal{B}_0 - 1) f - 2a(k_1 - k_2)\,.
\end{aligned}
$$

Proof. Upon dividing equations (2.326) by τ and restricting to the locus $\mathcal{L}$, equations (2.327) follow immediately.

Remembering $f = \log\tau$ and setting

$$
\begin{aligned}
\mathcal{A}_1 &:= -\mathcal{B}_{-1}\,, & \mathcal{B}_1 &:= \frac{1}{2}\left(\mathcal{B}_{-1} - \frac{\partial}{\partial a}\right),\\
\mathcal{A}_2 &:= -\left(\mathcal{B}_0 - a\frac{\partial}{\partial a}\right), & \mathcal{B}_2 &:= \frac{1}{2}\left(\mathcal{B}_0 - a\frac{\partial}{\partial a} - \frac{\partial}{\partial \beta}\right),
\end{aligned}
\qquad (2.329)
$$

we may recast (2.326) as (compare with (2.193))

$$
\mathcal{A}_k f = \mathcal{V}_k f + v_k, \quad \mathcal{B}_k f = \mathcal{W}_k f + w_k, \quad k = 1, 2\,, \qquad (2.330)
$$

where (compare with (2.191)) we note that

$$
\mathcal{V}_k|_{\mathcal{L}} = \frac{\partial}{\partial t_k}\,, \quad \mathcal{W}_k|_{\mathcal{L}} = \frac{\partial}{\partial s_k^+}\,, \qquad k = 1, 2\,. \qquad (2.331)
$$

To prove (2.328) we will copy the argument of Sect. 2.3.6 (see (2.195)). Indeed, compute

$$
\begin{aligned}
\mathcal{B}_1\mathcal{A}_1 f|_{\mathcal{L}} &= \mathcal{B}_1(\mathcal{V}_1 f + v_1) = \mathcal{B}_1\mathcal{V}_1 f|_{\mathcal{L}} + \mathcal{B}_1(v_1)|_{\mathcal{L}}\\
&\overset{(*)}{=} \mathcal{V}_1\mathcal{B}_1 f|_{\mathcal{L}} + \mathcal{B}_1(v_1)|_{\mathcal{L}}\\
&\overset{(^*_*)}{=} \frac{\partial}{\partial t_1}(\mathcal{W}_1 f + w_1)|_{\mathcal{L}} + \mathcal{B}_1(v_1)|_{\mathcal{L}}\\
&= \frac{\partial}{\partial t_1}\left(\frac{\partial}{\partial s_1^+} + \cdots\right) f|_{\mathcal{L}} + \left.\frac{\partial w_1}{\partial t_1}\right|_{\mathcal{L}} + \mathcal{B}_1(v_1)|_{\mathcal{L}}\,,
\end{aligned}
\qquad (2.332)
$$

where we used in $(*)$ that $\left.[\mathcal{B}_1, \mathcal{V}_1]\right|_{\mathcal{L}} = 0$ and in $(^*_*)$ that $\mathcal{V}_1|_{\mathcal{L}} = \partial/\partial t_1$, and so from (2.332) we must compute

$$
\begin{aligned}
\left.\frac{\partial}{\partial t_1}\mathcal{W}_1\right|_{\mathcal{L}} = \frac{\partial^2}{\partial t_1 \partial s_1^+}\,, \quad \left.\frac{\partial w_1}{\partial t_1}\right|_{\mathcal{L}} = \frac{1}{2}(k_1 + k_2)\,,\\
\mathcal{B}_1(v_1)|_{\mathcal{L}} = \frac{1}{2}(k_1 - k_2)\,.
\end{aligned}
\qquad (2.333)
$$

Now from (2.332) and (2.333), we find

$$
\mathcal{B}_1\mathcal{A}_1 f|_{\mathcal{L}} = \left.\frac{\partial^2 f}{\partial t_1 \partial s_1^+}\right|_{\mathcal{L}} + \frac{1}{2}(k_1 + k_2) + \frac{1}{2}(k_1 - k_2) = \frac{\partial^2 f}{\partial t_2 \partial s_1^+} + k_1\,,
$$

and so

$$\left.\frac{\partial^2 f}{\partial t_1 \partial s_1^+}\right|_{\mathcal{L}} = \mathcal{B}_1 \mathcal{A}_1 f|_{\mathcal{L}} - k_1 = -\mathcal{B}_{-1}\frac{1}{2}\left(\mathcal{B}_1 - \frac{\partial}{\partial a}\right) f|_{\mathcal{L}} - k_1 ,$$

which is just the first equation in (2.328). The crucial point in the calculation being (2.331) and $[\mathcal{B}_1, \mathcal{V}_1]|_{\mathcal{L}} = 0$. The other two formulas in (2.328) are done in precisely the same fashion, using the crucial facts (2.331) and $[\mathcal{B}_2, \mathcal{V}_1]|_{\mathcal{L}} = [\mathcal{A}_2, \mathcal{W}_1]|_{\mathcal{L}} = 0$ and the analogs of (2.333).

2.5.6 A PDE for the Gaussian Ensemble with External Source and the Pearcey PDE

From now on set: $k_1 = k_2 := k$ and restrict to the locus $\mathcal{L}$. From (2.315) and (2.316) we have the

3-KP *relations*:

$$\frac{\partial}{\partial t_1} g = \frac{\partial^2 f}{\partial t_2 \partial s_1^+} \Big/ \frac{\partial^2 f}{\partial t_1 \partial s_1^+} , \qquad -\frac{\partial}{\partial s_1^+} g = \frac{\partial^2 f}{\partial t_1 \partial s_2^+} \Big/ \frac{\partial^2 f}{\partial t_1 \partial s_1^+} , \tag{2.334}$$

with

$$f := \log \tau_{k,k} , \quad g := \log(\tau_{k+1,k}/\tau_{k-1,k}) ,$$

while from (2.327), we find

Virasoro relations on $\mathcal{L}$:

$$\frac{\partial g}{\partial t_1} = -\mathcal{B}_{-1} g + 2a , \quad \frac{\partial g}{\partial s^+} = \frac{1}{2}\left(\mathcal{B}_{-1} - \frac{\partial}{\partial a}\right) g - a . \tag{2.335}$$

Eliminating $\partial g/\partial t_1$, $\partial g/\partial s_1^+$ from (2.334) using (2.335) and then further eliminating $\mathcal{A}_1 g := -\mathcal{B}_- g$ and $\mathcal{B}_1 g := \frac{1}{2}(\mathcal{B}_{-1} - \partial/\partial a) g$ using $\mathcal{A}_1 \mathcal{B}_1 g = \mathcal{B}_1 \mathcal{A}_1 g$ yields

$$\begin{aligned} &\mathcal{B}_{-1}\left(\frac{\partial^2/\partial t_2 \partial s_1^+ - 2\,\partial^2/\partial t_1 \partial s_2^+)f}{\partial^2 f/\partial t_1 \partial s_1^+}\right) \\ &\qquad\qquad = \frac{\partial}{\partial a}\left(\frac{(\partial^2/\partial t_2 \partial s_1^+ - 2a\,\partial^2/\partial t_1 \partial s_1^+)f}{\partial^2 f/\partial t_1 \partial s_1^+}\right) , \end{aligned} \tag{2.336}$$

while from (2.328) we find the

Virasoro relations on $\mathcal{L}$:

$$\begin{aligned} 2\frac{\partial^2}{\partial t_1 \partial s_1^+} f &= \mathcal{B}_{-1}\left(\frac{\partial}{\partial a} - \mathcal{B}_{-1}\right) f - 2k =: F^+ , \\ 2\left(\frac{\partial^2}{\partial t_2 \partial s_1^+} - 2\frac{\partial^2}{\partial t_1 \partial s_2^+}\right) f &= H_1^+ - 2\mathcal{B}_{-1}\frac{\partial f}{\partial \mathcal{B}} , \\ 2\left(\frac{\partial^2}{\partial t_2 \partial s_1^+} - 2a\frac{\partial^2}{\partial t_1 \partial s_2^+}\right) f &= H_2^+ , \end{aligned} \tag{2.337}$$

where the precise formulas for H_i^+ will be given later. Substituting (2.337) into (2.336) and clearing the denominator yields[15]

$$\left\{\mathcal{B}_{-1}\frac{\partial f}{\partial \beta}, F^+\right\}_{\mathcal{B}_{-1}} = \left\{H_1^+, \frac{1}{2}F^+\right\}_{\mathcal{B}_{-1}} - \left\{H_2^+, \frac{1}{2}F^+\right\}_{\partial/\partial a}, \tag{2.338}$$

and by the involution: $a \to -a$, $\beta \to -\beta$, which by (2.302), clearly fixes $f = \log \tau_{k,k}$ on $t = s^+ = s^- = 0$, we find ($H_i^- = H^+_{i|a\to -a}$)

$$-\left\{\mathcal{B}_{-1}\frac{\partial f}{\partial \beta}, F^-\right\}_{\mathcal{B}_{-1}} = \left\{H_1^-, \frac{1}{2}F^-\right\}_{\mathcal{B}_{-1}} - \left\{H_2^-, \frac{1}{2}F^-\right\}_{-\partial/\partial a}. \tag{2.339}$$

These 2 relations (2.338) and (2.339) yield a linear system for:

$$\mathcal{B}_{-1}\frac{\partial f}{\partial \beta}, \quad \mathcal{B}_{-1}^2\frac{\partial f}{\partial \beta}.$$

Solving the system yields:

$$\mathcal{B}_{-1}\frac{\partial f}{\partial \beta} = R_1, \quad \mathcal{B}_{-1}^2\frac{\partial f}{\partial \beta} = R_2,$$

and so

$$\mathcal{B}_{-1}R_1(f) = R_2(f), \quad f = \log \tau_{k,k}(0,0,0,E)\big|_{\beta=0}.$$

Since

$$P(\mathrm{spec}(M) \subset E) = \left.\frac{\tau_{k,k}(0,0,0,E)}{\tau_{k,k}(0,0,0,\mathbb{R}}\right|_{\beta=0},$$

with

$$\tau_{k,k}(0,0,0,\mathbb{R})\big|_{\beta=0} = \left(\prod_0^{k-1} j!\right)^2 2^{k^2}(-2\pi)^k \exp(ka^2)a^{k^2},$$

we find the following theorem:

Theorem 2.5.7 (Adler–van Moerbeke [7]).

For $E = \bigcup_1^r [b_{2i-1}, b_{2i}]$, $A = \mathrm{diag}(\overbrace{-a,\dots,-a}^{k}, \overbrace{a,\dots,a}^{k})$,

$$P(a; b_1,\dots,b_{2r}) = \frac{\int_{\mathcal{H}_{2k(E)}} \exp\left(\mathrm{Tr}(-\frac{1}{2}M^2 + AM)\right) \mathrm{d}M}{\int_{\mathcal{H}_{2k(\mathbb{R})}} \exp\left(\mathrm{Tr}(-\frac{1}{2}M^2 + AM)\right) \mathrm{d}M} \tag{2.340}$$

satisfies a nonlinear 4th order PDE *in* $a, b_1, \dots b_r$:

$$(F^+\mathcal{B}_{-1}G^- + F^-\mathcal{B}_{-1}G^+)(F^+\mathcal{B}_{-1}F^- - F^-\mathcal{B}_{-1}F^+) \\ - (F^+G^- + F^-G^+)(F^+\mathcal{B}_{-1}^2F^- - F^-\mathcal{B}_{-1}^2F^+) = 0, \tag{2.341}$$

[15] $\{f,g\}_X := gXf - fXg$.

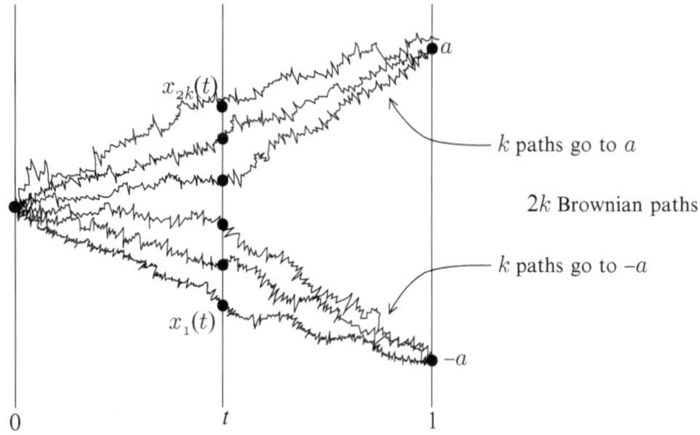

Fig. 2.6.

where

$$\mathcal{B}_{-1} = \sum_1^{2r} \frac{\partial}{\partial b_i}\,, \quad \mathcal{B}_0 = \sum_1^{2r} b_i \frac{\partial}{\partial b_i}\,,$$

$$\begin{aligned}
F^+ &:= -2k + \mathcal{B}_{-1}\left(\frac{\partial}{\partial a} - \mathcal{B}_{-1}\right)\log P\,,\\
G^+ &:= \{H_1^+, F^+\}_{\mathcal{B}_{-1}} - \{H_2^+, F^+\}_{\partial/\partial a}\,,\\
H_1^+ &:= \frac{\partial}{\partial a}\left(\mathcal{B}_0 - a\frac{\partial}{\partial a} - a\mathcal{B}_{-1} + 4\frac{\partial}{\partial a}\right)\log P + \mathcal{B}_0\mathcal{B}_{-1}\log P + 4ak + 4\frac{k^2}{a}\,,\\
H_2^+ &:= \frac{\partial}{\partial a}\left(\mathcal{B}_0 - a\frac{\partial}{\partial a} - a\mathcal{B}_{-1}\right)\log P + (2a\mathcal{B}_{-1}^2 - \mathcal{B}_0\mathcal{B}_{-1} + 2\mathcal{B}_{-1})\log P\,,
\end{aligned} \tag{2.342}$$

$$F^- = F^+|_{a\to -a} \quad \textit{and} \quad G^- = G^+|_{a\to -a}.$$ [16]

We now show how Theorem 2.5.7 implies Theorem 2.5.3. Indeed, remember our picture of $2k$ Brownian paths diverging at $t = \frac{1}{2}$.

Also, remembering the equivalence (2.270) between GUE with external source and the above Brownian motion, and (2.286), we find

$$\begin{aligned}
P_{k,k}^a(t; b_1, \ldots, b_{2r}) &:= \mathrm{Prob}_{k,k}^a(\text{all } x_i(t) \subset E)\\
&= P\left(\sqrt{\frac{2t}{1-t}}a; \sqrt{\frac{2}{t(1-t)}}(b_1, \ldots, b_r)\right)\\
&= \det(I - \widetilde{K}_{2k}^{E^c})\,,
\end{aligned} \tag{2.343}$$

where the *function* $P(*;*)$ is that of (2.340).

[16] Note $P(a; b_1, \ldots b_{2r}) = P(-a; b_1, \ldots, b_r)$.

Letting the number of particles $2k \to \infty$ and looking about the location $x = 0$ at time $t = \frac{1}{2}$ with a microscope, and slowing down time as follows:

$$k = \frac{1}{z^4}, \quad \pm a = \pm \frac{1}{z^2}, \quad b_i = u_i z, \quad t = \tfrac{1}{2} + \tau z^2, \qquad z \to 0, \tag{2.344}$$

which is just the Pearcey scaling (2.285), we find by Theorem 2.5.2 and (2.343), that

$$\begin{aligned} &\mathrm{Prob}^{1/z^2}_{(1/z^4,1/z^4)}\Big(\text{all } x_i(\tfrac{1}{2} + \tau z^2) \in \bigcup_1^r [z u_{2i-1}, z u_{2i}]\Big) \\ &\quad = P\bigg(\sqrt{\frac{2}{1-t}}a; \sqrt{\frac{2}{t(1-t)}}(b_1, \dots, b_{2r})\bigg)\bigg|_{\substack{a=1/z^2, b_i = u_i z \\ t = 1/2 + \tau z^2}} \\ &\quad = \det(I - K_\tau^P I_{\widetilde{E}^c}) + O(z) \\ &\quad =: Q(\tau; u_1, \dots, u_{2r}) + O(z), \end{aligned} \tag{2.345}$$

where K_τ^P is the Pearcey kernel (2.287) and $\widetilde{E} = \bigcup_1^r [u_{2i-1}, u_{2i}]$. Taking account of

$$\begin{aligned} P^a_{k,k}(t; b_1, \dots, b_{2r}) &= P\bigg(\sqrt{\frac{2t}{1-t}}a; \sqrt{\frac{2}{t(1-t)}}(b_1, \dots, b_r\bigg) \\ &=: P(A, B_1, \dots, B_{2r}), \end{aligned} \tag{2.346}$$

where the function $P(*, *)$ is that of (2.340), and the scaling (2.344) of the Pearcey process, we subject the equation (2.341) to both the change of coordinates involved in the equation (2.346) and the Pearcey scaling (2.344) simultaneously:

$$\begin{aligned} 0 &= \{(F^+\mathcal{B}_{-1}G^- + F^-\mathcal{B}_{-1}G^+)(F^+\mathcal{B}_{-1}F^- - F^-\mathcal{B}_{-1}F^+) \\ &\quad -(F^+G^- + F^-G^+)(F^+\mathcal{B}^2_{-1}F^- - F^-\mathcal{B}^2_{-1}F^+)\}\Big|_{\substack{A=(\sqrt{2}/z^2)\sqrt{(1/2+\tau z^2)/(1/2-\tau z^2)}, \\ B_i = u_i z\sqrt{2}/\sqrt{(1/4-\tau^2 z^4)}}} \\ &= \frac{1}{z^{17}}(\text{PDE in } \tau \text{ and } u \text{ for } \log P^a_{k,k}(t; b_1, \dots, b_{2r})|_{\text{scaling}}) + O\Big(\frac{1}{z^{15}}\Big) \\ &= \frac{1}{z^{17}}(\text{same PDE for } \log Q(\tau, u_1, \dots, u_{2r})) + O\Big(\frac{1}{z^{16}}\Big); \end{aligned}$$

the first step is accomplished by the chain rule and the latter step by (2.345), yielding Theorem 2.5.3 for $F = \log Q$.

A Hirota Symbol Residue Identity

Lemma A.1. *We have the following formal residue identity*

$$\frac{1}{2\pi\mathrm{i}}\oint_{\infty} f(t'-[z^{-1}],s',u')g(t''+[z^{-1}],s'',u'')\exp\left(\sum_1^{\infty}(t_i'-t_i'')z^i\right)z^r\,\mathrm{d}z$$
$$=\sum_{j\geq 0} s_{j-1-r}(-2a)s_j(\tilde{\partial}_t)\exp\left(\sum_1^{\infty}\left(a_l\frac{\partial}{\partial t_l}+b_l\frac{\partial}{\partial s_l}+c_l\frac{\partial}{\partial u_l}\right)\right)g\circ f\,, \quad (2.347)$$

where

$$\begin{aligned} t'&=t-a\,, & s'&=s-b\,, & u'&=u-c\,,\\ t''&=t+a\,, & s''&=s+b\,, & u''&=u+c\,, \end{aligned} \quad (2.348)$$

$$\tilde{\partial}_t=\left(\frac{\partial}{\partial t_1},\frac{1}{2}\frac{\partial}{\partial t_2},\frac{1}{3}\frac{\partial}{\partial t_3},\dots\right),\quad \exp\left(\sum_1^{\infty}t_iz^i\right)=\sum_0^{\infty}s_i(t)z^i\,, \quad (2.349)$$

and the Hirota symbol

$$p(\partial_t,\partial_s,\partial_u)g\circ f$$
$$:=p(\partial_{t'},\partial_{s'},\partial_{u'})g(t+t',s+s',u+u')f(t-t',s-s',u-u')\Big|_{\substack{t'=0\\ s'=0\\ u'=0}}. \quad (2.350)$$

Proof. By definition,

$$\frac{1}{2\pi\mathrm{i}}\oint_{\infty}\sum_{i=-\infty}^{i=\infty}a_iz^i\,\mathrm{d}z=a_{-1}\,, \quad (2.351)$$

and so by Taylor's Theorem, following [28] compute:

$$\oint_{\infty} f(t'-[z^{-1}],s',u')g(t''+[z^{-1}],s'',u'')\exp\left(\sum_1^{\infty}(t_i'-t_i'')z^i\right)z^r\frac{\mathrm{d}z}{2\pi\mathrm{i}}$$
$$=\oint_{\infty} f(t-a-[z^{-1}],s-b,u-c)g(t+a+[z^{-1}],s+b,u+c)$$
$$\times\exp\left(-2\sum_1^{\infty}a_iz^i\right)z^r\frac{\mathrm{d}z}{2\pi\mathrm{i}}$$
$$=\oint_{\infty}\exp\left(\sum_1^{\infty}\frac{1}{iz^i}\frac{\partial}{\partial a_i}\right)f(t-a,s-b,u-c)g(t+a,s+b,u+c)$$
$$\times\exp\left(-2\sum_1^{\infty}a_iz^i\right)z^r\frac{\mathrm{d}z}{2\pi\mathrm{i}}$$
$$=\oint_{\infty}\sum_1^{\infty}z^{-j}s_j(\tilde{\partial}_a)f(t-a,s-b,u-c)g(t+a,s+b,u+c)$$
$$\times\sum_{l=0}^{\infty}z^{l+r}s_l(-2a)\frac{\mathrm{d}z}{2\pi\mathrm{i}}$$

(picking out the residue term)

$$= \sum_{j=0}^{\infty} s_{j-1-r}(-2a) s_j(\tilde{\partial}_a) f(t-a, s-b, u-c) g(t+a, s+b, u+c)$$

$$= \sum_{j=0}^{\infty} s_{j-1-r}(-2a) s_j(\tilde{\partial}_a) \exp\left(\sum_1^{\infty}\left(a_l \frac{\partial}{\partial t_l'} + b_l \frac{\partial}{\partial s_l'} + c_l \frac{\partial}{\partial u_l'}\right)\right)$$
$$\times f(t-t', s-s', u-u') g(t+t', s+s', u+u')$$
$$\text{at } t' = s' = u' = 0$$

$$= \sum_{j=0}^{\infty} s_{j-1-r}(-2a) s_j(\tilde{\partial}_{t'}) \exp\left(\sum_1^{\infty}\left(a_l \frac{\partial}{\partial t_l'} + b_l \frac{\partial}{\partial s_l'} + c_l \frac{\partial}{\partial u_l'}\right)\right)$$
$$\times g(t+t', s+s', u+u') f(t-t', s-s', u-u')$$
$$\text{at } t' = s' = u' = 0$$

$$= \sum_{j=0}^{\infty} s_{j-1-r}(-2a) s_j(\tilde{\partial}_t) \exp\left(\sum_1^{\infty}\left(a_l \frac{\partial}{\partial t_l} + b_l \frac{\partial}{\partial s_l} + c_l \frac{\partial}{\partial u_l}\right)\right) g(t) \circ f(t) ,$$

completing the proof.

Proof of (2.28): To deduce (2.28) from (2.27), observe that since t, t' are arbitrary in (2.27), when we make the change of coordinates (2.348), a becomes arbitrary and we then apply Lemma A.1, with s and u absent, $r = 0$ and $f = g = \tau$, to deduce (2.28).

Proof of (2.166): To deduce (2.167) from (2.166), apply Lemma A.1 to the l.h.s. of (2.166), setting $f = \tau_n$, $g = \tau_{m+1}$, $r = n - m - 1$, where no u, u' is present, and when we make the change of coordinates (2.348), since t, t', s, s' are arbitrary, so is a and b, while in the r.h.s. of (2.166), we first need to make the change of coordinates $z \to z^{-1}$, so $z^{n-m-1}\,dz \to z^{m-n-1}\,dz$ (taking account of the switch in orientation) and then we apply Lemma A.1, with $f = \tau_{n+1}$, $g = \tau_m$, $r = m - n - 1$, and so deduce (2.167) from (2.166).

References

1. M. Adler, P. van Moerbeke: Adv. Math. **108**, 140 (1994)
2. M. Adler, P. van Moerbeke: Comm. Pure Appl. Math. **54**, 153 (2001)
3. M. Adler, P. van Moerbeke: Comm. Pure Appl. Math. **237**, 397 (2003)
4. M. Adler, P. van Moerbeke: Ann. of Math. (2) **149**, 921 (1999)
5. M. Adler, P. van Moerbeke: Ann. of Math. (2) **153**, 149 (2001)
6. M. Adler, P. van Moerbeke: Ann. Probab. **33**, 1326 (2005)
7. M. Adler, P. van Moerbeke: arXiv:math-pr/0509047 (2005)
8. M. Adler, P. van Moerbeke: Duke Math. J. **80**, 863 (1995) adler:ref
9. M. Adler, P. van Moerbeke: Comm. Pure Appl. Math. **50**, 241 (1997)
10. M. Adler, P. van Moerbeke: Commun. Math. Phys. **203**, 185 (1999)
11. M. Adler, P. van Moerbeke: Duke Math. J. **112**, 1 (2002)

12. M. Adler, P. van Moerbeke: PDE*s for n time correlation of the Pearcey process and the n coupled Gaussian ensemble with external source* (preprint 2006)
13. M. Adler, P. van Moerbeke, P. Vanhaecke: arXiv:math-ph/0601020 (2006)
14. M. Adler, P. van Moerbeke, P. Vanhaecke: *Algebraic Integrability, Painlevé Geometry and Lie Algebras* (Springer 2004)
15. M. Adler, P. van Moerbeke, P. Vanhaecke: *Moment matrices and multi-component* KP*, with applications to random matrix theory* (preprint 2006)
16. M. Adler, T. Shiota, P. van Moerbeke: Duke Math. J. **94**, 379 (1998)
17. M. Adler, T. Shiota, P. van Moerbeke: Comm. Math. Phys. **171**, 547 (1995)
18. M. Adler, T. Shiota, P. van Moerbeke: Phys. Lett. A **208**, 67 (1995)
19. M. Abramowitz, I. Stegun: *Handbook of Mathematical Functions* (Dover 1972)
20. D. Aldous, P. Diaconis: Bull. Amer. Math. Soc. (N.S.) **36**, 413 (1999)
21. A. Aptekarev, P. Bleher, A. Kuijlaars: arXiv:math-ph/0408041 (2004)
22. J. Baik, P. Deift, K. Johansson: J. Amer. Math. Soc. **12**, 1119 (1999)
23. D. Bessis, Cl. Itzykson, J.-B. Zuber: Adv. in Appl. Math. **1**, 109 (1980)
24. P. Bleher, A. Kuijlaars: Internat. Math. Res. Notices **3**, 109 (2004)
25. P. Bleher, A. Kuijlaars: Comm. Math. Phys. **252**, 43 (2004)
26. A. Borodin: Duke Math. J. **117**, 489 (2003)
27. E. Brézin, S. Hikami: Phys. Rev. E **58**, 7176 (1998)
28. E. Date, M. Jimbo, M. Kashiwara T. Miwa: Transformation groups for soliton equations. In: *Nonlinear Integrable Systems – Classical Theory and Quantum Theory* (World Sci. Publishing, Singapore, 1983) pp 39–119
29. P. Deift: Orthogonal polynomials and random matrices: a Riemann-Hilbert approach. In. *Courant Lecture Notes in Mathematics*, vol 3 (Courant Institute of Mathematical Sciences, New York; Amer. Math. Soc., Providence, RI, 1999)
30. F.J. Dyson: J. Math. Phys. **3**, 1191 (1962)
31. F.J. Dyson: J. Math. Phys. **3**, 140 (1962); 157 (1962); 166 (1962)
32. B. Eynard, M.L. Mehta: J. Phys. A **31**, 4449 (1998)
33. W. Feller: *An Introduction to Probability Theory and Its Applications*, vol. I–II (Wiley 1966)
34. M. Gaudin: Nuclear Phys. **25**, 447 (1961)
35. S. Helgason: *Differential Geometry, Lie Groups, and Symmetric Spaces* (Academic Press, Boston 1978)
36. M. Jimbo, T. Miwa, Y. Mori, M. Sato: Proc. Japan Acad. Ser. A Math. Sci. **55**, 317 (1979)
37. M. Jimbo, T. Miwa, Y. Mori, M. Sato: Phys. D **1**, 80 (1980)
38. K. Johansson: Duke Math. J. **91**, 151 (1998)
39. K. Johansson: arXiv:math pr/0306216 (2003)
40. V.G. Kac: *Infinite-Dimensional Lie Algebras* 3rd ed (Academic Press, Boston 1995)
41. V.G. Kac, A.K. Raina: Bombay lectures on highest weight representations of infinite-dimensional Lie algebras. In: *Advanced Series in Mathematical Physics*, vol 2 (World Scientific Publishing Co., Inc., Teaneck, NJ, 1987)
42. S. Karlin, J. McGregor: Pacific J. Math. **9**, 1141 (1959)
43. E.M. McMillan: A problem in the stability of periodic systems. In: *Topics in Modern Physics, A Tribute to E.U. Condon*, vol 424, ed by W.E. Britten, H. Odabasi (Colorado Ass. Univ. Press, Boulder 1971) pp 219–244
44. M.L. Mehta: *Random Matrices*, 2nd edn (Academic Press, Boston 1991)
45. L.A. Pastur: Teoret. Mat. Fiz. **10**, 102 (1972)

46. M.M. Prähofer, H. Spohn: J. Stat. Phys. **108**, 1071 (2002)
47. E.M. Rains: Topics in probability on compact Lie groups. Ph.D. Thesis, Harvard University, Cambridge (1995)
48. C.A. Tracy, H. Widom: Introduction to random matrices. In: *Geometric and Quantum Aspects of Integrable Systems*, vol 424, ed by G.F. Helminck (Lecture Notes in Phys., Springer, Berlin 1993) pp 103–130
49. C.A. Tracy, H. Widom: Commun. Math. Phys. **159**, 151 (1994)
50. C.A. Tracy, H. Widom: Commun. Math. Phys. **161**, 289 (1994)
51. C.A. Tracy, H. Widom: Commun. Math. Phys. **207**, 665 (1999)
52. C.A. Tracy, H. Widom: Probab. Theory Related Fields **119**, 350 (2001)
53. C.A. Tracy, H. Widom: arXiv:math-pr/0309082 (2003)
54. C.A. Tracy, H. Widom: arXiv:math-pr/0412005 (2004)
55. K. Ueno, K. Takasaki: Adv. Stud. Pure Math. **4**, 1 (1984)
56. P. van Moerbeke: Integrable foundations of string theory. In: *Lectures on Integrable Systems*, ed by O. Babelon, P. Cartier, Y. Kosmann-Schwarzbach (World Sci. Publishing, River Edge, NJ, 1994) pp 163–267
57. H. Widom: arXiv:math-pr/0308157 (2003)
58. E.P. Wigner: Proc. Camb. Phil. Soc. **47**, 790 (1951)

Part II

Random Matrices and Applications

3

Integral Operators in Random Matrix Theory

Harold Widom

Mathematics Department, 194 Baskin Engineering, University of California, Santa Cruz, CA 95064, USA, widom@math.ucsc.edu

3.1 Hilbert–Schmidt and Trace Class Operators. Trace and Determinant. Fredholm Determinants of Integral Operators

In this section we present some of the basic operator theory required before the applications to random matrix theory can be made.

We shall see first how to extend the trace and determinant, defined on finite-dimensional spaces, to appropriate classes of operators on a Hilbert space $\mathcal{H}$. We begin by introducing two classes of compact operators on $\mathcal{H}$, the *Hilbert–Schmidt operators* and the *trace class operators.*

The *Hilbert–Schmidt norm* is defined by $\|A\|_2 = \sqrt{\sum \|Ae_i\|^2}$ where $\{e_i\}$ is an orthonormal (ON) basis for $\mathcal{H}$. This definition is independent of the basis used and $\|A^*\|_2 = \|A\|_2$; both of these statements follow from the fact that if $\{f_j\}$ is another ON basis then

$$\sum_i \|Ae_i\|^2 = \sum_{i,j} |(Ae_i, f_j)|^2 = \sum_{i,j} |(e_i, A^* f_j)|^2 = \sum_j \|A^* f_j\|^2 .$$

J. Harnad (ed.), *Random Matrices, Random Processes and Integrable Systems*, CRM Series in Mathematical Physics, DOI 10.1007/978-1-4419-9514-8_3,

The Hilbert–Schmidt operators are those with finite Hilbert–Schmidt norm, and the set of Hilbert–Schmidt operators is denoted by $\mathcal{S}_2$. That the Hilbert–Schmidt norm is a norm in the usual sense, in particular that it satisfies the triangle inequality, follows readily from the fact that it is the l^2 norm of the sequence $\|Ae_i\|$.

The *trace norm* is defined by $\|A\|_1 = \sup\{\sum |(Ae_i, f_i)|: \{e_i\}, \{f_i\} \text{ are ON}\}$. Here the identity $\|A^*\|_1 = \|A\|_1$ is obvious. The trace class operators are those with finite trace norm and the set of trace class operators is denoted by $\mathcal{S}_1$. (That the trace norm is a norm is immediate.)

An important inequality is

$$\|BC\|_1 \leq \|B\|_2 \|C\|_2 \,. \tag{3.1}$$

To see this, let $\{e_i\}$ and $\{f_i\}$ be ON sets. Then

$$\begin{aligned}\sum |(BCe_i, f_i)| &= \sum |(Ce_i, B^* f_i)| \leq \sum \|(Ce_i\| \|B^* f_i)\| \\ &\leq \Big\{\sum \|(Ce_i\|^2\Big\}^{1/2} \Big\{\sum \|(B^* f_i\|^2\Big\}^{1/2} \leq \|C\|_2 \|B\|_2 \,.\end{aligned}$$

Taking the sup over all $\{e_i\}$ and $\{f_i\}$ gives (3.1). (The reason for inequality at the end rather than equality is that in the definition of Hilbert–Schmidt norms $\{e_i\}$ and $\{f_i\}$ have to be ON bases whereas in the definition of trace norm they are merely ON sets.)

The *polar decomposition* of an operator A is the representation $A = UP$, where $P = \sqrt{A^*A}$ (the positive operator square root constructed via the spectral theorem) and U is an isometry from $P(\mathcal{H})$ to $A(\mathcal{H})$. If A is compact so is P and the *singular numbers* $s_1 \geq s_2 \geq \cdots$ of A are defined to be the eigenvalues of P. Observe that

$$s_1 = \|P\| = \|A\| \,.$$

Taking e_i to be an ON basis of eigenvectors of P we see that

$$\|A\|_2^2 = \sum_i (UPe_i, UPe_i) = \sum s_i^2 (Ue_i, Ue_i) = \sum s_i^2 \,,$$

where we have used the fact that U is an isometry on the image of P.

The trace norm is also expressible in terms of the singular numbers. First, we have $\|A\|_1 \leq \|U\sqrt{P}\|_2 \|\sqrt{P}\|_2$ by (3.1). This product equals $\sum s_i$ since the singular numbers of both $U\sqrt{P}$ and $\sqrt{P}$ are the $\sqrt{s_i}$. To get the opposite inequality take the e_i as before and $f_i = Ue_i$. Then

$$\sum (Ae_i, f_i) = \sum (UPe_i, Ue_i) = \sum s_i (Ue_i, Ue_i) = \sum s_i \,.$$

Thus $\|A\|_1 \geq \sum s_i$, and so $\|A\|_1 = \sum s_i$.

So for $p = 1, 2$ we have $\mathcal{S}_p = \{A\colon \{s_n(A)\} \in l^p\}$. This is the definition of $\mathcal{S}_p$ in general and $\|A\|_p$ equals the l^p norm of the sequence $\{s_i(A)\}$. That this is actually a norm in general, that it satisfies the triangle inequality, is true but nontrivial. We shall only use the spaces $\mathcal{S}_1$ and $\mathcal{S}_2$, where we know this to be true.

With the e_i as before, let E_n be the projection on the space spanned by $e_1, \dots, e_n$. Then $A - AE_n = UP(I - E_n)$, and the eigenvalues of $P(I - E_n)$ are zero and $s_{n+1}, s_{n+2}, \dots$. Thus these are the singular numbers of $A - AE_n$ and it follows that AE_n converges as $n \to \infty$ to A in $\mathcal{S}_p$ if $A \in \mathcal{S}_p$. In particular the set of finite-rank operators, which we denote by $\mathcal{R}$, is dense in $\mathcal{S}_p$.

This also follows from the following useful characterization of s_n (sometimes used as its definition): If $\mathcal{R}_n$ is the set of operators of rank at most n then

$$s_n(A) = \mathrm{dist}(A, \mathcal{R}_{n-1}) \,.$$

(Here the distance refers to the operator norm.) To see this note first that since AE_{n-1} has rank at most $n-1$ the distance on the right is at most $\|A - AE_{n-1}\| = \|P(I - E_{n-1})\| = s_n(A)$. To obtain the opposite inequality, assume that B is an operator with rank at most $n-1$. It is easy to see that there is a unit vector of the form $x = \sum_{i=1}^n a_i e_i$ such that $Bx = 0$. Then

$$\|A - B\| \geq \|Ax\| = \left\| U \sum_{i=1}^n a_i s_i e_i \right\| = \left\{ \sum_{i=1}^n |a_i|^2 s_i^2 \right\}^{1/2} \geq s_n \,.$$

Thus the distance in question is at least $s_n(A)$.

From the above characterization of $s_i(A)$ we immediately get the inequalities

$$s_n(AB) \leq s_n(A)\|B\| \,, \quad s_n(A + B) \leq s_n(A) + \|B\| \,. \tag{3.2}$$

In particular the first one gives

$$\|AB\|_p \leq \|A\|_p \|B\| \,.$$

Taking adjoints gives the inequality $\|AB\|_p \leq \|A\| \, \|B\|_p$.

We are now ready to extend the definition of trace to the trace class operators (whence its name, of course) on $\mathcal{H}$.

Suppose at first that our Hilbert space is finite-dimensional, so the trace is defined and has the usual properties. If $\mathcal{M}$ is a subspace of $\mathcal{H}$ such that $\mathcal{M} \supset A(\mathcal{H})$ and $A(\mathcal{M}^\perp) = 0$ then $\mathrm{tr}\, A = \mathrm{tr}(A|_{\mathcal{M}})$. This follows from the fact that $\mathrm{tr}\, A = \sum (Ae_i, e_i)$ where $\{e_i\}$ is an ON basis of $\mathcal{M}$ augmented by and ON basis of $\mathcal{M}^\perp$.

Going back to our generally infinite-dimensional Hilbert space $\mathcal{H}$, we first define the trace on $\mathcal{R}$, the finite-rank operators. Let $\mathcal{M}$ be any finite-dimensional subspace such that $\mathcal{M} \supset A(\mathcal{H})$ and $A(\mathcal{M}^\perp) = 0$. (This is the same as $\mathcal{M} \supset A(\mathcal{H}) + A^*(\mathcal{H})$, so there are plenty of such subspaces.) It follows from the remark of the last paragraph that $\mathrm{tr}(A|_{\mathcal{M}})$ is independent of the particular $\mathcal{M}$ chosen, since for any two such $\mathcal{M}$ there is one containing them both. We define $\mathrm{tr}\, A$ to be $\mathrm{tr}(A|_{\mathcal{M}})$ for any such $\mathcal{M}$.

The trace, as defined on $\mathcal{R}$, is additive (if $\mathcal{M}_1$ is chosen for A_1 and $\mathcal{M}_2$ is chosen for A_2 then $\mathcal{M}_1 + \mathcal{M}_2$ may be chosen for A_1, A_2 and $A_1 + A_2$ simultaneously) and preserves multiplication by scalars. Moreover, if $\{e_i\}$ is an ON basis for $\mathcal{M}$ we have

$$|\operatorname{tr} A| = \left|\sum (Ae_i, e_i)\right| \le \|A\|_1 ,$$

by the definition of trace norm. In short, the trace is a continuous linear functional on the dense subspace $\mathcal{R}$ of $\mathcal{S}_1$. This allows us to extend the trace to a continuous linear functional on all of $\mathcal{S}_1$, which was our goal.

If $\{e_i\}$ is any ON basis for $\mathcal{H}$ then the expected formula

$$\operatorname{tr} A = \sum (Ae_i, e_i) \tag{3.3}$$

holds. The reason is that if $\mathcal{M}_n$ is the space spanned by $e_1, \ldots, e_n$ and E_n is the projection onto this space then $E_n A E_n \to A$ in trace norm and so

$$\operatorname{tr} A = \lim_{n\to\infty} \operatorname{tr} E_n A E_n = \lim_{n\to\infty} \operatorname{tr}(E_n A E_n|_{\mathcal{M}_n}) = \lim_{n\to\infty} \sum_{i=1}^{n} (Ae_i, e_i) .$$

Moving on to the determinant we can similarly define $\det(I + A) = \det\big((I + A)|_{\mathcal{M}}\big)$ for any $A \in \mathcal{R}$. If $\mathcal{H}$ is finite dimensional we have $|\det A| = \det P = \prod s_i(A)$, and so, using the second inequality of (3.2),

$$\begin{aligned} |\det(I + A)| = \prod s_i(I + A) &\le \prod (1 + s_i(A)) \\ &\le \exp\Big(\sum s_i(A)\Big) = \exp(\|A\|_1) . \end{aligned} \tag{3.4}$$

This inequality holds for all $A \in \mathcal{R}$. Since the determinant is not linear this will not be enough to extend its definition. To extend $\det(I+A)$ to all $A \in \mathcal{S}_1$ we use the inequality

$$|\det(I + B) - \det(I + A)| \le \|B - A\|_1 \exp(\|A\|_1 + \|B - A\|_1 + 1) .$$

This follows from the fact that the difference may be written

$$\int_0^1 \frac{\mathrm{d}}{\mathrm{d}t} \det(I{+}(1{-}t)A{+}tB)\,\mathrm{d}t = \int_0^1 \mathrm{d}t\,\frac{1}{2\pi\mathrm{i}} \int_{|u-t|=R} \frac{\det(I + (1-u)A + uB)}{(u-t)^2}\,du$$

where $R = 1/\|B - A\|_1$. One then uses (3.4) to bound the determinant in the integrand and the inequality follows. And from the inequality it follows that the function $A \to \det(I + A)$ is locally uniformly continuous on $\mathcal{R}$ with trace norm. This is enough to extend it continuously to all of $\mathcal{S}_1$, which was our second goal.

The finite-dimensional identity

$$\det(I+A)(I+B) = \det(I+A)\det(I+B) \tag{3.5}$$

gives immediately the same identity for $A, B \in \mathcal{R}$ and then by continuity it follows for all $A, B \in \mathcal{S}_1$. This shows in particular that if $I + A$ is invertible then $\det(I + A) \neq 0$.[1]

Our definitions of trace and determinant make no mention of eigenvalues. The connection is given by the following important and highly nontrivial.

Theorem of Lidskiĭ. *Let λ_i be the nonzero eigenvalues of $A \in \mathcal{S}_1$ repeated according to algebraic multiplicity.*[2] *Then $\sum |\lambda_i| \leq \|A\|_1$ and*

$$\operatorname{tr} A = \sum \lambda_i\,, \quad \det(I+A) = \prod (1+\lambda_i)\,.$$

We only outline the proof. The first statement follows from the finite-dimensional inequality $\sum |\lambda_i| \leq \|A\|_1$ (which follows from the Jordan form plus the definition of trace norm) by a limit argument.

Write $D(z) = \det(I + zA)$. This is an entire function of z whose only possible zeros are at the $-\lambda_i^{-1}$. In fact, the multiplicity of the zero of $D(z)$ at $-\lambda_i^{-1}$ is precisely the algebraic multiplicity of λ_i. In the finite-dimensional case this follows from the Jordan form again. In general using a *Riesz projection* one can write

$$I + zA = (I + zF)(I + zB)$$

where F is finite rank with λ_i an eigenvalue with the same multiplicity as for A, and B does not have λ_i as an eigenvalue.[3] It follows from this and (3.5) that $D(z)$ has a zero of exactly this multiplicity at $-\lambda_i^{-1}$.

Thus

$$D(z) = \mathrm{e}^{g(z)} \prod (1 + z\lambda_i)$$

for some entire function g satisfying $g(0) = 0$. One shows (using approximation by finite-rank operators) that there is an estimate

$$D(z) = O(\mathrm{e}^{\varepsilon|z|})$$

for all $\varepsilon > 0$ and from this and the first statement of the theorem one can deduce that $g(z) = o(|z|)$ as $z \to \infty$. This implies that g is a constant, which must be 0, and this gives

$$D(z) = \prod (1 + z\lambda_i)\,. \tag{3.6}$$

---

[1] The converse is also true. If $\det(I + A) \neq 0$ then -1 is not an eigenvalue of A, by Lidskiĭ's theorem stated next, and since A is compact this implies that $I + A$ is invertible.

[2] The algebraic multiplicity of λ is the dimension of its generalized eigenspace, the set of vectors x such that $(A - \lambda I)^n x = 0$ for some n.

[3] The Riesz projection is $E = (2\pi \mathrm{i})^{-1} \int_C (tI - A)^{-1}\, \mathrm{d}t$ where C is a little countour enclosing λ_i. The operators $F = AE$ and $B = A(I - E)$ have the stated properties.

Setting $z = 1$ gives the determinant identity.

Finally, the finite-dimensional result

$$D'(z) = D(z)\,\mathrm{tr}(A(I+zA)^{-1})\,, \tag{3.7}$$

which holds whenever $I + zA$ is invertible, carries over by approximation to all $A \in \mathcal{S}_1$. In particular $\mathrm{tr}\,A = D'(0)/D(0)$ and by (3.6) this equals $\sum \lambda_i$.

One consequence of Lidskiĭ's theorem is that

$$\mathrm{tr}\,AB = \mathrm{tr}\,BA\,, \quad \det(I+AB) = \det(I+BA)$$

for any operators A and B such that both products are trace class. This is a consequence of the easy fact that the two products have the same nonzero eigenvalues with the same algebraic multiplicities. (The weaker fact that these relations hold when both operators are Hilbert–Schmidt, or one is trace class and the other bounded, follows by approximation from the finite-dimensional result.)

All this has been abstract. Now we specialize to operators on Hilbert spaces $L^2(\mu)$ where μ is an arbitrary measure. There first question we ask is, what are the Hilbert–Schmidt opetarors on $L^2(\mu)$? The answer is that they are precisely the integral operators K with kernel $K(x,y)$ belonging to $L^2(\mu\times\mu)$.

To show that all such integral operators are Hilbert–Schmidt, let $\{e_i\}$ be an ON basis for $L^2(\mu)$. Then the doubly-indexed family of functions $\{e_i(x)e_j(y)^*\}$ is an ON basis for $L^2(\mu\times\mu)$.[4] The inner product of $K(x,y)$ with this vector equals

$$\iint e_i(x)^* K(x,y) e_j(y)\,\mathrm{d}\mu(y)\,\mathrm{d}\mu(x) = (Ke_j, e_i)\,.$$

Hence

$$\|K(x,y)\|^2_{L^2} = \sum_{i,j} |(Ke_j, e_i)|^2 = \sum_i \|Ke_j\|^2 = \|K\|_2^2\,. \tag{3.8}$$

Thus K is Hilbert–Schmidt and its Hilbert–Schmidt norm is the L^2 norm of the kernel.

Conversely, suppose K is a Hilbert–Schmidt operator. Then

$$\sum_{i,j=1}^{\infty} |(Ke_j, e_i)|^2 = \sum_{j=1}^{\infty} \|Ke_j\|^2 < \infty.$$

It follows that the series

$$\sum_{i,j=1}^{\infty} (Ke_j, e_i) e_i(x) e_j(y)^*$$

converges in $L^2(\mu\times\mu)$ to some function $K(x,y)$. If we denote by K' the integral operator with kernel $K(x,y)$ then $(K'e_j, e_i) = (Ke_j, e_i)$ for all i, j, so $K = K'$.

---

[4] We use the asterisc here to denote complex conjugate.

To determine when an operator, even an integral operator, is trace class is not so simple. We already have a sufficient condition, which is immediate from (3.1): If $K = K_1 K_2$ where K_1 and K_2 are Hilbert–Schmidt then K is trace class. In this case we also have the formula

$$\operatorname{tr} K = \iint K_1(x,y) K_2(y,x)\, \mathrm{d}\mu(y)\, \mathrm{d}\mu(x) \; . \tag{3.9}$$

To see this assume first that the two kernels are basis functions, so

$$K_1(x,y) = e_i(x) e_j(y)^* \; , \quad K_2(x,y) = e_k(x) e_l(y)^* \; .$$

Orthonormality of the e_i gives

$$K(x,y) = e_i(x) e_l(y)^* \delta_{jk} \; ,$$

and using (3.3) with the same basis functions gives $\operatorname{tr} K = \delta_{il}\delta_{jk}$. This is equal to the right side of (3.9).

Using the fact that both sides of (3.9) are bilinear in K_1 and K_2 we deduce that the identity holds when they are finite linear combinations of the basis functions $e_i(x)e_j(y)^*$. To extend this to general Hilbert–Schmidt K_i we use the density of these finite-rank kernels in $L^2(\mu \times \mu)$, the fact (which follows from (3.1)) that multiplication is continuous from $\mathcal{S}_2 \times \mathcal{S}_2$ to $\mathcal{S}_1$, and that the left side of (3.9) is continuous on $\mathcal{S}_1$ and the right side continuous on $\mathcal{S}_2 \times \mathcal{S}_2$.

Observe that (3.9) may be stated in this case as

$$\operatorname{tr} K = \int K(x,x)\, \mathrm{d}\mu(x) \; . \tag{3.10}$$

But one should be warned that this cannot be true for all trace class integral operators. For changing a kernel $K(x,y)$ on the diagonal $x = y$ does not change the integral operator but it certainly can change the right side above.

A useful consequence of what we have shown is that for a kernel acting on $L^2(-\infty,\infty)$ with Lebesgue measure we have

$$\|K\|_1 \le \|K(x,y)\|_2 + |J| \|K_x(x,y)\|_2$$

if K vanishes for x outside a finite interval J, and similarly with x and y interchanged. To see this we observe that for any x_0, $x \in J$

$$K(x,y) = \chi_J(x) K(x_0,y) + \int_J \varphi(x,z) K_x(z,y)\, \mathrm{d}z$$

where

$$\varphi(x,z) = \begin{cases} 1 & \text{if } x_0 < z < x, \\ -1 & \text{if } x < z < x_0, \\ 0 & \text{otherwise.} \end{cases}$$

The integral represents a product of two operators, the first with Hilbert–Schmidt norm at most $|J|$ and the second with Hilbert–Schmidt norm $\|K_x(x,y)\|_2$. The first term is a rank one operator of norm (and so trace norm) $|J|^{1/2}\|K(x_0,\ldots)\|_2$, and for some x_0 the second factor is at most $|J|^{-1/2} \times \|K(x,y)\|_2$. This gives the stated inequality. An easy computation, using (3.9) to compute the trace of the product, shows that that the formula (3.10) holds here also.

This extends to higher dimensions. In particular an operator with smooth kernel on a bounded domain is trace class, and (3.10) holds.

The classical Fredholm determinant formula for $D(z)$,

$$\det(I+zK) = \sum_{n=0}^{\infty} \frac{z^n}{n!} \int \cdots \int \det\big(K(x_i,x_j)\big)\, \mathrm{d}\mu(x_1)\cdots \mathrm{d}\mu(x_n)\,, \tag{3.11}$$

holds for any trace class operator K for which $\operatorname{tr} K = \int K(x,x)\,\mathrm{d}\mu(x)$. To see this we successively differentiate (3.7) and then set $z=0$ to see that the expansion about $z=0$ takes the form

$$\det(I+zK) = \sum_{n=0}^{\infty} z^n P_n(\operatorname{tr} K, \operatorname{tr} K^2, \ldots, \operatorname{tr} K^n)$$

with P_n a polynomial independent of K. If we expand the determinants in Fredholm's formula (3.11) we see that the right side is of the same form

$$\sum_{n=0}^{\infty} z^n Q_n(\operatorname{tr} K, \operatorname{tr} K^2, \ldots, \operatorname{tr} K^n)$$

with polynomials Q_n. The reason is that each permutation in the expansion of the determinant is a product of disjoint cycles. A cycle of length j contributes an integral

$$\int \cdots \int K(x_1,x_2)\cdots K(x_j,x_1)\, \mathrm{d}\mu(x_1)\cdots \mathrm{d}\mu(x_j)\,.$$

When $j>1$ this is equal to $\operatorname{tr} K^j$ by (3.10), and when $j=1$ this is equal to $\operatorname{tr} K$ by assumption.

The coefficients of z^n in the two sums depend only on the eigenvalues λ_i of K. Therefore to establish equality in general it suffices to show that for any numbers λ_i satisfying $\sum |\lambda_i| < \infty$ there is a trace class kernel K with these as eigenvalues such that the two sides agree. When μ is counting measure and K the diagonal matrix with diagonal entries λ_i then

$$\det(I+zK) = \prod_{i=1}^{\infty} (1+z\lambda_i)\,,$$

while the right side of (3.11) equals

$$\sum_{n=0}^{\infty} z^n \sum_{i_1<\cdots<i_n} \lambda_{i_1}\cdots\lambda_{i_n} \,.$$

These are the same.

Finally we derive the identity, generalizing (3.7),

$$\frac{\mathrm{d}}{\mathrm{d}z}\log\det\bigl(I+A(z)\bigr)=\mathrm{tr}\bigl(A'(z)(I+A(z)^{-1})\bigr) \tag{3.12}$$

for any differentiable family of trace class operators, valid whenever $I+A(z)$ is invertible. For h sufficiently small

$$\begin{aligned}\frac{\det(I+A(z+h))}{\det(I+A(z))} &= \det\Bigl[\bigl(I+A(z+h)\bigr)\bigl(I+A(z)\bigr)^{-1}\Bigr]\\ &= \det\Bigl[I+\bigl(A(z+h)-A(z)\bigr)\bigl(I+A(z)\bigr)^{-1}\Bigr]\,.\end{aligned}$$

From the differentiability of $A(z)$ we can write

$$A(z+h)-A(z)=hA'(z)+o_1(h)\,,$$

where $o_1(h)$ denotes a family of operators with trace norm $o(1)$ as $h\to 0$. Therefore the last operator in brackets is

$$I+hA'(z)\bigl(I+A(z)\bigr)^{-1}+o_1(h)\,.$$

The operator $I+hA'(z)\bigl(I+A(z)\bigr)^{-1}$ is invertible for small h with uniformly bounded norm, so the above may be factored as

$$\Bigl[I+hA'(z)\bigl(I+A(z)\bigr)^{-1}\Bigr]\bigl[I+o_1(h)\bigr]\,.$$

The second factor above has determinant $1+o(h)$, by the contuity of the determinant in $\mathcal{S}_1$. By (3.7), with z replaced by h and evaluated at $h=0$, the first factor has determinant $1+h\,\mathrm{tr}\bigl(A'(z)\bigl(I+A(z)\bigr)^{-1}\bigr)+o(h)$. So the determinant of the product has the same form and we have shown that

$$\frac{\det\bigl(I+A(z+h)\bigr)}{\det\bigl(I+A(z)\bigr)}=1+h\,\mathrm{tr}\bigl(A'(z)\bigl(I+A(z)\bigr)^{-1}\bigr)+o(h)\,.$$

This gives (3.12).

Except for the proof of Lidskiĭ's theorem this section is self-contained.The proof of Lidskiĭ's theorem we outlined follows [7, 4.7.14]. More thorough treatments of the $\mathcal{S}_p$ classes and trace and determinant may be found in [2,4], and a completely different development of the determinant is in [8].

3.2 Correlation Functions and Kernels of Integral Operators. Spacing Distributions as Operator Determinants. The Sine and Airy Kernels

We consider ensembles of $N \times N$ Hermitian matrices such that probability density of eigenvalues is of form

$$P(x_1, \ldots, x_N) = c_N \prod_{i<j} |x_i - x_j|^\beta \prod_i w(x_i) ,$$

where w is a weight function which is rapidly decreasing at $\pm\infty$. If F is a symmetric function then

$$\mathrm{E}\big(F(\lambda_1, \cdots, \lambda_N)\big) = \int \cdots \int P(x_1, \ldots, x_N) F(x_1, \ldots, x_N) \mathrm{d}x_1 \cdots \mathrm{d}x_N ,$$

where, as usual, E stands for expected value.

The cases of greatest interest are $\beta = 2, 1$ and 4 corresponding respectively to the unitary, orthogonal and symplectic ensembles.

We define $E(n, J)$ to be the probability that the set J contains exactly n eigenvalues. The probability density that there are eigenvalues at r and s and none in between equals

$$-\frac{\partial^2}{\partial r \partial s} E\big(0, (r, s)\big) .$$

(Thus $E(0, J)$ is sometimes referred to as the *spacing probability.*)

The probability density that there are eigenvalues near $x_1, \ldots, x_n$ is the *n-point correlation function* given by

$$R_n(x_1, \ldots, x_n) = \frac{N!}{(N-n)!} \int \cdots \int P(x_1, \ldots, x_N)\, \mathrm{d}x_{n+1} \cdots \mathrm{d}x_N . \tag{3.13}$$

For $\beta = 2$ we will find a kernel $K(x, y)$ in terms of which these are expressible.

We start with the fact that

$$\mathrm{E}\Big(\prod (1 + f(\lambda_i))\Big) = c_N \int \cdots \int \prod_{i<j} (x_i - x_j)^2 \prod_i \big[w(x_i)\big(1 + f(x_i)\big)\big]\, \mathrm{d}x ,$$

with c_N the normalizing constant such that the right side equals 1 when $f = 0$. (In the special case $f(x) = -\chi_J(x)$ we obtain $E(0, J)$.)

The first product in the integrand is the square of a Vandermonde determinant. There is a general identity

$$\int \cdots \int \det \varphi_i(x_j) \det \psi_i(x_j) \mathrm{d}\nu(x_1) \cdots \mathrm{d}\nu(x_N) = N! \det\Big(\int \varphi_i(x) \psi_j(x)\, d\nu(x)\Big) .$$

Taking $\varphi_i(x) = \psi_i(x) = x^i$, $\mathrm{d}\nu(x) = w(x)\,\mathrm{d}x$, we get

$$\mathrm{E}\Big(\prod(1+f(\lambda_i))\Big) = c'_N \det\Big(\int x^{i+j}(1+f(x))w(x)\,\mathrm{d}x\Big)\ (i,j=0,\dots,N-1)\,.$$

Replacing x^i by any $p_i(x)$, a polyomial of degree i, amounts to performing row and column operations. It follows that if we set

$$\varphi_i(x) = p_i(x)\sqrt{w(x)}\,.$$

Then the above becomes

$$\mathrm{E}\Big(\prod(1+f(\lambda_i))\Big) = c''_N \det\Big(\int \varphi_i(x)\varphi_j(x)\,\mathrm{d}x + \int \varphi_i(x)\varphi_j(x)f(x)\,\mathrm{d}x\Big)$$

with a different constant c''_N.

If we're clever we take the p_i to be the polynomials ON with respect to w, so the φ_i are ON with respect to Lebesque measure and the first summand in the determinant is δ_{ij}. Then taking $f=0$ gives $c''_N = 1$. This is the usual way of doing things. But let's not be clever, take the p_i arbitrary and set

$$M = (m_{ij}) = \Big(\int \varphi_i(x)\varphi_j(x)\,\mathrm{d}x\Big)\,,\quad M^{-1} = (\mu_{ij})\,,\quad \psi_i = \sum_j \mu_{ij}\varphi_j\,.$$

Then by factoring out M on the left we get

$$\mathrm{E}\Big(\prod(1+f(\lambda_i))\Big) = c'''_N \det\Big(\delta_{ij} + \int \psi_i(x)\varphi_j(x)f(x)\,\mathrm{d}x\Big)\,,$$

where $c'''_N = (\det M)c''_N$. Now taking $f=0$ gives $c'''_N = 1$.

Here is where operator determinants come in. There will be two Hilbert spaces. One is $L^2(\mathbf{R})$ and the other is $\mathbf{R}^N$, which consists of functions on the set $\{1,\cdots,N\}$. Let us define $A\colon L^2(\mathbf{R}) \to \mathbf{R}^N$ and $B\colon \mathbf{R}^N \to L^2(\mathbf{R})$ by the kernels

$$A(i,x) = \psi_i(x)f(x)\,,\quad B(x,j) = \varphi_j(x)\,.$$

Then $AB\colon \mathbf{R}^N \to \mathbf{R}^N$ has kernel

$$AB(i,j) = \int \psi_i(x)\varphi_j(x)f(x)\,\mathrm{d}x\,,$$

and $BA\colon L^2(\mathbf{R}) \to L^2(\mathbf{R})$ has kernel

$$BA(x,y) = \sum_{i=0}^{N-1} \varphi_i(x)\psi_i(y)f(y)\,,$$

The identity $\det(I+AB) = \det(I+BA)$ from Part I, which holds even if the operators go from one space to a different one,[5] gives

[5] Since our operators are of finite rank there is no need at this stage to use the full definition of determinants on infinite-dimensional spaces.

$$\mathrm{E}\Big(\prod(1+f(\lambda_i))\Big) = \det(I+BA)\,.$$

Thus, if we set

$$K(x,y) = \sum_{i=0}^{N-1} \varphi_i(x)\psi_i(y) = \sum_{i,j=0}^{N-1} \varphi_i(x)\mu_{ij}\varphi_j(y)\,, \tag{3.14}$$

then we have obtained

$$\mathrm{E}\Big(\prod(1+f(\lambda_i))\Big) = \det(I+Kf)\,. \tag{3.15}$$

Here K is the integral operator with kernel $K(x,y)$ and f denotes multiplication by the function f.

There is an abstract characterization of the operator K. If we define

$$\mathcal{H} = \{p\sqrt{w}\colon p \text{ is a polynomial of degree less than } N\}\,,$$

then K is the orthogonal projection operator from $L^2(\mathbf{R})$ to $\mathcal{H}$. To see this observe that the φ_i span $\mathcal{H}$. If a function g is orthogonal to all the φ_i then clearly $Kg = 0$. Thus $K(\mathcal{H}^\perp) = \{0\}$. If g is one of the φ_i, say $g = \varphi_k$, then

$$Kg = \sum_{i,j} \varphi_i\mu_{ij}(\varphi_j,\varphi_k) = \sum_{i,j}\varphi_i\mu_{ij}m_{jk} = \sum_i \varphi_i\delta_{ik} = \varphi_k = g\,.$$

Thus K acts as the identity operator on $\mathcal{H}$. Hence K is the orthogonal projection from $L^2(\mathbf{R})$ to $\mathcal{H}$ as claimed.

Taking $f = -\chi_J$ in (3.15) gives

$$E(0,J) = \det(I - K\chi_J)\,.$$

More generally, $E(n,J)$ is the coefficient of $(\lambda+1)^n$ in the expansion of

$$\int\cdots\int P(x)\prod_i[(\lambda+1)\chi_J(x_i) + \chi_{J^c}(x_i)]\,\mathrm{d}x \\ = \int\cdots\int P(x)\prod_i[1+\lambda\chi_J(x_i)]\,\mathrm{d}x = \det(I+\lambda K\chi_J)\,.$$

Thus

$$E(n,J) = \frac{1}{n!}\frac{\partial^n}{\partial\lambda^n}\det(I+\lambda K\chi_J)|_{\lambda=-1}\,.$$

Yet more generally, a similar expression for $E(n_1,\ldots,n_m;J_1,\ldots,J_m)$, the probability that J_i contains exactly n_i eigenvalues ($i = 1,\ldots,m$), can be found.

The n-point correlation function $R_n(y_1,\ldots,y_n)$ given by (3.13) equals the coefficient of $z_1\cdots z_n$ in the expansion of

$$\iint P(x) \prod_{i=1}^{N} \left[1 + \sum_{r=1}^{n} z_r \delta(x_i - y_r) \right] \mathrm{d}x \,.$$

To evaluate this integral by using the AB, BA trick as above we replace dx by the discrete measure $\sum_{r=1}^{n} \delta(x - y_r)$ and define f by $f(y_r) = z_r$. We then find that the integral equals

$$\det(\delta_{rs} + K(y_r, y_s) z_s)_{r,s=1,\dots,n} \,.$$

Taking the coefficient of $z_1 \cdots z_n$ gives

$$R_n(y_1, \dots, y_n) = \det\bigl(K(y_r, y_s)\bigr)_{r,s=1,\dots,n} \,.$$

For β equal to 1 and 4 the scalar kernel K is replaced by a matrix kernels. We indicate how this comes about when $\beta = 4$. We take the case when N is even and replace N by $2N$. Then we use the fact that

$$\prod_{i<j} (x_i - x_j)^4 = \det(x_j^i i x_j^{i-1}) \quad i = 0, \dots 2N, \; j = 1, \dots, N$$

and the general identity

$$\int \cdots \int \det\bigl(\varphi_i(x_j) \psi_i(x_j)\bigr) \, \mathrm{d}\nu(x_1) \cdots \mathrm{d}\nu(x_N)$$
$$= (2N)! \, \mathrm{Pf}\biggl(\int \bigl(\varphi_i(x)\psi_j(x) - \psi_i(x)\varphi_j(x)\bigr) \, \mathrm{d}\nu(x) \biggr) \,.$$

Here Pf denotes the Pfaffinan of the antisymmetric matrix on the right. Its square equals the determinant of the matrix. It follows that

$$\biggl(\mathrm{E}\Bigl(\prod (1 + f(\lambda_i)) \Bigr) \biggr)^2$$
$$= c_N \det\biggl(\int (j-i) x^{i+j-1} \bigl(1 + f(x)\bigr) w(x) \, \mathrm{d}x \biggr)_{i,j=0,\dots,2N-1} \,,$$

for some normalizing constant c_N.

As before we replace the sequence $\{x^i\}$ by any sequence $\{p_i(x)\}$ of polynomials of exact degree i, and set $\varphi_i = p_i w^{1/2}$. Except for another constant factor depending only on N, the above equals

$$\det\biggl(\int \bigl(\varphi_i(x)\varphi_j'(x) - \varphi_i'(x)\varphi_j(x)\bigr) \bigl(1 + f(x)\bigr) \, \mathrm{d}x \biggr) \,,$$

the extra terms arising from the differentiation of $w(x)^{1/2}$ having cancelled. We write this as

$$M + \biggl(\int \bigl(\varphi_i(x)\varphi_j'(x) - \varphi_i'(x)\varphi_j(x)\bigr) f(x) \, \mathrm{d}x \biggr)$$

where M is the matrix of integrals $\int \big(\psi_i(x)\psi_j'(x) - \psi_i'(x)\psi_i(x)\big)\,dx$.

Next we factor out M on the left. Its determinant is another constant depending only on N. If as before we set

$$M^{-1} = (\mu_{ij})\,, \quad \psi_i(x) = \sum_j \mu_{ij}\varphi_j(x)\,,$$

then the resulting matrix is

$$I + \left(\int \big(\psi_i(x)\varphi_j'(x) - \psi_i'(x)\varphi_j(x)\big)\, f(x)\,dx\right)\,,$$

Now $\psi_i(x)\varphi_j'(x) - \psi_i'(x)\varphi_j(x)$, the multiplier of $f(x)$ in the integrand, is not a single product but a sum of products. To use the AB, BA trick we want it to be a single product. In fact we can write it as one, as a matrix product

$$\big(\psi_i(x) \; - \psi_i'(x)\big)\begin{pmatrix}\varphi_j'(x)\\ \varphi_j(x)\end{pmatrix}\,.$$

Thus if we set

$$A(i,x) = f(x)\big(\psi_i(x) \; - \psi_i'(x)\big)\,, \quad B(x,i) = \begin{pmatrix}\varphi_i'(x)\\ \varphi_i(x)\end{pmatrix}\,,$$

then the above matrix is $I+AB$. In this case BA is the integral operator with matrix kernel $K(x,y)f(y)$ where

$$K(x,y) = \begin{pmatrix}\sum \varphi_i'(x)\psi_i(y) & -\sum \varphi_i'(x)\psi_i'(y)\\ \sum \varphi_i(x)\psi_i(y) & -\sum \varphi_i(x)\psi_i'(y)\end{pmatrix}\,.$$

The sums here are over $i = 0,\ldots,2N-1$. In terms of these matrix kernels there are then formulas for spacing probabilities (in terms of operator determinants) and correlation functions (in terms of block matrices).

This is the matrix kernel for $\beta = 4$. For $\beta = 1$ another identity yields another 2×2 matrix kernel.

We mentioned that it is usual in the case $\beta = 2$ to take the p_i to be the polynomials orthonormal with respect to the weight function w, so the formula (3.14) for the kernel simplifies to

$$\sum_{i=0}^{N-1} \varphi_i(x)\varphi_i(y)\,.$$

Analogously, when β equals 4 and 1, it has been usual to take the p_i to be so-called *skew-orthogonal polynomials*, and the formulas for the enteies in the matrix kernels take a simpler form. But (we won't go into detail) there are advantages to not doing this and using the abstract characterizations of the kernels instead.

The usefulness of the operator determinant representation becomes apparent when we let $N \to \infty$, since rather than having determinants of order N we have operators whose kernels $K_N(x, y)$ have N as a parameter. If these kernels converge in trace norm then the determinants for K_N converge to the corresponding determinant for the limit kernel. For the Gaussian unitary ensemble $w(x) = e^{-x^2}$ and

$$\frac{1}{\sqrt{2N}} K_N\left(z + \frac{x}{\sqrt{2N}}, z + \frac{y}{\sqrt{2N}}\right) \to \frac{1}{\pi} \frac{\sin(x-y)}{x-y} = K_{\mathrm{sine}}(x, y)$$

in trace norm on any bounded set (bulk scaling near z) and

$$\frac{1}{2^{1/2} N^{1/6}} K_N\left(\sqrt{2N} + \frac{x}{2^{1/2} N^{1/6}}, \sqrt{2N} + \frac{y}{2^{1/2} N^{1/6}}\right)$$
$$\to \frac{\mathrm{Ai}(x)\,\mathrm{Ai}'(y) - \mathrm{Ai}'(x) A(y)}{x - y} = K_{\mathrm{Airy}}(x, y)$$

in trace norm on any set which is bounded below, where Ai is the Airy function (edge scaling). Thus scaling limits of spacing distributions are equal to operator determinants for the *sine kernel* K_{sine} and the Airy *kernel* K_{Airy}. In particular we obtain the limiting distribution for the largest eigenvalue,

$$\lim_{N\to\infty} \Pr\left(\lambda_{\max} \le \sqrt{2N} + \frac{s}{2^{1/2} N^{1/6}}\right) = \det(I - K_{\mathrm{Airy}} \chi_{(s\infty)}) \,.$$

Universality theorems say that the same limiting kernels K_{sine} and K_{Airy} arise from bulk and edge scaling for large classes of random matrix ensembles.

The classical reference for random matrices is [5], which now has a successor [6]. This section followed the treatment in [10]. Bulk scaling of GUE can be found in [5,6] and edge-scaling in [1,3].

3.3 Differential Equations for Distribution Functions Arising in Random Matrix Theory. Representations in Terms of Painlevé Functions

Probably the most important distribution functions arising in random matrix theory which are representable as operator determinants are

$$F_2(s) = \det(I - K_{\mathrm{Airy}} \chi_{(s,\infty)}) \,, \tag{3.16}$$

and the analogous limiting distribution functions for the Gaussian orthogonal and symplectic ensembles, for which the kernels are matrix-valued. Their special importance lies in the fact that they arise also outside of random matrix theory in many other probabilistic models, in number theory, and elsewhere.

We shall show here that we have the formula

$$F_2(s) = e^{-\int_s^\infty (t-s)q(t)^2\,\mathrm{d}t}\,, \tag{3.17}$$

where q satisfies the Painlevé II equation

$$q''(s) = sq(s) + 2q(s)^3\,.$$

This, together with the asymptotics $q(s) \sim \mathrm{Ai}(s)$ as $s \to +\infty$, determines q uniquely. There is a whole class of kernels for which the analogous determinants are expressible in terms of solutions of differential equations (more precisely, systems of equations). Fortunately this most important one has the easiest derivation and we present it in detail. The actual proof is quite short but will require some preparation.

We think of our operators as acting on $L^2(J)$ where J is some fixed interval (s_0, ∞). We will want to differentiate the determinant on the right in (3.16) using formula (3.12). The formula in this case would read

$$\frac{\mathrm{d}}{\mathrm{d}s}\log\det\bigl(I - K(s)\bigr) = -\,\mathrm{tr}\left[\bigl(I - K(s)\bigr)^{-1}K'(s)\right], \tag{3.18}$$

where $K(s) = K_{\mathrm{Airy}}\chi_{(s,\infty)}$. But the operators $K(s)$ are not differentiable functions of s. Indeed, $K'(s)$ does not make sense as an operator on L^2.

So before we do anything else we have to arrange for (3.18) to make sense and to be correct. What we do is introduce the Sobolev space

$$H^1 = \{f\colon f, f' \in L^2\}$$

with Hilbert space norm $\|f\|_{H^1} = (\|f\|_2^2 + \|f'\|_2^2)^{1/2}$ and let our operators act on this space instead of L^2. That the operator determinants are the same for the two underlying spaces follows from the fact that $K(s)(L^2) \subset H^1$, which implies that the algebraic multiplicity of any nonzero eigenvalue of $K(s)$ is the same whether it acts on L^2 or H^1. Instead of the Airy kernel we shall consider at first, more generally, any kernel $K(x, y)$ which satisfies the following conditions:

If we define the function K^y by $K^y(x) = K(x, y)$ then

1. $K^y \in H^1$ for each $y \in J$, and $y \to K^y$ is a continuous function from J to H^1;
2. $\int_J \|K^y\|_{H^1}\,dy < \infty$.

(Condition 1 holds if $\|K^y\|_{H^1}$ and $\|(\partial K/\partial y)^y\|_{H^1}$ are locally bounded functions of y. These are easily seen to hold for the Airy kernel.) We shall show under these conditions that $K(s) = K\chi_{(s,\infty)}$ is a differentiable family of trace class operators on H^1.

What is nice about H^1 is that the point evaluations $f \to f(y)$ comprise a bounded and continuous family of continuous linear functionals on H^1. Continuity is immediate from

$$|f(y_2) - f(y_1)| \le \int_{y_1}^{y_2} |f'(y)|\,\mathrm{d}y \le \sqrt{y_2 - y_1}\,\|f\|_{H^1}\,.$$

If I is any interval of length 1 then

$$\min_{y\in I} | f(y) \leq \left\{ \int_I |f(y)|^2 \, \mathrm{d}y \right\}^{1/2} \leq \|f\|_{H^1} ,$$

and from this and the last inequality we deduce that

$$|f(y)| \leq 2\|f\|_{H^1}$$

for all y. (Take $y_2 = y$ we find an appropriate y_1 in an interval of length 1 containing y.) Thus the family of linear functionals is bounded.

It follows from the above-stated facts and Condition 1 that if $E(y)\colon H^1 \to H^1$ is defined by

$$E(y)f = f(y)K^y , \tag{3.19}$$

then $y \to E(y)$ is a family of operators which is continuous in the operator norm and satisfies $\|E(y)\| \leq 2\|K^y\|_{H^1}$. Each $E(y)$ is of rank one, so this is actually a continuous family of trace class operators on H^1, and the trace norm is the same as the operator norm. By Condition 2 we have

$$\int_J \|E(y)\|_1 \, \mathrm{d}y \leq 2 \int_J \|K^y\|_{H^1} \, \mathrm{d}y < \infty .$$

The identity

$$(K(s)f)(x) = \int_s^\infty K(x,y)f(y) \, \mathrm{d}y = \int_s^\infty (E(y)f)(x) \, \mathrm{d}y$$

shows that $K(s) = \int_s^\infty E(y) \, dy$. Hence $K(s)$ is trace class and $s \to K(s)$ is a differentiable function of s with derivative $-E(s)$.[6]

Thus (3.18) holds whenever $I - K(s)$ is invertible and gives

$$\frac{\mathrm{d}}{\mathrm{d}s} \log\det\bigl(I - K(s)\bigr) = \mathrm{tr}\bigl[\bigl(I - K(s)\bigr)^{-1} E(s)\bigr] .$$

For a more concrete expression for the right side above we denote by $R(x,y)$ the kernel of the operator $R = \bigl(I - K(s)\bigr)^{-1}K$. (This is a modified resolvent of $K(s)$; the actual resolvent is $\bigl(I - K(s)\bigr)^{-1}K(s)$.) The operator on the right has rank one and takes a function f to $R(x,s)f(s)$. Thus its only nonzero eigenvalue is $R(s,s)$, corresponding to the eigenfunction $R(x,s)$, and so this is its trace. This establishes the formula

---

[6] We use here the fact that the integral of trace class operators is trace class if their trace norms belong to L^1. This follows from the easily established fact that for any operator A

$$\|A\|_1 = \sup\{|\,\mathrm{tr}\, AB| : B \in \mathcal{R}, \|B\| \leq 1\} .$$

$$\frac{\mathrm{d}}{\mathrm{d}s} \log \det\bigl(I - K(s)\bigr) = R(s,s) \ , \tag{3.20}$$

which was our goal.

In order to apply this with $K = K_{\text{Airy}}$ we observe that $F_2(s)$ is a nonnegative monotonic function, being the limit of distribution functions, and so it is nonzero. This is equivalent to the invertibility of $I - K(s)$ in this case and so we may apply (3.20) to obtain the formula

$$\frac{\mathrm{d}}{\mathrm{d}s} \log F_2(s) = R(s,s) \ , \tag{3.21}$$

where $R(x,y)$ is the modified resolvent kernel for $K_{\text{Airy}}\chi_{(s,\infty)}$. For simplicity of notation we shall now write this operator as $K\chi$.

What will in the end enable us to establish (3.17) is that K satisfies two commutator relations which will yield corresponding relations for the resolvent kernel. The commutators will be with the operators M, multiplication by the independent variable x, and $D = d/dx$. But our Sobolev space H^1 is not closed under M or D, so we shall change our space to one which is closed under the actions of these operations. Before we do this we observe that if we set

$$\rho = I + R\chi$$

then

$$(I - K\chi)^{-1} = \rho \ , \tag{3.22}$$

in other words,

$$(I + R\chi)(I - K\chi) = (I - K\chi)(I + R\chi) = I \ .$$

After subtracting I from all terms this becomes identities involving the two kernels $K(x,y)\chi(y)$ and $R(x,y)\chi(y)$. From this it follows that (3.22) holds no matter what space the operators with these kernels act on, as long as it is closed under the these operations.

One space we might take that serves our purpose is

$$\mathcal{S} = \{f \in C^\infty : \text{each } f^{(n)}(x) \text{ is rapidly decreasing at } +\infty\} \ .$$

This is clearly closed under M and D, and is closed under $K\chi$ and $R\chi$ because of the rapid decrease of the Airy function at $+\infty$.

The two commutator relations are

$$[M, K\chi] = \mathrm{Ai} \otimes \mathrm{Ai}'\,\chi - \mathrm{Ai}' \otimes \mathrm{Ai}\,\chi \ , \tag{3.23}$$

$$[D, K\chi] = -\,\mathrm{Ai} \otimes \mathrm{Ai}\,\chi - K\delta \ . \tag{3.24}$$

Here $u \otimes v$ denotes the operator with kernel $u(x)v(y)$ and δ denotes multiplication by $\delta(y-s)$. (Of course δ does not act on $\mathcal{S}$, but $K\delta$ does. This operator is the same as $E(s)$ given by (3.19) but the δ notation is more transparent).

The first relation is immediate from the formula for the Airy kernel. For the second we use integration by parts and the fact that for the Airy kernel

$$(\partial_x + \partial_y)K(x,y) = -\operatorname{Ai}(x)\operatorname{Ai}(y)\,. \tag{3.25}$$

This follows using the differential equation $\operatorname{Ai}''(x) = x\operatorname{Ai}(x)$ satisfied by the Airy function.

In fact we shall not use (3.23), but instead

$$[D^2 - M, K\chi] = KD\delta + K\delta D\,. \tag{3.26}$$

To obtain this we apply the general identity $[D^2, K\chi] = D[D, K\chi] + [D, K\chi]D$. From (3.24) we obtain, first,

$$D[D, K\chi] = -\operatorname{Ai}'\otimes\operatorname{Ai}\chi + DK\delta\,,$$

and then using (3.25),

$$D[D, K\chi] = -\operatorname{Ai}'\otimes\operatorname{Ai}\chi + KD\delta - \operatorname{Ai}\otimes\operatorname{Ai}\delta\,.$$

From (3.24) again we obtain

$$[D, K\chi]D = \operatorname{Ai}\otimes(\operatorname{Ai}'\chi + \operatorname{Ai}\delta) + K\delta D\,.$$

Adding these two and using (3.23) we obtain (3.26).

We first derive two identities that are consequences of the commutators (3.24) and (3.26). We define

$$r = R(s,s)\,,\quad r_x = R_x(s,s)\,,\quad r_y = R_y(s,s)\,,$$

(where $R_x = \partial R/\partial x$ and $R_y = \partial R/\partial y$), and

$$Q = \rho\operatorname{Ai}\,,\quad q = Q(s)\,.$$

(Recall that $\rho = (I - K\chi)^{-1}$.) With this definition of q we shall derive the identity (3.17) and show that q satisfies the Painlevé equation. (The asymptotic behavior at $+\infty$ is almost immediate.)

The identities arising from the commutators are

$$r_x + r_y = -q^2 + r^2\,,\quad Q''(s) - sq = rQ'(s) - r_y q\,. \tag{3.27}$$

To establish these we need commutators, not with $K\chi$, but with $R = \rho K$. The commutator $[D, \rho]$ is obtained by multiplying (3.24) on the left and right by ρ. This gives

$$[D, \rho] = -(\rho\operatorname{Ai}\otimes\operatorname{Ai}\chi)\rho + R\delta\rho\,,$$

and from this we obtain

$$[D, R] = [D, \rho K] = \rho[D, K] + [D, \rho]K = -(\rho\operatorname{Ai}\otimes\operatorname{Ai})(I + \chi R) + R\delta R\,.$$

Notice that $I + \chi R = (I - \chi K)^{-1}$, and K is symmetric. Using these facts, and the general relation $(u \otimes v)T = u \otimes (T^t v)$, we see that the first term on the right equals $-\rho\,\mathrm{Ai} \otimes \rho\,\mathrm{Ai} = -Q \otimes Q$. Since R is symmetric the last term equals $R^s \otimes R^s$, in previous notation. We have shown that

$$[D, R] = -Q \otimes Q + R^s \otimes R^s \,.$$

The left side has kernel $R_x(x, y) + R_y(x, y)$, so setting $x = y = s$ in the kernels gives the first identity of (3.27).

The second commutator (3.26) gives similarly

$$[D^2 - M, \rho] = R(D\delta + \delta D)\rho \,.$$

Applying this to Ai and using $(D^2 - M)\,\mathrm{Ai} = 0$ give

$$Q''(x) - xQ(x) = -R_y(x, y)q + R(x, s)Q'(s) \,.$$

Setting $x = s$ gives the second identity of (3.27).

Since $\mathrm{d}\log F_2/\mathrm{d}s = r$ (this is (3.21)), to establish (3.17) we compute $\mathrm{d}r/\mathrm{d}s$. Now $R(x, y)$ depends also on s, since $K\chi$ does, and the chain rule gives

$$\frac{\mathrm{d}r}{\mathrm{d}s} = \frac{\mathrm{d}R}{\mathrm{d}s}(s, s) + R_x(s, s) + R_y(s, s) \,, \tag{3.28}$$

so we must evaluate the first factor. In fact

$$\frac{\mathrm{d}R}{\mathrm{d}s} = (I - K\chi)^{-1}\frac{\mathrm{d}K\chi}{\mathrm{d}s}(I - K\chi)^{-1}K \,.$$

The factor $\mathrm{d}(K\chi)/\mathrm{d}s$ is what we cumputed earlier to be $-E(s)$, which in the present notation is $-K\delta$. This gives

$$\frac{\mathrm{d}R}{\mathrm{d}s} = -R\delta R = -R^s \otimes R^s \,.$$

Setting $x = y = s$ in the kernels we obtain

$$\frac{\mathrm{d}R}{\mathrm{d}s}(s, s) = -r^2 \,.$$

Combining this with (3.28) and the first identity of (3.27) gives

$$\frac{\mathrm{d}r}{\mathrm{d}s} = \frac{\mathrm{d}R}{\mathrm{d}s}(s, s) + R_x(s, s) + R_y(s, s) \,, \quad r' = -q^2 \,. \tag{3.29}$$

(Here $' = \mathrm{d}/\mathrm{d}s$.) Thus $\mathrm{d}^2 \log F_2/\mathrm{d}s^2 = -q^2$, and (3.17) follows since both F_2 and q decrease rapidly at $+\infty$.

Next we derive the equation for q.

As with the computation of $r' = \mathrm{d}r/\mathrm{d}s$, the chain rule gives

$$q'(s) = Q'(s) - (R\delta\rho\,\mathrm{Ai})(s) = Q'(s) - rq \,, \tag{3.30}$$

and then (from $\mathrm{d}Q'/\mathrm{d}s = -R_x \delta\rho\, \mathrm{Ai}$)

$$q''(s) = Q''(s) - r_x q - (rq)' \,. \tag{3.31}$$

Combining (3.30) and (3.31) with the second identity of (3.27) gives

$$q'' = sq + (r^2 - r' - r_x - r_y)q \,.$$

By (3.29) and the first identity of (3.27) this equals $sq + 2q^3$. This completes the proof.

Systems of differential equations for distribution functions arising in random matrix theory are derived in [9]. The proof presented here for F_2 is a distillation of a different derivation from [11].

References

1. M.J. Bowick, E. Brézin: Phys. Lett. B **268**, 21 (1991)
2. N. Dunford, J. Schwartz: *Linear Operators. II. Spectral Theory* (Interscience, New York 1963)
3. P.J. Forrester: Nucl. Phys. B **402**, 709 (1993)
4. I.C. Gohberg, M.G. Krein: *Introduction to the Theory of Linear Nonselfadjoint Operators*, vol. 18 (Amer. Math. Soc., Providence 1967)
5. M.L. Mehta: *Random Matrices*, 2nd. ed. (Academic Press, San Diego 1991)
6. M.L. Mehta: *Random Matrices*, 3rd. ed. (Academic Press, San Diego 2004)
7. A. Pietsch: *Eigenvalues and s-Numbers* (Cambridge University Press, Cambridge 1987)
8. B. Simon: Adv. Math. **24**, 244 (1977)
9. C.A. Tracy, H. Widom: Commun. Math. Phys. **163**, 48 (1994)
10. C.A. Tracy, H. Widom: J. Stat. Phys. **92**, 809 (1998)
11. C.A. Tracy, H. Widom: Commun. Math. Phys. **252**, 7 (2004)

4

Lectures on Random Matrix Models

The Riemann–Hilbert Approach

Pavel M. Bleher

Department of Mathematical Sciences, Indiana University–Purdue University Indianapolis, 402 N. Blackford St., Indianapolis, IN 46202, USA
bleher@math.iupui.edu

Summary. This is a review of the Riemann–Hilbert approach to the large N asymptotics in random matrix models and its applications. We discuss the following topics: random matrix models and orthogonal polynomials, the Riemann–Hilbert approach to the large N asymptotics of orthogonal polynomials and its applications to the problem of universality in random matrix models, the double scaling limits, the large N asymptotics of the partition function, and random matrix models with external source.

Introduction

This article is a review of the Riemann–Hilbert approach to random matrix models. It is based on a series of 5 lectures given by the author at the miniprogram on "Random Matrices and their Applications" at the Centre de recherches mathématiques, Université de Montréal, in June 2005. The review contains 5 lectures:

Lecture 1. Random matrix models and orthogonal polynomials.
Lecture 2. Large N asymptotics of orthogonal polynomials. The Riemann–Hilbert approach.

J. Harnad (ed.), *Random Matrices, Random Processes and Integrable Systems*, CRM Series in Mathematical Physics, DOI 10.1007/978-1-4419-9514-8_4,

Lecture 3. Double scaling limit in a random matrix model.
Lecture 4. Large N asymptotics of the partition function of random matrix models.
Lecture 5. Random matrix models with external source.

The author would like to thank John Harnad for his invitation to give the series of lectures at the miniprogram. The lectures are based on the joint works of the author with several coauthors: Alexander Its, Arno Kuijlaars, Alexander Aptekarev, and Bertrand Eynard. The author is grateful to his coauthors for an enjoyable collaboration.

4.1 Random Matrix Models and Orthogonal Polynomials

The first lecture gives an introduction to random matrix models and their relations to orthogonal polynomials.

4.1.1 Unitary Ensembles of Random Matrices

4.1.1.1 Unitary Ensemble with Polynomial Interaction

Let $M = (M_{jk})_{j,k=1}^N$ be a random Hermitian matrix, $M_{kj} = \overline{M_{jk}}$, with respect to the probability distribution

$$\mathrm{d}\mu_N(M) = \frac{1}{Z_N}\,\mathrm{e}^{-N\,\mathrm{Tr}\,V(M)}\,\mathrm{d}M\,, \tag{4.1}$$

where

$$V(M) = \sum_{i=1}^{p} t_j M^j\,, \quad p = 2p_0\,,\ t_p > 0\,, \tag{4.2}$$

is a polynomial,

$$\mathrm{d}M = \prod_{j=1}^{N} \mathrm{d}M_{jj} \prod_{j\neq k}^{N} \mathrm{d}\,\mathrm{Re}\,M_{jk}\mathrm{d}\,\mathrm{Im}\,M_{jk}\,, \tag{4.3}$$

the Lebesgue measure, and

$$Z_N = \int_{\mathcal{H}_N} \mathrm{e}^{-N\,\mathrm{Tr}\,V(M)}\,\mathrm{d}M\,, \tag{4.4}$$

the partition function. The distribution $\mu_N(dM)$ is invariant with respect to any unitary conjugation,

$$M \to U^{-1}MU\,, \quad U \in \mathrm{U}(N)\,, \tag{4.5}$$

hence the name of the ensemble.

4.1.1.2 Gaussian Unitary Ensemble

For $V(M) = M^2$, the measure μ_N is the probability distribution of the Gaussian unitary ensemble (GUE). In this case,

$$\operatorname{Tr} V(M) = \operatorname{Tr} M^2 = \sum_{j,k=1}^{N} M_{kj} M_{jk} = \sum_{j=1}^{N} M_{jj}^2 + 2 \sum_{j>k} |M_{jk}|^2 , \tag{4.6}$$

hence

$$\mu_N^{\mathrm{GUE}}(\mathrm{d}M) = \frac{1}{Z_N^{\mathrm{GUE}}} \prod_{j=1}^{N} \exp(-N M_{jj}^2) \prod_{j>k} \exp(-2N|M_{jk}|^2)\, \mathrm{d}M , \tag{4.7}$$

so that the matrix elements in GUE are independent Gaussian random variables. The partition function of GUE is evaluated as

$$\begin{aligned} Z_N^{\mathrm{GUE}} &= \int_{\mathcal{H}_N} \prod_{j=1}^{N} \exp(-N M_{jj}^2) \prod_{j>k} \exp(-2N|M_{jk}|^2)\, \mathrm{d}M \\ &= \left(\frac{\pi}{N}\right)^{N/2} \left(\frac{\pi}{2N}\right)^{N(N-1)/2} = \left(\frac{\pi}{N}\right)^{N^2/2} \left(\frac{1}{2}\right)^{N(N-1)/2} . \end{aligned} \tag{4.8}$$

If $V(M)$ is not quadratic then the matrix elements M_{jk} are dependent.

4.1.1.3 Topological Large N Expansion

Consider the *free energy* of the unitary ensemble of random matrices,

$$F_N^0 = -N^{-2} \ln Z_N = -N^{-2} \ln \int_{\mathcal{H}_N} \mathrm{e}^{-N \operatorname{Tr} V(M)}\, \mathrm{d}M , \tag{4.9}$$

and the *normalized free energy*,

$$F_N = -N^{-2} \ln \frac{Z_N}{Z_N^{\mathrm{GUE}}} = -N^{-2} \ln \frac{\int_{\mathcal{H}_N} \exp(-N \operatorname{Tr} V(M))\, \mathrm{d}M}{\int_{\mathcal{H}_N} \exp(-N \operatorname{Tr} M^2)\, \mathrm{d}M} . \tag{4.10}$$

The normalized free energy can be expessed as

$$F_N = -N^{-2} \ln \langle \exp(-N \operatorname{Tr} V_1(M)) \rangle \tag{4.11}$$

where $V_1(M) = V(M) - M^2$ and

$$\langle f(M) \rangle = \frac{\int_{\mathcal{H}_N} f(M) \exp(-N \operatorname{Tr}(M^2))\, \mathrm{d}M}{\int_{\mathcal{H}_N} \exp(-N \operatorname{Tr}(M^2))\, \mathrm{d}M} , \tag{4.12}$$

the mathematical expectation of $f(M)$ with respect to GUE. Suppose that

$$V(M) = M^2 + t_3 M^3 + \cdots + t_p M^p \,. \tag{4.13}$$

Then (4.10) reduces to

$$F_N = -N^{-2} \ln\big\langle \exp\big(-N \operatorname{Tr}(t_3 M^3 + \cdots + t_p M^p)\big)\big\rangle \,. \tag{4.14}$$

F_N can be expanded into the asymptotic series in negative powers of N^2,

$$F_N \sim F + \sum_{g=1}^{\infty} \frac{F^{(2g)}}{N^{2g}} \,, \tag{4.15}$$

which is called the *topological large* N *expansion.* The Feynman diagrams representing $F^{(2g)}$ are realized on a two-dimensional Riemann closed manifold of genus g, and, therefore, F_N serves as a generating function for enumeration of graphs on Riemannian manifolds, see, e.g., the works [14, 33, 57, 65, 66, 77]. This in turn leads to a fascinating relation between the matrix integrals and the quantum gravity, see, e.g., the works [56, 101], and others.

4.1.1.4 Ensemble of Eigenvalues

The Weyl integral formula implies, see, e.g., [85], that the distribution of eigenvalues of M with respect to the ensemble μ_N is given as

$$\mathrm{d}\mu_N(\lambda) = \frac{1}{\widetilde{Z}_N} \prod_{j>k} (\lambda_j - \lambda_k)^2 \prod_{j=1}^{N} \exp\big(-NV(\lambda_j)\big)\, \mathrm{d}\lambda \,, \tag{4.16}$$

where

$$\widetilde{Z}_N = \int \prod_{j>k} (\lambda_j - \lambda_k)^2 \prod_{j=1}^{N} \exp\big(-NV(\lambda_j)\big)\, \mathrm{d}\lambda \,, \quad \mathrm{d}\lambda = \mathrm{d}\lambda_1 \cdots \mathrm{d}\lambda_N \,. \tag{4.17}$$

Respectively, for GUE,

$$\mathrm{d}\mu_N^{\mathrm{GUE}}(\lambda) = \frac{1}{Z_N^{\mathrm{GUE}}} \prod_{j>k} (\lambda_j - \lambda_k)^2 \prod_{j=1}^{N} \exp(-N\lambda_j^2)\, \mathrm{d}\lambda \,, \tag{4.18}$$

where

$$\widetilde{Z}_N^{\mathrm{GUE}} = \int \prod_{j>k} (\lambda_j - \lambda_k)^2 \prod_{j=1}^{N} \exp(-N\lambda_j^2)\, \mathrm{d}\lambda \,. \tag{4.19}$$

The constant $\widetilde{Z}_N^{\mathrm{GUE}}$ is a Selberg integral, and its exact value is

$$\widetilde{Z}_N^{\mathrm{GUE}} = \frac{(2\pi)^{N/2}}{(2N)^{N^2/2}} \prod_{n=1}^{N} n! \tag{4.20}$$

see, e.g., [85]. The partition functions Z_N and $\widetilde{Z}_N$ are related as follows:

$$\frac{\widetilde{Z}_N}{Z_N} = \frac{\widetilde{Z}_N^{\mathrm{GUE}}}{Z_N^{\mathrm{GUE}}} = \frac{1}{\pi^{N(N-1)/2}} \prod_{n=1}^{N} n! \tag{4.21}$$

One of the main problems is to evaluate the large N asymptotics of the partition function $\widetilde{Z}_N$ and of the correlations between eigenvalues.

The m-point correlation function is given as

$$R_{mN}(x_1, \dots, x_m) = \frac{N!}{(N-m)!} \int_{\mathbb{R}^{N-m}} p_N(x_1, \dots, x_N)\, \mathrm{d}x_{m+1} \cdots \mathrm{d}x_N\,, \tag{4.22}$$

where

$$p_N(x_1, \dots, x_N) = \widetilde{Z}_N^{-1} \prod_{j>k} (x_j - x_k)^2 \prod_{j=1}^{N} \exp\bigl(-NV(x_j)\bigr)\,. \tag{4.23}$$

The Dyson determinantal formula for correlation functions, see, e.g., [85], is

$$R_{mN}(x_1, \dots, x_m) = \det\bigl(K_N(x_k, x_l)\bigr)_{k,l=1}^{m}\,, \tag{4.24}$$

where

$$K_N(x, y) = \sum_{n=0}^{N-1} \psi_n(x)\psi_n(y) \tag{4.25}$$

and

$$\psi_n(x) = \frac{1}{h_n^{1/2}} P_n(x) \mathrm{e}^{-NV(x)/2}, \tag{4.26}$$

where $P_n(x) = x^n + a_{n-1}x^{n-1} + \cdots$ are monic orthogonal polynomials,

$$\int_{-\infty}^{\infty} P_n(x) P_m(x) \mathrm{e}^{-NV(x)}\, \mathrm{d}x = h_n \delta_{nm}\,. \tag{4.27}$$

Observe that the functions $\psi_n(x)$, $n = 0, 1, 2, \dots$, form an orthonormal basis in $L^2(\mathbb{R}^1)$, and K_N is the kernel of the projection operator onto the N-dimensional space generated by the first N functions ψ_n, $n = 0, \dots, N-1$. The kernel K_N can be expressed in terms of ψ_{N-1}, ψ_N only, due to the Christoffel–Darboux formula. Consider first recurrence and differential equations for the functions ψ_n.

4.1.1.5 Recurrence Equations and Discrete String Equations for Orthogonal Polynomials

The orthogonal polynomials satisfy the *three-term recurrence relation*, see, e.g., [95],

$$\begin{aligned} xP_n(x) &= P_{n+1}(x) + \beta_n P_n(x) + \gamma_n^2 P_{n-1}(x)\,, \\ \gamma_n &= \left(\frac{h_n}{h_{n-1}}\right)^{1/2} > 0\,,\ n \geq 1\,; \quad \gamma_0 = 0\,. \end{aligned} \tag{4.28}$$

For the functions ψ_n it reads as

$$x\psi_n(x) = \gamma_{n+1}\psi_{n+1}(x) + \beta_n\psi_n(x) + \gamma_n\psi_{n-1}(x)\,. \tag{4.29}$$

This allows the following calculation:

$$\begin{aligned} &(x-y)\sum_{n=0}^{N-1}\psi_n(x)\psi_n(y) \\ &\quad= \sum_{n=0}^{N-1}\Big[\big(\gamma_{n+1}\psi_{n+1}(x) + \beta_n\psi_n(x) + \gamma_n\psi_{n-1}(x)\big)\psi_n(y) \\ &\qquad\qquad - \psi_n(x)\big(\gamma_{n+1}\psi_{n+1}(y) + \beta_n\psi_n(y) + \gamma_n\psi_{n-1}(y)\big)\Big] \\ &\quad= \gamma_N[\psi_N(x)\psi_{N-1}(y) - \psi_{N-1}(x)\psi_N(y)] \end{aligned} \tag{4.30}$$

(telescopic sum), hence

$$K_N(x,y) = \sum_{n=0}^{N-1}\psi_n(x)\psi_n(y) = \gamma_N\frac{\psi_N(x)\psi_{N-1}(y) - \psi_{N-1}(x)\psi_N(y)}{x-y}\,. \tag{4.31}$$

which is the *Christoffel–Darboux formula.* For the density function we obtain that

$$p_N(x) = \frac{Q_N(x,x)}{N} = \frac{\gamma_N}{N}[\psi_N'(x)\psi_{N-1}(x) - \psi_{N-1}'(x)\psi_N(x)]\,. \tag{4.32}$$

Consider a matrix Q of the operator of multiplication by x, $f(x) \to xf(x)$, in the basis $\{\psi_n(x)\}$. Then by (4.29), Q is the symmetric tridiagonal Jacobi matrix,

$$Q = \begin{pmatrix} \beta_0 & \gamma_1 & 0 & 0 & \cdots \\ \gamma_1 & \beta_1 & \gamma_2 & 0 & \cdots \\ 0 & \gamma_2 & \beta_2 & \gamma_3 & \cdots \\ 0 & 0 & \gamma_3 & \beta_3 & \cdots \\ \vdots & \vdots & \vdots & \vdots & \ddots \end{pmatrix}\,. \tag{4.33}$$

Let $P = (P_{nm})_{n,m=0,1,2,\ldots}$ be a matrix of the operator $f(z) \to f'(z)$ in the basis $\psi_n(z)$, $n = 0, 1, 2, \ldots$, so that

$$P_{nm} = \int_{-\infty}^{\infty}\psi_n(x)\psi_m'(x)\,\mathrm{d}x\,. \tag{4.34}$$

Then $P_{mn} = -P_{nm}$ and

$$\begin{aligned}\psi_n'(z) &= -\frac{NV'(z)}{2}\psi_n(z) + \frac{P_n'(z)}{\sqrt{h_n}}\mathrm{e}^{-NV(z)/2} \\ &= -\frac{NV'(z)}{2}\psi_n(z) + \frac{n}{\gamma_n}\psi_{n-1}(z) + \cdots\,,\end{aligned} \tag{4.35}$$

hence

$$\begin{gathered}\left[P + \frac{NV'(Q)}{2}\right]_{nn} = 0\,, \quad \left[P + \frac{NV'(Q)}{2}\right]_{n,n+1} = 0\,, \\ \left[P + \frac{NV'(Q)}{2}\right]_{n,n-1} = \frac{n}{\gamma_n}\,.\end{gathered} \tag{4.36}$$

Since $P_{nn} = 0$, we obtain that

$$[V'(Q)]_{nn} = 0\,. \tag{4.37}$$

In addition,

$$\left[-P + \frac{NV'(Q)}{2}\right]_{n,n-1} = 0\,, \quad \left[P + \frac{NV'(Q)}{2}\right]_{n,n-1} = \frac{n}{\gamma_n}\,, \tag{4.38}$$

hence

$$\gamma_n[V'(Q)]_{n,n-1} = \frac{n}{N}\,. \tag{4.39}$$

Thus, we have the *discrete string equations* for the recurrence coefficients,

$$\begin{cases}\gamma_n[V'(Q)]_{n,n-1} = \dfrac{n}{N}\,, \\ [V'(Q)]_{nn} = 0\,.\end{cases} \tag{4.40}$$

The string equations can be brought to a variational form.

Proposition 4.1.1. *Define the infinite Hamiltonian,*

$$H(\gamma, \beta) = N\,\mathrm{Tr}\,V(Q) - \sum_{n=1}^{\infty} n \ln \gamma_n^2\,, \qquad \gamma = (\gamma_0, \gamma_1, \dots)\,,\ \beta = (\beta_0, \beta_1, \dots)\,. \tag{4.41}$$

Then equations (4.40) *can be written as*

$$\frac{\partial H}{\partial \gamma_n} = 0\,, \quad \frac{\partial H}{\partial \beta_n} = 0\,; \quad n \geq 1\,, \tag{4.42}$$

which are the Euler–Lagrange equations for the Hamiltonian H.

Proof. We have that

$$\frac{\partial H}{\partial \gamma_n} = N\,\mathrm{Tr}\left(V'(Q)\frac{\partial Q}{\partial \gamma_n}\right) - \frac{2n}{\gamma_n} = 2N[V'(Q)]_{n,n-1} - \frac{2n}{\gamma_n}\,, \tag{4.43}$$

and

$$\frac{\partial H}{\partial \beta_n} = N\,\mathrm{Tr}\left(V'(Q)\frac{\partial Q}{\partial \beta_n}\right) = N[V'(Q)]_{nn}\,, \tag{4.44}$$

hence (4.40) are equivalent to (4.42). □

Example. The even quartic model,

$$V(M) = \frac{t}{2}M^2 + \frac{g}{4}M^4 \,. \tag{4.45}$$

In this case, since V is even, $\beta_n = 0$, and we have one string equation,

$$\gamma_n^2(t + g\gamma_{n-1}^2 + g\gamma_n^2 + g\gamma_{n+1}^2) = \frac{n}{N} \,, \tag{4.46}$$

with the initial conditions: $\gamma_0 = 0$ and

$$\gamma_1 = \frac{\int_{-\infty}^{\infty} z^2 \mathrm{e}^{-NV(z)}\,\mathrm{d}z}{\int_{-\infty}^{\infty} \mathrm{e}^{-NV(z)}\,\mathrm{d}z} \,. \tag{4.47}$$

The Hamiltonian is

$$H(\gamma) = \sum_{n=1}^{\infty} \left[\frac{N}{2}\gamma_n^2(2t + g\gamma_{n-1}^2 + g\gamma_n^2 + g\gamma_{n+1}^2) - n\ln\gamma_n^2 \right] . \tag{4.48}$$

The minimization of the functional H is a useful procedure for a numerical solution of the string equations, see [18,36]. The problem with the initial value problem for the string equations, with the initial values $\gamma_0 = 0$ and (4.47), is that it is very unstable, while the minimization of H with $\gamma_0 = 0$ and some boundary conditions at $n = N$, say $\gamma_N = 0$, works very well. In fact, the boundary condition at $n = N$ creates a narrow boundary layer near $n = N$, and it does not affect significally the main part of the graph of γ_n^2. Figure 4.1

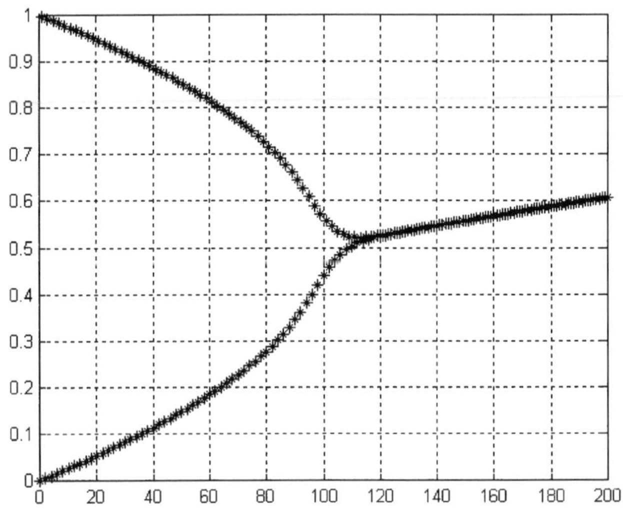

Fig. 4.1. A computer solution, $y = \gamma_n^2$, of the string equation for the quartic model: $g = 1$, $t = -1$, $N = 400$

presents a computer solution, $y = \gamma_n^2$, of the string equation for the quartic model: $g = 1$, $t = -1$, $N = 400$. For this solution, as shown in [17, 18], there is a critical value, $\lambda_c = \frac{1}{4}$, so that for any $\varepsilon > 0$, as $N \to \infty$,

$$\gamma_n^2 = R\left(\frac{n}{N}\right) + O(N^{-1}) , \quad \text{if } \frac{n}{N} \geq \lambda_c + \varepsilon , \tag{4.49}$$

and

$$\gamma_n^2 = \begin{cases} R\left(\frac{n}{N}\right) + O(N^{-1}) , & n = 2k+1 , \\ L\left(\frac{n}{N}\right) + O(N^{-1}) , & n = 2k , \end{cases} \quad \text{if } \frac{n}{N} \leq \lambda_c - \varepsilon . \tag{4.50}$$

The functions R for $\lambda \geq \lambda_c$ and R, L for $\lambda \leq \lambda_c$ can be found from string equation (4.46):

$$R(\lambda) = \frac{1 + \sqrt{1 + 12\lambda}}{6} , \quad \lambda > \lambda_c , \tag{4.51}$$

and

$$R(\lambda), L(\lambda) = \frac{1 \pm \sqrt{1 - 4\lambda}}{2} , \quad \lambda < \lambda_c . \tag{4.52}$$

We will discuss below how to justify asymptotics (4.49), (4.50), and their extension for a general V.

4.1.1.6 Differential Equations for the ψ-Functions

Define

$$\boldsymbol{\Psi}_n(z) = \begin{pmatrix} \psi_n(z) \\ \psi_{n-1}(z) \end{pmatrix} . \tag{4.53}$$

Then

$$\boldsymbol{\Psi}_n'(z) = N A_n(z) \boldsymbol{\Psi}_n(z) , \tag{4.54}$$

where

$$A_n(z) = \begin{pmatrix} -V'(z)/2 - \gamma_n u_n(z) & \gamma_n v_n(z) \\ -\gamma_n v_{n-1}(z) & V'(z)/2 + \gamma_n u_n(z) \end{pmatrix} \tag{4.55}$$

and

$$u_n(z) = [W(Q,z)]_{n,n-1} , \quad v_n(z) = [W(Q,z)]_{nn} , \tag{4.56}$$

where

$$W(Q,z) = \frac{V'(Q) - V'(z)}{Q - z} . \tag{4.57}$$

Observe that $\operatorname{Tr} A_n(z) = 0$.

Example. Even quartic model, $V(M) = (t/2)M^2 + (g/4)M^4$. Matrix $A_n(z)$:

$$A_n(z) = \begin{pmatrix} -\frac{1}{2}(tz + gz^3) - g\gamma_n^2 z & \gamma_n(gz^2 + \theta_n) \\ -\gamma_n(gz^2 + \theta_{n-1}) & \frac{1}{2}(tz + gz^3) + g\gamma_n^2 z \end{pmatrix} \tag{4.58}$$

where

$$\theta_n = t + g\gamma_n^2 + g\gamma_{n+1}^2 . \tag{4.59}$$

4.1.1.7 Lax Pair for the Discrete String Equations

Three-term recurrence relation (4.28) can be written as

$$\boldsymbol{\Psi}_{n+1}(z) = U_n(z)\boldsymbol{\Psi}_n(z)\,, \tag{4.60}$$

where

$$U_n(z) = \begin{pmatrix} \gamma_{n+1}^{-1}(z-\beta_n) & -\gamma_{n+1}^{-1}\gamma_n \\ 1 & 0 \end{pmatrix}. \tag{4.61}$$

Differential equation (4.54) and recurrence equation (4.60) form the Lax pair for discrete string equations (4.40). This means that the compatibility condition of (4.54) and (4.60),

$$U_n' = N(A_{n+1}U_n - U_nA_n)\,, \tag{4.62}$$

when written for the matrix elements, implies (4.40).

4.1.2 The Riemann–Hilbert Problem for Orthogonal Polynomials

4.1.2.1 Adjoint Functions

Introduce the *adjoint functions* to $P_n(z)$ as

$$Q_n(z) = \frac{1}{2\pi \mathrm{i}} \int_{-\infty}^{\infty} \frac{P_n(u)\, w(u)\, \mathrm{d}u}{u-z}\,, \qquad z \in \mathbb{C} \setminus \mathbb{R}\,, \tag{4.63}$$

where

$$w(z) = \mathrm{e}^{-NV(z)} \tag{4.64}$$

is the weight for the orthogonal polynomials P_n. Define

$$Q_{n\pm}(x) = \lim_{\substack{z\to x\\ \pm \operatorname{Im} z>0}} Q_n(z)\,, \qquad -\infty < x < \infty\,. \tag{4.65}$$

Then the well-known formula for the jump of the Cauchy type integral gives that

$$Q_{n+}(x) - Q_{n-}(x) = w(x)P_n(x)\,. \tag{4.66}$$

The asymptotics of $Q_n(z)$ as $z \to \infty$, $z \in \mathbb{C}$, is given as

$$\begin{aligned} Q_n(z) &= \frac{1}{2\pi \mathrm{i}} \int_{-\infty}^{\infty} \frac{w(u)P_n(u)\, \mathrm{d}u}{u-z} \\ &\cong -\frac{1}{2\pi \mathrm{i}} \int_{-\infty}^{\infty} w(u)P_n(u) \sum_{j=0}^{\infty} \frac{u^j}{z^{j+1}}\, \mathrm{d}u = -\frac{h_n}{2\pi \mathrm{i} z^{n+1}} + \sum_{j=n+2}^{\infty} \frac{\alpha_j}{z^j}\,, \end{aligned} \tag{4.67}$$

(due to the orthogonality, the first n terms cancel out). The sign $\cong$ in (4.67) means an asymptotic expansion, so that for any $k \geq n+2$, there exists a constant $C_k > 0$ such that for all $z \in \mathbb{C}$,

$$Q_n(z) - \left(-\frac{h_n}{2\pi i z^{n+1}} + \sum_{j=n+2}^{k} \frac{\alpha_j}{z^j}\right) \le \frac{C_k}{(1+|z|)^{k+1}} \,. \tag{4.68}$$

There can be some doubts about the uniformity of this asymptotics near the real axis, but we assume that the weight $w(z)$ is analytic in a strip $\{z : |\operatorname{Im} z| \le a\}$, $a > 0$, hence the contour of integration in (4.67) can be shifted, and (4.68) holds uniformly in the complex plane.

4.1.2.2 The Riemann–Hilbert Problem

Introduce now the matrix-valued function,

$$Y_n(z) = \begin{pmatrix} P_n(z) & Q_n(z) \\ CP_{n-1}(z) & CQ_{n-1}(z) \end{pmatrix} , \tag{4.69}$$

where the constant,

$$C = -\frac{2\pi i}{h_{n-1}} , \tag{4.70}$$

is chosen in such a way that

$$CQ_{n-1}(z) \cong \frac{1}{z^n} + \cdots , \tag{4.71}$$

see (4.67). The function Y_n solves the following Riemann–Hilbert problem (RHP):

(1) $Y_n(z)$ is analytic on $\mathbb{C}^+ \equiv \{\operatorname{Im} z \ge 0\}$ and $\mathbb{C}^- \equiv \{\operatorname{Im} z \le 0\}$ (two-valued on $\mathbb{R} = \mathbb{C}^+ \cap \mathbb{C}^-$).

(2) For any real x,

$$Y_{n+}(x) = Y_{n-}(x) j_Y(x), \quad j_Y(x) = \begin{pmatrix} 1 & w(x) \\ 0 & 1 \end{pmatrix} . \tag{4.72}$$

(3) As $z \to \infty$,

$$Y_n(z) \cong \left(I + \sum_{k=1}^{\infty} \frac{Y_k}{z^k}\right) \begin{pmatrix} z^n & 0 \\ 0 & z^{-n} \end{pmatrix} \tag{4.73}$$

where Y_k, $k = 1, 2, \ldots$, are some constant 2×2 matrices.

Observe that (4.72) follows from (4.66), while (4.73) from (4.71). The RHP (1)–(3) has some nice properties.

First of all, (4.69) is the only solution of the RHP. Let us sketch a proof of the uniqueness. It follows from (4.72), that

$$\det Y_{n+}(x) = \det Y_{n-}(x) , \tag{4.74}$$

hence $\det Y_n(z)$ has no jump at the real axis, and hence $Y_n(z)$ is an entire function. At infinity, by (4.73),

$$\det Y_n(z) \cong 1 + \cdots \tag{4.75}$$

hence

$$\det Y_n(z) \equiv 1 , \tag{4.76}$$

by the Liouville theorem. In particular, $Y_n(z)$ is invertible for any z. Suppose that $\widetilde{Y}_n$ is another solution of the RHP. Then $X_n = \widetilde{Y}_n Y_n^{-1}$ satisfies

$$\begin{aligned} X_{n+}(x) &= \widetilde{Y}_{n+}(x) Y_{n+}(x)^{-1} \\ &= \widetilde{Y}_{n-}(x) j_Y(x) j_Y(x)^{-1} Y_{n-}^{-1}(x) = X_{n-}(x) , \end{aligned} \tag{4.77}$$

hence X_n is an entire matrix-valued function. At infinity, by (4.73),

$$X_n(z) \cong I + \cdots \tag{4.78}$$

hence $X_n(z) \equiv I$, by the Liouville theorem. This implies that $\widetilde{Y}_n = Y_n$, the uniqueness.

The recurrence coefficients for the orthogonal polynomials can be found as

$$\gamma_n^2 = [Y_1]_{21}[Y_1]_{12} , \tag{4.79}$$

and

$$\beta_{n-1} = \frac{[Y_2]_{21}}{[Y_1]_{21}} - [Y_1]_{11} . \tag{4.80}$$

Indeed, from (4.69), (4.73),

$$Y_n(z)\begin{pmatrix} z^{-n} & 0 \\ 0 & z^n \end{pmatrix} = \begin{pmatrix} z^{-n}P_n(z) & z^n Q_n(z) \\ Cz^{-n}P_{n-1}(z) & z^n C Q_{n-1}(z) \end{pmatrix} \cong I + \sum_{k=1}^{\infty} \frac{Y_k}{z^k} , \tag{4.81}$$

hence by (4.67), (4.70), and (4.28),

$$[Y_1]_{21}[Y_1]_{12} = \left(-\frac{2\pi \mathrm{i}}{h_{n-1}}\right)\left(-\frac{h_n}{2\pi \mathrm{i}}\right) = \frac{h_n}{h_{n-1}} = \gamma_n^2 , \tag{4.82}$$

which proves (4.79). Also,

$$\frac{[Y_2]_{21}}{[Y_1]_{21}} - [Y_1]_{11} = p_{n-1,n-2} - p_{n,n-1} , \tag{4.83}$$

where

$$P_n(z) = \sum_{j=0}^{n} p_{nj} z^j . \tag{4.84}$$

From (4.28) we obtain that

$$p_{n-1,n-2} - p_{n,n-1} = \beta_{n-1} , \tag{4.85}$$

hence (4.80) follows. The normalizing constant h_n can be found as

$$h_n = -2\pi \mathrm{i}[Y_1]_{12}\,, \quad h_{n-1} = -\frac{2\pi \mathrm{i}}{[Y_1]_{21}}\,. \tag{4.86}$$

The reproducing kernel $K_N(x,y)$ of the eigenvalue correlation functions, see (4.24), is expressed in terms of $Y_{N+}(x)$ as follows:

$$K_N(x,y) = \exp\left(-\frac{NV(x)}{2}\right)\exp\left(-\frac{NV(y)}{2}\right) \times \frac{1}{2\pi \mathrm{i}(x-y)}\,(0\ 1)\,Y_{N+}^{-1}(y)Y_{N+}(x)\begin{pmatrix}1\\0\end{pmatrix}\,. \tag{4.87}$$

Indeed, by (4.31),

$$\begin{aligned} K_N(x,y) &= \gamma_N \frac{\psi_N(x)\psi_{N-1}(y) - \psi_{N-1}(x)\psi_N(y)}{x-y} \\ &= \exp\left(-\frac{NV(x)}{2}\right)\exp\left(-\frac{NV(y)}{2}\right) \\ &\quad \times \frac{\gamma_N}{\sqrt{h_N h_{N-1}}}\,\frac{P_N(x)P_{N-1}(y) - P_{N-1}(x)P_N(y)}{x-y}\,. \end{aligned} \tag{4.88}$$

From (4.69), (4.70), and (4.76), we obtain that

$$(0\ 1)\,Y_{N+}^{-1}(y)Y_{N+}(x)\begin{pmatrix}1\\0\end{pmatrix} = \frac{2\pi \mathrm{i}}{h_{N-1}}[P_N(x)P_{N-1}(y) - P_{N-1}(x)P_N(y)]\,. \tag{4.89}$$

Also,

$$\frac{\gamma_N}{\sqrt{h_N h_{N-1}}} = \frac{1}{h_{N-1}}\,, \tag{4.90}$$

hence (4.87) follows.

4.1.3 Distribution of Eigenvalues and Equilibrium Measure

4.1.3.1 Heuristics

We begin with some heuristic considerations to explain why we expect that the limiting distribution of eigenvalues solves a variational problem. Let us rewrite (4.17) as

$$\mathrm{d}\mu_N(\lambda) = \widetilde{Z}_N^{-1} \mathrm{e}^{-H_N(\lambda)}\,\mathrm{d}\lambda\,, \tag{4.91}$$

where

$$H_N(\lambda) = -\sum_{j\neq k} \ln|\lambda_j - \lambda_k| + N\sum_{j=1}^{N} V(\lambda_j)\,. \tag{4.92}$$

Given λ, introduce the probability measure on $\mathbb{R}^1$,

$$d\nu_\lambda(x) = N^{-1} \sum_{j=1}^{N} \delta(x - \lambda_j)\, dx \,. \tag{4.93}$$

Then (4.92) can be rewritten as

$$H_N(\lambda) = N^2 \left[- \iint_{x \neq y} \ln|x-y|\, d\nu_\lambda(x)\, d\nu_\lambda(y) + \int V(x)\, d\nu_\lambda(x) \right] . \tag{4.94}$$

Let ν be an arbitrary probability measure on $\mathbb{R}^1$. Set

$$I_V(\nu) = - \iint_{x \neq y} \ln|x-y|\, d\nu(x)\, d\nu(y) + \int V(x)\, d\nu(x) \,. \tag{4.95}$$

Then (4.91) reads

$$d\mu_N(\lambda) = \widetilde{Z}_N^{-1} \exp\bigl(-N^2 I_V(\nu_\lambda)\bigr)\, d\lambda \,. \tag{4.96}$$

Because of the factor N^2 in the exponent, we expect that for large N the measure μ_N is concentrated near the minimum of the functional I_V, i.e., near the *equilibrium measure* ν_V.

4.1.3.2 Equilibrium Measure

Consider the minimization problem

$$E_V = \inf_{\nu \in M_1(\mathbb{R})} I_V(\nu) \,, \tag{4.97}$$

where

$$M_1(\mathbb{R}) = \left\{ \nu : \nu \geq 0,\ \int_{\mathbb{R}} d\nu = 1 \right\} , \tag{4.98}$$

the set of probability measures on the line.

Proposition 4.1.2 (See [51]). *The infinum of $I_V(\nu)$ is attained uniquely at a measure $\nu = \nu_V$, which is called an equilibrium measure. The measure ν_V is absolutely continuous, and it is supported by a finite union of intervals, $J = \bigcup_{j=1}^q [a_j, b_j]$. On the support, its density has the form*

$$p_V(x) = \frac{d\nu_V}{dx}(x) = \frac{1}{2\pi i} h(x) R_+^{1/2}(x) \,, \quad R(x) = \prod_{j=1}^{q} (x - a_j)(x - b_j) \,. \tag{4.99}$$

Here $R^{1/2}(x)$ is the branch with cuts on J, which is positive for large positive x, and $R_+^{1/2}(x)$ is the value of $R^{1/2}(x)$ on the upper part of the cut. The function $h(x)$ is a polynomial, which is the polynomial part of the function $V'(x)/R^{1/2}(x)$ at infinity, i.e.,

$$\frac{V'(x)}{R^{1/2}(x)} = h(x) + O(x^{-1}) \,. \tag{4.100}$$

In particular, $\deg h = \deg V - 1 - q$.

There is a useful formula for the equilibrium density [51]:

$$\frac{\mathrm{d}\nu_V(x)}{\mathrm{d}x} = \frac{1}{\pi}\sqrt{q(x)}\,, \tag{4.101}$$

where

$$q(x) = -\left(\frac{V'(x)}{2}\right)^2 + \int \frac{V'(x)-V'(y)}{x-y}\,\mathrm{d}\nu_V(y)\,. \tag{4.102}$$

This, in fact, is an equation on q, since the right-hand side contains an integration with respect to ν_V. Nevertheless, if V is a polynomial of degree $p = 2p_0$, then (4.102) determines uniquely more then a half of the coefficients of the polynomial q,

$$q(x) = -\left(\frac{V'(x)}{2}\right)^2 - O(x^{p-2}). \tag{4.103}$$

Example. If $V(x)$ is *convex* then ν_V is regular (see Sect. 4.1.3.3), and the support of ν_V consists of a single interval, see, e.g., [82]. For the Gaussian ensemble, $V(x) = x^2$, hence, by (4.103), $q(x) = a^2 - x^2$. Since

$$\int_{-a}^{a} \frac{1}{\pi}\sqrt{a^2-x^2}\,\mathrm{d}x = 1\,,$$

we find that $a = \sqrt{2}$, hence

$$p_V(x) = \frac{1}{\pi}\sqrt{2-x^2}\,, \quad |x| \le \sqrt{2}\,, \tag{4.104}$$

the Wigner semicircle law.

4.1.3.3 The Euler–Lagrange Variational Conditions

A nice and important property of minimization problem (4.97) is that the minimizer is uniquely determined by the Euler-Lagrange variational conditions: for some real constant l,

$$2\int_{\mathbb{R}} \log|x-y|\,\mathrm{d}\nu(y) - V(x) = l\,, \quad \text{for } x \in J\,, \tag{4.105}$$

$$2\int_{\mathbb{R}} \log|x-y|\,\mathrm{d}\nu(y) - V(x) \le l\,, \quad \text{for } x \in \mathbb{R}\setminus J\,, \tag{4.106}$$

see [51].

Definition (See [52]). The equilibrium measure,

$$\mathrm{d}\nu_V(x) = \frac{1}{2\pi \mathrm{i}} h(x) R_+^{1/2}(x)\,\mathrm{d}x \tag{4.107}$$

is called *regular* (otherwise *singular*) if

(1) $h(x) \ne 0$ on the (closed) set J.
(2) Inequality (4.106) is strict,

$$2\int \log|x-y|\,\mathrm{d}\nu_V(y) - V(x) < l\,, \quad \text{for } x \in \mathbb{R}\setminus J\,. \tag{4.108}$$

4.1.3.4 Construction of the Equilibrium Measure: Equations on the End-Points

The strategy to construct the equilibrium measure is the following: first we find the end-points of the support, and then we use (4.100) to find $h(x)$ and hence the density. The number q of cuts is not, in general, known, and we try different q's. Consider the resolvent,

$$\omega(z) = \int_J \frac{\mathrm{d}\nu_V(x)}{z-x}\,, \qquad z \in \mathbb{C} \setminus J\,. \tag{4.109}$$

The Euler–Lagrange variational condition implies that

$$\omega(z) = \frac{V'(z)}{2} - \frac{h(z)R^{1/2}(z)}{2}\,. \tag{4.110}$$

Observe that as $z \to \infty$,

$$\omega(z) = \frac{1}{z} + \frac{m_1}{z^2} + \cdots\,, \qquad m_k = \int_J x^k \rho(x)\,\mathrm{d}x\,. \tag{4.111}$$

The equation

$$\frac{V'(z)}{2} - \frac{h(z)R^{1/2}(z)}{2} = \frac{1}{z} + O(z^{-2})\,. \tag{4.112}$$

gives $q+1$ equation on $a_1, b_1, \ldots, a_q, b_q$, if we substitute formula (4.100) for h. Remaining $q-1$ equation are

$$\int_{b_j}^{a_{j+1}} h(x)R^{1/2}(x)\,\mathrm{d}x = 0\,, \quad j = 1, \ldots, q-1\,, \tag{4.113}$$

which follow from (4.110) and (4.105).

Example. Even quartic model, $V(M) = (t/2)M^2 + \frac{1}{4}M^4$. For $t \geq t_c = -2$, the support of the equilibrium distribution consists of one interval $[-a, a]$ where

$$a = \left(\frac{-2t + 2(t^2+12)^{1/2}}{3}\right)^{1/2} \tag{4.114}$$

and

$$p_V(x) = \frac{1}{\pi}\left(c + \tfrac{1}{2}x^2\right)\sqrt{a^2 - x^2} \tag{4.115}$$

where

$$c = \frac{t + \left((t^2/4) + 3\right)^{1/2}}{3}\,. \tag{4.116}$$

In particular, for $t = -2$,

$$p_V(x) = \frac{1}{2\pi} x^2 \sqrt{4 - x^2}\,. \tag{4.117}$$

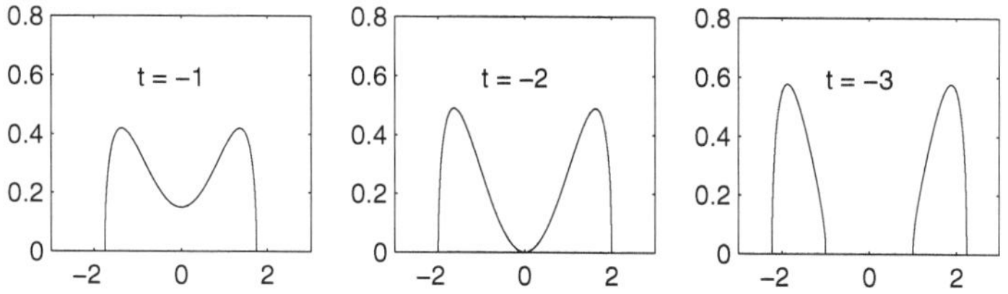

Fig. 4.2. The density function, $p_V(x)$, for the even quartic potential, $V(M) = (t/2)M^2 + \frac{1}{4}M^4$, for $t = -1, -2, -3$

For $t < -2$, the support consists of two intervals, $[-a, -b]$ and $[b, a]$, where

$$a = \sqrt{2-t}, \quad b = \sqrt{-2-t}\,, \tag{4.118}$$

and

$$p_V(x) = \frac{1}{2\pi}|x|\sqrt{(a^2-x^2)(x^2-b^2)}\,. \tag{4.119}$$

Figure 4.2 shows the density function for the even quartic potential, for $t = -1, -2, -3$.

4.2 Large N Asymptotics of Orthogonal Polynomials. The Riemann–Hilbert Approach

In this lecture we present the Riemann–Hilbert approach to the large N asymptotics of orthogonal polynomials. The central point of this approach is a construction of an asymptotic solution to the RHP, as $N \to \infty$. We call such a solution, a parametrix. In the original paper of Bleher and Its [17] the RH approach was developed for an even quartic polynomial $V(M)$ via a semiclassical solution of the differential equation for orthogonal polynomials. Then, in a series of papers, Deift, Kriecherbauer, McLaughlin, Venakides, and Zhou [51–53] developed the RH approach for a general real analytic V, with some conditions on the growth at infinity. The DKMVZ-approach is based on the Deift–Zhou steepest descent method, see [54]. In this lecture we present the main steps of the DKMVZ-approach. For the sake of simplicity, we will assume that V is regular. In this approach a sequence of transformations of the RHP is constructed, which reduces the RHP to a simple RHP which can be solved by a series of perturbation theory. This sequence of transformations gives the parametrix of the RHP in different regions on complex plane. The motivation for the first transformation comes from the Heine formula for orthogonal polynomials.

4.2.1 Heine's Formula for Orthogonal Polynomials

The Heine formula, see, e.g., [95], gives the Nth orthogonal polynomial as the matrix integral,

$$P_N(z) = \langle \det(z-M) \rangle \equiv Z_N^{-1} \int_{\mathcal{H}_N} \det(z-M) \exp\bigl(-N \operatorname{Tr} V(M)\bigr)\, \mathrm{d}M \,. \tag{4.120}$$

In the ensemble of eigenvalues,

$$\begin{aligned} P_N(z) &= \left\langle \prod_{j=1}^{N} (z-\lambda_j) \right\rangle \\ &\equiv \widetilde{Z}_N^{-1} \int \prod_{j=1}^{N} (z-\lambda_j) \prod_{j>k} (\lambda_j - \lambda_k)^2 \prod_{j=1}^{N} \exp\bigl(-NV(\lambda_j)\bigr)\, \mathrm{d}\lambda \,. \end{aligned} \tag{4.121}$$

Since ν_λ is close to the equilibrium measure ν for typical λ, we expect that

$$N^{-1} \log \left\langle \prod_{j=1}^{N} (z-\lambda_j) \right\rangle \approx \int_J \log(z-x)\, \mathrm{d}\nu_V(x) \,,$$

hence by the Heine formula,

$$N^{-1} \log P_N(z) \approx \int_J \log(z-x)\, \mathrm{d}\nu_V(x) \,. \tag{4.122}$$

This gives a *heuristic* semiclassical approximation for the orthogonal polynomial,

$$P_N(z) \approx \exp\left[N \int_J \log(z-x)\, \mathrm{d}\nu_V(x) \right] , \tag{4.123}$$

and it motivates the introduction of the "g-function."

4.2.1.1 g-Function

Define the g-function as

$$g(z) = \int_J \log(z-x)\, \mathrm{d}\nu_V(x) \,, \quad z \in \mathbb{C} \setminus (-\infty, b_q] \,, \tag{4.124}$$

where we take the principal branch for logarithm.

Properties of $g(z)$

(1) $g(z)$ is analytic in $\mathbb{C} \setminus (-\infty, b_q]$.
(2) As $z \to \infty$

$$g(z) = \log z - \sum_{j=1}^{\infty} \frac{g_j}{z^j} \,, \quad g_j = \int_J \frac{x^j}{j}\, \mathrm{d}\nu_V(x) \,. \tag{4.125}$$

(3) By (4.109), (4.110),

$$g'(z) = \omega(z) = \frac{V'(z)}{2} - \frac{h(z) R^{1/2}(z)}{2} \,. \tag{4.126}$$

(4) By (4.105),

$$g_+(x) + g_-(x) = V(x) + l\,, \quad x \in J\,. \tag{4.127}$$

(5) By (4.108),

$$g_+(x) + g_-(x) < V(x) + l\,, \quad x \in \mathbb{R} \setminus J\,. \tag{4.128}$$

(6) Equation (4.124) implies that the function

$$G(x) \equiv g_+(x) - g_-(x) \tag{4.129}$$

is pure imaginary for all real x, and $G(x)$ is constant in each component of $\mathbb{R} \setminus J$,

$$G(x) = \mathrm{i}\Omega_j \quad \text{for } b_j < x < a_{j+1},\, 1 \le j \le q-1\,, \tag{4.130}$$

where

$$\Omega_j = 2\pi \sum_{k=j+1}^{q} \int_{a_k}^{b_k} p_V(x)\,\mathrm{d}x\,, \quad 1 \le j \le q-1\,. \tag{4.131}$$

(7) Also,

$$G(x) = i\Omega_j - 2\pi\mathrm{i} \int_{b_j}^{x} p_V(s)\,\mathrm{d}s \quad \text{for } a_j < x < b_j,\, 1 \le j \le q\,, \tag{4.132}$$

where we set $\Omega_q = 0$.

Observe that from (4.132) and (4.99) we obtain that $G(x)$ is analytic on (a_j, b_j), and

$$\left.\frac{\mathrm{d}G(x+\mathrm{i}y)}{\mathrm{d}y}\right|_{y=0} = 2\pi p_V(x) > 0\,, \quad x \in (a_j, b_j)\,,\; 1 \le j \le q\,. \tag{4.133}$$

From (4.127) we have also that

$$G(x) = 2g_+(x) - V(x) - l = -[2g_-(x) - V(x) - l]\,, \quad x \in J\,. \tag{4.134}$$

4.2.2 First Transformation of the RH Problem

Our goal is to construct an asymptotic solution to RHP (4.72), (4.73) for $Y_N(z)$, as $N \to \infty$. In our construction we will assume that the equilibrium measure ν_V is regular. By (4.123) we expect that

$$P_N(z) \approx \mathrm{e}^{Ng(z)}\,, \tag{4.135}$$

therefore, we make the following substitution in the RHP:

$$Y_N(z) = \exp\left(\frac{Nl}{2}\sigma_3\right) T_N(z) \exp\left(N\left[g(z) - \frac{l}{2}\right]\sigma_3\right), \quad \sigma + 3 = \begin{pmatrix} 1 & 0 \\ 0 & -1 \end{pmatrix}. \tag{4.136}$$

Then $T_N(z)$ solves the following RH problem:

(1) $T_N(z)$ is analytic in $\mathbb{C}\setminus\mathbb{R}$.
(2) $T_{N+}(x)=T_{N-}(x)j_T(x)$ for $x\in\mathbb{R}$, where

$$j_T(x)=\begin{pmatrix}\exp(-N[g_+(x)-g_-(x)]) & \exp(N[g_+(x)+g_-(x)-V(x)-l]) \\ 0 & \exp(N[g_+(x)-g_-(x)])\end{pmatrix}. \tag{4.137}$$

(3) $T_N(z)=I+O(z^{-1})$, as $z\to\infty$.

The above properties of $g(z)$ ensure the following properties of the jump matrix j_T:

(1) $j_T(x)$ is exponentially close to the identity matrix on $(-\infty,a_1)\cup(b_q,\infty)$. Namely,

$$j_T(x)=\begin{pmatrix}1 & O(\mathrm{e}^{-Nc(x)}) \\ 0 & 1\end{pmatrix}, \quad x\in(-\infty,a_1)\cup(b_q,\infty), \tag{4.138}$$

where $c(x)>0$ is a continuous function such that

$$\lim_{x\to\pm\infty}\frac{c(x)}{\ln|x|}=\infty, \quad \lim_{x\to a_1}c(x)=\lim_{x\to b_q}c(x)=0. \tag{4.139}$$

(2) For $1\le j\le q-1$,

$$j_T(x)=\begin{pmatrix}\exp(-\mathrm{i}N\Omega_j) & O(\mathrm{e}^{-Nc(x)}) \\ 0 & \exp(\mathrm{i}N\Omega_j)\end{pmatrix}, \quad x\in(b_j,a_{j+1}), \tag{4.140}$$

where $c(x)>0$ is a continuous function such that

$$\lim_{x\to b_j}c(x)=\lim_{x\to a_{j+1}}c(x)=0. \tag{4.141}$$

(3) On J,

$$j_T(x)=\begin{pmatrix}\mathrm{e}^{-NG(x)} & 1 \\ 0 & \mathrm{e}^{NG(x)}\end{pmatrix}. \tag{4.142}$$

The latter matrix can be factorized as follows:

$$\begin{pmatrix}\mathrm{e}^{-NG(x)} & 1 \\ 0 & \mathrm{e}^{NG(x)}\end{pmatrix}=\begin{pmatrix}1 & 0 \\ \mathrm{e}^{NG(x)} & 1\end{pmatrix}\begin{pmatrix}0 & 1 \\ -1 & 0\end{pmatrix}\begin{pmatrix}1 & 0 \\ \mathrm{e}^{-NG(x)} & 1\end{pmatrix} \equiv j_-(x)j_Mj_+(x), \tag{4.143}$$

This leads to the second transformation of the RHP.

4.2.3 Second Transformation of the RHP: Opening of Lenses

The function $e^{-NG(x)}$ is analytic on each open interval (a_j, b_j). Observe that $|e^{-NG(x)}| = 1$ for real $x \in (a_j, b_j)$, and $e^{-NG(z)}$ is exponentially decaying for $\operatorname{Im} z > 0$. More precisely, by (4.133), there exists $y_0 > 0$ such that $e^{-NG(z)}$ satisfies the estimate,

$$|e^{-NG(z)}| \le e^{-Nc(z)}\,, \qquad z \in R_j^+ = \{z = x + iy : a_j < x < b_j, 0 < y < y_0\}\,, \tag{4.144}$$

where $c(z) > 0$ is a continuous function in R_j^+. Observe that $c(z) \to 0$ as $\operatorname{Im} z \to 0$. In addition, $|e^{NG(z)}| = |e^{-NG(\bar{z})}|$, hence

$$|e^{NG(z)}| \le e^{-Nc(z)}, \qquad z \in R_j^- = \{z = x + iy : a_j < x < b_j, 0 < -y < y_0\}\,, \tag{4.145}$$

where $c(z) = c(\bar{z}) > 0$. Consider a C^∞ curve γ_j^+ from a_j to b_j such that

$$\gamma_j^+ = \{x + iy : y = f_j(x)\}\,, \tag{4.146}$$

where $f_j(x)$ is a C^∞ function on $[a_j, b_j]$ such that

$$\begin{aligned} f_j(a_j) = f_j(b_j) = 0\,; \quad f_j'(a_j) = -f'(b_j) = \sqrt{3}\,; \\ 0 < f_j(x) < y_0\,,\ a_j < x < b_j\,. \end{aligned} \tag{4.147}$$

Consider the conjugate curve,

$$\gamma_j^- = \overline{\gamma_j^+} = \{x - iy : y = f_j(x)\}\,, \tag{4.148}$$

see Fig. 4.3. The region bounded by the interval $[a_j, b_j]$ and γ_j^+ (γ_j^-) is called the *upper* (*lower*) *lens*, $\mathcal{L}_j^\pm$, respectively. Define for $j = 1, \ldots, q$,

$$S_N(z) = \begin{cases} T_N(z) j_+^{-1}(z)\,, & \text{if } z \text{ is in the upper lens, } z \in \mathcal{L}_j^+\,, \\ T_N(z) j_-(z)\,, & \text{if } z \text{ is in the lower lens, } z \in \mathcal{L}_j^-\,, \\ T_N(z) & \text{otherwise}\,, \end{cases} \tag{4.149}$$

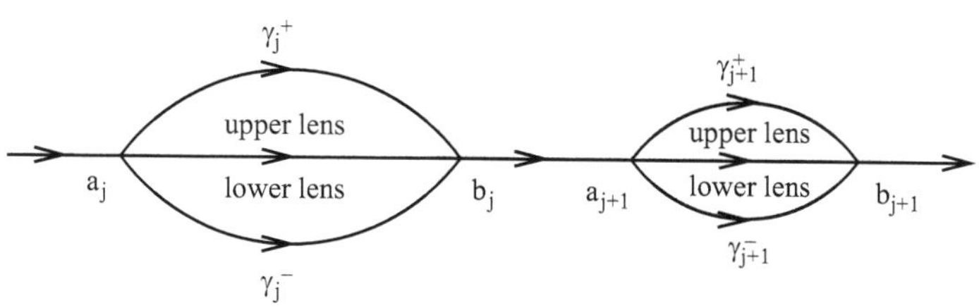

Fig. 4.3. The lenses

where

$$j_{\pm}(z) = \begin{pmatrix} 1 & 0 \\ \mathrm{e}^{\mp NG(z)} & 1 \end{pmatrix}. \tag{4.150}$$

Then $S_N(z)$ solves the following RH problem:

(1) $S_N(z)$ is analytic in $\mathbb{C} \setminus (\mathbb{R} \cup \Gamma)$, $\Gamma = \gamma_1^+ \cup \gamma_1^- \cup \cdots \cup \gamma_q^+ \cup \gamma_q^-$.
(2)

$$S_{N+}(z) = S_{N-}(z) j_S(z)\,, \quad z \in \mathbb{R} \cup \gamma\,, \tag{4.151}$$

where the jump matrix $j_S(z)$ has the following properties:
(a) $j_S(z) = \left(\begin{smallmatrix} 0 & 1 \\ -1 & 0 \end{smallmatrix}\right)$ for $z \in J$.
(b) $j_S(z) = j_T(z) = \begin{pmatrix} \exp(-\mathrm{i}N\Omega_j) & O(\exp(-c(x)N)) \\ 0 & \exp(\mathrm{i}N\Omega_j) \end{pmatrix}$ for $z \in (b_j, a_{j+1})$, $j = 1, \dots, q-1$, and $j_S(z) = j_T(z) = \left(\begin{smallmatrix} 1 & O(\exp(-c(x)N)) \\ 0 & 1 \end{smallmatrix}\right)$ for $z \in (-\infty, a_1) \cup (b_q, \infty)$.
(c) $j_S(z) = j_{\pm}(z) = \begin{pmatrix} 1 & 0 \\ O(\mathrm{e}^{-c(z)N}) & 1 \end{pmatrix}$ for $z \in \gamma_j^{\pm}$, $j = 1, \dots, q$, where $c(z) > 0$ is a continuous function such that $c(z) \to 0$ as $z \to a_j, b_j$.

(3) $S_N(z) = I + O(z^{-1})$, as $z \to \infty$.

We expect, and this will be justified later, that as $N \to \infty$, $S_N(z)$ converges to a solution of the model RHP, in which we drop the $O(\mathrm{e}^{-cN})$-terms in the jump matrix $j_S(z)$. Let us consider the model RHP.

4.2.4 Model RHP

We are looking for $M(z)$ that solves the following model RHP:

(1) $M(z)$ is analytic in $\mathbb{C} \setminus [a_1, b_q]$,
(2)

$$M_+(z) = M_-(z) j_M(z)\,, \quad z \in [a_1, b_q]\,, \tag{4.152}$$

where the jump matrix $j_M(z)$ is given by the following formulas:
(a) $j_M(z) = \left(\begin{smallmatrix} 0 & 1 \\ -1 & 0 \end{smallmatrix}\right)$ for $z \in J$
(b) $j_M(z) = \begin{pmatrix} \exp(-\mathrm{i}N\Omega_j) & 0 \\ 0 & \exp(\mathrm{i}N\Omega_j) \end{pmatrix}$ for $z \in [b_j, a_{j+1}]$, $j = 1, \dots, q-1$.

(3) $M(z) = I + O(z^{-1})$, as $z \to \infty$.

We will construct a solution to the model RHP by following the work [52].

4.2.4.1 Solution of the Model RHP. One-Cut Case

Assume that J consist of a single interval $[a, b]$. Then the model RH problem reduces to the following:

(1) $M(z)$ is analytic in $\mathbb{C} \setminus [a, b]$,
(2) $M_+(z) = M_-(z) \left(\begin{smallmatrix} 0 & 1 \\ -1 & 0 \end{smallmatrix}\right)$ for $z \in [a, b]$.
(3) $M(z) = I + O(z^{-1})$, as $z \to \infty$.

This RHP can be reduced to a pair of scalar RH problems. We have that

$$\begin{pmatrix} 0 & 1 \\ -1 & 0 \end{pmatrix} = \frac{1}{2}\begin{pmatrix} 1 & 1 \\ \mathrm{i} & -\mathrm{i} \end{pmatrix}\begin{pmatrix} \mathrm{i} & 0 \\ 0 & -\mathrm{i} \end{pmatrix}\begin{pmatrix} 1 & -\mathrm{i} \\ 1 & \mathrm{i} \end{pmatrix} \tag{4.153}$$

Let

$$\widetilde{M}(z) = \frac{1}{2}\begin{pmatrix} 1 & -i \\ 1 & i \end{pmatrix} M(z) \begin{pmatrix} 1 & 1 \\ \mathrm{i} & -\mathrm{i} \end{pmatrix}. \tag{4.154}$$

Then $\widetilde{M}(z)$ solves the following RHP:

(1) $\widetilde{M}(z)$ is analytic in $\mathbb{C} \setminus [a,b]$,

(2) $\widetilde{M}_+(z) = \widetilde{M}_-(z)\begin{pmatrix} \mathrm{i} & 0 \\ 0 & -\mathrm{i} \end{pmatrix}$ for $z \in [a,b]$.

(3) $\widetilde{M}(z) = I + O(z^{-1})$, as $z \to \infty$.

This is a pair of scalar RH problems, which can be solved by the Cauchy integral:

$$\begin{aligned}
\widetilde{M}(z) &= \begin{pmatrix} \exp\left(\dfrac{1}{2\pi\mathrm{i}}\displaystyle\int_a^b \frac{\log \mathrm{i}}{s-z}\,\mathrm{d}s\right) & 0 \\ 0 & \exp\left(\dfrac{1}{2\pi\mathrm{i}}\displaystyle\int_a^b \frac{\log(-\mathrm{i})}{s-z}\,\mathrm{d}s\right) \end{pmatrix} \\
&= \begin{pmatrix} \exp\left(\dfrac{1}{4}\log\left(\dfrac{z-b}{z-a}\right)\right) & 0 \\ 0 & \exp\left(-\dfrac{1}{4}\log\left(\dfrac{z-b}{z-a}\right)\right) \end{pmatrix} \\
&= \begin{pmatrix} \gamma^{-1} & 0 \\ 0 & \gamma \end{pmatrix},
\end{aligned} \tag{4.155}$$

where

$$\gamma(z) = \left(\frac{z-a}{z-b}\right)^{1/4} \tag{4.156}$$

with cut on $[a,b]$ and the branch such that $\gamma(\infty) = 1$. Thus,

$$\begin{aligned}
M(z) &= \frac{1}{2}\begin{pmatrix} 1 & 1 \\ \mathrm{i} & -\mathrm{i} \end{pmatrix}\begin{pmatrix} \gamma^{-1} & 0 \\ 0 & \gamma \end{pmatrix}\begin{pmatrix} 1 & 1 \\ i & -\mathrm{i} \end{pmatrix}^{-1} \\
&= \begin{pmatrix} \dfrac{\gamma(z)+\gamma^{-1}(z)}{2} & \dfrac{\gamma(z)-\gamma^{-1}(z)}{-2\mathrm{i}} \\ \dfrac{\gamma(z)-\gamma^{-1}(z)}{2\mathrm{i}} & \dfrac{\gamma(z)+\gamma^{-1}(z)}{2} \end{pmatrix}, \\
&\det M(z) = 1 .
\end{aligned} \tag{4.157}$$

At infinity we have

$$\gamma(z) = 1 + \frac{b-a}{4z} + O(z^{-2}), \tag{4.158}$$

hence

$$M(z) = I + \frac{1}{z}\begin{pmatrix} 0 & \dfrac{b-a}{-4\mathrm{i}} \\ \dfrac{b-a}{4\mathrm{i}} & 0 \end{pmatrix} + O(z^{-2})\,. \tag{4.159}$$

4.2.4.2 Solution of the Model RHP. Multicut Case

This will be done in three steps.

Step 1. Consider the auxiliary RHP,

(1) $Q(z)$ is analytic in $\mathbb{C} \setminus J$, $J = \bigcup_{j=1}^{q} [a_j, b_j]$,
(2) $Q_+(z) = Q_-(z) \left(\begin{smallmatrix} 0 & 1 \\ -1 & 0 \end{smallmatrix}\right)$ for $z \in J$.
(3) $Q(z) = I + O(z^{-1})$, as $z \to \infty$.

Then, similar to the one-cut case, this RHP is reduced to two scalar RHPs, and the solution is

$$Q(z) = \begin{pmatrix} \dfrac{\gamma(z)+\gamma^{-1}(z)}{2} & \dfrac{\gamma(z)-\gamma^{-1}(z)}{-2\mathrm{i}} \\ \dfrac{\gamma(z)-\gamma^{-1}(z)}{2\mathrm{i}} & \dfrac{\gamma(z)+\gamma^{-1}(z)}{2} \end{pmatrix}, \tag{4.160}$$

where

$$\gamma(z) = \prod_{j=1}^{q} \left(\frac{z-a_j}{z-b_j}\right)^{1/4}, \quad \gamma(\infty) = 1\,, \tag{4.161}$$

with cuts on J. At infinity we have

$$\gamma(z) = 1 + \sum_{j=1}^{q} \frac{b_j - a_j}{4z} + O(z^{-2})\,, \tag{4.162}$$

hence

$$Q(z) - I + \frac{1}{z}\begin{pmatrix} 0 & \displaystyle\sum_{j=1}^{q} \frac{b_j - a_j}{-4\mathrm{i}} \\ \displaystyle\sum_{j=1}^{q} \frac{b_j - a_j}{4\mathrm{i}} & 0 \end{pmatrix} + O(z^{-2})\,. \tag{4.163}$$

In what follows, we will modify this solution to satisfy part (b) in jump matrix in (4.152). This requires some Riemannian geometry and the Riemann theta function.

Step 2. Let X be the two-sheeted Riemannian surface of the genus

$$g = q - 1\,, \tag{4.164}$$

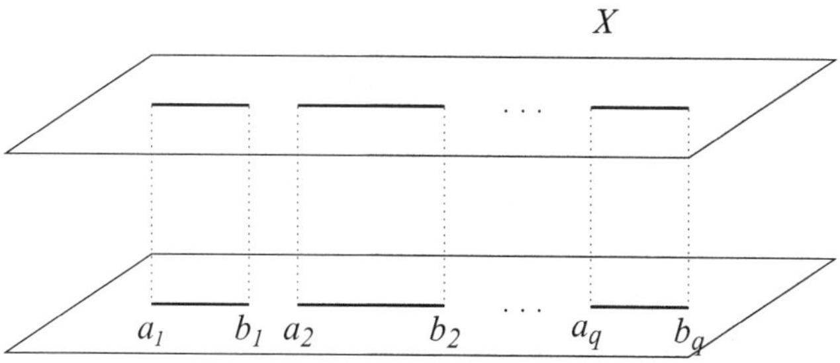

Fig. 4.4. The Riemannian surface associated to $\sqrt{R(z)}$

associated to $\sqrt{R(z)}$, where

$$R(z) = \prod_{j=1}^{q} (z - a_j)(z - b_j) \,, \tag{4.165}$$

with cuts on the intervals (a_j, b_j), $j = 1, \ldots, q$, see Fig. 4.4. We fix the first sheet of X by the condition that on this sheet,

$$\sqrt{R(x)} > 0 \,, \quad x > b_q \,. \tag{4.166}$$

We would like to introduce $2g$ cycles on X, forming a homology basis. To that end, consider, for $j = 1, \ldots, g$, a cycle A_j on X, which goes around the interval (b_j, a_{j+1}) in the negative direction, such that the part of A_j in the upper half-plane, $A_j^+ \equiv A_j \cup \{z : \operatorname{Im} z \geq 0\}$, lies on the first sheet of X, and the one in the lower half-plane, $A_j^- \equiv A_j \cup \{z : \operatorname{Im} z \leq 0\}$, lies on the second sheet, $j = 1, \ldots, g$. In addition, consider a cycle B_j on X, which goes around the interval (a_1, b_j) on the first sheet in the negative direction, see Fig. 4.5. Then the cycles $(A_1, \ldots, A_g, B_1, \ldots, B_g)$ form a canonical homology basis for X.

Consider the linear space Ω of holomorphic one-forms on X,

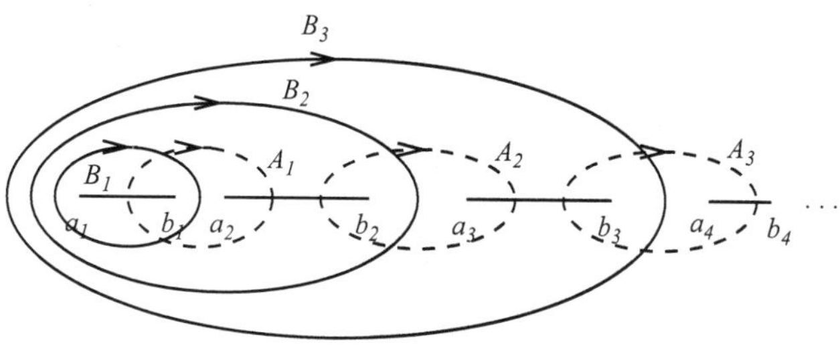

Fig. 4.5. The basis of cycles on X

$$\Omega = \left\{ \omega = \sum_{j=0}^{q-2} \frac{c_j z^j \, \mathrm{d}z}{\sqrt{R(z)}} \right\} . \tag{4.167}$$

The dimension of Ω is equal to g. Consider the basis in Ω,

$$\omega = (\omega_1, \dots, \omega_g) ,$$

with normalization

$$\int_{A_j} \omega_k = \delta_{jk} , \quad j, k = 1, \dots, g . \tag{4.168}$$

Such a basis exists and it is unique, see [67]. Observe that the numbers

$$m_{jk} = \int_{A_j} \frac{z^k \, \mathrm{d}z}{\sqrt{R(z)}} = 2 \int_{b_j}^{a_{j+1}} \frac{x^k \, \mathrm{d}x}{\sqrt{R(x)}} , \quad 1 \le j \le g , 1 \le k \le g-1 , \tag{4.169}$$

are real. This implies that the basis ω is real, i.e., the one-forms,

$$\omega_j = \sum_{k=0}^{g-1} \frac{c_{jk} z^k \, \mathrm{d}z}{\sqrt{R(z)}} , \tag{4.170}$$

have real coefficients c_{jk}.

Define the associated Riemann matrix of B-periods,

$$\tau = (\tau_{jk}) , \quad \tau_{jk} = \int_{B_j} \omega_k , \ j, k = 1, \dots, g . \tag{4.171}$$

Since $\sqrt{R(x)}$ is pure imaginary on (a_j, b_j), the numbers τ_{jk} are pure imaginary. It is known, see, e.g., [67], that the matrix τ is symmetric and $(-\mathrm{i}\tau)$ is positive definite.

The Riemann theta function with the matrix τ is defined as

$$\theta(s) = \sum_{m \in \mathbb{Z}^g} \exp\bigl(2\pi \mathrm{i}(m, s) + \pi \mathrm{i}(m, \tau m)\bigr) , \qquad s \in \mathbb{C}^g \ ; \ (m, s) = \sum_{j=1}^{g} m_j s_j . \tag{4.172}$$

The quadratic form $i(m, \tau m)$ is negative definite, hence the series is absolutely convergent and $\theta(s)$ is analytic in $\mathbb{C}^g$. The theta function is an even function,

$$\theta(-s) = \theta(s) , \tag{4.173}$$

and it has the following periodicity properties:

$$\theta(s + e_j) = \theta(s) \ ; \quad \theta(s + \tau_j) = \exp(-2\pi \mathrm{i} s_j - \pi \mathrm{i} \tau_{jj}) \theta(s) , \tag{4.174}$$

where $e_j = (0, \dots, 0, 1, 0, \dots, 0)$ is the jth basis vector in $\mathbb{C}^g$, and $\tau_j = \tau e_j$. This implies that the function

$$f(s) = \frac{\theta(s+d+c)}{\theta(s+d)}, \tag{4.175}$$

where $c, d \in \mathbb{C}^g$ are arbitrary constant vectors, has the periodicity properties,

$$f(s+e_j) = f(s)\,; \quad f(s+\tau_j) = \exp(-2\pi i c_j) f(s)\,. \tag{4.176}$$

Consider now the theta function associated with the Riemann surface X. It is defined as follows. Introduce the vector function,

$$u(z) = \int_{b_q}^{z} \omega\,, \quad z \in \mathbb{C} \setminus (a_1, b_q)\,, \tag{4.177}$$

where ω is the basis of holomorphic one-forms, determined by (4.168). The contour of integration in (4.177) lies in $\mathbb{C} \setminus (a_1, b_q)$, on the first sheet of X. We will consider $u(z)$ as a function with values in $\mathbb{C}^g/\mathbb{Z}^g$,

$$u\colon \mathbb{C} \setminus (a_1, b_q) \to \mathbb{C}^g/\mathbb{Z}^g\,. \tag{4.178}$$

On $[a_1, b_q]$ the function $u(z)$ is two-valued. From (4.171) we have that

$$u_+(x) - u_-(x) = \tau_j\,,\ x \in [b_j, a_{j+1}]\,; \quad 1 \le j \le q-1\,. \tag{4.179}$$

Since $\sqrt{R(x)}_- = -\sqrt{R(x)}_+$ on $[a_j, b_j]$, we have that the function $u_+(x) + u_-(x)$ is constant on $[a_j, b_j]$. It follows from (4.168) that, $\mod \mathbb{Z}^q$,

$$u_+(b_j) + u_-(b_j) = u_+(a_{j+1}) + u_-(a_{j+1})\,, \quad 1 \le j \le q-1\,. \tag{4.180}$$

Since $u_+(b_q) = u_-(b_q) = 0$, we obtain that

$$u_+(x) + u_-(x) = 0\,, \quad x \in J = \bigcup_{j=1}^{q} [a_j, b_j]\,. \tag{4.181}$$

Define

$$f_1(z) = \frac{\theta(u(z)+d+c)}{\theta(u(z)+d)}\,,\ f_2(z) = \frac{\theta(-u(z)+d+c)}{\theta(-u(z)+d)}\,, \quad z \in \mathbb{C} \setminus (a_1, b_q)\,, \tag{4.182}$$

where $c, d \in \mathbb{C}^g$ are arbitrary constant vectors. Then from (4.179) and (4.176) we obtain that for $1 \le j \le q-1$,

$$f_1(x+\mathrm{i}0) = \exp(-2\pi \mathrm{i} c_j) f_1(x-\mathrm{i}0)\,,\ f_2(x+\mathrm{i}0) = \exp(2\pi \mathrm{i} c_j) f_2(x-\mathrm{i}0)\,, \quad x \in (b_j, a_{j+1})\,, \tag{4.183}$$

and from (4.181) that

$$f_1(x+\mathrm{i}0) = f_2(x-\mathrm{i}0)\,,\ f_2(x+i0) = f_1(x-i0)\,, \quad x \in J\,. \tag{4.184}$$

Let us take

$$c = \frac{n\Omega}{2\pi}, \quad \Omega = (\Omega_1, \dots, \Omega_g), \tag{4.185}$$

and define the matrix-valued function,

$$F(z) = \begin{pmatrix} \frac{\theta(u(z)+d_1+c)}{\theta(u(z)+d_1)} & \frac{\theta(-u(z)+d_1+c)}{\theta(-u(z)+d_1)} \\ \frac{\theta(u(z)+d_2+c)}{\theta(u(z)+d_2)} & \frac{theta(-u(z)+d_2+c)}{\theta(-u(z)+d_2)} \end{pmatrix} \tag{4.186}$$

where $d_1, d_2 \in \mathbb{C}^g$ are arbitrary constant vectors. Then from (4.184), we obtain that

$$\begin{aligned} F_+(x) &= F_-(x) \begin{pmatrix} \exp(-\mathrm{i}N\Omega_j) & 0 \\ 0 & \exp(\mathrm{i}N\Omega_j) \end{pmatrix}, \\ &\qquad x \in (b_j, a_{j+1}) ;\ j = 1, \dots, q-1, \\ F_+(x) &= F_-(x) \begin{pmatrix} 0 & 1 \\ 1 & 0 \end{pmatrix}, \qquad x \in J . \end{aligned} \tag{4.187}$$

Step 3. Let us combine formulas (4.160) and (4.186), and let us set

$$\begin{aligned} M(z) &= F(\infty)^{-1} \\ &\times \begin{pmatrix} \frac{\gamma(z)+\gamma^{-1}(z)}{2} \frac{\theta(u(z)+d_1+c)}{\theta(u(z)+d_1)} & \frac{\gamma(z)-\gamma^{-1}(z)}{-2\mathrm{i}} \frac{\theta(-u(z)+d_1+c)}{\theta(-u(z)+d_1)} \\ \frac{\gamma(z)-\gamma^{-1}(z)}{2\mathrm{i}} \frac{\theta(u(z)+d_2+c)}{\theta(u(z)+d_2)} & \frac{\gamma(z)+\gamma^{-1}(z)}{2} \frac{\theta(-u(z)+d_2+c)}{\theta(-u(z)+d_2)} \end{pmatrix}, \end{aligned} \tag{4.188}$$

where

$$F(\infty) = \begin{pmatrix} \frac{\theta(u(\infty)+d_1+c)}{\theta(u(\infty)+d_1)} & 0 \\ 0 & \frac{\theta(-u(\infty)+d_2+c)}{\theta(-u(\infty)+d_2)} \end{pmatrix} \tag{4.189}$$

Then $M(z)$ has the following jumps:

$$\begin{aligned} M_+(x) &= M_-(x) \begin{pmatrix} \exp(-\mathrm{i}N\Omega_j) & 0 \\ 0 & \exp(\mathrm{i}N\Omega_j) \end{pmatrix}, \\ &\qquad x \in (b_j, a_{j+1}) ;\ j = 1, \dots, q-1, \\ M_+(x) &= M_-(x) \begin{pmatrix} 0 & 1 \\ -1 & 0 \end{pmatrix}, \qquad x \in J , \end{aligned} \tag{4.190}$$

which fits perfectly to the model RHP, and $M(\infty) = I$. It remains to find d_1, d_2 such that $M(z)$ is analytic at the zeros of $\theta(\pm u(z) + d_{1,2})$. These zeros can be cancelled by the zeros of the functions $\gamma(z) \pm \gamma^{-1}(z)$. Let us consider the latter zeros.

The zeros of $\gamma(z) \pm \gamma^{-1}(z)$ are the ones of $\gamma^2(z) \pm 1$, and hence of $\gamma^4(z) - 1$. By (4.161), the equation $\gamma^4(z) - 1 = 0$ reads

$$p(z) \equiv \prod_{j=1}^{q} \frac{z - a_j}{z - b_j} = 1 \,. \tag{4.191}$$

It is easy to see that

$$p(b_j + 0) = \infty \,, \; p(a_{j+1}) = 0 \,, \quad 1 \le j \le q - 1 \,, \tag{4.192}$$

hence equation (4.191) has a solution x_j on each interval (b_j, a_{j+1}),

$$p(x_j) = 1 \,, \; b_j < x_j < a_{j+1} \,; \quad 1 \le j \le q - 1 \,. \tag{4.193}$$

Since (4.191) has $(q-1)$ finite solutions, the numbers $\{x_j : 1 \le j \le q-1\}$ are all the solutions of (4.191). The function $\gamma(z)$, defined by (4.161), with cuts on J, is positive on $\mathbb{R} \setminus J$, hence

$$\gamma(x_j) = 1 \,. \tag{4.194}$$

Thus, we have $(q-1)$ zeros of $\gamma(z) - \gamma^{-1}(z)$ and no zeros of $\gamma(z) + \gamma^{-1}(z)$ on the sheet of $\gamma(z)$ under consideration.

Let us consider the zeros of the function $\theta(u(z)-d)$. The vector of Riemann constants is given by the formula

$$K = -\sum_{j=1}^{q-1} u(b_j) \,. \tag{4.195}$$

Define

$$d = -K + \sum_{j=1}^{q-1} u(z_j) \,. \tag{4.196}$$

Then

$$\theta(u(x_j) - d) = 0 \,, \quad 1 \le j \le q - 1 \,, \tag{4.197}$$

see [52], and $\{x_j : 1 \le j \le q-1\}$ are all the zeros of the function $\theta(u(z) - d)$. In addition, the function $\theta(u(z) + d)$ has no zeros at all on the upper sheet of X. In fact, all the zeros of $\theta(u(z) + d)$ lie on the lower sheet, above the same points $\{x_j : 1 \le j \le q-1\}$. Therefore, we set in (4.188),

$$d_1 = d \,, \quad d_2 = -d \,, \tag{4.198}$$

so that

$$M(z) = F(\infty)^{-1} \times \begin{pmatrix} \dfrac{\gamma(z)+\gamma^{-1}(z)}{2} \dfrac{\theta(u(z)+d+c)}{\theta(u(z)+d)} & \dfrac{\gamma(z)-\gamma^{-1}(z)}{-2\mathrm{i}} \dfrac{\theta(-u(z)+d+c)}{\theta(-u(z)+d)} \\ \dfrac{\gamma(z)-\gamma^{-1}(z)}{2\mathrm{i}} \dfrac{\theta(u(z)-d+c)}{\theta(u(z)-d)} & \dfrac{\gamma(z)+\gamma^{-1}(z)}{2} \dfrac{\theta(-u(z)-d+c)}{\theta(-u(z)-d)} \end{pmatrix}, \tag{4.199}$$

where

$$F(\infty) = \begin{pmatrix} \dfrac{\theta(u(\infty)+d+c)}{\theta(u(\infty)+d)} & 0 \\ 0 & \dfrac{\theta(-u(\infty)-d+c)}{\theta(-u(\infty)-d)} \end{pmatrix}. \tag{4.200}$$

This gives the required solution of the model RHP. As $z \to \infty$,

$$M(z) = I + \frac{M_1}{z} + O(z^{-2}), \tag{4.201}$$

where

$$M_1 = \begin{pmatrix} \left(\dfrac{\nabla\theta(u+d+c)}{\theta(u+d+c)} - \dfrac{\nabla\theta(u+d)}{\theta(u(\infty)+d)}, u'(\infty)\right) & \dfrac{\theta(-u+d+c)\theta(u+d)}{\theta(u+d+c)\theta(-u+d)} \displaystyle\sum_{j=1}^{q} \dfrac{(b_j-a_j)}{-4\mathrm{i}} \\ \dfrac{\theta(u-d+c)\theta(-u-d)}{\theta(-u-d+c)\theta(u-d)} \displaystyle\sum_{j=1}^{q} \dfrac{(b_j-a_j)}{4\mathrm{i}} & \left(\dfrac{\nabla\theta(u(\infty)+d-c)}{\theta(u(\infty)+d-c)} - \dfrac{\nabla\theta(u+d)}{\theta(u+d)}, u'(\infty)\right) \end{pmatrix},$$

$$u = u(\infty). \tag{4.202}$$

4.2.5 Construction of a Parametrix at Edge Points

We consider small disks $D(a_j, r)$, $D(b_j, r)$, $1 \le j \le q$, of radius $r > 0$, centered at the edge points,

$$D(a, r) \equiv \{z : |z - a| \le r\},$$

and we look for a local parametrix $U_N(z)$, defined on the union of these disks, such that

- $U_N(z)$ is analytic on $D \setminus (\mathbb{R} \cup \Gamma)$, where

$$D = \bigcup_{j=1}^{q} \big(D(a_j, r) \cup D(b_j, r)\big). \tag{4.203}$$

- $$U_{N+}(z) = U_{N-}(z) j_S(z), \quad z \in (\mathbb{R} \cup \Gamma) \cap D, \tag{4.204}$$

- as $N \to \infty$,

$$U_N(z) = \left(I + O\left(\frac{1}{N}\right)\right) M(z) \quad \text{uniformly for } z \in \partial D\,. \tag{4.205}$$

We consider here the edge point b_j, $1 \le j \le q$, in detail. From (4.127)) and (4.132), we obtain that

$$2g_+(x) = V(x) + l + \mathrm{i}\Omega_j - 2\pi\mathrm{i}\int_{b_j}^{x} p_V(s)\,\mathrm{d}s\,, \quad a_j < x < b_j\,, \tag{4.206}$$

hence

$$[2g_+(x) - V(x)] - [2g_+(b_j) - V(b_j)] = -2\pi\mathrm{i}\int_{b_j}^{x} p_V(s)\,\mathrm{d}s\,. \tag{4.207}$$

By using formula (4.99), we obtain that

$$[2g_+(b_j) - V(b_j)] - [2g_+(x) - V(x)] = \int_{b_j}^{x} h(s) R_+^{1/2}(s)\,\mathrm{d}s\,,$$
$$R(z) = \prod_{j=1}^{q}(x - a_j)(x - b_j)\,. \tag{4.208}$$

Since both $g(z)$ and $R^{1/2}(z)$ are analytic in the upper half-plane, we can extend this equation to the upper half-plane,

$$[2g_+(b_j) - V(b_j)] - [2g(z) - V(z)] = \int_{b_j}^{z} h(s) R^{1/2}(s)\,\mathrm{d}s\,, \tag{4.209}$$

where the contour of integration lies in the upper half-plane. Observe that

$$\int_{b_j}^{z} h(s) R^{1/2}(s)\,\mathrm{d}s = c(z - b_j)^{3/2} + O\big((z - b_j)^{5/2}\big) \tag{4.210}$$

as $z \to b_j$, where $c > 0$. Then it follows that

$$\beta(z) = \left\{\tfrac{3}{4}[2g_+(b_j) - V(b_j)] - [2g(z) - V(z)]\right\}^{2/3} \tag{4.211}$$

is analytic at b_j, real-valued on the real axis near b_j and $\beta'(b_j) > 0$. So β is a conformal map from $D(b_j, r)$ to a convex neighborhood of the origin, if r is sufficiently small (which we assume to be the case). We take Γ near b_j such that

$$\beta\big(\Gamma \cap D(b_j, r)\big) \subset \{z \mid \arg(z) = \pm 2\pi/3\}\,.$$

Then Γ and $\mathbb{R}$ divide the disk $D(b_j, r)$ into four regions numbered I, II, III, and IV, such that $0 < \arg\beta(z) < 2\pi/3$, $2\pi/3 < \arg\beta(z) < \pi$,

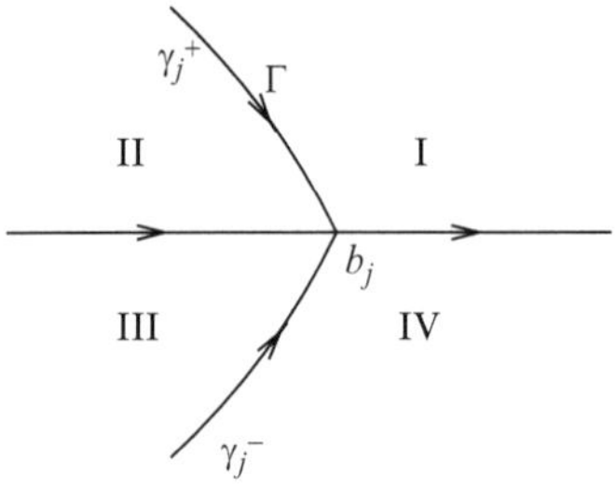

Fig. 4.6. Partition of a neighborhood of the edge point

$-\pi < \arg\beta(z) < -2\pi/3$, and $-2\pi/3 < \arg\beta(z) < 0$ for z in regions I, II, III, and IV, respectively, see Fig. 4.6.

Recall that the jumps j_S near b_q are given as

$$
\begin{aligned}
j_S &= \begin{pmatrix} 0 & 1 \\ -1 & 0 \end{pmatrix} && \text{on } [b_j - r, b_j) \\
j_S &= \begin{pmatrix} 1 & 0 \\ \mathrm{e}^{-NG(z)} & 1 \end{pmatrix} && \text{on } \gamma_j^+ \\
j_S &= \begin{pmatrix} 1 & 0 \\ \mathrm{e}^{NG(z)} & 1 \end{pmatrix} && \text{on } \gamma_j^- \\
j_S &= \begin{pmatrix} \exp(-N[g_+(z) - g_-(z)]) & \exp(N(g_+(z) + g_-(z) - V(z) - l)) \\ 0 & \exp(N[g_+(z) - g_-(z)]) \end{pmatrix} \\
& && \text{on } (b_j, b_j + r] \,.
\end{aligned}
\tag{4.212}
$$

We look for $U_N(z)$ in the form,

$$
U_N(z) = Q_N(z) \exp\left(-N \left[g(z) - \frac{V(z)}{2} - \frac{l}{2} \right] \sigma_3 \right) . \tag{4.213}
$$

Then the jump condition on $U_N(z)$, (4.204), is transformed to the jump condition on $Q_N(Z)$,

$$
Q_{N+}(z) = Q_{N-}(z) j_Q(z) \,, \tag{4.214}
$$

where

$$
\begin{aligned}
j_Q(z) = \exp\left(-N \left[g_-(z) - \frac{V(z)}{2} - \frac{l}{2} \right] \sigma_3 \right) j_S(z) \\
\times \exp\left(N \left[g_+(z) - \frac{V(z)}{2} - \frac{l}{2} \right] \sigma_3 \right) .
\end{aligned}
\tag{4.215}
$$

From (4.212), (4.127) and (4.134) we obtain that

$$\begin{aligned} j_Q &= \begin{pmatrix} 0 & 1 \\ -1 & 0 \end{pmatrix} \qquad \text{on } [b_j - r, b_j)] \,, \\ j_Q &= \begin{pmatrix} 1 & 0 \\ 1 & 1 \end{pmatrix} \qquad \text{on } \gamma_j^+ \,, \\ j_Q &= \begin{pmatrix} 1 & 0 \\ 1 & 1 \end{pmatrix} \qquad \text{on } \gamma_j^- \,, \\ j_Q &= \begin{pmatrix} 1 & 1 \\ 0 & 1 \end{pmatrix} \qquad \text{on } (b_j, b_j + r] \,. \end{aligned} \tag{4.216}$$

We construct $Q_N(z)$ with the help of the Airy function. The Airy function $\mathrm{Ai}(z)$ solves the equation $y'' = zy$ and for any $\varepsilon > 0$, in the sector $\pi + \varepsilon \leq \arg z \leq \pi - \varepsilon$, it has the asymptotics as $z \to \infty$,

$$\mathrm{Ai}(z) = \frac{1}{2\sqrt{\pi} z^{1/4}} \exp(-\tfrac{2}{3} z^{3/2})\left(1 + O(z^{-3/2})\right) . \tag{4.217}$$

The functions $\mathrm{Ai}(\omega z)$, $\mathrm{Ai}(\omega^2 z)$, where $\omega = \mathrm{e}^{2\pi \mathrm{i}/3}$, also solve the equation $y'' = zy$, and we have the linear relation,

$$\mathrm{Ai}(z) + \omega\, \mathrm{Ai}(\omega z) + \omega^2\, \mathrm{Ai}(\omega^2 z) = 0 . \tag{4.218}$$

We write

$$y_0(z) = \mathrm{Ai}(z) \,, \quad y_1(z) = \omega\, \mathrm{Ai}(\omega z) \,, \quad y_2(z) = \omega^2\, \mathrm{Ai}(\omega^2 z) \,, \tag{4.219}$$

and we use these functions to define

$$\Phi(z) = \begin{cases} \begin{pmatrix} y_0(z) & -y_2(z) \\ y_0'(z) & -y_2'(z) \end{pmatrix}, & \text{for } 0 < \arg z < 2\pi/3 \,, \\ \begin{pmatrix} -y_1(z) & -y_2(z) \\ -y_1'(z) & -y_2'(z) \end{pmatrix}, & \text{for } 2\pi/3 < \arg z < \pi \,, \\ \begin{pmatrix} -y_2(z) & y_1(z) \\ -y_2'(z) & y_1'(z) \end{pmatrix}, & \text{for } -\pi < \arg z < -2\pi/3 \,, \\ \begin{pmatrix} y_0(z) & y_1(z) \\ y_0'(z) & y_1'(z) \end{pmatrix}, & \text{for } -2\pi/3 < \arg z < 0 \,. \end{cases} \tag{4.220}$$

Observe that (4.218) reads

$$y_0(z) + y_1(z) + y_2(z) = 1 \,, \tag{4.221}$$

and it implies that on the discontinuity rays,

$$\Phi_+(z) = \Phi_-(z) j_Q(z) \,, \quad \arg z = 0, \pm\frac{2\pi}{3}, \pi \,. \tag{4.222}$$

Now we set

$$Q_N(z) = E_N(z) \Phi\left(N^{2/3} \beta(z)\right) , \tag{4.223}$$

so that

$$U_N(z) = E_N(z)\Phi(N^{2/3}\beta(z)) \exp\left(-N\left[g(z) - \frac{V(z)}{2} - \frac{l}{2}\right]\sigma_3\right), \tag{4.224}$$

where E_N is an analytic prefactor that takes care of the matching condition (4.205). Since $\Phi(z)$ has the jumps j_Q, we obtain that $U_N(z)$ has the jumps j_S, so that it satisfies jump condition (4.204). The analytic prefactor E_N is explicitly given by the formula,

$$E_N(z) = M(z)\Theta_N(z)L_N(z)^{-1}, \tag{4.225}$$

where $M(z)$ is the solution of the model RHP,

$$\Theta_N(z) = \exp\left(\pm\frac{N\Omega_j}{2}\sigma_3\right), \quad \pm \operatorname{Im} z \geq 0\,. \tag{4.226}$$

and

$$L_N(z) = \frac{1}{2\sqrt{\pi}} \begin{pmatrix} N^{-1/6}\beta^{-1/4}(z) & 0 \\ 0 & N^{1/6}\beta^{1/4}(z) \end{pmatrix} \begin{pmatrix} 1 & \mathrm{i} \\ -1 & \mathrm{i} \end{pmatrix} \tag{4.227}$$

where for $\beta^{1/4}(z)$ we take a branch which is positive for $z \in (b_j, b_j + r]$, with a cut on $[b_j - r, b_j)$. To prove the analyticity of $E_N(z)$, observe that

$$[M(x)\Theta_N(x)]_+ = [M(x)\Theta_N(x)]_- j_1(x)\,, \quad b_j - r \leq x \leq b_j + r\,, \tag{4.228}$$

where

$$j_1(x) = \exp\left(\frac{N\Omega_j}{2}\sigma_3\right) j_M(x) \exp\left(\frac{N\Omega_j}{2}\sigma_3\right). \tag{4.229}$$

From (4.152) we obtain that

$$\begin{aligned} j_1(x) &= \begin{pmatrix} 0 & 1 \\ -1 & 0 \end{pmatrix}, \qquad b_j - r \leq x < b_j\,, \\ j_1(x) &= I\,, \qquad b_j < x \leq b_j + r\,. \end{aligned} \tag{4.230}$$

From (4.227),

$$L_{N+}(x) = L_{N-}(x) j_2(x)\,, \quad b_j - r \leq x \leq b_j + r\,, \tag{4.231}$$

where $j_2(x) = I$ for $b_j < x \leq b_j + r$, and

$$j_2(x) = \begin{pmatrix} 1 & \mathrm{i} \\ -1 & \mathrm{i} \end{pmatrix}^{-1} \begin{pmatrix} -\mathrm{i} & 0 \\ 0 & \mathrm{i} \end{pmatrix} \begin{pmatrix} 1 & \mathrm{i} \\ -1 & \mathrm{i} \end{pmatrix} = \begin{pmatrix} 0 & 1 \\ -1 & 0 \end{pmatrix}, \quad b_j - r \leq x < b_j\,, \tag{4.232}$$

so that $j_2(x) = j_1(x)$, $b_j - r \leq x \leq b_j + r$. Therefore, $E_N(z)$ has no jump on $b_j - r \leq x \leq b_j + r$. Since the entries of both M and L have at most fourth-root singularities at b_j, the function $E_N(z)$ has a removable singularity at $z = b_j$, hence it is analytic in $D(b_j, r)$.

Let us prove matching condition (4.205). Consider first z in domain I on Fig. 4.6. From (4.217) we obtain that for $0 \le \arg z \le 2\pi/3$,

$$\begin{aligned} y_0(z) &= \frac{1}{2\sqrt{\pi} z^{1/4}} \exp(-\tfrac{2}{3} z^{3/2})(1 + O(z^{-3/2})) , \\ -y_2(z) &= \frac{\mathrm{i}}{2\sqrt{\pi} z^{1/4}} \exp(\tfrac{2}{3} z^{3/2})(1 + O(z^{-3/2})) , \end{aligned} \tag{4.233}$$

hence for z in domain I,

$$\begin{aligned} \Phi\big(N^{2/3}\beta(z)\big) = \frac{1}{2\sqrt{\pi}} N^{-\sigma_3/6} \beta(z)^{-\sigma_3/4} \begin{pmatrix} 1 & \mathrm{i} \\ -1 & \mathrm{i} \end{pmatrix} (I + O(N^{-1})) \\ \times \exp(-\tfrac{2}{3} N \beta(z)^{3/2} \sigma_3) . \end{aligned} \tag{4.234}$$

From (4.211),

$$\tfrac{2}{3}\beta(z)^{3/2} = \tfrac{1}{2}\{[2g_+(b_j) - V(b_j)] - [2g(z) - V(z)]\} , \tag{4.235}$$

and from (4.206),

$$2g_+(b_j) - V(b_j) = l + i\Omega_j , \tag{4.236}$$

hence

$$\frac{2}{3}\beta(z)^{3/2} = -g(z) + \frac{V(z)}{2} + \frac{l}{2} + \frac{i\Omega_j}{2} . \tag{4.237}$$

Therefore, from (4.224) and (4.234) we obtain that

$$\begin{aligned} U_N(z) = E_N(z) \frac{1}{2\sqrt{\pi}} N^{-\sigma_3/6} \beta(z)^{-\sigma_3/4} \begin{pmatrix} 1 & \mathrm{i} \\ -1 & \mathrm{i} \end{pmatrix} \\ \times \big(I + O(N^{-1})\big) \exp(-\mathrm{i} N \Omega_j/2) \end{aligned} \tag{4.238}$$

Then, from (4.225) and (4.227),

$$\begin{aligned} U_N(z) &= M(z) \exp\Big(\frac{\mathrm{i} N \Omega_j}{2}\Big) L_N(z)^{-1} \frac{1}{2\sqrt{\pi}} N^{-\sigma_3/6} \beta(z)^{-\sigma_3/4} \begin{pmatrix} 1 & \mathrm{i} \\ -1 & \mathrm{i} \end{pmatrix} \\ &\qquad \times \big(I + O(N^{-1})\big) \exp\Big(-\frac{\mathrm{i} N \Omega_j}{2}\Big) \\ &= M(z) \exp\Big(\frac{\mathrm{i} N \Omega_j}{2}\Big) \big(I + O(N^{-1})\big) \exp\Big(-\frac{\mathrm{i} N \Omega_j}{2}\Big) \\ &= M(z)\big(I + O(N^{-1})\big) , \end{aligned} \tag{4.239}$$

which proves (4.205) for z in region I. Similar calculations can be done for regions II, III, and IV.

4.2.6 Third and Final Transformation of the RHP

In the third and final transformation we put

$$\begin{aligned} R_N(z) &= S_N(z)M(z)^{-1} && \text{for } z \text{ outside the disks } D(a_j,r),\ D(b_j,r)\,, \\ & && 1\le j\le q\,, \\ R_N(z) &= S_N(z)U_N(z)^{-1} && \text{for } z \text{ inside the disks}\,. \end{aligned} \tag{4.240}$$

Then $R_N(z)$ is analytic on $\mathbb{C}\setminus\Gamma_R$, where Γ_R consists of the circles $\partial D(a_j,r)$, $\partial D(b_j,r)$, $1\le j\le q$, the parts of Γ outside of the disks $D(a_j,r)$, $D(b_j,r)$, $1\le j\le q$, and the real intervals $(-\infty,a_1-r),(b_1+r,a_2-r),\dots,(b_{q-1},a_q)$, (b_q+r,∞), see Fig. 4.7. There are the jump relations,

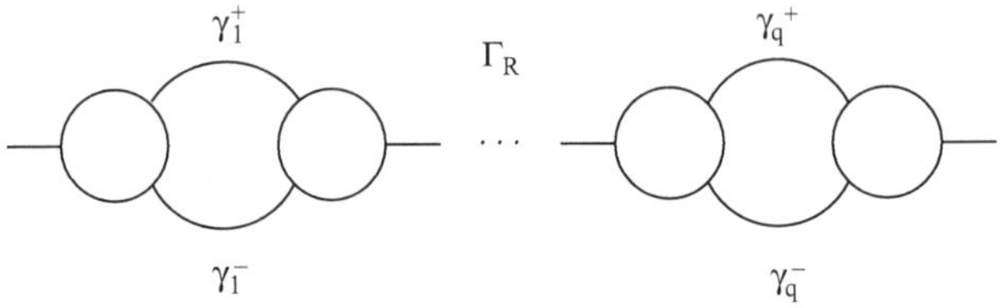

Fig. 4.7. The contour Γ_R for $R_N(z)$

$$R_{N+}(z) = R_{N-}(z)j_R(z)\,, \tag{4.241}$$

where

$$\begin{aligned} j_R(z) &= M(z)U_N(z)^{-1} && \text{on the circles}\,, \\ & && \text{oriented counterclockwise}\,, \\ j_R(z) &= M(z)j_S(z)M(z)^{-1} && \text{on the remaining parts of } \Gamma_R\,. \end{aligned} \tag{4.242}$$

We have that

$$\begin{aligned} j_R(z) &= I+O(N^{-1}) && \text{uniformly on the circles}\,, \\ j_R(z) &= I+O(\mathrm{e}^{-c(z)N}) && \text{for some } c(z)>0\,, \\ & && \text{on the remaining parts of } \Gamma_R\,. \end{aligned} \tag{4.243}$$

In addition, as $x\to\infty$, we have estimate (4.139) on $c(x)$. As $z\to\infty$, we have

$$R_N(z)\cong I+\sum_{j=1}^{\infty}\frac{R_j}{z^j}\,. \tag{4.244}$$

Thus, $R_N(z)$ solves the following RHP:

(1) $R_N(z)$ is analytic in $\mathbb{C}\setminus\Gamma_R$. and it is two-valued on Γ_R.
(2) On Γ_R, $R_N(z)$ satisfies jump condition (4.241), where the jump matrix $j_R(z)$ satisfies estimates (4.243).
(3) As $z\to\infty$, $R_N(z)$ has asymptotic expansion (4.244).

This RHP can be solved by a perturbation theory series.

4.2.7 Solution of the RHP for $R_N(z)$

Set

$$j_R^0(z) = j_R(z) - I. \tag{4.245}$$

Then by (4.243),

$$\begin{aligned} j_R^0(z) &= O(N^{-1}) && \text{uniformly on the circles ,} \\ j_R^0(z) &= O(\mathrm{e}^{-c(z)N}) && \text{for some } c(z) > 0 \text{ ,} \\ & && \quad \text{on the remaining parts of } \Gamma_R \text{ ,} \end{aligned} \tag{4.246}$$

where $c(x)$ satisfies (4.139) as $x \to \infty$. We can apply the following general result.

Proposition 4.2.1. *Assume that $v(z)$, $z \in \Gamma_R$, solves the equation*

$$v(z) = I - \frac{1}{2\pi \mathrm{i}} \int_{\Gamma_R} \frac{v(u) j_R^0(u)}{z_- - u} \, \mathrm{d}u \, , \quad z \in \Gamma_R \, , \tag{4.247}$$

where z_- means the value of the integral on the minus side of Γ_R. Then

$$R(z) = I - \frac{1}{2\pi \mathrm{i}} \int_{\Gamma_R} \frac{v(u) j_R^0(u)}{z - u} \, \mathrm{d}u \, , \quad z \in \mathbb{C} \setminus \Gamma_R \, , \tag{4.248}$$

solves the following RH problem:

(i) *$R(z)$ is analytic on $\mathbb{C} \setminus \Gamma_R$.*
(ii) *$R_+(z) = R_-(z) j_R(z)$, $z \in \Gamma_R$.*
(iii) *$R(z) = I + O(z^{-1})$, $z \to \infty$.*

Proof. From (4.247), (4.248),

$$R_-(z) = v(z) \, , \quad z \in \Gamma_R \, . \tag{4.249}$$

By the jump property of the Cauchy transform,

$$R_+(z) - R_-(z) = v(z) j_R^0(z) = R_-(z) j_R^0(z) \, , \tag{4.250}$$

hence $R_+(z) = R_-(z) j_R(z)$. From (4.248), $R(z) = I + O(z^{-1})$. Proposition 4.2.1 is proved. □

Equation (4.247) can be solved by perturbation theory, so that

$$v(z) = I + \sum_{k=1}^{\infty} v_k(z) \, , \tag{4.251}$$

where for $k \geq 1$,

$$v_k(z) = -\frac{1}{2\pi \mathrm{i}} \int_{\Gamma_R} \frac{v_{k-1}(u) j_R^0(u)}{z_- - u} \, \mathrm{d}u \, , \quad z \in \Gamma_R \, , \tag{4.252}$$

and $v_0(z) = I$. Series (4.251) is estimated from above by a convergent geometric series, so it is absolutely convergent. From (4.246) we obtain that there exists $C > 0$ such that

$$|v_k(z)| \le \frac{C^k}{N^k(1+|z|)} \,. \tag{4.253}$$

Observe that

$$v_1(z) = -\frac{1}{2\pi \mathrm{i}} \int_{\Gamma_R} \frac{j_R^0(u)}{z_- - u}\, du\,, \quad z \in \Gamma_R \,. \tag{4.254}$$

We apply this solution to find $R_N(z)$. The function $R_N(z)$ is given then as

$$R_N(z) = I + \sum_{k=1}^{\infty} R_{Nk}(z)\,, \tag{4.255}$$

where

$$R_{Nk}(z) = -\frac{1}{2\pi \mathrm{i}} \int_{\Gamma_R} \frac{v_{k-1}(u) j_R^0(u)}{z - u}\, du \,. \tag{4.256}$$

In particular,

$$R_{N1}(z) = -\frac{1}{2\pi \mathrm{i}} \int_{\Gamma_R} \frac{j_R^0(u)}{z - u}\, du \,. \tag{4.257}$$

From (4.253) we obtain that there exists $C_0 > 0$ such that

$$|R_{Nk}(z)| \le \frac{C_0 C^k}{N^k(1+|z|)} \,. \tag{4.258}$$

Hence from (4.255) we obtain that there exists $C_1 > 0$ such that for $k \ge 0$,

$$R_N(z) = I + \sum_{j=1}^{k} R_{Nj}(z) + \varepsilon_{Nk}(z)\,, \quad |\varepsilon_{Nk}(z)| \le \frac{C_1 C^k}{N^{k+1}(1+|z|)} \,. \tag{4.259}$$

In particular,

$$R_N(z) = I + O\left(\frac{1}{N(|z|+1)}\right) \quad \text{as } N \to \infty\,, \tag{4.260}$$

uniformly for $z \in \mathbb{C} \setminus \Gamma_R$.

4.2.8 Asymptotics of the Recurrent Coefficients

Let us summarize the large N asymptotics of orthogonal polynomials. From (4.240) and (4.260) we obtain that

$$\begin{aligned} S_N(z) &= \left(I + O\left(\frac{1}{N(|z|+1)}\right)\right) M(z)\,, \qquad z \in \mathbb{C} \setminus D\,,\\ S_N(z) &= \left(I + O\left(\frac{1}{N(|z|+1)}\right)\right) U_N(z)\,, \qquad z \in D; \end{aligned} \tag{4.261}$$

$$D = \bigcup_{j=1}^{q} [D(a_j, r) \cup D(b_j, r)].$$

From (4.149) we have that

$$T_N(z) = \begin{cases} S_N(z) \begin{pmatrix} 1 & 0 \\ e^{-NG(z)} & 1 \end{pmatrix}, & z \in \mathcal{L}^+ = \bigcup_{j=1}^{q} \mathcal{L}_j^+\,, \\ S_N(z) \begin{pmatrix} 1 & 0 \\ -e^{NG(z)} & 1 \end{pmatrix}, & z \in \mathcal{L}^- = \bigcup_{j=1}^{q} \mathcal{L}_j^-\,, \\ S_N(z)\,, & z \in \mathbb{C} \setminus (\mathcal{L}^+ \cup \mathcal{L}^-)\,. \end{cases} \tag{4.262}$$

Finally, from (4.136) we obtain that

$$Y_N(z) = \begin{cases} \exp\left(\frac{Nl}{2}\sigma_3\right)\left(I + O\left(\frac{1}{N(|z|+1)}\right)\right) M(z) \begin{pmatrix} 1 & 0 \\ \pm e^{\mp NG(z)} & 1 \end{pmatrix} \\ \quad \times \exp\left(N\left[g(z) - \frac{l}{2}\right]\sigma_3\right), \qquad z \in \mathcal{L}^{\pm} \setminus D\,, \\ \exp\left(\frac{Nl}{2}\sigma_3\right)\left(I + O\left(\frac{1}{N(|z|+1)}\right)\right) U_N(z) \\ \quad \times \exp\left(N\left[g(z) - \frac{l}{2}\right]\sigma_3\right), \qquad z \in D\,, \\ \exp\left(\frac{Nl}{2}\sigma_3\right)\left(I + O\left(\frac{1}{N(|z|+1)}\right)\right) M(z) \\ \quad \times \exp\left(N\left[g(z) - \frac{l}{2}\right]\sigma_3\right), \qquad z \in \mathbb{C} \setminus (D \cup \mathcal{L}^+ \cup \mathcal{L}^-)\,. \end{cases} \tag{4.263}$$

This gives the large N asymptotics of the orthogonal polynomials and their adjoint functions on the complex plane. Formulas (4.79) and (4.80) give then the large N asymptotics of the recurrent coefficients. Let us consider γ_N^2.

From (4.136) we obtain that for large z,

$$
\begin{aligned}
I+\frac{Y_1}{z}+\frac{Y_2}{z^2}+\cdots \\
&= Y_N(z)z^{-N\sigma_3} \\
&= \exp\left(\frac{Nl}{2}\sigma_3\right)T(z)\exp\left(N\left[g(z)-\frac{l}{2}-\log z\right)]\sigma_3\right) \\
&= \exp\left(\frac{Nl}{2}\sigma_3\right)\left(I+\frac{T_1}{z}+\frac{T_2}{z^2}+\cdots\right)\exp\left(N\left[g(z)-\frac{l}{2}-\log z\right]\sigma_3\right),
\end{aligned}
\tag{4.264}
$$

hence

$$
[Y_1]_{12}=\mathrm{e}^{Nl}[T_1]_{12},\quad [Y_1]_{21}=\mathrm{e}^{-Nl}[T_1]_{21} \tag{4.265}
$$

and

$$
\gamma_N^2=[Y_1]_{12}[Y_1]_{21}=[T_1]_{12}[T_1]_{21}\,. \tag{4.266}
$$

From (4.261), (4.262) we obtain further that

$$
\gamma_N^2=[M_1]_{12}[M_1]_{21}+O(N^{-1})\,, \tag{4.267}
$$

and from (4.202),

$$
\begin{aligned}
[M_1]_{12} &= \frac{\theta(-u(\infty)+d+c)\theta(u(\infty)+d)}{\theta(u(\infty)+d+c)\theta(-u(\infty)+d)}\sum_{j=1}^{q}\frac{(b_j-a_j)}{-4\mathrm{i}}\,, \\
[M_1]_{21} &= \frac{\theta(u(\infty)-d+c)\theta(-u(\infty)-d)}{\theta(-u(\infty)-d+c)\theta(u(\infty)-d)}\sum_{j=1}^{q}\frac{(b_j-a_j)}{4\mathrm{i}}\,,
\end{aligned}
\tag{4.268}
$$

hence

$$
\begin{aligned}
\gamma_N^2 &= \left[\frac{1}{4}\sum_{j=1}^{q}(b_j-a_j)\right]^2 \\
&\times\frac{\theta^2(u(\infty)+d)\theta(u(\infty)+(N\Omega/2\pi)-d)\theta(-u(\infty)+(N\Omega/2\pi)+d)}{\theta^2(u(\infty)-d)\theta(-u(\infty)+(N\Omega/2\pi)-d)\theta(u(\infty)+(N\Omega/2\pi)+d)} \\
&\qquad +O(N^{-1})\,,
\end{aligned}
\tag{4.269}
$$

where d is defined in (4.196). Consider now β_{N-1}.

From (4.264) we obtain that

$$
[Y_1]_{11}=[T_1]_{11}+Ng_1\,,\quad [Y_2]_{21}=\mathrm{e}^{-Nl}([T_2]_{21}+[T_1]_{21}Ng_1)\,, \tag{4.270}
$$

hence

$$
\beta_{N-1}=\frac{[Y_2]_{21}}{[Y_1]_{21}}-[Y_1]_{11}=\frac{[T_2]_{21}}{[T_1]_{21}}-[T_1]_{11}\,, \tag{4.271}
$$

and by (4.261), (4.262),

$$\beta_{N-1} = \frac{[M_2]_{21}}{[M_1]_{21}} - [M_1]_{11} + O(N^{-1}) \,. \tag{4.272}$$

From (4.199) we find that

$$\begin{aligned}
&\frac{[M_2]_{21}}{[M_1]_{21}} \\
&= \frac{\sum_{j=1}^q (b_j^2 - a_j^2)}{2\sum_{j=1}^q (b_j - a_j)} + \frac{\theta(u(\infty) - d)}{\theta(u(\infty) - d + c)} \left(\nabla_u \frac{\theta(u - d + c)}{\theta(u - d)} \bigg|_{u=u(\infty)}, u'(\infty) \right) \\
&= \frac{\sum_{j=1}^q (b_j^2 - a_j^2)}{2\sum_{j=1}^q (b_j - a_j)} + \left(\frac{\nabla\theta(u(\infty) - d + c)}{\theta(u(\infty) - d + c)} - \frac{\nabla\theta(u(\infty) - d)}{\theta(u(\infty) - d)}, u'(\infty) \right) , \\
&\qquad [M_1]_{11} = \frac{\theta(u(\infty) + d)}{\theta(u(\infty) + d + c)} \left(\nabla_u \frac{\theta(u + d + c)}{\theta(u + d)} \bigg|_{u=u(\infty)}, u'(\infty) \right) \\
&\qquad\qquad = \left(\frac{\nabla\theta(u(\infty) + d + c)}{\theta(u(\infty) + d + c)} - \frac{\nabla\theta(u(\infty) + d)}{\theta(u(\infty) + d)}, u'(\infty) \right) .
\end{aligned} \tag{4.273}$$

Hence,

$$\begin{aligned}
\beta_{N-1} &= \frac{\sum_{j=1}^q (b_j^2 - a_j^2)}{2\sum_{j=1}^q (b_j - a_j)} \\
&\quad + \left(\frac{\nabla\theta(u(\infty) + (N\Omega/2\pi) - d)}{\theta(u(\infty) + (N\Omega/2\pi) - d)} - \frac{\nabla\theta(u(\infty) + (N\Omega/2\pi) + d)}{\theta(u(\infty) + (N\Omega/2\pi) + d)} \right. \\
&\qquad\qquad \left. + \frac{\nabla\theta(u(\infty) + d)}{\theta(u(\infty) + d)} - \frac{\nabla\theta(u(\infty) - d)}{\theta(u(\infty) - d)}, u'(\infty) \right) \\
&\qquad\qquad + O(N^{-1}) \,.
\end{aligned} \tag{4.274}$$

This formula can be also written in the shorter form,

$$\begin{aligned}
\beta_{N-1} = \frac{\sum_{j=1}^q (b_j^2 - a_j^2)}{2\sum_{j=1}^q (b_j - a_j)} + \frac{\mathrm{d}}{\mathrm{d}z} \left[\log \frac{\theta(u(z) + (N\Omega/2\pi) - d)\theta(u(z) + d)}{\theta(u(z) + (N\Omega/2\pi) + d)\theta(u(z) - d)} \right] \bigg|_{z=\infty} \\
+ O(N^{-1}) \,.
\end{aligned} \tag{4.275}$$

In the one-cut case, $q = 1$, $a_1 = a$, $b_1 = b$, formulas (4.269), (4.274) simplify to

$$\gamma_N = \frac{b - a}{4} + O(N^{-1}) \,, \quad \beta_{N-1} = \frac{a + b}{2} + O(N^{-1}) \,. \tag{4.276}$$

Formula (4.269) is obtained in [52]. Formula (4.274) slightly differs from the formula for β_{N-1} in [52]: the first term, including a_js, b_js, is missing in [52].

4.2.9 Universality in the Random Matrix Model

By applying asymptotics (4.263) to reproducing kernel (4.87), we obtain the asymptotics of the eigenvalue correlation functions. First we consider the eigenvalue correlation functions in the bulk of the spectrum. Let us fix a point $x_0 \in \operatorname{Int} J = \bigcup_{j=1}^q (a_j, b_j)$. Then the density $p_V(x_0) > 0$. We have the following universal scaling limit of the reproducing kernel at x_0:

Theorem 4.2.1. *As* $N \to \infty$,

$$\lim_{N\to\infty} \frac{1}{Np_V(x_0)} K_N\left(x_0 + \frac{u}{Np_V(x_0)}, x_0 + \frac{v}{Np_V(x_0)}\right) = \frac{\sin[\pi(u-v)]}{\pi(u-v)} . \tag{4.277}$$

Proof. Assume that for some $1 \le j \le q$ and for some $\varepsilon > 0$, we have $\{x_0, x, y\} \in (a_j + \varepsilon, b_j - \varepsilon)$. By (4.87) and (4.136),

$$\begin{aligned} K_N(x,y) &= \frac{e^{-NV(x)/2} e^{-NV(y)/2}}{2\pi i(x-y)} (0\ 1)\, Y_{N+}^{-1}(y) Y_{N+}(x) \begin{pmatrix} 1 \\ 0 \end{pmatrix} \\ &= \frac{e^{-NV(x)/2} e^{-NV(y)/2}}{2\pi i(x-y)} \left(0\ \exp(N[g_+(y) - l'/2])\right) T_{N+}^{-1}(y) \\ &\qquad \times T_{N+}(x) \begin{pmatrix} \exp(N[g_+(x) - l/2]) \\ 0 \end{pmatrix} . \end{aligned} \tag{4.278}$$

Now, from (4.149) we obtain that

$$\begin{aligned} &K_N(x,y) \\ &\quad = \frac{e^{-NV(x)/2} e^{-NV(y)/2}}{2\pi i(x-y)} \left(0\ \exp(N[g_+(y) - l/2])\right) j_+(y) S_{N+}^{-1}(y) \\ &\qquad \times S_{N+}(x) j_+^{-1}(x) \begin{pmatrix} \exp(N[g_+(x) - l/2]) \\ 0 \end{pmatrix} \\ &\quad = \frac{e^{-NV(x)/2} e^{-NV(y)/2}}{2\pi i(x-y)} \\ &\qquad \times \left(\exp(N[-G(y) + g_+(y) - l/2])\ \exp(N[g_+(y) - l/2])\right) \\ &\qquad \times S_{N+}^{-1}(y) S_{N+}(x) \begin{pmatrix} \exp(N[g_+(x) - l/2]) \\ -\exp(N[G(x) + g_+(x) - l/2]) \end{pmatrix} . \end{aligned} \tag{4.279}$$

By (4.134),

$$-\frac{V(x)}{2} + g_+(x) - \frac{l}{2} = \frac{G(x)}{2} , \tag{4.280}$$

hence

$$\begin{aligned} K_N(x,y) &= \frac{1}{2\pi i(x-y)} \left(e^{NG(y)/2}\ e^{-NG(y)/2}\right) S_{N+}^{-1}(y) \\ &\qquad \times S_{N+}(x) \begin{pmatrix} e^{-NG(x)/2} \\ -e^{NG(x)/2} \end{pmatrix} . \end{aligned} \tag{4.281}$$

By (4.240),

$$S_{N+}(x) = R_N(x)M_+(x) \,. \tag{4.282}$$

Observe that $M_+(x)$ and $R_N(x)$ are analytic on $(a_j+\varepsilon, b_j-\varepsilon)$ and $R_N(x)$ satisfies estimate (4.260). This implies that as $x-y\to 0$,

$$S_{N+}^{-1}(y)S_{N+}(x) = I + O(x-y) \,, \tag{4.283}$$

uniformly in N. Since the function $G(x)$ is pure imaginary for real x, we obtain from (4.281) and (4.283) that

$$K_N(x,y) = \frac{1}{2\pi\mathrm{i}(x-y)}[\mathrm{e}^{-N(G(x)-G(y))/2} - \mathrm{e}^{N(G(x)-G(y))/2}] + O(1) \,. \tag{4.284}$$

By (4.132),

$$-\frac{N[G(x)-G(y)]}{2} = \pi\mathrm{i}N\int_y^x p_V(s)\,\mathrm{d}s = \pi\mathrm{i}Np_V(\xi)(x-y) \,, \qquad \xi\in[x,y] \,, \tag{4.285}$$

hence

$$K_N(x,y) = \frac{\sin[\pi Np_V(\xi)(x-y)]}{\pi(x-y)} + O(1) \,. \tag{4.286}$$

Let

$$x = x_0 + \frac{u}{Np_V(x_0)} \,, \quad y = x_0 + \frac{v}{Np_V(x_0)} \,, \tag{4.287}$$

where u and v are bounded. Then

$$\frac{1}{Np_V(x_0)}K_N(x,y) = \frac{\sin[\pi(u-v)]}{\pi(u-v)} + O(N^{-1}) \,, \tag{4.288}$$

which implies (4.277). □

Consider now the scaling limit at an edge point. Since the density p_V is zero at the edge point, we have to expect a different scaling of the eigenvalues. We have the following universal scaling limit of the reproducing kernel at the edge point:

Theorem 4.2.2. *If $x_0 = b_j$ for some $1\le j\le q$, then for some $c>0$, as $N\to\infty$,*

$$\lim_{N\to\infty}\frac{1}{(Nc)^{2/3}}K_N\left(x_0+\frac{u}{(Nc)^{2/3}}, x_0+\frac{v}{(Nc)^{2/3}}\right) = \frac{\mathrm{Ai}(u)\,\mathrm{Ai}'(v)-\mathrm{Ai}'(u)\,\mathrm{Ai}(v)}{u-v} \,. \tag{4.289}$$

Similarly, if $x_0=a_j$ for some $1\le j\le q$, then for some $c>0$, as $N\to\infty$,

$$\lim_{N\to\infty}\frac{1}{(Nc)^{2/3}}K_N\left(x_0-\frac{u}{(Nc)^{2/3}}, x_0-\frac{v}{(Nc)^{2/3}}\right) = \frac{\mathrm{Ai}(u)\,\mathrm{Ai}'(v)-\mathrm{Ai}'(u)\,\mathrm{Ai}(v)}{u-v} \,. \tag{4.290}$$

The proof is similar to the proof of Thm. 4.2.1, and we leave it to the reader.

4.3 Double Scaling Limit in a Random Matrix Model

4.3.1 Ansatz of the Double Scaling Limit

This lecture is based on the paper [18]. We consider the double-well quartic matrix model,

$$\mu_N(M) = Z_N^{-1} \exp(-N \operatorname{Tr} V(M)) \, \mathrm{d}M \tag{4.291}$$

(unitary ensemble), with

$$V(M) = \frac{tM^2}{2} + \frac{M^4}{4}, \quad t < 0 . \tag{4.292}$$

The critical point is $t_c = -2$, and the equilibrium measure is one-cut for $t > -2$ and two-cut for $t < -2$, see Fig. 4.2.

The corresponding monic orthogonal polynomials $P_n(z) = z^n + \cdots$ satisfy the orthogonality condition,

$$\int_{-\infty}^{\infty} P_n(z) P_m(z) \mathrm{e}^{-NV(z)} \, \mathrm{d}z = h_n \delta_{nm} . \tag{4.293}$$

and the recurrence relation,

$$zP_n(z) = P_{n+1}(z) + R_n P_{n-1}(z) . \tag{4.294}$$

The string equation has the form,

$$R_n(t + R_{n-1} + R_n + R_{n+1}) = \frac{n}{N} . \tag{4.295}$$

For any fixed $\varepsilon > 0$, the recurrent coefficients R_n have the scaling asymptotics as $N \to \infty$:

$$R_n = a\left(\frac{n}{N}\right) + (-1)^n b\left(\frac{n}{N}\right) + O(N^{-1}), \quad \varepsilon \le \frac{n}{N} \le \lambda_c - \varepsilon , \tag{4.296}$$

and

$$R_n = a\left(\frac{n}{N}\right) + O(N^{-1}), \quad \varepsilon^{-1} \ge \frac{n}{N} \ge \lambda_c + \varepsilon , \tag{4.297}$$

where

$$\lambda_c = \frac{t^2}{4} . \tag{4.298}$$

The scaling functions are:

$$a(\lambda) = -\frac{t}{2}, \quad b(\lambda) = \frac{\sqrt{t^2 - 4\lambda}}{2}, \quad \lambda < \lambda_c; , \tag{4.299}$$

and

$$a(\lambda) = \frac{-t + \sqrt{t^2 + 3g\lambda}}{6}, \quad \lambda > \lambda_c . \tag{4.300}$$

Our goal is to obtain the large N asymptotics of the recurrent coefficients R_n, when n/N is near the critical value λ_c. At this point we will assume that t is an arbitrary (bounded) negative number. In the end we will be interested in the case when t close to (-2). Let us give first some heuristic arguments for the critical asymptotics of R_n.

We consider $N \to \infty$ with the following scaling behavior of n/N:

$$\frac{n}{N} = \lambda_c + c_0 N^{-2/3} y, \quad c_0 = \left(\frac{t^2}{2}\right)^{1/3}, \tag{4.301}$$

where $y \in (-\infty, \infty)$ is a parameter. This limit is called the double scaling limit. We make the following Ansatz of the double scaling limit of the recurrent coefficient:

$$R_n = -\frac{t}{2} + N^{-1/3}(-1)^n c_1 u(y) + N^{-2/3} c_2 v(y) + O(N^{-1}), \tag{4.302}$$

where

$$c_1 = \left(2|t|\right)^{1/3}, \quad c_2 = \frac{1}{2}\left(\frac{1}{2|t|}\right)^{1/3}. \tag{4.303}$$

The choice of the constants c_0, c_1, c_2 secures that when we substitute this Ansatz into the left-hand side of string equation (4.295), we obtain that

$$\begin{aligned} &R_n(t + R_{n-1} + R_n + R_{n+1}) \\ &\quad = \frac{n}{N} + N^{-2/3} c_0 (v - 2u^2 - y) + N^{-1}(-1)^n (u'' - uv) + \cdots . \end{aligned} \tag{4.304}$$

By equating the coefficients at $N^{-2/3}$ and N^{-1} to 0, we arrive at the equations,

$$v = y + 2u^2 \tag{4.305}$$

and

$$u'' = yu + 2u^3, \tag{4.306}$$

the Painlevé II equation. The gluing of double scaling asymptotics (4.302) with (4.296) and (4.297) suggests the boundary conditions:

$$u \sim C\sqrt{-y}, \ y \to -\infty; \quad u \to 0, \ y \to \infty . \tag{4.307}$$

This selects uniquely the critical, Hastings–McLeod solution to the Painlevé II equation. Thus, in Ansatz (4.302) $u(y)$ is the Hastings–McLeod solution to Painlevé II and $v(y)$ is given by (4.305). The central question is how to prove Ansatz((4.302). This will be done with the help of the Riemann–Hilbert approach.

We consider the functions $\psi_n(z)$, $n = 0, 1, \dots$, defined in (4.26), and their adjoint functions,

$$\varphi_n(z) = \mathrm{e}^{NV(z)/2} \frac{1}{2\pi} \int_{-\infty}^{\infty} \frac{\mathrm{e}^{-NV(u)/2} \psi_n(u)\,\mathrm{d}u}{u - z}. \tag{4.308}$$

We define the Psi-matrix as

$$\Psi_n(z) = \begin{pmatrix} \psi_n(z) & \varphi_n(z) \\ \psi_{n-1}(z) & \varphi_{n-1}(z) \end{pmatrix}. \tag{4.309}$$

The Psi-matrix solves the Lax pair equations:

$$\Psi_n'(z) = N A_n(z) \Psi_n(z), \tag{4.310}$$

$$\Psi_{n+1}(z) = U_n(z) \Psi_n(z). \tag{4.311}$$

In the case under consideration, the matrix A_n is given by formula (4.58), with $g = 1$:

$$A_n(z) = \begin{pmatrix} -(tz/2 + z^3/2 + zR_n) & R_n^{1/2}(z^2 + \theta_n) \\ -R_n^{1/2}(z^2 + \theta_{n-1}) & tz/2 + z^3/2 + zR_n \end{pmatrix},$$

$$\theta_n = t + R_n + R_{n+1}. \tag{4.312}$$

Observe that (4.310) is a system of two differential equations of the first order. It can be reduced to the Schrödinger equation,

$$-\eta'' + N^2 U \eta = 0, \quad \eta \equiv \frac{\psi_n}{\sqrt{a_{11}}}, \tag{4.313}$$

where a_{ij} are the matrix elements of $A_n(z)$, and

$$U = -\det A_n + N^{-1}\left[(a_{11})' - a_{11}\frac{(a_{12})'}{a_{12}}\right] - N^{-2}\left[\frac{(a_{12})''}{2a_{12}} - \frac{3((a_{12})')^2}{4(a_{12})^2}\right], \tag{4.314}$$

see [17, 18].

The function $\Psi_n(z)$ solves the following RHP:

(1) $\Psi_n(z)$ is analytic on $\{\operatorname{Im} z \geq 0\}$ and on $\{\operatorname{Im} z \leq 0\}$ (two-valued on $\{\operatorname{Im} z = 0\}$).
(2) $\Psi_{n+}(x) = \Psi_{n-}(x) \left(\begin{smallmatrix} 1 & -\mathrm{i} \\ 0 & 1 \end{smallmatrix}\right)$, $x \in \mathbb{R}$.
(3) As $z \to \infty$,

$$\Psi_n(z) \sim \left(\sum_{k=0}^{\infty} \frac{\Gamma_k}{z^k}\right) \exp\left(-\left(\frac{NV(z)}{2} - n \ln z + \lambda_n\right)\sigma_3\right), \quad z \to \infty, \tag{4.315}$$

where Γ_k, $k = 0, 1, 2, \ldots$, are some constant 2×2 matrices, with

$$\Gamma_0 = \begin{pmatrix} 1 & 0 \\ 0 & R_n^{-1/2} \end{pmatrix}, \quad \Gamma_1 = \begin{pmatrix} 0 & 1 \\ R_n^{1/2} & 0 \end{pmatrix}, \tag{4.316}$$

$\lambda_n = (\ln h_n)/2$, and $\sigma_3 = \left(\begin{smallmatrix} 1 & 0 \\ 0 & -1 \end{smallmatrix}\right)$ is the Pauli matrix.

Observe that the RHP implies that $\det \Psi_n(z)$ is an entire function such that $\det \Psi_n(\infty) = R_n^{-1/2}$, hence

$$\det \Psi_n(z) = R_n^{-1/2}, \quad z \in \mathbb{C}. \tag{4.317}$$

We will construct a parametrix, an approximate solution to the RHP. To that end we use (4.310). We substitute Ansatz (4.302) into the matrix elements of A_n and we solve (4.310) in the semiclassical approximation, as $N \to \infty$. First we determine the turning points, the zeros of $\det A_n(z)$. From (4.312) we obtain that

$$\det A_n(z) = a_n(z) \equiv -\frac{tz^4}{2} - \frac{z^6}{4} + \left(\frac{n}{N} - \lambda_c\right) z^2 + R_n \theta_n \theta_{n+1}, \tag{4.318}$$

Ansatz (4.301), (4.302) implies that

$$\frac{n}{N} - \lambda_c = c_0 N^{-2/3} y, \quad \theta_n = 2c_2 N^{-2/3} v(y) + O(N^{-1}). \tag{4.319}$$

hence

$$\det A_n(z) = -\frac{tz^4}{2} - \frac{z^6}{4} + c_0 N^{-2/3} y z^2 - 2t[c_2 v(y)]^2 N^{-4/3} + O(N^{-5/3}). \tag{4.320}$$

We see from this formula that there are 4 zeros of $\det A_n$, approaching the origin as $N \to \infty$, and 2 zeros, approaching the points $\pm z_0$, $z_0 = \sqrt{-2t}$. Accordingly, we partition the complex plane into 4 domains:

(1) a neighborhood of the origin, the critical domain Ω^{CP},
(2) 2 neighborhoods of the simple turning points, the turning point domains $\Omega_{1,2}^{\mathrm{TP}}$,
(3) the rest of the complex plane, the WKB domain Ω^{WKB}.

We furthermore partition Ω^{WKB} into three domains: $\Omega_{1,2}^{\mathrm{WKB}}$ and $\Omega_{\infty}^{\mathrm{WKB}}$, see Fig. 4.8.

4.3.2 Construction of the Parametrix in Ω^{WKB}

In $\Omega_{\infty}^{\mathrm{WKB}}$ we define the parametrix by the formula,

$$\Psi^{\mathrm{WKB}}(z) = C_0 T(z) \exp(-(N\xi(z) + C_1)\sigma_3), \tag{4.321}$$

where $C_0 \neq 0$, C_1 are some constants (parameters of the solution). To introduce $T(z)$ and $\xi(z)$, we need some notations. We set

$$R_n^0 = -\frac{t}{2} + N^{-1/3}(-1)^n c_1 u(y) + N^{-2/3} c_2 v(y)\,, \tag{4.322}$$

as an approximation to R_n, and

$$A_n^0(z) = \begin{pmatrix} -(tz/2 + z^3/2 + zR_n^0) & (R_n^0)^{1/2}(z^2 + \theta_n^0) \\ -(R_n^0)^{1/2}(z^2 + \theta_{n-1}^0) & tz/2 + z^3/2 + zR_n^0 \end{pmatrix}, \quad \theta_n^0 = t + R_n^0 + R_{n+1}^0\,, \tag{4.323}$$

as an approximation to $A_n(z)$. We set

$$\begin{aligned} a_n^0(z) = -\frac{tz^4}{2} - \frac{z^6}{4} + \left(\frac{n}{N} - \lambda_{\rm c}\right) z^2 + N^{-4/3}(-t)^{1/3} 2^{-5/3}[v(y)^2 - 4w(y)^2] \\ - N^{-5/3}(-1)^n(-2t)^{-1/3} w(y)]\,, \quad w(y) = u'(y)\,, \end{aligned} \tag{4.324}$$

as an approximation to $\det A_n(z)$. Finally, we set

$$U^0 = -a_n^0(z) + N^{-1}\left[(a_{11}^0)' - a_{11}^0 \frac{(a_{12}^0)'}{a_{12}^0}\right], \tag{4.325}$$

as an approximation to the potential U in (4.314). With these notations,

$$\xi(z) = \int_{z_N}^{z} \mu(u)\,\mathrm{d}u\,, \quad \mu(z) = \sqrt{U^0(z)}\,, \tag{4.326}$$

where z_N is the zero of $U^0(z)$ which approaches z_0 as $N \to \infty$. Also,

$$T(z) = \left(\frac{a_{12}^0(z)}{\mu(z)}\right)^{1/2} \begin{pmatrix} 1 & 0 \\ -\frac{a_{11}^0(z)}{a_{12}^0(z)} & \frac{\mu(z)}{a_{12}^0(z)} \end{pmatrix}, \quad \det T(z) = 1\,. \tag{4.327}$$

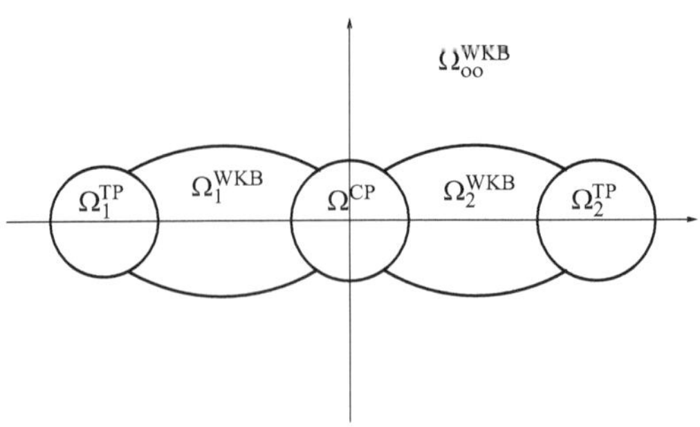

Fig. 4.8. The partition of the complex plane

From (4.321) we obtain the following asymptotics as $z \to \infty$:

$$\Psi^{\mathrm{WKB}}(z) = \sqrt{2}\, C_0 \left(I + z^{-1} (R_n^0)^{1/2} \begin{pmatrix} 0 & 1 \\ 1 & 0 \end{pmatrix} + O(z^{-2}) \right) \\ \times \exp\left(-\left(\frac{NV(z)}{2} - n \ln z + \lambda_n^0 + C_1 \right) \sigma_3 \right), \quad z \to \infty, \qquad (4.328)$$

where

$$\lambda_n^0 = \lim_{z\to\infty} \left[N\xi(z) - \left(\frac{NV(z)}{2} - n \ln z \right) \right]. \qquad (4.329)$$

The existence of the latter limit follows from (4.326).

In the domains $\Omega_{1,2}^{\mathrm{WKB}}$ we define

$$\Psi^{\mathrm{WKB}}(z) = \Psi_{\mathrm{a}}^{\mathrm{WKB}}(z) S_{\pm}, \quad \pm \operatorname{Im} z \geq 0, \qquad (4.330)$$

where $\Psi_{\mathrm{a}}^{\mathrm{WKB}}(z)$ is the analytic continuation of $\Psi^{\mathrm{WKB}}(z)$ from $\Omega_{\infty}^{\mathrm{WKB}}$ to $\Omega_{1,2}^{\mathrm{WKB}}$, from the upper half-plane, and

$$S_+ = \begin{pmatrix} 1 & 0 \\ \mathrm{i} & 1 \end{pmatrix} = S_- \begin{pmatrix} 1 & -\mathrm{i} \\ 0 & 1 \end{pmatrix}, \quad S_- = \begin{pmatrix} 1 & \mathrm{i} \\ \mathrm{i} & 0 \end{pmatrix}. \qquad (4.331)$$

Observe that $\Psi^{\mathrm{WKB}}(z)$ has jumps:

$$\Psi_+^{\mathrm{WKB}}(z) = \left(I + O(\mathrm{e}^{-cN}) \right) \Psi_-^{\mathrm{WKB}}(z), \quad z \in \partial\Omega_\infty^{\mathrm{WKB}} \cap (\partial\Omega_1^{\mathrm{WKB}} \cup \partial\Omega_2^{\mathrm{WKB}}), \qquad (4.332)$$

and

$$\Psi_+^{\mathrm{WKB}}(z) = \Psi_-^{\mathrm{WKB}}(z) \begin{pmatrix} 1 & -\mathrm{i} \\ 0 & 1 \end{pmatrix}, \quad z \in \mathbb{R} \cap (\Omega_1^{\mathrm{WKB}} \cup \Omega_2^{\mathrm{WKB}}). \qquad (4.333)$$

4.3.3 Construction of the Parametrix near the Turning Points

In Ω_2^{TP} we define the parametrix with the help of the Airy matrix-valued functions,

$$Y_{1,2}(z) = \begin{pmatrix} y_0(z) & y_{1,2}(z) \\ y_0'(z) & y_{1,2}'(z) \end{pmatrix}, \qquad (4.334)$$

where

$$y_0(z) = \mathrm{Ai}(z), \quad y_1(z) = \mathrm{e}^{-\pi\mathrm{i}/6} \,\mathrm{Ai}(\mathrm{e}^{-2\pi I/3} z), \\ y_2(z) = \mathrm{e}^{\pi\mathrm{i}/6} \,\mathrm{Ai}(\mathrm{e}^{2\pi\mathrm{i}/3} z). \qquad (4.335)$$

Let us remind that $\mathrm{Ai}(z)$ is a solution to the Airy equation $y'' = zy$, which has the following asymptotics as $z \to \infty$:

$$\mathrm{Ai}(z) = \frac{1}{2\sqrt{\pi} z^{1/4}} \exp\left(-\frac{2z^{3/2}}{3} + O(|z|^{-3/2}) \right), \\ -\pi + \varepsilon \leq \arg z \leq \pi - \varepsilon. \qquad (4.336)$$

The functions $y_j(z)$ satisfy the relation

$$y_1(z) - y_2(z) = -\mathrm{i}y_0(z) \,. \tag{4.337}$$

We define the parametrix in Ω_2^{TP} by the formula,

$$\Psi^{\mathrm{TP}}(z) = W(z)N^{\sigma_3/6}Y_{1,2}\big(N^{2/3}w(z)\big)\,, \quad \pm \operatorname{Im} z \geq 0\,, \tag{4.338}$$

where

$$w(z) = \left[\tfrac{3}{2}\xi(z)\right]^{2/3}, \tag{4.339}$$

with $\xi(z)$ defined in (4.326) above. Observe that $w(z)$ is analytic in Ω_2^{TP}. The matrix-valued function $W(z)$ is also analytic in Ω_2^{TP}, and it is found from the following condition of the matching $\Psi^{\mathrm{TP}}(z)$ to $\Psi^{\mathrm{WKB}}(z)$ on $\partial\Omega_2^{\mathrm{TP}}$:

$$\Psi^{\mathrm{TP}}(z) = \big(I + O(N^{-1})\big)\Psi^{\mathrm{WKB}}(z)\,, \quad z \in \partial\Omega_2^{\mathrm{TP}}\,, \tag{4.340}$$

see [17, 18]. A similar construction of the parametrix is used in the domain Ω_1^{TP}.

4.3.4 Construction of the Parametrix near the Critical Point

4.3.4.1 Model Solution

The crucial question is, what should be an Ansatz for the parametrix in the critical domain Ω^{CP}? To answer this question, let us construct a normal form of system of differential equations (4.310) at the origin. If we substitute Ansatz (4.302) into the matrix elements of $A_n(z)$, change

$$\Psi(z) = \Phi(CN^{1/3}z)\,, \quad C = \frac{(2t)^{1/6}}{2}\,, \tag{4.341}$$

and keep the leading terms, as $N \to \infty$, then we obtain the model equation (normal form),

$$\Phi'(s) = A(s)\Phi(s)\,, \tag{4.342}$$

where

$$A(s) = \begin{pmatrix} (-1)^n 4u(y)s & 4s^2 + (-1)^n 2w(y) + v(y) \\ -4s^2 + (-1)^n 2w(y) - v(y)) & -(-1)^n 4u(y)s \end{pmatrix}, \tag{4.343}$$

and $w(y) = u'(y)$. In fact, this is one of the equations of the Lax pair for the Hastings–Mcleod solution to Painlevé II. Equation (4.342) possesses three special solutions, $\boldsymbol{\Phi}_j$, $j = 0, 1, 2$, which are characterized by their asymptotics as $|s| \to \infty$:

$$\boldsymbol{\Phi}_0(s) = \begin{pmatrix} \Phi^1(s) \\ \Phi^2(s) \end{pmatrix} \sim \begin{pmatrix} \cos\left(\frac{4}{3}s^3 + ys - \frac{\pi n}{2}\right) \\ -\sin\left(\frac{4}{3}s^3 + ys - \frac{\pi n}{2}\right) \end{pmatrix} ,$$

$$|\arg s| \leq \frac{\pi}{3} - \varepsilon , \tag{4.344}$$

$$\boldsymbol{\Phi}_1(s) = \overline{\boldsymbol{\Phi}_2(\bar{s})} \sim \begin{pmatrix} (-\mathrm{i})^{n+1} \\ (-\mathrm{i})^n \end{pmatrix} \exp\bigl(\mathrm{i}(\tfrac{4}{3}\,s^3 + ys)\bigr) ,$$

$$-\frac{\pi}{3} + \varepsilon < \arg s < \frac{4\pi}{3} - \varepsilon .$$

The functions $\Phi^{1,2}(s)$ are real for real s and

$$\Phi^1(-s) = (-1)^n \Phi^1(s) , \quad \Phi^2(-s) = -(-1)^n \Phi^2(s) . \tag{4.345}$$

We define the 2×2 matrix-valued function on $\mathbb{C}$,

$$\Phi(s) = \bigl(\boldsymbol{\Phi}_0(s), \boldsymbol{\Phi}_{1,2}(s)\bigr) , \quad \pm \operatorname{Im} s \geq 0 , \tag{4.346}$$

which is two-valued on $\mathbb{R}$, and

$$\Phi_+(s) = \Phi_-(s) \begin{pmatrix} 1 & -\mathrm{i} \\ 0 & 1 \end{pmatrix} , \quad s \in \mathbb{R} . \tag{4.347}$$

The Ansatz for the parametrix in the critical domain is

$$\Psi_n^{\mathrm{CP}}(z) = \widetilde{C}_0 V(z) \Phi\bigl(N^{1/3}\zeta(z)\bigr) , \tag{4.348}$$

where $\widetilde{C}_0$ is a constant, a parameter of the solution, $\zeta(z)$ is an analytic scalar function such that $\zeta'(z) \neq 0$ and $V(z)$ is an analytic matrix-valued function. We now describe a procedure of choosing $\zeta(z)$ and $V(z)$. The essence of the RH approach is that we don't need to justify this procedure. We need only that $\zeta(z)$ and $V(z)$ are analytic in Ω^{CP}, and that on $\partial\Omega^{\mathrm{CP}}$, Ansatz (4.348) fits to the WKB Ansatz.

4.3.4.2 Construction of $\zeta(z)$. Step 1

To find $\zeta(z)$ and $V(z)$, let us substitute Ansatz (4.348) into equation (4.54). This gives

$$V(z)\bigl[\zeta'(z) N^{-2/3} A\bigl(N^{1/3}\zeta(z)\bigr)\bigr] V^{-1}(z) = A_n(z) - N^{-1} V'(z) V^{-1}(z) . \tag{4.349}$$

Let us drop the term of the order of N^{-1} on the right:

$$V(z)\bigl[\zeta'(z) N^{-2/3} A\bigl(N^{1/3}\zeta(z)\bigr)\bigr] V^{-1}(z) = A_n(z) , \tag{4.350}$$

and take the determinant of the both sides. This gives an equation on ζ only,

$$[\zeta'(z)]^2 f(\zeta(z)) = a_n(z)\,, \tag{4.351}$$

where

$$\begin{aligned} f(\zeta) &= N^{-2/3}\det A(N^{1/3}\zeta) \\ &= 16\zeta^4 + 8N^{-2/3}y\zeta^2 + N^{-4/3}[v^2(y) - 4w^2(y)] \end{aligned} \tag{4.352}$$

and

$$a_n(z) = \det A_n(z) = -\frac{tz^4}{2} - \frac{z^6}{4} + \left(\frac{n}{N} - \lambda_{\rm c}\right)z^2 + R_n\theta_n\theta_{n+1}\,, \tag{4.353}$$

where

$$\theta_n = t + R_n + R_{n+1}\,. \tag{4.354}$$

Equation (4.302) implies that

$$\theta_n = 2c_2N^{-2/3}v(y) + O(N^{-1})\,. \tag{4.355}$$

At this stage we drop all terms of the order of N^{-1}, and, therefore, we can simplify f and a_n to

$$f(\zeta) = 16\zeta^4 + 8N^{-2/3}y\zeta^2 \tag{4.356}$$

and

$$a_n(z) = -\frac{tz^4}{2} - \frac{z^6}{4} + \left(\frac{n}{N} - \lambda_{\rm c}\right)z^2\,. \tag{4.357}$$

Equation (4.351) is separable and we are looking for an analytic solution. To construct an analytic solution, let us make the change of variables,

$$z = CN^{-1/3}s\,, \quad \zeta = N^{-1/3}\sigma\,. \tag{4.358}$$

Then equation (4.351) becomes

$$[\sigma'(s)]^2 f_0(\sigma(s)) = a_0(s)\,, \tag{4.359}$$

where

$$\begin{aligned} f_0(\sigma) &= 16\sigma^4 + 8y\sigma^2\,, \\ a_0(s) &= 16s^4 + 8c_0^{-1}N^{2/3}\left(\frac{n}{N} - \lambda_{\rm c}\right)s^2 - N^{-2/3}cs^6\,. \end{aligned} \tag{4.360}$$

When we substitute (4.303) for y, we obtain that

$$a_0(s) = 16s^4 + 8ys^2 - N^{-2/3}cs^6\,. \tag{4.361}$$

When $y = 0$, (4.359) is easy to solve: by taking the square root of the both sides, we obtain that

$$\sigma^2\sigma' = s^2\left(1 - \frac{1}{16}N^{-2/3}cs^2\right)^{1/2}\,, \tag{4.362}$$

hence

$$\sigma(s) = \left[\int_0^s t^2 \left(1 - \frac{1}{16} N^{-2/3} ct^2 \right)^{1/2} \mathrm{d}t \right]^{1/3} \tag{4.363}$$

is an analytic solution to (4.359) in the disk $|s| \leq \varepsilon N^{1/3}$, for some $\varepsilon > 0$. This gives an analytic solution $\zeta(z) = N^{-1/3} \sigma(C^{-1} N^{1/3} z)$ to (4.353) in the disk $|z| \leq C\varepsilon$.

When $y \neq 0$, the situation is more complicated, and in fact, (4.359) has no analytic solution in the disk $|s| \leq \varepsilon N^{1/3}$. Consider, for instance, $y > 0$. By taking the square root of the both sides of (4.359), we obtain that

$$\sigma \left(\sigma^2 + \frac{y}{2} \right)^{1/2} \sigma' = s \left(s^2 + \frac{y}{2} - \frac{1}{16} N^{-2/3} cs^4 \right)^{1/2} , \tag{4.364}$$

The left-hand side has simple zeros at $\pm\sigma_0 = \pm \mathrm{i}\sqrt{y/2}$, and the right-hand side has simple zeros at $\pm s_0$, where $s_0 = \sigma_0 + O(N^{-2/3})$. The necessary and sufficient condition for the existence of an analytic solution to (4.359) in the disk $|s| \leq \varepsilon N^{1/3}$ is the equality of the periods,

$$\begin{aligned} P_1 &\equiv \int_{-\sigma_0}^{\sigma_0} \sigma \left(\sigma^2 + \frac{y}{2} \right)^{1/2} \mathrm{d}\sigma \\ &= P_2 \equiv \int_{-s_0}^{s_0} s \left(s^2 + \frac{y}{2} - \frac{1}{16} N^{-2/3} cs^4 \right)^{1/2} \mathrm{d}s , \end{aligned} \tag{4.365}$$

and, in fact, $P_1 \neq P_2$. To make the periods equal, we slightly change (4.303) as follows:

$$y = c_0^{-1} N^{2/3} \left(\frac{n}{N} - \lambda_c \right) + \alpha , \tag{4.366}$$

where α is a parameter. Then

$$P_2 = P_2(\alpha) \equiv \int_{-s_0(\alpha)}^{s_0(\alpha)} s \left(s^2 + \frac{y - \alpha}{2} - \frac{1}{16} N^{-2/3} cs^4 \right)^{1/2} \mathrm{d}s . \tag{4.367}$$

It is easy to check that $P_2'(0) \neq 0$, and therefore, there exists an $\alpha = O(N^{-2/3})$ such that $P_1 = P_2$. This gives an analytic solution $\sigma(s)$, and hence an analytic $\zeta(z)$.

4.3.4.3 Construction of $V(z)$

Next, we find a matrix-valued function $V(z)$ from (4.350). Both $V(z)$ and $V^{-1}(z)$ should be analytic at the origin. We have the following lemma.

Lemma 4.3.1. *Let $B = (b_{ij})$ and $D = (d_{ij})$ be two 2×2 matrices such that*

$$\operatorname{Tr} B = \operatorname{Tr} D = 0 , \quad \det B = \det D . \tag{4.368}$$

Then the equation $VB = DV$ has the following two explicit solutions:

$$V_1 = \begin{pmatrix} d_{12} & 0 \\ b_{11} - d_{11} & b_{12} \end{pmatrix}, \quad V_2 = \begin{pmatrix} b_{21} & d_{11} - b_{11} \\ 0 & d_{21} \end{pmatrix}. \tag{4.369}$$

We would like to apply Lemma 4.3.1 to

$$B = \zeta'(z)N^{-2/3}A\big(N^{1/3}\zeta(z)\big), \quad D = A_n(z). \tag{4.370}$$

The problem is that we need an analytic matrix valued function $V(z)$ which is invertible in a fixed neighborhood of the origin, but neither V_1 nor V_2 are invertible there. Nevertheless, we can find a linear combination of V_1 and V_2 (plus some negligibly small terms) which is analytic and invertible. Namely, we take

$$V(z) = \frac{1}{\sqrt{\det W(z)}} W(z) \tag{4.371}$$

where

$$W(z) = \begin{pmatrix} d_{12}(z) - b_{21}(z) - \alpha_{11} & b_{11}(z) - d_{11}(z) - \alpha_{12}z \\ b_{11}(z) - d_{11}(z) - \alpha_{21}z & b_{12}(z) - d_{21}(z) - \alpha_{22} \end{pmatrix}, \tag{4.372}$$

and the numbers $\alpha_{ij} = O(N^{-1})$ are chosen in such a way that the matrix elements of W vanish at the same points $\pm z_0$, $z_0 = O(N^{-1/3})$, on the complex plane. Then $V(z)$ is analytic in a disk $|z| < \varepsilon$, $\varepsilon > 0$.

4.3.4.4 Construction of $\zeta(z)$. Step 2

The accuracy of $\zeta(z)$, which is obtained from (4.351), is not sufficient for the required fit on $|z| = \varepsilon$, of Ansatz (4.348) to the WKB Ansatz. Therefore, we correct $\zeta(z)$ slightly by taking into account the term $-N^{-1}V'(z)V^{-1}(z)$ in (4.349). We have to solve the equation,

$$[\zeta'(z)]^2 N^{-4/3} \det A\big(N^{1/3}\zeta(z)\big) = \det[A_n(z) - N^{-1}V'(z)V^{-1}(z)]. \tag{4.373}$$

By change of variables (4.358), it reduces to

$$[\sigma'(s)]^2 f_1\big(\sigma(s)\big) = a_1(s), \tag{4.374}$$

where

$$\begin{aligned} f_1(\sigma) &= 16\sigma^4 + 8y\sigma^2 + [v^2(y) - 4w^2(y)]; \\ a_1(s) &= 16s^4 + 8(y-\alpha)s^2 + [v^2(y) - 4w^2(y)] + r_N(s), \\ r_N(s) &= O(N^{-2/3}). \end{aligned} \tag{4.375}$$

The function $f_1(\sigma)$ has 4 zeros, $\pm\sigma_j$ $j = 1, 2$. The function $a_1(s)$ is a small perturbation of $f_1(s)$, and it has 4 zeros, $\pm s_j$, such that $|s_j - \sigma_j| \to 0$ as

$N \to \infty$. Equation (4.374) has an analytic solution in the disk of radius $\varepsilon N^{2/3}$ if and only if the periods are equal,

$$P_{1j} \equiv \int_{-\sigma_j}^{\sigma_0} \sqrt{f_1(\sigma)}\,\mathrm{d}\sigma = P_{2j} \equiv \int_{-s_j}^{s_j} \sqrt{a_1(s)}\,\mathrm{d}s\,, \quad j = 1,2\,. \tag{4.376}$$

To secure the equality of periods we include $a_1(s)$ into the 2-parameter family of functions,

$$a_1(s) = 16s^4 + 8(y-\alpha)s^2 + [v^2(y) - 4w^2(y)] + r_N(s) + \beta\,, \tag{4.377}$$

where $-\infty < \alpha, \beta < \infty$ are the parameters. A direct calculation gives that

$$\det \begin{pmatrix} \dfrac{\partial P_{21}}{\partial \alpha} & \dfrac{\partial P_{21}}{\partial \beta} \\ \dfrac{\partial P_{22}}{\partial \alpha} & \dfrac{\partial P_{22}}{\partial \beta} \end{pmatrix} \neq 0\,, \tag{4.378}$$

see [18], hence, by the implicit function theorem, there exist $\alpha, \beta = O(N^{-2/3})$, which solve equations of periods (4.376). This gives an analytic function $\zeta(z)$, and hence, by (4.348), an analytic $\Psi^{\mathrm{CP}}(z)$.

The function $\Psi^{\mathrm{CP}}(z)$ matches the WKB-parametrix $\Psi^{\mathrm{WKB}}(z)$ on the boundary of Ω^{CP}. Namely, we have the following lemma.

Lemma 4.3.2 (See [18]). *If we take $\widetilde{C}_0 = C_0$ and $C_1 = -\frac{1}{4}\ln R_n^0$ then*

$$\Psi^{\mathrm{CP}}(z) = (I + O(N^{-1}))\Psi^{\mathrm{WKB}}(z)\,, \quad z \in \partial\Omega^{\mathrm{CP}}\,. \tag{4.379}$$

We omit the proof of the lemma, because it is rather straightforward, although technical. We refer the reader to the paper [18] for the details.

4.3.4.5 Proof of the Double Scaling Asymptotics

Let us summarize the construction of the parametrix in different domains. We define the parametrix Ψ_n^0 as

$$\Psi_n^0(z) = \begin{cases} \Psi^{\mathrm{WKB}}(z)\,, & z \in \Omega^{\mathrm{WKB}} = \Omega_\infty^{\mathrm{WKB}} \cup \Omega_1^{\mathrm{WKB}} \cup \Omega_2^{\mathrm{WKB}}\,, \\ \Psi^{\mathrm{TP}}(z)\,, & z \in \Omega_1^{\mathrm{TP}} \cup \Omega_2^{\mathrm{TP}}\,, \\ \Psi^{\mathrm{CP}}(z)\,, & z \in \Omega^{\mathrm{CP}}\,, \end{cases} \tag{4.380}$$

where $\Psi^{\mathrm{WKB}}(z)$ is given by (4.321), (4.330), $\Psi^{\mathrm{TP}}(z)$ by (4.338), and Ψ^{CP} by (4.348). Consider the quotient,

$$X(z) = \Psi_n(z)[\Psi_n^0(z)]^{-1}\,. \tag{4.381}$$

$X(z)$ has jumps on the contour Γ, depicted on Fig. 4.9, such that

$$X_+(z) = X_-(z)[I + O(N^{-1}(1+|z|)^{-2})]\,, \quad z \in \Gamma\,. \tag{4.382}$$

From (4.315) and (4.328) we obtain that

$$X(z) = X_0 + \frac{X_1}{z} + O(z^{-2})\,, \quad z \to \infty\,, \tag{4.383}$$

where

$$X_0 = \frac{1}{\sqrt{2}} C_0^{-1} \Gamma_0 \exp\big((C_1 + \lambda_n^0 - \lambda_n)\sigma_3\big) \tag{4.384}$$

and

$$X_1 = \frac{1}{\sqrt{2}} C_0^{-1} \big[\Gamma_1 \exp\big((C_1+\lambda_n^0-\lambda_n)\sigma_3\big) - \Gamma_0 \exp\big((C_1+\lambda_n^0-\lambda_n)\sigma_3\big)(R_n^0)^{1/2}\sigma_1\big]\,,$$
$$\sigma_1 = \begin{pmatrix} 0 & 1 \\ 1 & 0 \end{pmatrix}. \tag{4.385}$$

The RHP shares a remarkable property of well-posedness, see, e.g., [9, 42, 84, 102]. Namely, (4.382), (4.383) imply that

$$X_0^{-1} X(z) = I + O(N^{-1}(1+|z|)^{-1})\,, \quad z \in \mathbb{C}\,. \tag{4.386}$$

This in turn implies that

$$X_0^{-1} X_1 = O(N^{-1})\,, \tag{4.387}$$

or, equivalently,

$$\exp(-(C_1 + \lambda_n^0 - \lambda_n)\sigma_3) \Gamma_0^{-1} \Gamma_1 \exp\big((C_1 + \lambda_n^0 - \lambda_n)\sigma_3\big) - (R_n^0)^{1/2}\sigma_1 = O(N^{-1})\,. \tag{4.388}$$

By (4.316),

$$\exp(-(C_1 + \lambda_n^0 - \lambda_n)\sigma_3) \Gamma_0^{-1} \Gamma_1 \exp\big((C_1 + \lambda_n^0 - \lambda_n)\sigma_3\big) = \begin{pmatrix} 0 & \exp\big(-2(C_1 + \lambda_n^0 - \lambda_n)\big) \\ R_n \exp\big(2(C_1 + \lambda_n^0 - \lambda_n)\big) & 0 \end{pmatrix}, \tag{4.389}$$

hence (4.388) reduces to the system of equations,

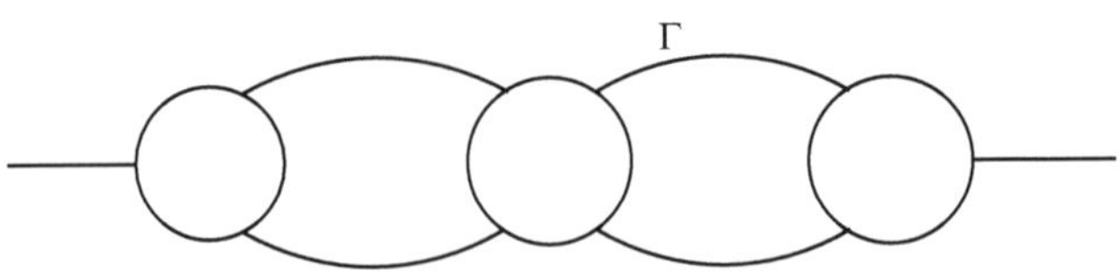

Fig. 4.9. The contour Γ

$$\begin{cases} \exp(-2(C_1+\lambda_n^0-\lambda_n)) = (R_n^0)^{1/2} + O(N^{-1}) \,, \\ R_n \exp(2(C_1+\lambda_n^0-\lambda_n)) = (R_n^0)^{1/2} + O(N^{-1}) \,. \end{cases} \tag{4.390}$$

By multiplying these equations, we obtain that

$$R_n = R_n^0 + O(N^{-1}) \,. \tag{4.391}$$

This proves Ansatz (4.302). Since $C_1 = -\frac{1}{4}\ln R_n^0$, we obtain from (4.391) that

$$\exp(2(\lambda_n - \lambda_n^0)) = 1 + O(N^{-1}) \,, \tag{4.392}$$

or equivalently,

$$h_n = \exp\left(2N \int_{z_N}^{\infty} \mu(u)\, du\right)(1 + O(N^{-1})) \,, \tag{4.393}$$

where

$$\int_{z_N}^{\infty} \mu(u)\, du \equiv \lim_{z\to\infty} \left[\int_{z_N}^{z} \mu(u)\, du - \left(\frac{V(z)}{2} - \frac{n \ln z}{N}\right)\right] . \tag{4.394}$$

Thus, we have the following theorem.

Theorem 4.3.1 (See [18]). *The recurrent coefficient R_n under the scaling* (4.301) *has asymptotics* (4.302). *The normalizing coefficient h_n has asymptotics* (4.393).

Equations (4.381) and (4.386) imply that

$$\Psi_n(z) = X_0\big[I + O\big(N^{-1}(1+|z|)^{-1}\big)\big]\Psi_n^0(z) \,, \quad z \in \mathbb{C} \,. \tag{4.395}$$

The number C_0 is a free parameter. Let us take $C_0 = 1$. From (4.384) and (4.391) we obtain that

$$X_0 = \frac{(R_n^0)^{1/4}}{\sqrt{2}}(1 + O(N^{-1})) \,, \tag{4.396}$$

hence

$$\Psi_n(z) = \frac{(R_n^0)^{1/4}}{\sqrt{2}}\Psi_n^0(z)[I + O(N^{-1})] \,, \quad z \in \mathbb{C} \,. \tag{4.397}$$

This gives the large N asymptotics of $\Psi_n(z)$ under scaling (4.301), as well as the asymptotics of the correlation functions. In particular, the asymptotics near the origin is described as follows.

Theorem 4.3.2 (See [18]). *Let $\boldsymbol{\Phi}_0(z;y) = \begin{pmatrix} \Phi^1(z;y) \\ \Phi^2(z;y) \end{pmatrix}$ be the solution for $n = 0$ to system* (4.342), *with the asymptotics at infinity as in* (4.344). *Then the following double scaling limit holds*:

$$\lim_{N\to\infty}\frac{1}{(cN^{1/3})^{m-1}}K_{Nm}\left(\frac{u_1}{cN^{1/3}},\dots,\frac{u_m}{cN^{1/3}};t_c+c_0yN^{-2/3}\right)$$
$$=\det\big(Q_c(u_i,u_j;y)\big)_{i,j=1,\dots,m}, \quad (4.398)$$

where $c=\zeta'(0)>0$, *and*

$$Q_c(u,v;y)=\frac{\Phi^1(u;y)\Phi^2(v;y)-\Phi^1(v;y)\Phi^2(u;y)}{\pi(u-v)}. \quad (4.399)$$

Let us mention here some further developments of Thms. 4.3.1, 4.3.2. They are extended to a general interaction $V(M)$ in the paper [40] of Claeys and Kuijlaars. The double scaling limit of the random matrix ensemble of the form $Z_N^{-1}|\det M|^{2\alpha}\mathrm{e}^{-N\operatorname{Tr}V(M)}\,\mathrm{d}M$, where $\alpha>-\frac{1}{2}$, is considered in the papers [41] of Claeys, Kuijlaars, and Vahlessen, and [75] of Its, Kuijlaars, and Östensson. In this case the double scaling limit is described in terms of a critical solution to the general Painlevé II equation $q''=sq+2q^3-\alpha$. The papers, [40,41,75] use the RH approach and the Deift–Zhou steepest descent method, discussed in Sect. 4.2 above. The double scaling limit of higher-order and Painlevé II hierachies is studied in the papers [15,92], and others. There are many physical papers on the double scaling limit related to the Painlevé I equation, see e.g., [34,56,59,70,71,101], and others. A rigorous RH approach to the Painlevé I double scaling limit is initiated in the paper [68] of Fokas, Its, and Kitaev. It is continued in the recent paper of Duits and Kuijlaars [61], who develop the RH approach and the Deift–Zhou steepest descent method to orthogonal polynomials on contours in complex plane with the exponential quartic weight, $\exp\big(-N(z^2/2+tz^4/4)\big)$, where $t<0$. Their results cover both the one-cut case $-1/12<t<0$ and the Painlevé I double scaling limit at $t=-1/12$.

4.4 Large N Asymptotics of the Partition Function of Random Matrix Models

4.4.1 Partition Function

The central object of our analysis is the partition function of a random matrix model,

$$Z_N=\int_{-\infty}^{\infty}\cdots\int_{-\infty}^{\infty}\prod_{1\le j<k\le N}(z_j-z_k)^2\exp\left(-N\sum_{j=1}^{N}V(z_j)\right)\mathrm{d}z_1\cdots\mathrm{d}z_N$$
$$=N!\prod_{n=0}^{N-1}h_n, \quad (4.400)$$

where $V(z)$ is a polynomial,

$$V(z) = \sum_{j=1}^{2d} v_j z^j , \quad v_{2d} > 0 , \tag{4.401}$$

and h_n are the normalization constants of the orthogonal polynomials on the line with respect to the weight $e^{-NV(z)}$,

$$\int_{-\infty}^{\infty} P_n(z)P_m(z)e^{-NV(z)}\,dz = h_n\delta_{nm} , \tag{4.402}$$

where $P_n(z) = z^n + \cdots$. We will be interested in the asymptotic expansion of the free energy,

$$F_N = -\frac{1}{N^2}\ln Z_N, \tag{4.403}$$

as $N \to \infty$.

Our approach will be based on the deformation τ_t of $V(z)$ to z^2,

$$\tau_t\colon V(z) \to V(z;t) = (1-t^{-1})z^2 + V(t^{-1/2}z) , \tag{4.404}$$

$1 \le t < \infty$, so that

$$\tau_1 V(z) = V(z) , \quad \tau_\infty V(z) = z^2 . \tag{4.405}$$

Observe that

$$\tau_t\tau_s = \tau_{ts} . \tag{4.406}$$

We start with the following proposition, which describe the deformation equations for h_n and the recurrent coefficients of orthogonal polynomials under the change of the coefficients of $V(z)$.

Proposition 4.4.1. *We have that*

$$\begin{aligned}
\frac{1}{N}\frac{\partial \ln h_n}{\partial v_k} &= -[Q^k]_{nn} , \\
\frac{1}{N}\frac{\partial \gamma_n}{\partial v_k} &= \frac{\gamma_n}{2}([Q^k]_{n-1,n-1} - [Q^k]_{nn}) , \\
\frac{1}{N}\frac{\partial \beta_n}{\partial v_k} &= \gamma_n[Q^k]_{n,n-1} - \gamma_{n+1}[Q^k]_{n+1,n} ,
\end{aligned} \tag{4.407}$$

where Q is the Jacobi matrix defined in (4.33).

The proof of Prop. 4.4.1 is given in [19]. It uses some results of the works of Eynard and Bertola, Eynard, Harnad [10]. For even V, it was proved earlier by Fokas, Its, Kitaev [68].

We will be especially interested in the derivatives with respect to v_2. For $k = 2$, Prop. 4.4.1 gives that

$$\begin{aligned}\frac{1}{N}\frac{\partial \ln h_n}{\partial v_2} &= -\gamma_n^2 - \beta_n^2 - \gamma_{n+1}^2\,,\\ \frac{1}{N}\frac{\partial \gamma_n}{\partial v_2} &= \frac{\gamma_n}{2}(\gamma_{n-1}^2 + \beta_{n-1}^2 - \gamma_{n+1}^2 - \beta_n^2)\,,\\ \frac{1}{N}\frac{\partial \beta_n}{\partial v_2} &= \gamma_n^2\beta_{n-1} + \gamma_n^2\beta_n - \gamma_{n+1}^2\beta_n - \gamma_{n+1}^2\beta_{n+1}\,.\end{aligned} \tag{4.408}$$

Next we describe the v_2-deformation of Z_N.

Proposition 4.4.2 (See [19]). *We have the following relation:*

$$\frac{1}{N^2}\frac{\partial^2 \ln Z_N}{\partial v_2^2} = \gamma_N^2(\gamma_{N-1}^2 + \gamma_{N+1}^2 + \beta_N^2 + 2\beta_N\beta_{N-1} + \beta_{N-1}^2)\,. \tag{4.409}$$

Observe that the equation is local in N. For $V(z) = v_2z^2 + v_4z^4$, Prop. 4.4.2 was obtained earlier by Its, Fokas, Kitaev [74]. Proposition 4.4.2 can be applied to deformation (4.404). Let $\gamma_n(\tau)$, $\beta_n(\tau)$ be the recurrence coefficients for orthogonal polynomials with respect to the weight $\mathrm{e}^{-NV(z;\tau)}$, and let

$$Z_N^{\mathrm{Gauss}} = \int_{-\infty}^{\infty}\cdots\int_{-\infty}^{\infty}\prod_{1\le j<k\le N}(z_j - z_k)^2\exp\left(-N\sum_{j=1}^N z_j^2\right)\mathrm{d}z_1\cdots\mathrm{d}z_N \tag{4.410}$$

be the partition function for the Gaussian ensemble. It can be evaluated explicitly, and the corresponding free energy has the form,

$$F_N^{\mathrm{Gauss}} = -\frac{1}{N^2}\ln\left(\frac{(2\pi)^{N/2}}{(2N)^{N^2/2}}\prod_{n=1}^N n!\right). \tag{4.411}$$

By integrating twice formula (4.409), we obtain the following result:

Proposition 4.4.3.

$$\begin{aligned}F_N(t) = F_N^{\mathrm{Gauss}} + \int_t^\infty \frac{t-\tau}{\tau^2}\{&\gamma_N^2(\tau)[\gamma_{N-1}^2(\tau) + \gamma_{N+1}^2(\tau) + \beta_N^2(\tau)\\ &+ 2\beta_N(\tau)\beta_{N-1}(\tau) + \beta_{N-1}^2(\tau)] - \tfrac{1}{2}\}\,\mathrm{d}\tau\,.\end{aligned} \tag{4.412}$$

4.4.2 Analyticity of the Free Energy for Regular V

The basic question of statistical physics is the existence of the free energy in the thermodynamic limit and the analyticity of the limiting free energy with respect to various parameters. The values of the parameters at which the free energy is not analytic are the critical points. When applied to the "gas" of eigenvalues, this question refers to the existence of the limit,

$$F = \lim_{N\to\infty} F_N\,, \tag{4.413}$$

and to the analyticity of F with respect to the coefficients v_j of the polynomial V. The existence of limit (4.413) is proven under general conditions on V, not only polynomials, see the work of Johansson [78] and references therein. In fact, F is the energy E_V of minimization problem (4.97), (4.98), so that

$$F = I_V(\nu_V) , \tag{4.414}$$

where ν_V is the equilibrium measure. The following theorem establishes the analyticity of F for regular V. We call V regular, if the corresponding equilibrium measure ν_V is regular, as defined in (4.107), (4.108). We call V, q-cut regular, it the measure ν_V is regular and its support consists of q intervals.

Theorem 4.4.1 (See [19]). *Suppose that $V(z)$ is a q-cut regular polynomial of degree $2d$. Then for any $p \le 2d$, there exists $t_1 > 0$ such that for any $t \in [-t_1, t_1]$,*

(1) *the polynomial $V(z) + tz^p$ is q-cut regular.*
(2) *The end-points of the support intervals, $[a_i(t), b_i(t)]$, $i = 1, \dots, q$, of the equilibrium measure for $[V(z) + tz^p]$ are analytic in t.*
(3) *The free energy $F(t)$ is analytic in t.*

Proof. Consider for $j = 0, \dots, q$, the quantities

$$T_j(a, b; t) = \frac{1}{2\pi \mathrm{i}} \oint_\Gamma \frac{V'(z;t) z^j}{\sqrt{R(z)}} \, \mathrm{d}z \,, \quad R(z) = \prod_{i=1}^{q} (z - a_i)(z - b_i) \,, \tag{4.415}$$

where Γ is a contour around $[a_1, b_q]$. Consider also for $k = 1, \dots, q-1$, the quantities

$$N_k(a, b; t) = \frac{1}{2\pi \mathrm{i}} \oint_{\Gamma_k} h(z;t) \sqrt{R(z)} \, \mathrm{d}z \,, \tag{4.416}$$

where Γ_k is a contour around $[b_k, a_{k+1}]$. Then, as shown by Kuijlaars and McLaughlin in [82], the Jacobian of the map $\{[a_i, b_i] : i = 1, \dots, q\} \to \{T_j, N_k\}$ is nonzero at the solution, $\{[a_i(t), b_i(t)] : i = 1, \dots, q\}$, to the equations $\{T_j = 2\delta_{jq}, N_k = 0\}$. By the implicit function theorem, this implies the analyticity of $[a_i(t), b_i(t)]$ in t. □

4.4.3 Topological Expansion

In the paper [65], Ercolani and McLaughlin proves topological expansion (4.15) for polynomial V of form (4.13), with small values of the parameters t_j. Their proof is based on a construction of the asymptotic large N expansion of the parametrix for the RHP. Another proof of topological expansion (4.15) is given in the paper [19]. Here we outline the main steps of the proof of [19]. We start with a preliminary result, which follows easily from the results of [52].

Proposition 4.4.4. *Suppose $V(x)$ is one-cut regular. Then there exists $\varepsilon > 0$ such that for all n in the interval*

$$1-\varepsilon \le \frac{n}{N} \le 1+\varepsilon\,,$$

the recurrence coefficients admit the uniform asymptotic representation,

$$\gamma_n = \gamma\left(\frac{n}{N}\right) + O(N^{-1})\,, \quad \beta_n = \beta\left(\frac{n}{N}\right) + O(N^{-1})\,. \tag{4.417}$$

The functions $\gamma(s)$, $\beta(s)$ are expressed as

$$\gamma(s) = \frac{b(s)-a(s)}{4}\,, \quad \beta(s) = \frac{a(s)+b(s)}{2}\,, \tag{4.418}$$

where $[a(s), b(s)]$ is the support of the equilibrium measure for the polynomial $s^{-1}V(x)$.

The next theorem gives an asymptotic expansion for the recurrence coefficients.

Theorem 4.4.2. *Suppose that $V(x)$ is a one-cut regular polynomial. Then there exists $\varepsilon > 0$ such that for all n in the interval*

$$1-\varepsilon \le \frac{n}{N} \le 1+\varepsilon\,,$$

the recurrence coefficients admit the following uniform asymptotic expansion as $N \to \infty$:

$$\begin{aligned} \gamma_n &\sim \gamma\left(\frac{n}{N}\right) + \sum_{k=1}^{\infty} N^{-2k} f_{2k}\left(\frac{n}{N}\right), \\ \beta_n &\sim \beta\left(\frac{n+\frac{1}{2}}{N}\right) + \sum_{k=1}^{\infty} N^{-2k} g_{2k}\left(\frac{n+\frac{1}{2}}{N}\right), \end{aligned} \tag{4.419}$$

where $f_{2k}(s)$, $g_{2k}(s)$, $k \ge 1$, are analytic functions on $[1-\varepsilon, 1+\varepsilon]$.

Sketch of the proof of the theorem. The Riemann–Hilbert approach gives an asymptotic expansion in powers of N^{-1}. We want to show that the odd coefficients vanish. To prove this, we use induction in the number of the coefficient and the invariance of the string equations,

$$\gamma_n [V'(Q)]_{n,n-1} = \frac{n}{N}\,, \quad [V'(Q)]_{nn} = 0\,, \tag{4.420}$$

with respect to the change of variables

$$\{\gamma_j \to \gamma_{2n-j},\ \beta_j \to \beta_{2n-j-1} : j = 0, 1, 2, \dots\}\,. \tag{4.421}$$

This gives the cancellation of the odd coefficients, which proves the theorem. □

The main condition, under which the topological expansion is proved in [19], is the following:

Hypothesis R. For all $t \geq 1$ the polynomial $\tau_t V(z)$ is one-cut regular.

Theorem 4.4.3 (See [19]). *If a polynomial $V(z)$ satisfies Hypothesis* R, *then its free energy admits the asymptotic expansion,*

$$F_N - F_N^{\mathrm{Gauss}} \sim F + N^{-2}F^{(2)} + N^{-4}F^{(4)} + \cdots . \tag{4.422}$$

The leading term of the asymptotic expansion is:

$$F = \int_1^\infty \frac{1-\tau}{\tau^2}[2\gamma^4(\tau) + 4\gamma^2(\tau)\beta^2(\tau) - \tfrac{1}{2}]\,\mathrm{d}\tau , \tag{4.423}$$

where

$$\gamma(\tau) = \frac{b(\tau) - a(\tau)}{4} , \quad \beta(\tau) = \frac{a(\tau) + b(\tau)}{2} , \tag{4.424}$$

and $[a(\tau), b(\tau)]$ is the support of the equilibrium measure for the polynomial $V(z;\tau)$.

To prove this theorem, we substitute asymptotic expansions (4.419) into formula (4.412) and check that the odd powers of N^{-1} cancel out. See [19].

4.4.4 One-Sided Analyticity at a Critical Point

According to the definition, see (4.107)–(4.108), the equilibrium measure is singular in the following three cases:

(1) $h(c) = 0$ where $c \in (a_j, b_j)$, for some $1 \leq j \leq q$,
(2) $h(a_j) = 0$ or $h(b_j) = 0$, for some $1 \leq j \leq q$,
(3) for some $c \notin J$,

$$2\int \log|c - y|\,\mathrm{d}\nu_V(y) - V(c) = l . \tag{4.425}$$

More complicated cases appear as a combination of these three basic ones. The cases (1) and (3) are generic for a one-parameter family of polynomials. The case (1) means a split of the support interval (a_j, b_j) into two intervals. A typical illustration of this case is given in Fig. 4.2. Case (3) means a birth of a new support interval at the point c. Case (2) is generic for a two-parameter family of polynomials.

Introduce the following hypothesis.

Hypothesis S$_q$. $V(z;t)$, $t \in [0, t_0]$, $t_0 > 0$, is a one-parameter family of real polynomials such that

(i) $V(z;t)$ is q-cut regular for $0 < t \leq t_0$,

(ii) $V(z;0)$ is q-cut singular and $h(a_i) \neq 0$, $h(b_i) \neq 0$, $i = 1, \dots, q$, where $\bigcup_{i=1}^{q} [a_i, b_i]$ is the support of the equilibrium measure for $V(z;0)$.

We have the following result, see [19].

Theorem 4.4.4. *Suppose $V(z;t)$ satisfies Hypothesis S_q. Then the end-points $a_i(t), b_i(t)$ of the equilibrium measure for $V(z;t)$, the density function, and the free energy are analytic, as functions of t, at $t = 0$.*

The proof of this theorem is an extension of the proof of Thm. 4.4.1, and it is also based on the work of Kuijlaars and McLaughlin [82]. Theorem shows that the free energy can be analytically continued through the critical point, $t = 0$, if conditions (i) and (ii) are fulfilled. If $h(a_j) = 0$ or $h(b_j) = 0$, then the free energy is expected to have an algebraic singularity at $t = 0$, but this problem has not been studied yet in details. As concerns the split of the support interval, this case was studied for a nonsymmetric quartic polynomial in the paper of Bleher and Eynard [15]. To describe the result of [15], consider the singular quartic polynomial,

$$V_c'(x) = \frac{1}{T_c}(x^3 - 4c_1 x^2 + 2c_2 x + 8c_1)\,, \quad T_c = 1 + 4c_1^2\,;\ V_c(0) = 0\,, \tag{4.426}$$

where we denote

$$c_k = \cos k\pi\epsilon\,. \tag{4.427}$$

It corresponds to the critical density

$$\rho_c(x) = \frac{1}{2\pi T_c}(x - 2c_1)^2\sqrt{4 - x^2}\,. \tag{4.428}$$

Observe that $0 < \epsilon < 1$ is a parameter of the problem which determines the location of the critical point,

$$-2 < 2c_1 = 2\cos\pi\epsilon < 2\,. \tag{4.429}$$

We include V_c into the one-parameter family of quartic polynomials, $\{V(x;T) : T > 0\}$, where

$$V'(x;T) = \frac{1}{T}(x^3 - 4c_1 x^2 + 2c_2 x + 8c_1)\,; \quad V(0;T) = 0\,. \tag{4.430}$$

Let $F(T)$ be the free energy corresponding to $V(x;T)$.

Theorem 4.4.5. *The function $F(T)$ can be analytically continued through $T = T_c$ both from $T \geq T_c$ and from $T \leq T_c$. At $T = T_c$, $F(T)$ is continuous, as well as its first two derivatives, but the third derivative jumps.*

This corresponds to the third-order phase transition. Earlier the third-order phase transition was observed in a circular ensemble of random matrices by Gross and Witten [72].

4.4.5 Double Scaling Limit of the Free Energy

Consider an even quartic critical potential,

$$V(z) = \tfrac{1}{4}z^4 - z^2 , \tag{4.431}$$

and its deformation,

$$\tau_t V(z) \equiv V(z;t) = \frac{1}{4t^2}z^4 + \left(1 - \frac{2}{t}\right)z^2 . \tag{4.432}$$

Introduce the scaling,

$$t = 1 + N^{-2/3}2^{-2/3}x . \tag{4.433}$$

The Tracy–Widom distribution function defined by the formula

$$F_{\mathrm{TW}}(x) = \exp\left[\int_x^\infty (x-y)u^2(y)\,\mathrm{d}y\right] , \tag{4.434}$$

where $u(y)$ is the Hastings–McLeod solution to the Painlevé II, see (4.306), (4.307).

Theorem 4.4.6 (See [19]). *For every* $\varepsilon > 0$,

$$\begin{aligned} F_N(t) - F_N^{\mathrm{Gauss}} \\ &= F_N^{\mathrm{reg}}(t) - N^{-2}\log F_{\mathrm{TW}}\big((t-1)2^{2/3}N^{2/3}\big) + O(N^{-7/3+\varepsilon}) , \end{aligned} \tag{4.435}$$

as $N \to \infty$ *and* $|(t-1)N^{2/3}| < C$, *where*

$$F_N^{\mathrm{reg}}(t) \equiv F(t) + N^{-2}F^{(2)}(t) \tag{4.436}$$

is the sum of the first two terms of the topological expansion.

4.5 Random Matrix Model with External Source

4.5.1 Random Matrix Model with External Source and Multiple Orthogonal Polynomials

We consider the Hermitian random matrix ensemble with an external source,

$$\mathrm{d}\mu_n(M) = \frac{1}{Z_n}\exp\big(-n\operatorname{Tr}(V(M) - AM)\big)\,\mathrm{d}M , \tag{4.437}$$

where

$$Z_n = \int \exp\big(-n\operatorname{Tr}(V(M) - AM)\big)\,\mathrm{d}M , \tag{4.438}$$

and A is a fixed Hermitian matrix. Without loss of generality we may assume that A is diagonal,

$$A = \mathrm{diag}(a_1, \dots, a_n)\,, \quad a_1 < \cdots < a_n\,. \tag{4.439}$$

Define the monic polynomial

$$P_n(z) = \int \det(z - M)\, \mathrm{d}\mu_n(M)\,. \tag{4.440}$$

Proposition 4.5.1. (a) *There is a constant* $\widetilde{Z}_n$ *such that*

$$P_n(z) = \frac{1}{\widetilde{Z}_n} \int \prod_{j=1}^{n} (z - \lambda_j) \prod_{j=1}^{n} \exp\big(-(V(\lambda_j) - a_j \lambda_j)\big) \Delta(\lambda)\, \mathrm{d}\lambda\,, \tag{4.441}$$

where

$$\Delta(\lambda) = \prod_{i>j} (\lambda_i - \lambda_j) \tag{4.442}$$

and $\mathrm{d}\lambda = \mathrm{d}\lambda_1\, \mathrm{d}\lambda_2 \cdots \mathrm{d}\lambda_n$.

(b) *Let*

$$m_{jk} = \int_{-\infty}^{\infty} x^k \exp\big(-(V(x) - a_j x)\big)\, \mathrm{d}x\,. \tag{4.443}$$

Then we have the determinantal formula

$$P_n(z) = \frac{1}{\widetilde{Z}_n} \begin{vmatrix} m_{10} & m_{11} & \cdots & m_{1n} \\ \vdots & \vdots & \ddots & \vdots \\ m_{n0} & m_{n1} & \cdots & m_{nn} \\ 1 & z & \cdots & z^n \end{vmatrix}\,. \tag{4.444}$$

(c) *For* $j = 1, \dots, n$,

$$\int_{-\infty}^{\infty} P_n(x) \exp\big(-(V(x) - a_j x)\big)\, \mathrm{d}x = 0\,, \tag{4.445}$$

and these equations each uniquely determine the monic polynomial P_n.

Proposition 4.5.1 can be extended to the case of multiple a_js as follows.

Proposition 4.5.2. *Suppose* A *has distinct eigenvalues* a_i, $i = 1, \dots, p$, *with respective multiplicities* n_i, *so that* $n_1 + \cdots + n_p = n$. *Let* $n^{(i)} = n_1 + \cdots + n_i$ *and* $n^{(0)} = 0$. *Define*

$$w_j(x) = x^{d_j - 1} \exp\big(-(V(x) - a_i x)\big), \quad j = 1, \dots, n, \tag{4.446}$$

where $i = i_j$ *is such that* $n^{(i-1)} < j \le n^{(i)}$ *and* $d_j = j - n^{(i-1)}$. *Then the following hold.*

(a) *There is a constant* $\widetilde{Z}_n > 0$ *such that*

$$P_n(z) = \frac{1}{\widetilde{Z}_n} \int \prod_{j=1}^{n} (z - \lambda_j) \prod_{j=1}^{n} w_j(\lambda_j) \Delta(\lambda)\, \mathrm{d}\lambda\,. \tag{4.447}$$

(b) *Let*

$$m_{jk} = \int_{-\infty}^{\infty} x^k w_j(x)\,\mathrm{d}x\ . \tag{4.448}$$

Then we have the determinantal formula

$$P_n(z) = \frac{1}{\widetilde{Z}_n} \begin{vmatrix} m_{10} & m_{11} & \cdots & m_{1n} \\ \vdots & \vdots & \ddots & \vdots \\ m_{n0} & m_{n1} & \cdots & m_{nn} \\ 1 & z & \cdots & z^n \end{vmatrix}\ . \tag{4.449}$$

(c) *For* $i = 1, \ldots, p$,

$$\int_{-\infty}^{\infty} P_n(x) x^j \exp\bigl(-(V(x) - a_i x)\bigr)\,\mathrm{d}x = 0\ , \quad j = 0, \ldots, n_i - 1\ , \tag{4.450}$$

and these equations uniquely determine the monic polynomial P_n.

The relations (4.450) can be viewed as multiple orthogonality conditions for the polynomial P_n. There are p weights $\exp\bigl(-(V(x) - a_j x)\bigr)$, $j = 1, \ldots, p$, and for each weight there are a number of orthogonality conditions, so that the total number of them is n. This point of view is especially useful in the case when A has only a small number of distinct eigenvalues.

4.5.1.1 Determinantal Formula for Eigenvalue Correlation Functions

P. Zinn-Justin proved in [103, 104] a determinantal formula for the eigenvalue correlation functions of the random matrix model with external source. We relate the determinantal kernel to the multiple orthogonal polynomials.

Let Σ_n be the collection of functions

$$\Sigma_n := \{x^j \exp(a_i x) \mid i = 1, \ldots, p,\ j = 0, \ldots, n_i - 1\}. \tag{4.451}$$

We start with a lemma.

Lemma 4.5.1. *There exists a unique function* Q_{n-1} *in the linear span of* Σ_n *such that*

$$\int_{-\infty}^{\infty} x^j Q_{n-1}(x) \mathrm{e}^{-V(x)}\,\mathrm{d}x = 0\ , \tag{4.452}$$

$j = 0, \ldots, n-2$, *and*

$$\int_{-\infty}^{\infty} x^{n-1} Q_{n-1}(x) \mathrm{e}^{-V(x)}\,\mathrm{d}x = 1\ . \tag{4.453}$$

Consider $P_0, \dots, P_n$, a sequence of multiple orthogonal polynomials such that $\deg P_k = k$, with an increasing sequence of the multiplicity vectors, so that $k_i \le l_i$, $i = 1, \dots, p$, when $k \le l$. Consider the biorthogonal system of functions, $\{Q_k(x) : k = 0, \dots, n-1\}$,

$$\int_{-\infty}^{\infty} P_j(x) Q_k(x) \mathrm{e}^{-V(x)}\, \mathrm{d}x = \delta_{jk} , \tag{4.454}$$

for $j, k = 0, \dots, n-1$. Define the kernel

$$K_n(x,y) = \exp\left(-\frac{1}{2}(V(x)+V(y))\right) \sum_{k=0}^{n-1} P_k(x) Q_k(y) . \tag{4.455}$$

Theorem 4.5.1. *The* m*-point correlation function of eigenvalues has the determinantal form*

$$R_m(\lambda_1, \dots, \lambda_m) = \det(K_n(\lambda_j, \lambda_k))_{1 \le j,k \le m} . \tag{4.456}$$

4.5.1.2 Christoffel–Darboux Formula

We will assume that there are only two distinct eigenvalues, $a_1 = a$ and $a_2 = -a$, with multiplicities n_1 and n_2, respectively. We redenote P_n by P_{n_1,n_2}. Set

$$h_{n_1,n_2}^{(j)} = \int_{-\infty}^{\infty} P_{n_1,n_2}(x) x^{n_j} w_j(x)\, \mathrm{d}x , \tag{4.457}$$

$j = 1, 2$, which are non-zero numbers.

Theorem 4.5.2. *With the notation introduced above,*

$$\begin{aligned} (x-y) \exp\left(\frac{1}{2}(V(x)+V(y))\right) K_n(x,y) \\ = P_{n_1,n_2}(x) Q_{n_1,n_2}(y) - \frac{h_{n_1,n_2}^{(1)}}{h_{n_1-1,n_2}^{(1)}} P_{n_1-1,n_2}(x) Q_{n_1+1,n_2}(y) \\ - \frac{h_{n_1,n_2}^{(2)}}{h_{n_1,n_2-1}^{(2)}} P_{n_1,n_2-1}(x) Q_{n_1,n_2+1}(y) . \end{aligned} \tag{4.458}$$

4.5.1.3 Riemann–Hilbert Problem

The Rieman–Hilbert problem is to find $Y \colon \mathbb{C} \setminus \mathbb{R} \to \mathbb{C}^{3\times 3}$ such that

- Y is analytic on $\mathbb{C} \setminus \mathbb{R}$,

- for $x \in \mathbb{R}$, we have

$$Y_+(x) = Y_-(x) \begin{pmatrix} 1 & w_1(x) & w_2(x) \\ 0 & 1 & 0 \\ 0 & 0 & 1 \end{pmatrix} \tag{4.459}$$

where $Y_+(x)$ ($Y_-(x)$) denotes the limit of $Y(z)$ as $z \to x$ from the upper (lower) half-plane,
- as $z \to \infty$, we have

$$Y(z) = \left(I + O\left(\frac{1}{z}\right)\right) \begin{pmatrix} z^n & 0 & 0 \\ 0 & z^{-n_1} & 0 \\ 0 & 0 & z^{-n_2} \end{pmatrix} \tag{4.460}$$

where I denotes the 3×3 identity matrix.

Proposition 4.5.3. *There exists a unique solution to the RH problem,*

$$Y = \begin{pmatrix} P_{n_1,n_2} & \mathrm{C}(P_{n_1,n_2} w_1) & \mathrm{C}(P_{n_1,n_2} w_2) \\ c_1 P_{n_1-1,n_2} & c_1\,\mathrm{C}(P_{n_1-1,n_2} w_1) & c_1\,\mathrm{C}(P_{n_1-1,n_2} w_2) \\ c_2 P_{n_1,n_2-1} & c_2\,\mathrm{C}(P_{n_1,n_2-1} w_1) & c_2\,\mathrm{C}(P_{n_1,n_2-1} w_2) \end{pmatrix} \tag{4.461}$$

with constants

$$c_1 = -2\pi \mathrm{i}\big(h^{(1)}_{n_1-1,n_2}\big)^{-1}, \quad c_2 = -2\pi \mathrm{i}\big(h^{(2)}_{n_1,n_2-1}\big)^{-1}, \tag{4.462}$$

and where $\mathrm{C} f$ *denotes the Cauchy transform of* f, *i.e.,*

$$\mathrm{C} f(z) = \frac{1}{2\pi \mathrm{i}} \int_{\mathbb{R}} \frac{f(s)}{s-z}\, \mathrm{d}s \,. \tag{4.463}$$

The Christoffel–Darboux formula, (4.458), can be expressed in terms of the solution of RH problem:

$$K_n(x,y) = \exp\left(-\frac{1}{2}(V(x)+V(y))\right) \times \frac{\exp(a_1 y)[Y^{-1}(y)Y(x)]_{21} + \exp(a_2 y)[Y^{-1}(y)Y(x)]_{31}}{2\pi \mathrm{i}(x-y)} \,. \tag{4.464}$$

4.5.1.4 Recurrence and Differential Equations

The recurrence and differential equations are nicely formulated in terms of the function

$$\Psi(z) = \begin{pmatrix} 1 & 0 & 0 \\ 0 & c_1^{-1} & 0 \\ 0 & 0 & c_2^{-1} \end{pmatrix} Y(z) \begin{pmatrix} w(z) & 0 & 0 \\ 0 & \mathrm{e}^{-Naz} & 0 \\ 0 & 0 & \mathrm{e}^{Naz} \end{pmatrix}, \tag{4.465}$$

where

$$w(z) = \mathrm{e}^{-NV(z)} \,. \tag{4.466}$$

The function Ψ solves the following RH problem:

- Ψ is analytic on $\mathbb{C}\setminus\mathbb{R}$,
- for $x\in\mathbb{R}$,

$$\Psi_+(x)=\Psi_-(x)\begin{pmatrix}1&1&1\\0&1&0\\0&0&1\end{pmatrix}, \tag{4.467}$$

- as $z\to\infty$,

$$\Psi(z)\sim\left(I+\frac{\Psi^{(1)}_{n_1,n_2}}{z}+\cdots\right)\begin{pmatrix}z^n w&0&0\\0&c_1^{-1}z^{-n_1}\mathrm{e}^{-Naz}&0\\0&0&c_2^{-1}z^{-n_2}\mathrm{e}^{Naz}\end{pmatrix}. \tag{4.468}$$

The recurrence equation for Ψ has the form:

$$\Psi_{n_1+1,n_2}(z)=U_{n_1,n_2}(z)\Psi_{n_1,n_2}(z)\,, \tag{4.469}$$

where

$$U_{n_1,n_2}(z)=\begin{pmatrix}z-b_{n_1,n_2}&-c_{n_1,n_2}&-d_{n_1,n_2}\\1&0&0\\1&0&e_{n_1,n_2}\end{pmatrix} \tag{4.470}$$

and

$$\begin{gathered}c_{n_1,n_2}=\frac{h^{(1)}_{n_1,n_2}}{h^{(1)}_{n_1-1,n_2}}\neq 0\,,\quad d_{n_1,n_2}=\frac{h^{(2)}_{n_1,n_2}}{h^{(2)}_{n_1,n_2-1}}\neq 0\,,\\ e_{n_1,n_2}=\frac{h^{(2)}_{n_1+1,n_2-1}}{h^{(2)}_{n_1,n_2-1}}\neq 0\,.\end{gathered} \tag{4.471}$$

Respectively, the recurrence equations for the multiple orthogonal polynomials are

$$\begin{aligned}P_{n_1+1,n_2}(z)=(z-b_{n_1,n_2})P_{n_1,n_2}(z)-c_{n_1,n_2}P_{n_1-1,n_2}(z)\\ -d_{n_1,n_2}P_{n_1,n_2-1}(z)\,,\end{aligned} \tag{4.472}$$

and

$$P_{n_1+1,n_2-1}(z)=P_{n_1,n_2}(z)+e_{n_1,n_2}P_{n_1,n_2-1}(z)\,. \tag{4.473}$$

The differential equation for Ψ is

$$\Psi'_{n_1,n_2}(z)=NA_{n_1,n_2}(z)\Psi_{n_1,n_2}(z)\,, \tag{4.474}$$

where

$$\begin{aligned}A_{n_1,n_2}(z)=-\left[\left(I+\frac{\Psi^{(1)}_{n_1,n_2}}{z}+\cdots\right)\begin{pmatrix}V'(z)&0&0\\0&0&0\\0&0&0\end{pmatrix}\left(I+\frac{\Psi^{(1)}_{n_1,n_2}}{z}+\cdots\right)^{-1}\right]_{\mathrm{pol}}\\ +\begin{pmatrix}0&0&0\\0&-a&0\\0&0&a\end{pmatrix},\end{aligned} \tag{4.475}$$

where $[f(z)]_{\text{pol}}$ means the polynomial part of $f(z)$ at infinity.

For the Gaussian model, $V(x) = x^2/2$, the recurrence equation reduces to

$$\Psi_{n_1+1,n_2} = \begin{pmatrix} z-a & -n_1/n & -n_2/n \\ 1 & 0 & 0 \\ 1 & 0 & -2a \end{pmatrix} \Psi_{n_1,n_2}\,, \tag{4.476}$$

where $n = n_1 + n_2$, and the differential equation reads

$$\Psi'_{n_1,n_2} = n \begin{pmatrix} -z & n_1/n & n_2/n \\ -1 & -a & 0 \\ -1 & 0 & a \end{pmatrix} \Psi_{n_1,n_2}\,. \tag{4.477}$$

In what follows, we will restrict ourselves to the case when n is even and

$$n_1 = n_2 = \frac{n}{2}\,, \tag{4.478}$$

so that

$$A = \operatorname{diag}(\underbrace{-a,\ldots,-a}_{n/2},\underbrace{a,\ldots,a}_{n/2}). \tag{4.479}$$

4.5.2 Gaussian Matrix Model with External Source and Non-Intersecting Brownian Bridges

Consider n independent Brownian motions (Brownian bridges) $x_j(t)$, $j = 1, \ldots, n$, on the line, starting at the origin at time $t = 0$, half ending at $x = 1$ and half at $x = -1$ at time $t = 1$, and conditioned not to intersect for $t \in (0,1)$. Then at any time $t \in (0,1)$ the positions of n non-intersecting Brownian bridges are distributed as the scaled eigenvalues,

$$\lambda_j = \frac{x_j}{\sqrt{t(1-t)}}\,,$$

of a Gaussian random matrix with the external source

$$a(t) = \sqrt{\frac{t}{1-t}}\,.$$

Figure 4.10 gives an illustration of the non-intersecting Brownian bridges. See also the paper [99] of Tracy and Widom on non-intersecting Brownian excursions.

In the Gaussian model the value $a = 1$ is critical, and we will discuss its large n asymptotics in the three cases:

(1) $a > 1$, two cuts,
(2) $a < 1$, one cut,
(3) $a = 1$, double scaling limit.

In the picture of the non-intersecting Brownian bridges this transforms to a critical time $t = \frac{1}{2}$, and there are two cuts for $t > \frac{1}{2}$, one cut for $t < \frac{1}{2}$, and the double scaling limit appears in a scaled neighborhood of $t = \frac{1}{2}$.

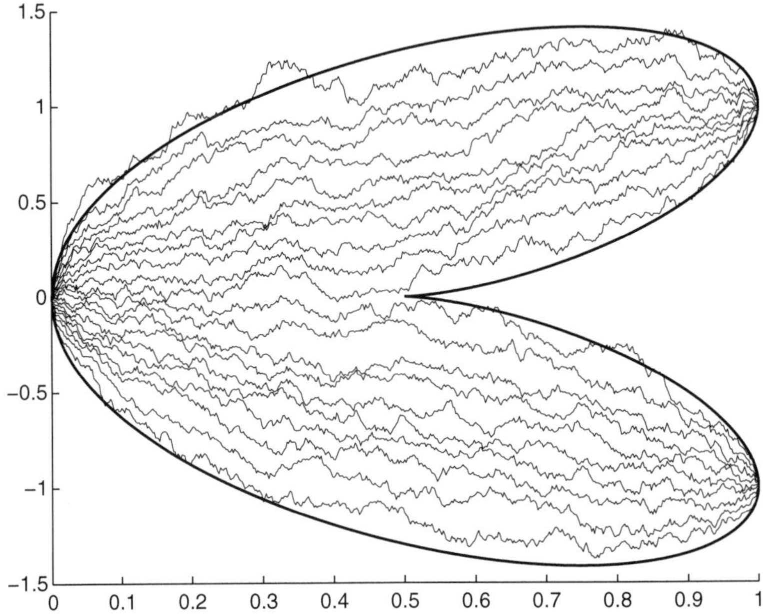

Fig. 4.10. Non-intersecting Brownian paths that start at one point and end at two points. At any intermediate time the positions of the paths are distributed as the eigenvalues of a Gaussian random matrix ensemble with external source. As their number increases the paths fill out a region whose boundary has a cusp

4.5.3 Gaussian Model with External Source. Main Results

First we describe the limiting mean density of eigenvalues. The limiting mean density follows from earlier work of Pastur [89]. It is based on an analysis of the equation (Pastur equation)

$$\xi^3 - z\xi^2 + (1-a^2)\xi + a^2 z = 0\,, \tag{4.480}$$

which yields an algebraic function $\xi(z)$ defined on a three-sheeted Riemann surface. The restrictions of $\xi(z)$ to the three sheets are denoted by $\xi_j(z)$, $j = 1, 2, 3$. There are four real branch points if $a > 1$ which determine two real intervals. The two intervals come together for $a = 1$, and for $0 < a < 1$, there are two real branch points, and two purely imaginary branch points. Figure 4.11 depicts the structure of the Riemann surface $\xi(z)$ for $a > 1$, $a = 1$, and $a < 1$.

In all cases we have that the limiting mean eigenvalue density $\rho(x) = \rho(x; a)$ is given by

$$\rho(x; a) = \frac{1}{\pi} \operatorname{Im} \xi_{1+}(x)\,, \quad x \in \mathbb{R}\,, \tag{4.481}$$

where $\xi_{1+}(x)$ denotes the limiting value of $\xi_1(z)$ as $z \to x$ with $\operatorname{Im} z > 0$. For $a = 1$ the limiting mean eigenvalue density vanishes at $x = 0$ and $\rho(x;a) \sim |x|^{1/3}$ as $x \to 0$.

We note that this behavior at the closing (or opening) of a gap is markedly different from the behavior that occurs in the usual unitary random matrix ensembles $Z_n^{-1} \mathrm{e}^{-n \operatorname{Tr} V(M)} \, \mathrm{d}M$ where a closing of the gap in the spectrum typically leads to a limiting mean eigenvalue density ρ that satisfies $\rho(x) \sim (x - x^*)^2$ as $x \to x^*$ if the gap closes at $x = x^*$. In that case the local eigenvalue correlations can be described in terms of ψ-functions associated with the Painlevé II equation, see above and [18, 40]. The phase transition for the model under consideration is different, and it cannot be realized in a unitary random matrix ensemble.

Theorem 4.5.3. *The limiting mean density of eigenvalues*

$$\rho(x) = \lim_{n \to \infty} \frac{1}{n} K_n(x,x) \tag{4.482}$$

exists for every $a > 0$. It satisfies

$$\rho(x) = \frac{1}{\pi} |\operatorname{Im} \xi(x)| \,, \tag{4.483}$$

where $\xi = \xi(x)$ is a solution of the cubic equation,

$$\xi^3 - x\xi^2 - (a^2 - 1)\xi + xa^2 = 0 \,. \tag{4.484}$$

The support of ρ consists of those $x \in \mathbb{R}$ for which (4.484) has a non-real solution.

(a) *For $0 < a < 1$, the support of ρ consists of one interval $[-z_1, z_1]$, and ρ is real analytic and positive on $(-z_1, z_1)$, and it vanishes like a square root at the edge points $\pm z_1$, i.e., there exists a constant $\rho_1 > 0$ such that*

$$\rho(x) = \frac{\rho_1}{\pi} |x \mp z_1|^{1/2} \big(1 + o(1)\big) \quad \text{as } x \to \pm z_1,\ x \in (-z_1, z_1) \,. \tag{4.485}$$

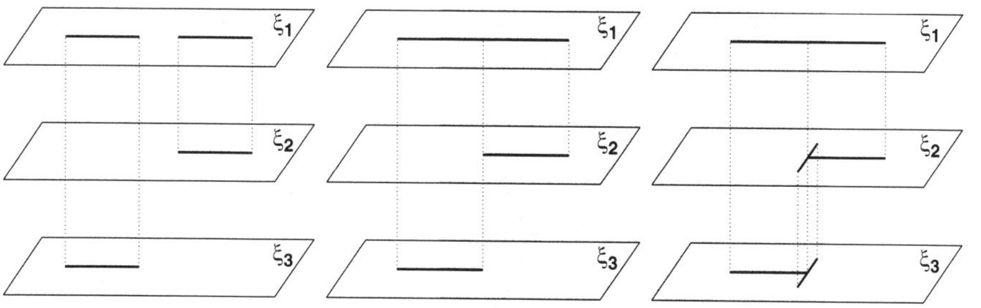

Fig. 4.11. The structure of the Riemann surface for (4.480) for the values $a > 1$ (left), $a = 1$ (middle) and $a < 1$ (right). In all cases the eigenvalues of M accumulate on the interval(s) of the first sheet with a density given by (4.481)

(b) *For $a = 1$, the support of ρ consists of one interval $[-z_1, z_1]$, and ρ is real analytic and positive on $(-z_1, 0) \cup (0, z_1)$, it vanishes like a square root at the edge points $\pm z_1$, and it vanishes like a third root at 0, i.e., there exists a constant $c > 0$ such that*

$$\rho(x) = c|x|^{1/3}\big(1 + o(1)\big)\,, \quad \text{as } x \to 0\,. \tag{4.486}$$

(c) *For $a > 1$, the support of ρ consists of two disjoint intervals $[-z_1, -z_2] \cup [z_2, z_1]$ with $0 < z_2 < z_1$, ρ is real analytic and positive on $(-z_1, -z_2) \cup (z_2, z_1)$, and it vanishes like a square root at the edge points $\pm z_1$, $\pm z_2$.*

To describe the universality of local eigenvalue correlations in the large n limit, we use a rescaled version of the kernel K_n:

$$\widehat{K}_n(x, y) = \exp\Big(n\big(h(x) - h(y)\big)\Big)K_n(x, y) \tag{4.487}$$

for some function h. The rescaling is allowed because it does not affect the correlation functions R_m, which are expressed as determinants of the kernel. The function h has the following form on $(-z_1, -z_2) \cup (z_2, z_1)$:

$$h(x) = -\frac{1}{4}x^2 + \operatorname{Re}\int_{z_1}^{x} \xi_{1+}(s)\,\mathrm{d}s\,, \tag{4.488}$$

where ξ_1 is a solution of the Pastur equation. The local eigenvalue correlations in the bulk of the spectrum in the large n limit are described by the sine kernel. The bulk of the spectrum is the open interval $(-z_1, z_1)$ for $a < 1$, and the union of the two open intervals, $(-z_1, -z_2)$ and (z_2, z_1), for $a \geq 1$ ($z_2 = 0$ for $a = 1$).

We have the following result:

Theorem 4.5.4. *For every x_0 in the bulk of the spectrum we have that*

$$\lim_{n\to\infty} \frac{1}{n\rho(x_0)}\widehat{K}_n\left(x_0 + \frac{u}{n\rho(x_0)}, x_0 + \frac{v}{n\rho(x_0)}\right) = \frac{\sin \pi(u - v)}{\pi(u - v)}\,.$$

At the edge of the spectrum the local eigenvalue correlations are described in the large n limit by the Airy kernel:

Theorem 4.5.5. *For every $u, v \in \mathbb{R}$ we have*

$$\lim_{n\to\infty} \frac{1}{(\rho_1 n)^{2/3}}\widehat{K}_n\left(z_1 + \frac{u}{(\rho_1 n)^{2/3}}, z_1 + \frac{v}{(\rho_1 n)^{2/3}}\right) = \frac{\operatorname{Ai}(u)\operatorname{Ai}'(v) - \operatorname{Ai}'(u)\operatorname{Ai}(v)}{u - v}\,.$$

A similar limit holds near the edge point $-z_1$ and also near the edge points $\pm z_2$ if $a > 1$.

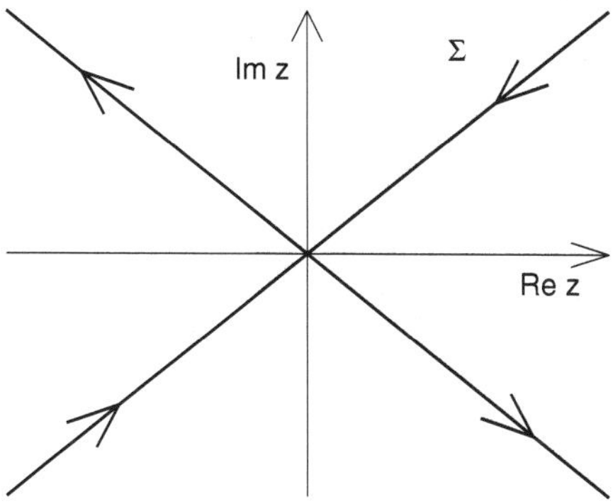

Fig. 4.12. The contour Σ that appears in the definition of $q(y)$

As is usual in a critical case, there is a family of limiting kernels that arise when a changes with n and $a \to 1$ as $n \to \infty$ in a critical way. These kernels are constructed out of Pearcey integrals and therefore they are called Pearcey kernels. The Pearcey kernels were first described by Brézin and Hikami [30,31]. A detailed proof of the following result was recently given by Tracy and Widom [98].

Theorem 4.5.6. *We have for every fixed* $b \in \mathbb{R}$,

$$\lim_{n\to\infty} \frac{1}{n^{3/4}} K_n\left(\frac{x}{n^{3/4}}, \frac{y}{n^{3/4}}; 1 + \frac{b}{2\sqrt{n}}\right) = K^{\mathrm{cusp}}(x, y; b) \tag{4.489}$$

where K^{cusp} *is the Pearcey kernel*

$$K^{\mathrm{cusp}}(x, y; b) = \frac{p(x)q''(y) - p'(x)q'(y) + p''(x)q(y) - bp(x)q(y)}{x - y} \tag{4.490}$$

with

$$\begin{aligned} p(x) &= \frac{1}{2\pi} \int_{-\infty}^{\infty} \exp\left(-\frac{1}{4}s^4 - \frac{b}{2}s^2 + \mathrm{i}sx\right) \mathrm{d}s\,, \\ q(y) &= \frac{1}{2\pi} \int_{\Sigma} \exp\left(\frac{1}{4}t^4 + \frac{b}{2}t^2 + \mathrm{i}ty\right) \mathrm{d}t\,. \end{aligned} \tag{4.491}$$

The contour Σ *consists of the four rays* $\arg y = \pm\pi/4, \pm 3\pi/4$, *with the orientation shown in Fig.* 4.12.

The functions (4.491) are called Pearcey integrals [91]. They are solutions of the third-order differential equations $p'''(x) = xp(x) + bp'(x)$ and $q'''(y) = -yq(y) + bq'(y)$, respectively.

Theorem 4.5.6 implies that local eigenvalue statistics of eigenvalues near 0 are expressed in terms of the Pearcey kernel. For example we have the following corollary of Thm. 4.5.6.

Corollary 4.5.1. *The probability that a matrix of the ensemble* (4.437), (4.479), *with* $a=1+bn^{-1/2}/2$ *has no eigenvalues in the interval* $[cn^{-3/4}, dn^{-3/4}]$ *converges, as* $n \to \infty$, *to the Fredholm determinant of the integral operator with kernel* $K^{\mathrm{cusp}}(x,y;b)$ *acting on* $L^2(c,d)$.

Similar expressions hold for the probability to have one, two, three,..., eigenvalues in an $O(n^{-3/4})$ neighborhood of $x=0$.

Tracy and Widom [98] and Adler and van Moerbeke [3] gave differential equations for the gap probabilities associated with the Pearcey kernel and with the more general Pearcey process which arises from considering the non-intersecting Brownian motion model at several times near the critical time. See also [88] where the Pearcey process appears in a combinatorial model on random partitions.

Brézin and Hikami and also Tracy and Widom used a double integral representation for the kernel in order to establish Thm. 4.5.6. We will describe the approach of [23], based on the Deift–Zhou steepest descent method for the Riemann–Hilbert problem for multiple Hermite polynomials. This method is less direct than the steepest descent method for integrals. However, an approach based on the Riemann–Hilbert problem may be applicable to more general situations, where an integral representation is not available. This is the case, for example, for the general (non-Gaussian) unitary random matrix ensemble with external source, (4.437), with a general potential V.

The proof of the theorems above is based on the construction of a parametrix of the RHP, and we will describe this construction for the cases $a>1$, $a<1$, and $a=1$.

4.5.4 Construction of a Parametrix in the Case $a > 1$

Consider the Riemann surface given by (4.480)) for $a>1$, see the left surface on Fig. 4.11. There are three roots to this equation, which behave at infinity as

$$\xi_1(z) = z - \frac{1}{z} + O\left(\frac{1}{z^3}\right), \quad \xi_{2,3}(z) = \pm a + \frac{1}{2z} + O\left(\frac{1}{z^2}\right). \tag{4.492}$$

We need the integrals of the ξ-functions,

$$\lambda_k(z) = \int^z \xi_k(s)\,\mathrm{d}s\,, \quad k=1,2,3\,, \tag{4.493}$$

which we take so that λ_1 and λ_2 are analytic on $\mathbb{C}\setminus(-\infty, z_1]$ and λ_3 is analytic on $\mathbb{C}\setminus(-\infty,-z_2]$. Then, as $z\to\infty$,

$$\begin{aligned}\lambda_1(z) &= \frac{z^2}{2} - \ln z + l_1 + O\left(\frac{1}{z^2}\right),\\ \lambda_{2,3}(z) &= \pm az + \frac{1}{2}\ln z + l_{2,3} + O\left(\frac{1}{z}\right),\end{aligned} \tag{4.494}$$

where l_1, l_2, l_3 are some constants, which we choose as follows. We choose l_1 and l_2 such that

$$\lambda_1(z_1) = \lambda_2(z_1) = 0 \,, \tag{4.495}$$

and then l_3 such that

$$\lambda_3(-z_2) = \lambda_{1+}(-z_2) = \lambda_{1-}(-z_2) - \pi \mathrm{i} \,. \tag{4.496}$$

First Transformation of the RH Problem

Using the functions λ_j and the constants l_j, $j = 1, 2, 3$, we define

$$\begin{aligned} T(z) &= \operatorname{diag}\big(\exp(-nl_1), \exp(-nl_2), \exp(-nl_3)\big) Y(z) \\ &\quad \times \operatorname{diag}\Big(\exp\big(n(\lambda_1(z) - \tfrac{1}{2}z^2)\big), \exp\big(n(\lambda_2(z) - az)\big), \exp\big(n(\lambda_3(z) + az)\big)\Big) \,. \end{aligned} \tag{4.497}$$

Then $T_+(x) = T_-(x) j_T(x)$, $x \in \mathbb{R}$, where for $x \in [z_2, z_1]$,

$$j_T = \begin{pmatrix} \exp(n(\lambda_1 - \lambda_2)_+) & 1 & \exp\big(n(\lambda_3 - \lambda_{1-})\big) \\ 0 & \exp(n(\lambda_1 - \lambda_2)_-) & 0 \\ 0 & 0 & 1 \end{pmatrix} \tag{4.498}$$

and for $x \in [-z_1, -z_2]$,

$$j_T = \begin{pmatrix} \exp(n(\lambda_1 - \lambda_3)_+) & \exp\big(n(\lambda_{2+} - \lambda_{1-})\big) & 1 \\ 0 & 1 & 0 \\ 0 & 0 & \exp(n(\lambda_1 - \lambda_3)_-) \end{pmatrix} \,. \tag{4.499}$$

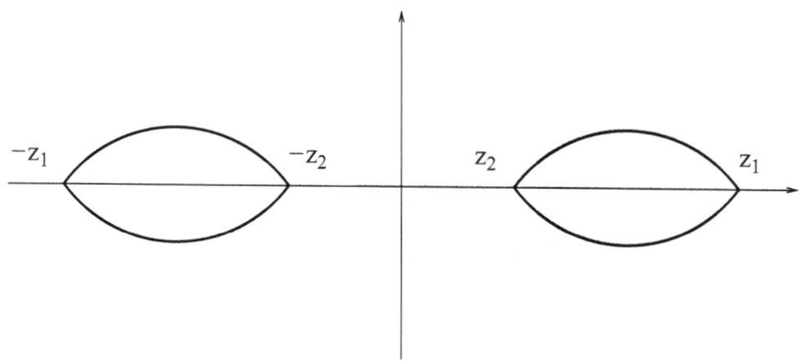

Fig. 4.13. The lenses for $a > 1$

Second Transformation of the RH Problem: Opening of Lenses

The lens structure is shown on Fig. 4.13. Set in the right lens,

$$S(z)=\begin{cases} T(z)\begin{pmatrix} 1 & 0 & 0 \\ -\exp(n(\lambda_1(z)-\lambda_2(z))) & 1 & -\exp(n(\lambda_3(z)-\lambda_2(z))) \\ 0 & 0 & 1 \end{pmatrix} \\ \qquad\qquad \text{in the upper lens region}\,, \\ T(z)\begin{pmatrix} 1 & 0 & 0 \\ \exp(n(\lambda_1(z)-\lambda_2(z))) & 1 & -\exp(n(\lambda_3(z)-\lambda_2(z))) \\ 0 & 0 & 1 \end{pmatrix} \\ \qquad\qquad \text{in the lower lens region}\,, \end{cases} \tag{4.500}$$

and, respectively, in the left lens,

$$S(z)=\begin{cases} T(z)\begin{pmatrix} 1 & 0 & 0 \\ 0 & 1 & 0 \\ -\exp(n(\lambda_1(z)-\lambda_3(z))) & -\exp(n(\lambda_2(z)-\lambda_3(z))) & 1 \end{pmatrix} \\ \qquad\qquad \text{in the upper lens region}\,, \\ T(z)\begin{pmatrix} 1 & 0 & 0 \\ 0 & 1 & 0 \\ \exp(n(\lambda_1(z)-\lambda_3(z))) & -\exp(n(\lambda_2(z)-\lambda_3(z))) & 1 \end{pmatrix} \\ \qquad\qquad \text{in the lower lens region}\,. \end{cases} \tag{4.501}$$

Then

$$S_+(x)=S_-(x)j_S(x)\,;\quad j_S(x)=\begin{pmatrix} 0 & 1 & 0 \\ -1 & 0 & 0 \\ 0 & 0 & 1 \end{pmatrix}\,,\ x\in[z_2,z_1]\,, \tag{4.502}$$

and

$$S_+(x)=S_-(x)j_S(x)\,;\quad j_S(x)=\begin{pmatrix} 0 & 0 & 1 \\ 0 & 1 & 0 \\ -1 & 0 & 0 \end{pmatrix}\,,\ x\in[-z_1,-z_2]\,. \tag{4.503}$$

In addition, S has jumps on the boundary of the lenses, which are exponentially small away of the points $\pm z_{1,2}$. The RH problem for S is approximated by the model RH problem.

Model RH Problem

- M is analytic on $\mathbb{C}\setminus([-z_1,-z_2]\cup[z_2,z_1])$,
- $$M_+(x)=M_-(x)j_S(x)\,,\quad x\in(-z_1,-z_2)\cup(z_2,z_1)\,, \tag{4.504}$$

- as $z \to \infty$,

$$M(z) = I + O\left(\frac{1}{z}\right), \tag{4.505}$$

where the jump matrix is

$$j_S(x) = \begin{cases} \begin{pmatrix} 0 & 1 & 0 \\ -1 & 0 & 0 \\ 0 & 0 & 1 \end{pmatrix}, & x \in (z_2, z_1) \\ \begin{pmatrix} 0 & 0 & 1 \\ 0 & 1 & 0 \\ -1 & 0 & 0 \end{pmatrix}, & x \in (-z_1, -z_2). \end{cases} \tag{4.506}$$

Solution to the model RH problem has the form:

$$M(z) = A(z)B(z)C(z), \tag{4.507}$$

where

$$A(z) = \operatorname{diag}\left(1, -\frac{\mathrm{i}}{\sqrt{2}}, -\frac{\mathrm{i}}{\sqrt{2}}\right),$$
$$B(z) = \begin{pmatrix} \xi_1^2(z) - a^2 & \xi_2^2(z) - a^2 & \xi_3^2(z) - a^2 \\ \xi_1(z) + a & \xi_2(z) + a & \xi_3(z) + a \\ \xi_1(z) - a & \xi_2(z) - a & \xi_3(z) - a \end{pmatrix} \tag{4.508}$$

and

$$C(z) = \operatorname{diag}\left(\frac{1}{\sqrt{Q(\xi_1(z))}}, \frac{1}{\sqrt{Q(\xi_2(z))}}, \frac{1}{\sqrt{Q(\xi_3(z))}}\right) \tag{4.509}$$

where

$$Q(z) = z^4 - (1 + 2a^2)z^2 + (a^2 - 1)a^2. \tag{4.510}$$

Parametrix at Edge Points

We consider small disks $D(\pm z_j, r)$ with radius $r > 0$ and centered at the edge points, and look for a local parametrix P defined on the union of the four disks such that

- P is analytic on $D(\pm z_j, r) \setminus (\mathbb{R} \cup \Gamma)$,
-

$$P_+(z) = P_-(z) j_S(z), \quad z \in (\mathbb{R} \cup \Gamma) \cap D(\pm z_j, r), \tag{4.511}$$

- as $n \to \infty$,

$$P(z) = \left(I + O\left(\frac{1}{n}\right)\right) M(z) \quad \text{uniformly for } z \in \partial D(\pm z_j, r). \tag{4.512}$$

We consider here the edge point z_1 in detail. We note that as $z \to z_1$,

$$\begin{aligned}\lambda_1(z) &= q(z-z_1) + \frac{2\rho_1}{3}(z-z_1)^{3/2} + O(z-z_1)^2 \\ \lambda_2(z) &= q(z-z_1) - \frac{2\rho_1}{3}(z-z_1)^{3/2} + O(z-z_1)^2\end{aligned} \tag{4.513}$$

so that

$$\lambda_1(z) - \lambda_2(z) = \frac{4\rho_1}{3}(z-z_1)^{3/2} + O(z-z_1)^{5/2} \tag{4.514}$$

as $z \to z_1$. Then it follows that

$$\beta(z) = \left[\tfrac{3}{4}(\lambda_1(z) - \lambda_2(z))\right]^{2/3} \tag{4.515}$$

is analytic at z_1, real-valued on the real axis near z_1 and $\beta'(z_1) = \rho_1^{2/3} > 0$. So β is a conformal map from $D(z_1, r)$ to a convex neighborhood of the origin, if r is sufficiently small (which we assume to be the case). We take Γ near z_1 such that

$$\beta(\Gamma \cap D(z_1, r)) \subset \{z \mid \arg(z) = \pm 2\pi/3\}\,.$$

Then Γ and $\mathbb{R}$ divide the disk $D(z_1, r)$ into four regions numbered I, II, III, and IV, such that $0 < \arg\beta(z) < 2\pi/3$, $2\pi/3 < \arg\beta(z) < \pi$, $-\pi < \arg\beta(z) < -2\pi/3$, and $-2\pi/3 < \arg\beta(z) < 0$ for z in regions I, II, III, and IV, respectively, see Fig. 4.14.

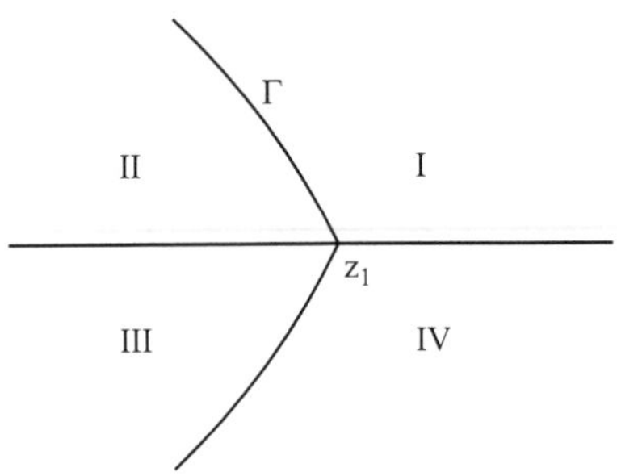

Fig. 4.14. Partition of a neighborhood of the edge point

Recall that the jumps j_S near z_1 are given as

$$
\begin{aligned}
j_S &= \begin{pmatrix} 0 & 1 & 0 \\ -1 & 0 & 0 \\ 0 & 0 & 1 \end{pmatrix} \qquad \text{on } [z_1 - r, z_1) \\
j_S &= \begin{pmatrix} 1 & 0 & 0 \\ \exp(n(\lambda_1 - \lambda_2)) & 1 & \exp(n(\lambda_3 - \lambda_2)) \\ 0 & 0 & 1 \end{pmatrix} \\
&\qquad \text{on the upper boundary of the lens in } D(z_1, r) \\
j_S &= \begin{pmatrix} 1 & 0 & 0 \\ \exp(n(\lambda_1 - \lambda_2)) & 1 & -\exp(n(\lambda_3 - \lambda_2)) \\ 0 & 0 & 1 \end{pmatrix} \\
&\qquad \text{on the lower boundary of the lens in } D(z_1, r) \\
j_S &= \begin{pmatrix} 1 & \exp(n(\lambda_2 - \lambda_1)) & \exp(n(\lambda_3 - \lambda_1)) \\ 0 & 1 & 0 \\ 0 & 0 & 1 \end{pmatrix} \\
&\qquad \text{on } (z_1, z_1 + r]\,.
\end{aligned}
\tag{4.516}
$$

We write

$$
\widetilde{P} = \begin{cases} P \begin{pmatrix} 1 & 0 & 0 \\ 0 & 1 & -\exp(n(\lambda_3 - \lambda_2)) \\ 0 & 0 & 1 \end{pmatrix} & \text{in regions I and IV} \\ P & \text{in regions II and III.} \end{cases}
\tag{4.517}
$$

Then the jumps for $\widetilde{P}$ are $\widetilde{P}_+ = \widetilde{P}_- j_{\widetilde{P}}$ where

$$
\begin{aligned}
j_{\widetilde{P}} &= \begin{pmatrix} 0 & 1 & 0 \\ -1 & 0 & 0 \\ 0 & 0 & 1 \end{pmatrix} \qquad \text{on } [z_1 - r, z_1) \\
j_{\widetilde{P}} &= \begin{pmatrix} 1 & 0 & 0 \\ e^{n(\lambda_1 - \lambda_2)} & 1 & 0 \\ 0 & 0 & 1 \end{pmatrix} \\
&\qquad \text{on the upper side of the lens in } D(z_1, r) \\
j_{\widetilde{P}} &= \begin{pmatrix} 1 & 0 & 0 \\ e^{n(\lambda_1 - \lambda_2)} & 1 & 0 \\ 0 & 0 & 1 \end{pmatrix} \\
&\qquad \text{on the lower side of the lens in } D(z_1, r) \\
j_{\widetilde{P}} &= \begin{pmatrix} 1 & e^{n(\lambda_2 - \lambda_1)} & 0 \\ 0 & 1 & 0 \\ 0 & 0 & 1 \end{pmatrix} \\
&\qquad \text{on } (z_1, z_1 + r]\,.
\end{aligned}
\tag{4.518}
$$

We also need the matching condition

$$\widetilde{P}(z) = \left(I + O\left(\frac{1}{n}\right)\right)M(z) \quad \text{uniformly for } z \in \partial D(z_1, r)\,. \tag{4.519}$$

The RH problem for $\widetilde{P}$ is essentially a 2×2 problem, since the jumps (4.518) are non-trivial only in the upper 2×2 block. A solution can be constructed in a standard way out of Airy functions. The Airy function $\mathrm{Ai}(z)$ solves the equation $y'' = zy$ and for any $\varepsilon > 0$, in the sector $\pi + \varepsilon \le \arg z \le \pi - \varepsilon$, it has the asymptotics as $z \to \infty$,

$$\mathrm{Ai}(z) = \frac{1}{2\sqrt{\pi}z^{1/4}} \exp(-\tfrac{2}{3}z^{3/2})\big(1 + O(z^{-3/2})\big)\,. \tag{4.520}$$

The functions $\mathrm{Ai}(\omega z)$, $\mathrm{Ai}(\omega^2 z)$, where $\omega = e^{\frac{2\pi i}{3}}$, also solve the equation $y'' = zy$, and we have the linear relation,

$$\mathrm{Ai}(z) + \omega\,\mathrm{Ai}(\omega z) + \omega^2\,\mathrm{Ai}(\omega^2 z) = 0\,. \tag{4.521}$$

Write

$$y_0(z) = \mathrm{Ai}(z)\,, \quad y_1(z) = \omega\,\mathrm{Ai}(\omega z)\,, \quad y_2(z) = \omega^2\,\mathrm{Ai}(\omega^2 z)\,, \tag{4.522}$$

and we use these functions to define

$$\Phi(z) = \begin{cases} \begin{pmatrix} y_0(z) & -y_2(z) & 0 \\ y_0'(z) & -y_2'(z) & 0 \\ 0 & 0 & 1 \end{pmatrix}, & \text{for } 0 < \arg z < 2\pi/3\,, \\ \begin{pmatrix} -y_1(z) & -y_2(z) & 0 \\ -y_1'(z) & -y_2'(z) & 0 \\ 0 & 0 & 1 \end{pmatrix}, & \text{for } 2\pi/3 < \arg z < \pi\,, \\ \begin{pmatrix} -y_2(z) & y_1(z) & 0 \\ -y_2'(z) & y_1'(z) & 0 \\ 0 & 0 & 1 \end{pmatrix}, & \text{for } -\pi < \arg z < -2\pi/3\,, \\ \begin{pmatrix} y_0(z) & y_1(z) & 0 \\ y_0'(z) & y_1'(z) & 0 \\ 0 & 0 & 1 \end{pmatrix}, & \text{for } -2\pi/3 < \arg z < 0\,. \end{cases} \tag{4.523}$$

Then

$$\begin{aligned} \widetilde{P}(z) &= E_n(z)\Phi\big(n^{2/3}\beta(z)\big) \\ &\times \mathrm{diag}\Big(\exp\Big(\tfrac{1}{2}n\big(\lambda_1(z) - \lambda_2(z)\big)\Big), \exp\Big(-\tfrac{1}{2}n\big(\lambda_1(z) - \lambda_2(z)\big)\Big), 1\Big) \end{aligned} \tag{4.524}$$

where E_n is an analytic prefactor that takes care of the matching condition (4.519). Explicitly, E_n is given by

$$E_n = \sqrt{\pi}M \begin{pmatrix} 1 & -1 & 0 \\ -\mathrm{i} & -\mathrm{i} & 0 \\ 0 & 0 & 1 \end{pmatrix} \begin{pmatrix} n^{1/6}\beta^{1/4} & 0 & 0 \\ 0 & n^{-1/6}\beta^{-1/4} & 0 \\ 0 & 0 & 1 \end{pmatrix}. \tag{4.525}$$

A similar construction works for a parametrix P around the other edge points.

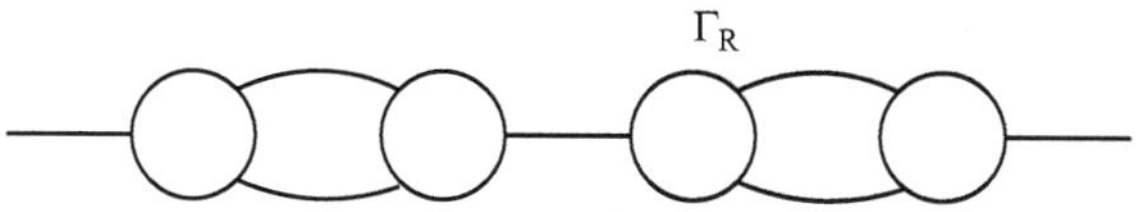

Fig. 4.15. The contour Γ_R for R

Third Transformation

In the third and final transformation we put

$$\begin{aligned} R(z) &= S(z)M(z)^{-1} && \text{for } z \text{ outside the disks } D(\pm z_j, r)\ ,\ j=1,2 \\ R(z) &= S(z)P(z)^{-1} && \text{for } z \text{ inside the disks}\ . \end{aligned} \tag{4.526}$$

Then R is analytic on $\mathbb{C} \setminus \Gamma_R$, where Γ_R consists of the four circles $\partial D(\pm z_j, r)$, $j = 1, 2$, the parts of Γ outside the four disks, and the real intervals $(-\infty, -z_1 - r)$, $(-z_2 + r, z_2 - r)$, $(z_1 + r, \infty)$, see Fig. 4.15. There are jump relations

$$R_+ = R_- j_R \tag{4.527}$$

where

$$\begin{aligned} j_R &= MP^{-1} && \text{on the circles, oriented counterclockwise} \\ j_R &= M j_S M^{-1} && \text{on the remaining parts of } \Gamma_R\ . \end{aligned} \tag{4.528}$$

We have that $j_R = I + O(1/n)$ uniformly on the circles, and $j_R = I + O(\mathrm{e}^{-cn})$ for some $c > 0$ as $n \to \infty$, uniformly on the remaining parts of Γ_R. So we can conclude

$$j_R(z) = I + O\left(\frac{1}{n}\right) \quad \text{as } n \to \infty, \text{ uniformly on } \Gamma_R\ . \tag{4.529}$$

As $z \to \infty$, we have

$$R(z) = I + O(1/z)\ . \tag{4.530}$$

From (4.527), (4.529), (4.530) and the fact that we can deform the contours in any desired direction, it follows that

$$R(z) = I + O\left(\frac{1}{n(|z|+1)}\right) \quad \text{as } n \to \infty\ . \tag{4.531}$$

uniformly for $z \in \mathbb{C} \setminus \Gamma_R$, see [47,52,53,81].

4.5.5 Construction of a Parametrix in the Case $a < 1$

4.5.5.1 λ-functions

Consider the Riemann surface given by (4.480) for $a < 1$, see the right surface on Fig. 4.11. There are three roots to this equation, which behave at infinity as in (4.492). We need the integrals of the ξ-functions, which we define as

$$\begin{aligned}\lambda_1(z)=\int_{z_1}^{z}\xi_1(s)\,\mathrm{d}s\,,\quad \lambda_2(z)=\int_{z_1}^{z}\xi_2(s)\,\mathrm{d}s\,,\\ \lambda_3(z)=\int_{-z_{1+}}^{z}\xi_3(s)\,\mathrm{d}s+\lambda_{1-}(-z_1)\,,\end{aligned}\tag{4.532}$$

The path of integration for λ_3 lies in $\mathbb{C}\setminus((-\infty,0]\cup[-\mathrm{i}z_2,\mathrm{i}z_2])$, and it starts at the point $-z_1$ on the upper side of the cut. All three λ-functions are defined on their respective sheets of the Riemann surface with an additional cut along the negative real axis. Thus λ_1, λ_2, λ_3 are defined and analytic on $\mathbb{C}\setminus(-\infty,z_1]$, $\mathbb{C}\setminus((-\infty,z_1]\cup[-\mathrm{i}z_2,\mathrm{i}z_2])$, and $\mathbb{C}\setminus((-\infty,0]\cup[-\mathrm{i}z_2,\mathrm{i}z_2])$, respectively. Their behavior at infinity is

$$\begin{aligned}\lambda_1(z)&=\tfrac{1}{2}z^2-\log z+\ell_1+O(1/z)\\ \lambda_2(z)&=az+\tfrac{1}{2}\log z+\ell_2+O(1/z)\\ \lambda_3(z)&=-az+\tfrac{1}{2}\log z+\ell_3+O(1/z)\end{aligned}\tag{4.533}$$

for certain constants ℓ_j, $j=1,2,3$. The λ_j's satisfy the following jump relations

$$\begin{aligned}\lambda_{1\mp}&=\lambda_{2\pm} && \text{on } (0,z_1)\,,\\ \lambda_{1-}&=\lambda_{3+} && \text{on } (-z_1,0)\,,\\ \lambda_{1+}&=\lambda_{3-}-\pi\mathrm{i} && \text{on } (-z_1,0)\,,\\ \lambda_{2\mp}&=\lambda_{3\pm} && \text{on } (0,\mathrm{i}z_2)\,,\\ \lambda_{2\mp}&=\lambda_{3\pm}-\pi\mathrm{i} && \text{on } (-\mathrm{i}z_2,0)\,,\\ \lambda_{1+}&=\lambda_{1-}-2\pi\mathrm{i} && \text{on } (-\infty,-z_1)\,,\\ \lambda_{2+}&=\lambda_{2-}+\pi\mathrm{i} && \text{on } (-\infty,0)\,,\\ \lambda_{3+}&=\lambda_{3-}+\pi\mathrm{i} && \text{on } (-\infty,-z_1)\,,\end{aligned}\tag{4.534}$$

where the segment $(-\mathrm{i}z_2,\mathrm{i}z_2)$ is oriented upwards.

4.5.5.2 First Transformation $Y\mapsto U$

We define for $z\in\mathbb{C}\setminus(\mathbb{R}\cup[-iz_2,iz_2])$,

$$\begin{aligned}U(z)&=\operatorname{diag}\big(\exp(-n\ell_1),\exp(-n\ell_2),\exp(-n\ell_3)\big)Y(z)\\ &\quad\times\operatorname{diag}\Big(\exp\big(n(\lambda_1(z)-\tfrac{1}{2}z^2)\big),\exp\big(n(\lambda_2(z)-az)\big),\exp\big(n(\lambda_3(z)+az)\big)\Big).\end{aligned}\tag{4.535}$$

This coincides with the first transformation for $a>1$. Then U solves the following RH problem.

- $U\colon\mathbb{C}\setminus(\mathbb{R}\cup[-iz_2,iz_2])\to\mathbb{C}^{3\times3}$ is analytic.
- U satisfies the jumps

$$U_+ = U_- \begin{pmatrix} \exp(n(\lambda_{1+} - \lambda_{1-})) & \exp(n(\lambda_{2+} - \lambda_{1-})) & \exp(n(\lambda_{3+} - \lambda_{1-})) \\ 0 & \exp(n(\lambda_{2+} - \lambda_{2-})) & 0 \\ 0 & 0 & \exp(n(\lambda_{3+} - \lambda_{3-})) \end{pmatrix} \quad \text{on } \mathbb{R}\,, \tag{4.536}$$

and

$$U_+ = U_- \begin{pmatrix} 1 & 0 & 0 \\ 0 & \exp(n(\lambda_{2+} - \lambda_{2-})) & 0 \\ 0 & 0 & \exp(n(\lambda_{3+} - \lambda_{3-})) \end{pmatrix} \quad \text{on } [-\mathrm{i}z_2, \mathrm{i}z_2]\,. \tag{4.537}$$

- $U(z) = I + O(1/z)$ as $z \to \infty$.

4.5.5.3 Second Transformation $U \mapsto T$: Global Opening of a Lens on $[-\mathrm{i}z_2, \mathrm{i}z_2]$

The second transformation is the opening of a lens on the interval $[-\mathrm{i}z_2, \mathrm{i}z_2]$. We consider a contour Σ, which goes first from $-\mathrm{i}z_2$ to $\mathrm{i}z_2$ around the point z_1, and then from $\mathrm{i}z_2$ to $-\mathrm{i}z_2$ around the point $-z_1$, see Fig. 4.16, and such that for $z \in \Sigma$,

$$\pm\big(\operatorname{Re}\lambda_2(z) - \operatorname{Re}\lambda_3(z)\big) > 0\,, \quad \pm\operatorname{Re} z > 0\,. \tag{4.538}$$

Observe that inside the curvilinear quadrilateral marked by a solid line on Fig. 4.16, $\pm\big(\operatorname{Re}\lambda_2(z)-\operatorname{Re}\lambda_3(z)\big) < 0$, hence the contour Σ has to stay outside of this quadrilateral. We set $T = U$ outside Σ, and inside Σ we set

$$\begin{aligned} T &= U \begin{pmatrix} 1 & 0 & 0 \\ 0 & 1 & 0 \\ 0 & -\exp(n(\lambda_2 - \lambda_3)) & 1 \end{pmatrix} \quad \text{for } \operatorname{Re} z < 0 \text{ inside } \Sigma\,, \\ T &= U \begin{pmatrix} 1 & 0 & 0 \\ 0 & 1 & -\exp(n(\lambda_3 - \lambda_2)) \\ 0 & 0 & 1 \end{pmatrix} \quad \text{for } \operatorname{Re} z > 0 \text{ inside } \Sigma\,. \end{aligned} \tag{4.539}$$

4.5.5.4 Third Transformation $T \mapsto S$: Opening of a Lens on $[-z_1, z_1]$

We open a lens on $[-z_1, z_1]$, inside of Σ, see Fig. 4.17, and we define $S = T$ outside of the lens and

$$S = T \begin{pmatrix} 1 & 0 & 0 \\ 0 & 1 & 0 \\ -\exp(n(\lambda_1 - \lambda_3)) & 0 & 1 \end{pmatrix}$$
in upper part of the lens in left half-plane ,

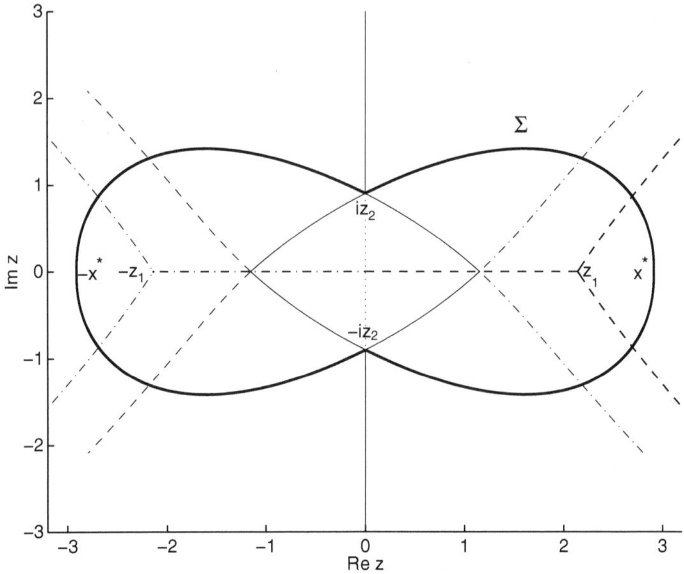

Fig. 4.16. Contour Σ which is such that $\operatorname{Re}\lambda_2 < \operatorname{Re}\lambda_3$ on the part of Σ in the left half-plane and $\operatorname{Re}\lambda_2 > \operatorname{Re}\lambda_3$ on the part of Σ in the right half-plane

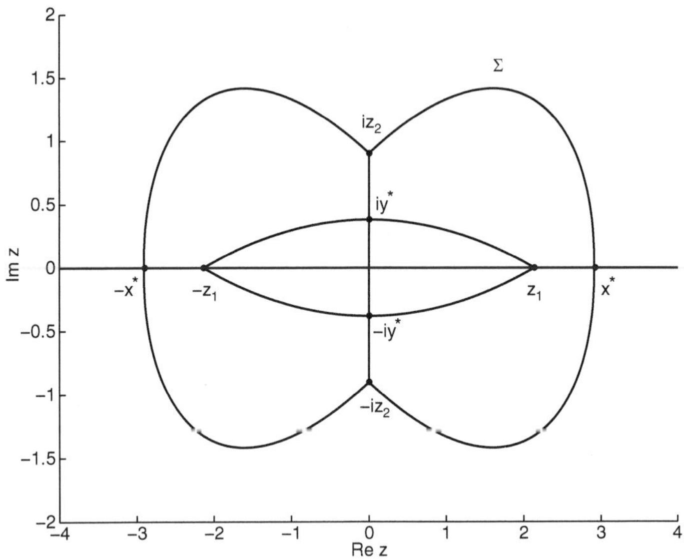

Fig. 4.17. Opening of a lens around $[-z_1, z_1]$

$$S = T \begin{pmatrix} 1 & 0 & 0 \\ 0 & 1 & 0 \\ \exp(n(\lambda_1 - \lambda_3)) & 0 & 1 \end{pmatrix}$$

in lower part of the lens in left half-plane ,

$$S = T \begin{pmatrix} 1 & 0 & 0 \\ -\exp(n(\lambda_1 - \lambda_2)) & 1 & 0 \\ 0 & 0 & 1 \end{pmatrix}$$
in upper part of the lens in right half-plane ,

$$S = T \begin{pmatrix} 1 & 0 & 0 \\ \exp(n(\lambda_1 - \lambda_2)) & 1 & 0 \\ 0 & 0 & 1 \end{pmatrix}$$
in lower part of the lens in right half-plane . (4.540)

Then S satisfies the following RH problem:

- S is analytic outside the real line, the vertical segment $[-\mathrm{i}z_2, \mathrm{i}z_2]$, the curve Σ, and the upper and lower lips of the lens around $[-z_1, z_1]$.
- S satisfies the following jumps on the real line

$$S_+ = S_- \begin{pmatrix} 1 & \exp(n(\lambda_{2+} - \lambda_{1-})) & \exp(n(\lambda_{3+} - \lambda_{1-})) \\ 0 & 1 & 0 \\ 0 & 0 & 1 \end{pmatrix} \quad \text{on } (-\infty, -x^*] \quad (4.541)$$

$$S_+ = S_- \begin{pmatrix} 1 & 0 & \exp(n(\lambda_{3+} - \lambda_{1-})) \\ 0 & 1 & 0 \\ 0 & 0 & 1 \end{pmatrix} \quad \text{on } (-x^*, -z_1] \quad (4.542)$$

$$S_+ = S_- \begin{pmatrix} 0 & 0 & 1 \\ 0 & 1 & 0 \\ -1 & 0 & 0 \end{pmatrix} \quad \text{on } (-z_1, 0) \quad (4.543)$$

$$S_+ = S_- \begin{pmatrix} 0 & 1 & 0 \\ -1 & 0 & 0 \\ 0 & 0 & 1 \end{pmatrix} \quad \text{on } (0, z_1) \quad (4.544)$$

$$S_+ = S_- \begin{pmatrix} 1 & \exp(n(\lambda_2 - \lambda_1)) & 0 \\ 0 & 1 & 0 \\ 0 & 0 & 1 \end{pmatrix} \quad \text{on } [z_1, x^*) \quad (4.545)$$

$$S_+ = S_- \begin{pmatrix} 1 & \exp(n(\lambda_2 - \lambda_1)) & \exp(n(\lambda_3 - \lambda_1)) \\ 0 & 1 & 0 \\ 0 & 0 & 1 \end{pmatrix} \quad \text{on } [x^*, \infty) . \quad (4.546)$$

S has the following jumps on the segment $[-\mathrm{i}z_2, \mathrm{i}z_2]$,

$$S_+ = S_- \begin{pmatrix} 1 & 0 & 0 \\ 0 & 0 & 1 \\ 0 & -1 & \exp(n(\lambda_{3+} - \lambda_{3-})) \end{pmatrix} \quad \text{on } (-\mathrm{i}z_2, -\mathrm{i}y^*) \quad (4.547)$$

$$S_+ = S_- \begin{pmatrix} 1 & 0 & 0 \\ 0 & 0 & 1 \\ \exp(n(\lambda_1 - \lambda_{3-})) & -1 & \exp(n(\lambda_{3+} - \lambda_{3-})) \end{pmatrix}$$

$$\text{on } (-\mathrm{i}y^*, 0) \tag{4.548}$$

$$S_+ = S_- \begin{pmatrix} 1 & 0 & 0 \\ 0 & 0 & 1 \\ -\exp(n(\lambda_1 - \lambda_{3-})) & -1 & \exp(n(\lambda_{3+} - \lambda_{3-})) \end{pmatrix}$$

$$\text{on } (0, \mathrm{i}y^*) \tag{4.549}$$

$$S_+ = S_- \begin{pmatrix} 1 & 0 & 0 \\ 0 & 0 & 1 \\ 0 & -1 & \exp(n(\lambda_{3+} - \lambda_{3-})) \end{pmatrix} \quad \text{on } (\mathrm{i}y^*, \mathrm{i}z_2)\,. \tag{4.550}$$

The jumps on Σ are

$$S_+ = S_- \begin{pmatrix} 1 & 0 & 0 \\ 0 & 1 & 0 \\ 0 & \exp(n(\lambda_2 - \lambda_3)) & 1 \end{pmatrix} \quad \text{on } \{z \in \Sigma \mid \operatorname{Re} z < 0\} \tag{4.551}$$

$$S_+ = S_- \begin{pmatrix} 1 & 0 & 0 \\ 0 & 1 & \exp(n(\lambda_3 - \lambda_2)) \\ 0 & 0 & 1 \end{pmatrix} \quad \text{on } \{z \in \Sigma \mid \operatorname{Re} z > 0\}\,. \tag{4.552}$$

Finally, on the upper and lower lips of the lens, we find jumps

$$S_+ = S_- \begin{pmatrix} 1 & 0 & 0 \\ 0 & 1 & 0 \\ \exp(n(\lambda_1 - \lambda_3)) & 0 & 1 \end{pmatrix}$$

$$\text{on the lips of the lens in the left half-plane} \tag{4.553}$$

$$S_+ = S_- \begin{pmatrix} 1 & 0 & 0 \\ \exp(n(\lambda_1 - \lambda_2)) & 1 & 0 \\ 0 & 0 & 1 \end{pmatrix}$$

$$\text{on the lips of the lens in the right half-plane}\,. \tag{4.554}$$

- $S(z) = I + O(1/z)$ as $z \to \infty$.

As $n \to \infty$, the jump matrices have limits. Most of the limits are the identity matrix, except for the jumps on $(-z_1, z_1)$, see (4.543) and (4.544), and on $(-\mathrm{i}z_2, \mathrm{i}z_2)$, see (4.547)–(4.550). The limiting model RH problem can be solved explicitly. The solution is similar to the case $a > 1$, and it is given by formulas (4.507)–(4.510), with cuts of the function $\sqrt{P(z)}$ on the intervals $[-z_1, z_1]$ and $[-\mathrm{i}z_2, \mathrm{i}z_2]$.

4.5.5.5 Local Parametrix at the Branch Points for $a < 1$

Near the branch points the model solution M will not be a good approximation to S. We need a local analysis near each of the branch points. In a small circle around each of the branch points, the parametrix P should have the same jumps as S, and on the boundary of the circle P should match with M in the sense that

$$P(z) = M(z)\big(I + O(1/n)\big) \tag{4.555}$$

uniformly for z on the boundary of the circle.

The construction of P near the real branch points $\pm z_1$ makes use of Airy functions and it is the same as the one given above for the case $a > 1$. The parametrix near the imaginary branch points $\pm \mathrm{i}z_2$ is also constructed with Airy functions. We give the construction near $\mathrm{i}z_2$. There are three contours, parts of Σ, meeting at $\mathrm{i}z_2$: left, right and vertical, see Fig. 4.17. We want an analytic P in a neigborhood of $\mathrm{i}z_2$ with jumps

$$\begin{aligned}
P_+ &= P_- \begin{pmatrix} 1 & 0 & 0 \\ 0 & 1 & 0 \\ 0 & \exp(n(\lambda_2 - \lambda_3)) & 1 \end{pmatrix} && \text{on left contour} \\
P_+ &= P_- \begin{pmatrix} 1 & 0 & 0 \\ 0 & 1 & \exp(n(\lambda_3 - \lambda_2)) \\ 0 & 0 & 1 \end{pmatrix} && \text{on right contour} \\
P_+ &= P_- \begin{pmatrix} 1 & 0 & 0 \\ 0 & 0 & 1 \\ 0 & -1 & \exp(n(\lambda_{3+} - \lambda_{3-})) \end{pmatrix} && \text{on vertical part}\,.
\end{aligned} \tag{4.556}$$

In addition we need the matching condition (4.555). Except for the matching condition (4.555), the problem is a 2×2 problem. We define

$$f(z) = \left[\tfrac{3}{4}(\lambda_2 - \lambda_3)(z)\right]^{2/3} \tag{4.557}$$

such that

$$\arg f(z) = \pi/3\,, \quad \text{for } z = \mathrm{i}y\,,\ \ y > z_2\,.$$

Then $s = f(z)$ is a conformal map, which maps $[0, \mathrm{i}z_2]$ into the ray $\arg s = -2\pi/3$, and which maps the parts of Σ near $\mathrm{i}z_2$ in the right and left half-planes into the rays $\arg s = 0$ and $\arg s = 2\pi/3$, respectively. The local parametrix has the form,

$$P(z) = E(z)\Phi\big(n^{2/3}f(z)\big) \begin{pmatrix} 1 & 0 & 0 \\ 0 & \exp\big(\tfrac{1}{2}n(\lambda_2 - \lambda_3)\big) & 0 \\ 0 & 0 & \exp\big(-\tfrac{1}{2}n(\lambda_2 - \lambda_3)\big) \end{pmatrix} \tag{4.558}$$

where E is analytic. The model matrix-valued function Φ is defined as

$$\begin{aligned}
\Phi &= \begin{pmatrix} 1 & 0 & 0 \\ 0 & y_0 & -y_2 \\ 0 & y_0' & -y_2' \end{pmatrix} && \text{for } 0 < \arg s < 2\pi/3\,, \\
\Phi &= \begin{pmatrix} 1 & 0 & 0 \\ 0 & y_0 & y_1 \\ 0 & y_0' & y_1' \end{pmatrix} && \text{for } -2\pi/3 < \arg s < 0\,, \\
\Phi &= \begin{pmatrix} 1 & 0 & 0 \\ 0 & -y_1 & -y_2 \\ 0 & -y_1' & -y_2' \end{pmatrix} && \text{for } 2\pi/3 < \arg s < 4\pi/3\,,
\end{aligned} \tag{4.559}$$

where $y_0(s) = \mathrm{Ai}(s)$, $y_1(s) = \omega\,\mathrm{Ai}(\omega s)$, $y_2(s) = \omega^2\,\mathrm{Ai}(\omega^2 s)$ with $\omega = 2\pi/3$ and Ai the standard Airy function. In order to achieve the matching (4.555) we define the prefactor E as

$$E = ML^{-1} \tag{4.560}$$

with

$$L = \frac{1}{2\sqrt{\pi}} \begin{pmatrix} 1 & 0 & 0 \\ 0 & n^{-1/6} f^{-1/4} & 0 \\ 0 & 0 & n^{1/6} f^{1/4} \end{pmatrix} \begin{pmatrix} 1 & 0 & 0 \\ 0 & 1 & \mathrm{i} \\ 0 & -1 & \mathrm{i} \end{pmatrix} \tag{4.561}$$

where $f^{1/4}$ has a branch cut along the vertical segment $[0, \mathrm{i}z_2]$ and it is real and positive where f is real and positive. The matching condition (4.555) now follows from the asymptotics of the Airy function and its derivative.

A similar construction gives the parametrix in the neighborhood of $-\mathrm{i}z_2$.

4.5.5.6 Fourth Transformation $S \mapsto R$

Having constructed N and P, we define the final transformation by

$$\begin{aligned} R(z) &= S(z)M(z)^{-1} \qquad \text{away from the branch points}\,, \\ R(z) &= S(z)P(z)^{-1} \qquad \text{near the branch points}\,. \end{aligned} \tag{4.562}$$

Since jumps of S and N coincide on the interval $(-z_1, z_1)$ and the jumps of S and P coincide inside the disks around the branch points, we obtain that R is analytic outside a system of contours as shown in Fig. 4.18.

On the circles around the branch points there is a jump

$$R_+ = R_-\big(I + O(1/n)\big)\,, \tag{4.563}$$

which follows from the matching condition (4.555). On the remaining contours, the jump is

$$R_+ = R_-\big(I + O(\mathrm{e}^{-cn})\big) \tag{4.564}$$

for some $c > 0$. Since we also have the asymptotic condition $R(z) = I + O(1/z)$ as $z \to \infty$, we may conclude as that

$$R(z) = I + O\left(\frac{1}{n(|z|+1)}\right) \quad \text{as } n \to \infty\,, \tag{4.565}$$

uniformly for $z \in \mathbb{C}$.

4.5.6 Double Scaling Limit at $a = 1$

This section is based on the paper [23].

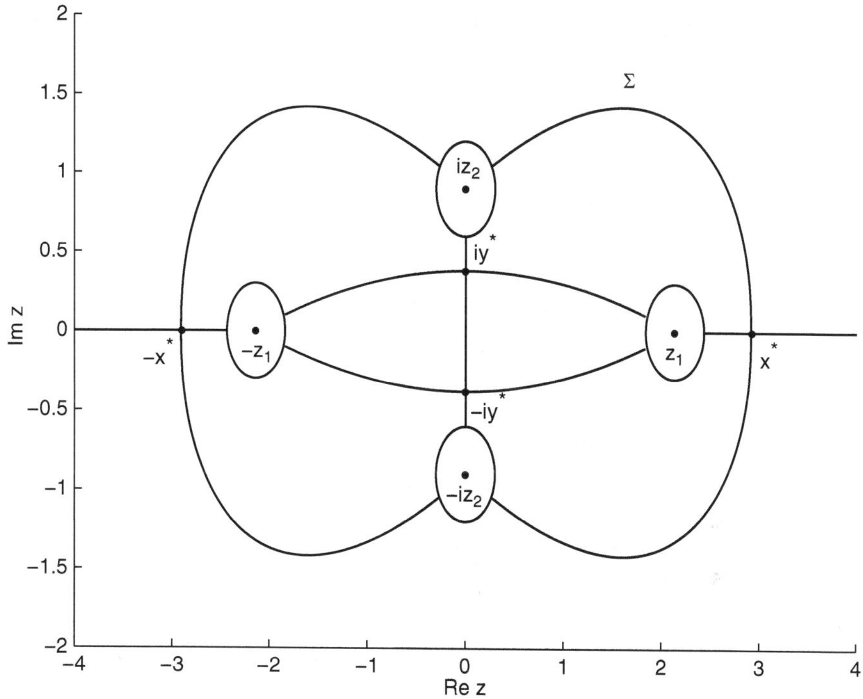

Fig. 4.18. R has jumps on this system of contours

4.5.6.1 Modified Pastur Equation

The analysis in for the cases $a > 1$ and $0 < a < 1$ was based on the Pastur equation (4.480), and it would be natural to use (4.480) also in the case $a = 1$. Indeed, that is what we tried to do, and we found that it works for $a \equiv 1$, but in the double scaling regime $a = 1 + b/(2\sqrt{n})$ with $b \neq 0$, it led to problems that we were unable to resolve in a satisfactory way. A crucial feature of our present approach is a *modification* of (4.480) when a is close to 1, but different from 1. At $x = 0$ we wish to have a *double branch point* for all values of a so that the structure of the Riemann surface is as in the middle figure of Fig. 4.11 for all a.

For $c > 0$, we consider the Riemann surface for the equation

$$z = \frac{w^3}{w^2 - c^2} \tag{4.566}$$

where w is a new auxiliary variable. The Riemann surface has branch points at $z^* = 3\sqrt{3}/2c$, $-z^*$ and a double branch point at 0. There are three inverse functions w_k, $k = 1, 2, 3$, that behave as $z \to \infty$ as

$$\begin{aligned} w_1(z) &= z - \frac{c^2}{z} + O\left(\frac{1}{z^3}\right) \\ w_2(z) &= c + \frac{c^2}{2z} + O\left(\frac{1}{z^2}\right) \\ w_3(z) &= -c + \frac{c^2}{2z} + O\left(\frac{1}{z^2}\right) \end{aligned} \tag{4.567}$$

and which are defined and analytic on $\mathbb{C} \setminus [-z^*, z^*]$, $\mathbb{C} \setminus [0, z^*]$ and $\mathbb{C} \setminus [-z^*, 0]$, respectively.

Then we define the modified ξ-functions

$$\xi_k = w_k + \frac{p}{w_k}, \quad \text{for } k = 1, 2, 3, \tag{4.568}$$

which we also consider on their respective Riemann sheets. In what follows we take

$$c = \frac{a + \sqrt{a^2 + 8}}{4} \quad \text{and} \quad p = c^2 - 1. \tag{4.569}$$

Note that $a = 1$ corresponds to $c = 1$ and $p = 0$. In that case the functions coincide with the solutions of (4.480) that we used in our earlier works. From (4.566), 4.568), and (4.569) we obtain the modified Pastur equation

$$\xi^3 - z\xi^2 + (1 - a^2)\xi + a^2 z + \frac{(c^2 - 1)^3}{c^2 z} = 0, \tag{4.570}$$

where c is given by (4.569). This equation has three solutions, with the following behavior at infinity:

$$\begin{aligned} \xi_1(z) &= z - \frac{1}{z} + O\left(\frac{1}{z^3}\right), \\ \xi_{2,3}(z) &= \pm a + \frac{1}{2z} + O\left(\frac{1}{z^2}\right), \end{aligned} \tag{4.571}$$

and the cuts as in the middle figure of Fig. 4.11. At zero the functions ξ_k have the asymptotics,

$$\xi_k(z) = \begin{cases} -\omega^{2k} z^{1/3} f_2(z) - \omega^k z^{-1/3} g_2(z) + \dfrac{z}{3} & \text{for } \operatorname{Im} z > 0, \\ -\omega^k z^{1/3} f_2(z) - \omega^{2k} z^{-1/3} g_2(z) + \dfrac{z}{3} & \text{for } \operatorname{Im} z < 0, \end{cases} \tag{4.572}$$

where the functions $f_2(z), g_2(z)$ are analytic at the origin and real for real z, with

$$f_2(0) = c^{2/3} + \frac{1}{3} c^{-4/3}(c^2 - 1), \quad g_2(0) = c^{-2/3}(c^2 - 1). \tag{4.573}$$

We define then the functions λ_k as

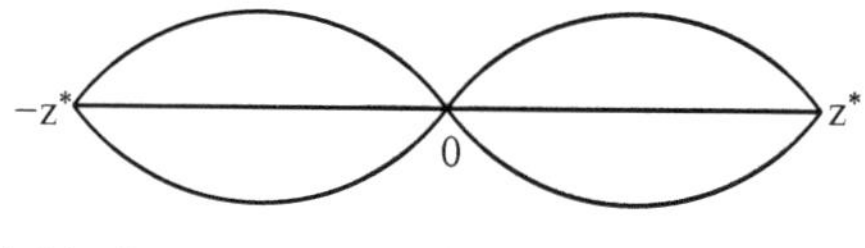

Fig. 4.19. Lens structure for the double scaling limit

$$\lambda_k(z) = \int_{0+}^{z} \xi_k(s)\,\mathrm{d}s \tag{4.574}$$

where the path of integration starts at 0 on the upper side of the cut and is fully contained (except for the initial point) in $\mathbb{C} \setminus (-\infty, z^*]$, and we define the first transformation of the RHP by the same formula (4.497) as for the case $a > 1$. For what follows, observe that the λ-functions have the following asymptotics at the origin:

$$\lambda_k(z) = \begin{cases} -\frac{3}{4}\omega^{2k} z^{4/3} f_3(z) - \frac{1}{2}\omega^k z^{2/3} g_3(z) + \frac{z^2}{6} & \text{for } \operatorname{Im} z > 0\,, \\ \lambda_{k-}(0) - \frac{3}{4}\omega^k z^{4/3} f_3(z) - \frac{1}{2}\omega^{2k} z^{2/3} g_3(z) + \frac{z^2}{6} & \\ & \text{for } \operatorname{Im} z < 0\,, \end{cases} \tag{4.575}$$

where the function f_3 and g_3 are analytic at the origin and

$$\begin{aligned} f_3(0) &= f_2(0) = c^{2/3} + \tfrac{1}{3}c^{-4/3}(c^2-1)\,, \\ g_3(0) &= 3g_2(0) = 3c^{-2/3}(c^2-1)\,, \end{aligned} \tag{4.576}$$

The second transformation, the opening of lenses is given by formulas (4.500), (4.501). The lens structure is shown on Fig. 4.19. The model solution is defined as

$$M(z) = \begin{pmatrix} M_1(w_1(z)) & M_1(w_2(z)) & M_1(w_3(z)) \\ M_2(w_1(z)) & M_2(w_2(z)) & M_2(w_3(z)) \\ M_3(w_1(z)) & M_3(w_2(z)) & M_3(w_3(z)) \end{pmatrix} \tag{4.577}$$

where M_1, M_2, M_3 are the three scalar valued functions

$$\begin{gathered} M_1(w) = \frac{w^2-c^2}{w\sqrt{w^2-3c^2}}\,, \quad M_2(w) = \frac{-\mathrm{i}}{\sqrt{2}}\frac{w+c}{w\sqrt{w^2-3c^2}}\,, \\ M_3(w) = \frac{-\mathrm{i}}{\sqrt{2}}\frac{w-c}{w\sqrt{w^2-3c^2}};. \end{gathered} \tag{4.578}$$

The construction of a parametrix P at the edge points $\pm z^*$ can be done with Airy functions in the same way as for $a > 1$.

4.5.6.2 Parametrix at the Origin

The main issue is the construction of a parametrix at the origin, and this is where the Pearcey integrals come in. The Pearcey differential equation $p'''(\zeta) = \zeta p(\zeta) + bp'(\zeta)$ admits solutions of the form

$$p_j(\zeta) = \int_{\Gamma_j} \exp\left(-\frac{1}{4}s^4 - \frac{b}{2}s^2 + \mathrm{i}s\zeta\right) \mathrm{d}s \tag{4.579}$$

for $j = 0, 1, 2, 3, 4, 5$, where

$$\begin{aligned} \Gamma_0 &= (-\infty, \infty)\,, & \Gamma_1 &= (\mathrm{i}\infty, 0] \cup [0, \infty)\,, \\ \Gamma_2 &= (\mathrm{i}\infty, 0] \cup [0, -\infty)\,, & \Gamma_3 &= (-\mathrm{i}\infty, 0] \cup [0, -\infty)\,, \\ \Gamma_4 &= (-\mathrm{i}\infty, 0] \cup [0, \infty)\,, & \Gamma_5 &= (-\mathrm{i}\infty, \mathrm{i}\infty) \end{aligned} \tag{4.580}$$

or any other contours that are homotopic to them as for example given in Fig. 4.20. The formulas (4.580) also determine the orientation of the contours Γ_j.

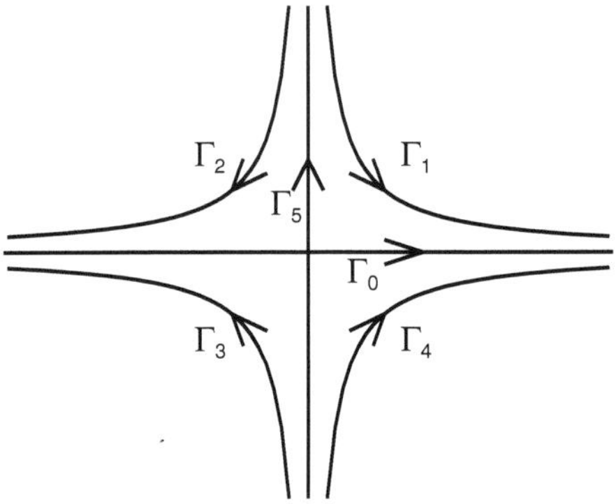

Fig. 4.20. The contours Γ_j, $j = 0, 1, \ldots, 5$, equivalent to those in (4.580), that are used in the definition of the Pearcey integrals p_j

Define $\Phi = \Phi(\zeta; b)$ in six sectors by

$$\Phi = \begin{pmatrix} -p_2 & p_1 & p_5 \\ -p_2' & p_1' & p_5' \\ -p_2'' & p_1'' & p_5'' \end{pmatrix} \quad \text{for } 0 < \arg\zeta < \pi/4 \tag{4.581}$$

$$\Phi = \begin{pmatrix} p_0 & p_1 & p_4 \\ p_0' & p_1' & p_4' \\ p_0'' & p_1'' & p_4'' \end{pmatrix} \quad \text{for } \pi/4 < \arg\zeta < 3\pi/4 \tag{4.582}$$

$$\Phi = \begin{pmatrix} -p_3 & -p_5 & p_4 \\ -p_3' & -p_5' & p_4' \\ -p_3'' & -p_5'' & p_4'' \end{pmatrix} \quad \text{for } 3\pi/4 < \arg\zeta < \pi \tag{4.583}$$

$$\Phi = \begin{pmatrix} p_4 & -p_5 & p_3 \\ p_4' & -p_5' & p_3' \\ p_4'' & -p_5'' & p_3'' \end{pmatrix} \quad \text{for } -\pi < \arg\zeta < -3\pi/4 \tag{4.584}$$

$$\Phi = \begin{pmatrix} p_0 & p_2 & p_3 \\ p_0' & p_2' & p_3' \\ p_0'' & p_2'' & p_3'' \end{pmatrix} \quad \text{for } -3\pi/4 < \arg\zeta < -\pi/4 \tag{4.585}$$

$$\Phi = \begin{pmatrix} p_1 & p_2 & p_5 \\ p_1' & p_2' & p_5' \\ p_1'' & p_2'' & p_5'' \end{pmatrix} \qquad \text{for } -\pi/4 < \arg\zeta < 0\,. \tag{4.586}$$

We define the local parametrix Q in the form

$$Q(z) = E(z)\Phi\bigl(n^{3/4}\zeta(z); n^{1/2}b(z)\bigr)\exp\bigl(n\Lambda(z)\bigr)\exp(-nz^2/6)\,,$$
$$\Lambda = \operatorname{diag}(\lambda_1, \lambda_2, \lambda_3)\,, \tag{4.587}$$

where E is an analytic prefactor, and

$$\zeta(z) = \zeta(z;a) = z[f_3(z;a)]^{3/4} \tag{4.588}$$

and

$$b(z) = b(z;a) = \frac{g_3(z;a)}{f_3(z;a)^{1/2}}\,. \tag{4.589}$$

The functions f_3, g_3 appear in (4.575), in the asymptotics of the λ-functions. In (4.588) and (4.589) the branch of the fractional powers is chosen which is real and positive for real values of z near 0. The prefactor $E(z)$ is defined as

$$E(z) = -\sqrt{\frac{3}{2\pi}}\exp(-nb(z)^2/8)M(z)K(\zeta(z))^{-1}\begin{pmatrix} n^{1/4} & 0 & 0 \\ 0 & 1 & 0 \\ 0 & 0 & n^{-1/4} \end{pmatrix}\,, \tag{4.590}$$

where

$$K(\zeta) = \begin{cases} \begin{pmatrix} \zeta^{-1/3} & 0 & 0 \\ 0 & 1 & 0 \\ 0 & 0 & \zeta^{1/3} \end{pmatrix}\begin{pmatrix} -\omega & \omega^2 & 1 \\ -1 & 1 & 1 \\ -\omega^2 & \omega & 1 \end{pmatrix} & \text{for } \operatorname{Im}\zeta > 0\,, \\ \begin{pmatrix} \zeta^{-1/3} & 0 & 0 \\ 0 & 1 & 0 \\ 0 & 0 & \zeta^{1/3} \end{pmatrix}\begin{pmatrix} \omega^2 & \omega & 1 \\ 1 & 1 & 1 \\ \omega & \omega^2 & 1 \end{pmatrix} & \text{for } \operatorname{Im}\zeta < 0\,. \end{cases} \tag{4.591}$$

4.5.6.3 Final Transformation

We fix $b \in \mathbb{R}$ and let $a = 1 + b/(2\sqrt{n})$ and we define

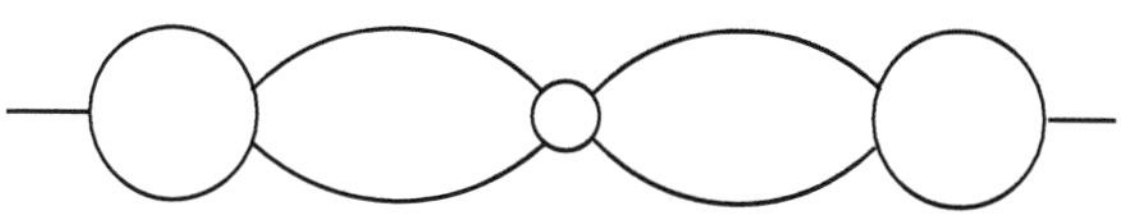

Fig. 4.21. The contour Σ_R. The matrix-valued function R is analytic on $\mathbb{C} \setminus \Sigma_R$. The disk around 0 has radius $n^{-1/4}$ and is shrinking as $n \to \infty$. The disks are oriented counterclockwise and the remaining parts of Σ_R are oriented from left to right

$$R(z) = \begin{cases} S(z)M(z)^{-1}\,, & \text{for } z \in \mathbb{C} \setminus \Sigma_S \text{ outside the disks } D(0, n^{-1/4}) \text{ and } D(\pm 3\sqrt{3}/2, r)\,, \\ S(z)P(z)^{-1}\,, & \text{for } z \in D(\pm 3\sqrt{3}/2, r) \setminus \Sigma_S\,, \\ S(z)Q(z)^{-1}\,, & \text{for } z \in D(0, n^{-1/4}) \setminus \Sigma_S\,. \end{cases} \tag{4.592}$$

Then $R(z)$ is analytic inside the disks and also across the real interval between the disks. Thus, $R(z)$ is analytic outside the contour Σ_R shown in Fig. 4.21. On the contour Σ_R the function $R(z)$ has jumps, so that $R_+(z) = R_-(z)j_R(z)$, where

$$j_R(z) = I + O(n^{-1}) \qquad \text{uniformly for } \left|z \mp \frac{3\sqrt{3}}{2}\right| = r\,, \tag{4.593}$$

$$j_R(z) = I + O(n^{-1/6}) \quad \text{uniformly for } |z| = n^{-1/4}\,, \tag{4.594}$$

and there exists $c > 0$ so that

$$j_R(z) = I + O\left(\frac{e^{-cn^{2/3}}}{1+|z|^2}\right) \quad \text{uniformly for } z \text{ on the remaining parts of } \Sigma_R\,. \tag{4.595}$$

Also, as $z \to \infty$, we have $R(z) = I + O(1/z)$. This implies that

$$R(z) = I + O\left(\frac{n^{-1/6}}{1+|z|}\right) \tag{4.596}$$

uniformly for $z \in \mathbb{C} \setminus \Sigma_R$.

4.5.7 Concluding Remarks

The Riemann–Hilbert approach is a new powerful method for random matrix models and orthogonal polynomials. In this paper we reviewed the approach and some of its applications. Let us mention some recent developments. The RH approach to orthogonal polynomials with complex exponential weights is considered in the recent work of Bertola and Mo [13]. The RHP for discrete orthogonal polynomials and its applications is developed in the monograph [8] of Baik, Kriecherbauer, McLaughlin and Miller. Applications of random matrix models to the exact solution of the six-vertex model with domain wall boundary conditions are considered in the works of Zinn-Justin [105] and Colomo and Pronko [43]. The RH approach to the six-vertex model with domain wall boundary conditions is developed in the work of Bleher and Fokin [16]. The RHP for a two matrix model is considered in the works of Bertola, Eynard, Harnad [11] and Kuijlaars and McLaughlin [83]. The universality results for the scaling limit of correlation functions for orthogonal and symplectic ensembles of random matrices are obtained in the works of Stojanovic [94], Deift and Gioev [48, 49], Costin, Deift and Gioev [44], Deift, Gioev, Kriecherbauer, and Vanlessen [50].

Acknowledgements

The author is supported in part by the National Science Foundation (NSF) Grant DMS-0652005.

References

1. M. Adler, P. van Moerbeke: Ann. Math. **153** 149, (2001)
2. M. Adler, P. van Moerbeke: Ann. Probab. **33**, 1326 (2005)
3. M. Adler, P. van Moerbeke: arXiv:math.PR/0509047
4. A.I. Aptekarev: J. Comput. Math. Appl. **99**, 423 (1998)
5. A.I. Aptekarev, P.M. Bleher, A.B.J. Kuijlaars: Commun. Math. Phys. **259**, 367 (2005)
6. A.I. Aptekarev, A. Branquinho, and W. Van Assche: Trans. Amer. Math. Soc. **355**, 3887 (2003)
7. J. Baik, P. Deift, K. Johansson: J. Amer. Math. Soc. **12**, 1119 (1999)
8. J. Baik, T. Kriecherbauer, K.T.-R. McLaughlin, P.D. Miller: *Discrete Orthogonal Polynomials* (Princeton Univ. Press, Princeton NJ 2006)
9. R. Beals, P.A. Deift, C. Tomei: *Direct and Inverse Scattering on the Line* (Amer. Math. Soc., Providence, RI 1988)
10. M. Bertola, B. Eynard, J. Harnad: J. Phys. A **36**, 3067 (2003)
11. M. Bertola, B. Eynard, J. Harnad: Commun. Math. Phys. **243**, 193 (2003)
12. M. Bertola, B. Eynard, J. Harnad: Commun. Math. Phys. **263**, 401 (2006)
13. M. Bertola, M.Y. Mo: arXiv:math-ph/0605043
14. D. Bessis, C. Itzykson, J.-B. Zuber: Adv. Applied Math. **1**, 109 (1980)
15. P.M. Bleher, B. Eynard: J. Phys. A **36**, 3085 (2003)
16. P.M. Bleher, V. Fokin: Commun. Math. Phys. **268**, 223 (2006)
17. P.M. Bleher, A.R. Its: Ann. Math. **150** 185, (1999)
18. P.M. Bleher, A.R. Its: Commun. Pure Appl. Math., **56**, 433 (2003)
19. P.M. Bleher, A.Its: Ann. Inst. Fourier **55**, 1943 (2005)
20. P.M. Bleher, A.B.J. Kuijlaars: Int. Math. Res. Not. **2004**, 109 (2004)
21. P.M. Bleher, A.B.J. Kuijlaars, Commun. Math. Phys. **252**, 43 (2004)
22. P.M. Bleher, A.B.J. Kuijlaars: Ann. Inst. Fourier **55**, 2001 (2005)
23. P.M. Bleher, A.B.J. Kuijlaars, Commun. Math. Phys. **270**, 481 (2006)
24. G. Bonnet, F. David, B. Eynard: J. Phys. A **33**, 6739 (2000)
25. A. Boutet de Monvel, L.A. Pastur, M. Shcherbina: J. Statist. Phys. **79**, 585 (1995)
26. M.J. Bowick, E. Brézin: Phys. Letts. B **268**, 21 (1991)
27. E. Brézin, S. Hikami: Phys. Rev. E **55**, 4067 (1997)
28. E. Brézin, S. Hikami: Nucl. Phys. B **479**, 697 (1996)
29. E. Brézin, S. Hikami: Phys. Rev. E **56**, 264 (1997)
30. E. Brézin, S. Hikami: Phys. Rev. E **57**, 4140 (1998)
31. E. Brézin, S. Hikami: Phys. Rev. E **58**, 7176 (1998)
32. E. Brézin, S. Hikami, A. Zee: Phys. Rev. E **51**, 5442 (1995)
33. E. Brézin, C. Itzykson, G. Parisi, J.-B. Zuber: Commun. Math. Phys. **59**, 35 (1978)
34. E. Brézin, V. A. Kazakov: Phys. Lett. B **236**, 144 (1990)
35. E. Brézin, A. Zee: Nucl. Phys. B **402**, 613 (1993)

36. R.C. Brower, N. Deo, S. Jain, C.-I. Tan: Nucl. Phys. B **405**, 166 (1993)
37. Y. Chen, T. Grava: J. Phys. A **35**, L45 (2002)
38. Y. Chen, M. Ismail: J. Phys. A **30**, 7817 (1997)
39. Y. Chen, M. Ismail: J. Phys. A **30**, 6633 (1997)
40. T. Claeys, A.B.J. Kuijlaars: Commun. Pure Appl. Math. **59**, 1573 (2006)
41. T. Claeys, A.B.J. Kuijlaars, M. Vanlessen: arXiv:math-ph/0508062
42. K.F. Clancey, I. Gohberg: Factorization of matrix functions and singular integral operators (Birkhäuser, Basel Boston 1981).
43. F. Colomo, A.G. Pronko: J. Stat. Mech. **2005**, P01005 (2005)
44. O. Costin, P. Deift, D. Gioev: arXiv:math-ph/0610063
45. C. Crnkovic, M. Douglas, G. Moore: Int. J. Mod. Phys. A **7**, 7693 (1992)
46. E. Daems, A.B.J. Kuijlaars: J. Approx. Theory **130** 190 (2004)
47. P. Deift: *Orthogonal Polynomials and Random Matrices: a Riemann–Hilbert Approach* (Amer. Math. Soc., Providence RI 1999)
48. P. Deift, D. Gioev: arXiv:math-ph/0411057
49. P. Deift, D. Gioev: arXiv:math-ph/0507023
50. P. Deift, D. Gioev, T. Kriecherbauer, M. Vanlessen: arXiv:math-ph/0612007
51. P. Deift, T. Kriecherbauer, K.T.-R. McLaughlin: J. Approx. Theory **95**, 388 (1998)
52. P. Deift, T. Kriecherbauer, K.T.-R. McLaughlin, S. Venakides, X. Zhou: Commun. Pure Appl. Math. **52** 1335, (1999)
53. P. Deift, T. Kriecherbauer, K.T.-R. McLaughlin, S. Venakides, X. Zhou: Commun. Pure Appl. Math. **52**, 1491 (1999)
54. P. Deift, X. Zhou, Ann. Math. **137**, 295 (1993)
55. K. Demeterfi, N. Deo, S. Jain, C.-I. Tan: Phys. Rev. D, **42**, 4105 (1990)
56. P. Di Francesco, P. Ginsparg, and J. Zinn-Justin: 2D gravity and random matrices, Phys. Rep. **254** (1995)
57. P. Di Francesco: Matrix model combinatorics: applications to folding and coloring. In: *Random Matrices and Their Applications*, ed by P.M. Bleher, A.R. Its (Cambridge Univ. Press, Cambridge 2001) pp 111–170
58. M. Douglas, N. Seiberg, S. Shenker: Phys. Lett. B **244**, 381 (1990)
59. M. Douglas, S. Shenker: Nucl. Phys. B **335**, 635 (1990)
60. B.A. Dubrovin: Usp. Mat. Nauk **36**, 11 (1981) (Russian); English transl.: Russ. Math. Surveys **36**, 11 (1981)
61. M. Duits, A. Kuijlaars: Nonlinearity **19**, 2211 (2006)
62. F.J. Dyson: Commun. Math. Phys. **19**, 235 (1970)
63. B. Eynard: Nucl. Phys. B **506**, 633 (1997)
64. B. Eynard: arXiv:math-ph/0109018
65. N.M. Ercolani, K.D.T.-R. McLaughlin: Int. Math. Res. Not. **2003**, 755 (2003)
66. N.M. Ercolani, K.D.T.-R. McLaughlin, V.U. Pierce: arXiv:math-ph/0606010.
67. H.M. Farkas, I. Kra: *Riemann Surfaces*, 2nd edn (Springer, New York 1992)
68. A.S. Fokas, A. R. Its, A.V. Kitaev: Commun. Math. Phys. **147**, 395 (1992)
69. P.J. Forrester, NucL. Phys. B **402**, 709 (1993)
70. D. Gross, A. Migdal: Phys. Rev. Lett. **64** 127 (1990)
71. D. Gross, A. Migdal, Nucl. Phys. B **340** 333 (1990)
72. D. Gross, E. Witten: Phys. Rev. D **21**, 446 (1980)
73. T. Hollowood, L. Miramontes, A. Pasquinucci, C. Nappi, Nucl. Phys. B **373**, 247 (1992)
74. A.R. Its, A.V. Kitaev, A.S. Fokas, Zap. Nauch. Sem. LOMI **187** 3 (1991) (Russian); English transl, J. Math. Sci. **73**, 415 (1995)

75. A.R. Its, A.B.J. Kuijlaars, J. Östensson: arXiv:0704.1972.
76. A.R. Its, V. Novokshenov: *The Isomonodromic Deformation Method in the Theory of Painlevé Equations* (Springer, Berlin 1986)
77. C. Itzykson, J.-B. Zuber: J. Math. Phys. **21**, 411 (1980)
78. K. Johansson: Duke Math. J. **91**, 151 (1988)
79. A.V. Kitaev: Zap. Nauchn. Sem. Leningrad. Otdel. Mat. Inst. Steklov. (LOMI) **187**, 53 (1991) (Russian); English transl.: J. Math. Sci. **73**, 446 (1995)
80. T. Kriecherbauer, K.T.-R. McLaughlin: Internat. Math. Res. Notices **1999**, 299 (1999)
81. A.B.J. Kuijlaars: Riemann–Hilbert analysis for orthogonal polynomials. In: *Orthogonal Polynomials and Special Functions*, ed by E. Koelink, W. Van Assche (Springer, Berlin 2003) pp 167–210
82. A.B.J. Kuijlaars, K.T.-R. McLaughlin: Commun. Pure Appl. Math. **53**, 736 (2000)
83. A.B.J. Kuijlaars, K.T.-R. McLaughlin: J. Comput. Appl. Math. **178**, 313 (2005)
84. G.S. Litvinchuk, I.M. Spitkovskii: *Factorization of Measurable Matrix Functions* (Birkhäuser, Basel 1987).
85. M.L. Mehta: *Random Matrices*, 2nd edn (Academic Press, Boston, 1991)
86. G. Moore: Progr. Theor. Phys. Suppl. **102**, 255 (1990)
87. G. Moore: Commun. Math. Phys. **133**, 261 (1990)
88. A. Okounkov, N. Reshetikhin: arXiv:math.CO/0503508
89. L.A. Pastur: Teoret. Mat. Fiz. **10**, 102 (1972) (Russian); English transl.: Theor. Math. Phys. **10**, 67 (1972)
90. L.A. Pastur, M. Shcherbina: J. Stat. Phys. **86**, 109 (1997)
91. T. Pearcey: Philos. Mag. **37**, 311 (1946)
92. V. Periwal, D. Shevitz: Nucl. Phys. B **333**, 731 (1990)
93. E.B. Saff, V. Totik: *Logarithmic Potentials with External Field* (Springer, Berlin 1997)
94. A. Stojanovic: Math. Phys. Anal. Geom. **3** 339 (2000); Errata: **7**, 347 (2004)
95. G. Szegő: *Orthogonal Polynomials*, 4th edn (Amer. Math. Soc., Providence RI 1975)
96. C.A. Tracy, H. Widom: Introduction to random matrices. In: *Geometric and Quantum Aspects of Integrable Systems* (Springer, Berlin, 1993) pp 103–130
97. C.A. Tracy, H. Widom: Commun. Math. Phys. **159**, 151 (1994)
98. C.A. Tracy, H. Widom: Commun. Math. Phys. **263**, 381 (2006)
99. C.A. Tracy, H. Widom: arXiv:math.PR/0607321
100. W. Van Assche, J.S. Geronimo, A.B.J. Kuijlaars: Riemann–Hilbert problems for multiple orthogonal polynomials. In: *Special Functions* 2000: *Current Perspective and Future Directions*, ed by J. Bustoz, M.E.H. Ismail, S.K. Suslov (Kluwer Acad. Publ., Dordrecht, 2001) pp 23–59
101. E. Witten: Two-dimensional gravity and intersection theory on moduli spaces. In: *Surveys in Differential Geometry* (Lehigh Univ., Bethlehem PA, 1991) pp 243–310
102. X. Zhou: SIAM J. Math. Anal. **20**, 966 (1989)
103. P. Zinn-Justin: Nucl. Phys. B **497**, 725 (1997)
104. P. Zinn-Justin: Commun. Math. Phys. **194**, 631 (1998)
105. P. Zinn-Justin: Phys. Rev. E **62**, 3411 (2000)

5

Large N Asymptotics in Random Matrices

The Riemann–Hilbert Approach

Alexander R. Its

Department of Mathematics Sciences, Indiana University–Purdue University Indianapolis, Indianapolis, IN 46202-3216, USA itsa@math.iupui.edu

5.1 The RH Representation of the Orthogonal Polynomials and Matrix Models

5.1.1 Introduction

5.1.1.1 Hermitian Matrix Model

The Hermitian matrix model is defined as the ensemble $\mathcal{H}_N$ of random Hermitian $N \times N$ matrices $M = (M_{ij})_{i,j=1}^{N}$ with the probability distribution

$$\mu_N(\mathrm{d}M) = \widehat{Z}_N^{-1} \exp\bigl(-N \operatorname{Tr} V(M)\bigr)\, \mathrm{d}M \,. \tag{5.1}$$

Here the (Haar) measure $\mathrm{d}M$ is the Lebesgue measure on $\mathcal{H}_N \equiv \mathbb{R}^{N^2}$, i.e.,

$$\mathrm{d}M = \prod_{j} \mathrm{d}M_{jj} \prod_{j<k} \mathrm{d}M_{jk}^{\mathrm{R}} \mathrm{d}M_{jk}^{\mathrm{I}} \,, \quad M_{jk} = M_{jk}^{\mathrm{R}} + \mathrm{i}M_{jk}^{\mathrm{I}} \,.$$

The exponent $V(M)$ is a polynomial of even degree with a positive leading coefficient,

J. Harnad (ed.), *Random Matrices, Random Processes and Integrable Systems*, CRM Series in Mathematical Physics, DOI 10.1007/978-1-4419-9514-8_5,

$$V(z) = \sum_{j=1}^{2m} t_j z^j , \quad t_{2m} > 0 ,$$

and the normalization constant $\widehat{Z}_N$, which is also called *partition function*, is given by the equation,

$$\widehat{Z}_N = \int_{\mathcal{H}_N} \exp(-N \operatorname{Tr} V(M)) \, \mathrm{d}M ,$$

so that,

$$\int_{\mathcal{H}_N} \mu_N(\mathrm{d}M) = 1 .$$

The model is also called a *unitary ensemble*. The use of the word "unitary" refers to the invariance properties of the ensemble under unitary conjugation. The special case when $V(M) = M^2$ is called the *Gaussian Unitary Ensemble* (GUE). (we refer to the book [40] as a basic reference for random matrices; see also the more recent survey [45] and monograph [12]).

5.1.1.2 Eigenvalue Statistics

Let

$$z_0(M) < \cdots < z_N(M)$$

be the ordered eigenvalues of M. It is a basic fact (see e.g. [45] or [12]) that the measure (5.1) induces a probability measure on the eigenvalues of M, which is given by the expression

$$\frac{1}{Z_N} \prod_{1 \le j < k \le N} (z_j - z_k)^2 \exp\Big(-N \sum_{j=1}^{N} V(z_j) \Big) \, \mathrm{d}z_0 \ldots \mathrm{d}z_N , \tag{5.2}$$

where the reduced partition function Z_N is represented by the multiple integral

$$Z_N = \int_{-\infty}^{\infty} \cdots \int_{-\infty}^{\infty} \prod_{1 \le j < k \le N} (z_j - z_k)^2 \exp\Big(-N \sum_{j=1}^{N} V(z_j) \Big) \, \mathrm{d}z_0 \ldots \mathrm{d}z_N . \tag{5.3}$$

The principal object of interest in the random matrix theory is the *m-point correlation function* $K_{Nm}(z_0 \cdots z_m)$ which is defined by the relation

$$\begin{aligned} K_{Nm}(z_0 \cdots z_m) \, \mathrm{d}z_0 \ldots \mathrm{d}z_m \\ &= \text{the joint probability to find the } k\text{th eigenvalue in the} \\ &\quad \text{interval } [z_k, z_k + \mathrm{d}z_k],\ k = 1, \ldots, m. \end{aligned}$$

The principal issue is the *universality properties* of the random matrix ensembles. More specifically, this means the analysis of the m-point correlation function $K_{Nm}(z_0 \cdots z_m)$ and other related distribution functions of the theory, as the size N of the matrices approaches infinity and the proof that the limiting distribution functions are independent on the choice of the measure (i.e., the polynomial $V(M)$). In view of (5.2), (5.3), this analysis suggests the asymptotic evaluation of the multiple integrals of the form

$$\int_{-\infty}^{\infty} \cdots \int_{-\infty}^{\infty} \prod_{1 \le j < k \le N} (z_j - z_k)^2 \exp\left(-N \sum_{j=1}^{N} V(z_j)\right) \prod_{j=1}^{N} f_j(z_j) \,\mathrm{d}z_0 \ldots \mathrm{d}z_N, \tag{5.4}$$

where $f_j(z)$ are given functions, as the number of integrations goes to infinity.

The fundamental difficulties occure when one moves beyond the GUE case. Indeed, in the latter case the integrals (5.4), for the relevant to the theory choices of $f_j(z)$, can be evaluated (reduced to single integrals) via the classical Selberg's formula. It is not known if there is any direct analog of Selberg's formula for arbitrary polynomial $V(z)$. This makes the analysis of (5.4), and most importantly its asymptotic evaluation as $N \to \infty$, an extremely challenging enterprise. Until very recently, only the leading terms of the large N asymptotics of Z_N had been known for generic $V(z)$ (see [43] for more details). This result in fact is highly nontrivial, and it has been of great importance for the development of the Riemann–Hilbert method which is the subject of these lectures.

During the last ten years, a considerable progress has been achieved in random matrix theory which has allowed solutions to a number of the long-standing problems related to the universality in the "beyond GUE" cases (see, e.g., [2, 4, 19, 41]). To a large extend, these developments are due to the deep and profound connections of the random matrices to the theory of integrable systems whose different aspects were first discovered in the early 90s in the physical papers [7,21,31], and in the mathematical papers [45,46]. In turn, this discovery was based on the remarkable works of Mehta, Gauden and Dyson of 60s–70s that linked the random matrix theory to orthogonal polynomials.

5.1.1.3 Connection to Orthogonal Polynomials

Let $\{P_n(z)\}_{n=0}^{\infty}$ be a system of monic orthogonal polynomials on the line, with the exponential weight $\omega(z) = \mathrm{e}^{-NV(z)}$ generated by the same polynomial function $V(z)$ as in the measure μ_N of the Hermitian matrix model (5.1). In other words, let the collection of polynomials $\{P_n(z)\}_{n=0}^{\infty}$ be defined by the relations,

$$\int_{-\infty}^{\infty} P_n(z) P_m(z) \mathrm{e}^{-NV(z)} \,\mathrm{d}z = h_n \delta_{nm}\,, \quad P_n = z^n + \cdots\,. \tag{5.5}$$

Put also ,

$$\psi_n(z) = \frac{1}{\sqrt{h_n}} P_n(z) \mathrm{e}^{-NV(z)/2}.$$

The remarkable formula found by Dyson expresses the m-point correlation function $K_{Nm}(z_0, \dots, z_m)$ in terms of the Christoffel–Darboux kernel corresponding to $P_n(\lambda)$ as follows (see [22]; see also [12, 40, 45]).

$$K_{Nm}(z_0 \cdots z_m) = \det(K_N(z_i, z_j))_{i,j=1}^{m} , \tag{5.6}$$

$$K_N(z, z') = \sqrt{\frac{h_{N+1}}{h_N}} \frac{\psi_{N+1}(z)\psi_N(z') - \psi_N(z)\psi_{N+1}(z')}{z - z'} . \tag{5.7}$$

Simultaneously, the partition function Z_N can be evaluated as the product of the norms h_n,

$$Z_N = N! \prod_{n=0}^{N-1} h_n . \tag{5.8}$$

Equations (5.6)–(5.8) translate the basic asymptotic questions of the random matrix theory to the asymptotic analysis of the orthogonal polynomials $P_n(\lambda)$ as $n, N \to \infty$ and $n/N = O(1)$. This does not though help immediately, since the asymptotic problem indicated constitutes one of the principal analytic challenges of the orthogonal polynomial theory itself. In fact, until very recently, one could only approach the problem by using either the classical Heine–Borel formula or the second order linear ODE, which both can be written for any exponential weight $\omega(\lambda)$ (see [10, 29]). The Heine–Borel formula gives for the polynomials $P_n(\lambda)$ an n-fold integral representation of the same type (5.4) or, equivalently, a representation in terms of $n \times n$ Hankel determinants (see, e.g., [12]). Therefore, for the non-Gaussian $V(z)$, i.e., for non-Hermite polynomials, we again face the "beyond GUE" problem, i.e., the impossibility to use the Selberg integral approach for the large n, N asymptotic analysis. Similarly, in the non-Hermite cases, one also loses the possibility to use directly the second order linear ODE mentioned above. Indeed, it is *only in the Gaussian case that the coefficients of this ODE are known a priori for all n.* In general case, they are expressed in terms of the recurrence coefficients of the three term recurrence relation satisfied by P_n which in turn are determined by a *nonlinear* difference equation – the so-called discrete string equation or Freud's equation (see, e.g., [4]). Hence one has to solve two problems simultaneously: to find semiclassical asymptotics for the recurrence coefficients and for the orthogonal polynomials. A way to overcome these difficulties is to use the *Riemann–Hilbert method*, which can be also thought of as a certain "non-commutative" analog of Selberg's integral.

5.1.1.4 The Riemann–Hilbert Method. A General introduction

The Riemann–Hilbert method reduces a particular problem at hand to the *Riemann–Hilbert problem* of analytic factorization of a given matrix-valued function $G(z)$ defined on an oriented contour Γ in the complex z-plane. More precisely, the Riemann–Hilbert problem determined by the pair (Γ, G) consits in finding an matrix-valued function $Y(z)$ with the following properties (we use notation $H(\Omega)$ for the set of analytic in Ω functions).

- $Y(z) \in H(\mathbb{C} \setminus \Gamma)$,
- $Y_+(z) = Y_-(z)G(z)$, $z \in \Gamma$, $\quad Y_\pm = \lim Y(z')$, $z' \to z \in \pm\text{side of } \Gamma$,
- $Y(z) \to I$ as $z \to \infty$.

The Riemann–Hilbert problem can be properly viewed as a "nonabelian" analog of contour integral representation which it reduces to in the abelian case, i.e., when $[G(z_1), G(z_2)] = 0$, $\forall z_1, z_2 \in \Gamma$ (we will say more on this matter and on the history of the method later on; see also [34]).

The main benefit of reducing originally nonlinear problems to the analytic factorization of given matrix functions arises in asymptotic analysis. In typical applications, the jump matrices $G(z)$ are characterized by oscillatory dependence on external large parameters, say space x and time t (in the case of the orthogonal polynomials, these are integers n, N and the coefficients t_j of the potential $V(z)$). The asymptotic evaluation of the solution $Y(z, x, t)$ of the Riemann–Hilbert problem as $x, t \to \infty$ turns out to be in some (not all!) ways quite similar to the asymptotic evaluation of oscillatory contour integrals via the classical method of steepest descent. Indeed, after about 20 years of significant efforts by several authors starting from the 1973 works of Shabat, Manakov, and Ablowitz and Newell (see [17] for a detailed historical review), the development of the relevant scheme of asymptotic analysis of integrable systems finally culminated in the *nonlinear steepest descent method* for oscillatory Riemann–Hilbert problems, which was introduced in 1992 by Deift and Zhou in [13]. In complete analogy to the classical method, it examines the analytic structure of $G(z)$ in order to deform the contour Γ to contours where the oscillatory factors involved become exponentially small as $x, t \to \infty$, and hence the original Riemann–Hilbert problem reduces to a collection of local Riemann–Hilbert problems associated with the relevant saddle points. The noncommutativity of the matrix setting requires, however, developing of several totally new and rather sophisticated technical ideas, which, in particular, enable an explicit solution of the local Riemann–Hilbert problems. (for more details see the following lectures and the original papers of Deift and Zhou, and also the review article [17].) Remarkably, as, in particular, we will see later in this lectures, the final result of the analysis is as efficient as the asymptotic evaluation of the oscillatory integrals.

In the next sections we will present the Riemann–Hilbert setting for the orthogonal polynomials, which was first suggested and used in the series of papers [28, 29, 35, 36] in connection with the matrix model of 2D quantum gravity.

5.1.2 The RH Representation of the Orthogonal Polynomials

Let $\{P_n(z)\}$ be the collection of orthogonal polynomials (5.5). Put

$$Y(z) = \begin{pmatrix} P_n(z) & 1/(2\pi \mathrm{i}) \int_{-\infty}^{\infty} P_n(z')\mathrm{e}^{-NV(z')}/(z'-z)\mathrm{d}z' \\ (-2\pi\mathrm{i}/h_{n-1})P_{n-1}(z) & -1/h_{n-1}\int_{-\infty}^{\infty} P_{n-1}(z')\mathrm{e}^{-N(z')}/(z'-z)\,\mathrm{d}z' \end{pmatrix}$$

$$z \notin \mathbb{R}\,. \quad (5.9)$$

Theorem 5.1.1 ([29]). *The matrix valued function $Y(z)$ solves the following Riemann–Hilbert problem.*

(1) $Y(z) \in H(\mathbb{C} \setminus \mathbb{R})$ and $Y(z)|_{\mathbb{C}_\pm} \in C(\bar{\mathbb{C}}_\pm)$.
(2) $Y_+(z) = Y_-(z) \left(\begin{smallmatrix} 1 & \mathrm{e}^{-NV(z)} \\ 0 & 1 \end{smallmatrix}\right)$, $z \in \mathbb{R}$.
(3) $Y(z) \left(\begin{smallmatrix} z^{-n} & 0 \\ 0 & z^{n} \end{smallmatrix}\right) \equiv Y(z) z^{n\sigma_3} \mapsto I$, $z \to \infty$.

Here, σ_3 denotes the Pauli matrix,

$$\sigma_3 = \begin{pmatrix} 1 & 0 \\ 0 & -1 \end{pmatrix},$$

$\mathbb{C}_+$ and $\mathbb{C}_-$ stand for the upper and lower half-planes, respectively, $\overline{\mathbb{C}}_+$ and $\overline{\mathbb{C}}_-$ denote the respective closures, and $Y_\pm(z)$ are the boundary values of $Y(z)$ on the real line,

$$Y_\pm(z) = \lim_{\substack{z' \to z \\ \pm \operatorname{Im} z' > 0}} Y(z').$$

The asymptotic condition (3) means the uniform estimate

$$|Y(z) \cdot z^{n\sigma_3} - I| \le \frac{C}{|z|}, \quad |z| > 1\,, \tag{5.10}$$

for some positive constant C.

Proof. The first part of statement (1) is a direct corollary of the standard properties of Cauchy type integrals. The continuity of the function $Y(z)$ up to the real line, from above and from below, follows from the possibilty to bend, up and down, the contour of integrations (i.e., the real line) in the Cauchy integrals in the right hand side of equation (5.9)). Indeed, assuming that $\operatorname{Im} z' > 0$, we can rewrite the integral representation for $Y_{12}(z')$ in the form,

$$Y_{12}(z') = \frac{1}{2\pi\mathrm{i}} \int_{-\infty}^{\infty} \frac{P_n(z'')\mathrm{e}^{-NV(z'')}}{z'' - z'}\,\mathrm{d}z''$$
$$= \frac{1}{2\pi\mathrm{i}} \int_{\Gamma_+} \frac{P_n(z'')\mathrm{e}^{-NV(z'')}}{z'' - z'}\,\mathrm{d}z''\,,$$

where the contour Γ_+ is depicted in Fig. 5.1. Hence, the limit value $Y_{12+}(z) \equiv \lim_{z' \to z, \operatorname{Im} z' > 0} Y(z')$, $z \in \mathbb{R}$ exists and, moreover, is given by the equation,

$$Y_{12+}(z) = \frac{1}{2\pi\mathrm{i}} \int_{\Gamma_+} \frac{P_n(z')\mathrm{e}^{-NV(z')}}{z' - z}\,\mathrm{d}z'\,. \tag{5.11}$$

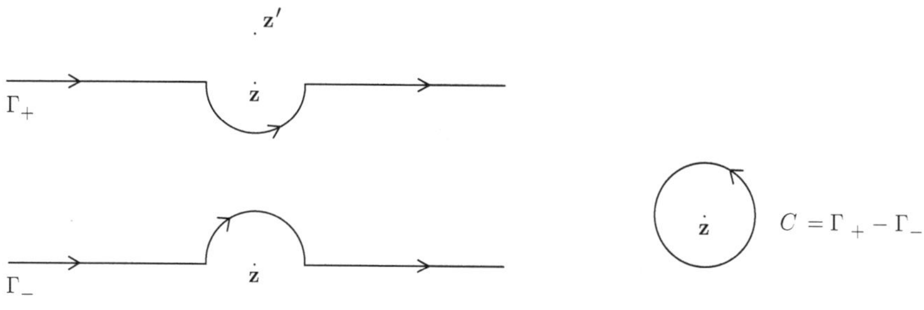

Fig. 5.1. The contours $\Gamma_\pm$ and C

Similarly,

$$Y_{12-}(z) = \frac{1}{2\pi\mathrm{i}} \int_{\Gamma_-} \frac{P_n(z')\mathrm{e}^{-NV(z')}}{z'-z}\,\mathrm{d}z' , \tag{5.12}$$

with the contour Γ_- depicted in Fig. 5.1 as well.

In addition, from (5.11) and (5.12) it follows that

$$Y_{12+}(z) - Y_{12-}(z) = \frac{1}{2\pi\mathrm{i}} \int_{C} \frac{P_n(z')\mathrm{e}^{-NV(z')}}{z'-z}\,\mathrm{d}z',$$

where $C \equiv \Gamma_+ - \Gamma_-$ is a circle centered at z (see again Fig. 5.1) and hence, by residue theorem,

$$Y_{12+}(z) - Y_{12-}(z) = P_n(z)\mathrm{e}^{-NV(z)} \tag{5.13}$$

Similarly,

$$Y_{22+}(z) - Y_{22-}(z) = -\frac{2\pi\mathrm{i}}{h_{n-1}} P_{n-1}(z)\mathrm{e}^{-NV(z)} \tag{5.14}$$

It is also obvious that

$$Y_{11+}(z) = Y_{11-}(z), \quad Y_{21+}(z) = Y_{21-}(z). \tag{5.15}$$

The matrix jump equation (2) of the theorem is equivalent to the scalar equations (5.13)–(5.15) and, therefore, the only statement left is the asymptotic relation (3). We also note that so far we *have not* used the orthogonality condition (5.5). To prove (3) we need first the following simple general fact.

Lemma 5.1.1. *Let $P_n(z) = z^n + \cdots$, be an arbitrary monic polynomial of degree n. Then the following asymptotic relation takes place.*

$$\int_{-\infty}^{\infty} \frac{P_n(z')\mathrm{e}^{-NV(z')}}{z'-z}\,\mathrm{d}z' \cong \sum_{k=1}^{\infty} \frac{c_k}{z^k}, \quad z \to \infty, \tag{5.16}$$

where the coefficients c_k are given by the equations,

$$c_k = -\int_{-\infty}^{\infty} P_n(z) \cdot z^{k-1} \mathrm{e}^{-NV(z)} \, \mathrm{d}z. \tag{5.17}$$

The asymptotics (5.16) *means that for any* $q \in N$ *there is a positive constant* C_q *such that,*

$$\left| \int_{-\infty}^{\infty} \frac{P_n(z')\mathrm{e}^{-NV(z')}}{z' - z} \, \mathrm{d}z' - \sum_{k=1}^{q} \frac{c_k}{z^k} \right| \le \frac{C_q}{|z|^{q+1}} \quad \forall \, |z| \ge 1 \, . \tag{5.18}$$

Proof of the lemma. Let Ω_{ψ_0} denote the domain,

$$z : \psi_0 \le |\arg z| \equiv |\psi| \le \pi - \psi_0 \, , \quad 0 < \psi_0 < \frac{\pi}{2} \, ,$$

depicted in Fig. 5.2.

Then, assuming that $z \in \Omega_{\psi_0}$, we have that

$$\int_{-\infty}^{\infty} \frac{P_n(z')}{z' - z} \mathrm{e}^{-NV(z')} \, \mathrm{d}z' = -\sum_{k=1}^{q} \frac{c_k}{z^k} - \frac{1}{z^{q+1}} \int_{-\infty}^{\infty} \frac{P_n(z')(z')^q \mathrm{e}^{-NV(z')}}{1 - z'/z} \, \mathrm{d}z' \, .$$

Observe also that

$$|z'| < \frac{1}{2}|z| \implies \left| 1 - \frac{z'}{z} \right| \ge 1 - \left| \frac{z'}{z} \right| \ge \frac{1}{2} \, ,$$

while

$$|z'| > \frac{1}{2}|z| \implies \left| 1 - \frac{z'}{z} \right| = \sqrt{\left(1 - \left| \frac{z'}{z} \right| \cos \psi \right)^2 + \left| \frac{z'}{z} \right|^2 \sin^2 \psi} \ge \tfrac{1}{2} |\sin \psi_0| .$$

Therefore, there exists the constant C_{q,ψ_0} such that

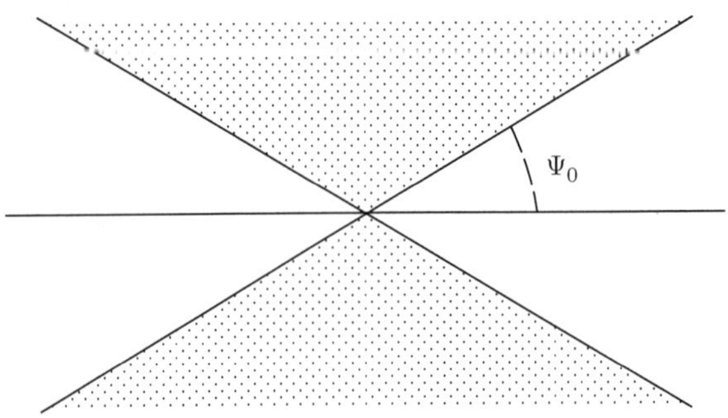

Fig. 5.2. The domain Ω_{ψ_0}

$$\left| \int_{-\infty}^{\infty} \frac{P_n(z')(z')^q e^{-NV(z')}}{1 - z'/z} \, dz' \right| \leq C_{q,\psi_0} \, .$$

This leads to the estimate (5.18) for all $|z| > 1$ and $z \in \Omega_{\psi_0}$. By rotating slightly the contour $\mathbb{R}$ we exclude dependence on ψ_0 and extend the estimate to all $|z| > 1$. □

From Lemma 5.1.1 and orthogonality of $P_n(z)$ it follows that

$$Y_{12}(z) = O(z^{-n-1})$$
$$Y_{22}(z) = z^{-n} + O(z^{-n-1}) \, .$$

Therefore,

$$Y(z) = \begin{pmatrix} z^n + O(z^{n-1}) & O(z^{-n-1}) \\ O(z^{n-1}) & z^{-n} + O(z^{-n-1}) \end{pmatrix}$$
$$\Longrightarrow Y(z) \cdot z^{-n\sigma_3} = \begin{pmatrix} 1 + O(1/z) & O(1/z) \\ O(1/z) & 1 + O(1/z) \end{pmatrix} = I + O\left(\frac{1}{z}\right) .$$

This proves (3). □

Theorem 5.1.2. *$Y(z)$ is defined by conditions* (1)–(3) *uniquely.*

Proof. First we notice that (1)–(3) implies that

$$\det Y(z) \equiv 1.$$

Indeed, the scalar function $d(z) \equiv \det Y(z)$ satisfies the following conditions,

(1d) $d(z) \in H(\mathbb{C} \setminus \mathbb{R})$ and $d(z)|_{\mathbb{C}_\pm} \in C(\overline{\mathbb{C}}_\pm)$.
(2d) $d_+(z) = d_-(z) \det \left(\begin{smallmatrix} 1 & e^{-NV(z)} \\ 0 & 1 \end{smallmatrix} \right) = d_-(z)$, $z \in \mathbb{R}$.
(3d) $d(z) \mapsto 1$, $z \to \infty$.

In virtue of Liouville's theorem, conditions (1d)–(3d) imply that $d(z) \equiv 1$. Assume now that $\widetilde{Y}(z)$ is another solution of the Riemann–Hilbert problem (1)–(3) and consider the matrix ratio.

$$X(z) := \widetilde{Y}(z) Y^{-1}(z) \, .$$

The matrix function $X(z)$ is holomorphioc in $\mathbb{C} \setminus \mathbb{R}$ and its restrictions to $\mathbb{C}_\pm$ are continuous up to the real axis. Since both functions, $\widetilde{Y}(z)$ and $Y(z)$ has the same jump matrix across tha real axis, their ratio $X(z)$ has no jump across the real axis, i.e., on the real axis,

$$X_+(z) = X_-(z).$$

Taking into account that

$$X(z) \mapsto I, \quad z \to \infty,$$

and using again Liouville's theorem, we conclude that $X(z) \equiv I$. This proves the theorem. □

Remark. Thm. 5.1.1 and Lemma 5.1.1 imply, in particular, that normalization condition (3) can be in fact extended to the whole asymptotic series,

$$Y(z)z^{-n\sigma_3} \cong I + \sum_{k=1}^{\infty} \frac{m_k}{z^k}, \quad z \to \infty. \tag{5.19}$$

Thm.s 5.1.1 and 5.1.2 yields the following *Riemann–Hilbert representation* of the orthogonal polynomials $P_n(z)$ and their norms h_n in terms of the solution $Y(z) \equiv Y(z;n)$ of the Riemann–Hilbert problem (1)–(3).

$$P_n(z) = Y_{11}(z)\,, \quad h_n = -2\pi \mathrm{i}(m_1)_{12}\,, \quad h_{n-1}^{-1} = \frac{\mathrm{i}}{2\pi}(m_1)_{21}\,, \tag{5.20}$$

where the matrix $m_1 \equiv m_1(n)$ is the first matrix coefficient of the asymptotic series (5.19).

Equations (5.20) reduce the problem of the asymptotic analysis of the orthogonal polynomials $P_n(z)$ as n, $N \to \infty$ to the problem of the asymptotic analysis of the solution $Y(z)$ of the Riemann–Hilbert problem (1)–(3) as n, $N \to \infty$. Before we proceed with this analysis, we need to review some basic facts of the general Riemann–Hilbert theory.

5.1.3 Elements of the RH Theory

Let us present in more detail the general setting of a Riemann–Hilbert (factorization) problem.

The RH Data

Let Γ be an oriented contour on the complex z-plane. The contour Γ might have points of self-intersection, and it might have more than one connected component. Fig. 5.3 depicts typical contours appearing in applications.

The orientation defines the $+$ and the $-$ sides of Γ in the usual way. Suppose in addition that we are given a map G from Γ into the set of $p \times p$ invertible matrices,

$$G : \Gamma \to \mathrm{GL}(p, \mathbb{C}).$$

The Riemann–Hilbert problem determined by the pair (Γ, G) consists in finding an $p \times p$ matrix function $Y(z)$ with the following properties

(a) $Y(z) \in H(\mathbb{C} \setminus \Gamma)$
(b) $Y_+(z) = Y_-(z)G(z)$, $z \in \Gamma$
(c) $Y(z) \mapsto I$, $z \to \infty$.

We note that the orthogonal polynomial Riemann–Hilbert problem (1)–(3) can be easily reformulated as a problem of the type (a)–(c) by introducing the function,

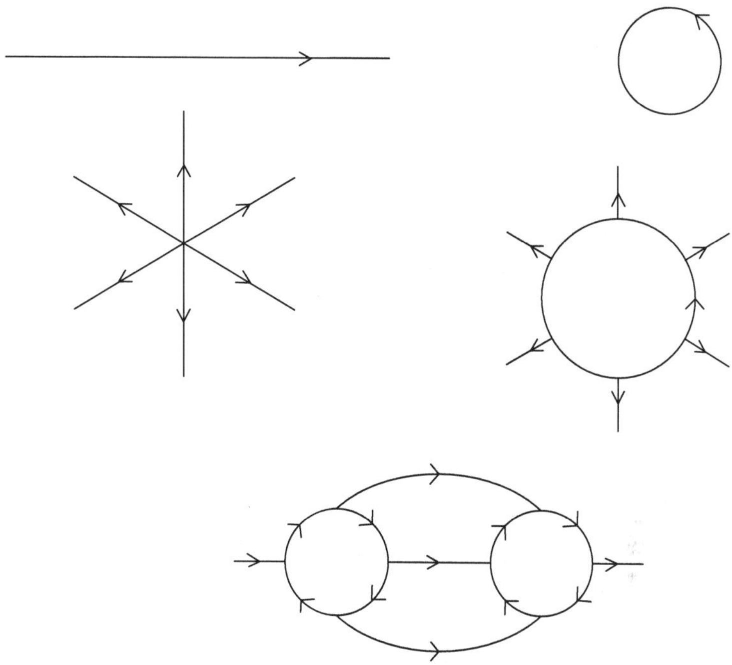

Fig. 5.3. Typical contours Γ

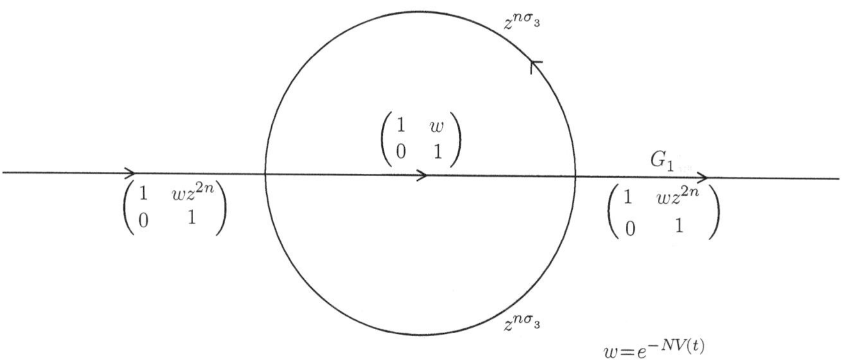

Fig. 5.4. The normalized form of the orthogonal polynomial RH problem

$$\widetilde{Y}(z) = \begin{cases} Y(z) & \text{for } |z| < 1 \,, \\ Y(z)z^{-n\sigma_3} & \text{for } |z| > 1 \,. \end{cases}$$

Indeed, in terms of $\widetilde{Y}(z)$ the problem (1)–(3) becomes the RH problem of the type (a)–(c) posed on the contour Γ which is the union of the real axes and the unit circle, and with the jump matrices indicated in Fig. 5.4.

Precise Sense of (a)–(b)

Let us make the following assumptions about the contours Γ and the jump matrix $G(z)$.

1. The contour Γ is a finite union of smooth curves in $\mathbb{C}P^1$ with the finite number of self-intersections and such that $\mathbb{C}P^1 \setminus \Gamma$ has a finite number of connected components.

2. The jump contour Γ can be represented as a finite union of smooth components in such a way that the restriction of the jump matrix $G(z)$ on any smooth component admits an analytic continuation in a neighborhood of the component.

3. $G(z) \to I$ along any infinite branch of Γ; moreover, the limit is taken exponentially fast.

4. $\det G(z) \equiv 1$.

5. Let A be a point of self-intersection, or a *node* point, of the contour Γ. Let Γ_k, $k = 1, \ldots, q$ denote the smooth components of Γ intersecting at A. We assume that Γ_k are numbered counterclockwise. Let $G_k(z)$ denote the restriction of the jump matrix $G(z)$ on the component Γ_k. By property 2, each $G_k(z)$ is analytic in a neghborhood of the point A; we denote this neighborhood Ω_A. Then, the following *cyclic*, or *trivial monodromy* condition, holds at every node point A:

$$G_1^{\pm 1}(z) G_2^{\pm 1}(z) \cdots G_q^{\pm 1}(z) = I, \quad \forall z \in \Omega_A, \tag{5.21}$$

where in $G_k^{\pm 1}$ the sign "+" is chosen if Γ_k is oriented outwards from A, and the sign "−" is chosen if Γ_k is oriented inwards toward A. We shall say that this property constitutes the *smoothness* of the jump matrix $G(z)$ (cf. [3]). These assumptions allow the following specification of the setting of the RH problem

(a) $Y\big|_{\Omega_k} \in H(\Omega_k) \cap C(\bar{\Omega}_k)$, $\mathbb{C} \setminus \Gamma = \bigcup \Omega_k$.
(b) $Y_+(z) = Y_-(z) G(z)$ $\forall z \in \Gamma$ as continuous functions.
(c) $|I - Y(z)| \leq C/|z|$ $|z| \geq 1$.

This is, of course, not the most general way to set a Riemann–Hilbert problem. Nevertheless, it well covers the needs of the random matrix theory. A more general setting requires an advanced L_p-theory of the Cauchy operators and the Riemann–Hilbert problems themselves and can be found in [3, 8, 15, 16, 39, 49].

We are going now to collect some of the basic facts concerning the Cauchy operators and the general Riemann–Hilbert theory. The proofs as well as more details can be found in the references [3, 8, 15, 16, 39, 49].

Plemelj–Sokhotskii Formulae

The most fundamental ingredient of the Riemann–Hilbert theory is the Plemelj–Sokhotskii formulae for the boundary values of the Cauchy type integrals, which we will formulate as the following theorem.

Theorem 5.1.3. *Let Γ be an oriented smooth closed contour, and $g(z)$ be a Hölder class (matrix-valued) function defined on Γ. Consider the* Cauchy type *integral,*

$$y(z) = \frac{1}{2\pi \mathrm{i}} \int_\Gamma \frac{g(z')}{z'-z} \, \mathrm{d}z' .$$

Denote $\Omega_\pm$ the connected components of the set $\mathbb{C} \setminus \Gamma$. Then the following statements are true.

- $y(z) \in H(\mathbb{C} \setminus \Gamma)$, $y(z)|_{\Omega_\pm} \in C(\bar{\Omega}_\pm)$, *and* $y(z) \to 0$ *as* $z \to \infty$
- *The boundary values, $y_\pm(z)$, of the function $y(z)$ satisfy the equations,*

$$y_\pm(z) = \pm \frac{1}{2} g(z) + \frac{1}{2\pi \mathrm{i}} \, \mathrm{v.p.} \int_\Gamma \frac{g(z')}{z'-z} \, \mathrm{d}z' , \quad z \in \Gamma , \tag{5.22}$$

where

$$\mathrm{v.p.} \int_\Gamma \frac{g(z')}{z'-z} \, \mathrm{d}z' \equiv \lim_{\epsilon \to 0} \int_{\Gamma_\epsilon} \frac{g(z')}{z'-z} \, \mathrm{d}z', \quad \Gamma_\epsilon = \Gamma \setminus (\Gamma \cap \{|z'-z| < \epsilon\}),$$

is the principal value *of the Cauchy integral $\int_\Gamma g(z')/(z'-z) \, \mathrm{d}z'$, $z \in \Gamma$.*
- *(corollary of* (5.22)*)*

$$y_+(z) - y_-(z) = g(z), \quad z \in \Gamma. \tag{5.23}$$

The proof of these classical facts can be found, e.g., in the monograph [30]. We also note that in our proof of Thm. 5.1.1 above, we have practically proven this lemma in the case when $g(z)$ is analytic in the neighborhood of Γ.

Thm. 5.1.3 admits, after proper modifications of its statements, the natural, although nontrivial generalizations to the cases of the piece-wise smooth contours, the contours with the end points and to the functions $g(z)$ belonging to the L_p spaces (see references above). We will discuss in more detail these generalizations when we need them. We shall call the pair "contour–function" and admissible pair, if, with proper modifications, the statements of Thm. 5.1.3 hold.

Corollary 5.1.1. *Given an admissible pair $(\Gamma, g(z))$ the* additive *Riemann–Hilbert problem,*

- $y(z) \in H(\mathbb{C} \setminus \Gamma)$
- $y_+(z) = y_-(z) + g(z)$, $z \in \Gamma$
- $y(z) \to 0$, $z \to \infty$.

admits an explicit solution in terms of the Cauchy integral,

$$y(z) = \frac{1}{2\pi \mathrm{i}} \int_\Gamma \frac{g(z')}{z'-z} \, \mathrm{d}z' . \tag{5.24}$$

Abelian Case. Integral Representations

Assume that the matrices $G(z_1)$ and $G(z_2)$ commute for all pairs $z_1, z_2 \in \Gamma$,

$$[G(z_1), G(z_2)] = 0\,, \quad \forall z_1, z_2 \in \Gamma\,.$$

For instance, this is true for the triangular matrices,

$$G(z) = \begin{pmatrix} 1 & g(z) \\ 0 & 1 \end{pmatrix}\,, \tag{5.25}$$

and, of course, in the scalar case, when $p = 1$. Looking for the solution of the corresponding RH problem from the same commutative subgroup, we can apply logarithm and transform the original, multiplicative jump relation into the additive one,

$$\ln Y_+(z) = \ln Y_-(z) + \ln G(z)\,.$$

Hence, we arrive at the following integral representation of the solution of the RH problem (a)–(c).

$$Y(z) = \exp\left\{\frac{1}{2\pi \mathrm{i}} \int_\Gamma \frac{\ln G(z')}{z' - z}\, \mathrm{d}z'\right\}\,. \tag{5.26}$$

In the triangular case (5.25) this formula simplifies to the expression,

$$Y(z) = \begin{pmatrix} 1 & 1/(2\pi \mathrm{i}) \int_\Gamma g(z')/(z' - z)\, \mathrm{d}z' \\ 0 & 1 \end{pmatrix}\,.$$

It should be noted that there is still a nontrivial technical matter of how to threat equation (5.26) in more general situation. Even in scalar case, there is a subtle question of a possibility for the problem to have a non-zero index, that is, if $\partial\Gamma = 0$ and $\Delta \ln G|_\Gamma \neq 0$. All the same, formula (5.26), after a suitable modification in the case of nonzero index (see, e.g., [30]), yields a contour-integral representation for the solution of the RH problem (a)–(c) in the Abelian case.

Non-Abelian Case. Integral Equations

In the general case, the solution $Y(z)$ of the Riemann–Hilbert problem determined by the pair $(\Gamma, G(z))$ can be always expressed in terms of its boundary values on the contour Γ with the help of the following Cauchy-type integrals.

$$Y(z) = I + \frac{1}{2\pi \mathrm{i}} \int_\Gamma \frac{Y_-(z')(G(z') - I)}{z' - z}\, \mathrm{d}z' \tag{5.27}$$

$$= I + \frac{1}{2\pi \mathrm{i}} \int_\Gamma \frac{Y_+(z')(I - G^{-1}(z'))}{z' - z} dz'\,, \quad z \notin \Gamma\,. \tag{5.28}$$

Note. These integrals have no singularities at the node points since the cyclic condition (5.21).

The proof of (5.27), (5.28) is achieved by observing that each of these equations is equivalent to the relation,

$$Y(z) = I + \frac{1}{2\pi \mathrm{i}} \int_{\Gamma} \frac{Y_+(z') - Y_-(z')}{z' - z} \, \mathrm{d}z' \,, \quad z \notin \Gamma \,, \tag{5.29}$$

which in turn can be shown to be true by application of the Cauchy formula and the normalization condition $Y(z) \to I$ as $z \to \infty$. We will demonstrate this procedure in the two particular examples of the contour Γ. Any other admissible contours can be treated in a similar way.

Case 1. Let Γ be a closed simple contour as indicated in Fig. 5.5. Observe that

$$\Gamma = \partial\Omega_+ = -\partial\Omega_- \,,$$

where $\Omega_\pm$ are connected components of $\mathbb{C} \setminus \Gamma$ – see Fig. 5.5. Assume, for instance, that $z \in \Omega_+$. Then by Cauchy theorem,

$$\frac{1}{2\pi \mathrm{i}} \int_{\Gamma} \frac{Y_+(z')}{z' - z} \, \mathrm{d}z' = \frac{1}{2\pi \mathrm{i}} \int_{\partial\Omega_+} \frac{Y_+(z')}{z' - z} \, \mathrm{d}z'$$
$$= \mathrm{res}\,|_{z'=z} \frac{Y(z')}{z' - z} = Y(z) \,,$$

and

$$\frac{1}{2\pi \mathrm{i}} \int_{\Gamma} \frac{Y_-(z')}{z' - z} \, \mathrm{d}z' = -\frac{1}{2\pi \mathrm{i}} \int_{\partial\Omega_-} \frac{Y_-(z')}{z' - z} \, \mathrm{d}z'$$
$$= -\,\mathrm{res}\,|_{z'=\infty} \frac{Y(z')}{z' - z} = I \,,$$

where in the last equation we have taken into account the normalization condition $Y(z) \to I$, $z \to \infty$ which in the closed contour case becomes just the relation $Y(\infty) = I$. Substitution of these two equations into (5.29) makes the latter an identity.

If $z \in \Omega_-$ we would have instead

$$\frac{1}{2\pi \mathrm{i}} \int_{\Gamma} \frac{Y_+(z')}{z' - z} \, \mathrm{d}z' = \frac{1}{2\pi \mathrm{i}} \int_{\partial\Omega_+} \frac{Y_+(z')}{z' - z} \, \mathrm{d}z' = 0$$

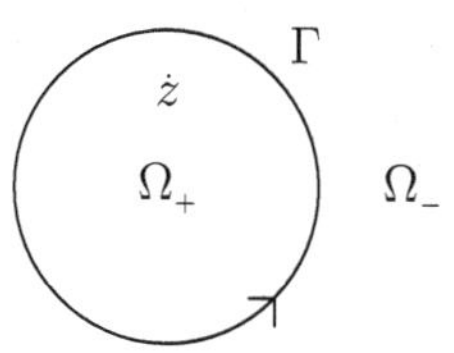

Fig. 5.5. The contour and the domains of Case 1

and

$$\frac{1}{2\pi \mathrm{i}}\int_{\Gamma}\frac{Y_-(z')}{z'-z}\,\mathrm{d}z' = -\frac{1}{2\pi \mathrm{i}}\int_{\partial\Omega_-}\frac{Y_-(z')}{z'-z}\,\mathrm{d}z'$$
$$= -\operatorname{res}|_{z'=z}\frac{Y(z')}{z'-z} - \operatorname{res}|_{z'=\infty}\frac{Y(z')}{z'-z} = -Y(z)+I\,,$$

and (5.29) again follows.

Case 2. Let Γ be an unbounded piece-wise smooth contour as indicated in Fig. 5.6. Denote the right-hand side of (5.27) as $\widetilde{Y}(z)$. Our aim is to show that $Y(z) = \widetilde{Y}(z)$. To this end we first observe that

$$\widetilde{Y}(z) = I + \frac{1}{2\pi \mathrm{i}}\int_{\Gamma}\frac{Y_-(z')(G(z)-I)}{z'-z}\,\mathrm{d}z' \equiv I + \frac{1}{2\pi \mathrm{i}}\int_{\Gamma}\frac{Y_+(z')-Y_-(z')}{z'-z}\,\mathrm{d}z$$
$$= \lim_{R\to\infty}\widetilde{Y}_R(z)\,,$$

where

$$\widetilde{Y}_R(z) = I + \frac{1}{2\pi \mathrm{i}}\int_{\Gamma_R}\frac{Y_+(z')-Y_-(z')}{z'-z}\,\mathrm{d}z,$$

and Γ_R is the R–cut-off of the contour Γ depicted in Fig. 5.7. Introducing the big circle $C^R = C_1^R + C_2^R$ (see Fig. 5.7 again), the formula for $\widetilde{Y}_R(z)$ can be rewritten as the relation,

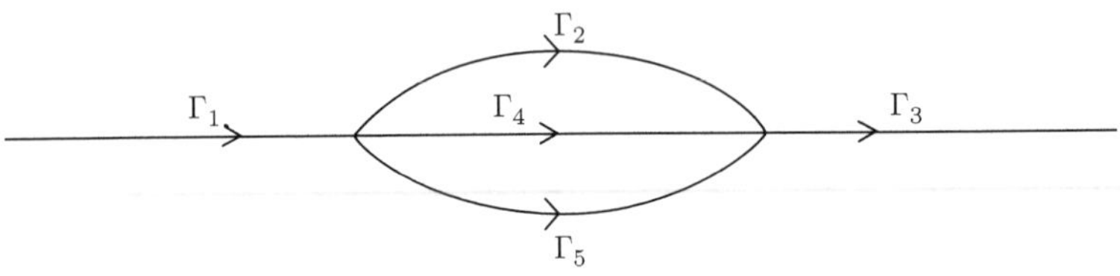

Fig. 5.6. The contour and the domains of Case 2

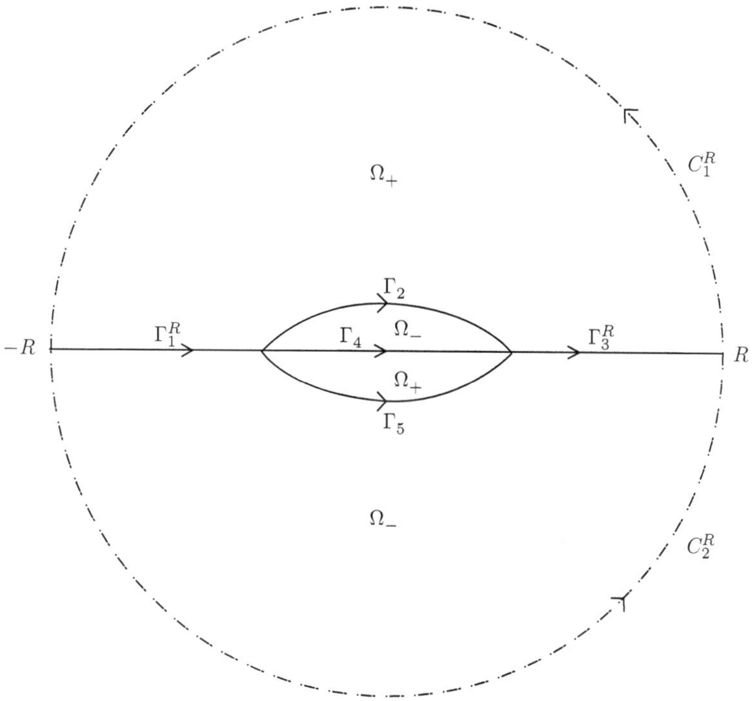

Fig. 5.7. The contour Γ_R

$$
\begin{aligned}
\widetilde{Y}_R(z) = I &+ \left\{ \frac{1}{2\pi \mathrm{i}} \int_{\Gamma_1^R} \frac{Y_+(z')}{z'-z} \,\mathrm{d}z' + \frac{1}{2\pi \mathrm{i}} \int_{\Gamma_2} \frac{Y_+(z')}{z'-z} \,\mathrm{d}z' \right. \\
&\qquad \left. + \frac{1}{2\pi \mathrm{i}} \int_{\Gamma_3^R} \frac{Y_+(z')}{z'-z} \,\mathrm{d}z' + \frac{1}{2\pi \mathrm{i}} \int_{C_1^R} \frac{Y(z')}{z'-z} \,\mathrm{d}z' \right\} \\
&+ \left\{ -\frac{1}{2\pi \mathrm{i}} \int_{\Gamma_2} \frac{Y_-(z')}{z'-z} \,\mathrm{d}z' + \frac{1}{2\pi \mathrm{i}} \int_{\Gamma_4} \frac{Y_+(z')}{z'-z} \,\mathrm{d}z' \right\} \\
&+ \left\{ -\frac{1}{2\pi \mathrm{i}} \int_{\Gamma_4} \frac{Y_-(z')}{z'-z} \,\mathrm{d}z' + \frac{1}{2\pi \mathrm{i}} \int_{\Gamma_5} \frac{Y_t(z')}{z'-z} \,\mathrm{d}z' \right\} \\
&+ \left\{ -\frac{1}{2\pi \mathrm{i}} \int_{\Gamma_1^R} \frac{Y_-(z')}{z'-z} \,\mathrm{d}z' - \frac{1}{2\pi \mathrm{i}} \int_{\Gamma_5} \frac{Y_-(z')}{z'-z} \,\mathrm{d}z' \right. \\
&\qquad \left. - \frac{1}{2\pi \mathrm{i}} \int_{\Gamma_3^R} \frac{Y_-(z')}{z'-z} \,\mathrm{d}z' + \frac{1}{2\pi \mathrm{i}} \int_{C_2^R} \frac{Y(z')}{z'-z} \,\mathrm{d}z' \right\} \\
&- \frac{1}{2\pi \mathrm{i}} \int_{C^R} \frac{Y(z')}{z'-z} ,
\end{aligned}
$$

which in turn is transformed to the equation

$$\widetilde{Y}_R(z) = I + \frac{1}{2\pi \mathrm{i}} \int_{\partial\Omega_+} \frac{Y_+(z')}{z'-z}\,\mathrm{d}z' + \frac{1}{2\pi \mathrm{i}} \int_{\partial\Omega_-} \frac{Y_+(z')}{z'-z}\,\mathrm{d}z' - \frac{1}{2\pi \mathrm{i}} \int_{C^R} \frac{Y(z')}{z'-z}\,.$$

In this last equation, the domains $\Omega_\pm$ are indicated in Fig. 5.7 and $Y_+(z)$ means the boundary value of $Y(z)$ taking from the inside of the respective domain. The normalization condition, $Y(z) \to I$, as $z \to \infty$, implies that the integral over the circle C^R tends to I. At the same time, every z, which is not on Γ, belongs either to Ω_+ or to Ω_- for all sufficiently large R. Therefore, in virtue of the Cauchy theorem the sum of the first two integrals equals $Y(z)$ and hence $\widetilde{Y}_R(z) = Y(z)$ for all sufficiently large R. This in turn implies that

$$\widetilde{Y}(z) = \lim_{R\to\infty} \widetilde{Y}_R(z) = Y(z),$$

as needed.

The arguments which has been just used can be in fact applied to a general admissible contour Γ. Indeed, any contour Γ, after perhaps re-orrientation of some of its parts and addition some new parts accompanied with the trivial jump matrices, can be made to so-call *full contour*. The latter means that $\mathbb{C} \setminus \Gamma = \Omega_+ \cup \Omega_-$ where $\Omega_\pm$ are disjoint, possibly multiply-connected open sets such that $\Gamma = \partial\Omega_+ = -\partial\Omega_-$. Therefore, (5.29) can be always re-written as

$$Y(z) = I + \frac{1}{2\pi \mathrm{i}} \int_{\partial\Omega_+} \frac{Y_+(z')}{z'-z}\,\mathrm{d}z' + \frac{1}{2\pi \mathrm{i}} \int_{\partial\Omega_-} \frac{Y_+(z')}{z'-z}\,\mathrm{d}z'\,, \tag{5.30}$$

where $Y_+(z)$ means the boundary value of $Y(z)$ from the inside of the respective domain. Introducing the R–cut-off of the contour Γ and the big circle C^R, we see that the right-hand side of (5.30) is equal to

$$I + \lim_{R\to\infty} \left\{ \frac{1}{2\pi \mathrm{i}} \int_{\partial\Omega_+^R} \frac{Y_+(z')}{z'-z}\,\mathrm{d}z' + \frac{1}{2\pi \mathrm{i}} \int_{\partial\Omega_-^R} \frac{Y_+(z')}{z'-z}\,\mathrm{d}z' - \frac{1}{2\pi \mathrm{i}} \int_{C^R} \frac{Y(z')}{z'-z} \right\},$$

which is in virtue of the Cauchy theorem and the normalization condition at $z = \infty$ coincide with $Y(z)$.

Corollary. *The solution of the RH problem admits the representation:*

$$Y(z) = I + \frac{1}{2\pi \mathrm{i}} \int_\Gamma \frac{\rho(z')(G(z') - I)}{z'-z}\,\mathrm{d}z' \quad z \notin \Gamma \tag{5.31}$$

where $\rho(z) \equiv Y_-(z)$ satisfies the integral equation

$$\begin{aligned} \rho(z) &= I + \frac{1}{2\pi \mathrm{i}} \int_\Gamma \frac{\rho(z')(G(z') - I)}{z'-z_-}\,\mathrm{d}z' \\ &\equiv I + (C_-[\rho(G-I)])(z), \quad z \in \Gamma, \\ (C_\pm f)(z) &= \lim_{\tilde{z}\to z_\pm} \int_\Gamma \frac{f(z')}{z'-\tilde{z}} \frac{\mathrm{d}z'}{2\pi \mathrm{i}}, \end{aligned} \tag{5.32}$$

and $\tilde{z} \to z_\pm$ denotes the non-tangential limit from the $\pm$ side of Γ, respectively.

5.1.3.1 L_2-boundedness of $C_\pm$ and a small-norm theorem

Theorem 5.1.4. *Let $f \in L^p(\Gamma, |dz|)$, $1 \le p < \infty$. Then the following statements are true.*

- *$(C_\pm f)(z)$, as defined in (5.32), exists for $z \in \Gamma$ a.e.*
- *Assume, in addition, that $1 < p < \infty$. Then*

$$\|C_\pm f\|\|_{L^p(\Gamma)} \le c_p \|f\|\|_{L^p(\Gamma)} \tag{5.33}$$

for some constant c_p. In other words, the Cauchy operators C_+ and C_- are bounded *in the space $L^p(\Gamma)$ for all $1 < p < \infty$ and, in particular, they are bounded in $L^2(\Gamma)$.*
- *As operators in $L^p(\Gamma)$, $1 < p < \infty$, the Cauchy operators $C_\pm$ satisfy the relation (the Plemelj–Sokhotskii formulae for L^p),*

$$C_\pm = \pm\tfrac{1}{2}\mathbf{1} - \tfrac{1}{2}H, \tag{5.34}$$

and (hence)

$$C_+ - C_- = \mathbf{1}. \tag{5.35}$$

Here, $\mathbf{1}$ denotes the identical operator in L^p and H is the Hilbert transform,

$$(Hf)(z) = \text{v.p.} \int_\Gamma \frac{f(z')}{z - z'} \frac{\mathrm{d}z'}{\pi \mathrm{i}}.$$

This theorem summarizes the basic functional analytic properties of the Cauchy operators $C_\pm$ which we will need. We refer the reader to the monographs [8, 39] for the proofs of the theorem's statements. The most important for us is the L^2 -boundedness of the Cauchy operators. The next theorem is a direct corollary of this crucial property, and it plays a central technical role in the asymptotic aspects of the Riemann–Hilbert method.

Theorem 5.1.5. *Assume that the jump matrix $G(z)$ depends on an extra parameter t, and that the matrix $G(z)$ is close to the unit matrix, as $t \to \infty$. More precisely, we assume that*

$$\|G - I\|_{L^2(\Gamma)\cap L^\infty(\Gamma)} < \frac{C}{t^\epsilon}, \quad t \ge t_0\,, \ \epsilon > 0\,. \tag{5.36}$$

Then, for sufficiently large t, the Riemann–Hilbert problem determined by the pair (Γ, G) is uniquely solvable, and its solution $Y(z) \equiv Y(z,t)$ satisfies the following uniform estimate.

$$\|Y(z,t) - I\| < \frac{C}{(1 + |z|^{1/2})t^\epsilon}, \quad z \in \mathbb{K}\,, \ t \ge t_1 \ge t_0\,, \tag{5.37}$$

where $\mathbb{K}$ is a closed subset of $\mathbb{C}P^1 \setminus \Gamma$ satisfying,

$$\frac{\operatorname{dist}\{z;\Gamma\}}{1+|z|} \ge c(\mathbb{K}), \quad \forall z \in \mathbb{K}.$$

If, in addition,

$$\|G - I\|_{L^1(\Gamma)} < \frac{C}{t^{\epsilon}}, \tag{5.38}$$

then $|z|^{1/2}$ in the right-hand side of (5.37) can be replaced by $|z|$:

$$\|Y(z,t) - I\| < \frac{C}{(1+|z|)t^{\epsilon}}, \quad z \in \mathbb{K}, \; t \ge t_1 \ge t_0, \tag{5.39}$$

Proof. We shall use the integral equation (5.32). which can be rewritten as the following equation in $L^2(\Gamma)$:

$$\rho_0(z,t) = F(z,t) + \mathcal{K}[\rho_0](z,t), \quad z \in \Gamma, \tag{5.40}$$

where

$$\rho_0(z,t) := \rho(z,t) - I,$$

$$F(z,t) = \frac{1}{2\pi \mathrm{i}} \int_{\Gamma} \frac{G(z',t) - I}{z' - z_-} \, \mathrm{d}z' \equiv C_-(G-I)(z,t),$$

$$\mathcal{K}[\rho_0](z,t) = \frac{1}{2\pi \mathrm{i}} \int_{\Gamma} \frac{\rho_0(z',t)(G(z',t) - I)}{z' - z_-} \, \mathrm{d}z' \equiv C_-(\rho_0(G-I))(z,t).$$

Due to estimate (5.36) on $G - I$ and the *boundedness* of the Cauchy operator C_- in the space $L^2(\Gamma)$, it follows that F and $\mathcal{K}$ satisfy the estimates

$$\|F\|_{L^2} \le \|C_-\|_{L^2} \cdot \|G - I\|_{L^2} < \frac{C}{t^{\epsilon}},$$

$$\|\mathcal{K}\|_{L^2} \le \|C_-\|_{L^2} \cdot \|G - I\|_{L^\infty} < \frac{C}{t^{\epsilon}},$$

This proves that the integral operator $\mathcal{K}$ is a contraction, thus the solution ρ_0 of (5.40) exists, and, moreover, it satisfies

$$\|\rho_0\|_{L^2} \le 2\|F\|_{L^2} \le \frac{C}{t^{\epsilon}}, \quad t \ge t_1 \ge t_0.$$

This in turn implies the solvability of the Riemann–Hilbert problem. In addition, using the integral representation (5.31) we find

$$\begin{aligned}
\|Y(z,t) - I\| &\le \left\| \int_{\Gamma} \frac{I - G(z',t)}{z' - z} \, \mathrm{d}z' \right\| + \left\| \int_{\Gamma} \frac{\rho_0(z',t)\big(I - G(z',t)\big)}{z' - z} \, \mathrm{d}z' \right\| \\
&\le \|I - G\|_{L^2} \sqrt{\int_{\Gamma} \frac{\mathrm{d}z'}{|z' - z|^2}} + \frac{1}{\operatorname{dist}\{z, \Gamma\}} \|I - G\|_{L^2} \|\rho_0\|_{L^2} \\
&\le \frac{C}{(1+|z|^{1/2})t^{\epsilon}}, \quad t \ge t_1 \ge t_0,
\end{aligned}$$

for z belonging to a subset $\mathbb{K}$. This proves the estimate (5.37). If in addition we have (5.38), then the above estimation of $\|Y(z,t)-I\|$ can be modified as

$$\|Y(z,t)-I\| \leq \frac{1}{\mathrm{dist}\{z,\Gamma\}}\big(\|I-G\|_{L^1} + \|I-G\|_{L^2}\|\rho_0\|_{L^2}\big)\,,$$

and the improved estimate (5.39) follows. The theorem is proven.

It should be noticed that in all our applications, whenever we arrive at a jump matrix which is close to identity it will always be accompanied with the both, $L^2 \cap L^\infty$ and L^1 - norm inequalities (5.36) and (5.38)). That is, we shall always have estimate (5.39) for the Riemann–Hibert problems whose jump matrices are close to identity.

Two Final Remarks

Remark 5.1.1. The following formal arguments show that the Riemann–Hilbert representation of the orthogonal polynomials (as well as similar representations in the general theory of integrable systems) can be treated as a *non-Abelian* analog of the integral representations. Indeed, given any contour integral,

$$\int_\Gamma w(z)\,\mathrm{d}z\,,$$

it can be expressed as the following limit.

$$\int_\Gamma w(z)\,\mathrm{d}z = -2\pi\mathrm{i}\lim_{z\to\infty} zY_{12}(z)\,,$$

where $Y(z)$ is the solution of the following 2×2 *Abelian* Riemann–Hilbert problem,

- $Y(z) \in H(\mathbb{C}\setminus\Gamma)$;
- $Y_+(z) = Y_-(z)\left(\begin{smallmatrix}1 & w(z)\\ 0 & 1\end{smallmatrix}\right)$;
- $Y(z) \mapsto I$, $z\to\infty$.

Remark 5.1.2. Let us introduce the function

$$\Psi(z) = Y(z)\exp\biggl(-\frac{NV(z)\sigma_3}{2}\biggr)\,.$$

Then in terms of this function, the orthogonal polynomial Riemann–Hilbert problem (1)–(3) can be re-written as follows,

(1′) $\Psi(z) \in H(\mathbb{C}\setminus\mathbb{R})$
(2′) $\Psi_+(z) = \Psi_-(z)\left(\begin{smallmatrix}1 & 1\\ 0 & 1\end{smallmatrix}\right)$, $z\in\mathbb{R}$.
(3′) $\Psi(z) = \big(I + O(1/z)\big)\exp(-NV(z)\sigma_3/2 + n\ln z\sigma_3)$, $z\to\infty$.

The principal feature of this RH problem is that its jump matrix is constant with respect to all the parameters involved. By standard arguments of the Soliton theory, based on the Liouville theorem, this implies that the function $\Psi(z) \equiv \Psi(z, n, t_j)$ satisfies the following system of linear differential-difference equations.

$$\left\{\begin{aligned} &\Psi(z, n+1) = U(z)\Psi(z, n) \\ &\frac{\partial \Psi}{\partial z} = A(z)\Psi(z) \\ &\frac{\partial \Psi}{\partial t_j} = V_j(z)\Psi(z) \end{aligned}\right\} \qquad \text{“Lax-pair”}$$

Here, $U(z)$, $A(z)$, and $V_k(z)$ are polynomial on z (for their exact expressions in terms of the corresponding h_n see [10, 29]). The first two equations constitute the Lax pair for the relevant Freud equation, i.e., the nonlinear difference equation on the recursive coefficients, $R_n = h_n/h_{n-1}$. The compatability conditions of the third equations with the different j generate the KP-type hierarchy of the integrable PDEs; the compatability condition of the second and the third equations produces the Schlesinger-type systems and generate the Virasoro-type constraints; the compatability condition of the first and the third equations is related to the Toda-type hierarchy and vertex operators. For more on this very important algebraic ramifications of the random matrix theory see [1, 47, 48] and the lectures of M. Adler and P. van Moerbeke.

The linear system above allows to treat the RH problem (1)–(3) as an *inverse monodromy problem* for the second, i.e., z-equation, and use the isomonodromy method [37] from Soliton Theory for the asymptotic analysis of the orthogonal polynomials. This was first done in [28, 35]. At that time though there were some lack of mathematical rigor in our analysis. These gaps were filled later on with the advance of the NSDM of Deift and Zhou for oscillatory RH problems.

There are two closely related though rather different in a number of key points ways of the asymptotic analysis of the problem (1)–(3). The first one has been developed in [4, 5], and it represents an advance version of the original isomonodromy approach of [28, 35]. It uses an a priori information about the behavior of the parameters h_n obtained from the formal analysis of the Lax pair; indeed, this approach goes back to the 1970s pioneering works of Zakharov and Manakov on the asymptotic analysis of integrable systems (see survey [17]). The second method has been suggested by P. Deift, T. Kriecherbauer, K. T-R McLaughlin, S. Venakides, and X. Zhou in [19], and it constitutes an extension of the original Deift-Zhou approach to the RH problems of the types (1)–(3). The very important advantage of the DKMVZ-scheme is that it does not use any a priori info.

In these lectures, we will present the second, i.e., the DKMVZ approach, considering a quartic potential $V(z) = t_1z^2 + t_2z^4$ as a case study. This will allow us to keep all the calculations on the very elementary level. General case and many more on the matter will be presented in the lectures of Pavel Bleher and Ken McLaughlin. Few remarks about the first approach, i.e., about the method of [4, 5], will be made at the end of Section 5.4.3.

5.2 The Asymptotic Analysis of the RH Problem. The DKMVZ Method

5.2.1 A Naive Approach

But for the normalization condition (3), the RH problem (1)–(3) looks almost as an Abelian problem, so that one might be tempted to make the "obvious" transformation,

$$Y(z) \mapsto \Phi(z) := Y(z)z^{-n\sigma_3} \tag{5.41}$$

and try to make use of the integral representation (5.26). Indeed, with the help of change (5.41), the conditions (1)–(3) transforms to the conditions ,

(1″) $\Phi(z) \in H(\mathbb{C} \setminus \mathbb{R})$
(2″) $\Phi_+(z) = \Phi_-(z) \begin{pmatrix} 1 & z^{2n}e^{-NV(z)} \\ 0 & 1 \end{pmatrix}$, $z \in \mathbb{R}$.
(3″) $\Phi(z) \mapsto I$, $z \to \infty$.

which look quite "Abelian." *But* the price is an extra singularity – the pole of order n which, as the transformation (5.41) suggests, the function $\Phi(z)$ must have at $z = 0$. In other words, the properties (1″)–(3″) has to be supplemented by the forth property,

(4″) $\Phi(z)z^{n\sigma_3} = O(1)$, $z \to 0$,

which means we have gained nothing compare to the original setting (1)–(3) of the RH problem.

5.2.2 The g-Function

Rewrite our "naive" formula (5.41) as

$$Y(z) = \Phi(z)e^{ng(z)\sigma_3} \quad g(z) = \ln z\ , \tag{5.42}$$

and notice that in order to regularize the normalization condition at $z = \infty$ we do not actually need $g(z) \equiv \ln z$. In fact, what is necessary is to have the following properties of the function $g(z)$.

(a) $g(z) = \ln z + O(1/z)$, $z \to \infty$.
(b) $g(z) \in H(\mathbb{C} \setminus \mathbb{R})$.
(c) $|g(z)| < C_R$, $|z| < R$, $\forall R > 0$.

Of course, there are infinitely many ways to achieve (a), (b), (c). Indeed, any function of the form

$$g(z) = \int \ln(z - s)\rho(s)\,\mathrm{d}s, \qquad \int \rho(s)\,\mathrm{d}s = 1 \tag{5.43}$$

would do. Let us then try to figure out what else we would like to gain from the transformation (5.42). To this end let us write down the RH problem in terms of $\Phi(z)$.

$$\Phi_+(z) = \Phi_-(z) \begin{pmatrix} e^{n(g_- - g_+)} & e^{n(g_- + g_+ - \frac{1}{\lambda}V)} \\ 0 & e^{n(g_+ - g_-)} \end{pmatrix} \quad \lambda = \frac{n}{N} . \tag{5.44}$$

Suppose now that the real line $\mathbb{R}$ can be divided into the two subsets,

$$\mathbb{R} = J \cup J^{c} \quad J = [a, b] ,$$

in such a way that the boundary values $g_\pm$ of the g-function exhibit the following behavior.

(d) For $z \in J^{c}$,

$$g_+ - g_- \equiv 2\pi i m_{1,2}$$

for some integers $m_{1,2}$ (m_1 and m_2 correspond to (b, ∞) and $(-\infty, a)$, respectively), and

$$\mathrm{Re}\left(g_+ + g_- - \frac{1}{\lambda} V \right) < 0 .$$

(e) For $z \in J$,

$$g_+ + g_- - \frac{1}{\lambda} V \equiv 0 ,$$

and

$$\mathrm{Re}(g_+ - g_-) \equiv 0 , \quad \frac{\mathrm{d}}{\mathrm{d}y} \mathrm{Re}(g_+ - g_-)(z + \mathrm{i}y)|_{y=0} > 0 , \quad y \in \mathbb{R} .$$

(f) The function

$$h \equiv g_+ - g_-$$

admits the analytic continuation in the lens-shaped region, $\Omega = \Omega_1 \cup \Omega_2$, around the interval J as indicated in Fig. 5.8.

Note that, in virtue of the first identity in (e), condition (f) is satisfied automatically for any polynomial $V(z)$. In fact, it is valid for any real analytic $V(z)$.

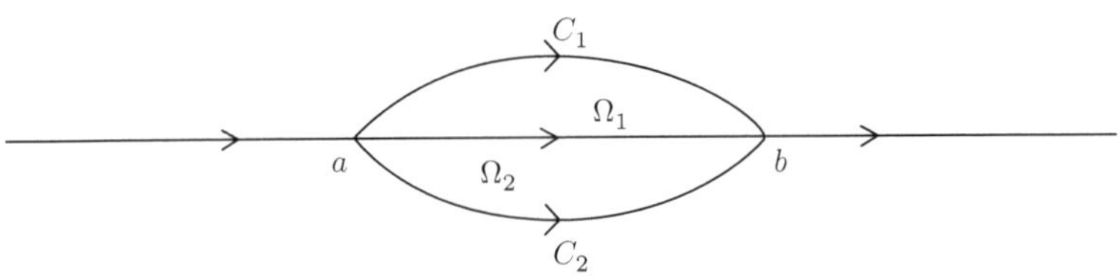

Fig. 5.8. The lens-shaped region Ω and the contour Γ

Postponing the discussion of the existence of such a function $g(z)$, let us see what these conditions give us.

Denote $G_\Phi \equiv G_\Phi(z)$ the Φ-jump matrix. Then, in view of (5.44), property (d) implies that

$$G_\Phi = \begin{pmatrix} 1 & \mathrm{e}^{n(g_+ + g_- - 1/\lambda V)} \\ 0 & 1 \end{pmatrix} \sim I \tag{5.45}$$

as $n \to \infty$ and $z \in J^c$, while from the first condition in (e), we conclude that

$$G_\Phi = \begin{pmatrix} \mathrm{e}^{-nh(z)} & 1 \\ 0 & \mathrm{e}^{nh(z)} \end{pmatrix}$$

as $z \in J$. We also observe that

$$\begin{pmatrix} \mathrm{e}^{-nh(z)} & 1 \\ 0 & \mathrm{e}^{nh(z)} \end{pmatrix} = \begin{pmatrix} 1 & 0 \\ \mathrm{e}^{nh} & 1 \end{pmatrix} \begin{pmatrix} 0 & 1 \\ -1 & 0 \end{pmatrix} \begin{pmatrix} 1 & 0 \\ \mathrm{e}^{-nh} & 1 \end{pmatrix} . \tag{5.46}$$

Recall also that the function $h(z)$ admits the analytical continuation in the region Ω around the interval (a, b); moreover, in view of the second condition in (e), it satisfies there the inequalities,

$$\operatorname{Re} h(z) \begin{cases} > 0\,, & \text{if } \operatorname{Im} z > 0 \\ < 0\,, & \text{if } \operatorname{Im} z < 0\,. \end{cases} \tag{5.47}$$

Asymptotic relation (5.45), identity (5.46) and inequalities (5.47) suggest the final transformation of the original RH problem – the so-called *opening of the lenses.* This is a key procedure of the DKMVZ scheme and it is described as follows.

Let $\Omega = \Omega_1 \cup \Omega_2$ be the lens-shaped region around the interval J introduced above and depicted in Fig. 5.8. Define the function $\widehat{\Phi}(z)$ by the equations,

- for z outside the domain Ω,

$$\widehat{\Phi}(z) = \Phi(z)\,;$$

- for z within the domain Ω_1 (the upper lens),

$$\widehat{\Phi}(z) = \Phi(z) \begin{pmatrix} 1 & 0 \\ -\mathrm{e}^{-nh(z)} & 1 \end{pmatrix}\,;$$

- for z within the domain Ω_2 (the lower lens),

$$\widehat{\Phi}(z) = \Phi(z) \begin{pmatrix} 1 & 0 \\ \mathrm{e}^{nh(z)} & 1 \end{pmatrix}\,.$$

With the passing,

$$Y(z) \to \Phi(z) \to \widehat{\Phi}(z),$$

the original RH problem (1)–(3) transforms to the RH problem posed on the contour Γ consisting of the real axes and the curves C_1 and C_2 which form the boundary of the domain Ω,

$$\partial\Omega = C_1 - C_2$$

(see Fig. 5.8). For the function $\widehat{\Phi}(z)$ we have,

(1‴) $\widehat{\Phi}(z) \in H(\mathbb{C} \setminus \Gamma)$
(2‴) $\widehat{\Phi}_+(z) = \widehat{\Phi}_-(z)G_{\widehat{\Phi}}(z)$, $z \in \Gamma$.
(3‴) $\widehat{\Phi}(z) \mapsto I$, $z \to \infty$.

The jump matrix $G_{\widehat{\Phi}}(z)$, by virtue of the definition of $\widehat{\Phi}(z)$ is given by the relations,

- for z real and outside of the interval $[a, b]$,

$$G_{\widehat{\Phi}}(z) = \begin{pmatrix} 1 & \mathrm{e}^{n(g_+ + g_- - 1/\lambda V)} \\ 0 & 1 \end{pmatrix} ;$$

- for z on the curve C_1,

$$G_{\widehat{\Phi}}(z) = \begin{pmatrix} 1 & 0 \\ \mathrm{e}^{-nh} & 1 \end{pmatrix} ;$$

- for z on the curve C_2,

$$G_{\widehat{\Phi}}(z) = \begin{pmatrix} 1 & 0 \\ \mathrm{e}^{nh} & 1 \end{pmatrix} ;$$

- for z on the interval $[a, b]$,

$$G_{\widehat{\Phi}}(z) = \begin{pmatrix} 0 & 1 \\ -1 & 0 \end{pmatrix} .$$

For $z \in \mathbb{R} \setminus [a, b]$, the matrix $G_{\widehat{\Phi}}(z)$ coincide with the jump matrix $G_{\Phi}(z)$ and hence, in virtue of the property (d) of the g-function, the asymptotic relation (cf. (5.45)),

$$G_{\widehat{\Phi}}(z) \sim I\,, \quad n \to \infty,$$

holds. Simultaneously, the inequalities (5.47) implies that

$$G_{\widehat{\Phi}}(z) \sim I\,, \quad n \to \infty\,,$$

for all $z \in C_1 \cup C_2$, $z \neq a, b$. Therefore, one could expect that the $\widehat{\Phi}$-RH problem (1‴)—(3‴) reduces asymptotically to the following RH problem with the *constant* jump matrix:

$$\widehat{\Phi}(z) \sim \Phi^{\infty}(z), \quad n \to \infty, \tag{5.48}$$

(1^∞) $\Phi^\infty(z) \in H(\mathbb{C} \setminus [a,b])$.
(2^∞) $\Phi^\infty_+(z) = \Phi^\infty_-(z) \left(\begin{smallmatrix} 0 & 1 \\ -1 & 0 \end{smallmatrix}\right)$, $z \in (a,b)$.
(3^∞) $\Phi^\infty(\infty) = I$.

The important fact is that the RH problem $1^\infty - 3^\infty$ is an Abelian one and hence can be solved explicitly. Indeed, diagonalizing the jump matrix from 2^∞ we arrive at the relation

$$\begin{pmatrix} 0 & 1 \\ -1 & 0 \end{pmatrix} = \begin{pmatrix} 1 & 1 \\ i & -i \end{pmatrix} \begin{pmatrix} i & 0 \\ 0 & -i \end{pmatrix} \begin{pmatrix} 1 & 1 \\ i & -i \end{pmatrix}^{-1} .$$

This equation suggests the substitution,

$$\Phi^\infty(z) = \begin{pmatrix} 1 & 1 \\ i & -i \end{pmatrix} X^\infty(z) \begin{pmatrix} 1 & 1 \\ i & -i \end{pmatrix}^{-1} , \tag{5.49}$$

which, in turn, transforms the problem $1^\infty - 3^\infty$ to the RH problem with the diagonal jump matrix,

($1^{\infty\prime}$) $X^\infty(z) \in H(\mathbb{C} \setminus [a,b])$;
($2^{\infty\prime}$) $X^\infty_+ = X^\infty_- \left(\begin{smallmatrix} i & 0 \\ 0 & -i \end{smallmatrix}\right)$, $z \in (a,b)$;
($3^{\infty\prime}$) $X^\infty(z)(\infty) = I$.

The solution of this problem is easy:

$$X^\infty(z) = \begin{pmatrix} \beta(z) & 0 \\ 0 & \beta^{-1}(z) \end{pmatrix} , \quad \beta(z) = \left(\frac{z-b}{z-a}\right)^{1/4} .$$

Here, the function $\beta(z)$ is defined on the z-plane cut along the interval $[a,b]$, and its branch is fixed by the condition,

$$\beta(\infty) = 1.$$

For the original function $\Phi^\infty(z)$ this yields the representation,

$$\Phi^\infty(z) = \begin{pmatrix} (\beta(z) + \beta^{-1}(z))/2 & (\beta(z) - \beta^{-1}(z))/(2\mathrm{i}) \\ -(\beta(z) - \beta^{-1}(z))/(2\mathrm{i}) & (\beta(z) + \beta^{-1}(z))/2 \end{pmatrix} \tag{5.50}$$

Remark. Equation (5.50) can be obtained directly from the general formula (5.26) for the solutions of Abelian RH problems.

Now, we have to address the following two issues.

I. Existence and construction of g-function.
II. Justification of the asymptotic statement (5.48).

5.2.3 Construction of the g-Function

First let us provide ourselves with a little bit of flexibility. Observe that the normalization condition at $z=\infty$ allows to generalize (5.42) as follows:

$$Y(z)=\mathrm{e}^{(nl/2)\sigma_3}\Phi(z)\mathrm{e}^{n(g(z)-l/2)\sigma_3}\,,\quad l=\text{constant}\,. \tag{5.51}$$

This implies the following modification of the Φ-jump matrix (cf. (5.44)).

$$G_\Phi=\begin{pmatrix}\mathrm{e}^{-nh} & \mathrm{e}^{n(g_++g_--(1/\lambda)V-l)}\\ 0 & \mathrm{e}^{nh}\end{pmatrix}\,,\quad h\equiv g_+-g_-.$$

Hence, the conditions (d), (e) on the g-function can be relaxed. Namely, the properties of the difference g_+-g_- stay the same, while for the sum g_++g_- we now only need that

$$\left.\begin{aligned}&\operatorname{Re}\left(g_++g_--\frac{1}{\lambda}V-l\right)<0 &&\text{on } J^{\mathrm{c}}\\ &g_++g_--\frac{1}{\lambda}V-l=0 &&\text{on } J\end{aligned}\right\}\quad\text{for some constant } l\,.$$

Let us try to find such $g(z)$. To this end, it is easier to work with $g'(z)$, so that for $g'(z)$ we have:

$$g'(z)\in H(\mathbb{C}\setminus[a,b])$$

$$g'_++g'_-=\frac{1}{\lambda}V'\quad z\in(a,b) \tag{5.52}$$

$$g'(z)=\frac{1}{z}+\cdots\,,\quad z\to\infty \tag{5.53}$$

(Note: a, b are also unknown!).

Put

$$\varphi(z)=\frac{g'(z)}{\sqrt{(z-a)(z-b)}},$$

where $\sqrt{(z-a)(z-b)}$ is defined on the z-plane cut along the interval $[a,b]$ and fixed by the relation,

$$\sqrt{(z-a)(z-b)}\sim z\,,\quad z\to\infty\,.$$

Then, (5.52) transforms into the additive RH problem,

$$\varphi_+-\varphi_-=\frac{1}{\lambda}\frac{V'(z)}{\sqrt{(z-a)(z-b)}_+}\,,$$

which, by virtue of the Plemelj–Sokhotskii formulae, yields the equation,

$$\varphi(z)=\frac{1}{2\pi\mathrm{i}\lambda}\int_a^b\frac{V'(s)}{\sqrt{(s-a)(s-b)}_+}\frac{\mathrm{d}s}{s-z}$$

and hence

$$g'(z) = \frac{\sqrt{(z-a)(z-b)}}{2\pi \mathrm{i}\lambda} \int_a^b \frac{V'(s)}{\sqrt{(s-a)(s-b)}_+} \frac{\mathrm{d}s}{s-z} \,. \tag{5.54}$$

Taking into account the asymptotic condition (5.53), we also obtain the system on a, b:

$$\begin{cases} \displaystyle\int_a^b \frac{V'(s)}{\sqrt{(s-a)(s-b)}_+} \,\mathrm{d}s = 0 \\ \displaystyle\frac{\mathrm{i}}{2\pi\lambda} \int_a^b \frac{sV'(s)}{\sqrt{(s-a)(s-b)}_+} \,\mathrm{d}s = 1. \end{cases} \tag{5.55}$$

Equations (5.54) and (5.55) produces the needed g-function, *provided* the inequality part of the conditions (d)–(e) satisfies, i.e., provided

$$\begin{aligned} &\mathrm{Re}\left(g_+ + g_- - \frac{1}{\lambda}V - l\right) < 0 && \text{on } J^{\mathrm{c}} \\ &\frac{\mathrm{d}}{\mathrm{d}y} \mathrm{Re}(g_+ - g_-)(z + \mathrm{i}y)|_{y=0} > 0 :, \ y \in \mathbb{R} && \text{on } J \,. \end{aligned} \tag{5.56}$$

In what follows, we will show how the conditions (5.55), (5.56) can be realized for quartic potential,

$$V(z) = \frac{t}{2}z^2 + \frac{\kappa}{4}z^4 \,, \quad \kappa > 0 \,. \tag{5.57}$$

The evenness of $V(z)$ suggests the a priori choice

$$b = -a \equiv z_0 > 0 \,.$$

Then, the first equation in (5.55) is valid automatically. Instead of the direct analysis of the second relation in (5.55), let us transform expression (5.54) to an equivalent one, which can be obtain by the actual evaluation of the integral in (5.54). Indeed, we have

$$\begin{aligned} \int_{-z_0}^{z_0} & \frac{V'(s)}{\sqrt{(s^2 - z_0^2)}_+} \frac{\mathrm{d}s}{s-z} \\ &= \frac{1}{2} \int_{C_0} \frac{V'(s)}{\sqrt{s^2 - z_0^2}} \frac{\mathrm{d}s}{s-z} \\ &= \pi \mathrm{i} \,\mathrm{res}_{s=z} \frac{V'(s)}{\sqrt{s^2 - z_0^2}(s-z)} + \pi \mathrm{i} \,\mathrm{res}_{s=\infty} \frac{V'(s)}{\sqrt{s^2 - z_0^2}(s-z)} \\ &= \pi \mathrm{i} \frac{V'(z)}{\sqrt{z^2 - z_0^2}} - \pi \mathrm{i}\left(\kappa z^2 + \frac{\kappa}{2}z_0^2 + t\right) . \end{aligned}$$

Here, C_0 is the clos]ed contour around the interval $[-z_0, z_0]$ such that the both points z and ∞ lie to the left of C_0. Hence,

$$g'(z) = \frac{1}{2\lambda}V'(z) - \frac{1}{2\lambda}\left(t + \kappa z^2 + \frac{\kappa}{2}z_0^2\right)\sqrt{z^2 - z_0^2} \tag{5.58}$$

Finally, z_0 is determined from the condition,

$$g'(z) \mapsto \frac{1}{z} + \cdots, \qquad z \to \infty,$$

which implies the equations,

$$\frac{3\kappa}{16\lambda}z_0^4 + \frac{t}{4\lambda}z_0^2 = 1$$
$$z_0^4 + \frac{4t}{3\kappa}z_0^2 - \frac{16\lambda}{3\kappa} = 0.$$

Solving these equations, we obtain

$$z_0^2 = -\frac{2t}{3\kappa} + \sqrt{\frac{4t^2}{g\kappa^2} + \frac{16\lambda}{3\kappa}}$$
$$= \frac{-2t + 2\sqrt{t^2 + 12x\lambda}}{3\kappa}.$$

Our (almost) final result (for $V(z) = (t/2)z^2 + (\kappa/4)z^4$) then is:

$$g(z) = -\frac{1}{\lambda}\int_{z_0}^{z}(b_0 + b_2 s^2)\sqrt{s^2 - z_0^2\,\mathrm{d}s} + \frac{1}{2\lambda}V(z) + \frac{l}{2}, \tag{5.59}$$

where

$$z_0 = \left(\frac{-2t + \sqrt{4t^2 + 48\kappa\lambda}}{3\kappa}\right)^{1/2},$$
$$b_2 = \frac{\kappa}{2}, \quad b_0 = \frac{2t + \sqrt{t^2 + 12\kappa\lambda}}{6},$$

and the constant l is determined from the condition

$$g(z) = \ln z + O\left(\frac{1}{z}\right), \qquad z \to \infty.$$

Remark 5.2.1. A similar evaluation of the integral in (5.54) for an arbitrary polynomial potential $V(z)$ yields the following representation for the "one interval" g-function in the general case,

$$g'(z) = \frac{1}{2\lambda}V'(z) + Q(z)\sqrt{(z-a)(z-b)},$$

where $Q(z)$ is a polynomial.

The last thing we have to perform is to check the inequality conditions (5.56) and the inclusion $1/(2\pi \mathrm{i})(g_+ - g_-) \in \mathbb{Z}$ on J^c. We first notice that

$$g_+ + g_- - \frac{1}{\lambda}V - l = \left\{ \begin{array}{ll} -\dfrac{2}{\lambda}\displaystyle\int_{z_0}^{z} (b_0 + b_2 s^2)\sqrt{s^2 - z_0^2}\,\mathrm{d}s & z \in (z_0, +\infty) \\ -\dfrac{2}{\lambda}\displaystyle\int_{-z_0}^{z} (b_0 + b_2 s^2)\sqrt{s^2 - z_0^2}\,\mathrm{d}s & z \in (-\infty, -z_0) \end{array} \right\} < 0$$

$$\text{for all } x > 0\,,\ t \in \mathbb{R}\,,$$

and also

$$g_+ - g_- = 0 \quad \text{on } (z_0, \infty)$$

while

$$g_+ - g_- = -\int_{C_0} \mathrm{d}g(z) = -\operatorname{res}_{z=\infty} \mathrm{d}g(z) = 2\pi \mathrm{i} \quad \text{on } (-\infty, -z_0)\,.$$

Hence, the first inequality in (5.56) and the inclusion $n/(2\pi \mathrm{i})(g_+ - g_-) \in \mathbb{Z}$ on J^c, are satisfied. Assume now that $z \in (-z_0, z_0)$. We have that

$$h(z) = g_+(z) - g_-(z) = -\frac{2\mathrm{i}}{\lambda}\int_{z_0}^{z} (b_0 + b_2 s^2)\sqrt{z_0^2 - s^2}\,\mathrm{d}s\,,$$

and

$$\frac{\mathrm{d}}{\mathrm{d}y}\operatorname{Re} h(z + \mathrm{i}y)\bigg|_{y=0} = \frac{2}{\lambda}(b_0 + b_2 z^2)\sqrt{z_0^2 - z^2} > 0\,,$$

for all $z \in (-z_0, z_0)$, *provided that*

$$b_0 > 0\,, \quad \text{that is, } 2t + \sqrt{t^2 + 12\kappa\lambda} > 0\,. \tag{5.60}$$

Condition (5.60) is always satisfies in the case $t > 0$. For negative t, one needs,

$$\lambda > \frac{t^2}{4\kappa}\,, \tag{5.61}$$

in order to secure (5.60). Inequality (5.61) constitutes the one-cut condition for the quartic potential, $t < 0$.

Conclusion. If

$$\begin{gathered} V(z) = \frac{tz^2}{2} + \frac{\kappa}{4}z^4 \\ \kappa > 0 \quad \text{and} \quad t > 0\,, \quad \text{or} \quad t < 0\,, \quad \lambda > \frac{t^2}{4\kappa} \end{gathered} \tag{5.62}$$

then a one-cut g function exists and is given by (5.59).

Remark 5.2.2. If $t < 0$ and $\lambda < t^2/(4\kappa)$, then the determining g-function condition, i.e., condition (e), can be realized if the set J is taken to be the union of two intervals,

$$J = (-z_2, -z_1) \cup (z_1, z_2) \,, \quad 0 < z_0 < z_2$$

(the symmetric choice of J reflects again the evenness of $V(z)$). Indeed, assuming this form of J, the basic g-function relation, i.e., the scalar RH problem,

$$g_+ + g_- - \frac{1}{\lambda} V - l = 0 \quad \text{on } J \,,$$

would yield the following replacement of the (5.58).

$$g'(z) = \frac{1}{2\lambda} V'(z) - \frac{\kappa}{2\lambda} z \sqrt{(z^2 - z_1^2)(z^2 - z_2^2)} \tag{5.63}$$

The asymptotic condition, $g'(z) \sim 1/z$, $z \to \infty$, allows, as before, to determine the end points of the set J. In fact, we have that

$$z_{1,2} = \left(\frac{-t \mp 2\sqrt{\lambda\kappa}}{\kappa} \right)^{1/2} .$$

With the suitable choice of the constant of integration $l/2$, the g-function is now given by the formula,

$$g(z) = -\frac{\kappa}{2\lambda} \int_{z_2}^{z} s \sqrt{(s^2 - z_1^2)(s^2 - z_2^2)} \, \mathrm{d}s + \frac{1}{2\lambda} V(z) + \frac{l}{2} \,.$$

One can check that this function satisfies all the conditions (a)–(f) (the (d) and (e) are "l"-modified) except the first relation in (d) is replaced now by the equations,

$$\begin{aligned} g_+ - g_- &\equiv 0 \,, && \text{on } (z_2, +\infty) \,, \\ g_+ - g_- &\equiv \pi \mathrm{i} \,, && \text{on } (-z_1, z_1) \,, \\ g_+ - g_- &\equiv 2\pi \mathrm{i} \,, && \text{on } (-\infty, -z_2) \,. \end{aligned}$$

This leads to the necessity to introduce in the model Φ^∞-RH problem an extra jump across the interval $(-z_1, z_1)$ where the jump matrix equals $(-1)^n I$. The model RH problem is still Abelian, and it admits an elementary solution (see also [4] where this model problem is implicitly discussed).

Remark 5.2.3. For arbitrary real analytic $V(z)$, the set J is always a finite union of disjoint intervals (see [18]; see also the lectures of Pavel Bleher in this volume). Similar to the two-cut case outlined in previous remark, the first relation of condition (d) is replaced by the identities,

$$g_+ - g_- \equiv i\alpha_k, \quad \text{on } k\text{th connected component of } J^c \,.$$

There are no general constraints on the values of the real constants α_k, so that generically the model Φ^∞-RH problem is not Abelian. Nevertheless, it still can be solved explicitly in terms of the Riemann theta-functions with the help of the algebraic geometric methods of Soliton theory (see [19]; see also the lectures of Ken McLaughlin and Pavel Bleher in this volume)

Remark 5.2.4. The g-function is closely related to the another fundamental object of the random matrix theory. Indeed, the corresponding measure (see (5.43)), $d\mu_{\mathrm{eq}}(s) \equiv \rho(s)\,\mathrm{d}s$, is the so-called *equilibrium measure*, which minimizes the functional,

$$I_V(\mu) = -\iint_{\mathbb{R}^2} \ln|s-u|\,\mathrm{d}\mu(s)\,\mathrm{d}\mu(u) + \int_{\mathbb{R}^1} V(s)\,\mathrm{d}\mu(s),$$

defined on the Borel probability measures ($\int_{\mathbb{R}^1} \mathrm{d}\mu(s) = 1$) on the real lines. Simultaneously, $\mathrm{d}\mu_{\mathrm{eq}}(s)$ is the limiting eigenvalue density and the limiting density of the zeros of the associated orthogonal polynomials. We refer the reader to [18, 43] for more details (see also the lectures of Ken McLaughlin and Pavel Bleher in this volume).

5.3 The Parametrix at the End Points. The Conclusion of the Asymptotic Analysis

The existence of the g-function alone does not prove the asymptotic relation (5.48). Indeed, the asymptotic conditions,

$$G_{\widehat{\Phi}}(z)|_{C_1\cup C_2} \sim I, \quad G_{\widehat{\Phi}}(z)|_{\mathbb{R}\setminus[-z_0,z_0]} \sim I,$$

which we used to justify (5.48) are not valid at the end points $\pm z_0$. Therefore, the jump matrix $G_{\widehat{\Phi}}(z)$ is *not* L^∞-close to $G_{\Phi^\infty}(z)$ and the general Theorem 5.1.5 of Lecture 5.1 is not applicable. The parametrix solutions near the end points are needed.

5.3.1 The Model Problem Near $z = z_0$

Let $\mathcal{B}$ be a small neighborhood of the point z_0. Observe that

$$\begin{aligned} g_+ + g_- - \frac{1}{\lambda}V - l &= -\frac{2}{\lambda}\int_{z_0}^{z} (b_0 + b_2 s^2)\sqrt{s^2 - z_0^2}\,\mathrm{d}s \\ &= -c_0(z-z_0)^{\frac{3}{2}} + \cdots\,, \end{aligned} \tag{5.64}$$

as $z \in \mathcal{B}$, $z > z_0$. Simultaneously,

$$h = -\frac{2}{\lambda}\int_{z_0}^{z} (b_0 + b_2 s^2)\sqrt{s^2 - z_0^2}\,\mathrm{d}s = -c_0(z-z_0)^{\frac{3}{2}} + \cdots\,, \tag{5.65}$$

as $z \in C_1 \cap \mathcal{B}$ and

$$h = \frac{2}{\lambda} \int_{z_0}^{z} (b_0 + b_2 s^2) \sqrt{s^2 - z_0^2}\, \mathrm{d}s = c_0 (z - z_0)^{\frac{3}{2}} + \cdots , \tag{5.66}$$

as $z \in C_2 \cap \mathcal{B}$. Here,

$$c_0 = \frac{4}{3\lambda}(b_0 + b_2 z_0^2)\sqrt{2z_0} = \frac{2}{3\lambda}\sqrt{2z_0 t^2 + 24 z_0 \kappa},$$

and the function $(z - z_0)^{3/2}$ is defined on the z-plane cut along $(-\infty, z_0]$ and fixed by the condition,

$$(z - z_0)^{3/2} > 0, \quad z_0 > 0.$$

Equations (5.64)–(5.66) suggest the introduction of the following local variable,

$$\begin{aligned} w(z) &= \left(\frac{3n}{4}\right)^{2/3} \left(\frac{2}{\lambda}\int_{z_0}^{z} (b_0 + b_2 s^2)\sqrt{s^2 - z_0^2}\, \mathrm{d}s\right)^{2/3} \\ &\equiv \left(\frac{3n}{4}\right)^{2/3} \left(-2g(z) + \frac{1}{\lambda} V(z) + l\right)^{2/3} . \end{aligned} \tag{5.67}$$

Since $w(z) \sim n^{2/3} c_1 (z - z_0)$ with $c_1 > 0$, it follows that equation (5.67) indeed define in the neighborhood $\mathcal{B}$ a holomorphic change of variable:

$$\mathcal{B} \mapsto D_n(0) \equiv \{w : |w| < n^{2/3} c_1 \rho\} , \quad 0 < \rho < \frac{z_0}{2} . \tag{5.68}$$

The action of the map $z \to w$ on the part of the contour Γ of the $\widehat{\Phi}$ - RH problem (1‴)–(3‴), which is inside of the neighborhood $\mathcal{B}$, is indicated in Fig. 5.9.

Observe that the jump matrix $G_{\widehat{\Phi}}$ inside the neighborhood $\mathcal{B}$ can be written down in the form,

$$G_{\widehat{\Phi}}(z) = \exp(-\tfrac{2}{3} w^{3/2}(z)\sigma_3) S \exp(\tfrac{2}{3} w^{3/2}(z)\sigma_3) ,$$

where the piecewise constant matrix S is given by the equations,

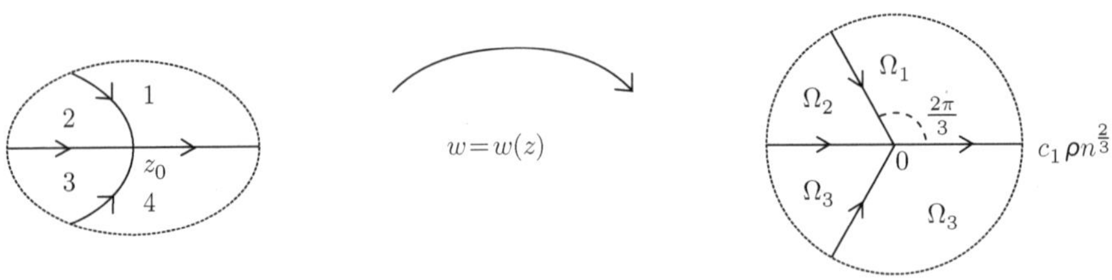

Fig. 5.9. The map $z \to w$

$$S = \begin{pmatrix} 1 & 1 \\ 0 & 1 \end{pmatrix} \equiv S_1 , \qquad z \in [z_0, +\infty) \cap \mathcal{B} ,$$

$$S = \begin{pmatrix} 1 & 0 \\ 1 & 1 \end{pmatrix} \equiv S_2 , \qquad z \in C_1 \cap \mathcal{B} ,$$

$$S = \begin{pmatrix} 0 & 1 \\ -1 & 0 \end{pmatrix} \equiv S_3 , \quad z \in (-\infty, -z_0] \cap \mathcal{B} ,$$

$$S = \begin{pmatrix} 1 & 0 \\ 1 & 1 \end{pmatrix} \equiv S_4 , \qquad z \in C_2 \cap \mathcal{B} .$$

Therefore, the map, $z \to w$, transforms the $\mathcal{B}$ - part of the $\widehat{\Phi}$-RH problem $(1''')$–$(3''')$ into the following model RH problem in the w-plane.

(1^0) $Y^0(w) \in H(\mathbb{C} \setminus \Gamma_w)$
(2^0) $Y^0_+(w) = Y^0_-(w) \exp(-\frac{2}{3}w^{3/2}\sigma_3) S \exp(\frac{2}{3}w^{3/2}\sigma_3)$, $w \in \Gamma_w$
(3^0) $Y^0(w) = w^{-\sigma_3/4} \begin{pmatrix} 1/2 & \mathrm{i}/2 \\ -1/2 & \mathrm{i}/2 \end{pmatrix} \big(I + O(1/w)\big)$, $w \to \infty$.

Here, the contour Γ_w is the union of the four rays,

$$\Gamma_w = \bigcup_{k=1} \Gamma_k ,$$

which are depicted in Fig. 5.10. The branch of the function $w^{1/2}$ is define on the w-plane cut along $(-\infty, 0]$ and fixed by the condition $w^{1/2} > 0$ as $w > 0$.

The normalization condition (3^0), whose appearance has not been explained yet, comes from the fact that we want the "interior" function $Y^0(w(z))$

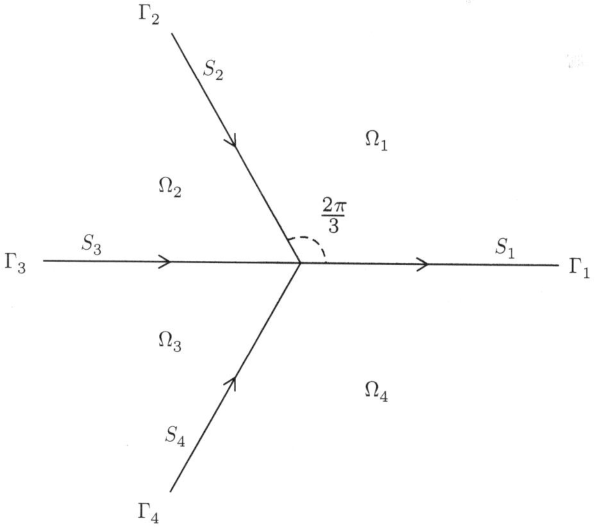

Fig. 5.10. The model RH problem at the end point z_0

to match asymptotically, as $n \to \infty$, the "exterior" function $\Phi^\infty(z)$ at the boundary of $\mathcal{B}$. In other words, to specify the behavior of $Y^0(w)$ as $w \to \infty$, we must look at the behavior of $\Phi^\infty(z)$ at $z = z_0$. To this end, we notice that the function $\Phi^\infty(z)$ admits the following factorization (cf. (5.49)):

$$\begin{aligned}\Phi^\infty(z) &= \begin{pmatrix} 1 & -1 \\ -\mathrm{i} & -\mathrm{i} \end{pmatrix} \begin{pmatrix} \beta^{-1}(z) & 0 \\ 0 & \beta(z) \end{pmatrix} \begin{pmatrix} 1/2 & \mathrm{i}/2 \\ -1/2 & \mathrm{i}/2 \end{pmatrix} \\ &= \begin{pmatrix} 1 & -1 \\ -\mathrm{i} & -\mathrm{i} \end{pmatrix} \begin{pmatrix} (\beta/w^{1/4})^{-1} & 0 \\ 0 & \beta/w^{1/4} \end{pmatrix} w^{-\sigma_3/4} \begin{pmatrix} 1/2 & \mathrm{i}/2 \\ -1/2 & \mathrm{i}/2 \end{pmatrix} . \end{aligned} \tag{5.69}$$

Since,

$$\frac{\beta(z)}{w^{1/4}(z)} = (2z_0 n^{2/3} c_1)^{-1/4} + \cdots \equiv \sum_{k=0}^{\infty} b_k (z - z_0)^k \,, \quad b_0 > 0 \,, \ |z - z_0| < z_0 \,,$$

the matrix function,

$$E(z) = \begin{pmatrix} 1 & -1 \\ -\mathrm{i} & -\mathrm{i} \end{pmatrix} \begin{pmatrix} (\beta/w^{1/4})^{-1} & 0 \\ 0 & \beta/w^{1/4} \end{pmatrix} \equiv \begin{pmatrix} \beta^{-1} & -\beta \\ -\mathrm{i}\beta^{-1} & -\mathrm{i}\beta \end{pmatrix} w^{\sigma_3/4}(z) \,, \tag{5.70}$$

is holomorphic at $z = z_0$. Therefore we have that

$$\Phi^\infty(z) = E(z) w^{-\sigma_3/4} \begin{pmatrix} 1/2 & \mathrm{i}/2 \\ -1/2 & \mathrm{i}/2 \end{pmatrix} , \tag{5.71}$$

where the left matrix multiplier, $E(z)$, is holomorphic and invertible in the neighborhood of z_0. Equation (5.71) explains the choice of the normalization condition at $w = \infty$ we made in the model problem (1^0)–(3^0). The holomorphic factor $E(z)$ has no relevance to the setting of the Riemann–Hilbert problem in the w-plane; it will be restored latter on, when we start actually assembling the parametrix for $\widehat{\Phi}(z)$ in $\mathcal{B}$ (see (5.75) below).

5.3.2 Solution of the Model Problem

Put

$$\Psi^0(w) := Y^0(w) \exp(-\tfrac{2}{3} w^{3/2} \sigma_3) \,.$$

The jump condition (2^0) and the asymptotic condition (3^0) in terms of the function $\Psi^0(w)$ become

(a) $\Psi^0_+(w) = \Psi^0_-(w) S$, $w \in \Gamma_w$,

and

(b) $\Psi^0(w) = w^{-\sigma_3/4} \begin{pmatrix} 1/2 & \mathrm{i}/2 \\ -1/2 & \mathrm{i}/2 \end{pmatrix} (I + O(1/w)) \exp(-\frac{2}{3} w^{3/2} \sigma_3)$, $w \to \infty$, respectively. Moreover, if $\mathcal{M}_0$ denote the operator of analytic continuation around the point $w = 0$, we see that

$$\begin{aligned}\mathcal{M}_0[\Psi^0] &= \Psi^0 S_1^{-1} S_4 S_3 S_2 = \Psi^0 \begin{pmatrix}1 & -1\\ 0 & 1\end{pmatrix}\begin{pmatrix}1 & 0\\ 1 & 1\end{pmatrix}\begin{pmatrix}0 & 1\\ -1 & 0\end{pmatrix}\begin{pmatrix}1 & 0\\ 1 & 1\end{pmatrix}\\ &= \Psi^0 I = \Psi^0 \,. \end{aligned} \tag{5.72}$$

Let Ω_k, $k = 1, 2, 3, 4$ be the connected components of $\mathbb{C} \setminus \Gamma_w$ (see Fig. 5.10). Then, monodromy relation (5.72) means that

(c) $\Psi^0|_{\Omega_k}$ admits the analytic continuation to the whole $\mathbb{C}$ as an entire function.

It is also worth noticing, that

(d) $\det \Psi^0(w) \equiv 2\mathrm{i} \neq 0$.

We are now ready to proceed with one of the most fundamental though quite simple methodological ingredients of the classical monodromy theory of linear systems as well as of the modern theory of integrable systems. That is, we are going to consider the "logarithmic derivative" of the matrix valued function $\Psi^0(w)$,

$$A(w) := \frac{\mathrm{d}\Psi^0}{\mathrm{d}w}(\Psi^0)^{-1} \,.$$

Since along each ray Γ_k the Ψ-jump matrix, $S \equiv S_k$, is constant, the matrix valued function $A(w)$ has no jumps across Γ. Indeed, for $z \in \Gamma_k$, we have that,

$$\begin{aligned} A_+ = \frac{\mathrm{d}\Psi^0_+}{\mathrm{d}w}(\Psi^0_+)^{-1} &= \frac{\mathrm{d}(\Psi^0_- S_k)}{\mathrm{d}w}(\Psi^0_- S_k)^{-1}\\ &= \frac{\mathrm{d}\Psi^0_-}{\mathrm{d}w} S_k S_k^{-1}(\Psi^0_-)^{-1} = \frac{\mathrm{d}\Psi^0_-}{\mathrm{d}w}(\Psi^0_-)^{-1} = A_- \,. \end{aligned}$$

Together with (d), this means that $A(w) \in H(\mathbb{C} \setminus \{0\})$. From property (c), it follows that $w = 0$ is a removable singularity and hence we conclude that $A(w)$ is an entire function. Simultaneously, from the asymptotic relation (b) we have:

$$\begin{aligned}A(w) &= \frac{\mathrm{d}\Psi^0}{\mathrm{d}w}(\Psi^0)^{-1}\\ &= w^{-\sigma_3/4}\begin{pmatrix}1/2 & \mathrm{i}/2\\ -1/2 & \mathrm{i}/2\end{pmatrix}\left(I+O\left(\frac{1}{w}\right)\right)(-w^{1/2}\sigma_3)\\ &\qquad\times\left(I+O\left(\frac{1}{w}\right)\right)\begin{pmatrix}1 & -1\\ -\mathrm{i} & -\mathrm{i}\end{pmatrix}w^{\sigma_3/4}+O\left(\frac{1}{w}\right)\\ &= -w^{1/2}w^{-\sigma_3/4}\left[\begin{pmatrix}1/2 & \mathrm{i}/2\\ -1/2 & \mathrm{i}/2\end{pmatrix}\begin{pmatrix}1 & -1\\ \mathrm{i} & \mathrm{i}\end{pmatrix}+O\left(\frac{1}{w}\right)\right]w^{\sigma_3/4}+O\left(\frac{1}{w}\right)\\ &= -w^{1/2}w^{-\sigma_3/4}\left(\begin{pmatrix}0 & -1\\ -1 & 0\end{pmatrix}+O\left(\frac{1}{w}\right)\right)w^{\sigma_3/4}+O\left(\frac{1}{w}\right)\\ &= w^{1/2}\left[\begin{pmatrix}0 & 1/\sqrt{w}\\ \sqrt{w} & 0\end{pmatrix}+\begin{pmatrix}0 & 0\\ d/\sqrt{w} & 0\end{pmatrix}+O\left(\frac{1}{w}\right)\right]+O\left(\frac{1}{w}\right)\\ &= \begin{pmatrix}0 & 1\\ w+d & 0\end{pmatrix}+O\left(\frac{1}{\sqrt{w}}\right), \qquad w\to\infty\,.\end{aligned}$$

The last equation, together with already established fact of the entirety of $A(w)$, implies that

$$A(w)=\begin{pmatrix}0 & 1\\ w+d & 0\end{pmatrix}, \qquad \forall w\,.$$

In other words, we arrive at the following important statement: *The (unique) solution of the Riemann–Hilbert problem* (a)–(b) *satisfies (if exists) the following differential equation*:

$$\frac{\mathrm{d}\Psi^0}{\mathrm{d}w}=\begin{pmatrix}0 & 1\\ w+d & 0\end{pmatrix}\Psi^0. \tag{5.73}$$

Here, d is some constant. In fact, in a few lines, we will show that d must be zero.

Equation (5.73) is a matrix form of the classical Airy equation and hence is solvable in terms of contour integrals. Indeed, denoting

$$y(w)=\left(\Psi^0(w)\right)_{1j}, \quad j=1 \text{ or } 2\,,$$

we derive from (5.73) that

$$y''=(w+d)y\,, \quad \left(\Psi^0(w)\right)_{2j}=y'(w)\,.$$

The first relation is the Airy equation with the shifted argument. Its general solution has the exponential behavior as $w\to\infty$ whose characteristic exponents are

$$\pm\tfrac{2}{3}(w+d)^{3/2}=\pm(\tfrac{2}{3}w^{3/2}+dw^{1/2})+O(w^{-1/2})\,.$$

Therefore, to reproduce the characteristic exponents of the asymptotic condition (b) we must assume that in (5.73)

$$d = 0 \, .$$

The above heuristic arguments *together with the known asymptotic behavior of the Airy functions in the whole neighborhood of the infinity* motivate the following *explicit* construction for the solution of the model RH problem (1^0) - (3^0).

$$Y^0(w) := \Psi^0(w) \exp(\tfrac{2}{3} w^{3/2} \sigma_3) \, , \quad \Psi^0(w) = \Psi_{\mathrm{Ai}}(w) C \, , \tag{5.74}$$

where $\Psi_{\mathrm{Ai}}(w)$ and C are given by the explicit formulae,

$$\Psi_{\mathrm{Ai}}(w) = \begin{pmatrix} y_0(w) & \mathrm{i} y_1(w) \\ y_0'(w) & \mathrm{i} y_1'(w) \end{pmatrix}$$

$$y_0(w) = \mathrm{Ai}(w) := \frac{1}{2\pi \mathrm{i}} \int_{-\mathrm{i}\infty}^{\mathrm{i}\infty} \exp(\tfrac{1}{3} z^3 - wz) \, \mathrm{d}z$$

$$y_1(w) = \mathrm{e}^{-\pi \mathrm{i}/6} \, \mathrm{Ai}(\mathrm{e}^{-2\pi \mathrm{i}/3} w) \, ,$$

and

$$C = \sqrt{\pi} \begin{cases} I & w \in \Omega_1 \\ \begin{pmatrix} 1 & 0 \\ -1 & 1 \end{pmatrix} & w \in \Omega_2 \\ \begin{pmatrix} 0 & -1 \\ 1 & 1 \end{pmatrix} & w \in \Omega_3 \\ \begin{pmatrix} 1 & -1 \\ 0 & 1 \end{pmatrix} & w \in \Omega_4 \, . \end{cases}$$

The indicated choice of the matrix C in the domains Ω_k is due to the known asymptotic behavior of the Airy functions in the neighborhood of the infinity. In fact, the matrix C is made out of the relevant Stokes multipliers of the Airy functions in such a way which guarantees that the matrix product $\Psi_{\mathrm{Ai}}(w) C$ satisfies the asymptotic condition (b) at $w = \infty$. Therefore, in order to prove that (5.74) indeed defines the solution of the RH problem (1^0)–(3^0), we only need to check the jump relations (a).

The validity of (a) on the rays Γ_1 and Γ_2 is obvious. On the ray Γ_3 we have,

$$\begin{aligned} \Psi_+^0 = \Psi^0|_{\Omega_2} = \Psi_{\mathrm{Ai}} \begin{pmatrix} 1 & 0 \\ -1 & 1 \end{pmatrix} &= \Psi^0|_{\Omega_3} \begin{pmatrix} 1 & 1 \\ -1 & 0 \end{pmatrix} \begin{pmatrix} 1 & 0 \\ -1 & 1 \end{pmatrix} \\ &= \Psi_-^0 \begin{pmatrix} 1 & 1 \\ -1 & 0 \end{pmatrix} \begin{pmatrix} 1 & 0 \\ -1 & 1 \end{pmatrix} = \Psi_-^0 \begin{pmatrix} 0 & 1 \\ -1 & 0 \end{pmatrix} \equiv \Psi_-^0 S_3 \, . \end{aligned}$$

Similarly, on the ray Γ_4,

$$\begin{aligned} \Psi_+^0 = \Psi^0|_{\Omega_3} = \Psi_{\mathrm{Ai}} \begin{pmatrix} 0 & -1 \\ 1 & 1 \end{pmatrix} &= \Psi^0|_{\Omega_4} \begin{pmatrix} 1 & 1 \\ 0 & 1 \end{pmatrix} \begin{pmatrix} 0 & -1 \\ 1 & 1 \end{pmatrix} \\ &= \Psi_-^0 \begin{pmatrix} 1 & 1 \\ 0 & 1 \end{pmatrix} \begin{pmatrix} 0 & -1 \\ 1 & 1 \end{pmatrix} = \Psi_-^0 \begin{pmatrix} 1 & 0 \\ 1 & 1 \end{pmatrix} \equiv \Psi_-^0 S_4 \, . \end{aligned}$$

This completes the proof of the representation (5.74) for the solution of the model Riemann–Hilbert problem (1^0)–(3^0).

Remark. From the known error-terms for Airy integral we conclude that (b) could be improved as:

$$\Psi^0(w) = w^{-\sigma_3/4}\begin{pmatrix} 1/2 & \mathrm{i}/2 \\ -1/2 & \mathrm{i}/2 \end{pmatrix}\left(I + O\left(\frac{1}{w^{3/2}}\right)\right)\exp(-\tfrac{2}{3}w^{3/2}\sigma_3) \quad w \to \infty\,.$$

5.3.3 The Final Formula for the Parametrix

Recalling the holomorphic left multiplier $E(z)$ in (5.71) we define in the neighborhood,

$$\mathcal{B} = w^{-1}\big(D_n(0)\big)\,,$$

the parametrix for the solution of our basic $\widehat{\Phi}$-RH problem $(1''')$–$(3''')$ by the equation,

$$\Phi^0(z) = E(z)Y^0(w(z)) \equiv E(z)\Psi_{\mathrm{Ai}}(w(z))C\exp(\tfrac{2}{3}w^{3/2}(z)\sigma_3)\,. \tag{5.75}$$

In this formulas, it is assumed that (cf. Fig. 5.9)

$$C_1 \cap \mathcal{B} = w^{-1}\big(\Gamma_1 \cap D_n(0)\big)\,, \quad C_2 \cap \mathcal{B} = w^{-1}\big(\Gamma_4 \cap D_n(0)\big)\,.$$

The neighborhood $\mathcal{B}$ is splited into four regions (see again Fig. 5.9)

$$\mathcal{B}_k = w^{-1}\big(\Omega_k \cap D_n(0)\big)\,, \quad k = 1, \dots, 4$$

and the matrix C in (5.75) satisfies the relations,

$$C = \sqrt{\pi}\begin{cases} I & z \in \mathcal{B}_1 \\ \begin{pmatrix} 1 & 0 \\ -1 & 1 \end{pmatrix} & z \in \mathcal{B}_2 \\ \begin{pmatrix} 0 & -1 \\ 1 & 1 \end{pmatrix} & z \in \mathcal{B}_3 \\ \begin{pmatrix} 1 & -1 \\ 0 & 1 \end{pmatrix} & z \in \mathcal{B}_4 \end{cases}$$

The following are two principal properties of the parametrix $\Phi^0(z)$.

1. Inside $\mathcal{B}$, the jumps of $\Phi^0(z)$ are *exactly* the same as of $\widehat{\Phi}(z)$.
2. On the boundary of $\mathcal{B}$, $\partial\mathcal{B}$ we have:

$$\Phi^0(z) = \Phi^\infty(z)\left(I + O\left(\frac{1}{n}\right)\right) = \left(I + O\left(\frac{1}{n}\right)\right)\Phi^\infty(z) \quad z \in \partial\mathcal{B}\,,\ n \to \infty\,.$$

The second statement follows from formula (5.71) for $\Phi^\infty(z)$, asymptotic estimate (3^0) for the model function $Y^0(w)$ (with the improved by $O(w^{-3/2})$ error term) and the fact that for $z \in \partial\mathcal{B}$ we have that

$$\frac{1}{w(z)} = O\left(\frac{1}{n^{2/3}}\right), \quad n \to \infty .$$

Finally we notice that, due to the evennes of the potential $V(z)$, both the solution of the basic $\widehat{\Phi}$-RH problem and the model function $\Phi^{\infty}(z)$ satisfy the symmetry relation,

$$\widehat{\Phi}(z) = \sigma_3 \widehat{\Phi}(-z) \sigma_3 , \quad \Phi^{\infty}(z) = \sigma_3 \Phi^{\infty}(-z) \sigma_3 . \tag{5.76}$$

Therefore, we can avoid a separate analysis of the end point $-z_0$ and produce a relevant parametrix for the solution $\widehat{\Phi}(z)$ at this point by the expression,

$$\sigma_3 \Phi^0(-z) \sigma_3 .$$

In the case of general potential $V(z)$, when there is no symmetry relation which would conviniently connect different end points of the support J of the equilibrium measure, one has to construct the relevant parametrix to each of the end points separately, following the patern we have developed above for the point z_0.

5.3.4 The Conclusion of the Asymptotic Analysis

Denote

$$-\mathcal{B} := \{z : -z \in \mathcal{B}\} ,$$

and define the function,

$$\widehat{\Phi}^{\mathrm{as}}(z) = \begin{cases} \Phi^{\infty}(z) & z \in \mathbb{C} \setminus \big(\mathcal{B} \cup (-\mathcal{B})\big) \\ \Phi^0(z) & z \in \mathcal{B} \\ \sigma_3 \Phi^0(-z) \sigma_3 & z \in -\mathcal{B} . \end{cases} \tag{5.77}$$

We claim that the exact solution of the $\widehat{\Phi}$-RH problem (1‴)–(3‴) satisfies the following uniform estimate,

$$\widehat{\Phi}(z) = \left(I + O\left(\frac{1}{(1+|z|)n}\right)\right) \widehat{\Phi}^{\mathrm{as}}(z) , \quad n \to \infty , \ z \in \mathbb{C} . \tag{5.78}$$

In order to prove (5.78) we consider the matrix ratio

$$X(z) := \widehat{\Phi}(z) \big(\widehat{\Phi}^{\mathrm{as}}(z)\big)^{-1} . \tag{5.79}$$

This function solves the Riemann–Hilbert problem posed on the contour Γ_X depicted in Fig. 5.11. The important feature of this problem is the absence of any jumps inside the domains $\mathcal{B}$ and $-\mathcal{B}$ and across the part of the real line between the boundary curves $C_r \equiv \partial \mathcal{B}$ and $C_l \equiv \partial(-\mathcal{B})$. Indeed, the cancelling out of the jumps of $\widehat{\Phi}(z)$ and $\widehat{\Phi}^{\mathrm{as}}(z)$ inside $\mathcal{B}$, $-\mathcal{B}$ and across the indicated part of the real line follows directly from the basic jump properties of the model functions $\Phi^{\infty}(z)$ and $\Phi^{0}(z)$, and from the symmetry relation (5.76).

The X-jump matrix, $G_X(z)$, satisfies the following *uniform* estimates as $n \to \infty$.

$$G_X(z) \equiv \Phi^{\infty}(z) \begin{pmatrix} 1 & \exp\bigl(n(g_+ + g_- - (1/\lambda)V - l)\bigr) \\ 0 & 1 \end{pmatrix} \left(\Phi^{0}(z)\right)^{-1}$$

$$= I + O\bigl(\exp(-cnz^2)\bigr) , \quad c > 0 , \ z \in \Gamma_l \cup \Gamma_r ,$$

$$G_X(z) \equiv \Phi^{\infty}(z) \begin{pmatrix} 1 & 0 \\ \mathrm{e}^{\mp nh} & 1 \end{pmatrix} \left(\Phi^{\infty}(z)\right)^{-1}$$

$$= I + O(\mathrm{e}^{-cn}) , \quad c > 0 , \ z \in C_{u,d} ,$$

$$G_X(z) \equiv \Phi^{0}(z)\bigl(\Phi^{\infty}(z)\bigr)^{-1} = I + O\Bigl(\frac{1}{n}\Bigr) , \qquad z \in C_r ,$$

$$G_X(z) \equiv \sigma_3 G_X(-z)\sigma_3 = I + O\Bigl(\frac{1}{n}\Bigr) , \qquad z \in C_l .$$

The above estimates show that, in our final RH transformations, we have achieved the needed closeness of the jump-matrix to the identity matrix. Indeed, we have

$$\|I - G_X\|_{L_2 \cap L_\infty} \le \frac{C}{n} . \tag{5.80}$$

We can now apply the general Theorem 5.1.5 and deduce the inequality,

$$\|X(z) - I\| < \frac{C}{(1 + |z|)n} , \qquad z \in \mathbb{K} , \ n > n_0 , \tag{5.81}$$

where $\mathbb{K}$ is a closed subset of $\mathbb{C}P^1 \setminus \Gamma_X$ satisfying,

$$\frac{\mathrm{dist}\{z; \Gamma\}}{1 + |z|} \ge c(\mathbb{K}) , \quad \forall z \in \mathbb{K} .$$

Observe, that the choice of the contour Γ_X is not rigid; one can always slightly deform the contour Γ_X and slightly rotate its infinite parts. Therefore, the

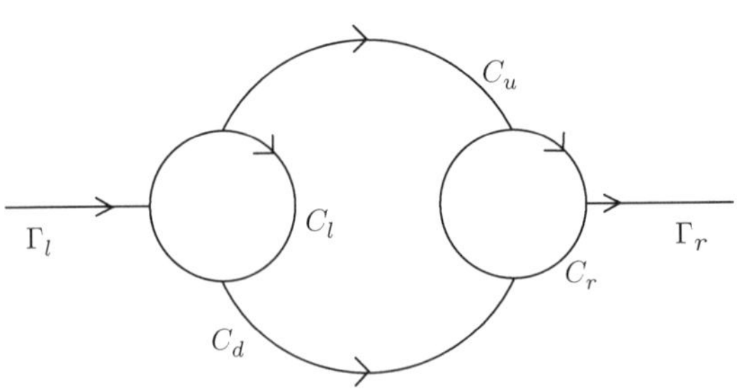

Fig. 5.11. The final, X-RH problem

domain of validity of the estimate (5.81) can be made the whole z-plane. This completes the proof of the asymptotic equation (5.78).

Remark. It can be checked directly, using the above expressions of the jump matrix $G_X(z)$ in terms of the functions $\Phi^{\infty}(z)$ and $\Phi^0(z)$, that the smoothness condition (5.21) is satisfied at each node point of the contour Γ_X. This also can be deduced from the a priori continuity of the function $X(z)$, defined by (5.79), in the closure of every connected component of the set $\mathbb{C} \setminus \Gamma_X$.

The asymptotic formula (5.78) provides us with the uniform asymptotics, as $n, N \to \infty$, for all complex z, of the orthogonal polynomials $P_n(z)$ corresponding to the quartic weight $\mathrm{e}^{-NV(z)}$, $V(z) = tz^2/2 + \kappa z^4/4$ under the one-cut condition (5.62) on the parameters κ and t. For instance, the oscillatory part of the asymptotics reads as follows.

$$P_n(x) \exp\left(-\frac{NV(x)}{2} - \frac{nl}{2}\right) = \mathrm{Re}\left(\frac{\beta_+(x) + \beta_+^{-1}(x)}{2} \mathrm{e}^{-iN\psi(x)}\right) + O\left(\frac{1}{n}\right), \quad |x| < z_0 , \tag{5.82}$$

where

$$\beta_+(x) = \mathrm{e}^{\mathrm{i}\pi/4} \left(\frac{z_0 - x}{z_0 + x}\right)^{1/4}, \quad \psi(x) = \int_{z_0}^{x} (b_0 + b_2 u^2)\sqrt{z_0^2 - u^2}\, \mathrm{d}u ,$$

and the parameters (functions of λ!) l, z_0, b_0, and b_2 are defined in (5.59). Also, for the norm h_n and the recurrence coefficient $R_n = h_n/h_{n-1}$ we obtain, with the help of (5.20), the asymptotic representations,

$$h_n = \pi z_0 \mathrm{e}^{nl}\left(1 + O\left(\frac{1}{n}\right)\right), \tag{5.83}$$

and

$$R_n = \frac{z_0^2}{4} + O\left(\frac{1}{n}\right). \tag{5.84}$$

In conclusion, we notice that estimate (5.80) allows us to iterate the singular integral equation corresponding to the X-RH problem and, in principal, obtain the complete asymptotic series for all the above objects. It should be mentioned that although the Riemann–Hilbert iterating procedure rigorously establishes the existence of the complete asymptotic series mentioned, its use for obtaining compact formulae for the coefficients of these expansions is very involved. The better way to get the higher corrections to the estimates (5.82)–(5.84) is to use the relevant Freud difference equation for R_n (see the end of the first lecture) in conjunction with the *already found* leading term of the asymptotics and the *already established* existence of the whole series (see, e.g., [6]). There exists also a very beautiful formal procedure [24], based on the so called "loop equation" (see, for example, survey [23]) which allows to obtain compact expressions for the terms of any order in the expensions of h_n and R_n and, in fact, in the expansions of the partition function Z_N. This procedure has been recently made rigorous (again, with the help of the RH approach) in the work [25].

5.4 The Critical Case. The Double Scaling Limit and the Second Painlevé Equation

In this lecture, we shall discuss the asymptotic solution of the $\widehat{\Phi}$-RH problem when $t < 0$ and the ratio $\lambda \equiv n/N$ is close to or coincide with its critical value,

$$\lambda_c = \frac{t^2}{4\kappa} .$$

More precisely, we shall assume that

$$\frac{n}{N} \equiv \lambda = \lambda_c + cxN^{-2/3} . \tag{5.85}$$

Here, x is a scaling variable which is assumed to be bounded and c is a normalization constant which we choose to be,

$$c = \left(\frac{t^2}{2\kappa}\right)^{1/3} .$$

This choice of c will be motivated later on. The principal feature of the double scaling limit (5.85) is the value $\frac{2}{3}$ of its critical exponent. From the asymptotic analysis which will follow, we will see that this is precisely the exponent which guarantees the existence of the nontrivial scaling limit in the random matrix model under consideration.

The condition (5.85) implies that now we have (cf. (5.60))

$$\frac{\mathrm{d}}{\mathrm{d}y} \operatorname{Re} h|_{z=0} \sim 0 ,$$

which yields the replacement of the previous two-lenses jump contour Γ by the four-lenses contour Γ_c depicted in Fig. 5.12 and hence the need to construct a parametrix to the solution $\widehat{\Phi}$ in the neighborhood of the point $z = 0$.

5.4.1 The Parametrix at $z = 0$

To find out what the relevant model RH problem in the neighborhood of $z = 0$ should be, we shall analyse the function $h(z)$ near the point $z = 0$. First, we notice that under condition (5.85) the end point z_0 and the coefficient b_0, defined in (5.59), satisfy the estimates,

$$z_0^2 = z_c^2 - \frac{2c}{t}xN^{-2/3} + O(N^{-4/3}), \quad z_c = \sqrt{\frac{2|t|}{\kappa}}, \tag{5.86}$$

and

$$b_0 = -\frac{\kappa c}{2t}xN^{-2/3} + O(N^{-4/3}). \tag{5.87}$$

The function $h(z)$, which is participating in the $\widehat{\Phi}$ jump matrix as indicated in Fig. 5.12, is given by the formula,

$$\begin{aligned} h(z) &= -\frac{2}{\lambda}\int_{z_0}^{z}(b_0 + b_2u^2)\sqrt{u^2 - z_0^2}\,\mathrm{d}u \\ &= \mathrm{i}\pi - \frac{2}{\lambda}\int_0^z (b_0 + b_2u^2)\sqrt{u^2 - z_0^2}\,\mathrm{d}u, \end{aligned} \tag{5.88}$$

where

$$\pi \equiv -\mathrm{i}h(0) = -\frac{2}{\lambda}\int_{z_0}^{0}(b_0 + b_2u^2)\sqrt{z_0^2 - u^2}\,\mathrm{d}u = \pi,$$

and, unlike the previous formulas for $g(z)$ and $h(z)$, the function $\sqrt{z^2 - z_0^2}$ is now defined in the domain $\mathbb{C}\setminus(-\infty, z_0)\cup(z_0, +\infty)$ and fixed by the condition $-\mathrm{i}\sqrt{z^2 - z_0^2}|_{z=0} > 0$. From (5.88) and with the help of (5.86) and (5.87) we arrive at the following estimate for $h(z)$.

$$h(z) = \mathrm{i}\pi - \frac{2\mathrm{i}}{\lambda}\big(D_0(z) + xN^{-2/3}D_1(z)\big) + O(N^{-4/3}), \tag{5.89}$$

where

$$D_0(z) = \frac{\kappa}{2}\int_0^z u^2\sqrt{z_c^2 - u^2}\,\mathrm{d}u, \quad D_1(z) = c\int_0^z \frac{\mathrm{d}u}{\sqrt{z_c^2 - u^2}}.$$

Observe that near $z = 0$,

$$D_0(z) = \frac{\kappa z_c}{6}z^3 - \frac{\kappa}{20z_c}z^5 + \cdots, \tag{5.90}$$

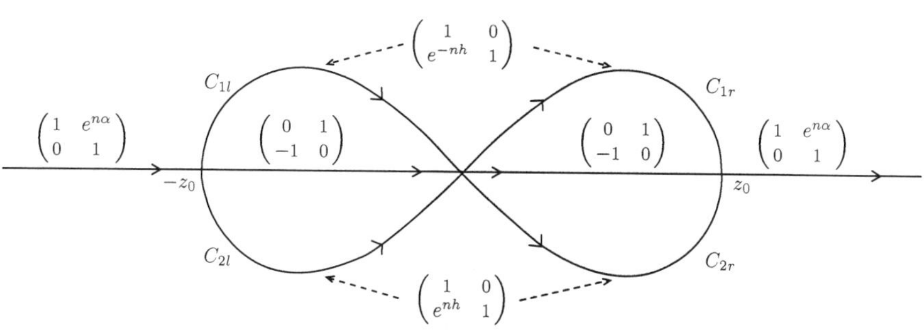

Fig. 5.12. The contour Γ_c and the $\widehat{\Phi}$-RH problem in the critical case

and

$$D_1(z) = C^{-1}z - \frac{1}{6Cz_c^2}z^3 + \cdots , \quad C = \frac{2^{5/6}}{(|t|\kappa)^{1/6}} . \tag{5.91}$$

Hence, with some constants a_1, a_2, we have that

$$D_0(z) + xN^{-2/3}D_1(z) = a_1z^3 + a_2xN^{-2/3}z + \cdots .$$

This representation suggests to introduce, in the neighborhood of the point $z = 0$, the local variable $\zeta = \zeta(z)$, in such a way that the following relation satisfies,

$$\frac{4}{3}\zeta^3(z) + xN^{-2/3}\zeta(z) = D_0(z) + xN^{-2/3}D_1(z) + O(N^{-4/3}) . \tag{5.92}$$

By solving this equation perturbatively, we find that, within the indicated error,

$$\zeta(z) = \zeta_0(z) + xN^{-2/3}\zeta_1(z) , \tag{5.93}$$

where

$$\zeta_0(z) = [\tfrac{3}{4}D_0(z)]^{1/3} ,$$

and

$$\zeta_1(z) = \frac{D_1(z) - \zeta_0(z)}{4\zeta^2(z)} .$$

The important point is that both these functions are holomorphic and invertible in a neighborhood of $z = 0$. Indeed, from (5.90) we obtain at once that

$$\zeta_0(z) = C^{-1}z - \frac{1}{10Cz_c^2}z^3 + \cdots ,$$

where C is exactly the same constant as in (5.91). This, in particular, means that $\zeta_1(z)$ is indeed holomorphic at $z = 0$, and in fact,

$$\zeta_1(z) = -\frac{C}{60z_c^2}z + \cdots .$$

The above estimates show that, for sufficiently large N, the function $\zeta(z)$ is conformal in some (fixed) neighborhood $\mathcal{B}_0$ of the point $z = 0$. This, in turn, allows us to introduce the following local variable in $\mathcal{B}_0$,

$$\xi(z) = N^{1/3}\zeta(z) . \tag{5.94}$$

Equation (5.94) defines in the neighborhood $\mathcal{B}_0$ a local change of variable (cf. (5.67))

$$\mathcal{B}_0 \to D_n(0) \equiv \{\xi : |\xi| < N^{1/3}\rho\} , \quad 0 < \rho < \frac{z_c}{2C} .$$

The action of the map $z \to \xi$ on the part of the contour Γ_c which is inside the neighborhood $\mathcal{B}_0$ is depicted in Fig. 5.13.

By virtue of its definition, the function $\xi(z)$ satisfies the asymptotic relation (cf. (5.89)),

$$-n(h - \mathrm{i}\pi) = \frac{8\mathrm{i}}{3}\xi^3 + 2\mathrm{i}x\xi + O(N^{-1/3})\,. \tag{5.95}$$

Therefore, the jump matrix $G_{\widehat{\Phi}}$ inside the neighborhood $\mathcal{B}_0$ can be uniformly estimated as

$$G_{\widehat{\Phi}}(z) = G^0_{\widehat{\Phi}}(z) + O(N^{-1/3})\,, \tag{5.96}$$

where the matrix function $G^0_{\widehat{\Phi}}(z)$ is defined by the equations,

$$G^0_{\widehat{\Phi}}(z) = \begin{cases} \begin{pmatrix} 1 & 0 \\ \exp(-\mathrm{i}n\pi + (8\mathrm{i}/3)\xi^3(z) + 2\mathrm{i}x\xi(z)) & 1 \end{pmatrix} & z \in \Gamma_c \cap \mathcal{B}_0\,,\ \operatorname{Im} z > 0 \\ \begin{pmatrix} 1 & 0 \\ \exp(\mathrm{i}n\pi - (8\mathrm{i}/3)\xi^3(z) - 2\mathrm{i}x\xi(z)) & 1 \end{pmatrix} & z \in \Gamma_c \cap \mathcal{B}_0\,,\ \operatorname{Im} z < 0 \\ \begin{pmatrix} 0 & 1 \\ -1 & 0 \end{pmatrix} & z \in \Gamma_c \cap \mathcal{B}_0\,,\ \operatorname{Im} z = 0 \end{cases}$$

Therefore, the map $z \to \xi$ transforms the $\mathcal{B}_0$-part of the critical $\widehat{\Phi}$-RH problem into the following model RH problem in the ξ-plane.

(1^c) $Y^c(\xi) \in H(\mathbb{C} \setminus \Gamma^0_c)$
(2^c) $Y^c_+(\xi) = Y^c_-(\xi) G^0(\xi)$,

$$G^0(\xi) = \begin{cases} \begin{pmatrix} 1 & 0 \\ \exp(-\mathrm{i}n\pi + (8\mathrm{i}/3)\xi^3 + 2\mathrm{i}x\xi) & 1 \end{pmatrix} & \xi \in \Gamma^0_{c2} \cup \Gamma^0_{c3} \\ \begin{pmatrix} 1 & 0 \\ \exp(\mathrm{i}n\pi - (8\mathrm{i}/3)\xi^3 - 2\mathrm{i}x\xi) & 1 \end{pmatrix} & \xi \in \Gamma^0_{c5} \cup \Gamma^0_{c6} \\ \begin{pmatrix} 0 & 1 \\ -1 & 0 \end{pmatrix} & \xi \in \Gamma^0_{c1} \cup \Gamma^0_{c4} \end{cases}$$

(3^c) $Y^c(\xi) = \Lambda\big(I + O(1/\xi)\big)$, $\xi \to \infty$,

$$\Lambda = \begin{cases} I & \operatorname{Im}\xi > 0 \\ \begin{pmatrix} 0 & -1 \\ 1 & 0 \end{pmatrix} & \operatorname{Im}\xi < 0\,. \end{cases}$$

Here, the contour Γ^0_c is the union of the six rays,

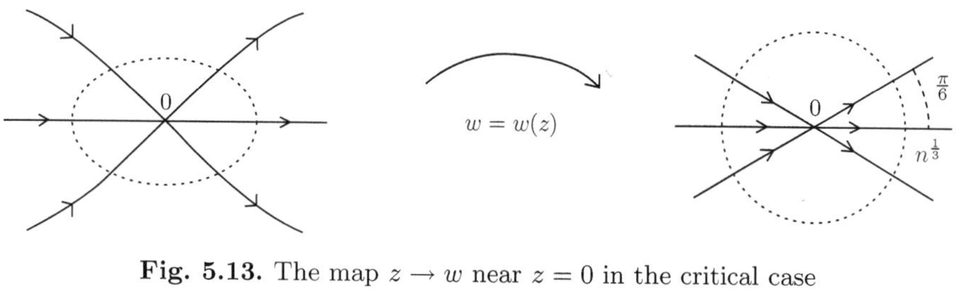

Fig. 5.13. The map $z \to w$ near $z = 0$ in the critical case

$$\Gamma_c^0 = \bigcup_{k=1}^{6} \Gamma_{ck}^0 \,,$$

which are depicted in Fig. 5.14.

The normalization condition (3^c) comes, as in the case of the end-point parametrix, from the fact that we want the "interior" function $Y^c(w(z))$ to match asymptotically, as $n \to \infty$, the "exterior" function $\Phi^\infty(z)$ at the boundary of $\mathcal{B}_0$. In other words, to specify the behavior of $Y^c(w)$ as $w \to \infty$, we must look at the behavior of $\Phi^\infty(z)$ at $z = 0$. To this end, we notice that the factorization (5.69) of the function $\Phi^\infty(z)$, which have already been used in the setting of the end point RH problem, can be rewritten as

$$\begin{aligned}\Phi^\infty(z) &= \begin{pmatrix} 1 & -1 \\ -\mathrm{i} & -\mathrm{i} \end{pmatrix} \begin{pmatrix} \beta_0^{-1}(z) & 0 \\ 0 & \beta_0(z) \end{pmatrix} \begin{pmatrix} 1/2 & \mathrm{i}/2 \\ -1/2 & \mathrm{i}/2 \end{pmatrix} \begin{cases} I & \operatorname{Im} z > 0 \\ \begin{pmatrix} 0 & -1 \\ 1 & 0 \end{pmatrix} & \operatorname{Im} z < 0 \end{cases} \\ &\equiv E_0(z) \begin{cases} I & \operatorname{Im} z > 0 \\ \begin{pmatrix} 0 & -1 \\ 1 & 0 \end{pmatrix} & \operatorname{Im} z < 0 \,. \end{cases}\end{aligned} \tag{5.97}$$

Here, the function $\beta_0(z)$ defines by the same formula as the function $\beta(z)$ before, but on the z-plane cut along $(-\infty, -z_0) \cup (z_0, +\infty)$, so that the both functions – $\beta_0(z)$ and $E_0(z)$ are holomorphic (and $E_0(z)$ invertible) at $z = 0$. Formula (5.97) explains the normalization condition (3^c).

Although there are no explicit contour integral representations for the solution $Y^c(\xi) \equiv Y^c(\xi, x)$ of the model RH problem (1^c)–(3^c), the solution exists for all real x and is analytic in x. This important fact follows from the modern theory of Painlevé equations where the RH problem (1^c)–(3^c) appears in connection with the second Painlevé equation. We will discuss this issue in some detail in Section 5.4.3.

Similar to the analysis done at the end points, we shall now define the parametrix for the solution of the $\widehat{\Phi}$-RH problem at the point $z = 0$ by the equation (cf. (5.75)),

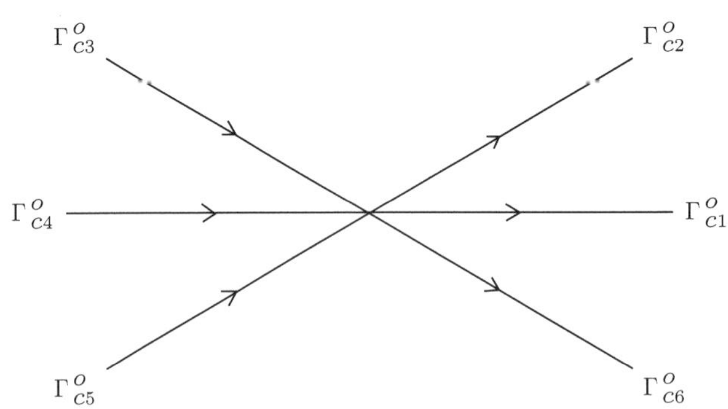

Fig. 5.14. The contour Γ_c^0

$$\Phi^c(z) = E_0(z)Y^c\big(\xi(z)\big)\,, \quad z \in \mathcal{B}_0\,. \tag{5.98}$$

We also notice that at the end points $\pm z_0$ we can use the old parametrics, i.e., $\Phi^0(z)$, at the point z_0, and $\sigma_3\Phi^0(-z)\sigma_3$, at the point $-z_0$. The function $\Phi^0(z)$ is defined in (5.75).

5.4.2 The Conclusion of the Asymptotic Analysis in the Critical Case

Instead of the solution $\widehat{\Phi}(z)$ directly, it is more convenient to approximate its Ψ-form, i.e., the function

$$\widehat{\Psi}(z) = \widehat{\Phi}(z)\exp(ng_0(z)\sigma_3)\,, \tag{5.99}$$

where

$$g_0(z) = -\frac{1}{\lambda}\int_{z_0}^{z}(b_0 + b_2u^2)\sqrt{u^2 - z_0^2}\,\mathrm{d}u = g(z) - \frac{1}{2\lambda}V(z) - \frac{l}{2}\,. \tag{5.100}$$

It is important to emphasize that, unlike (5.88) defining the function $h(z)$, in (5.100) the function $\sqrt{z^2 - z_0^2}$ is defined on the z-plane cut along the interval $[-z_0, z_0]$. In particular, this means that

$$g_0(z) = \tfrac{1}{2}h(z)\,, \quad \operatorname{Im} z > 0\,,$$

while,

$$g_0(z) = -\tfrac{1}{2}h(z)\,, \quad \operatorname{Im} z < 0.$$

Also, we note that,

$$g_0(z) = -\frac{1}{2\lambda}V(z) - \frac{l}{2} + \ln z + O\left(\frac{1}{z}\right), \quad z \to \infty\,.$$

In terms of the function $\widehat{\Psi}(z)$, the basic RH problem is formulated as follows.

(1'''') $\widehat{\Psi}(z) \in H(\mathbb{C} \setminus \Gamma_c)$
(2'''') $\widehat{\Psi}_+(z) = \widehat{\Psi}_-(z)G_{\widehat{\Psi}}(z)$, $z \in \Gamma_c$.
(3'''') $\widehat{\Psi}(z) = \big(I + O(1/z)\big)\exp(-NV(z)\sigma_3 + n\ln z\sigma_3 - (l/2)\sigma_3)$, $z \to \infty$.

The piecewise jump matrix $G_{\widehat{\Psi}}(z)$ is given by the relations,

- for z real and outside of the interval $[-z_0, z_0]$,

$$G_{\widehat{\Psi}}(z) = \begin{pmatrix}1 & 1\\ 0 & 1\end{pmatrix}\,;$$

- for z on the curves C_{1r}, C_{1l}, C_{2r}, and C_{2l}

$$G_{\widehat{\Phi}}(z) = \begin{pmatrix}1 & 0\\ 1 & 1\end{pmatrix}\,;$$

- for z on the interval $[-z_0, z_0]$,

$$G_{\widehat{\Phi}}(z) = \begin{pmatrix} 0 & 1 \\ -1 & 0 \end{pmatrix} .$$

It is also worth noticing that, in terms of the function $g_0(z)$, the estimate (5.95) reads,

$$ng_0(z) = \begin{cases} in\pi/2 - (4\mathrm{i}/3)\xi^3(z) - \mathrm{i}x\xi(z) + O(N^{-1/3}) , \\ \qquad\qquad z \in \mathcal{B}_0 , \ \operatorname{Im} z > 0 \\ -in\pi/2 + (4\mathrm{i}/3)\xi^3(z) + \mathrm{i}x\xi(z) + O(N^{-1/3}) , \\ \qquad\qquad z \in \mathcal{B}_0 , \ \operatorname{Im} z < 0 . \end{cases} \tag{5.101}$$

Put

$$\Psi^{\infty}(z) := \Phi^{\infty}(z)\exp(ng_0(z)\sigma_3) ,$$

$$\Psi^{c}(z) := \Phi^{c}(z) \begin{cases} \exp\left((in\pi/2 - (4\mathrm{i}/3)\xi^3(z) - \mathrm{i}x\xi(z))\sigma_3\right) , \\ \qquad\qquad z \in \mathcal{B}_0 , \ \operatorname{Im} z > 0 \\ \exp\bigl(-(in\pi/2 - (4\mathrm{i}/3)\xi^3(z) - \mathrm{i}x\xi(z))\sigma_3\bigr) , \\ \qquad\qquad z \in \mathcal{B}_0 , \ \operatorname{Im} z < 0 , \end{cases}$$

$$\Psi^{0}(z) := \Phi^{0}(z)\exp\bigl(-\tfrac{2}{3}w^{3/2}(z)\bigr) ,$$

and define the function (cf. (5.77)),

$$\widehat{\Psi}^{\mathrm{as}}(z) = \begin{cases} \Psi^{\infty}(z) & z \in \mathbb{C} \setminus \bigl(\mathcal{B}_0 \cup \mathcal{B} \cup (-\mathcal{B})\bigr) \\ \Psi^{c}(z) & z \in \mathcal{B}_0 \\ \Psi^{0}(z) & z \in \mathcal{B} \\ (-1)^n \sigma_3 \Psi^{0}(-z)\sigma_3 & z \in -\mathcal{B} . \end{cases} \tag{5.102}$$

Our aim is to prove that the exact solution $\widehat{\Psi}(z)$ of the RH problem (1'''')–(3'''') satisfies the estimate (cf. (5.78)),

$$\widehat{\Psi}(z) - \left(I + O\left(\frac{1}{(1+|z|)N^{1/3}}\right)\right)\widehat{\Psi}^{\mathrm{as}}(z) ,$$

$$N \to \infty , \ \frac{n}{N} = \lambda_c + cxN^{-2/3} , \ z \in \mathbb{C} . \tag{5.103}$$

Similar to the previous, one-cut case, in order to prove (5.103) we consider the matrix ratio (cf. (5.79)),

$$X^{c}(z) := \widehat{\Psi}(z)\bigl(\widehat{\Psi}^{\mathrm{as}}(z)\bigr)^{-1} . \tag{5.104}$$

This function solves the Riemann–Hilbert problem posed on the contour Γ_{X^c} depicted in Fig. 5.15. Similar to the one-cut X-RH problem, the important feature of this problem is the absence of any jumps inside the domains $\mathcal{B}_0$, $\mathcal{B}$ and $-\mathcal{B}$ and across the part of the real line between the boundary curves $C_r \equiv \partial\mathcal{B}$, $C_0 \equiv \partial\mathcal{B}_0$, and $C_l \equiv \partial(-\mathcal{B})$. As before, the cancelling out of the jumps of $\widehat{\Psi}(z)$ and $\widehat{\Psi}^{as}(z)$ inside $\mathcal{B}$, $\mathcal{B}_0$, and $-\mathcal{B}$ and across the indicated part of the real line follows directly from the basic jump properties of the model functions $\Psi^{\infty}(z)$ and $\Psi^{0}(z)$. The only difference, comparing to the one-cut considerations, is that one has to take into account that in terms of Ψ -functions, the symmetry relation (5.76) should be replaced by the relation,

$$\widehat{\Psi}(z) = (-1)^n \sigma_3 \widehat{\Psi}(-z)\sigma_3.$$

This modification is a consequence of the symmetry equation,

$$g_0(-z) = g_0(z) - \pi i,$$

which in turn can be easily checked directly.

On the "one-cut" parts of the contour Γ_{X^c},the X^c-jump matrix, $G_{X^c}(z)$, satisfies the previous "one-cut" uniform estimates as $N \to \infty$. That is:

$$\begin{aligned} G_{X^c}(z) &\equiv \Phi^{\infty}(z)\begin{pmatrix} 1 & \exp\bigl(n(g_+ + g_- - (1/\lambda)V - l)\bigr) \\ 0 & 1 \end{pmatrix}\bigl(\Phi^0(z)\bigr)^{-1} \\ &= I + O\bigl(\mathrm{e}^{-cNz^2}\bigr), \quad c>0,\ z \in \Gamma_l \cup \Gamma_r, \\ G_{X^c}(z) &\equiv \Phi^{\infty}(z)\begin{pmatrix} 1 & 0 \\ \mathrm{e}^{\mp nh} & 1 \end{pmatrix}(\Phi^{\infty}(z))^{-1} = I + O\bigl(\mathrm{e}^{-cN}\bigr), \\ &\qquad c>0, \quad z \in C_{ul} \cup C_{ur}\ (\mathrm{e}^{-nh}),\ z \in C_{dl} \cup C_{dr}\ (\mathrm{e}^{nh}), \end{aligned}$$

and

$$G_{X^c}(z) \equiv \Phi^0(z)\bigl(\Phi^{\infty}(z)\bigr)^{-1} = I + O\Bigl(\frac{1}{N}\Bigr), \quad z \in C_r \cup C_l\,.$$

In order to estimate the jump accross the curve $C_0 = \partial\mathcal{B}_0$ we notice that, as follows from (5.101) (and the asymptotic normalization condition (3^c) for the model function $Y^c(\xi)$),

$$\Psi^c(z) = \left(I + O\Bigl(\frac{1}{N^{1/3}}\Bigr)\right)\Psi^{\infty}(z), \quad N \to \infty, \tag{5.105}$$

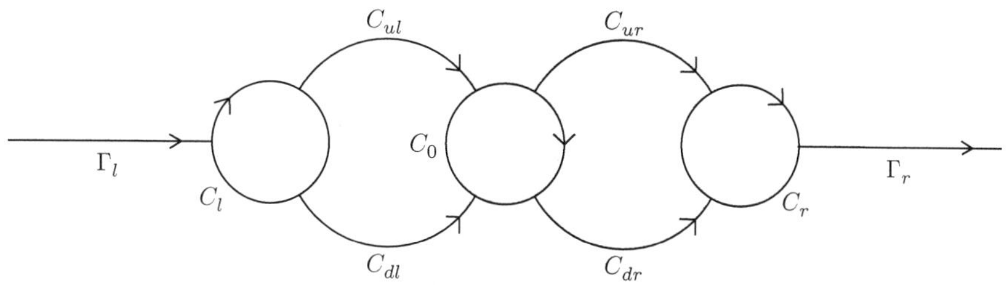

Fig. 5.15. The final, X_c-RH problem

uniformly on C_0 (in fact, in some neighborhood of C_0). Therefore, for the jump matrix G_{X^c} on the curve C_0 we have,

$$G_{X^c}(z) = \bigl(X^c_-(z)\bigr)^{-1} X^c_+(z) = \widehat{\Psi}^{\mathrm{as}}_-(z)\bigl(\widehat{\Psi}^{\mathrm{as}}_+(z)\bigr)^{-1}$$
$$= \Psi^c(z)\bigl(\Psi^\infty(z)\bigr)^{-1} = \left(I + O\left(\frac{1}{N^{1/3}}\right)\right). \tag{5.106}$$

From the above estimates we arrive at the key inequality (cf. (5.80)),

$$\|I - G_{X^c}\|_{L_2 \cap L_\infty} \leq \frac{C}{N^{1/3}}\,. \tag{5.107}$$

By exactly the same arguments as in the one-cut case (see the proof of (5.78) in Section 5.3.4), we derive from (5.107) the estimate,

$$X^c(z) = I + O\left(\frac{1}{(1+|z|)N^{1/3}}\right), \quad N \to \infty\,, \ z \in \mathbb{C}\,, \tag{5.108}$$

and (5.103) follows.

The improvement of the error term in (5.103) can be achieved, in principle, by iterating the relevant singular integral equations. In Section 5.4.4 we will show how, with the help of iterations, one can calculate explicitly the order $O(N^{-1/3})$ correction in (5.103). However (see also the end of Section 5.3.4), to achieve the $O(N^{-1})$ accuracy in the estimation of $\widehat{\Psi}(z)$, the corresponding multiple integral representation becomes extremely cumbersome. Indeed, in paper [5], where the first rigorous evaluation of the double scaling limit (5.85) has been performed, an alternative approach to the asymptotic solution of the $\widehat{\Phi}$-RH problem is used. This approach is based on a rather nontrivial modification of both the model solution $\Psi^\infty(z)$ and the model solution $\Psi^c(z)$. These modifications are based on the prior heuristic study of the Freud equation associated to the quartic potential $V(z)$ for the recurrsion coefficients R_n followed by the formal WKB-type analysis of the associated Lax pair. This analysis uses also the Lax pair associated to the model RH problem (1^c)-(3^c) and produces an improved pair of model functions $\Psi^\infty(z)$ and $\Psi^c(z)$ for which the asymptotic relation (5.105) holds with the accuracy $O(N^{-1})$. We refer the reader to [5] and to the lectures of Pavel Bleher in this volume for more details about this construction.

Remark. The asymptotic solution of the Riemann–Hilbert problem corresponding to a generic real analytic critical potential $V(z)$ (the corresponding equilibrium measure vanishes quadratically at an interior point of the support) has been performed in [9]. The analysis of [9] is similar to the one presented above except for the parametrix at the critical point, where the authors of [9] suggest a rather non-trivial modification of the local variable so that the relation, similar to (5.101), is satisfied *exactly.*

5.4.3 Analysis of the RH Problem (1^c)–(3^c). The Second Painlevé Equation

PutF

$$\widetilde{Y}^c(\xi) = \begin{cases} Y^c(\xi) & \operatorname{Im} z > 0 \\ \\ Y^c(\xi) \left(\begin{smallmatrix} 0 & 1 \\ -1 & 0 \end{smallmatrix}\right) & \operatorname{Im} z < 0 \,. \end{cases} \tag{5.109}$$

This simple transformation eliminates the jump accross the real line and, simultaneously, simplifies the asymptotic condition at $\xi = \infty$. Indeed, in terms of $\widetilde{Y}^c(\xi)$ the RH Problem (1^c)–(3^c) transforms to the problem,

($1^{\tilde{c}}$) $\widetilde{Y}^c(\xi) \in H(\mathbb{C} \setminus \Gamma^p)$
($2^{\tilde{c}}$) $\widetilde{Y}^c_+(\xi) = \widetilde{Y}^c_-(\xi)\widetilde{G}^0(\xi)$,

$$\widetilde{G}^0(\xi) = \begin{cases} \begin{pmatrix} 1 & 0 \\ \exp(-in\pi + (8\mathrm{i}/3)\xi^3 + 2\mathrm{i}x\xi) & 1 \end{pmatrix}, & \xi \in \Gamma^p_1 \cup \Gamma^p_2 \\ \begin{pmatrix} 1 & -\exp(in\pi - (8\mathrm{i}/3)\xi^3 - 2\mathrm{i}x\xi) \\ 0 & 1 \end{pmatrix}, & \xi \in \Gamma^p_3 \cup \Gamma^p_4 \end{cases}$$

($3^{\tilde{c}}$) $\widetilde{Y}^c(\xi) = \bigl(I + O(1/\xi)\bigr)$, $\xi \to \infty$,

Here, the contour Γ^p is the union of the four rays,

$$\Gamma^p = \bigcup_{k=1}^{4} \Gamma^p_k \equiv \Gamma^0_c \setminus \Gamma^0_{c1} \cup \Gamma^0_{c4} \,,$$

which are depicted in Fig. 5.16. Let us also make one more trivial gauge transformation,

$$\widetilde{Y}^c(\xi) \to Y^p(\xi) = \mathrm{e}^{-\frac{in\pi}{2}\sigma_3}\widetilde{Y}^c(\xi)\mathrm{e}^{\frac{in\pi}{2}\sigma_3}, \tag{5.110}$$

which brings the model RH problem (1^c)- (3^c) to the following universal form,

(1^p) $Y^p(\xi) \in H(\mathbb{C} \setminus \Gamma^p)$
(2^p) $Y^p_+(\xi) = Y^p_-(\xi) \exp(-(4\mathrm{i}/3)\xi^3\sigma_3 - \mathrm{i}x\xi\sigma_3) S \exp(4\mathrm{i}/3)\xi^3\sigma_3 + \mathrm{i}x\xi\sigma_3)$, $\xi \in \Gamma^p$

$$S = \begin{cases} \left(\begin{smallmatrix} 1 & 0 \\ 1 & 1 \end{smallmatrix}\right) \equiv S_1 = S_2; , & \xi \in \Gamma^p_1 \cup \Gamma^p_2 \\ \left(\begin{smallmatrix} 1 & -1 \\ 0 & 1 \end{smallmatrix}\right) \equiv S_3 = S_4 \,, & \xi \in \Gamma^p_3 \cup \Gamma^p_4 \end{cases} \tag{5.111}$$

(3^p) $Y^p(\xi) = \bigl(I + O(1/\xi)\bigr)$, $\xi \to \infty$,

The problem (1^p)–(3^p) is depicted in Fig. 5.16, and it plays in the analysis of the parametrix at the internal (double) zero of the equilibrium measure (the point $z = 0$) precisely the same role as the universal Airy RH-problem[1] (1^0)–(3^0) plays in the analysis of the parametrix at the end point of the support of

[1] Note that, although we are using the same letter, the matrices S_k in (2^p) are *not* the matrices S_k in (2^0).

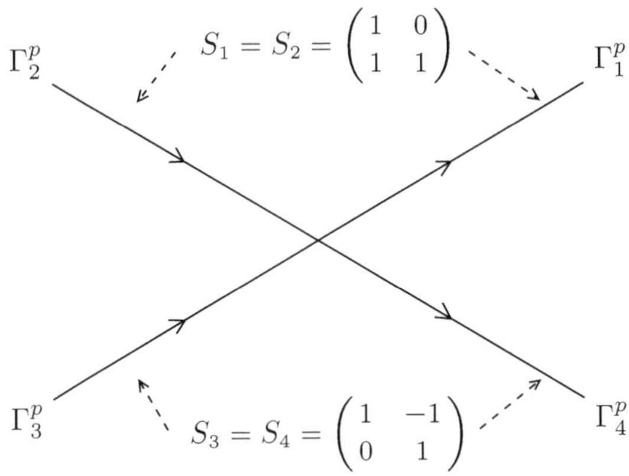

Fig. 5.16. The contour Γ^p and the Painlevé II RH problem

the equilibrium measure (the point $z = z_0$). However, the important difference is that we do not posses any explicit representation, in terms of contour integrals of elementary functions, for the solution $Y^p(\xi)$ of the problem (1^p)–(3^p). Therefore, the proof of the existence of the function $Y^p(\xi) \equiv Y^p(\xi, x)$ and its smooth dependence on the real parameter x is much more difficult than in the case of the Airy-solvable problem (1^0)–(3^0). Indeed, the existence of the solution $Y^p(\xi)$ and its smooth dependence on the real x is a nontrivial fact of the modern theory of Painlevé equations, whose appearance in context of the problem (1^p) –(3^p) we are going to explain now.

Put

$$\Psi^p(\xi, x) := Y^p(\xi, x) \exp\left(-\frac{4i}{3}\xi^3\sigma_3 - ix\xi\sigma_3\right) .$$

The jump condition (2^p) and the asymptotic condition (3^p) in terms of the function $\Psi^p(\xi)$ become

(a) $\Psi^p_+(\xi) = \Psi^p_-(\xi)S$, $\xi \in \Gamma^p$

and

(b) $\Psi^p(\xi) = (I + m_1^p/\xi + \cdots) \exp(-(4\mathrm{i}/3)\xi^3\sigma_3 - \mathrm{i}x\xi\sigma_3)$, $\xi \to \infty$,

respectively. Moreover, similar to the Airy case, we have the cyclic relation,

$$S_4^{-1} S_3 S_2 S_1^{-1} = I .$$

Therefore, we can now proceed with considering the following, this time two logarithmic derivatives,

$$A(\xi) = \frac{\partial \Psi^p}{\partial \xi}(\Psi^p)^{-1} ,$$

and

$$U(\xi) = \frac{\partial \Psi^p}{\partial x}(\Psi^p)^{-1} .$$

Repeating the same arguments as we used when deriving (5.73), which are based on manipulations with the asymptotic expansion (b) and the exploitation of the Liouville theorem, we establish that both the matrices $A(\xi)$ and $U(\xi)$ are in fact polynomials in ξ. Indeed we find that,

$$A(\xi) = -4\mathrm{i}\xi^2\sigma_3 + 4\xi u\sigma_1 - 2u_x\sigma_2 - (\mathrm{i}x + 2iu^2)\sigma_3 \, , \tag{5.112}$$

and

$$U(\xi) = -\mathrm{i}\xi\sigma_3 + u\sigma_1 \, , \tag{5.113}$$

where,

$$\sigma_1 = \begin{pmatrix} 0 & 1 \\ 1 & 0 \end{pmatrix} , \quad \sigma_1 = \begin{pmatrix} 0 & -\mathrm{i} \\ 1 & 0 \end{pmatrix} , \quad u_x \equiv \frac{\mathrm{d}u}{\mathrm{d}x} ,$$

and the function $u(x)$ is defined by the coefficient $m_1 \equiv m_1(x)$ of the series (b) according to the equation,

$$u(x) = 2\mathrm{i}\big(m_1^p(x)\big)_{12} \, . \tag{5.114}$$

In other words, the matrix valued function $\Psi^p(\xi, x)$ satisfies the following system of linear differential equations (cf. (5.73)),

$$\begin{cases} \dfrac{\partial \Psi^p}{\partial \xi} = A(\xi)\Psi^p \\ \dfrac{\partial \Psi^p}{\partial x} = U(\xi)\Psi^p, \end{cases} \tag{5.115}$$

with the coefficient matrices $A(\xi)$ and $U(\xi)$ defined by the relations (5.112) and (5.113), respectively.

Consider the compatibility condition of system (5.115):

$$\Psi^p_{x\xi} = \Psi^p_{\xi x} \iff A_x - U_\xi = [U, A] \, . \tag{5.116}$$

The matrix equation (5.116) is satisfied identically with respect to ξ. By a simple straightforward calculations, one can check that the matrix relation (5.116) is equivalent to a single scalar differential equation for the function $u(x)$,

$$u_{xx} = 2u^3 + xu \, . \tag{5.117}$$

The latter equation is a particular case of the second Painlevé equation (see, e.g., [33]).

In the terminology of integrable systems, the linear equations (5.115) form a *Lax pair* for the nonlinear ODE (5.117)) – the Flaschka–Newell Lax pair [26], and the matrix relation (5.116) is its *zero curvature or Lax* representation. The restrictions of the function $\Psi^p(\xi)$ to the connected components of the domain $\mathbb{C}\setminus\Gamma^p$ constitute the canonical solutions of the first, i.e., ξ-equation of the Lax pair (5.115), and the matrices S_k from (5.111) are its *Stokes matrices*. The Stokes matrices of the ξ-equation form a complete set of the first integrals of the Painlevé equation (5.117). The particular choice of Stokes matrices indicated in (5.111) corresponds, as we will show shortly, to a selection of the so-called Hastings–McLeod solution of the Painlevé II equation (5.117).

Unlike the Airy case of the problem (1^0)–(3^0), (5.115) do not provide us directly with the solution of the problem (1^p)–(3^p). However, they can be used to prove the existence of the solution and its smooth dependence on the real x. The proof is as follows.

Hastings and McLeod have shown in [32] that there exists a unique solution of (5.117) smooth for all real x, which satisfies the asymptotic condition,

$$u(x) \sim \frac{1}{2\sqrt{\pi}} x^{-1/4} \exp(-\tfrac{2}{3}x^{3/2}) , \quad x \to +\infty . \tag{5.118}$$

They also showed that on the other end of the real line the solution $u(x)$ has a power like behavior,

$$u(x) \sim \sqrt{-\frac{x}{2}} , \quad x \to -\infty . \tag{5.119}$$

Simultaneously, one can (quite easily) solve asymptotically as $x \to +\infty$ the singular integral equation (5.27) corresponding to the problem (1^p)–(3^p) (see [26]; see also [37, 38, 44]), or one can (again quite easily) solve asymptotically the problem itself directly – with the help of the nonlinear steepest descent method (see [14]; see also [27]). Evaluating then, using (5.114), the asymptotics of the corresponding function $u(x)$, one discovers that it is exactly the asymptotics (5.118). Hence, the choice (5.111) of the Stokes data for the ξ-equation indeed corresponds to the Hastings–Mcleaod Painlevé transcendents. The independence of the Stokes matrices of x and the global existence of the solution $u(x)$ imply that the canonical solutions of the ξ-equation provide the solution of the Riemann–Hilbert problem (1^p)–(3^p) for all real x.

Independent of [32] a proof of the solvability of the problem (1^p)–(3^p), as well as the meromorphic extension of its solution on the whole x-plane together with the detail description of the asymptotic behavior of the function $u(x)$ everywhere on the x-plane, can be obtained (though not very easily, this time) via the advanced Riemann–Hilbert theory of Painlevé equations developed during the last 20 years. We refer the reader to the recent monograph [27] for more on this issue and for the history of the subject.

5.4.4 The Painlevé Asymptotics of the Recurrence Coefficients

It follows from (5.20) that the recurrence coefficients R_n are given by the formula

$$R_n = (m_1)_{12}(m_1)_{21} , \tag{5.120}$$

where m_1 is the matrix coefficient of the z^{-1} term of the series (5.19). Hence we need the asymptotics of the coefficient m_1.

For large z,

$$\widehat{\Psi}^{as}(z) = \Psi^{\infty}(z) ,$$

which together with (5.104) implies that, for large z,

$$\widehat{\Psi}(z) = X^c(z)\Psi^{\infty}(z) ,$$

or

$$\widehat{\Phi}(z) = X^c(z)\Phi^{\infty}(z) .$$

Recalling definition (5.51) of the function $\Phi(z)$, we rewrite the last relation as an equation for the original function $Y(z)$,

$$Y(z) = \mathrm{e}^{nl/2} X^c(z)\Phi^{\infty}(z)\mathrm{e}^{n(g(z)-l/2)} . \tag{5.121}$$

Define g_1, m_1^{∞}, and $m_1^{X^c}$ as the coefficients of the asymptotic series as $z \to \infty$,

$$g(z) = \ln z + \frac{g_1}{z} + \cdots ,$$

$$\Phi^{\infty}(z) = I + \frac{m_1^{\infty}}{z} + \cdots ,$$

and

$$X^c(z) = I + \frac{m_1^{X^c}}{z} + \cdots ,$$

respectively. We have that,

$$m_1 = ng_1 + \exp\biggl(\frac{nl}{2}\sigma_3\biggr)(m_1^{\infty} + m_1^{X^c})\exp\biggl(-\frac{nl}{2}\sigma_3\biggr) . \tag{5.122}$$

The coefficients g_1 and m_1^{∞} are elementary objects and can be easily found from the explicit formulae for $g(z)$ and $\Phi^{\infty}(z)$. Let us discuss the evaluation of the matrix $m_1^{X^c}$.

From the singular integral equation (5.27) corresponding to the X^c-RH problem, i.e., from the equation,

$$X^c(z) = I + \frac{1}{2\pi \mathrm{i}}\int_{\Gamma_{X^c}} \frac{X^c_-(z')(G_{X^c}(z') - I)}{z' - z}\,\mathrm{d}z' , \tag{5.123}$$

we derive that

$$m_1^{X^c} = -\frac{1}{2\pi \mathrm{i}}\int_{\Gamma_{X^c}} X^c_-(z')(G_{X^c}(z') - I)\,\mathrm{d}z' . \tag{5.124}$$

Taking into account (5.108) and the fact that everywhere on Γ_{X^c} but C_0 the difference $G_{X^c}(z') - I$ is of order $O(N^{-1})$, we obtain from (5.124) the asymptotic formula,

$$m_1^{X^c} = -\frac{1}{2\pi \mathrm{i}}\int_{C_0} (G_{X^c}(z') - I)\,\mathrm{d}z' + O\biggl(\frac{1}{N^{2/3}}\biggr) . \tag{5.125}$$

To proceed we need the details about the jump matrix $G_{X^c}(z)$ on the curve C_0.

From (5.106) it follows that, for $z \in C_0$,

$$
\begin{aligned}
&G_{X^c}(z)\\
&= \Psi^c(z)\big(\Psi^\infty(z)\big)^{-1}\\
&= E_0(z)Y^c(\xi(z)) \exp\left[\operatorname{sign}(\operatorname{Im} z)\left(\frac{in\pi}{2}-\frac{4i}{3}\xi^3(z)-ix\xi(z)\right)\sigma_3 - ng_0(z)\sigma_3\right]\\
&\qquad\qquad\times \Lambda^{-1}E_0^{-1}(z)\\
&= E_0(z)\widetilde{Y}^c(\xi(z))\Lambda \exp\left[\operatorname{sign}(\operatorname{Im} z)\left(\frac{in\pi}{2}-\frac{4i}{3}\xi^3(z)-ix\xi(z)\right)\sigma_3 - ng_0(z)\sigma_3\right]\\
&\qquad\qquad\times \Lambda^{-1}E_0^{-1}(z)\\
&= E_0(z)\widetilde{Y}^c(\xi(z)) \exp\left(\frac{in\pi}{2}-\frac{4i}{3}\xi^3(z)-ix\xi(z)-n\tilde{g_0}(z)\sigma_3\right)E_0^{-1}(z)\,.
\end{aligned}
\tag{5.126}
$$

Here, the function $\tilde{g_0}(z)$ is given by the same integral (5.100)) as the function $g_0(z)$, but the root $\sqrt{z^2-z_0^2}$ is defined on the plane cut along $(-\infty,-z_0)\cup(z_0,+\infty)$, so that

$$
\tilde{g_0}(z) = \operatorname{sign}(\operatorname{Im} z) g_0(z)\,.
$$

We also note that the function $\tilde{g_0}(z)$, as well as the functions $\xi(z)$ and $E_0(z)$, is holomorphic in $\mathcal{B}_0$.

By a straightforward though slightly tiresome calculation we can refine the estimate (5.101) as follows,

$$
n\tilde{g_0}(z) = \frac{in\pi}{2}-\frac{4i}{3}\xi^3(z)-ix\xi(z)+ia(z)x^2N^{-1/3}+O(N^{-1})\,,\quad z\in\overline{\mathcal{B}}_0\,, \tag{5.127}
$$

where

$$
a(z) = \frac{D_1(z)-\zeta_0(z)}{3D_0(z)}D_1(z) - D_2(z)\,,
$$

and $D_0(z)$, $D_1(z)$, and $\zeta_0(z)$ are the elementary functions which have been defined in (5.89) and (5.93), respectively, when we introduced the local variable $\zeta(z)$, and $D_2(z)$ is the integral,

$$
D_2(z) = -\frac{c^2}{2t}\int_0^z \frac{du}{(z_0^2-u^2)^{3/2}}\,.
$$

Actually, these exact formulas for $a(z)$ are not really of importance for our current calculations. What will matters is that $a(z)$ is holomorphic at $z=0$ (cf. the holomorphicity at $z=0$ of the function $\zeta_1(z)$ from (5.93)).

Using (5.127) and the asymptotic expansion $(3^{\tilde{c}})$ of the function $\widetilde{Y}^c(\xi)$, we derive from (5.126) the following estimate, uniform for $z\in C_0$

$$G_{X^c}(z) = I + \frac{1}{\xi(z)} E_0(z)\tilde{m}_1^p E_0^{-1}(z) - \mathrm{i}x^2 a(z) E_0(z)\sigma_3 E_0^{-1}(z) N^{-1/3} + O(N^{-2/3}) \,, \quad N \to \infty \,, \; z \in C_0 \,, \tag{5.128}$$

where the matrix $\tilde{m}_1^p$ is the coefficient of the asymptotic series,

$$\widetilde{Y}^c(\xi) = I + \frac{\tilde{m}_1^p}{\xi} + \cdots \,, \qquad \xi \to \infty \,.$$

Note that the matrix function $\tilde{m}_1^p \equiv \tilde{m}_1^p(x)$ is related to the similar function $m_1^p(x)$ from (5.114) by the equation (see (5.110)),

$$\tilde{m}_1^p = \exp\left(\frac{\mathrm{i}n\pi}{2}\sigma_3\right) m_1^p \exp\left(-\frac{\mathrm{i}n\pi}{2}\sigma_3\right) . \tag{5.129}$$

We also recall that $\xi(z) = N^{1/3}\zeta(z)$ and hence

$$\frac{1}{\xi(z)} = O(N^{-1/3}) \,, \quad N \to \infty \,, \; z \in C_0 \,.$$

With the help of (5.128), we can now easily estimate the integral in the right-hand side of (5.125). Taking into account that the function $a(z)$ is analytic in $\mathcal{B}_0$ and hence the integral of $a(z)E_0(z)\sigma_3 E_0^{-1}(z)$ vanishes, we obtain

$$\begin{aligned} m_1^{X^c} &= \operatorname{res}|_{z=0}\left(\frac{1}{\xi(z)} E_0(z)\tilde{m}_1^p E_0^{-1}(z)\right) + O(N^{-2/3}) \\ &= CN^{-1/3} E_0(0)\tilde{m}_1^p E_0^{-1}(0) + O(N^{-2/3}) \,, \end{aligned} \tag{5.130}$$

where C is the constant defined by (5.91).

We are almost ready to produce the final estimate for the matrix $m_1^{X^c}$. One more step is needed, this time algebraic.

Put

$$\widetilde{Y}^p(\xi) = \sigma_1 Y^p(-\xi)\sigma_1 \,.$$

Since,

$$S_{1,2} = \sigma_1 S_{3,4}^{-1}\sigma_1 \,,$$

we immediately conclude that $\widetilde{Y}^p(\xi)$ solves the same Riemann–Hilbert problem, i.e., (1^p)–(3^p). Therefore, by the uniqueness theorem, the functions $\widetilde{Y}^p(\xi)$ and $Y^p(\xi)$ coincide. Hence, we arrive at the symmetry relation,

$$Y^p(\xi) = \sigma_1 Y^p(-\xi)\sigma_1 \,,$$

which in turn implies that

$$m_1^p = -\sigma_1 m_1^p \sigma_1 \,. \tag{5.131}$$

From (5.131), we come to the following structure of the matrix coefficient m_1^p:

$$(m_1^p)_{11} = -(m_1^p)_{22}\,, \quad (m_1^p)_{12} = -(m_1^p)_{21}\,.$$

Introducing notation, $v := (m_1^p)_{11}$, and recalling the relation (5.114) between the Hastings–McLeod Painlevé function $u(x)$ and the 12 entry of the matrix m_1^p, we obtain finally that the matrix m_1^p can be written as

$$m_1^p = \begin{pmatrix} v & -(i/2)u \\ (i/2)u & -v \end{pmatrix}, \tag{5.132}$$

and hence (cf. (5.129))

$$\tilde{m}_1^p = \begin{pmatrix} v & (-1)^{n+1}(\mathrm{i}/2)u \\ (-1)^n(\mathrm{i}/2)u & -v \end{pmatrix}. \tag{5.133}$$

As a matter of fact, the function $v(x)$ can be also expressed in terms of $u(x)$ and its derivatives, but we won't need this expression.

From the definition (5.97) of the matrix factor $E_0(z)$, we have that,

$$E_0(0) = \frac{1}{\sqrt{2}}\begin{pmatrix} 1 & 1 \\ -1 & 1 \end{pmatrix}.$$

The substitution of this equation and (5.133) into (5.130) concludes the asymptotic evaluation of the matrix $m_1^{X^c}$. Indeed, we obtain that

$$m_1^{X^c} = CN^{-1/3}\begin{pmatrix} 0 & -v - (\mathrm{i}/2)(-1)^n u \\ -v + (\mathrm{i}/2)(-1)^n u & 0 \end{pmatrix} + O(N^{-2/3})\,. \tag{5.134}$$

The last object we need in order to be able to evaluate R_n, is the matrix coefficient m_1^∞. We derive it easily from (5.50):

$$m_1^\infty = \begin{pmatrix} 0 & (\mathrm{i}/2)z_0 \\ -(\mathrm{i}/2)z_0 & 0 \end{pmatrix} = \begin{pmatrix} 0 & (\mathrm{i}/2)z_c \\ -(\mathrm{i}/2)z_c & 0 \end{pmatrix} + O(N^{-2/3})\,, \tag{5.135}$$

where z_c, we recall, is given in (5.86). Using (5.135) and (5.134) in (5.122) (note that we we do not need g_1, although it also can be evaluated by elementary means), we arrive at the estimates,

$$(m_1)_{12} = \mathrm{e}^{nl}\left(\frac{\mathrm{i}}{2}z_c + \frac{\mathrm{i}}{2}CN^{-1/3}(-1)^{n+1}u - CN^{-1/3}v + O(N^{-2/3})\right),$$

and

$$(m_1)_{21} = \mathrm{e}^{-nl}\left(-\frac{\mathrm{i}}{2}z_c - \frac{\mathrm{i}}{2}CN^{-1/3}(-1)^{n+1}u - CN^{-1/3}v + O(N^{-2/3})\right),$$

which in turn provide as with the asymptotic formula for R_n,

$$
\begin{aligned}
R_n = (m_1)_{12}(m_1)_{21} &= -\left(\frac{\mathrm{i}}{2}z_c + \frac{\mathrm{i}}{2}CN^{-1/3}(-1)^{n+1}u\right)^2 + O(N^{-2/3}) \\
&= \frac{z_c^2}{4} + \frac{z_c}{2}CN^{-1/3}(-1)^{n+1}u + O(N^{-2/3}) \\
&= \frac{|t|}{2\kappa} + N^{-1/3}C_1(-1)^{n+1}u(x) + O(N^{-2/3});, \quad C_1 = \left(\frac{2|t|}{\kappa^2}\right)^{1/3} .
\end{aligned}
\tag{5.136}
$$

Remark 5.4.1. As has already been indicated above (see the end of Section 5.3.4), from the Riemann–Hilbert analysis one can rigorously establish the existence of the full asymptotic expansions for the coefficients R_n. A direct substitution of this series in the difference Freud equation would allow then to evaluate, in principal, the corrections to (5.136) of any order. For instance, one obtains (see also [5]) that,

$$
\begin{aligned}
R_n = \frac{|t|}{2\kappa} + N^{-1/3}C_1(-1)^{n+1}u(x) + N^{-2/3}C_2\big(x + 2u^2(x)\big) + O(N^{-1}) \, , \\
C_2 = \frac{1}{2}\left(\frac{1}{2|t|\kappa}\right)^{1/3} .
\end{aligned}
\tag{5.137}
$$

It also should be said that asymptotics (5.137) were first suggested (via the formal analysis of the Freud equation) in physical papers by Douglas, Seiberg, and Shenker [20], Crnković and Moor [11], and Periwal and Shevitz [42].

Remark 5.4.2. Using estimate (5.128) in the integral equation (5.123), we can improve our main asymptotic formula (5.103). Indeed we have that

$$
\widehat{\Psi}(z) = \left(I + r(z) + O\left(\frac{1}{(1+|z|)N^{2/3}}\right)\right)\widehat{\Psi}^{\mathrm{as}}(z) \, , \tag{5.138}
$$

where

$$
r(z) = \frac{1}{z}CN^{-1/3}E_0(0)\tilde{m}_1^p E_0^{-1}(0) \, , \tag{5.139}
$$

if $z \notin \overline{\mathcal{B}}_0$, and

$$
\begin{aligned}
r(z) = I + \frac{1}{z}CN^{-1/3}E_0(0)\tilde{m}_1^p E_0^{-1}(0) - \frac{1}{\xi(z)}E_0(z)\tilde{m}_1^p E_0^{-1}(z) \\
+ \mathrm{i}x^2N^{-1/3}a(z)E_0(z)\sigma_3E_0^{-1}(z) \, ,
\end{aligned}
\tag{5.140}
$$

if $z \in \mathcal{B}_0$.

As it has already been mentioned in the end of Section 5.4.2, a further improvement of the estimate (5.103), if one wants compact formulas, needs an essential modification of the construction of the parametrix at $z = 0$ (see [5]).

References

1. M. Adler, T. Shiota, P. van Moerbeke: Phys. Lett. A **208**, 67 (1995)
2. J. Baik, P.A. Deift, K. Johansson: J. Amer. Math. Soc. **12**, 1119 (1999)
3. R. Beals, P.A. Deift, C. Tomei, *Direct and Inverse Scattering on the Line* (Amer. Math. Soc., Providence, RI 1988)
4. P.M. Bleher, A.R. Its: Ann. Math. **150**, 185 (1999)
5. P.M. Bleher, A.R. Its: Commun. Pure Appl. Math. **56**, 433 (2003)
6. P.M. Bleher, A.R. Its: Ann. Inst. Fourier **55**, 1943 (2005)
7. É. Brézin, V.A. Kazakov: Phys. Lett. B **236**, 144 (1990).
8. K.F. Clancey, I. Gohberg, *Factorization of Matrix Functions and Singular Integral Operators* (Birkhäuser, Basel 1981)
9. T. Claeys, A.B.J. Kuijlaars: Commun. Pure Appl. Math. **59** 1573 (2006)
10. Y. Chen, M.E. H. Ismail, J. Phys. A **30**, 7817 (1997)
11. Č. Crnković, G. Moore: Phys. Lett. B **257**, 322 (1991).
12. P.A. Deift: *Orthogonal Polynomials and Random Matrices: A Riemann–Hilbert Approach* (Courant Inst Math. Sci., New York 1999)
13. P.A. Deift, X. Zhou: Ann. Math. **137**, 295 (1993)
14. P.A. Deift, X. Zhou: Commun. Pure Appl. Math. **48**, 277 (1995)
15. P.A. Deift, X. Zhou: Int. Math. Res. Not. **2002**, 2121 (2002)
16. P.A. Deift, X. Zhou: Acta Math. **188**, 163 (2002)
17. P.A. Deift, A.R. Its, X. Zhou: Long-time asymptotics for integrable nonlinear wave equations. In: *Important Developments in Soliton Theory*, ed. by A.S. Fokas, V.E. Zakharov, (Springer, Berlin 1993) pp 181–204
18. P.A. Deift, T. Kriecherbauer, K.T.-R. McLaughlin: J. Approx. Theory **95**, 399 (1998)
19. P.A. Deift, T. Kriecherbauer, K.T.-R. McLaughlin, S. Venakides, X. Zhou: Commun. Pure Appl. Math. **52**, 1335 (1999)
20. M.R. Douglas, N. Seiberg, S.H. Shenker: Phys. Lett. B **244**, 381 (1990)
21. M.R. Douglas, S.H. Shenker: Nucl. Phys. B **335**, 635 (1990)
22. F.J. Dyson: Commun. Math. Phys. **19**, 235 (1970)
23. P. Di Francesco, P. Ginsparg, J. Zinn-Justin: Physics Rep. **254** (1995)
24. B. Eynard: arXiv:hep-th/0210047
25. N.M. Ercolani, K.D.T.-R. McLaughlin: arXiv:math-ph/0609048
26. H. Flaschka and A.C. Newell: Commun. Math. Phys. **76**, 65 (1980)
27. A.S. Fokas, A.R. Its, A.A. Kapaev, V.Yu. Novokshenov: *Painlevé Transcendents. The Riemann–Hilbert Approach* (Amer. Math. Soc., Providence, RI 2006)
28. A.S. Fokas, A.R. Its, A.V. Kitaev: Commun. Math. Phys. **142**, 313 (1991)
29. A.S. Fokas, A.R. Its, A.V. Kitaev: Commun. Math. Phys. **147**, 395 (1992)
30. F.D. Gakhov: *Boundary Value Problems*, 3rd edn (Nauka, Moscow 1977)
31. D.J. Gross, A.A. Migdal: Phys. Rev. Lett. **64**, 127 (1990)
32. S.P. Hastings, J.B. McLeod: Arch. Rational Mech. Anal. **73**, 31 (1980)
33. E.L. Ince, *Ordinary Differential Equations* (Dover, New York 1956)
34. A.R. Its, Notices Amer. Math. Soc. **50**, 1389 (2003)
35. A.R. Its, A.V. Kitaev, A.S. Fokas: Usp. Mat. Nauk **45**, 135 (1990); Russian Math. Surveys **45**, 155 (1990)
36. A.R. Its, A.V. Kitaev, A.S. Fokas: Zap. Nauch. Semin. LOMI **187**, 3 (1991); J. Math. Sci. **73**, 415 (1995)
37. A.R. Its and V.Yu. Novokshenov, *The Isomonodromy Deformation Method in the Theory of Painlevé Equations* (Springer, Berlin 1986)

38. A.A. Kapaev: Phys. Lett. A, **167**, 356 (1992)
39. G.S. Litvinchuk, I.M. Spitkovskii: *Factorization of Measurable Matrix Functions* (Birkhäuser, Basel 1987)
40. M.L. Mehta: *Random Matrices*, 2nd edn (Academic Press, Boston, 1991)
41. L. Pastur, M. Shcherbina: J. Stat. Phys. **86**, 109 (1997)
42. V. Periwal, D. Shevitz: Nucl. Phys. B **344**, 731 (1990)
43. E.B. Saff, V. Totik, *Logarithmic Potentials with External Fields* (Springer, Berlin 1997)
44. B.I. Suleimanov: Differentsial′nye Uravneniya **23**, 834 (1987); Differential Equations **23**, 569 (1987)
45. C.A. Tracy, H. Widom: Introduction to random matrices. In: *Geometric and Quantumww Aspects of Integrable Systems*, ed. by G.F. Helminck (Springer, Berlin 1993) pp 103–130
46. C.A. Tracy, H. Widom: Commun. Math. Phys. **163**, 33 (1994)
47. P. van Moerbeke: Astérisque **276**, 411 (2002)
48. P. van Moerbeke: Integrable lattices: random matrices and random permutations. In: *Random Matrices and Their Applications*, ed. by P.M. Bleher, A.R. Its (Cambridge Univ. Press, Cambridge 2001) pp 321–406
49. X. Zhou: SIAM J. Math. Anal. **20**, 966 (1989)

6

Formal Matrix Integrals and Combinatorics of Maps

B. Eynard

Service de Physique Théorique de Saclay, 91191 Gif-sur-Yvette Cedex, France
eynard@spht.saclay.cea.fr

Summary. This article is a short review on the relationship between convergent matrix integrals, formal matrix integrals, and combinatorics of maps.

6.1 Introduction

This article is a short review on the relationship between *convergent matrix integrals*, *formal matrix integrals*, and *combinatorics of maps*. We briefly summarize results developed over the last 30 years, as well as more recent discoveries.

We recall that formal matrix integrals are *identical* to combinatorial generating functions for maps, and that formal matrix integrals are in general very *different* from convergent matrix integrals. Both may coincide perturbatively (i.e., up to terms smaller than any negative power of N), only for some potentials which correspond to negative weights for the maps, and therefore not very interesting from the combinatorics point of view.

We also recall that both convergent and formal matrix integrals are solutions of the same set of loop equations, and that loop equations do not have a unique solution in general.

Finally, we give a list of the classical matrix models which have played an important role in physics in the past decades. Some of them are now well understood, some are still difficult challenges.

J. Harnad (ed.), *Random Matrices, Random Processes and Integrable Systems*, CRM Series in Mathematical Physics, DOI 10.1007/978-1-4419-9514-8_6,

Matrix integrals were first introduced by physicists [55], mostly in two ways:

- in nuclear physics, solid state physics, quantum chaos, *convergent matrix integrals* are studied for the eigenvalues statistical properties [6, 33, 48, 52]. Statistical properties of the spectrum of large random matrices show some amazing universal behaviors, and it is believed that they correspond to some kind of "central limit theorem" for nonindependent random variables. This domain is very active and rich, and many important recent progresses have been achieved by the mathematicians community. Universality was proved in many cases, in particular using the Riemann–Hilbert approach of Bleher–Its [5] and Deift Venakides Zhou Mac Laughlin [19], and also by large deviation methods [34, 35].
- in Quantum Chromodynamics, quantum gravity, string theory, conformal field theory, *formal matrix integrals* are studied for their combinatorial property of being generating functions of maps [20]. This fact was first discovered by t'Hooft in 1974 [49], then further developed mostly by BIPZ [12] as well as Ambjorn, David, Kazakov [18, 20, 32, 36, 40]. For a long time, physicist's papers have been ambiguous about the relationship between formal and convergent matrix integrals, and many people have been confused by those ill-defined matrix integrals. However, if one uses the word "formal matrix integral", many physicist's results of the 80's till now are perfectly rigorous, especially those using loop equations. Only results regarding convergency properties were non rigorous, but as far as combinatorics is concerned, *convergency is not an issue.*

The ambiguity in physicist's ill-defined matrix integrals started to become obvious when E. Kanzieper and V. Freilikher [42], and later Brezin and Deo in 1998 [11] tried to compare the topological expansion of a formal matrix integral derived from loop equations, and the asymptotics of the convergent integral found with the help of orthogonal polynomials. The two results did not match. The orthogonal polynomial's method showed clearly that the convergent matrix integrals had no large N power series expansion (it contained some $(-1)^N$). The origin of this puzzle has now been understood [9], and it comes from the fact that formal matrix integrals and convergent matrix integrals are different objects in general.

This short review is only about combinatoric properties of formal matrix integrals. Matrix models is a very vast topic, and many important applications, methods and points of view are not discussed here. In particular, critical limits (which include asymptotics of combinatoric properties of maps), the link with integrable systems, with conformal field theory, with algebraic geometry, with orthogonal polynomials, group theory, number theory, probabilities and many other aspects, are far beyond the scope of such a short review.

6.2 Formal Matrix Integrals

In this section we introduce the notion of formal matrix integrals of the form:

$$Z_{\text{formal}}(t) = \int_{\text{formal}} \mathrm{d}M_1\,\mathrm{d}M_2 \cdots \mathrm{d}M_p \\ \exp\left(-\frac{N}{t}\left(\sum_{i,j=1}^{p} \frac{C_{ij}}{2} \operatorname{Tr} M_i M_j - NV(M_1,\ldots,M_p)\right)\right) . \tag{6.1}$$

The idea is to formally expand the exponential $\exp\big((N^2/t)V\big)$ in powers of t^{-1}, and compute the Gaussian integral for each term. The result is a formal series in powers of t. So, let us define it precisely.

Definition 6.2.1. Q is an invariant noncommutative monomial of $M_1,\ldots,$ M_p, if $Q=1$ or if Q is of the form:

$$\overline{Q} = \prod_{r=1}^{R} \frac{1}{N} \operatorname{Tr}(W_r) , \tag{6.2}$$

where each W_r is an arbitrary word written with the alphabet $M_1,\ldots,M_p$. Q is the equivalence class of $\overline{Q}$ under permutations of the W_rs, and cyclic permutations of the letters of each W_r.

The degree of Q is the sum of lengths of all W_rs.

Invariant noncommutative polynomials of $M_1,\ldots,M_p$ are complex finite linear combinations of monomials:

$$V = \sum_Q t_Q Q , \quad t_Q \in \mathbb{C} . \tag{6.3}$$

The degree of a polynomial is the maximum degree of its monomials.

They are called invariant, because they are left unchanged if one conjugates all matrices $M_i \to U M_i U^{-1}$ with the same invertible matrix U.

Invariant polynomials form an algebra over $\mathbb{C}$.

Let $V(M_1,\ldots,M_p)$ be an arbitrary invariant polynomial of degree d in $M_1,\ldots,M_p$, which contains only monomials of degree at least 3.

Proposition 6.2.1. *Let C be a $p\times p$ symmetric positive definite matrix, then the following Gaussian integral*

$$A_k(t) \\ = \frac{\int_{H_N\times\cdots\times H_N} \mathrm{d}M_1\,\mathrm{d}M_2\cdots\mathrm{d}M_p\,(N^{2k}t^{-k}/k!)V^k \exp(-(N/2t)\operatorname{Tr}\sum_{i,j=1}^{p} C_{ij}M_iM_j)}{\int_{H_N\times\cdots\times H_N} \mathrm{d}M_1\,\mathrm{d}M_2\cdots\mathrm{d}M_p\,\exp(-(N/2t)\operatorname{Tr}\sum_{i,j=1}^{p} C_{ij}M_iM_j)} , \tag{6.4}$$

where $\mathrm{d}M_i$ *is the usual Lebesgue* ($U(N)$ *invariant*) *measure on the space of Hermitian matrices* H_N, *is absolutely convergent and has the following properties*:

- $A_k(t)$ *is a polynomial in* t, *of the form*:

$$A_k(t) = \sum_{k/2 \le j \le kd/2-k} A_{k,j} t^j \ . \tag{6.5}$$

- $A_k(t)$ *is a Laurent polynomial in* N.
- $A_0(t) = 1$.

Proof. $A_0 = 1$ is trivial. Let $d = \deg V$. Since V is made of monomials of degree at least 3 and at most d, then V^k is a sum of invariant monomials whose degree l is between $3k$ and dk. According to Wick's theorem, the Gaussian integral of a monomial of degree l is zero if l is odd, and it is proportional to $t^{l/2}$ if l is even. Since $3k \le l \le dk$ we have:

$$0 \le k/2 \le l/2 - k \le dk/2 - k \ . \tag{6.6}$$

Thus $A_k(t)$ is a finite linear combination of positive integer powers of t, i.e., it is a polynomial in t, of the form of (6.5).

The matrix size N's dependence comes in several ways. First there is the factor N^{2k}. The matrix size also appears in the matrix products (each matrix product is a sum over an index which runs from 1 to N), in the traces (it means the first and last index of a matrix product have to be identified, thus there is a Kroenecker's δ_{ij} of two indices). And after Gaussian integration over all matrix elements, the Wick's theorem pairings result in $N^{-l/2}$ times some product of Kroenecker's δ of pairs of indices (times some elements of the matrix C^{-1} which are independent of N). The matrix indices thus appear only in sums and $\delta' s$, and the result of the sum over indices is an integer power of N. Thus, each $A_k(t)$ is a finite sum (sum for each monomial of V^k, and the Gaussian integral of each monomial is a finite sum of Wick's pairings) of positive or negative powers of N, i.e., a Laurent polynomial in N. □

Definition 6.2.2. The formal matrix integral $Z_{\mathrm{formal}}(t)$ is defined as the formal power series:

$$Z_{\mathrm{formal}}(t) = \sum_j Z_j t^j \ , \quad Z_j = \sum_{k=0}^{2j} A_{k,j} \tag{6.7}$$

and each Z_i is a Laurent polynomial in N. Notice that $Z_0 = 1$.

By abuse of notation, $Z_{\mathrm{formal}}(t)$ is often written:

$$Z_{\mathrm{formal}}(t) = \frac{\int_{\mathrm{formal}} \mathrm{d}M_1\, \mathrm{d}M_2 \cdots \mathrm{d}M_p\, \exp(-(N/t)(\sum_{i,j=1}^p C_{ij}/2\, \mathrm{Tr}\, M_i M_j) - NV(M_1, \ldots, M_p))}{\int_{H_N \times \cdots \times H_N} \mathrm{d}M_1\, \mathrm{d}M_2 \cdots \mathrm{d}M_p\, \exp(-(N/2t)\, \mathrm{Tr} \sum_{i,j=1}^p C_{ij} M_i M_j)} \tag{6.8}$$

but it does not mean that it has anything to do with the corresponding convergent (if it converges) integral. In fact, the integral can be absolutely convergent only if $\deg(V)$ is even and if the t_Q corresponding to the highest-degree terms of V have a negative real part. But as we shall see below, the relevant case for combinatorics, corresponds to all t_Qs positive, and in that case, the formal integral is *never* a convergent one.

Definition 6.2.3. The formal free energy $F_{\text{formal}}(t)$ is defined as the formal log of Z_{formal}.

$$F_{\text{formal}}(t) = \ln\big(Z_{\text{formal}}(t)\big) = \sum_j F_j t^j \tag{6.9}$$

We have $F_0 = 0$. Each F_j is a Laurent polynomial in N.

6.2.1 Combinatorics of Maps

Recall that an invariant monomial is a product of terms, each term being the trace of a word in an alphabet of p letters. Thus, an invariant monomial is given by:

- the number R of traces, ($R-1$ is called the crossing number of Q),
- R words written in an alphabet of p letters.

The R words can be permuted together, and in each word the letters can be cyclically permuted. We label the invariant monomials by the equivalence classes of those permutations.

Another graphical way of coding invariant monomials is the following:

Definition 6.2.4. To each invariant monomial Q we associate biunivoquely a *Q-gon* (generalized polygon) as follows:

- to each word we associate an oriented polygon (in the usual sense), with as many edges as the length of the word, and whose edges carry a "color" between 1 and p, given by the corresponding letter in the word.
- the R words are glued together by their centers on their upper face (in accordance with the orientation), so that they make a surface with $R-1$ crossings.
- $R-1$ which is the number of traces minus one (i.e., one trace corresponds to a crossing number zero), is called the crossing number of the Q-gon.
- The degree $\deg(Q)$ of the Q-gon is the total number of edges (sum of lengths of all words).
- to Q-gon we associate a symmetry factor $s_Q = \#\text{Aut}(Q)$ which is the number of symmetries which leave Q unchanged.

An example is given in Fig. 6.1.

Notice that we allow a Q-gon to be made of polygons with possibly one or two sides. We will most often call the Q-gons polygons. The usual polygons are Q-gons with no crossing, i.e., $R=1$.

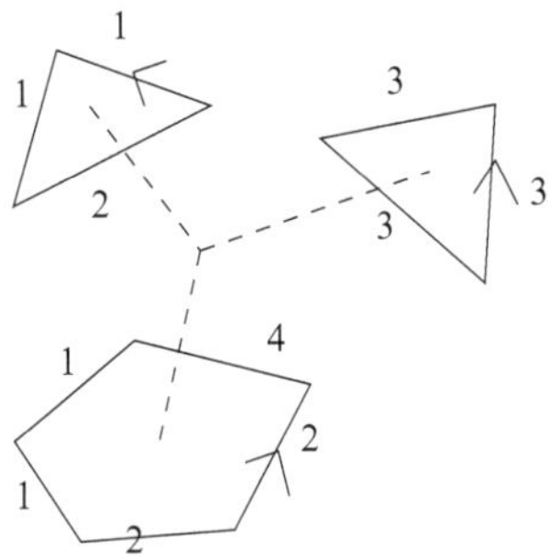

Fig. 6.1. The invariant monomial $Q = N^{-3}\,\mathrm{Tr}(M_1^2 M_2)\,\mathrm{Tr}(M_3^3)\,\mathrm{Tr}(M_2^2 M_4 M_1^2)$ of degree 11, is represented by 2 triangles and one pentagon glued together by their center. The dashed lines mean that the 3 centers should actually be seen as only one point. Its symmetry factor is $s_Q = 3$ because we can perform 3 rotations on the triangle of color (3,3,3).

Definition 6.2.5. Let $p \geq 1$ and $d \geq 3$ be given integers. Let $\mathcal{Q}_{d,p}$ be the set of all invariant monomials Q (or polygons) of degree $3 \leq \deg(Q) \leq d$.

$\mathcal{Q}_{d,p}$ is clearly a finite set.

Definition 6.2.6. Let $\mathcal{S}_{l,d,p}$ be the set of oriented discrete surfaces such that $\#\,\text{edges} - \#\,Q\text{-gons} + \#\,\text{crossings} = l$, and obtained by gluing together polygons (belonging to $\mathcal{Q}_{d,p}$) by their sides (following the orientation). The edges of the polygons carry colors among p possible colors (thus each edge of the surface, is at the border of two polygons, and has two colors, one on each side).

Let $\overline{\mathcal{S}}_{l,d,p}$ be the subset of $\mathcal{S}_{l,d,p}$ which contains only connected surfaces. Such surfaces are also called "maps."

Proposition 6.2.2. *$\mathcal{S}_{l,d,p}$ is a finite set.*

Proof. Indeed, since all polygons of $\mathcal{Q}_{d,p}$ have at least 3 sides, we have $\#\,\text{edges} \geq \frac{3}{2}\,\#\,\text{polygons}$, and thus $\#\,\text{edges} \leq 2l$ and $\#\,\text{polygons} \leq 4l/3$, and thus the number of discrete surfaces in $\mathcal{S}_l$, is finite. We can give a very large bound on $\#\,\mathcal{S}_{l,d,p}$. We have:

$$\#\,\mathcal{S}_{l,d,p} \leq (4dl/3)^p (2dl/3)!(4dl/3)! \leq (4dl/3)^p (2dl)!\,. \tag{6.10}$$

□

Theorem 6.2.1 (t'Hooft 1974 and followers). *If the potential V is an invariant polynomial given by*

$$V = \sum_{Q \in \mathcal{Q}_{d,p}} \frac{t_Q}{s_Q} Q \tag{6.11}$$

then:

$$Z_l = \sum_{S \in \mathcal{S}_{l,d,p}} \frac{1}{\# \operatorname{Aut}(S)} N^{\chi(S)} \prod_Q t_Q^{n_Q(S)} \prod_{i,j} \left((C^{-1})_{i,j}\right)^{E_{i,j}(S)/2} \tag{6.12}$$

$$F_l = \sum_{S \in \overline{\mathcal{S}}_{l,d,p}} \frac{1}{\# \operatorname{Aut}(S)} N^{\chi(S)} \prod_Q t_Q^{n_Q(S)} \prod_{i,j} \left((C^{-1})_{i,j}\right)^{E_{i,j}(S)/2} \tag{6.13}$$

where Z_l and F_l were defined in Def. 6.2.2, Def. 6.2.3, and where:

- *$n_Q(S)$ is the number of Q-gons in S,*
- *$E_{ij}(S)$ is the number of edges of S which glue two polygons whose sides have colors i and j,*
- *$\chi(S)$ is the Euler characteristic of S.*
- $\operatorname{Aut}(S)$ *is the group of automorphisms of S, i.e., the group of symmetries which leave S unchanged.* $\# \operatorname{Aut}(S)$ *is called the symmetry factor.*

In other words, Z_l is the formal generating function which enumerates discrete surfaces of $\mathcal{S}_{l,d,p}$, according to their Euler characteristics, their number of polygons, number of edges according to their color, number of crossings... F_l is the formal generating function for the same surfaces with the additional condition that they are connected.

An example is given in Fig. 6.2.

Proof. It is a mere application of Feynman's graphical representation of Wick's theorem[1].

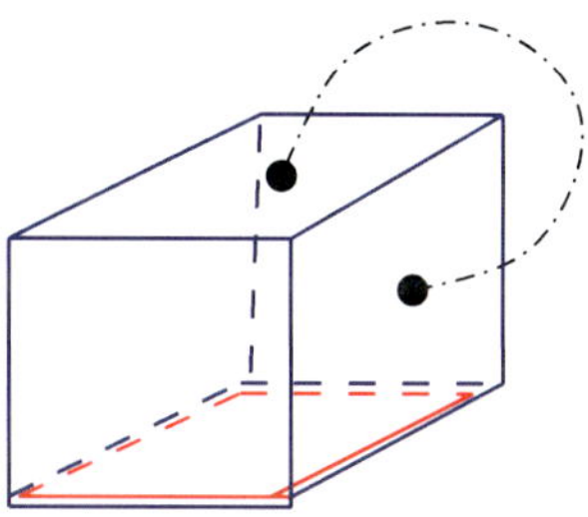

Fig. 6.2. If

$$V = \frac{t_4}{4} \frac{1}{N} \operatorname{Tr} M_1^4 + \frac{\tilde{t}_4}{4} \frac{1}{N} \operatorname{Tr} M_2^4 + \frac{t_{4,4}}{32N^2} (\operatorname{Tr} M_1^4)^2,$$

the above surface contributes to the term $N^0 t_4^3 \tilde{t}_4 t_{4,4} (C^{-1})_{1,1}^4 (C^{-1})_{1,2}^4$. Indeed M_1 is represented in blue, M_2 in red, so that $\operatorname{Tr} M_1^4$ corresponds to blue squares, $\operatorname{Tr} M_2^4$ corresponds to red squares, and $(\operatorname{Tr} M_1^4)^2$ corresponds to pairs of blue squares glued at their center. Its Euler characteristic is $\chi = 0$ (it is a torus), and this surface has no automorphism (other than identity), i.e., $\# \operatorname{Aut} = 1$. It corresponds to $l = 7$.

[1] Although Feynman's graphs are sometimes regarded as nonrigorous, let us emphasize that it is only when Feynman's graphs and Wick's theorem are used for functional integrals that they are nonrigorous. Here we have finite dimensional Gaussian integrals, and Wick's theorem and Feynman's graphs are perfectly rigorous.

The only nontrivial part is the power of N, because N enters not only in the weights but also in the variables of integrations. It was the great discovery of t'Hooft to recognize that the power of N is precisely the Euler characteristics. Let us explain it again.

First, V^k is decomposed into a combination of monomials:

$$V = \sum_Q \frac{t_Q}{s_Q} Q \,, \quad V^k = \sum_G T_{k,G}\, G \tag{6.14}$$

where G is a product of Qs, and

$$T_{k,G} = \sum_{Q_1 \cup Q_2 \cup \cdots \cup Q_k = G} t_{Q_1} t_{Q_2} \cdots t_{Q_k} \frac{1}{\prod_i s_{Q_i}} \,.$$

G is thus a collection of polygons, some of them glued by their centers. So far the polygons of G are not yet glued by their sides to form a surface. Remember that each Q carries a factor N^{-R} if Q is made of R traces.

Then, for each G, the Gaussian integral is computed by Wick's theorem and gives a sum of Wick's pairings, i.e., it is the sum over all possible ways of pairing two Ms, i.e., it is the sum over all possible ways of gluing together all polygons by their sides, i.e. corresponds to the sum over all surfaces S. The pairing $\langle M_{i\,ab} M_{j\,cd}\rangle$ of the (a,b) matrix element of M_i and the (c,d) matrix element of M_j gives a factor:

$$\langle M_{i\,ab} M_{j\,cd}\rangle = \frac{t}{N} (C^{-1})_{ij} \delta_{ad} \delta_{bc} \,. \tag{6.15}$$

The double line notation for pairings (see [20]) allows to see clearly that the sum over all indices is $N^{\#\,\text{vertices}(S)}$. The total power of N is thus:

$$2k - \sum_Q R_Q + \#\,\text{vertices} - \#\,\text{edges} - \#\,\text{polygons} \tag{6.16}$$

Now, notice that

$$\#\,\text{polygons} = \sum_Q R_Q = k + \sum_Q (R_Q - 1) = k + \#\,\text{crossings} \tag{6.17}$$

Thus the total power of N is:

$$\#\,\text{vertices} - \#\,\text{edges} + \#\,\text{polygons} - 2\,\#\,\text{crossings} = \chi \tag{6.18}$$

which is the Euler characteristic of S.

We leave to the reader to see in the literature how to find the symmetry factor. □

Corollary 6.2.1. *$N^{-2}F_l$ is a polynomial in N^{-2}:*

$$F_l = \sum_{g=0}^{l+1} N^{2-2g} F_{l,g} \,. \tag{6.19}$$

Proof. Indeed the Euler characteristic of connected graphs is of the form $\chi = 2-2g$ where g is a positive integer and is the genus (number of handles). The fact that it is a polynomial comes from the fact that F_l is a finite sum.

It is easy to find some bound on g. We have:

$$\begin{aligned} 2g-2 &= -\#\,\text{vertices} + \#\,\text{edges} - \#\,\text{polygons} + 2\,\#\,\text{crossings} \\ &= -\#\,\text{vertices} + 2l - \#\,\text{edges} + \#\,\text{polygons} \end{aligned} \tag{6.20}$$

and using $\#\,\text{edges} \geq \frac{3}{2}\,\#\,\text{polygons}$, we have:

$$2g-2 \leq -\#\,\text{vertices} + 2l - \tfrac{1}{2}\,\#\,\text{polygons} \leq 2l\,. \tag{6.21}$$

□

6.2.2 Topological Expansion

Definition 6.2.7. We define the genus g free energy as the formal series in t:

$$F^{(g)}(t) = -\sum_{l=0}^{\infty} t^{l+2-2g} F_{l,g} \tag{6.22}$$

$F^{(g)}$ is the generating function for connected oriented discrete surfaces of genus g.

Remark. The minus sign in front of $F^{(g)}$ is there for historical reasons, because in thermodynamics the free energy is $-\ln Z$.

There is no reason a priori to believe that $F^{(g)}(t)$ might have a finite radius of convergence in t. However, for many classical matrix models, it is proved that $\forall g$, $F^{(g)}$ has a finite radius of convergence because it can be expressed in terms of algebraic functions.

There is also no reason a priori to believe that the $F^{(g)}$s should be the limits of convergent matrix integrals. There are many works which prove that the number of terms in $F^{(g)}$ grows with g like $(\beta g)!$ for some β. If t and all t_Qs are positive (this is the interesting case for combinatorics and statistical physics because we want all surfaces to be counted with a positive weight), then $F^{(g)}$ is positive and grows like $(\beta g)!$, and therefore the sum of the topological series *cannot* be convergent (even under Borel resummation). For negative t, it is only an asymptotic series, and at least in some cases, it can be made convergent using Borel transforms.

6.3 Loop Equations

Loop equations is the name given to Schwinger–Dyson equations in the context of random matrices [53]. The reason for that name is explained below. Let us recall that Schwinger–Dyson equations are in fact nothing but integration by parts, or the fact that the integral of a total derivative vanishes.

In particular, in the context of convergent matrix integrals we have:

$$\begin{aligned}
0 = \sum_{i<j} \int_{H_N^p} & \mathrm{d}M_1\,\mathrm{d}M_2 \cdots \mathrm{d}M_p \left(\frac{\partial}{\partial\,\mathrm{Re}\,M_{k\,i,j}} - i \frac{\partial}{\partial\,\mathrm{Im}\,M_{k\,i,j}} \right) \\
& \times \Bigg(\left(G(M_1,\ldots,M_p) + G(M_1,\ldots,M_p)^{\dagger}\right)_{ij} \\
& \times \exp\Bigg(-\frac{N}{t} \Bigg(\sum_{i,j=1}^{p} \frac{C_{ij}}{2} \operatorname{Tr}(M_i M_j) - N V(M_1,\ldots,M_p) \Bigg)\Bigg)\Bigg) \\
+ \sum_i \int_{H_N^p} & \mathrm{d}M_1\,\mathrm{d}M_2 \cdots \mathrm{d}M_p \frac{\partial}{\partial M_{k\,i,i}} \\
& \times \Bigg((G(M_1,\ldots,M_p) + G(M_1,\ldots,M_p)^{\dagger})_{ii} \\
& \times \exp\Bigg(-\frac{N}{t} \Bigg(\sum_{i,j=1}^{p} \frac{C_{ij}}{2} \operatorname{Tr}(M_i M_j) - N V(M_1,\ldots,M_p) \Bigg)\Bigg)\Bigg)
\end{aligned} \tag{6.23}$$

where G is any noncommutative polynomial, and k is an integer between 1 and p.

Therefore, Schwinger–Dyson equations for matrix integrals give relationships between expectation values of invariant polynomials. Namely:

$$\frac{N}{t} \left\langle \operatorname{Tr}\left(\Big(\sum_j C_{kj} M_j - N D_k(V) \Big) G \right) \right\rangle = \langle K_k(G) \rangle \tag{6.24}$$

where D_k is the non commutative derivative with respect to M_k, and $K_k(G)$ is some invariant polynomial which can be computed by the following rules:

- Leibnitz rule

$$\begin{aligned}
& K_k\big(A(M_1,\ldots,M_k,\ldots,M_p) B(M_1,\ldots,M_k,\ldots,M_p)\big) \\
& \quad = \Big(K_k\big(A(M_1,\ldots,M_k,\ldots,M_p) B(M_1,\ldots,m_k,\ldots,M_p)\big) \Big)_{m_k \to M_k} \\
& \quad + \Big(K_k\big(A(M_1,\ldots,m_k,\ldots,M_p) B(M_1,\ldots,M_k,\ldots,M_p)\big) \Big)_{m_k \to M_k} ;
\end{aligned} \tag{6.25}$$

- split rule

$$\begin{aligned}
& K_k\big(A(M_1,\ldots,m_k,\ldots,M_p) M_k^l B(M_1,\ldots,m_k,\ldots,M_p)\big) \\
& \qquad = \sum_{j=0}^{l-1} \operatorname{Tr}(A(M_1,\ldots,m_k,\ldots,M_p) M_k^j) \\
& \qquad\qquad \times \operatorname{Tr}\big(M_k^{l-j-1} B(M_1,\ldots,m_k,\ldots,M_p)\big) ;
\end{aligned} \tag{6.26}$$

- merge rule

$$K_k\Big(A(M_1,\ldots,m_k,\ldots,M_p)\,\mathrm{Tr}\big(M_k^l B(M_1,\ldots,m_k,\ldots,M_p)\big)\Big) = \sum_{j=0}^{l-1} \mathrm{Tr}(A(M_1,\ldots,m_k,\ldots,M_p)M_k^j B(M_1,\ldots,m_k,\ldots,M_p)M_k^{l-j-1})\ ; \tag{6.27}$$

- no M_k rule

$$K_k(A(M_1,\ldots,m_k,\ldots,M_p)) = 0\ . \tag{6.28}$$

Since each rule allows to strictly decrease the partial degree in M_k, this set of rules allows to compute $K_k(G)$ for any G.

For any G and any k we get one loop equation of the form (6.24).

Definition 6.3.1. The formal expectation value of some invariant polynomial $P(M_1,\ldots,M_k)$ is the formal power series in t defined by:

$$A_{k,P}(t) = \frac{\int_{H_N\times\cdots\times H_N} \mathrm{d}M_1\,\mathrm{d}M_2\cdots\mathrm{d}M_p\,(N^{2k}\,t^{-k}/k!)PV^k \exp\big(-(N/2t)\,\mathrm{Tr}(\sum_{i,j=1}^p C_{ij}M_iM_j)\big)}{\int_{H_N\times\cdots\times H_N} \mathrm{d}M_1\,\mathrm{d}M_2\cdots\mathrm{d}M_p \exp\big(-(N/2t)\,\mathrm{Tr}(\sum_{i,j=1}^p C_{ij}M_iM_j)\big)}\ . \tag{6.29}$$

$A_{k,P}(t)$ is a polynomial in t, of the form:

$$A_{k,P}(t) = \sum_{\deg(P)/2+k/2\le j\le \deg(P)/2+kd/2-k} A_{k,P,j}t^j \tag{6.30}$$

and we define the following quantity

$$A_{P,j}(t) = \sum_{k=0}^{2j-\deg P} A_{k,P,j} \tag{6.31}$$

and the formal series

$$A_P(t) = \sum_{j=\deg P/2}^{\infty} A_{P,j}\,t^j \tag{6.32}$$

Again, each $A_{k,P}$, $A_{k,P,j}$, $A_{P,j}$ is a Laurent polynomial in N.

The formal expectation value of P is defined as:

$$\langle P(M_1,\ldots,M_k)\rangle_{\text{formal}} = \frac{A_P(t)}{Z_{\text{formal}}(t)} \tag{6.33}$$

where the division by Z_{formal} is to be taken in the sense of formal series, and it can be performed since $Z_{\text{formal}}(t) = \sum_{j=0}^{\infty} Z_jt^j$ with $Z_0 = 1$.

The formal expectation value is often written by abuse of notations

$$\langle P\rangle_{\text{formal}} = \frac{\int_{\text{formal}} \mathrm{d}M_1 \cdots \mathrm{d}M_p P \exp(-(N/t)(\sum_{i,j=1}^{p}(C_{ij}/2)\operatorname{Tr}(M_iM_j) - NV(M_1,\ldots,M_p)))}{\int_{\text{formal}} \mathrm{d}M_1 \cdots \mathrm{d}M_p \exp(-(N/t)(\sum_{i,j=1}^{p}(C_{ij}/2)\operatorname{Tr}(M_iM_j) - NV(M_1,\ldots,M_p)))} . \tag{6.34}$$

Theorem 6.3.1. *The formal expectation values of the formal matrix integral satisfy the same loop equations as the convergent matrix integral ones, i.e., they satisfy* (6.24) *for any k and G.*

Proof. It is clear that Gaussian integrals, and thus formal integrals satisfy (6.23). The remainder of the derivation of loop equations for convergent integrals is purely algebraic, and thus it holds for both convergent and formal integrals. □

On a combinatoric point of view, loop equations are the generalisation of Tutte's equations for counting planar maps [50, 51]. This is where the name "loop equations" comes from: indeed, similarly to Thm. 6.2.1, formal expectation values of traces are generating functions for open discrete surfaces with as many boundaries as the number of traces (and again the power of N is the Euler characteristic of the surface). The boundaries are "loops," and the combinatorical interpretation of Schwinger–Dyson equations is a recursion on the size of the boundaries, i.e. how to build discrete surfaces by gluing loops à la Tutte [50, 51].

Notice that in general, the loop equations *don't have a unique solution.* One can find a unique solution only with additional constraints not contained in the loop equations themselves. Thus, the fact that both convergent and formal matrix models obey the same loop equations does not mean they have to coincide. Many explicit examples where both do not coincide have been found in the literature. It is easy to see on a very simple example that Schwinger–Dyson equations can't have a unique solution: consider the Airy function $\int_\gamma \exp(t^3/3 - tx)\,\mathrm{d}t$ where γ is any path in the complex plane, going from ∞ to ∞ such that the integral is convergent. There are only two homologicaly independent choices of γ (one going from $+\infty$ to $\mathrm{e}^{2i\pi/3}\infty$ and one from $+\infty$ to $\mathrm{e}^{-2i\pi/3}\infty$). Schwinger–Dyson equations are: $\langle nt^{n-1} + t^{n+2} - xt^n\rangle = 0$ for all n. It is clear that loop equations are independent of the path γ, while their solution clearly depends on γ.

Theorem 6.3.2. *The formal matrix integral is the solution of loop equations with the following additional requirements:*

- *the expectation value of every monomial invariant is a formal power series in N^{-2}.*
- *the $t \to 0$ limit of the expectation value of any monomial invariant of degree ≥ 1 vanishes:*

$$\lim_{t\to 0}\langle Q\rangle = 0 \quad \textit{if } Q \neq 1. \tag{6.35}$$

Proof. The fact that all expectation values have a formal N^{-2} expansion follows from the construction. The fact that $\lim_{t\to 0}\langle Q\rangle = 0$ if $Q \neq 1$, follows from the fact that we are making a formal Taylor expansion at small t, near the minimum of the quadratic part, i.e., near $M_i = 0$, $i = 1, \dots, p$. □

However, even those requirements don't necessarily provide a unique solution to loop equations. Notice that there exist formal solutions of loop equations which satisfy the first point (there is a formal N^{-2} expansion), but not the second point ($\lim_{t\to 0}\langle Q\rangle = 0$). Those solutions are related to so-called "multicut" solutions, they also have a known combinatoric interpretation, but we don't consider them in this short review (see for instance [9, 26] for examples).

There is a conjecture about the relationship between different solutions of loop equations:

Conjecture 6.3.1. The convergent matrix integral (we assume V to be such that the integral is convergent, i.e. the highest t_Qs have negative real part)

$$Z_{\mathrm{conv}} = \int_{H_N^p} \mathrm{d}M_1\,\mathrm{d}M_2\cdots\mathrm{d}M_p\,\exp\left(-\frac{N}{t}\left(\sum_{i,j=1}^{p}\frac{C_{ij}}{2}\,\mathrm{Tr}(M_iM_j) - NV(M_1,\dots,M_p)\right)\right) \tag{6.36}$$

is a finite linear combination of convergent formal solutions of loop equations (i.e., a formal solution of loop equations $\ln Z = -\sum_{g=G}^{\infty} N^{2-2g}F^{(g)}$, such that the N^{-2} series is convergent.), i.e.,

$$Z_{\mathrm{conv}} = \sum_i c_i Z_i\,, \quad \ln Z_i = -\sum_{g=0}^{\infty} N^{2-2g}\,F_i^{(g)}\,. \tag{6.37}$$

Hint. It amounts to exchanging the large N and small t limits. First, notice that convergent matrix integrals are usualy defined on H_N^p, but can be defined on any "integration path" in the complexified of H_N^p, which is $M_N(\mathbb{C})^p$, as long as the integral is convergent. The homology space of such contours is of finite dimension (because there are a finite number of variables $p \times N^2$, and a finite number of possible sectors at ∞ because the integration measure is the exponential of a polynomial). Thus, the set of "generalized convergent matrix integrals" defined on arbitrary paths, is a finite-dimensional vector space which we note: Gen. The hermitian matrix integral defined on H_N^p is only one point in that vector space.

Second, notice that every such generalized convergent matrix integral satisfies the same set of loop equations, and that loop equations of type (6.23) are clearly linear in Z. Thus, the set of solutions of loop equations is a vector space which contains the vector space of generalized convergent matrix integrals.

Third, notice that formal integrals are solutions of loop equations, and therefore, formal integrals which are also convergent, belong to the vector space of generalized convergent matrix integrals.

Fourth, one can try to compute any generalized convergent matrix integral by small t saddle point approximation (at finite N). In that purpose, we need to identify the small t saddle points, i.e., a set of matrices $(\underline{M}_1, \dots, \underline{M}_p) \in M_N(\mathbb{C})^p$ such that $\forall i, j, k$ one has:

$$\frac{\partial}{\partial M_{k_{i,j}}} \left(\sum_{l,m} \operatorname{Tr}\left(\tfrac{1}{2} C_{l,m} M_l M_m\right) - V(M_1, \dots, M_p) \right)_{M_n = \underline{M}_n} = 0 \tag{6.38}$$

and such that

$$\operatorname{Im}\left(\sum_{l,m} \operatorname{Tr}\left(\tfrac{1}{2} C_{l,m} M_l M_m\right) - V(M_1, \dots, M_p) \right)$$
$$= \operatorname{Im}\left(\sum_{l,m} \operatorname{Tr}\left(\tfrac{1}{2} C_{l,m} \underline{M}_l \underline{M}_m\right) - V(\underline{M}_1, \dots, \underline{M}_p) \right) \tag{6.39}$$

and

$$\operatorname{Re}\left(\sum_{l,m} \operatorname{Tr}\left(\tfrac{1}{2} C_{l,m} M_l M_m\right) - V(M_1, \dots, M_p) \right)$$
$$\geq \operatorname{Re}\left(\sum_{l,m} \operatorname{Tr}\left(\tfrac{1}{2} C_{l,m} \underline{M}_l \underline{M}_m\right) - V(\underline{M}_1, \dots, \underline{M}_p) \right). \tag{6.40}$$

If such a saddle point exist, then it is possible to replace $\exp\big((N^2/t)V\big)$ by its Taylor series expansion in the integral and exchange the summation and the integration (because both the series and the integral are absolutely convergent, this is nothing but WKB expansion). This proves that saddle points are formal integrals and at the same time they are generalized convergent integrals, thus they are formal convergent integrals.

The conjecture thus amounts to claim that saddle points exist, and that there exist as many saddle points as the dimension of Gen, and that they are independent, so that the saddle points form a basis of Gen.

Notice that a linear combination of convergent formal solutions has no N^{-2} expansion in general, and thus the set of convergent formal integrals is not a vector space.

This conjecture is proved for the 1-matrix model with real potential [19,22], and for complex potentials it can be derived from Bertola–Man Yue [3] (indeed, the asymptotics of the partition function follow from those of the orthogonal polynomials). It is the physical intuition that it should hold for more general cases. It somehow corresponds to the small t saddle point method. Each saddle point has a WKB expansion, i.e., is a convergent formal solution, and the whole function is a sum over all saddle points. The coefficients of the linear combination reflect the homology class of the path (here H_N^p) on which Z is defined. This path has to be decomposed as a linear combination of steepest descent paths. The coefficients of that linear combination make the c_is of (6.37) (this is the generalisation of the idea introduced in [9]).

6.4 Examples

6.4.1 1-Matrix Model

$$Z_{\text{formal}} = \frac{\int \mathrm{d}M \, \exp(-(N/t)\, \mathrm{Tr}(\mathcal{V}(M)))}{\int \mathrm{d}M \, \exp(-(N/t)\, \mathrm{Tr}((C/2)M^2))} \tag{6.41}$$

where $\mathcal{V}(M) = (C/2)M^2 - V(M)$ is a polynomial in M, such that $V(M)$ contains only terms with degree ≥ 3.

For any α and γ, let us parametrise the potential $\mathcal{V}$ as:

$$\begin{cases} x(z) = \alpha + \gamma(z + 1/z) \,, \\ \mathcal{V}'(x(z)) = \sum_{j=0}^{d} v_j (z^j + z^{-j}) \,. \end{cases} \tag{6.42}$$

We determine α and γ by the following conditions:

$$v_0 = 0 \,, \quad v_1 = \frac{t}{\gamma} \,, \tag{6.43}$$

i.e., α and β are solutions of an algebraic equation and exist for almost any t, and they are algebraic functions of t. In general, they have a finite radius of convergence in t. We chose the branches of $\alpha(t)$ and $\gamma(t)$ which vanish at $t = 0$:

$$\alpha(t=0) = 0 \,, \quad \gamma(t=0) = 0 \,. \tag{6.44}$$

Then we define:

$$y(z) = \frac{1}{2} \sum_{j=1}^{d} v_j (z^j - z^{-j}) \,. \tag{6.45}$$

The curve $(x(z), y(z))$, $z \in \mathbb{C}$, is called the spectral curve. It is a genus zero hyperelliptical curve $y^2 = \text{Polynomial}(x)$. It has only two branch points solutions of $x'(z) = 0$ which correspond to $z = \pm 1$, i.e., $x = \alpha \pm 2\gamma$. y as a function of x has a cut along the segment $[\alpha - 2\gamma, \alpha + 2\gamma]$. Notice that we have:

$$\operatorname*{Res}_{z \to \infty} y \, \mathrm{d}x = t = - \operatorname*{Res}_{z \to 0} y \, \mathrm{d}x \tag{6.46}$$

$$\operatorname*{Res}_{z \to \infty} \mathcal{V}'(x) y \, \mathrm{d}x = 0 \,, \quad \operatorname*{Res}_{z \to \infty} x \mathcal{V}'(x) y \, \mathrm{d}x = t^2 \tag{6.47}$$

Then one has [2]:

$$
\begin{aligned}
F^{(0)} \\
&= \frac{1}{2}\left(\operatorname*{Res}_{z\to\infty} \mathcal{V}(x) y \,\mathrm{d}x - t \operatorname*{Res}_{z\to\infty} \mathcal{V}(x)\frac{\mathrm{d}z}{z} - \frac{3}{2}t^2 - t^2\ln\left(\frac{\gamma^2 C}{t}\right)\right) \\
&= \frac{1}{2}\left(-\sum_{j\geq 1} \frac{\gamma^2}{j}(v_{j+1}-v_{j-1})^2 - \frac{2\gamma t}{j}(-1)^j (v_{2j-1}-v_{2j+1}) - \frac{3}{2}t^2 - t^2 \ln\left(\frac{C\gamma^2}{t}\right)\right)
\end{aligned}
\tag{6.48}
$$

and: [13, 27]

$$
F^{(1)} = -\frac{1}{24}\ln\left(\frac{\gamma^2 y'(1)y'(-1)}{t^2}\right) . \tag{6.49}
$$

Expressions are known for all $F^{(g)}$s and we refer the reader to [14, 25]. Those expressions are detailed in the next section about the 2-matrix model, and one has to keep in mind that the 1-matrix model is a special case of the 2-matrix model with $\mathcal{V}_1(x) = \mathcal{V}(x) + x^2/2$ and $\mathcal{V}_2(y) = y^2/2$.

As a fore-taste, let us just show an expression for $F^{(2)}$:

$$
\begin{aligned}
&-2F^{(2)} \\
&= \operatorname*{Res}_{z_1\to\pm1}\operatorname*{Res}_{z_2\to\pm1}\operatorname*{Res}_{z_3\to\pm1} \Phi(z_1)E(z_1,z_2)E(1/z_1,z_3)\frac{1}{(z_2-1/z_2)^2}\frac{1}{(z_3-1/z_3)^2} \\
&\quad + 2\operatorname*{Res}_{z_1\to\pm1}\operatorname*{Res}_{z_2\to\pm1}\operatorname*{Res}_{z_3\to\pm1} \Phi(z_1)E(z_1,z_2)E(z_2,z_3)\frac{1}{(1/z_1-1/z_2)^2}\frac{1}{(z_3-1/z_3)^2} \\
&\quad + 2\operatorname*{Res}_{z_1\to\pm1}\operatorname*{Res}_{z_2\to\pm1}\operatorname*{Res}_{z_3\to\pm1} \Phi(z_1)E(z_1,z_2)E(z_2,z_3)\frac{1}{(1/z_1-1/z_3)^2}\frac{1}{(1/z_2-z_3)^2}
\end{aligned}
\tag{6.50}
$$

where the residue is first evaluated for z_3 then z_2 then z_1, and where:

$$
E(z,z') = \frac{1}{4\gamma}\frac{1}{z'(z-z')(z-1/z')y(z')} \tag{6.51}
$$

$$
\Phi(z) = -\frac{1}{4\gamma}\frac{1}{zy(z)(z-1/z)}\int_{1/z}^{z} y\,\mathrm{d}x . \tag{6.52}
$$

6.4.1.1 Example Triangulated Maps

Consider the particular case $V(x) = (t_3/3)x^3$.

Let $T = tt_3^2/C^3$, and a be a solution of:

$$
a - a^3 = 4T \tag{6.53}
$$

and consider the branch of $a(T)$ which goes to 1 when $T = 0$. If we parametarize:

$$
T = \frac{tt_3^2}{C^3} = \frac{\sin(3\phi)}{6\sqrt{3}} \tag{6.54}
$$

we have:

$$a = \frac{\cos(\pi/6 - \phi)}{\cos(\pi/6)} \,. \tag{6.55}$$

We have:

$$\alpha = \frac{C}{2t_3}(1-a) \,, \quad \gamma^2 = \frac{t}{aC} \tag{6.56}$$

$$v_0 = 0 \,, \quad v_1 = \frac{t}{\gamma} \,, \quad v_2 = -t_3\gamma^2 \tag{6.57}$$

which gives:

$$F^{(0)} = \frac{5}{12}t^2 - \left(\frac{t}{4C} + \frac{C}{6t}\right)\frac{1}{a} - \frac{t^2}{2}\ln a \,, \tag{6.58}$$

$$F^{(1)} = -\frac{1}{24}\ln\left(\frac{1 - 2a\sqrt{1-a^2}}{a^2}\right) \,. \tag{6.59}$$

The radius of convergence of $F^{(g)}$ is $|T| < 1/(6\sqrt{3})$ for all g.

6.4.1.2 Example Square Maps

Consider the particular case $V(x) = (t_4/4)x^4$, and write $T = tt_4/C^2$. Define:

$$b = \sqrt{1 - 12T} \tag{6.60}$$

We find

$$\alpha = 0 \,, \quad \gamma^2 = \frac{2t}{C(1+b)} \tag{6.61}$$

$$v_1 = \frac{t}{\gamma} \,, \quad v_2 = 0 \,, \quad v_3 = -t_4\gamma^3 \tag{6.62}$$

We find:

$$F^{(0)} = \frac{t^2}{2}\left(-\frac{(1-b)^2}{12(1+b)^2} + \frac{2(1-b)}{3(1+b)} + \ln\left(\frac{1+b}{2}\right)\right) \,, \tag{6.63}$$

$$F^{(1)} = -\frac{1}{12}\ln\left(\frac{2b}{1+b}\right) \,. \tag{6.64}$$

The radius of convergence of $F^{(g)}$ is $|T| < 1/12$ for all g.

6.4.2 2-Matrix Model

The 2-matrix model was introduced by Kazakov [37], as the Ising model on a random map.

$$Z_{\text{formal}} = \frac{\int \mathrm{d}M_1\,\mathrm{d}M_2\,\exp\bigl(-(N/t)\,\mathrm{Tr}(\mathcal{V}_1(M_1)+\mathcal{V}_2(M_2)-C_{12}M_1M_2)\bigr)}{\int \mathrm{d}M_1\,\mathrm{d}M_2\,\exp\bigl(-(N/t)\,\mathrm{Tr}((C_{11}/2)M_1^2+(C_{22}/2)M_2^2-M_1M_2)\bigr)} \tag{6.65}$$

where $\mathcal{V}_1(M_1) = (C_{11}/2)M_1^2 - V_1(M_1)$ is a polynomial in M_1, such that $V_1(M_1)$ contains only terms with degree ≥ 3, and where $\mathcal{V}_2(M_2) = (C_{22}/2)M_2^2 - V_2(M_2)$ is a polynomial in M_2, such that $V_2(M_2)$ contains only terms with degree ≥ 3, and we assume $C_{12} = 1$:

$$C = \begin{pmatrix} C_{11} & -1 \\ -1 & C_{22} \end{pmatrix} \tag{6.66}$$

Indeed, it generates surfaces made of polygons of two possible colors (call them polygons carrying a spin $+$ or $-$) glued by their sides (no crossing here). The weight of each surface depends on the number of neighbouring polygons with same spin or different spin, which is precisely the Ising model. If $C_{11} = C_{22}$ and $V_1 = V_2$, this is an Ising model without magnetic field, otherwise it is an Ising model with magnetic field.

Let us describe the solution.

Consider the following rational curve

$$\begin{cases} x(z) = \gamma z + \sum_{k=0}^{\deg V_2'} \alpha_k z^{-k} \\ y(z) = \gamma z^{-1} + \sum_{k=0}^{\deg V_1'} \beta_k z^k \end{cases} \tag{6.67}$$

where all coefficients γ, α_k, β_k are determined by:

$$y(z) - V_1'\bigl(x(z)\bigr) \underset{z\to\infty}{\sim} -\frac{t}{\gamma z} + O(z^{-2}) \tag{6.68}$$

$$x(z) - V_2'\bigl(y(z)\bigr) \underset{z\to 0}{\sim} -\frac{tz}{\gamma} + O(z^2) \tag{6.69}$$

The coefficients γ, α_k, β_k are algebraic functions of t, and we must choose the branch such that $\gamma \to 0$ at $t = 0$.

The curve $\bigl(x(z), y(z)\bigr)$, $z \in \mathbb{C}$, is called the spectral curve. It is a genus zero algebraic curve. There are $\deg V_2$ branch points solutions of $x'(z) = 0$.

Notice that we have:

$$\operatorname*{Res}_{z\to\infty} y\,\mathrm{d}x = t = -\operatorname*{Res}_{z\to 0} y\,\mathrm{d}x\,. \tag{6.70}$$

Then one has [2]:

$$F^{(0)} = \frac{1}{2}\Biggl(\operatorname*{Res}_{z\to\infty} \mathcal{V}_1(x)y\,\mathrm{d}x + \operatorname*{Res}_{z\to 0}\bigl(xy - \mathcal{V}_2(y)\bigr)y\,\mathrm{d}x - t\operatorname*{Res}_{z\to\infty} \mathcal{V}_1(x)\frac{\mathrm{d}z}{z} \\ - t\operatorname*{Res}_{z\to 0}\bigl(xy - \mathcal{V}_2(y)\bigr)\frac{\mathrm{d}z}{z} - \frac{3}{2}t^2 - t^2\ln\Bigl(\frac{\gamma^2 \det C}{t}\Bigr)\Biggr) \tag{6.71}$$

and [27]:

$$F^{(1)} = -\frac{1}{24}\ln\left(\frac{(\tilde{t}_{\deg(V_2)})^2}{t^2}\prod_{i=1}^{\deg V_2}\gamma y'(a_i)\right) \tag{6.72}$$

where a_i are the solutions of $x'(a_i) = 0$, and $\tilde{t}_{\deg(V_2)}$ is the leading coefficient of $\mathcal{V}_2$.

The other $F^{(g)}$s are found as follows [15]:

Let a_i, $i = 1, \ldots, \deg V_2$ be the branch points, i.e., the solutions of $x'(a_i) = 0$. If z is close to a branch-point a_i, we denote $\bar{z}$ the other point such that

$$z \to a_i\ , \quad x(\bar{z}) = x(z) \quad \text{and} \quad \bar{z} \to a_i\ . \tag{6.73}$$

Notice that $\bar{z}$ depends on the branch-point, i.e., $\bar{z}$ is not globally defined. We also define:

$$\Phi(z) = \int_{z_0}^{z} y\,\mathrm{d}x \tag{6.74}$$

$\Phi(z)$ depends on the base-point z_0 and on the path between z and z_0, but the calculations below don't depend on that.

We define recursively:

$$W_1^{(0)}(p) = 0 \tag{6.75}$$

$$W_2^{(0)}(p,q) = \frac{1}{(p-q)^2} \tag{6.76}$$

$$\begin{aligned} &W_{k+1}^{(g)}(p,p_1,\ldots,p_k)\\ &= -\frac{1}{2}\sum_i \operatorname*{Res}_{z\to a_i} \frac{(z-\bar{z})\,\mathrm{d}z}{(p-z)(p-\bar{z})(y(z)-y(\bar{z})x'(\bar{z})}\\ &\times\left(W_{k+2}^{(g-1)}(z,\bar{z},p_1,\ldots,p_k) + \sum_{h=0}^{g}\sum_{I\subset\{1,2,\ldots,k\}} W_{1+|I|}^{(h)}(z,p_I)W_{1+k-|I|}^{(g-h)}(\bar{z},p_{K/\{I\}})\right). \end{aligned} \tag{6.77}$$

This system is a triangular system, and all $W_k^{(g)}$ are well-defined in a finite number of steps $\leq g(g+1)/2 + k$.

Then we have [15]:

$$F^{(g)} = \frac{1}{2-2g}\sum_i \operatorname*{Res}_{z\to a_i} \Phi(z)W_1^{(g)}(z)\,\mathrm{d}z\ , \quad g > 2\ . \tag{6.78}$$

The 1-matrix case is a special case of this when the curve is hyperelliptical (in that case $\bar{z}$ is globally defined), it corresponds to $\deg V_2 = 2$.

6.4.3 Chain of Matrices

$$Z_{\text{formal}} = \frac{\int \mathrm{d}M_1 \cdots \mathrm{d}M_p \exp\left(-(N/t)\,\mathrm{Tr}(\sum_{i=1}^p \mathcal{V}_i(M_i) - \sum_{i=1}^{p-1} M_i M_{i+1})\right)}{\int \mathrm{d}M_1 \ldots \mathrm{d}M_p \exp\left(-(N/t)\,\mathrm{Tr}(\sum_{i=1}^p (C_{ii}/2)M_i^2 - \sum_{i=1}^{p-1} M_i M_{i+1})\right)} \tag{6.79}$$

where $\mathcal{V}_i(M_i) = (C_{ii}/2)M_i^2 - V_i(M_i)$ is a polynomial in M_i, such that $V_i(M_i)$ contains only terms with degree ≥ 3. The matrix C is:

$$C = \begin{pmatrix} C_{11} & -1 & & & \\ -1 & C_{22} & -1 & & \\ & & \ddots & & \\ & & -1 & C_{pp} & -1 \end{pmatrix} \tag{6.80}$$

Consider the following rational curve

$$x_i(z) = \sum_{k=-s_i}^{r_i} \alpha_{i,k} z^k \quad \forall i = 0, \ldots, p+1 \tag{6.81}$$

where all coefficients γ, α_k, β_k are determined by:

$$\begin{aligned} & x_{i+1} + x_{i-1} = \mathcal{V}'(x_i) \qquad\qquad \forall i = 1, \ldots, p \\ & x_0(z) \underset{z\to\infty}{\sim} \frac{t}{\gamma z} + O(z^{-2}) \\ & x_{p+1}(z) \underset{z\to 0}{\sim} \frac{tz}{\gamma} + O(z^2)\,. \end{aligned} \tag{6.82}$$

The coefficients γ, $\alpha_{i,k}$ are algebraic functions of t, and we must choose the branch such that $\gamma \to 0$ at $t = 0$.

The curve $\big(x_1(z), x_2(z)\big)$, $z \in \mathbb{C}$, is called the spectral curve. It is a genus zero algebraic curve.

Notice that we have $\forall i = 1, \ldots, p-1$:

$$\operatorname*{Res}_{z\to\infty} x_{i+1} dx_i = t = -\operatorname*{Res}_{z\to 0} x_{i+1} dx_i \tag{6.83}$$

Then one has [24]:

$$\begin{aligned} F^{(0)} = \frac{1}{2}\Bigg(& \sum_{i=1}^{p} \operatorname*{Res}_{z\to\infty} \big(\mathcal{V}_i(x_i) - \tfrac{1}{2} x_i \mathcal{V}_i'(x_i)\big) x_{i+1}\, \mathrm{d}x_i \\ & - t \sum_{i=1}^{p} \operatorname*{Res}_{z\to\infty} \big(\mathcal{V}_i(x_i) - \tfrac{1}{2} x_i \mathcal{V}_i'(x_i)\big) \frac{\mathrm{d}z}{z} - t^2 \ln\left(\frac{\gamma^2 \det C}{t}\right)\Bigg)\,. \end{aligned} \tag{6.84}$$

$F^{(1)}$ and the other $F^{(g)}$s have never been computed, but it is strongly believed that they should be given by the same formula as in the 2-matrix case.

6.4.4 Closed Chain of Matrices

$$Z_{\text{formal}} = \frac{\int \mathrm{d}M_1 \cdots \mathrm{d}M_p \, \exp\big(-(N/t)\,\mathrm{Tr}(\sum_{i=1}^{p} \mathcal{V}_i(M_i) - \sum_{i=1}^{p-1} M_i M_{i+1} - M_p M_1)\big)}{\int \mathrm{d}M_1 \cdots \mathrm{d}M_p \, \exp\big(-(N/t)\,\mathrm{Tr}(\sum_{i=1}^{p} (C_{ii}/2) M_i^2 - \sum_{i=1}^{p-1} M_i M_{i+1} - M_p M_1)\big)} \,. \tag{6.85}$$

It is the case where the matrix C of quadratic interactions has the form:

$$C = \begin{pmatrix} C_{11} & -1 & & & -1 \\ -1 & C_{22} & -1 & & \\ & & \ddots & & \\ -1 & & -1 & C_{pp} & -1 \end{pmatrix} \tag{6.86}$$

This model is yet unsolved, apart from very few cases ($p = 2$, $p = 3$ (Potts model), $p = 4$ with cubic potentials), and there are also results in the large p limit. This model is still a challenge.

6.4.5 $O(n)$ Model

$$Z_{\text{formal}} = \frac{\int \mathrm{d}M \, \mathrm{d}M_1 \cdots \mathrm{d}M_n \, \exp\big(-(\frac{N}{t})\,\mathrm{Tr}((\frac{C_M}{2}) M^2 + (\frac{C}{2}) \sum_{i=1}^{n} M_i^2 - V(M) - \sum_{i=1}^{n} M M_i^2)\big)}{\int \mathrm{d}M \, \mathrm{d}M_1 \cdots \mathrm{d}M_n \, \exp\big(-(\frac{N}{t})\mathrm{Tr}((\frac{C_M}{2}) M^2 + (\frac{C}{2}) \sum_{i=1}^{n} M_i^2)\big)} \tag{6.87}$$

where V contains at least cubic terms.

We write:

$$\mathcal{V}(x) = -V\left(-\left(x + \frac{C}{2}\right)\right) + \frac{C_M}{2}\left(x + \frac{C}{2}\right)^2 \,. \tag{6.88}$$

This model can easily be analytically continued for noninteger ns. Indeed, combinatoricaly, it is the generating function of a loop gas living on the maps. n is the "fugacity" of loops, i.e., the n dependence of the weight of each configuration is $n^{\#\,\text{loops}}$, and the C dependence is $C^{-\text{length of loops}}$. One often writes:

$$n = 2\cos(\nu\pi) \tag{6.89}$$

The $O(n)$ model was introduced by I. Kostov in [43] then in [21, 44], and it plays a very important role in many areas of physics, and lately, it has started to play a special role in string theory, as an effective model for the check of ADS-CFT correspondence.

The leading order 1-cut solution of this model is well known [28–30]. For any two distinct complex numbers a and b, define:

$$m = 1 - \frac{a^2}{b^2}\,, \qquad \tau = \frac{iK'(m)}{K(m)} \tag{6.90}$$

where $K(m) = K'(1-m)$ are the complete elliptical integrals ([1]).

We also consider the following elliptical function (defined on the torus $(1,\tau)$):

$$x(u) = ib\frac{\mathrm{cn}(2K(m)u, m)}{\mathrm{sn}(2K(m)u, m)} \tag{6.91}$$

where sn and cn are the elliptical sine and cosine functions.

Then we define the following function on the universal covering (i.e., on the complex plane):

$$G_\nu(u) = H\left(\mathrm{e}^{\mathrm{i}\nu\pi/2}\frac{\theta_1(u+\nu/2)}{\theta_1(u)} + \mathrm{e}^{-\mathrm{i}\nu\pi/2}\frac{\theta_1(u-\nu/2)}{\theta_1(u)}\right) \tag{6.92}$$

where H is a normalization constant such that:

$$\lim_{u\to 0} G_\nu(u)/x(u) = 1 \tag{6.93}$$

It satisfies:

$$\begin{aligned} G_\nu(u+\tau) + G_\nu(u-\tau) - nG_\nu(u) &= 0\,, \\ G(u+1) = G(u)\,, \quad G(u) &= G(-\tau-u)\,. \end{aligned} \tag{6.94}$$

We have:

$$\begin{aligned} &G_\nu(u)^2 + G_\nu(-u)^2 - nG_\nu(u)G_\nu(-u) \\ &\qquad = (2+n)(x^2(u) - e_\nu^2)\,, \quad e_\nu = x\left(\frac{\nu}{2}\right) \end{aligned} \tag{6.95}$$

$$\begin{aligned} &G_{1-\nu}(u)^2 + G_{1-\nu}(-u)^2 + nG_{1-\nu}(u)G_{1-\nu}(-u) \\ &\qquad = (2-n)(x^2(u) - e_{1-\nu}^2) \end{aligned} \tag{6.96}$$

$$\begin{aligned} &G_{1-\nu}(u)G_\nu(u) - G_{1-\nu}(-u)G_\nu(-u) + \frac{n}{2}G_{1-\nu}(-u)G_\nu(u) \\ &\qquad - \frac{n}{2}G_{1-\nu}(u)G_\nu(-u) = x(u)b\frac{2\sin(\nu\pi)}{\mathrm{sn}(\nu K)\,\mathrm{sn}((1-\nu)K)}\,. \end{aligned} \tag{6.97}$$

Then we define:

$$A(x^2) = \mathrm{Pol}\left(\frac{(2+n)(x^2-e_\nu^2)(\mathcal{V}'.G_{1-\nu})_+ - xbc(\mathcal{V}'.G_\nu)_-}{(x^2-b^2)^2}\right) \tag{6.98}$$

$$B(x^2) = \mathrm{Pol}\left(\frac{(2-n)(x^2-e_{1-\nu}^2)(1/x)(\mathcal{V}'.G_\nu)_- - bc(\mathcal{V}'.G_{1-\nu})_+}{(x^2-b^2)^2}\right) \tag{6.99}$$

where Pol means the polynomial part in x at large x (i.e., at $u \to 0$), and the subscript $+$ and $-$ mean the even and odd part, and where

$$c = \frac{2\sin(\nu\pi)}{\operatorname{sn}(\nu K)\operatorname{sn}((1-\nu)K)} \tag{6.100}$$

A and B are polynomials of x^2.

Then, a and b are completely determined by the 2 conditions:

$$\operatorname*{Res}_{u\to 0}\left(\frac{A\big(x^2(u)\big)G_{1-\nu}(u)_+}{x(u)} + B\big(x^2(u)\big)G_\nu(u)_-\right)\mathrm{d}x(u) = 0 \tag{6.101}$$

$$\operatorname*{Res}_{u\to 0}\big(A\big(x^2(u)\big)G_{1-\nu}(u)_- + x(u)B\big(x^2(u)\big)G_\nu(u)_+\big)\,\mathrm{d}x(u) = t\,. \tag{6.102}$$

Once we have determined a, b, m, A, B, we define the resolvent:

$$\omega(u) = A(x(u)^2)G_{1-\nu}(u) + x(u)B(x(u)^2)G_\nu(u) \tag{6.103}$$

which is the first term of the formal large N expansion of:

$$\frac{1}{N}\left\langle \operatorname{Tr}\frac{1}{x(u)-M}\right\rangle = \omega(u) + O(1/N^2) \tag{6.104}$$

and which satisfies:

$$\omega(u+\tau) + \omega(u-\tau) + n\omega(u) = \mathcal{V}'(x(u))\,, \tag{6.105}$$

$$\omega(u+1) = \omega(u)\,, \quad \omega(u) = \omega(-\tau-u)\,. \tag{6.106}$$

The free energy is then found by:

$$\frac{\partial F^{(0)}}{\partial t^2} = \left(1-\frac{n}{2}\right)\ln\big(a^2 g(m)\big) \tag{6.107}$$

where

$$\frac{g'}{g} = \frac{1}{m(1-m)\operatorname{sn}^2(\nu K(m), m)}\,. \tag{6.108}$$

All this is described in [23]. The special cases of n integer, and in general when ν is rational, can be computed with more details.

6.4.6 Potts Model

$$Z_{\text{formal}} = \frac{\int \mathrm{d}M_1\cdots \mathrm{d}M_Q\, \exp(-(N/t)\operatorname{Tr}(\sum_{i=1}^Q \frac{1}{2}M_i^2 + V(M_i) + (C/2)\sum_{i,j}^Q M_iM_j))}{\int \mathrm{d}M\,\mathrm{d}M_1\cdots \mathrm{d}M_n\, \exp(-(N/t)\operatorname{Tr}(\sum_{i=1}^Q \frac{1}{2}M_i^2 + (C/2)\sum_{i,j}^Q M_iM_j))} \tag{6.109}$$

or equivalently:

$$Z_{\text{formal}} = \frac{\int \mathrm{d}M\,\mathrm{d}M_1 \cdots \mathrm{d}M_Q \, \exp(-(N/t)\,\mathrm{Tr}(\sum_{i=1}^{Q} \frac{1}{2}M_i^2 + V(M_i) + (C/2)M^2 - C\sum_{i=1}^{Q} MM_i))}{\int \mathrm{d}M\,\mathrm{d}M_1 \cdots \mathrm{d}M_Q \, \exp(-(N/t)\,\mathrm{Tr}(\sum_{i=1}^{Q} \frac{1}{2}M_i^2 + (C/2)M^2 - C\sum_{i=1}^{Q} MM_i))} \tag{6.110}$$

The Random lattice Potts model was first introduced by V. Kazakov in [38, 39]. The loop equations have been written and solved to leading order, in particular in [10, 17, 57].

6.4.7 3-Color Model

$$Z_{\text{formal}} = \frac{\int \mathrm{d}M_1\,\mathrm{d}M_2\,\mathrm{d}M_3 \, \exp(-(N/t)\,\mathrm{Tr}(\frac{1}{2}(M_1^2+M_2^2+M_3^2) - g(M_1M_2M_3 + M_1M_3M_2)))}{\int \mathrm{d}M_1\,\mathrm{d}M_2\,\mathrm{d}M_3 \, \exp(-(N/t)\,\mathrm{Tr}(\frac{1}{2}(M_1^2+M_2^2+M_3^2)))} \tag{6.111}$$

The loop equations have been written and solved to leading order, in particular in [31, 45].

6.4.8 6-Vertex Model

$$Z_{\text{formal}} = \frac{\int \mathrm{d}M\,\mathrm{d}M^\dagger \, \exp(-(N/t)\,\mathrm{Tr}(MM^\dagger - M^2M^{2\dagger} + \cos\beta(MM^\dagger)^2))}{\int \mathrm{d}M\,\mathrm{d}M^\dagger \, \exp(-(N/t)\,\mathrm{Tr}(MM^\dagger))} \tag{6.112}$$

where M is a complex matrix. The loop equations have been written and solved to leading order, in particular in [46, 56].

6.4.9 ADE Models

Given a Dynkin diagram of A, D or E Lie algebra, and let A be its adjacency matrix ($A_{ij} = A_{ji} = \#$ links between i and j). We define:

$$Z_{\text{formal}} = \int \prod_i \mathrm{d}M_i \prod_{i<j} \mathrm{d}B_{ij} \, \exp\left(-\frac{N}{T}\,\mathrm{Tr}\left(\sum_i \frac{1}{2}M_i^2 - \frac{g}{3}M_i^3 + \frac{1}{2}\sum_{i,j} B_{ij}B_{ij}^t + \frac{K}{2}\sum_{i,j} A_{ij}B_{ij}B_{ij}^t M_i\right)\right) \tag{6.113}$$

where $B_{ji} = B_{ij}^t$ are complex matrices, and M_i are Hermitian matrices.

The loop equations have been written and solved to leading order, in particular in [44, 47].

6.4.10 ABAB Models

$$
Z_{\text{formal}} = \int \prod_{i=1}^{n_1} \mathrm{d}A_i \prod_{j=1}^{n_2} \mathrm{d}B_j \exp\left(-\frac{N}{T}\operatorname{Tr}\left(\sum_i \frac{A_i^2}{2} + \sum_i \frac{B_i^2}{2} + g\sum_{i,j} A_i B_j A_i B_j\right)\right). \tag{6.114}
$$

This model is yet unsolved, apart from few very special cases [41]. However, its solution would be of great interest in the understanding of Temperley–Lieb algebra.

6.5 Discussion

6.5.1 Summary of Some Known Results

We list here some properties which are rather well understood.

- The fact that formal matrix integrals are generating functions for counting discrete surfaces (also called maps) is well understood, as was explained in this review.
- The fact that formal matrix integrals i.e. generating functions of maps satisfy Schwinger–Dyson equations is well understood.
- The fact that formal matrix integrals and convergent integrals don't coincide in general is well understood. In the examples of the 1-matrix model, 2-matrix Ising model or chain of matrices, it is understood that they may coincide only in the "1-cut case", i.e. if the classical spectral curve has genus zero.
- The fact that the formal 1-matrix, 2-matrix or chain of matrices integrals are τ functions of some integrable hierarchies is well understood too.
- For the formal 1-matrix model and 2-matrix Ising model, all $F^{(g)}$s have been computed explicitly. The result is written in terms of residues of rational functions.
- For the chain of matrices, $F^{(0)}$ is known explicitly [24], and it is strongly believed that all $F^{(g)}$s are given by the same expression as for the 2-matrix model.
- Multi-cut formal matrix models are well studied too, and they can be rewritten in terms of multi-matrix models. For the 1 and 2 matrix models, the expressions of the $F^{(g)}$s are known explicitly (in terms of residues of meromorphic forms on higher genus spectral curves).

6.5.2 Some Open Problems

• The large N asymptotics of convergent matrix integrals, are not directly relevant for combinatorics, but are a very important question for other applications, in particular bi-orthogonal polynomials, universal statistical properties of eigenvalues of large random matrices, and many other applications. This question is highly nontrivial, and so far, it has been solved only for the 1-matrix model [5,19,22], and a few other cases (e.g., the 6-vertex model [8]). The method mostly used to prove the asymptotic formulae is the Riemann-Hilbert method [5,19], which consists in finding a Riemann–Hilbert problem for the actual matrix integral and for its conjectured asymptotics, and compare both. There is a hope that probabilists' methods like large deviations could be at leats as efficient.

• For the 2-matrix model and chain of matrices, the topological expansion of formal "mixed" expectation values (e.v. of invariant monomials whose words contain at least 2 different letters) has not yet been computed. This problem is a challenge in itself, and has applications to the understanding of "boundary conformal field theory". In terms of combinatorics, it corresponds to find generating functions for open surfaces with boundaries of prescribed colors.

• Many matrix models have not yet been solved, even to planar order. For instance the closed chain of matrices where $C_{ij} = \delta_{i,j+1} + \delta_{i,j-1} + C_i\delta_{ii}$ and $C_{p1} = C_{1p} = 1$. For the Potts model, 6-vertex model, 3-color model, $O(n)$ model, ADE models, only the planar resolvent is known. For the $A_nB_mA_nB_m$, almost nothing is known, although this model describes the combinatorics of Temperly–Lieb algebra.

• Limits of critical points are still to be understood and classified. Only a small number of critical points have been studied so far. They have been related to KdV or KP hierarchies. Critical points, i.e., radius of convergence of the t-formal power series, are in relationship with asymptotics numbers of large maps (through Borel's transform), and thus critical points allow to count maps made of a very large number of polygons, i.e., they can be interpreted as counting continuous surfaces (this was the idea of 2D quantum gravity in the 80s and early 90s). This is yet to be better understood [7,16].

• Extensions to other types of matrices (non-Hermitian) have been very little studied compared to the hermitian case, and much is still to be understood. For instance real symmetric matrices or quaternion-self-dual matrices have been studied from the begining [48], and they count nonorientable maps. Complex matrices, and normal complex matrices have played an increasing role in physics, because of their relationship with Laplacian growth problems [54], or some limits of string theory [4]. Complex matrices count maps with arrows on the edges (see the 6-vertex model [56]). Other types of matrix ensembles have been introduced in relationship with symmetric spaces [58], and it is not clear what they count.

• And there are so many applications of random matrices to physics, mathematics, biology, economics, to be investigated...

Acknowledgements

I would like to thank Philippe Di Francesco for his critical reading of the manuscript. This work is partly supported by the Enigma European network MRT-CT-2004-5652, by the ANR project Géométrie et intégrabilité en physique mathématique ANR-05-BLAN-0029-01, and by the Enrage European network MRTN-CT-2004-005616.

References

1. M. Abramowitz, I.A. Stegun: *Handbook of Mathematical Functions* (Dover, New York, 1964)
2. M. Bertola: J. High Energy Phys. **2003**, 062 (2003)
3. M. Bertola, Mo M. Y.: Int. Math. Res. Pap. **2005**, 565 (2005)
4. D. Berenstein, J. Maldacena, H. Nastase: J. High Energy Phys. **2002**, 013 (2002)
5. P. Bleher, A. Its: Ann. Math. **150**, 185 (1999)
6. P. Bleher, A. Its (eds): *Random Matrix Models and Their Applications* (Cambridge University Press, Cambridge, 2001).
7. P. Bleher, B. Eynard: J. Phys. A **36**, 3085 (2003)
8. P. Bleher, V. Fokin: Commun. Math. Phys. **268**, 223 (2006)
9. G. Bonnet, F. David, B. Eynard: J. Phys. A **33**, 6739 (2000)
10. G. Bonnet, B. Eynard: Phys. Lett. **B463**, 273 (1999)
11. É. Brézin, N. Deo: Phys. Rev. E **59**, 3901 (1999)
12. É. Brézin, C. Itzykson, G. Parisi, J. Zuber: Commun. Math. Phys. **59**, 35 (1978)
13. L. Chekhov: Theor. Math. Phys. **141**, 1640 (2004)
14. L. Chekhov, B. Eynard: J. High Energy Phys. **2006**, 014 (2006)
15. L. Chekhov, B. Eynard, N. Orantin: arXiv:math-ph/0603003.
16. G.M. Cicuta: arXiv:cond-mat/0012078.
17. J.M. Daul: arXiv:hep-th/9502014.
18. F. David: Nucl. Phys. **B257** 45 (1985)
19. P. Deift, T. Kriecherbauer, K.T.-R. McLaughlin, S. Venakides, Z. Zhou: Commun. Pure Appl. Math. **52**, 1335 (1999)
20. P. Di Francesco, P. Ginsparg, J. Zinn-Justin: Phys. Rep. **254**, 1 (1995)
21. B. Duplantier, I. Kostov: Phys. Rev. Lett. **61**, 1433 (1988)
22. N. Ercolani, K.T.-R. McLaughlin: Int. Math. Res. Not. **2003**, 755 (2003)
23. B. Eynard: Gravitation quantique bidimensionnelle et matrices aléatoires, PhD Thesis, Université Paris 6, Paris (1995)
24. B. Eynard, J. High Energy Phys. **2003**, 018 (2003)
25. B. Eynard: J. High Energy Phys. **2005**, 034 (2005)
26. B. Eynard: Le modèle à deux matrices — Polynômes biorthogonaux, problème de Riemann–Hilbert et géométrie algébrique, Habilitation Thesis, Université Paris 7, Paris (2005)
27. B. Eynard, D. Korotkin, A. Kokotov: Nucl. Phys. **B694**, 443 (2004)
28. B. Eynard, J. Zinn-Justin: Nucl. Phys. **B386**, 558 (1992)
29. B. Eynard, C. Kristjansen: Nucl. Phys. **B455**, 577 (1995)
30. B. Eynard, C. Kristjansen: Nucl. Phys. **B466**, 463 (1996)
31. B. Eynard, C. Kristjansen: Nucl. Phys. **B516**, 529 (1998)

32. D.J. Gross, T. Piran, S. Weinberg (eds): *Two-Dimensional Quantum Gravity and Random Surfaces* (World Scientific, Singapore, 1991).
33. T. Guhr, A. Müller-Groeling, H.A. Weidenmüller: Phys. Rep. **299**, 189 (1998)
34. A. Guionnet, E. Maurel-Segala: ALEA Lat. Am. J. Probab. Math. Stat. **1**, 241 (2006)
35. A. Guionnet, O. Zeitouni: J. Funct. Anal. **188**, 461 (2002)
36. V.A. Kazakov: Phys. Lett. **B150**, 282 (1985)
37. V.A. Kazakov: Phys Lett. **A119**, 140 (1986).
38. V.A. Kazakov: Nucl. Phys. B Proc. Suppl. **4**, 93 (1988)
39. V.A. Kazakov: Mod. Phys. Lett. **A4**, 1691 (1989)
40. V.A. Kazakov, A.A. Migdal, I.K. Kostov: Phys. Lett. **B157**, 295 (1985)
41. V.A. Kazakov, P. Zinn-Justin: Nucl. Phys. **B546**, 647 (1999)
42. E. Kanzieper, V. Freilikher, Phys. Rev. E **57**, 6604 (1998)
43. I.K. Kostov: Mod. Phys. Lett. **A4**, 217 (1989)
44. I.K. Kostov: Nucl.Phys. **B376**, 539 (1992)
45. I.K. Kostov: Phys. Lett. **B49**, 245 (2002)
46. I.K. Kostov: Nucl. Phys. **B575**, 513 (2000)
47. I.K. Kostov, Phys. Lett. **B297** 74 (1992).
48. M.L. Mehta, *Random Matrices*, 2nd edn, (Academic Press, New York, 1991).
49. G. 't Hooft, Nucl. Phys. **B72**, 461 (1974).
50. W.T. Tutte: Can. J. Math. **14**, 21 (1962)
51. W.T. Tutte: Can. J. Math. **15**, 249 (1963)
52. P. van Moerbeke: Integrable lattices: random matrices and random permutations. In: *Random Matrices and Their Applications* ed by. P. Bleher, A. Its (Cambridge University Press, Cambridge, 2001) pp 321–406
53. S.R. Wadia: Phys. Rev. D **24**, 970(1981)
54. P. Wiegmann, A. Zabrodin: arXiv:hep-th/0309253.
55. E.P. Wigner: Proc. Cambridge Philos. Soc. **47**, 790 (1951).
56. P. Zinn-Justin: Europhys. Lett. **50**, 15 (2000)
57. P. Zinn-Justin: J. Stat. Phys. **98**, 245 (2001)
58. M.R. Zirnbauer: J. Math. Phys. **37**, 4986 (1996)

7

Application of Random Matrix Theory to Multivariate Statistics

Momar Dieng[1] and Craig A. Tracy[2]

[1] Department of Mathematics, University of Arizona, Tucson, AZ 85721, USA
momar@math.arizona.edu

[2] Department of Mathematics, University of California, Davis, CA 95616, USA
tracy@math.ucdavis.edu

Summary. This is an expository account of the edge eigenvalue distributions in random matrix theory and their application in multivariate statistics. The emphasis is on the Painlevé representations of these distribution functions.

7.1 Multivariate Statistics

7.1.1 Wishart Distribution

The basic problem in statistics is testing the agreement between actual observations and an underlying probability model. Pearson in 1900 [27] introduced the famous χ^2 *test* where the sampling distribution approaches, as the sample

J. Harnad (ed.), *Random Matrices, Random Processes and Integrable Systems*, CRM Series in Mathematical Physics, DOI 10.1007/978-1-4419-9514-8_7,

size increases, to the χ^2 distribution. Recall that if X_j are independent and identically distributed standard normal random variables, $N(0,1)$, then the distribution of

$$\chi_n^2 := X_1^2 + \cdots + X_n^2 \tag{7.1}$$

has density

$$f_n(x) = \begin{cases} \frac{1}{2^{n/2}\Gamma(n/2)} x^{n/2-1} \mathrm{e}^{-x/2} & \text{for } x > 0, \\ 0 & \text{for } x \le 0, \end{cases} \tag{7.2}$$

where $\Gamma(x)$ is the gamma function.

In classical *multivariate statistics*[3] it is commonly assumed that the underlying distribution is the multivariate normal distribution. If X is a $p \times 1$-variate normal with $\mathbb{E}(X) = \mu$ and $p \times p$ covariance matrix $\Sigma = \operatorname{cov}(X) := \mathbb{E}\big((X-\mu)\otimes(X-\mu)\big)$,[4] denoted $N_p(\mu, \Sigma)$, then if $\Sigma > 0$ the density function of X is

$$f_X(x) = (2\pi)^{-p/2} (\det \Sigma)^{-1/2} \exp\big[-\tfrac{1}{2}\big(x-\mu, \Sigma^{-1}(x-\mu)\big)\big]\,, \quad x \in \mathbb{R}^p\,,$$

where $(\cdot,\cdot)$ is the standard inner product on $\mathbb{R}^p$.

It is convenient to introduce a matrix notation: If X is a $n \times p$ matrix (the *data matrix*) whose rows X_j are independent $N_p(\mu, \Sigma)$ random variables,

$$X = \begin{pmatrix} \longleftarrow & X_1 & \longrightarrow \\ \longleftarrow & X_2 & \longrightarrow \\ & \vdots & \\ \longleftarrow & X_n & \longrightarrow \end{pmatrix},$$

then we say X is $N_p(\mathbf{1}\otimes\mu, I_n \otimes \Sigma)$ where $\mathbf{1} = (1,1,\dots,1)$ and I_n is the $n \times n$ identity matrix. We now introduce the multivariate generalization of (7.1).

Definition 7.1.1. If $A = X^t X$, where the $n \times p$ matrix X is $N_p(0, I_n \otimes \Sigma)$, then A is said to have *Wishart distribution* with n degrees of freedom and covariance matrix Σ. We write A is $W_p(n, \Sigma)$.

To state the generalization of (7.2) we first introduce the *multivariate Gamma function.* If $\mathcal{S}_m^+$ is the space of $p \times p$ positive definite, symmetric matrices, then

$$\Gamma_p(a) := \int_{\mathcal{S}_p^+} \mathrm{e}^{-\operatorname{tr}(A)} (\det A)^{a-(p+1)/2}\, \mathrm{d}A$$

where $\operatorname{Re}(a) > (m-1)/2$ and $\mathrm{d}A$ is the product Lebesgue measure of the $\frac{1}{2}p(p+1)$ distinct elements of A. By introducing the matrix factorization

[3] There are many excellent textbooks on multivariate statistics, we mention Anderson [1], Muirhead [26], and for a shorter introduction, Bilodeau and Brenner [4].

[4] If u and v are vectors we denote by $u \otimes v$ the matrix with (i,j) matrix element $u_i v_j$.

$A = T^t T$ where T is upper-triangular with positive diagonal elements, one can evaluate this integral in terms of ordinary gamma functions, see [26, p. 62]. Note that $\Gamma_1(a)$ is the usual gamma function $\Gamma(a)$. The basic fact about the Wishart distributions is

Theorem 7.1.1 (Wishart [38]). *If A is $W_p(n, \Sigma)$ with $n \geq p$, then the density function of A is*

$$\frac{1}{2^{pn/2}\Gamma_p(n/2)(\det \Sigma)^{n/2}} \exp\bigl(-\tfrac{1}{2}\operatorname{tr}(\Sigma^{-1}A)\bigr)(\det A)^{(n-p-1)/2}\,. \tag{7.3}$$

For $p = 1$ and $\Sigma = 1$ (7.3) reduces to (7.2). The case $p = 2$ was obtained by Fisher in 1915 and for general p by Wishart in 1928 using geometrical arguments. Most modern proofs follow James [20]. The importance of the Wishart distribution lies in the fact that the *sample covariance matrix*, S, is $W_p(n, 1/n\Sigma)$ where

$$S := \frac{1}{n}\sum_{j=1}^{N}(X_i - \overline{X}) \otimes (X_j - \overline{X})\,, \quad N = n+1\,,$$

and X_j, $j = 1, \ldots, N$, are independent $N_p(\mu, \Sigma)$ random vectors, and $\overline{X} = (1/N)\sum_j X_j$.

Principle component analysis,[5] a multivariate data reduction technique, requires the eigenvalues of the sample covariance matrix; in particular, the largest eigenvalue (the largest principle component variance) is most important. The next major result gives the joint density for the eigenvalues of a Wishart matrix.

Theorem 7.1.2 (James [21]). *If A is $W_p(n, \Sigma)$ with $n \geq p$ the joint density function of the eigenvalues $l_1, \ldots, l_p$ of A is*

$$\frac{\pi^{p^2/2}2^{-pn/2}(\det \Sigma)^{-n/2}}{\Gamma_p(p/2)\Gamma_p(n/2)} \prod_{j=1}^{p} l_j^{(n-p-1)/2} \prod_{j<k}(l_j - l_k) \times \int_{\mathbb{O}(p)} \exp\bigl(-\tfrac{1}{2}\operatorname{tr}(\Sigma^{-1}HLH^t)\bigr)\,\mathrm{d}H \tag{7.4}$$

where $\mathbb{O}(p)$ is the orthogonal group of $p \times p$ matrices, $\mathrm{d}H$ is normalized Haar measure and L is the diagonal matrix $\operatorname{diag}(l_1, \ldots, l_p)$. *(We take $l_1 > l_2 > \cdots > l_p$.)*

Remark 7.1.1. The difficult part of this density function is the integral over the orthogonal group $\mathbb{O}(p)$. There is no known closed formula for this integral

[5] See, for example, [26, Chap. 9], and [22] for a discussion of some current issues in principle component analysis.

though James and Constantine (see [26, Chap. 7] for references) developed the theory of *zonal polynomials* which allow one to write infinite series expansions for this integral. However, these expansions converge slowly; and zonal polynomials themselves, lack explicit formulas such as are available for Schur polynomials. For complex Wishart matrices, the group integral is over the unitary group $\mathbb{U}(p)$; and this integral can be evaluated using the Harish-Chandra–Itzykson–Zuber integral [39].

There is one important case where the integral can be (trivially) evaluated.

Corollary 7.1.1. *If $\Sigma = I_p$, then the joint density* (7.4) *simplifies to*

$$\frac{\pi^{p^2/2} 2^{-pn/2} (\det \Sigma)^{-n/2}}{\Gamma_p(p/2)\Gamma_p(n/2)} \prod_{j=1}^{p} l_j^{(n-p-1)/2} \exp\Big(-\frac{1}{2}\sum_j l_j\Big) \prod_{j<k} (l_j - l_k) \,. \tag{7.5}$$

7.1.2 An Example with $\Sigma \neq cI_p$

This section uses the theory of zonal polynomials as can be found in Muirhead [26, Chap. 7] or Macdonald [23]. This section is not used in the remainder of the chapter. Let $\lambda = (\lambda_1, \dots, \lambda_p)$ be a partition into not more than p parts. We let $C_\lambda(Y)$ denote the zonal polynomial of Y corresponding to λ. It is a symmetric, homogeneous polynomial of degree $|\lambda|$ in the eigenvalues $y_1, \dots, y_p$ of Y. The normalization we adopt is defined by

$$(\operatorname{tr} Y)^k = (y_1 + \cdots + y_p)^k = \sum_{\substack{\lambda \vdash k \\ l(\lambda) \le p}} C_\lambda(Y) \,.$$

The fundamental integral formula for zonal polynomials is[6]

Theorem 7.1.3. *Let $X, Y \in \mathcal{S}_p^+$, then*

$$\int_{\mathbb{O}(p)} C_\lambda(XHYH^t)\, \mathrm{d}H = \frac{C_\lambda(X) C_\lambda(Y)}{C_\lambda(I_p)} \tag{7.6}$$

where $\mathrm{d}H$ is normalized Haar measure.

By expanding the exponential and using (7.6) it follows that

$$\int_{\mathbb{O}(p)} \exp\big(z \operatorname{tr}(XHYH^t)\big)\, \mathrm{d}H = \sum_{k \ge 0} \frac{z^k}{k!} \sum_{\substack{\lambda \vdash k \\ l(\lambda) \le p}} \frac{C_\lambda(X) C_\lambda(Y)}{C_\lambda(I_p)} \,. \tag{7.7}$$

We examine (7.7) for the special case ($|\rho| < 1$)

[6] See, for example, [26, Thm. 7.2.5].

$$\Sigma = (1-\rho)I_p + \rho\mathbf{1}\otimes\mathbf{1} = \begin{pmatrix} 1 & \rho & \rho & \cdots & \rho \\ \rho & 1 & \rho & \cdots & \rho \\ \vdots & & & & \vdots \\ \rho & \rho & \rho & \cdots & 1 \end{pmatrix} .$$

We have

$$\Sigma^{-1} = (1-\rho)^{-1}I_p - \frac{\rho}{(1-\rho)(1+(p-1)\rho)}\mathbf{1}\otimes\mathbf{1}$$

and

$$\det\Sigma = (1-\rho)^{p-1}(1+(p-1)\rho) .$$

For this choice of Σ, let $Y = \alpha\mathbf{1}\otimes\mathbf{1}$ where $\alpha = \rho/\big((2(1-\rho)(1+(p-1)\rho)\big)$, then

$$\begin{aligned}\int_{\mathbb{O}(p)} & \exp\big(-\tfrac{1}{2}\operatorname{tr}(\Sigma^{-1}HLH^t)\big)\,\mathrm{d}H \\ &= \exp\Big(-\frac{1}{2(1-\rho)}\sum_j\lambda_j\Big)\int_{\mathbb{O}(p)}\exp\big(\operatorname{tr}(YHLH^t)\big)\,\mathrm{d}H \\ &= \exp\Big(-\frac{1}{2(1-\rho)}\sum_j\lambda_j\Big)\sum_{k\ge 0}\frac{C_{(k)}(\alpha\mathbf{1}\otimes\mathbf{1})C_{(k)}(\Lambda)}{k!C_{(k)}(I_p)}\end{aligned}$$

where we used the fact that the only partition $\lambda \vdash k$ for which $C_\lambda(Y)$ is nonzero is $\lambda = (k)$. And for this partition, $C_{(k)}(Y) = \alpha^k p^k$. Define the symmetric functions g_n[7] by

$$\prod_{j\ge 1}(1-x_j y)^{-1/2} = \sum_{n\ge 0} g_n(x)y^n ,$$

then it is known that [23]

$$C_{(k)}(L) = \frac{2^{2k}(k!)^2}{(2k)!}g_k(L) .$$

Using the known value of $C_{(k)}(I_p)$ we find

$$\int_{\mathbb{O}(p)}\exp\big(-\tfrac{1}{2}\operatorname{tr}(\Sigma^{-1}HLH^t)\big)\,\mathrm{d}H = \exp\Big(-\frac{1}{2(1-\rho)}\sum_j\lambda_j\Big)\sum_{k\ge 0}\frac{(\alpha p)^k}{(\frac{1}{2}p)_k}g_k(L)$$

where $(a)_k = a(a+1)\cdots(a+k-1)$ is the Pochammer symbol.

[7] In the theory of zonal polynomials, the g_n are the analogue of the complete symmetric functions h_n.

7.2 Edge Distribution Functions

7.2.1 Summary of Fredholm Determinant Representations

In this section we define three Fredholm determinants from which the edge eigenvalue distributions, for the three symmetry classes orthogonal, unitary and symplectic, will ensue. This section follows [31, 33, 36]; see also, [15, 16].

In the unitary case ($\beta = 2$), define the trace class operator K_2 on $L^2(s, \infty)$ with *Airy kernel*

$$K_{\mathrm{Ai}}(x,y) := \frac{\mathrm{Ai}(x)\,\mathrm{Ai}'(y) - \mathrm{Ai}'(x)\,\mathrm{Ai}(y)}{x-y} = \int_0^\infty \mathrm{Ai}(x+z)\,\mathrm{Ai}(y+z)\,\mathrm{d}z \quad (7.8)$$

and associated Fredholm determinant, $0 \le \lambda \le 1$,

$$D_2(s,\lambda) = \det(I - \lambda K_2)\,. \quad (7.9)$$

Then we introduce the distribution functions

$$F_2(s) = F_2(s,1) = D_2(s,1)\,, \quad (7.10)$$

and for $m \ge 2$, the distribution functions $F_2(s,m)$ are defined recursively below by (7.25).

In the symplectic case ($\beta = 4$), we define the trace class operator K_4 on $L^2(s,\infty) \oplus L^2(s,\infty)$ with matrix kernel

$$K_4(x,y) := \frac{1}{2}\begin{pmatrix} S_4(x,y) & SD_4(x,y) \\ IS_4(x,y) & S_4(y,x) \end{pmatrix} \quad (7.11)$$

where

$$\begin{aligned} S_4(x,y) &= K_{\mathrm{Ai}}(x,y) - \frac{1}{2}\,\mathrm{Ai}(x,y)\int_y^\infty \mathrm{Ai}(z)\,\mathrm{d}z\,, \\ SD_4(x,y) &= -\partial_y K_{\mathrm{Ai}}(x,y) - \frac{1}{2}\,\mathrm{Ai}(x)\,\mathrm{Ai}(y)\,, \\ IS_4(x,y) &= -\int_x^\infty K_{\mathrm{Ai}}(z,y)\,dz + \frac{1}{2}\int_x^\infty \mathrm{Ai}(z)\,\mathrm{d}z \int_y^\infty \mathrm{Ai}(z)\,\mathrm{d}z\,, \end{aligned}$$

and the associated Fredholm determinant, $0 \le \lambda \le 1$,

$$D_4(s,\lambda) = \det(I - \lambda K_4)\,. \quad (7.12)$$

Then we introduce the distribution functions (note the square root)

$$F_4(s) = F_4(s,1) = \sqrt{D_4(s,1)}\,, \quad (7.13)$$

and for $m \ge 2$, the distribution functions $F_4(s,m)$ are defined recursively below by (7.27).

In the orthogonal case ($\beta = 1$), we introduce the matrix kernel

$$K_1(x,y) := \begin{pmatrix} S_1(x,y) & SD_1(x,y) \\ IS_1(x,y) - \varepsilon(x,y) & S_1(y,x) \end{pmatrix} \tag{7.14}$$

where

$$\begin{aligned} S_1(x,y) &= K_{\mathrm{Ai}}(x,y) - \frac{1}{2}\,\mathrm{Ai}(x)\left(1 - \int_y^\infty \mathrm{Ai}(z)\,\mathrm{d}z\right), \\ SD_1(x,y) &= -\partial_y K_{\mathrm{Ai}}(x,y) - \tfrac{1}{2}\,\mathrm{Ai}(x)\,\mathrm{Ai}(y), \\ IS_1(x,y) &= -\int_x^\infty K_{\mathrm{Ai}}(z,y)\,\mathrm{d}z + \frac{1}{2}\left(\int_y^x \mathrm{Ai}(z)\,\mathrm{d}z + \int_x^\infty \mathrm{Ai}(z)\,\mathrm{d}z \int_y^\infty \mathrm{Ai}(z)\,\mathrm{d}z\right), \\ \varepsilon(x-y) &= \tfrac{1}{2}\,\mathrm{sgn}(x-y). \end{aligned}$$

The operator K_1 on $L^2(s,\infty) \oplus L^2(s,\infty)$ with this matrix kernel is *not* trace class due to the presence of ε. As discussed in [36], one must use the weighted space $L^2(\rho) \oplus L^2(\rho^{-1})$, $\rho^{-1} \in L^1$. Now the determinant is the 2-determinant,

$$D_1(s,\lambda) = \det_2(I - \lambda K_1 \chi_J) \tag{7.15}$$

where χ_J is the characteristic function of the interval (s,∞). We introduce the distribution functions (again note the square root)

$$F_1(s) = F_1(s,1) = \sqrt{D_1(s,1)}\,, \tag{7.16}$$

and for $m \geq 2$, the distribution functions $F_1(s,m)$ are defined recursively below by (7.27). This is the first indication that the determinant $D_1(s,\lambda)$ might be more subtle than either $D_2(s,\lambda)$ or $D_4(s,\lambda)$.

7.2.2 Universality Theorems

Suppose A is $W_p(n, I_p)$ with eigenvalues $l_1 > \cdots > l_p$. We define scaling constants

$$\mu_{np} = \left(\sqrt{n-1} + \sqrt{p}\right)^2, \quad \sigma_{np} = \left(\sqrt{n-1} + \sqrt{p}\right)\left(\frac{1}{\sqrt{n-1}} + \frac{1}{\sqrt{p}}\right)^{1/3}.$$

The following theorem establishes, under the null hypothesis $\Sigma = I_p$, that the largest principal component variance, l_1, converges in law to F_1.

Theorem 7.2.1 (Johnstone, [22]). *If $n,p \to \infty$ such that $n/p \to \gamma$, $0 < \gamma < \infty$, then*

$$\frac{l_1 - \mu_{np}}{\sigma_{np}} \xrightarrow{\mathfrak{D}} F_1(s,1)\,.$$

Johnstone's theorem generalizes to the mth largest eigenvalue.

Theorem 7.2.2 (Soshnikov [29]). *If $n, p \to \infty$ such that $n/p \to \gamma$, $0 < \gamma < \infty$, then*

$$\frac{l_m - \mu_{np}}{\sigma_{np}} \xrightarrow{\mathfrak{D}} F_1(s, m), \quad m = 1, 2, \dots .$$

Soshnikov proved his result under the additional assumption $n - p = O(p^{1/3})$. We remark that a straightforward generalization of Johnstone's proof [22] together with results of Dieng [10] show this restriction can be removed. Subsequently, El Karoui [14] extended Thm. 7.2.2 to $0 < \gamma \leq \infty$. The extension to $\gamma = \infty$ is important for modern statistics where $p \gg n$ arises in applications.

Going further, Soshnikov lifted the Gaussian assumption, again establishing a F_1 universality theorem. In order to state the generalization precisely, let us redefine the $n \times p$ matrices $X = \{x_{i,j}\}$ such that $A = X^t X$ to satisfy

1. $\mathbb{E}(x_{ij}) = 0$, $\mathbb{E}(x_{ij}^2) = 1$.
2. The random variables x_{ij} have symmetric laws of distribution.
3. All even moments of x_{ij} are finite, and they decay at least as fast as a Gaussian at infinity: $\mathbb{E}(x_{ij}^{2m}) \leq (\text{const}\, m)^m$.
4. $n - p = O(p^{1/3})$.

With these assumptions,

Theorem 7.2.3 (Soshnikov [29]).

$$\frac{l_m - \mu_{np}}{\sigma_{np}} \xrightarrow{\mathfrak{D}} F_1(s, m), \quad m = 1, 2, \dots .$$

It is an important open problem to remove the restriction $n - p = O(p^{1/3})$.

For real symmetric matrices, Deift and Gioev [8], building on the work of Widom [37], proved F_1 universality when the Gaussian weight function $\exp(-x^2)$ is replaced by $\exp\bigl(-V(x)\bigr)$ where V is an even degree polynomial with positive leading coefficient.

Table 7.3 in Sect. 7.9 displays a comparison of the percentiles of the F_1 distribution with percentiles of empirical Wishart distributions. Here l_j denotes the jth largest eigenvalue in the Wishart Ensemble. The percentiles in the l_j columns were obtained by finding the ordinates corresponding to the F_1-percentiles listed in the first column, and computing the proportion of eigenvalues lying to the left of that ordinate in the empirical distributions for the l_j. The bold entries correspond to the levels of confidence commonly used in statistical applications. The reader should compare Table 7.3 to similar ones in [14, 22].

7.3 Painlevé Representations: A Summary

The Gaussian β-ensembles are probability spaces on N-tuples of random variables $\{l_1,\ldots,l_N\}$, with joint density functions P_β given by[8]

$$P_\beta(l_1,\ldots,l_N) = P_\beta^{(N)}(\boldsymbol{l}) = C_\beta^{(N)} \exp\left[-\sum_{j=1}^{N} l_j^2\right] \prod_{j<k} |l_j - l_k|^\beta . \tag{7.17}$$

The $C_\beta^{(N)}$ are normalization constants, given by

$$C_\beta^{(N)} = \pi^{-N/2} 2^{-N-\beta N(N-1)/4} \prod_{j=1}^{N} \frac{\Gamma(1+\gamma)\Gamma(1+\beta/2)}{\Gamma(1+\beta/2j)} . \tag{7.18}$$

By setting $\beta = 1, 2, 4$ we recover the (finite N) *Gaussian Orthogonal Ensemble* (GOE_N), *Gaussian Unitary Ensemble* (GUE_N), and *Gaussian Symplectic Ensemble* (GSE_N), respectively. For the remainder of the chapter we restrict to these three cases, and refer the reader to [12] for recent results on the general β case. Originally the l_j are eigenvalues of randomly chosen matrices from corresponding matrix ensembles, so we will henceforth refer to them as eigenvalues. With the eigenvalues ordered so that $l_j \geq l_{j+1}$, define

$$\hat{l}_m^{(N)} = \frac{l_m - \sqrt{2N}}{2^{-1/2} N^{-1/6}} , \tag{7.19}$$

to be the rescaled mth eigenvalue measured from edge of spectrum. For the largest eigenvalue in the β-ensembles (proved only in the $\beta = 1, 2, 4$ cases) we have

$$\hat{l}_1^{(N)} \xrightarrow{\mathfrak{D}} \hat{l}_1 , \tag{7.20}$$

whose law is given by the Tracy–Widom distributions.[9][8]

---

[8] In many places in the random matrix theory literature, the parameter β (times $\frac{1}{2}$) appears in front of the summation inside the exponential factor (7.17), in addition to being the power of the Vandermonde determinant. That convention originated in [24], and was justified by the alternative physical and very useful interpretation of (7.17) as a one-dimensional Coulomb gas model. In that language the potential $W = \frac{1}{2}\sum_i l_i^2 - \sum_{i<j} \ln|l_i - l_j|$ and $P_\beta^{(N)}(\boldsymbol{l}) = C\exp(-W/kT) = C\exp(-\beta W)$, so that $\beta = (kT)^{-1}$ plays the role of inverse temperature. However, by an appropriate choice of specialization in Selberg's integral, it is possible to remove the β in the exponential weight, at the cost of redefining the normalization constant $C_\beta^{(N)}$. We choose the latter convention in this work since we will not need the Coulomb gas analogy. Moreover, with computer simulations and statistical applications in mind, this will in our opinion make later choices of standard deviations, renormalizations, and scalings more transparent. It also allows us to dispose of the $\sqrt{2}$ that is often present in F_4.

Theorem 7.3.1 (Tracy, Widom [31, 33]).

$$F_2(s) = \mathbb{P}_2(\hat{l}_1 \le s) = \exp\left[-\int_s^\infty (x-s)q^2(x)\,\mathrm{d}x\right], \tag{7.21}$$

$$F_1(s) = \mathbb{P}_1(\hat{l}_1 \le s) = \left(F_2(s)\right)^{1/2} \exp\left[-\frac{1}{2}\int_s^\infty q(x)\,\mathrm{d}x\right], \tag{7.22}$$

$$F_4(s) = \mathbb{P}_4(\hat{l}_1 \le s) = \left(F_2(s)\right)^{1/2} \cosh\left[-\frac{1}{2}\int_s^\infty q(x)\,\mathrm{d}x\right]. \tag{7.23}$$

The function q is the unique (see [6,19]) solution to the Painlevé II equation

$$q'' = xq + 2q^3, \tag{7.24}$$

such that $q(x) \sim \mathrm{Ai}(x)$ as $x \to \infty$, where $\mathrm{Ai}(x)$ is the solution to the Airy equation which decays like $\frac{1}{2}\pi^{-1/2}x^{-1/4}\exp(-\frac{2}{3}x^{3/2})$ at $+\infty$. The density functions f_β corresponding to the F_β are graphed in Fig. 7.1.[10]

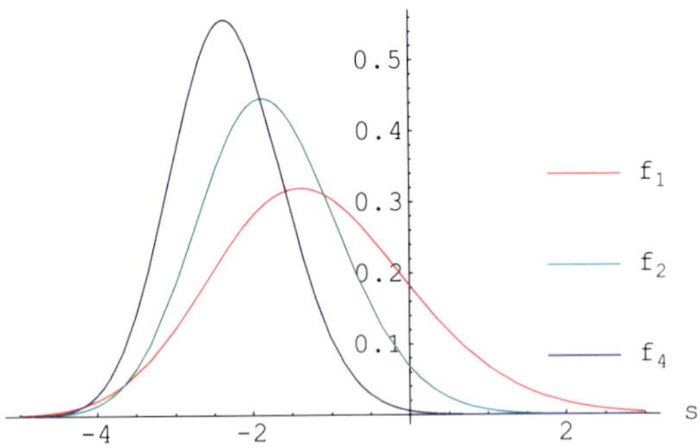

Fig. 7.1. Tracy–Widom density functions.

Let $F_2(s,m)$ denote the distribution for the mth largest eigenvalue in GUE. Tracy and Widom showed [31] that if we define $F_2(s,0) \equiv 0$, then

$$F_2(s,m+1) - F_2(s,m) = \frac{(-1)^m}{m!}\frac{\mathrm{d}^m}{\mathrm{d}\lambda^m} D_2(s,\lambda)\Big|_{\lambda=1}, \quad m \ge 0, \tag{7.25}$$

where (7.9) has the Painlevé representation

$$D_2(s,\lambda) = \exp\left[-\int_s^\infty (x-s)q^2(x,\lambda)\,\mathrm{d}x\right], \tag{7.26}$$

and $q(x,\lambda)$ is the solution to (7.24) such that $q(x,\lambda) \sim \sqrt{\lambda}\,\mathrm{Ai}(x)$ as $x \to \infty$. The same combinatorial argument used to obtain the recurrence (7.25) in the $\beta = 2$ case also works for the $\beta = 1, 4$ cases, leading to

[10] Actually, for $\beta = 4$, the density of $F_4(\sqrt{2}s)$ is graphed to agree with Mehta's original normalization [24] as well as with [33].

$$F_\beta(s, m+1) - F_\beta(s, m) = \frac{(-1)^m}{m!} \frac{\mathrm{d}^m}{\mathrm{d}\lambda^m} D_\beta^{1/2}(s, \lambda)\big|_{\lambda=1} , \qquad m \geq 0 , \ \beta = 1, 4 , \quad (7.27)$$

where $F_\beta(s, 0) \equiv 0$. Given the similarity in the arguments up to this point and comparing (7.26) to (7.21), it is natural to conjecture that $D_\beta(s, \lambda)$, $\beta = 1, 4$, can be obtained simply by replacing $q(x)$ by $q(x, \lambda)$ in (7.22) and (7.23).

That this is not the case for $\beta = 1$ was shown by Dieng [10, 11]. A hint that $\beta = 1$ is different comes from the following interlacing theorem.

Theorem 7.3.2 (Baik, Rains [3]). *In the appropriate scaling limit, the distribution of the largest eigenvalue in* GSE *corresponds to that of the second largest in* GOE*. More generally, the joint distribution of every second eigenvalue in the* GOE *coincides with the joint distribution of all the eigenvalues in the* GSE*, with an appropriate number of eigenvalues.*

This interlacing property between GOE and GSE had long been in the literature, and had in fact been noticed by Mehta and Dyson [25]. In this context, Forrester and Rains [17] classified all weight functions for which alternate eigenvalues taken from an orthogonal ensemble form a corresponding symplectic ensemble, and similarly those for which alternate eigenvalues taken from a union of two orthogonal ensembles form an unitary ensemble. The following theorem gives explicit formulas for $D_1(s, \lambda)$ and $D_4(s, \lambda)$; and hence, from (7.27), a recursive procedure to determine $F_1(\cdot, m)$ and $F_4(\cdot, m)$ for $m \geq 2$.

Theorem 7.3.3 (Dieng [10,11]). *In the edge scaling limit, the distributions for the mth largest eigenvalues in the* GOE *and* GSE *satisfy the recurrence* (7.27) *with*[11]8

$$D_1(s, \lambda) = D_2(s, \tilde{\lambda}) \frac{\lambda - 1 - \cosh \mu(s, \tilde{\lambda}) + \sqrt{\tilde{\lambda}} \sinh \mu(s, \tilde{\lambda})}{\lambda - 2} , \qquad (7.28)$$

$$D_4(s, \lambda) = D_2(s, \lambda) \cosh^2 \left(\frac{\mu(s, \lambda)}{2} \right) , \qquad (7.29)$$

where

$$\mu(s, \lambda) := \int_s^\infty q(x, \lambda) \, \mathrm{d}x, \quad \tilde{\lambda} := 2\lambda - \lambda^2 , \qquad (7.30)$$

and $q(x, \lambda)$ is the solution to (7.24) *such that $q(x, \lambda) \sim \sqrt{\lambda}\,\mathrm{Ai}(x)$ as $x \to \infty$.*

Note the appearance of $\tilde{\lambda}$ in the arguments on the right-hand side of (7.28). In Fig. 7.2 we compare the densities $f_1(s, m)$, $m = 1, \ldots, 4$, with finite N GOE simulations. This last theorem also provides a new proof of the Baik–Rains interlacing theorem.

Corollary 7.3.1 (Dieng [10, 11]).

$$F_4(s, m) = F_1(s, 2m) , \quad m \geq 1 . \qquad (7.31)$$

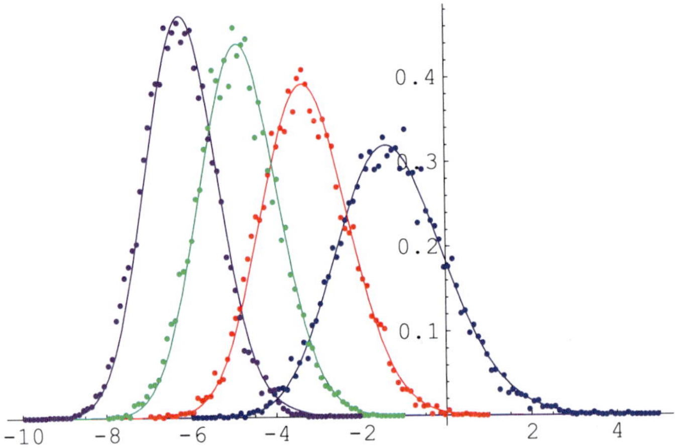

Fig. 7.2. 10^4 realizations of $10^3 \times 10^3$ GOE matrices; the solid curves are, from right to left, the theoretical limiting densities for the first through fourth largest eigenvalue.

The proofs of these theorems occupy the bulk of the remaining part of the chapter. In the last section, we present an efficient numerical scheme to compute $F_\beta(s,m)$ and the associated density functions $f_\beta(s,m)$. We implemented this scheme using MATLAB®,[12] and compared the results to simulated Wishart distributions.

7.4 Preliminaries

7.4.1 Determinant Matters

We gather in this short section more or less classical results for further reference.

Theorem 7.4.1.

$$\prod_{0\le j<k\le N} (x_j - x_k)^4 = \det(x_k^j j x_k^{j-1})_{\substack{j=0,\ldots,2N-1\\ k=1,\ldots,N}} \,.$$

Theorem 7.4.2. *If A, B are Hilbert–Schmidt operators on a general*[13] *Hilbert space $\mathcal{H}$, then*

$$\det(I + AB) = \det(I + BA)\,.$$

[12] MATLAB® is a registered trademark of The MathWorks, Inc., 3 Apple Hill Drive, Natick, MA 01760-2098; Phone: 508-647-7000; Fax: 508-647-7001. Copies of the code are available by contacting the first author.

[13] See [18] for proof.

Theorem 7.4.3 (de Bruijn [7]).

$$\int\cdots\int \det\bigl(\varphi_j(x_k)\bigr)_{1\le j,k\le N} \det\bigl(\psi_j(x_k)\bigr)_{1\le j,k\le N}\, \mathrm{d}\mu(x_1)\cdots \mathrm{d}\mu(x_N) = N!\det\Bigl(\int \varphi_j(x)\psi_k(x)\,\mathrm{d}\mu(x)\Bigr)_{1\le j,k\le N}, \tag{7.32}$$

$$\int\cdots\int_{x_1\le\cdots\le x_N} \det\bigl(\varphi_j(x_k)\bigr)_{j,k=1}^{N} \mathrm{d}\mu(x_1)\cdots \mathrm{d}\mu(x_N) = \mathrm{Pf}\Bigl(\iint \mathrm{sgn}(x-y)\varphi_j(x)\varphi_k(x)\,\mathrm{d}\mu(x)\,\mathrm{d}\mu(y)\Bigr)_{j,k=1}^{N}, \tag{7.33}$$

$$\int\cdots\int \det\bigl(\varphi_j(x_k)\psi_j(x_k)\bigr)_{\substack{1\le j\le 2N\\ 1\le k\le N}} \mathrm{d}\mu(x_1)\cdots \mathrm{d}\mu(x_N) = (2N)!\,\mathrm{Pf}\Bigl(\int \varphi_j(x)\psi_k(x)-\varphi_k(x)\psi_j(x)\,\mathrm{d}\mu(x)\Bigr)_{j,k=1}^{2N}, \tag{7.34}$$

where Pf *denotes the Pfaffian. The last two integral identities were discovered by de Bruijn* [7] *in an attempt to generalize the first one. The first and last are valid in general measure spaces. In the second identity, the space needs to be ordered. In the last identity, the left-hand side determinant is a* $2N\times 2N$ *determinant whose columns are alternating columns of the* φ_j *and* ψ_j *(i.e., the first four columns are* $\{\varphi_j(x_1)\}$, $\{\psi_j(x_1)\}$, $\{\varphi_j(x_2)\}$, $\{\psi_j(x_2)\}$, *respectively for* $j=1,\dots,2N$*), hence the notation, and asymmetry in indexing.*

A large portion of the foundational theory of random matrices, in the case of invariant measures, can be developed from Thms. 7.4.2 and 7.4.3 as was demonstrated in [34, 37].

7.4.2 Recursion Formula for the Eigenvalue Distributions

With the joint density function defined as in (7.17), let J denote the interval (t,∞), and $\chi=\chi_J(x)$ its characteristic function.[14] We denote by $\widetilde{\chi}=1-\chi$ the characteristic function of the complement of J, and define $\widetilde{\chi}_\lambda=1-\lambda\chi$. Furthermore, let $E_{\beta,N}(t,m)$ equal the probability that exactly the m largest eigenvalues of a matrix chosen at random from a (finite N) β-ensemble lie in J. We also define

$$G_{\beta,N}(t,\lambda)=\int\cdots\int_{x_i\in\mathbb{R}} \widetilde{\chi}_\lambda(x_1)\cdots\widetilde{\chi}_\lambda(x_N)P_\beta(x_1,\dots,x_N)\,\mathrm{d}x_1\cdots \mathrm{d}x_N\,. \tag{7.35}$$

For $\lambda=1$ this is just $E_{\beta,N}(t,0)$, the probability that no eigenvalues lie in (t,∞), or equivalently the probability that the largest eigenvalue is less than

[14] Much of what is said here is still valid if J is taken to be a finite union of open intervals in $\mathbb{R}$ (see [32]). However, since we will only be interested in edge eigenvalues we restrict ourselves to (t,∞) from here on.

t. In fact we will see in the following propositions that $G_{\beta,N}(t,\lambda)$ is in some sense a generating function for $E_{\beta,N}(t,m)$.

Proposition 7.4.1.

$$G_{\beta,N}(t,\lambda) = \sum_{k=0}^{N}(-\lambda)^k \binom{N}{k} \int \cdots \int_{x_i \in J} P_\beta(x_1,\ldots,x_N)\,\mathrm{d}x_1 \cdots \mathrm{d}x_N\,. \tag{7.36}$$

Proof. Using the definition of the $\widetilde{\chi}_\lambda(x_1)$ and multiplying out the integrand of (7.35) gives

$$\begin{aligned} &G_{\beta,N}(t,\lambda) \\ &\quad = \sum_{k=0}^{N}(-\lambda)^k \int \cdots \int_{x_i \in \mathbb{R}} e_k(\chi(x_1),\ldots,\chi(x_N)) P_\beta(x_1,\ldots,x_N)\,\mathrm{d}x_1 \cdots \mathrm{d}x_N\,, \end{aligned}$$

where, in the notation of [30], $e_k = m_{1^k}$ is the kth elementary symmetric function. Indeed each term in the summation arises from picking k of the $\lambda\chi$-terms, each of which comes with a negative sign, and $N-k$ of the 1's. This explains the coefficient $(-\lambda)^k$. Moreover, it follows that e_k contains $\binom{N}{k}$ terms. Now the integrand is symmetric under permutations of the x_i. Also if $x_i \notin J$, all corresponding terms in the symmetric function are 0, and they are 1 otherwise. Therefore we can restrict the integration to $x_i \in J$, remove the characteristic functions (hence the symmetric function), and introduce the binomial coefficient to account for the identical terms up to permutation. □

Proposition 7.4.2.

$$E_{\beta,N}(t,m) = \frac{(-1)^m}{m!}\frac{\mathrm{d}^m}{\mathrm{d}\lambda^m} G_{\beta,N}(t,\lambda)\bigg|_{\lambda=1}, \quad m \ge 0\,. \tag{7.37}$$

Proof. This is proved by induction. As noted above, $E_{\beta,N}(t,0) = G_{\beta,N}(t,1)$ so it holds for the degenerate case $m=0$. When $m=1$ we have

$$\begin{aligned} &-\frac{\mathrm{d}}{\mathrm{d}\lambda} G_{\beta,N}(t,\lambda)\bigg|_{\lambda=1} \\ &\quad = -\frac{\mathrm{d}}{\mathrm{d}\lambda} \int \cdots \int \widetilde{\chi}_\lambda(x_1)\cdots\widetilde{\chi}_\lambda(x_n) P_\beta^{(N)}(\boldsymbol{x})\,\mathrm{d}x_1\cdots dx_n\bigg|_{\lambda=1} \\ &\quad = -\sum_{j=1}^{N} - \int \cdots \int \widetilde{\chi}(x_1)\cdots\widetilde{\chi}(x_{j-1})\chi(x_j) \\ &\qquad\qquad \times \widetilde{\chi}(x_{j+1})\cdots\widetilde{\chi}(x_N) P_\beta^{(N)}(\boldsymbol{x})\,\mathrm{d}x_1\cdots\mathrm{d}x_N\,. \end{aligned}$$

The integrand is symmetric under permutations so we can make all terms look the same. There are $N = \binom{N}{1}$ of them so we get

$$
\begin{aligned}
&-\frac{\mathrm{d}}{\mathrm{d}\lambda}G_{\beta,N}(t,\lambda)\bigg|_{\lambda=1}\\
&\quad=\binom{N}{1}\int\cdots\int\chi(x_1)\widetilde{\chi}(x_2)\cdots\widetilde{\chi}(x_N)P_\beta^{(N)}(\boldsymbol{x})\,\mathrm{d}x_1\cdots\mathrm{d}x_N\bigg|_{\lambda=1}\\
&\quad=\binom{N}{1}\int\cdots\int\chi(x_1)\chi(x_2)\cdots\chi(x_N)P_\beta^{(N)}(\boldsymbol{x})\,\mathrm{d}x_1\cdots\mathrm{d}x_N=E_{\beta,N}(t,1)\,.
\end{aligned}
$$

When $m=2$ then

$$
\begin{aligned}
&\frac{1}{2}\left(-\frac{\mathrm{d}}{\mathrm{d}\lambda}\right)^2G_{\beta,N}(t,\lambda)\bigg|_{\lambda=1}\\
&\quad=\frac{N}{2}\sum_{j=2}^{N}\int\cdots\int\chi(x_1)\widetilde{\chi}(x_2)\cdots\widetilde{\chi}(x_{j-1})\chi(x_j)\\
&\qquad\qquad\times\widetilde{\chi}(x_{j+1})\cdots\widetilde{\chi}(x_N)P_\beta^{(N)}(\boldsymbol{x})\,\mathrm{d}x_1\cdots\mathrm{d}x_N\bigg|_{\lambda=1}\\
&\quad=\frac{N(N-1)}{2}\int\cdots\int\chi(x_1)\chi(x_2)\widetilde{\chi}(x_3)\cdots\widetilde{\chi}(x_N)P_\beta^{(N)}(\boldsymbol{x})\,\mathrm{d}x_1\cdots\mathrm{d}x_N\bigg|_{\lambda=1}\\
&\quad=\binom{N}{2}\int\cdots\int\chi(x_1)\chi(x_2)\widetilde{\chi}(x_3)\cdots\widetilde{\chi}(x_N)P_\beta^{(N)}(\boldsymbol{x})\,\mathrm{d}x_1\cdots\mathrm{d}x_N\bigg|_{\lambda=1}\\
&\quad=\binom{N}{2}\int\cdots\int\chi(x_1)\chi(x_2)\chi(x_3)\cdots\chi(x_N)P_\beta^{(N)}(\boldsymbol{x})\,\mathrm{d}x_1\cdots\mathrm{d}x_N\\
&\quad=E_{\beta,N}(t,2)\,,
\end{aligned}
$$

where we used the previous case to get the first equality, and again the invariance of the integrand under symmetry to get the second equality. By induction then,

$$
\begin{aligned}
&\frac{1}{m!}\left(-\frac{\mathrm{d}}{\mathrm{d}\lambda}\right)^m G_{\beta,N}(t,\lambda)\bigg|_{\lambda=1}\\
&\quad=\frac{N(N-1)\cdots(N-m+2)}{m!}\sum_{j=m}^{N}\int\cdots\int\chi(x_1)\widetilde{\chi}(x_2)\cdots\widetilde{\chi}(x_{j-1})\chi(x_j)\\
&\qquad\qquad\times\widetilde{\chi}(x_{j+1})\cdots\widetilde{\chi}(x_N)P_\beta^{(N)}(\boldsymbol{x})\,\mathrm{d}x_1\cdots\mathrm{d}x_N\\
&\quad=\frac{N(N-1)\cdots(N-m+1)}{m!}\int\cdots\int\chi(x_1)\cdots\chi(x_m)\widetilde{\chi}(x_{m+1})\cdots\widetilde{\chi}(x_N)\\
&\qquad\qquad\times P_\beta^{(N)}(\boldsymbol{x})\,\mathrm{d}x_1\cdots\mathrm{d}x_N\\
&\quad=\binom{N}{m}\int\cdots\int\chi(x_1)\cdots\chi(x_m)\widetilde{\chi}(x_{m+1})\cdots\widetilde{\chi}(x_N)P_\beta^{(N)}(\boldsymbol{x})\,\mathrm{d}x_1\cdots\mathrm{d}x_N\\
&\quad=E_{\beta,N}(t,m)\,.
\end{aligned}
$$

□

If we define $F_{\beta,N}(t,m)$ to be the distribution of the mth largest eigenvalue in the (finite N) β-ensemble, then the following probabilistic result is immediate from our definition of $E_{\beta,N}(t,m)$.

Corollary 7.4.1.

$$F_{\beta,N}(t,m+1) - F_{\beta,N}(t,m) = E_{\beta,N}(t,m)\,. \tag{7.38}$$

7.5 The Distribution of the mth Largest Eigenvalue in the GUE

7.5.1 The Distribution Function as a Fredholm Determinant

We follow [34] for the derivations that follow. The GUE case corresponds to the specialization $\beta = 2$ in (7.17) so that

$$G_{2,N}(t,\lambda) = C_2^{(N)} \int_{x_i\in\mathbb{R}} \cdots \int \prod_{j<k} (x_j - x_k)^2 \prod_j^N w(x_j) \prod_j^N (1 + f(x_j))\, \mathrm{d}x_1 \cdots \mathrm{d}x_N \tag{7.39}$$

where $w(x) = \exp(-x^2)$, $f(x) = -\lambda\chi_J(x)$, and $C_2^{(N)}$ depends only on N. In the steps that follow, additional constants depending solely on N (such as $N!$) which appear will be lumped into $C_2^{(N)}$. A probability argument will show that the resulting constant at the end of all calculations simply equals 1. Expressing the Vandermonde as a determinant

$$\prod_{1\le j<k\le N} (x_j - x_k) = \det(x_k^j)_{\substack{j=0,\dots,N\\ k=1,\dots,N}} \tag{7.40}$$

and using (7.32) with $\varphi_j(x) = \psi_j(x) = x^j$ and $d\mu(x) = w(x)\big(1 + f(x)\big)$ yields

$$G_{2,N}(t,\lambda) = C_2^{(N)} \det\left(\int_{\mathbb{R}} x^{j+k} w(x)\big(1+f(x)\big)\,\mathrm{d}x\right)_{j,k=0}^{N-1}. \tag{7.41}$$

Let $\{\varphi_j(x)\}$ be the sequence obtained by orthonormalizing the sequence $\{x^j w^{1/2}(x)\}$. It follows that

$$G_{2,N}(t,\lambda) = C_2^{(N)} \det\left(\int_{\mathbb{R}} \varphi_j(x)\varphi_k(x)\big(1+f(x)\big)\,\mathrm{d}x\right)_{j,k=0}^{,N-1} \tag{7.42}$$

$$= C_2^{(N)} \det\left(\delta_{j,k} + \int_{\mathbb{R}} \varphi_j(x)\varphi_k(x) f(x)\,\mathrm{d}x\right)_{j,k=0}^{N-1}. \tag{7.43}$$

The last expression is of the form $\det(I+AB)$ for $A\colon L^2(\mathbb{R}) \to \mathbb{C}^N$ with kernel $A(j,x) = \varphi_j(x)f(x)$ whereas $B\colon \mathbb{C}^N \to L^2(\mathbb{R})$ with kernel $B(x,j) = \varphi_j(x)$. Note that $AB\colon \mathbb{C}^N \to \mathbb{C}^N$ has kernel

$$AB(j,k) = \int_{\mathbb{R}} \varphi_j(x)\varphi_k(x) f(x)\,\mathrm{d}x \tag{7.44}$$

whereas $BA\colon L^2(\mathbb{R}) \to L^2(\mathbb{R})$ has kernel

$$BA(x,y) = \sum_{k=0}^{N-1} \varphi_k(x)\varphi_k(y) := K_{2,N}(x,y)\,. \tag{7.45}$$

From Thm. 7.4.2 it follows that

$$G_{2,N}(t,\lambda) = C_2^{(N)} \det(I + K_{2,N} f)\,, \tag{7.46}$$

where $K_{2,N}$ has kernel $K_{2,N}(x,y)$ and $K_{2,N}f$ acts on a function by first multiplying it by f and acting on the product with $K_{2,N}$. From (7.39) we see that setting $f = 0$ in the last identity yields $C_2^{(N)} = 1$. Thus the above simplifies to

$$G_{2,N}(t,\lambda) = \det(I + K_{2,N} f)\,. \tag{7.47}$$

7.5.2 Edge Scaling and Differential Equations

We specialize $w(x) = \exp(-x^2)$, $f(x) = -\lambda\chi_J(x)$, so that the $\{\varphi_j(x)\}$ are in fact the Hermite polynomials times the square root of the weight. Using the Plancherel–Rotach asymptotics of Hermite polynomials, it follows that in the *edge scaling limit,*

$$\lim_{N\to\infty} \frac{1}{2^{1/2}N^{1/6}} K_{N,2}\left(\sqrt{2N} + \frac{x}{2^{1/2}N^{1/6}}, \sqrt{2N} + \frac{y}{2^{1/2}N^{1/6}}\right) \times \chi_J\left(\sqrt{2N} + \frac{y}{2^{1/2}N^{1/6}}\right) \tag{7.48}$$

is $K_{\mathrm{Ai}}(x,y)$ as defined in (7.8). As operators, the convergence is in trace class norm to K_2. (A proof of this last fact can be found in [36].) For notational convenience, we denote the corresponding operator K_2 by K in the rest of this subsection. It is convenient to view K as the integral operator on $\mathbb{R}$ with kernel

$$K(x,y) = \frac{\varphi(x)\psi(y) - \psi(x)\varphi(y)}{x-y}\chi_J(y)\,, \tag{7.49}$$

where $\varphi(x) = \sqrt{\lambda}\,\mathrm{Ai}(x)$, $\psi(x) = \sqrt{\lambda}\,\mathrm{Ai}'(x)$ and J is (s,∞) with

$$t = \sqrt{2N} + \frac{s}{\sqrt{2}N^{1/6}}\,. \tag{7.50}$$

Note that although $K(x,y)$, φ and ψ are functions of λ as well, this dependence will not affect our calculations in what follows. Thus we omit it to avoid cumbersome notation. The Airy equation implies that φ and ψ satisfy the relations

$$\frac{\mathrm{d}}{\mathrm{d}x}\varphi = \psi\,, \quad \frac{\mathrm{d}}{\mathrm{d}x}\psi = x\varphi\,. \tag{7.51}$$

We define $D_{2,N}(s,\lambda)$ to be the Fredholm determinant $\det(I - K_{N,2})$. Thus in the edge scaling limit

$$\lim_{N\to\infty} D_{2,N}(s,\lambda) = D_2(s,\lambda) .$$

We define the operator

$$R = (I-K)^{-1}K , \tag{7.52}$$

whose kernel we denote $R(x,y)$. Incidentally, we shall use the notation $\doteq$ in reference to an operator to mean "has kernel." For example $R \doteq R(x,y)$. We also let M stand for the operator whose action is multiplication by x. It is well known that

$$\frac{\mathrm{d}}{\mathrm{d}s} \log\det(I-K) = -R(s,s) . \tag{7.53}$$

For functions f and g, we write $f \otimes g$ to denote the operator specified by

$$f \otimes g \doteq f(x)g(y) , \tag{7.54}$$

and define

$$Q(x,s) = Q(x) = \big((I-K)^{-1}\varphi\big)(x) , \tag{7.55}$$

$$P(x,s) = P(x) = \big((I-K)^{-1}\psi\big)(x) . \tag{7.56}$$

Then straightforward computation yields the following facts

$$\begin{aligned} [M,K] &= \varphi\otimes\psi - \psi\otimes\varphi, \\ [M,(I-K)^{-1}] &= (I-K)^{-1}[M,K](I-K)^{-1} = Q\otimes P - P\otimes Q . \end{aligned} \tag{7.57}$$

On the other hand if $(I-K)^{-1} \doteq \rho(x,y)$, then

$$\rho(x,y) = \delta(x-y) + R(x,y) , \tag{7.58}$$

and it follows that

$$[M,(I-K)^{-1}] \doteq (x-y)\rho(x,y) = (x-y)R(x,y) . \tag{7.59}$$

Equating the two representation for the kernel of $[M,(I-K)^{-1}]$ yields

$$R(x,y) = \frac{Q(x)P(y) - P(x)Q(y)}{x-y} . \tag{7.60}$$

Taking the limit $y \to x$ and defining $q(s) = Q(s,s)$, $p(s) = P(s,s)$, we obtain

$$R(s,s) = Q'(s,s)p(s) - P'(s,s)q(s) . \tag{7.61}$$

Let us now derive expressions for $Q'(x)$ and $P'(x)$. If we let the operator D stand for differentiation with respect to x,

$$\begin{aligned} Q'(x,s) &= D(I-K)^{-1}\varphi \\ &= (I-K)^{-1}D\varphi + [D,(I-K)^{-1}]\varphi \\ &= (I-K)^{-1}\psi + [D,(I-K)^{-1}]\varphi \\ &= P(x) + [D,(I-K)^{-1}]\varphi\,. \end{aligned} \tag{7.62}$$

We need the commutator

$$[D,(I-K)^{-1}] = (I-K)^{-1}[D,K](I-K)^{-1}\,. \tag{7.63}$$

Integration by parts shows

$$[D,K] \doteq \left(\frac{\partial K}{\partial x} + \frac{\partial K}{\partial y}\right) + K(x,s)\delta(y-s)\,. \tag{7.64}$$

The δ function comes from differentiating the characteristic function χ. Moreover,

$$\left(\frac{\partial K}{\partial x} + \frac{\partial K}{\partial y}\right) = \varphi(x)\varphi(y)\,. \tag{7.65}$$

Thus

$$[D,(I-K)^{-1}] \doteq -Q(x)Q(y) + R(x,s)\rho(s,y)\,. \tag{7.66}$$

(Recall $(I-K)^{-1} \doteq \rho(x,y)$.) We now use this in (7.62) to obtain

$$\begin{aligned} Q'(x,s) &= P(x) - Q(x)(Q,\varphi) + R(x,s)q(s) \\ &= P(x) - Q(x)u(s) + R(x,s)q(s)\,, \end{aligned}$$

where the inner product (Q,φ) is denoted by $u(s)$. Evaluating at $x=s$ gives

$$Q'(s,s) = p(s) - q(s)u(s) + R(s,s)q(s)\,. \tag{7.67}$$

We now apply the same procedure to compute P'.

$$\begin{aligned} P'(x,s) &= D(I-K)^{-1}\psi \\ &= (I-K)^{-1}D\psi + [D,(I-K)^{-1}]\psi \\ &= M(I-K)^{-1}\varphi + [(I-K)^{-1},M]\varphi + [D,(I-K)^{-1}]\psi \\ &= xQ(x) + (P\otimes Q - Q\otimes P)\varphi + (-Q\otimes Q)\psi + R(x,s)p(s) \\ &= xQ(x) + P(x)(Q,\varphi) - Q(x)(P,\varphi) - Q(x)(Q,\psi) + R(x,s)p(s) \\ &= xQ(x) - 2Q(x)v(s) + P(x)u(s) + R(x,s)p(s)\,. \end{aligned}$$

Here $v = (P,\varphi) = (\psi,Q)$. Setting $x=s$ we obtain

$$P'(s,s) = sq(s) + 2q(s)v(s) + p(s)u(s) + R(s,s)p(s)\,. \tag{7.68}$$

Using this and the expression for $Q'(s,s)$ in (7.61) gives

$$R(s,s) = p^2 - sq^2 + 2q^2v - 2pqu\,. \tag{7.69}$$

Using the chain rule, we have

$$\frac{\mathrm{d}q}{\mathrm{d}s} = \left(\frac{\partial}{\partial x} + \frac{\partial}{\partial s}\right) Q(x,s)\bigg|_{x=s} . \tag{7.70}$$

The first term is known. The partial with respect to s is

$$\frac{\partial Q(x,s)}{\partial s} = (I-K)^{-1}\frac{\partial K}{\partial s}(I-K)^{-1}\varphi = -R(x,s)q(s) ,$$

where we used the fact that

$$\frac{\partial K}{\partial s} \doteq -K(x,s)\delta(y-s) . \tag{7.71}$$

Adding the two partial derivatives and evaluating at $x = s$ gives

$$\frac{\mathrm{d}q}{\mathrm{d}s} = p - qu . \tag{7.72}$$

A similar calculation gives

$$\frac{\mathrm{d}p}{\mathrm{d}s} = sq - 2qv + pu . \tag{7.73}$$

We derive first-order differential equations for u and v by differentiating the inner products. Recall that

$$u(s) = \int_s^\infty \varphi(x)Q(x,s)\,\mathrm{d}x .$$

Thus

$$\begin{aligned}\frac{\mathrm{d}u}{\mathrm{d}s} &= -\varphi(s)q(s) + \int_s^\infty \varphi(x)\frac{\partial Q(x,s)}{\partial s}\,\mathrm{d}x \\ &= -\left(\varphi(s) + \int_s^\infty R(s,x)\varphi(x)\,\mathrm{d}x\right)q(s) = -(I-K)^{-1}\varphi(s)q(s) = -q^2 .\end{aligned}$$

Similarly,

$$\frac{\mathrm{d}v}{\mathrm{d}s} = -pq . \tag{7.74}$$

From the first-order differential equations for q, u and v it follows immediately that the derivative of $u^2 - 2v - q^2$ is zero. Examining the behavior near $s = \infty$ to check that the constant of integration is zero then gives

$$u^2 - 2v = q^2. \tag{7.75}$$

We now differentiate (7.72) with respect to s, use the first order differential equations for p and u, and then the first integral to deduce that q satisfies the Painlevé II equation (7.24). Checking the asymptotics of the Fredholm determinant $\det(I-K)$ for large s shows we want the solution with boundary condition

$$q(s,\lambda) \sim \sqrt{\lambda}\,\mathrm{Ai}(s) \quad \text{as } s \to \infty. \tag{7.76}$$

That a solution q exists and is unique follows from the representation of the Fredholm determinant in terms of it. Independent proofs of this, as well as the asymptotics as $s \to -\infty$ were given by [6, 9, 19]. Since $[D,(I-K)^{-1}] \doteq (\partial/\partial x + \partial/\partial y)R(x,y)$, (7.66) says

$$\left(\frac{\partial}{\partial x} + \frac{\partial}{\partial y}\right) R(x,y) = -Q(x)Q(y) + R(x,s)\rho(s,y)\,. \tag{7.77}$$

In computing $\partial Q(x,s)/\partial s$ we showed that

$$\frac{\partial}{\partial s}(I-K)^{-1} \doteq \frac{\partial}{\partial s} R(x,y) = -R(x,s)\rho(s,y)\,. \tag{7.78}$$

Adding these two expressions,

$$\left(\frac{\partial}{\partial x} + \frac{\partial}{\partial y} + \frac{\partial}{\partial s}\right) R(x,y) = -Q(x)Q(y)\,, \tag{7.79}$$

and then evaluating at $x = y = s$ gives

$$\frac{\mathrm{d}}{\mathrm{d}s} R(s,s) = -q^2\,. \tag{7.80}$$

Integration (and recalling (7.53)) gives,

$$\frac{\mathrm{d}}{\mathrm{d}s} \log\det(I-K) = -\int_s^\infty q^2(x,\lambda)\,\mathrm{d}x\,; \tag{7.81}$$

and hence,

$$\begin{aligned}\log\det(I-K) &= -\int_s^\infty \left(\int_y^\infty q^2(x,\lambda)\,\mathrm{d}x\right)\mathrm{d}y \\ &= -\int_s^\infty (x-s)q^2(x,\lambda)\,\mathrm{d}x\,. \end{aligned} \tag{7.82}$$

To summarize, we have shown that $D_2(s,\lambda)$ has the Painlevé representation (7.26) where q satisfies the Painlevé II equation (7.24) subject to the boundary condition (7.76).

7.6 The Distribution of the mth Largest Eigenvalue in the GSE

7.6.1 The Distribution Function as a Fredholm Determinant

The GSE corresponds case corresponds to the specialization $\beta = 4$ in (7.17) so that

$$G_{4,N}(t,\lambda) = C_4^{(N)} \int \cdots \int_{x_i\in\mathbb{R}} \prod_{j<k}(x_j - x_k)^4 \prod_j^N w(x_j) \prod_j^N (1+f(x_j))\,\mathrm{d}x_1\cdots\mathrm{d}x_N \quad (7.83)$$

where $w(x) = \exp(-x^2)$, $f(x) = -\lambda\chi_J(x)$, and $C_4^{(N)}$ depends only on N. As in the GUE case, we will absorb into $C_4^{(N)}$ any constants depending only on N that appear in the derivation. A simple argument at the end will show that the final constant is 1. These calculations follow [34]. By Thm. 7.4.1, $G_{4,N}(t,\lambda)$ is given by the integral

$$C_4^{(N)} \int \cdots \int_{x_i\in\mathbb{R}} \det(x_k^j, jx_k^{j-1})_{\substack{j=0,\ldots,2N-1\\ k=1,\ldots,N}} \prod_{i=1}^N w(x_i) \prod_{i=1}^N (1+f(x_i))\,\mathrm{d}x_1\cdots\mathrm{d}x_N$$

which, if we define $\varphi_j(x) = x^{j-1}w(x)\big(1+f(x)\big)$ and $\psi_j(x) = (j-1)x^{j-2}$ and use the linearity of the determinant, becomes

$$G_{4,N}(t,\lambda) = C_4^{(N)} \int \cdots \int_{x_i\in\mathbb{R}} \det\big(\varphi_j(x_k), \psi(x_k)\big)_{\substack{1\le j\le 2N\\ 1\le k\le N}}\,\mathrm{d}x_1\cdots\mathrm{d}x_N\,.$$

Now using (7.34), we obtain

$$\begin{aligned} G_{4,N}(t,\lambda) &= C_4^{(N)}\,\mathrm{Pf}\Big(\int \varphi_j(x)\psi_k x - \varphi_k(x)\psi_j(x)\,\mathrm{d}x\Big)_{j,k=1}^{2N} \\ &= C_4^{(N)}\,\mathrm{Pf}\Big(\int (k-j)x^{j+k-3}w(x)\big(1+f(x)\big)\,\mathrm{d}x\Big)_{j,k=1}^{2N} \\ &= C_4^{(N)}\,\mathrm{Pf}\Big(\int (k-j)x^{j+k-1}w(x)\big(1+f(x)\big)\,\mathrm{d}x\Big)_{j,k=0}^{2N-1}, \end{aligned}$$

where we let $k \to k+1$ and $j \to j+1$ in the last line. Remembering that the square of a Pfaffian is a determinant, we obtain

$$G_{4,N}^2(t,\lambda) = C_4^{(N)}\det\Big(\int (k-j)x^{j+k-1}w(x)\big(1+f(x)\big)\,\mathrm{d}x\Big)_{j,k=0}^{2N-1}.$$

Row operations on the matrix do not change the determinant, so we can replace $\{x^j\}$ by an arbitrary sequence $\{p_j(x)\}$ of polynomials of degree j obtained by adding rows to each other. Note that the general (j,k) element in the matrix can be written as

$$\left[\left(\frac{\mathrm{d}}{\mathrm{d}x}x^k\right)x^j - \left(\frac{\mathrm{d}}{\mathrm{d}x}x^j\right)x^k\right]w(x)\big(1+f(x)\big)\,.$$

Thus when we add rows to each other the polynomials we obtain will have the same general form (the derivatives factor). Therefore we can assume without loss of generality that $G_{4,N}^2(t,\lambda)$ equals

$$C_4^{(N)} \det\left(\int [p_j(x)p_k'(x) - p_j'(x)p_k(x)]w(x)\big(1+f(x)\big)\,\mathrm{d}x\right)_{j,k=0}^{2N-1},$$

where the sequence $\{p_j(x)\}$ of polynomials of degree j is arbitrary. Let $\psi_j = p_j w^{1/2}$ so that $p_j = \psi_j w^{-1/2}$. Substituting this into the above formula and simplifying, we obtain

$$\begin{aligned} G_{4,N}^2(t,\lambda) &= C_4^{(N)} \det\left[\int \Big(\big(\psi_j(x)\psi_k'(x) - \psi_k(x)\psi_j'(x)\big)\big(1+f(x)\big)\Big)\mathrm{d}x\right]_{j,k=0}^{2N-1} \\ &= C_4^{(N)} \det[M+L] = C_4^{(N)} \det[M]\det[I + M^{-1}L]\,, \end{aligned}$$

where M, L are matrices given by

$$\begin{aligned} M &= \left(\int \big(\psi_j(x)\psi_k'(x) - \psi_k(x)\psi_j'(x)\big)\,\mathrm{d}x\right)_{j,k=0}^{2N-1}, \\ L &= \left(\int \big(\psi_j(x)\psi_k'(x) - \psi_k(x)\psi_j'(x)\big) f(x)\,\mathrm{d}x\right)_{j,k=0}^{2N-1}. \end{aligned}$$

Note that $\det[M]$ is a constant which depends only on N so we can absorb it into $C_4^{(N)}$. Also if we denote

$$M^{-1} = \{\mu_{jk}\}_{j,k=0}^{2N-1}, \quad \eta_j = \sum_{k=0}^{2N-1} \mu_{jk}\psi_k(x)\,,$$

it follows that

$$M^{-1}\cdot N = \left\{\int \big(\eta_j(x)\psi_k'(x) - \eta_j'(x)\psi_k(x)\big) f(x)\,\mathrm{d}x\right\}_{j,k=0}^{2N-1}.$$

Let $A\colon L^2(\mathbb{R}) \times L^2(\mathbb{R}) \to \mathbb{C}^{2N}$ be the operator defined by the $2N \times 2$ matrix

$$A(x) = \begin{pmatrix} \eta_0(x) & -\eta_0'(x) \\ \eta_1(x) & -\eta_1'(x) \\ \vdots & \vdots \end{pmatrix}.$$

Thus if

$$g = \begin{pmatrix} g_0(x) \\ g_1(x) \end{pmatrix} \in L^2(\mathbb{R}) \times L^2(\mathbb{R})\,,$$

we have

$$Ag = A(x)g = \begin{pmatrix} \int (\eta_0 g_0 - \eta_0' g_1)\,\mathrm{d}x \\ \int (\eta_1 g_0 - \eta_1' g_1)\,\mathrm{d}x \\ \vdots \end{pmatrix} \in \mathbb{C}^{2N}.$$

Similarly we define $B\colon \mathbb{C}^{2n} \to L^2(\mathbb{R}) \times L^2(\mathbb{R})$ given by the $2 \times 2n$ matrix

$$B(x) = f \cdot \begin{pmatrix} \psi_0'(x) & \psi_1'(x) & \cdots \\ \psi_0(x) & \psi_1(x) & \cdots \end{pmatrix} .$$

Explicitly if

$$\alpha = \begin{pmatrix} \alpha_0 \\ \alpha_1 \\ \vdots \end{pmatrix} \in \mathbb{C}^{2N} ,$$

then

$$B\alpha = B(x) \cdot \alpha = \begin{pmatrix} f \sum_{i=0}^{2N-1} \alpha_i \psi_i' \\ f \sum_{i=0}^{2N-1} \alpha_i \psi_i \end{pmatrix} \in L^2 \times L^2 .$$

Observe that $M^{-1} \cdot L = AB\colon \mathbb{C}^{2n} \to \mathbb{C}^{2n}$. Indeed

$$\begin{aligned} AB\alpha &= \begin{pmatrix} \sum_{i=0}^{2N-1} \left[\int (\eta_0 \psi_i' - \eta_0' \psi_i) f \, \mathrm{d}x \right] \alpha_i \\ \sum_{i=0}^{2N-1} \left[\int (\eta_1 \psi_i' - \eta_1' \psi_i) f \, \mathrm{d}x \right] \alpha_i \\ \vdots \end{pmatrix} \\ &= \left\{ \int (\eta_j \psi_k' - \eta_j' \psi_k) f \, \mathrm{d}x \right\} \cdot \begin{pmatrix} \alpha_0 \\ \alpha_1 \\ \vdots \end{pmatrix} = (M^{-1} \cdot L)\alpha . \end{aligned}$$

Therefore, by (7.4.2)

$$G_{4,N}^2(t,\lambda) = C_4^{(N)} \det(I + M^{-1} \cdot L) = C_4^{(N)} \det(I + AB) = C_4^{(N)} \det(I + BA)$$

where $BA\colon L^2(\mathbb{R}) \to L^2(\mathbb{R})$. From our definition of A and B it follows that

$$\begin{aligned} BAg &= \begin{pmatrix} f \sum_{i=0}^{2n-1} \psi_i' \left(\int (\eta_i g_0 - \eta_i' g_1) \, \mathrm{d}x \right) \\ f \sum_{i=0}^{2N-1} \psi_i' \left(\int (\eta_i g_0 - \eta_i' g_1) \, \mathrm{d}x \right) \end{pmatrix} \\ &= f \begin{pmatrix} \int \sum_{i=0}^{2N-1} \psi_i'(x) \eta_i(y) g_0(y) \, \mathrm{d}y - \int \sum_{i=0}^{2N-1} \psi_i'(x) \eta_i'(y) g_1(y) \, \mathrm{d}y \\ \int \sum_{i=0}^{2N-1} \psi_i(x) \eta_i(y) g_0(y) \, \mathrm{d}y - \int \sum_{i=0}^{2N-1} \psi_i(x) \eta_i'(y) g_1(y) \, \mathrm{d}y \end{pmatrix} \\ &= f K_{4,N} g , \end{aligned}$$

where $K_{4,N}$ is the integral operator with matrix kernel

$$K_{4,N}(x,y) = \begin{pmatrix} \sum_{i=0}^{2N-1} \psi_i'(x) \eta_i(y) & -\sum_{i=0}^{2N-1} \psi_i'(x) \eta_i'(y) \\ \sum_{i=0}^{2N-1} \psi_i(x) \eta_i(y) & -\sum_{i=0}^{2N-1} \psi_i(x) \eta_i'(y) \end{pmatrix} .$$

Recall that $\eta_j(x) = \sum_{k=0}^{2N-1} \mu_{jk} \psi_k(x)$ so that

$$K_{4,N}(x,y) = \begin{pmatrix} \sum_{j,k=0}^{2N-1} \psi_j'(x) \mu_{jk} \psi_k(y) & -\sum_{j,k=0}^{2N-1} \psi_j'(x) \mu_{jk} \psi_k'(y) \\ \sum_{j,k=0}^{2N-1} \psi_j(x) \mu_{jk} \psi_k(y) & -\sum_{j,k=0}^{2N-1} \psi_j(x) \mu_{jk} \psi_k'(y) \end{pmatrix} .$$

Define ε to be the following integral operator

$$\varepsilon \doteq \varepsilon(x-y) = \begin{cases} \frac{1}{2} & \text{if } x > y, \\ -\frac{1}{2} & \text{if } x < y. \end{cases} \tag{7.84}$$

As before, let D denote the operator that acts by differentiation with respect to x. The fundamental theorem of calculus implies that $D\varepsilon = \varepsilon D = I$. We also define

$$S_N(x,y) = \sum_{j,k=0}^{2N-1} \psi_j'(x)\mu_{jk}\psi_k(y) \, .$$

Since M is antisymmetric,

$$\begin{aligned} S_N(y,x) &= \sum_{j,k=0}^{2N-1} \psi_j'(y)\mu_{jk}\psi_k(x) \\ &= -\sum_{j,k=0}^{2N-1} \psi_j'(y)\mu_{kj}\psi_k(x) = -\sum_{j,k=0}^{2N-1} \psi_j(y)\mu_{kj}\psi_k'(x) \, , \end{aligned}$$

after re-indexing. Note that

$$\varepsilon S_N(x,y) = \sum_{j,k=0}^{2N-1} \varepsilon D\psi_j(x)\mu_{jk}\psi_k(y) = \sum_{j,k=0}^{2N-1} \psi_j(x)\mu_{jk}\psi_k(y) \, ,$$

and

$$-\frac{\mathrm{d}}{\mathrm{d}y} S_N(x,y) = \sum_{j,k=0}^{2N-1} \psi_j'(x)\mu_{jk}\psi_k'(y) \, .$$

Thus we can now write succinctly

$$K_N(x,y) = \begin{pmatrix} S_N(x,y) & -\mathrm{d}S_N(x,y)/\mathrm{d}y \\ \varepsilon S_N(x,y) & S_N(y,x) \end{pmatrix} . \tag{7.85}$$

To summarize, we have shown that $G_{4,N}^2(t,\lambda) = C_4^{(N)} \det(I - K_{4,N} f)$. Setting $f \equiv 0$ on both sides (where the original definition of $G_{4,N}(t,\lambda)$ as an integral is used on the left) shows that $C_4^{(N)} = 1$. Thus

$$G_{4,N}(t,\lambda) = \sqrt{D_{4,N}(t,\lambda)} \, , \tag{7.86}$$

where we define

$$D_{4,N}(t,\lambda) = \det(I + K_{4,N} f) \, , \tag{7.87}$$

and $K_{4,N}$ is the integral operator with matrix kernel (7.85).

7.6.2 Gaussian Specialization

We would like to specialize the above results to the case of a Gaussian weight function

$$w(x) = \exp(x^2) \tag{7.88}$$

and indicator function

$$f(x) = -\lambda \chi_J\ , \quad J = (t, \infty)\ .$$

We want the matrix

$$M = \left\{ \int \left(\psi_j(x) \psi_k'(x) - \psi_k(x) \psi_j'(x) \right) \mathrm{d}x \right\}_{j,k=0}^{2N-1}$$

to be the direct sum of N copies of

$$Z = \begin{pmatrix} 0 & 1 \\ -1 & 0 \end{pmatrix}\ ,$$

so that the formulas are the simplest possible, since then μ_{jk} can only be 0 or ± 1. In that case M would be skew-symmetric so that $M^{-1} = -M$. In terms of the integrals defining the entries of M this means that we would like to have

$$\int \left(\psi_{2j}(x) \frac{\mathrm{d}}{\mathrm{d}x} \psi_{2k+1}(x) - \psi_{2k+1}(x) \frac{\mathrm{d}}{\mathrm{d}x} \psi_{2j}(x) \right) \mathrm{d}x = \delta_{j,k}\ ,$$
$$\int \left(\psi_{2j+1}(x) \frac{\mathrm{d}}{\mathrm{d}x} \psi_{2k}(x) - \psi_{2k}(x) \frac{\mathrm{d}}{\mathrm{d}x} \psi_{2j+1}(x) \right) \mathrm{d}x = -\delta_{j,k}$$

and otherwise

$$\int \left(\psi_j(x) \frac{\mathrm{d}}{\mathrm{d}x} \psi_k(x) - \psi_j(x) \frac{\mathrm{d}}{\mathrm{d}x} \psi_k(x) \right) \mathrm{d}x = 0\ .$$

It is easier to treat this last case if we replace it with three nonexclusive conditions

$$\int \left(\psi_{2j}(x) \frac{\mathrm{d}}{\mathrm{d}x} \psi_{2k}(x) - \psi_{2k}(x) \frac{\mathrm{d}}{\mathrm{d}x} \psi_{2j}(x) \right) \mathrm{d}x = 0\ ,$$
$$\int \left(\psi_{2j+1}(x) \frac{\mathrm{d}}{\mathrm{d}x} \psi_{2k+1}(x) - \psi_{2k+1}(x) \frac{\mathrm{d}}{\mathrm{d}x} \psi_{2j+1}(x) \right) \mathrm{d}x = 0\ ,$$

(so when the parity is the same for j, k, which takes care of diagonal entries, among others) and

$$\int \left(\psi_j(x) \frac{\mathrm{d}}{\mathrm{d}x} \psi_k(x) - \psi_j(x) \frac{\mathrm{d}}{\mathrm{d}x} \psi_k(x) \right) \mathrm{d}x = 0\ ,$$

whenever $|j-k|>1$, which targets entries outside of the tridiagonal. Define

$$\varphi_k(x) = \frac{1}{c_k} H_k(x) \exp(-x^2/2) \quad \text{for } c_k = \sqrt{2^k k! \sqrt{\pi}} \tag{7.89}$$

where the H_k are the usual Hermite polynomials defined by the orthogonality condition

$$\int_{\mathbb{R}} H_j(x) H_k(x) \exp(-x^2)\, \mathrm{d}x = c_j^2 \delta_{j,k} \ .$$

Then it follows that

$$\int_{\mathbb{R}} \varphi_j(x) \varphi_k(x)\, \mathrm{d}x = \delta_{j,k} \ .$$

Now let

$$\psi_{2j+1}(x) = \frac{1}{\sqrt{2}} \varphi_{2j+1}(x) \psi_{2j}(x) = -\frac{1}{\sqrt{2}} \varepsilon \varphi_{2j+1}(x) \ .$$

This definition satisfies our earlier requirement that $\psi_j = p_j w^{1/2}$ with w defined in (7.88). In particular we have in this case

$$p_{2j+1}(x) = \frac{1}{c_j \sqrt{2}} H_{2j+1}(x) \ .$$

Let ε as in (7.84), and D denote the operator that acts by differentiation with respect to x as before, so that $D\varepsilon = \varepsilon D = I$. It follows that

$$\begin{aligned} &\int_{\mathbb{R}} \left[\psi_{2j}(x) \frac{\mathrm{d}}{\mathrm{d}x} \psi_{2k+1}(x) - \psi_{2k+1}(x) \frac{\mathrm{d}}{\mathrm{d}x} \psi_{2j}(x) \right] \mathrm{d}x \\ &\qquad = \frac{1}{2} \int_{\mathbb{R}} \left[-\varepsilon \varphi_{2j+1}(x) \frac{\mathrm{d}}{\mathrm{d}x} \varphi_{2k+1}(x) + \varphi_{2k+1}(x) \frac{\mathrm{d}}{\mathrm{d}x} \varepsilon \varphi_{2j+1}(x) \right] \mathrm{d}x \\ &\qquad = \frac{1}{2} \int_{\mathbb{R}} \left[-\varepsilon \varphi_{2j+1}(x) \frac{\mathrm{d}}{\mathrm{d}x} \varphi_{2k+1}(x) + \varphi_{2k+1}(x) \varphi_{2j+1}(x) \right] \mathrm{d}x \ . \end{aligned}$$

We integrate the first term by parts and use the fact that

$$\frac{\mathrm{d}}{\mathrm{d}x} \varepsilon \varphi_j(x) = \varphi_j(x)$$

and also that φ_j vanishes at the boundary (i.e., $\varphi_j(\pm\infty) = 0$) to obtain

$$\begin{aligned} &\int_{\mathbb{R}} \left[\psi_{2j}(x) \frac{\mathrm{d}}{\mathrm{d}x} \psi_{2k+1}(x) - \psi_{2k+1}(x) \frac{\mathrm{d}}{\mathrm{d}x} \psi_{2j}(x) \right] \mathrm{d}x \\ &\qquad = \frac{1}{2} \int_{\mathbb{R}} \left[-\varepsilon \varphi_{2j+1}(x) \frac{\mathrm{d}}{\mathrm{d}x} \varphi_{2k+1}(x) + \varphi_{2k+1}(x) \varphi_{2j+1}(x) \right] \mathrm{d}x \\ &\qquad = \frac{1}{2} \int_{\mathbb{R}} [\varphi_{2j+1}(x) \varphi_{2k+1}(x) + \varphi_{2k+1}(x) \varphi_{2j+1}(x)]\, \mathrm{d}x \\ &\qquad = \frac{1}{2} \int_{\mathbb{R}} [\varphi_{2j+1}(x) \varphi_{2k+1}(x) + \varphi_{2k+1}(x) \varphi_{2j+1}(x)]\, \mathrm{d}x \\ &\qquad = \frac{1}{2} (\delta_{j,k} + \delta_{j,k}) = \delta_{j,k} \ , \end{aligned}$$

as desired. Similarly

$$\int_{\mathbb{R}} \left[\psi_{2j+1}(x) \frac{\mathrm{d}}{\mathrm{d}x} \psi_{2k}(x) - \psi_{2k}(x) \frac{\mathrm{d}}{\mathrm{d}x} \psi_{2j+1}(x) \right] \mathrm{d}x$$
$$= \frac{1}{2} \int_{\mathbb{R}} \left[-\varphi_{2j+1}(x) \frac{\mathrm{d}}{\mathrm{d}x} \varepsilon\varphi_{2k+1}(x) + \varepsilon\varphi_{2k+1}(x) \frac{\mathrm{d}}{\mathrm{d}x} \varphi_{2j+1}(x) \right] \mathrm{d}x = -\delta_{j,k} \, .$$

Moreover,

$$p_{2j+1}(x) = \frac{1}{c_j\sqrt{2}} H_{2j+1}(x)$$

is certainly an odd function, being the multiple of and odd Hermite polynomial. On the other hand, one easily checks that ε maps odd functions to even functions on $L^2(\mathbb{R})$. Therefore

$$p_{2j}(x) = -\frac{1}{c_j\sqrt{2}} \varepsilon H_{2j+1}(x)$$

is an even function, and it follows that

$$\int_{\mathbb{R}} \left[\psi_{2k}(x) \frac{\mathrm{d}}{\mathrm{d}x} \psi_{2j}(x) - \psi_{2j}(x) \frac{\mathrm{d}}{\mathrm{d}x} \psi_{2k}(x) \right] \mathrm{d}x$$
$$= \int_{\mathbb{R}} \left[p_{2j}(x) \frac{\mathrm{d}}{\mathrm{d}x} p_{2k}(x) - p_{2k}(x) \frac{\mathrm{d}}{\mathrm{d}x} p_{2j}(x) \right] w(x) \, \mathrm{d}x = 0 \, ,$$

since both terms in the integrand are odd functions, and the weight function is even. Similarly,

$$\int_{\mathbb{R}} \left[\psi_{2k+1}(x) \frac{\mathrm{d}}{\mathrm{d}x} \psi_{2j+1}(x) - \psi_{2j+1}(x) \frac{\mathrm{d}}{\mathrm{d}x} \psi_{2k+1}(x) \right] \mathrm{d}x$$
$$= \int_{\mathbb{R}} \left[p_{2j+1}(x) \frac{\mathrm{d}}{\mathrm{d}x} p_{2k+1}(x) - p_{2k+1}(x) \frac{\mathrm{d}}{\mathrm{d}x} p_{2j+1}(x) \right] w(x) \, \mathrm{d}x = 0 \, .$$

Finally it is easy to see that if $|j-k| > 1$ then

$$\int_{\mathbb{R}} \left[\psi_j(x) \frac{\mathrm{d}}{\mathrm{d}x} \psi_k(x) - \psi_j(x) \frac{\mathrm{d}}{\mathrm{d}x} \psi_k(x) \right] \mathrm{d}x = 0 \, .$$

Indeed both differentiation and the action of ε can only "shift" the indices by 1. Thus by orthogonality of the φ_j, this integral will always be 0. Hence by choosing

$$\psi_{2j+1}(x) = \frac{1}{\sqrt{2}} \varphi_{2j+1}(x), \quad \psi_{2j}(x) = -\frac{1}{\sqrt{2}} \varepsilon\varphi_{2j+1}(x) \, ,$$

we force the matrix

$$M = \left\{ \int_{\mathbb{R}} \left(\psi_j(x)\psi'_k(x) - \psi_k(x)\psi'_j(x) \right) \mathrm{d}x \right\}_{j,k=0}^{2n-1}$$

to be the direct sum of N copies of

$$Z = \begin{pmatrix} 0 & 1 \\ -1 & 0 \end{pmatrix} .$$

Hence $M^{-1} = -M$ where $M^{-1} = \{\mu_{j,k}\}_{j,k=0,2N-1}$. Moreover, with our above choice, $\mu_{j,k} = 0$ if j, k have the same parity or $|j-k| > 1$, and $\mu_{2k,2j+1} = \delta_{jk} = -\mu_{2j+1,2k}$ for $j,k = 0,\ldots,N-1$. Therefore

$$\begin{aligned} S_N(x,y) &= -\sum_{j,k=0}^{2N-1} \psi'_j(x)\mu_{jk}\psi_k(y) \\ &= -\sum_{j=0}^{N-1} \psi'_{2j}(x)\psi_{2j+1}(y) + \sum_{j=0}^{N-1} \psi'_{2j+1}(x)\psi_{2j}(y) \\ &= \frac{1}{2}\left[\sum_{j=0}^{N-1} \varphi_{2j+1}(x)\varphi_{2j+1}(y) - \sum_{j=0}^{N-1} \varphi'_{2j+1}(x)\varepsilon\varphi_{2j+1}(y) \right] . \end{aligned}$$

Recall that the H_j satisfy the differentiation formulas (see for example [2, p. 280])

$$H'_j(x) = 2xH_j(x) - H_{j-1}(x) j = 1,2,\ldots . \tag{7.90}$$

$$H'_j(x) = 2jH_{j-1}(x) \qquad j = 1,2,\ldots . \tag{7.91}$$

Combining (7.89) and (7.90) yields

$$\varphi'_j(x) = x\varphi_j(x) - \frac{c_{j+1}}{c_j}\varphi_{j+1}(x) . \tag{7.92}$$

Similarly, from (7.89) and (7.91) we have

$$\varphi'_j(x) = -x\varphi_j(x) + 2j\frac{c_{j-1}}{c_j}\varphi_{j-1}(x) . \tag{7.93}$$

Combining (7.92) and (7.93), we obtain

$$\varphi'_j(x) = \sqrt{\frac{j}{2}}\varphi_{j-1}(x) - \sqrt{\frac{j+1}{2}}\varphi_{j+1}(x) . \tag{7.94}$$

Let $\varphi = \left(\varphi_1\ \varphi_2\ \cdots\right)^t$ and $\varphi' = \left(\varphi'_1\ \varphi'_2\ \cdots\right)^t$. Then we can rewrite (7.94) as

$$\varphi' = A\varphi$$

where $A = \{a_{j,k}\}$ is the infinite antisymmetric tridiagonal matrix with $a_{j,j-1} = \sqrt{j/2}$. Hence,

$$\varphi_j'(x) = \sum_{k\geq 0} a_{jk}\varphi_k(x) .$$

Moreover, using the fact that $D\varepsilon = \varepsilon D = I$ we also have

$$\varphi_j(x) = \varepsilon\varphi_j'(x) = \varepsilon\left(\sum_{k\geq 0} a_{jk}\varphi_k(x)\right) = \sum_{k\geq 0} a_{jk}\varepsilon\varphi_k(x) .$$

Combining the above results, we have

$$\begin{aligned}\sum_{j=0}^{N-1} \varphi_{2j+1}'(x)\varepsilon\varphi_{2j+1}(y) &= \sum_{j=0}^{N-1}\sum_{k\geq 0} a_{2j+1,k}\varphi_k(x)\varepsilon\varphi_{2j+1}(x) \\ &= -\sum_{j=0}^{N-1}\sum_{k\geq 0} a_{k,2j+1}\varphi_k(x)\varepsilon\varphi_{2j+1}(x) .\end{aligned}$$

Note that $a_{k,2j+1} = 0$ unless $|k-(2j+1)| = 1$, that is unless k is even. Thus we can rewrite the sum as

$$\begin{aligned}&\sum_{j=0}^{N-1} \varphi_{2j+1}'(x)\varepsilon\varphi_{2j+1}(y) \\ &\qquad = -\sum_{\substack{k,j\geq 0 \\ k \text{ even} \\ k\leq 2N}} a_{k,j}\varphi_k(x)\varepsilon\varphi_j(y) - a_{2N,2N+1}\varphi_{2N}(x)\varepsilon\varphi_{2N+1}(y) \\ &\qquad = -\sum_{\substack{k\geq 0 \\ k \text{ even} \\ k\leq 2N}} \varphi_k(x)\sum_{j\geq 0} a_{k,j}\varepsilon\varphi_j(y) + a_{2N,2N+1}\varphi_{2N}(x)\varepsilon\varphi_{2N+1}(y)\end{aligned}$$

where the last term takes care of the fact that we are counting an extra term in the sum that was not present before. The sum over j on the right is just $\varphi_k(y)$, and $a_{2N,2N+1} = -\sqrt{(2N+1)/2}$. Therefore

$$\begin{aligned}\sum_{j=0}^{N-1} \varphi_{2j+1}'(x)\varepsilon\varphi_{2j+1}(y) &= \sum_{\substack{k\geq 0 \\ k \text{ even} \\ k\leq 2N}} \varphi_k(x)\varphi_k(y) - \sqrt{\frac{2N+1}{2}}\varphi_{2N}(x)\varepsilon\varphi_{2N+1}(y) \\ &= \sum_{j=0}^{N} \varphi_{2j}(x)\varphi_{2j}(y) - \sqrt{\frac{2N+1}{2}}\varphi_{2N}(x)\varepsilon\varphi_{2N+1}(y) .\end{aligned}$$

It follows that

$$S_N(x,y) = \frac{1}{2}\left[\sum_{j=0}^{2N} \varphi_j(x)\varphi_j(y) - \sqrt{\frac{2N+1}{2}}\varphi_{2N}(x)\varepsilon\varphi_{2N+1}(y)\right] .$$

We redefine

$$S_N(x,y) = \sum_{n=0}^{2N} \varphi_n(x)\varphi_n(y) = S_N(y,x) \tag{7.95}$$

so that the top left entry of $K_N(x,y)$ is

$$S_N(x,y) + \sqrt{\frac{2N+1}{2}}\varphi_{2N}(x)\varepsilon\varphi_{2N+1}(y)\,.$$

If S_N is the operator with kernel $S_N(x,y)$ then integration by parts gives

$$S_N D f = \int_{\mathbb{R}} S(x,y)\frac{\mathrm{d}}{\mathrm{d}y}f(y)\,\mathrm{d}y = \int_{\mathbb{R}}\left(-\frac{\mathrm{d}}{\mathrm{d}y}S_N(x,y)\right)f(y)\,\mathrm{d}y\,,$$

so that $-\mathrm{d}S_N(x,y)/\mathrm{d}y$ is in fact the kernel of $S_N D$. Therefore (7.86) now holds with $K_{4,N}$ being the integral operator with matrix kernel $K_{4,N}(x,y)$ whose (i,j)-entry $K_{4,N}^{(i,j)}(x,y)$ is given by

$$\begin{aligned}
K_{4,N}^{(1,1)}(x,y) &= \frac{1}{2}\left[S_N(x,y) + \sqrt{\frac{2N+1}{2}}\varphi_{2N}(x)\varepsilon\varphi_{2N+1}(y)\right],\\
K_{4,N}^{(1,2)}(x,y) &= \frac{1}{2}\left[SD_N(x,y) - \frac{\mathrm{d}}{\mathrm{d}y}\left(\sqrt{\frac{2N+1}{2}}\varphi_{2N}(x)\varepsilon\varphi_{2N+1}(y)\right)\right],\\
K_{4,N}^{(2,1)}(x,y) &= \frac{\varepsilon}{2}\left[S_N(x,y) + \sqrt{\frac{2N+1}{2}}\varphi_{2N}(x)\varepsilon\varphi_{2N+1}(y)\right],\\
K_{4,N}^{(2,2)}(x,y) &= \frac{1}{2}\left[S_N(x,y) + \sqrt{\frac{2N+1}{2}}\varepsilon\varphi_{2N+1}(x)\varphi_{2N}(y)\right].
\end{aligned}$$

We let $2N+1 \to N$ so that N is assumed to be odd from now on (this will not matter in the end since we will take $N \to \infty$). Therefore the $K_{4,N}^{(i,j)}(x,y)$ are given by

$$\begin{aligned}
K_{4,N}^{(1,1)}(x,y) &= \frac{1}{2}\left[S_N(x,y) + \sqrt{\frac{N}{2}}\varphi_{N-1}(x)\varepsilon\varphi_N(y)\right],\\
K_{4,N}^{(1,2)}(x,y) &= \frac{1}{2}\left[SD_N(x,y) - \sqrt{\frac{N}{2}}\varphi_{N-1}(x)\varphi_N(y)\right],\\
K_{4,N}^{(2,1)}(x,y) &= \frac{\varepsilon}{2}\left[S_N(x,y) + \sqrt{\frac{N}{2}}\varphi_{N-1}(x)\varepsilon\varphi_N(y)\right],\\
K_{4,N}^{(2,2)}(x,y) &= \frac{1}{2}\left[S_N(x,y) + \sqrt{\frac{N}{2}}\varepsilon\varphi_N(x)\varphi_{N-1}(y)\right],
\end{aligned}$$

where

$$S_N(x,y) = \sum_{n=0}^{N-1} \varphi_n(x)\varphi_n(y)\,.$$

Define

$$\varphi(x) = \left(\frac{N}{2}\right)^{1/4} \varphi_N(x), \quad \psi(x) = \left(\frac{N}{2}\right)^{1/4} \varphi_{N-1}(x) ,$$

so that

$$\begin{aligned}
K_{4,N}^{(1,1)}(x,y) &= \tfrac{1}{2}\chi(x)[S_N(x,y) + \psi(x)\varepsilon\varphi(y)]\chi(y) , \\
K_{4,N}^{(1,2)}(x,y) &= \tfrac{1}{2}\chi(x)[SD_N(x,y) - \psi(x)\varphi(y)]\chi(y), \\
K_{4,N}^{(2,1)}(x,y) &= \tfrac{1}{2}\chi(x)[\varepsilon S_N(x,y) + \varepsilon\psi(x)\varepsilon\varphi(y)]\chi(y) , \\
K_{4,N}^{(2,2)}(x,y) &= \tfrac{1}{2}\chi(x)[S_N(x,y) + \varepsilon\varphi(x)\psi(y)]\chi(y) .
\end{aligned}$$

Notice that

$$\begin{aligned}
\tfrac{1}{2}\chi(S + \psi\otimes\varepsilon\varphi)\chi &\doteq K_{4,N}^{(1,1)}(x,y) , \\
\tfrac{1}{2}\chi(SD - \psi\otimes\varphi)\chi &\doteq K_{4,N}^{(1,2)}(x,y) , \\
\tfrac{1}{2}\chi(\varepsilon S + \varepsilon\psi\otimes\varepsilon\varphi)\chi &\doteq K_{4,N}^{(2,1)}(x,y) , \\
\tfrac{1}{2}\chi(S + \varepsilon\varphi\otimes\varepsilon\psi)\chi &\doteq K_{4,N}^{(2,2)}(x,y) .
\end{aligned}$$

Therefore

$$K_{4,N} = \frac{1}{2}\chi \begin{pmatrix} S + \psi\otimes\varepsilon\varphi & SD - \psi\otimes\varphi \\ \varepsilon S + \varepsilon\psi\otimes\varepsilon\varphi & S + \varepsilon\varphi\otimes\psi \end{pmatrix} \chi . \tag{7.96}$$

Note that this is identical to the corresponding operator for $\beta = 4$ obtained by Tracy and Widom in [33], the only difference being that φ, ψ, and hence also S, are redefined to depend on λ. This will affect boundary conditions for the differential equations we will obtain later.

7.6.3 Edge Scaling

7.6.3.1 Reduction of the Determinant

We want to compute the Fredholm determinant (7.86) with $K_{4,N}$ given by (7.96) and $f = \chi_{(t,\infty)}$. This is the determinant of an operator on $L^2(J)\times L^2(J)$. Our first task will be to rewrite the determinant as that of an operator on $L^2(J)$. This part follows exactly the proof in [33]. To begin, note that

$$[S, D] = \varphi\otimes\psi + \psi\otimes\varphi \tag{7.97}$$

so that, using the fact that $D\varepsilon = \varepsilon D = I$,

$$\begin{aligned}
[\varepsilon, S] = \varepsilon S - S\varepsilon = \varepsilon SD\varepsilon - \varepsilon DS\varepsilon = \varepsilon[S,D]\varepsilon &= \varepsilon\varphi\otimes\psi\varepsilon + \varepsilon\psi\otimes\varphi\varepsilon \\
&= \varepsilon\varphi\otimes\varepsilon^t\psi + \varepsilon\psi\otimes\varepsilon^t\varphi = -\varepsilon\varphi\otimes\varepsilon\psi - \varepsilon\psi\otimes\varepsilon\varphi ,
\end{aligned} \tag{7.98}$$

where the last equality follows from the fact that $\varepsilon^t = -\varepsilon$. We thus have

$$D(\varepsilon S + \varepsilon\psi \otimes \varepsilon\varphi) = S + \psi \otimes \varepsilon\varphi\,, \quad D(\varepsilon SD - \varepsilon\psi \otimes \varphi) = SD - \psi \otimes \varphi\,.$$

The expressions on the right side are the top matrix entries in (7.96). Thus the first row of $K_{4,N}$ is, as a vector,

$$D(\varepsilon S + \varepsilon\psi \otimes \varepsilon\varphi, \varepsilon SD - \varepsilon\psi \otimes \varphi)\,.$$

Now (7.98) implies that

$$\varepsilon S + \varepsilon\psi \otimes \varepsilon\varphi = S\varepsilon - \varepsilon\varphi \otimes \varepsilon\psi\,.$$

Similarly (7.97) gives

$$\varepsilon[S, D] = \varepsilon\varphi \otimes \psi + \varepsilon\psi \otimes \varphi\,,$$

so that

$$\varepsilon SD - \varepsilon\psi \otimes \varphi = \varepsilon DS + \varepsilon\varphi \otimes \psi = S + \varepsilon\varphi \otimes \psi\,.$$

Using these expressions we can rewrite the first row of $K_{4,N}$ as

$$D(S\varepsilon - \varepsilon\varphi \otimes \varepsilon\psi, S + \varepsilon\varphi \otimes \psi)\,.$$

Now use (7.98) to show the second row of $K_{4,N}$ is

$$(S\varepsilon - \varepsilon\varphi \otimes \varepsilon\psi, S + \varepsilon\varphi \otimes \psi)\,.$$

Therefore,

$$\begin{aligned} K_{4,N} &= \chi \begin{pmatrix} D(S\varepsilon - \varepsilon\varphi \otimes \varepsilon\psi) & D(S + \varepsilon\varphi \otimes \psi) \\ S\varepsilon - \varepsilon\varphi \otimes \varepsilon\psi & S + \varepsilon\varphi \otimes \psi \end{pmatrix} \chi \\ &= \begin{pmatrix} \chi D & 0 \\ 0 & \chi \end{pmatrix} \begin{pmatrix} (S\varepsilon - \varepsilon\varphi \otimes \varepsilon\psi)\chi & (S + \varepsilon\varphi \otimes \psi)\chi \\ S\varepsilon - \varepsilon\varphi \otimes \varepsilon\psi)\chi & (S + \varepsilon\varphi \otimes \psi)\chi \end{pmatrix}. \end{aligned}$$

Since $K_{4,N}$ is of the form AB, we can use (7.4.2) and deduce that $D_{4,N}(s,\lambda)$ is unchanged if instead we take $K_{4,N}$ to be

$$\begin{aligned} K_{4,N} &= \begin{pmatrix} (S\varepsilon - \varepsilon\varphi \otimes \varepsilon\psi)\chi & (S + \varepsilon\varphi \otimes \psi)\chi \\ (S\varepsilon - \varepsilon\varphi \otimes \varepsilon\psi)\chi & (S + \varepsilon\varphi \otimes \psi)\chi \end{pmatrix} \begin{pmatrix} \chi D & 0 \\ 0 & \chi \end{pmatrix} \\ &= \begin{pmatrix} (S\varepsilon - \varepsilon\varphi \otimes \varepsilon\psi)\chi D & (S + \varepsilon\varphi \otimes \psi)\chi \\ (S\varepsilon - \varepsilon\varphi \otimes \varepsilon\psi)\chi D & (S + \varepsilon\varphi \otimes \psi)\chi \end{pmatrix}. \end{aligned}$$

Therefore

$$D_{4,N}(s,\lambda) = \det \begin{pmatrix} I - \frac{1}{2}(S\varepsilon - \varepsilon\varphi \otimes \varepsilon\psi)\lambda\chi D & -\frac{1}{2}(S + \varepsilon\varphi \otimes \psi)\lambda\chi \\ -\frac{1}{2}(S\varepsilon - \varepsilon\varphi \otimes \varepsilon\psi)\lambda\chi D & I - \frac{1}{2}(S + \varepsilon\varphi \otimes \psi)\lambda\chi \end{pmatrix}. \tag{7.99}$$

Now we perform row and column operations on the matrix to simplify it, which do not change the Fredholm determinant. Justification of these operations is given in [33]. We start by subtracting row 1 from row 2 to get

$$\begin{pmatrix} I - \frac{1}{2}(S\varepsilon - \varepsilon\varphi \otimes \varepsilon\psi)\lambda\chi D & -\frac{1}{2}(S + \varepsilon\varphi \otimes \psi)\lambda\chi \\ -I & I \end{pmatrix} .$$

Next, adding column 2 to column 1 yields

$$\begin{pmatrix} I - \frac{1}{2}(S\varepsilon - \varepsilon\varphi \otimes \varepsilon\psi)\lambda\chi D - \frac{1}{2}(S + \varepsilon\varphi \otimes \psi)\lambda\chi & -\frac{1}{2}(S + \varepsilon\varphi \otimes \psi)\lambda\chi \\ 0 & I \end{pmatrix} .$$

Thus the determinant we want equals the determinant of

$$I - \tfrac{1}{2}(S\varepsilon - \varepsilon\varphi \otimes \varepsilon\psi)\lambda\chi D - \tfrac{1}{2}(S + \varepsilon\varphi \otimes \psi)\lambda\chi \,. \tag{7.100}$$

So we have reduced the problem from the computation of the Fredholm determinant of an operator on $L^2(J) \times L^2(J)$, to that of an operator on $L^2(J)$.

7.6.3.2 Differential Equations

Next we want to write the operator in (7.100) in the form

$$(I - K_{2,N})\left(I - \sum_{i=1}^{L} \alpha_i \otimes \beta_i\right) , \tag{7.101}$$

where the α_i and β_i are functions in $L^2(J)$. In other words, we want to rewrite the determinant for the GSE case as a finite dimensional perturbation of the corresponding GUE determinant. The Fredholm determinant of the product is then the product of the determinants. The limiting form for the GUE part is already known, and we can just focus on finding a limiting form for the determinant of the finite dimensional piece. It is here that the proof must be modified from that in [33]. A little rearrangement of (7.100) yields (recall $\varepsilon^t = -\varepsilon$)

$$I - \frac{\lambda}{2}S\chi - \frac{\lambda}{2}S\varepsilon\chi D - \frac{\lambda}{2}\varepsilon\varphi \otimes \chi\psi - \frac{\lambda}{2}\varepsilon\varphi \otimes \psi\varepsilon\chi D \,.$$

Writing $\varepsilon[\chi, D] + \chi$ for $\varepsilon\chi D$ and simplifying gives

$$I - \lambda S\chi - \lambda\varepsilon\varphi \otimes \psi\chi - \frac{\lambda}{2}S\varepsilon[\chi, D] - \frac{\lambda}{2}\varepsilon\varphi \otimes \psi\varepsilon[\chi, D]$$

Let $\sqrt{\lambda}\varphi \to \varphi$, and $\sqrt{\lambda}\psi \to \psi$ so that $\lambda S \to S$ and (7.100) goes to

$$I - S\chi - \varepsilon\varphi \otimes \psi\chi - \tfrac{1}{2}S\varepsilon[\chi, D] - \tfrac{1}{2}\varepsilon\varphi \otimes \psi\varepsilon[\chi, D] \,.$$

Now we define $R := (I - S\chi)^{-1}S\chi = (I - S\chi)^{-1} - I$ (the resolvent operator of $S\chi$), whose kernel we denote by $R(x, y)$, and $Q_\varepsilon := (I - S\chi)^{-1}\varepsilon\varphi$. Then (7.100) factors into

$$A = (I - S\chi)B \,.$$

where B is

$$I - Q_\varepsilon \otimes \chi\psi - \tfrac{1}{2}(I + R)S\varepsilon[\chi, D] - \tfrac{1}{2}(Q_\varepsilon \otimes \psi)\varepsilon[\chi, D] .$$

Hence

$$D_{4,N}(t, \lambda) = D_{2,N}(t, \lambda) \det(B) .$$

In order to find $\det(B)$ we use the identity

$$\varepsilon[\chi, D] = \sum_{k=1}^{2m} (-1)^k \varepsilon_k \otimes \delta_k , \tag{7.102}$$

where ε_k and δ_k are the functions $\varepsilon(x - a_k)$ and $\delta(x - a_k)$ respectively, and the a_k are the endpoints of the (disjoint) intervals considered, $J = \cup_{k=1}^{m}(a_{2k-1}, a_{2k})$. In our case $m = 1$ and $a_1 = t$, $a_2 = \infty$. We also make use of the fact that

$$a \otimes b \times c \otimes d = (b, c) \times a \otimes d \tag{7.103}$$

where $(\cdot, \cdot)$ is the usual L^2-inner product. Therefore

$$(Q_\varepsilon \otimes \psi)\varepsilon[\chi, D] = \sum_{k=1}^{2} (-1)^k Q_\varepsilon \otimes \psi \times \varepsilon_k \otimes \delta_k = \sum_{k=1}^{2} (-1)^k (\psi, \varepsilon_k) Q_\varepsilon \otimes \delta_k .$$

It follows that

$$\frac{D_{4,N}(t, \lambda)}{D_{2,N}(t, \lambda)} \tag{7.104}$$

is the determinant of

$$I - Q_\varepsilon \otimes \chi\psi - \frac{1}{2} \sum_{k=1}^{2} (-1)^k [(S + RS)\varepsilon_k + (\psi, \varepsilon_k) Q_\varepsilon] \otimes \delta_k . \tag{7.105}$$

We now specialize to the case of one interval $J = (t, \infty)$, so $m = 1$, $a_1 = t$ and $a_2 = \infty$. We write $\varepsilon_t = \varepsilon_1$, and $\varepsilon_\infty = \varepsilon_2$, and similarly for δ_k. Writing out the terms in the summation and using the fact that

$$\varepsilon_\infty = -\tfrac{1}{2} , \tag{7.106}$$

yields

$$\begin{aligned} I - Q_\varepsilon \otimes \chi\psi &+ \tfrac{1}{2}[(S + RS)\varepsilon_t + (\psi, \varepsilon_t) Q_\varepsilon] \otimes \delta_t \\ &+ \tfrac{1}{4}[(S + RS)1 + (\psi, 1) Q_\varepsilon] \otimes \delta_\infty . \end{aligned} \tag{7.107}$$

Now we can use the formula

$$\det\left(I - \sum_{i=1}^{L} \alpha_i \otimes \beta_i\right) = \det\big(\delta_{jk} - (\alpha_j, \beta_k)\big)_{j,k=1}^{L} . \tag{7.108}$$

In order to simplify the notation in preparation for the computation of the various inner products, define

$$\begin{aligned} Q(x,\lambda,t) &:= (I - S\chi)^{-1}\varphi\,, & P(x,\lambda,t) &:= (I - S\chi)^{-1}\psi\,, \\ Q_\varepsilon(x,\lambda,t) &:= (I - S\chi)^{-1}\varepsilon\varphi\,, & P_\varepsilon(x,\lambda,t) &:= (I - S\chi)^{-1}\varepsilon\psi\,, \end{aligned} \tag{7.109}$$

$$\begin{aligned} q_N &:= Q(t,\lambda,t)\,, & p_N &:= P(t,\lambda,t)\,, \\ q_\varepsilon &:= Q_\varepsilon(t,\lambda,t)\,, & p_\varepsilon &:= P_\varepsilon(t,\lambda,t)\,, \\ u_\varepsilon &:= (Q,\chi\varepsilon\varphi) = (Q_\varepsilon,\chi\varphi)\,, & v_\varepsilon &:= (Q,\chi\varepsilon\psi) = (P_\varepsilon,\chi\psi)\,, \\ \tilde{v}_\varepsilon &:= (P,\chi\varepsilon\varphi) = (Q_\varepsilon,\chi\varphi)\,, & w_\varepsilon &:= (P,\chi\varepsilon\psi) = (P_\varepsilon,\chi\psi)\,, \end{aligned} \tag{7.110}$$

$$\begin{gathered} \mathcal{P}_4 := \int_{\mathbb{R}} \varepsilon_t(x) P(x,t)\,\mathrm{d}x\,, \quad \mathcal{Q}_4 := \int_{\mathbb{R}} \varepsilon_t(x) Q(x,t)\,\mathrm{d}x\,, \\ \mathcal{R}_4 := \int_{\mathbb{R}} \varepsilon_t(x) R(x,t)\,\mathrm{d}x\,, \end{gathered} \tag{7.111}$$

where we remind the reader that ε_t stands for the function $\varepsilon(x-t)$. Note that all quantities in (7.110) and (7.111) are functions of t and λ alone. Furthermore, let

$$c_\varphi = \varepsilon\varphi(\infty) = \frac{1}{2}\int_{-\infty}^{\infty} \varphi(x)\,\mathrm{d}x\,, \quad c_\psi = \varepsilon\psi(\infty) = \frac{1}{2}\int_{-\infty}^{\infty} \psi(x)\,\mathrm{d}x\,. \tag{7.112}$$

Recall from the previous section that when $\beta = 4$ we take N to be odd. It follows that φ and ψ are odd and even functions respectively. Thus when $\beta = 4$, $c_\varphi = 0$ while computation using known integrals for the Hermite polynomials gives

$$c_\psi = (\pi N)^{1/4} 2^{-3/4-N/2} \frac{(N!)^{1/2}}{(N/2)!}\sqrt{\lambda}\,. \tag{7.113}$$

Hence computation yields

$$\lim_{N\to\infty} c_\psi = \sqrt{\frac{\lambda}{2}}\,. \tag{7.114}$$

At $t = \infty$,

$$u_\varepsilon(\infty) = 0\,, \quad q_\varepsilon(\infty) = c_\varphi\,, \tag{7.115}$$

$$\mathcal{P}_4(\infty) = -c_\psi\,, \quad \mathcal{Q}_4(\infty) = -c_\varphi\,, \quad \mathcal{R}_4(\infty) = 0\,. \tag{7.116}$$

In (7.108), $L = 3$ and if we denote $a_4 = (\psi, \varepsilon_t)$, then we have explicitly

$$\begin{gathered} \alpha_1 = Q_\varepsilon\,, \quad \alpha_2 = -\tfrac{1}{2}[(S + RS)\varepsilon_t + a_4 Q_\varepsilon]\,, \quad \alpha_3 = -\tfrac{1}{4}[(S+RS)1 + (\psi,1)Q_\varepsilon]\,, \\ \beta_1 = \chi\psi\,, \quad \beta_2 = \delta_t\,, \quad \beta_3 = \delta_\infty\,. \end{gathered}$$

However notice that

$$((S+RS)\varepsilon_t, \delta_\infty) = (\varepsilon_t, \delta_\infty) = 0\,, \quad ((S+RS)1, \delta_\infty) = (1, R_\infty) = 0\,, \tag{7.117}$$

and $(Q_\varepsilon, \delta_\infty) = c_\varphi = 0$. Therefore the terms involving $\beta_3 = \delta_\infty$ are all 0 and we can discard them reducing our computation to that of a 2×2 determinant instead with

$$\alpha_1 = Q_\varepsilon \,, \quad \alpha_2 = -\tfrac{1}{2}[(S + RS)\varepsilon_t + a_4 Q_\varepsilon] \,, \quad \beta_1 = \chi\psi \,, \quad \beta_2 = \delta_t \,. \tag{7.118}$$

Hence

$$(\alpha_1, \beta_1) = \tilde{v}_\varepsilon, \quad (\alpha_1, \beta_2) = q_\varepsilon \,, \tag{7.119}$$

$$(\alpha_2, \beta_1) = -\tfrac{1}{2}(\mathcal{P}_4 - a_4 + a_4 \tilde{v}_\varepsilon) \,, \tag{7.120}$$

$$(\alpha_2, \beta_2) = -\tfrac{1}{2}(\mathcal{R}_4 + a_4 q_\varepsilon) \,. \tag{7.121}$$

We want the limit of the determinant

$$\det\big(\delta_{jk} - (\alpha_j, \beta_k)\big)_{1 \le j,k \le 2} \,, \tag{7.122}$$

as $N \to \infty$. In order to get our hands on the limits of the individual terms involved in the determinant, we will find differential equations for them first as in [33]. Adding $a_4/2$ times row 1 to row 2 shows that a_4 falls out of the determinant, so we will not need to find differential equations for it. Thus our determinant is now

$$\det \begin{pmatrix} 1 - \tilde{v}_\varepsilon & -q_\varepsilon \\ \frac{1}{2}\mathcal{P}_4 & 1 + \frac{1}{2}\mathcal{R}_4 \end{pmatrix} \,. \tag{7.123}$$

Proceeding as in [33] we find the following differential equations

$$\frac{\mathrm{d}}{\mathrm{d}t} u_\varepsilon = -q_N q_\varepsilon \,, \qquad \frac{\mathrm{d}}{\mathrm{d}t} q_\varepsilon = q_N - q_N \tilde{v}_\varepsilon - p_N u_\varepsilon \,, \tag{7.124}$$

$$\frac{\mathrm{d}}{\mathrm{d}t} \mathcal{Q}_4 = -q_N(\mathcal{R}_4 + 1) \,, \qquad \frac{\mathrm{d}}{\mathrm{d}t} \mathcal{P}_4 = -p_N(\mathcal{R}_4 + 1) \,, \tag{7.125}$$

$$\frac{\mathrm{d}}{\mathrm{d}t} \mathcal{R}_4 = -p_N \mathcal{Q}_4 - q_N \mathcal{P}_4 \,. \tag{7.126}$$

Now we change variable from t to s where $t = \tau(s) = \sqrt{2N} + s/(\sqrt{2}N^{1/6})$ and take the limit $N \to \infty$, denoting the limits of q_ε, $\mathcal{P}_4$, $\mathcal{Q}_4$, $\mathcal{R}_4$, and the common limit of u_ε and $\tilde{v}_\varepsilon$ respectively by $\bar{q}$, $\overline{\mathcal{P}}_4$, $\overline{\mathcal{Q}}_4$, $\overline{\mathcal{R}}_4$ and $\bar{u}$. Also $\overline{\mathcal{P}}_4$ and $\overline{\mathcal{Q}}_4$ differ by a constant, namely $\overline{\mathcal{Q}}_4 = \overline{\mathcal{P}}_4 + \sqrt{2}/2$. These limits hold uniformly for bounded s so we can interchange $\lim_{N\to\infty}$ and $\mathrm{d}/\mathrm{d}s$. Also $\lim_{N\to\infty} N^{-1/6} q_N = \lim_{N\to\infty} N^{-1/6} p_N = q$, where q is as in (7.26). We obtain the systems

$$\frac{\mathrm{d}}{\mathrm{d}s} \bar{u} = -\frac{1}{\sqrt{2}} q \bar{q} \,, \qquad \frac{\mathrm{d}}{\mathrm{d}s} \bar{q} = \frac{1}{\sqrt{2}} q (1 - 2\bar{u}) \,, \tag{7.127}$$

$$\frac{\mathrm{d}}{\mathrm{d}s} \overline{\mathcal{P}}_4 = -\frac{1}{\sqrt{2}} q (\overline{\mathcal{R}}_4 + 1) \,, \quad \frac{\mathrm{d}}{\mathrm{d}s} \overline{\mathcal{R}}_4 = -\frac{1}{\sqrt{2}} q \left(2\overline{\mathcal{P}}_4 + \sqrt{\frac{\lambda}{2}} \right) \,, \tag{7.128}$$

The change of variables $s \to \mu = \int_s^\infty q(x, \lambda)\, \mathrm{d}x$ transforms these systems into constant coefficient ordinary differential equations

$$\frac{\mathrm{d}}{\mathrm{d}\mu}\bar{u} = \frac{1}{\sqrt{2}}\bar{q}\,, \qquad \frac{\mathrm{d}}{\mathrm{d}\mu}\bar{q} = -\frac{1}{\sqrt{2}}(1-2\bar{u})\,, \tag{7.129}$$

$$\frac{\mathrm{d}}{\mathrm{d}\mu}\overline{\mathcal{P}}_4 = \frac{1}{\sqrt{2}}(\overline{\mathcal{R}}_4+1)\,, \quad \frac{\mathrm{d}}{\mathrm{d}\mu}\overline{\mathcal{R}}_4 = \frac{1}{\sqrt{2}}\left(2\overline{\mathcal{P}}_4+\sqrt{\frac{\lambda}{2}}\right). \tag{7.130}$$

Since $\lim_{s\to\infty}\mu = 0$, corresponding to the boundary values at $t=\infty$ which we found earlier for $\mathcal{P}_4, \mathcal{R}_4$, we now have initial values at $\mu = 0$. Therefore

$$\bar{u}(\mu=0) = \bar{q}(\mu=0) = 0\,, \tag{7.131}$$

$$\overline{\mathcal{P}}_4(\mu=0) = -\sqrt{\frac{\lambda}{2}}\,, \quad \overline{\mathcal{R}}_4(\mu=0) = 0\,. \tag{7.132}$$

We use this to solve the systems and get

$$\bar{q} = \frac{1}{2\sqrt{2}}(\mathrm{e}^{-\mu}-\mathrm{e}^{\mu})\,, \tag{7.133}$$

$$\bar{u} = \tfrac{1}{2}(1-\tfrac{1}{2}\mathrm{e}^{\mu}-\tfrac{1}{2}\mathrm{e}^{-\mu})\,, \tag{7.134}$$

$$\overline{\mathcal{P}}_4 = \frac{1}{2\sqrt{2}}\left(\frac{2-\sqrt{\lambda}}{2}\mathrm{e}^{\mu}-\frac{2+\sqrt{\lambda}}{2}\mathrm{e}^{-\mu}-\sqrt{\lambda}\right), \tag{7.135}$$

$$\overline{\mathcal{R}}_4 = \frac{2-\sqrt{\lambda}}{4}e^{\mu}+\frac{2+\sqrt{\lambda}}{4}\mathrm{e}^{-\mu}-1\,. \tag{7.136}$$

Substituting these expressions into the determinant gives (7.29), namely

$$D_4(s,\lambda) = D_2(s,\lambda)\cosh^2\left(\frac{\mu(s,\lambda)}{2}\right), \tag{7.137}$$

where $D_\beta = \lim_{N\to\infty} D_{\beta,N}$. Note that even though there are λ-terms in (7.135) and (7.136), these do not appear in the final result (7.137), making it similar to the GUE case where the main conceptual difference between the $m=1$ (largest eigenvalue) case and the general m is the dependence of the function q on λ. The right hand side of the above formula clearly reduces to the $\beta=4$ Tracy–Widom distribution when we set $\lambda=1$. Note that where we have $D_4(s,\lambda)$ above, Tracy and Widom (and hence many RMT references) write $D_4(s/\sqrt{2},\lambda)$ instead. Tracy and Widom applied the change of variable $s\to s/\sqrt{2}$ in their derivation in [33] so as to agree with Mehta's form of the $\beta=4$ joint eigenvalue density, which has $-2x^2$ in the exponential in the weight function, instead of $-x^2$ in our case. To switch back to the other convention, one just needs to substitute in the argument $s/\sqrt{2}$ for s everywhere in our results. At this point this is just a cosmetic discrepancy, and it does not change anything in our derivations since all the differentiations are done with respect to λ anyway. It *does* change conventions for rescaling data while doing numerical work though.

7.7 The Distribution of the mth Largest Eigenvalue in the GOE

7.7.1 The Distribution Function as a Fredholm Determinant

The GOE corresponds case corresponds to the specialization $\beta = 1$ in (7.17) so that

$$G_{1,N}(t,\lambda) = C_1^{(N)} \int_{x_i\in\mathbb{R}} \cdots \int \prod_{j<k} |x_j - x_k| \prod_j^N w(x_j) \prod_j^N (1 + f(x_j))\,\mathrm{d}x_1 \cdots \mathrm{d}x_N \quad (7.138)$$

where $w(x) = \exp(-x^2)$, $f(x) = -\lambda\chi_J(x)$, and $C_1^{(N)}$ depends only on N. As in the GSE case, we will lump into $C_1^{(N)}$ any constants depending only on N that appear in the derivation. A simple argument at the end will show that the final constant is 1. These calculations more or less faithfully follow and expand on [34]. We want to use (7.33), which requires an ordered space. Note that the above integrand is symmetric under permutations, so the integral is $n!$ times the same integral over ordered pairs $x_1 \le \cdots \le x_N$. So we can rewrite (7.138) as

$$(N!) \int_{x_1\le\cdots\le x_N\in\mathbb{R}} \cdots \int \prod_{j<k} (x_k - x_j) \prod_{i=1}^N w(x_k) \prod_{i=1}^N (1 + f(x_k))\,\mathrm{d}x_1 \cdots \mathrm{d}x_N\,,$$

where we can remove the absolute values since the ordering insures that $(x_j - x_i) \ge 0$ for $i < j$. Recall that the Vandermonde determinant is

$$\varDelta_N(x) = \det(x_k^{j-1})_{j,k=1}^N = (-1)^{N(N-1)/2} \prod_{j<k} (x_j - x_k)\,.$$

Therefore what we have inside the integrand above is, up to sign

$$\det\Big(x_k^{j-1} w(x_k)\big(1 + f(x_k)\big)\Big)_{j,k=1}^N\,.$$

Note that the sign depends only on N. Now we can use (7.33) with

$$\varphi_j(x) = x^{j-1} w(x)\big(1 + f(x)\big)\,.$$

In using (7.33) we square both sides so that the right-hand side is now a determinant instead of a Pfaffian. Therefore $G_{1,N}^2(t,\lambda)$ equals

$$C_1^{(N)} \det\bigg(\iint \operatorname{sgn}(x-y) x^{j-1} y^{k-1} \big(1 + f(x)\big) w(x) w(y)\,\mathrm{d}x\,\mathrm{d}y\bigg)_{j,k=1}^N\,.$$

Shifting indices, we can write it as

$$C_1^{(N)} \det\left(\iint \operatorname{sgn}(x-y)x^j y^k\big(1+f(x)\big)w(x)w(y)\,\mathrm{d}x\,\mathrm{d}y\right)_{j,k=1}^{N-1} \tag{7.139}$$

where $C_1^{(N)}$ is a constant depending only on N, and is such that the right side is 1 if $f \equiv 0$. Indeed this would correspond to the probability that $\lambda_p^{\mathrm{GOE}(N)} < \infty$, or equivalently to the case where the excluded set J is empty. We can replace x^j and y^k by any arbitrary polynomials $p_j(x)$ and $p_k(x)$, of degree j and k respectively, which are obtained by row operations on the matrix. Indeed such operations would not change the determinant. We also replace $\operatorname{sgn}(x-y)$ by $\varepsilon(x-y) = \frac{1}{2}\operatorname{sgn}(x-y)$ which just produces a factor of 2 that we absorb in $C_1^{(N)}$. Thus $G_{1,N}^2(t,\lambda)$ now equals

$$C_1^{(N)} \det\left(\iint \varepsilon(x-y)p_j(x)p_k(y)\big(1+f(x)\big) \times \big(1+f(y)\big)w(x)w(y)\,\mathrm{d}x\,\mathrm{d}y\right)_{j,k=0}^{N-1}. \tag{7.140}$$

Let $\psi_j(x) = p_j(x)w(x)$ so the above integral becomes

$$C_1^{(N)} \det\left(\iint \varepsilon(x-y)\psi_j(x)\psi_k(y) \times \big(1+f(x)+f(y)+f(x)f(y)\big)\,\mathrm{d}x\,\mathrm{d}y\right)_{j,k=0}^{N-1}. \tag{7.141}$$

Partially multiplying out the term we obtain

$$C_1^{(N)} \det\left(\iint \varepsilon(x-y)\psi_j(x)\psi_k(y)\,\mathrm{d}x\,\mathrm{d}y + \iint \varepsilon(x-y)\psi_j(x)\psi_k(y) \times \big(f(x)+f(y)+f(x)f(y)\big)\,\mathrm{d}x\,\mathrm{d}y\right)_{j,k=0}^{N-1}. \tag{7.142}$$

Define

$$M = \left(\iint \varepsilon(x-y)\psi_j(x)\psi_k(y)\,\mathrm{d}x\,\mathrm{d}y\right)_{j,k=0}^{N-1}, \tag{7.143}$$

so that $G_{1,N}^2(t,\lambda)$ is now

$$C_1^{(N)} \det\left(M + \iint \varepsilon(x-y)\psi_j(x)\psi_k(y) \times \big(f(x)+f(y)+f(x)f(y)\big)\,\mathrm{d}x\,\mathrm{d}y\right)_{j,k=0}^{N-1}.$$

Let ε be the operator defined in (7.84). We can use operator notation to simplify the expression for $G_{1,N}^2(t,\lambda)$ a great deal by rewriting the double integrals as single integrals. Indeed

$$\iint \varepsilon(x-y)\psi_j(x)\psi_k(y)f(x)\,\mathrm{d}x\,\mathrm{d}y = \int f(x)\psi_j(x) \int \varepsilon(x-y)\psi_k(y)\,\mathrm{d}y\,\mathrm{d}x$$
$$= \int f\psi_j \varepsilon\psi_k\,\mathrm{d}x\,.$$

Similarly,

$$\begin{aligned}\iint \varepsilon(x-y)\psi_j(x)\psi_k(y)f(y)\,\mathrm{d}x\,\mathrm{d}y &= -\iint \varepsilon(y-x)\psi_j(x)\psi_k(y)f(y)\,\mathrm{d}x\,\mathrm{d}y \\ &= -\int f(y)\psi_k(y)\int \varepsilon(y-x)\psi_j(x)\,\mathrm{d}x\,\mathrm{d}y \\ &= -\int f(x)\psi_k(x)\int \varepsilon(x-y)\psi_j(y)\,\mathrm{d}y\,\mathrm{d}x \\ &= -\int f\psi_k\varepsilon\psi_j\,\mathrm{d}x\,.\end{aligned}$$

Finally,

$$\begin{aligned}&\iint \varepsilon(x-y)\psi_j(x)\psi_k(y)f(x)f(y)\,\mathrm{d}x\,\mathrm{d}y \\ &\quad = -\iint \varepsilon(y-x)\psi_j(x)\psi_k(y)f(x)f(y)\,\mathrm{d}x\,\mathrm{d}y \\ &\quad = -\int f(y)\psi_k(y)\int \varepsilon(y-x)f(x)\psi_j(x)\,\mathrm{d}x\,\mathrm{d}y \\ &\quad = -\int f(x)\psi_k(x)\int \varepsilon(x-y)f(y)\psi_j(y)\,\mathrm{d}y\,\mathrm{d}x \\ &\quad = -\int f\psi_k\varepsilon(f\psi_j)\,\mathrm{d}x\,.\end{aligned} \tag{7.144}$$

It follows that

$$\begin{aligned}&G^2_{1,N}(t,\lambda) \\ &\quad = C_1^{(N)} \det\left(M + \int [f\psi_j\varepsilon\psi_k - f\psi_k\varepsilon\psi_j - f\psi_k\varepsilon(f\psi_j)]\,\mathrm{d}x\right)_{j,k=0}^{N-1}.\end{aligned} \tag{7.145}$$

If we let $M^{-1} = (\mu_{jk})_{j,k=0}^{N-1}$, and factor $\det(M)$ out, then $G^2_{1,N}(t,\lambda)$ equals

$$\begin{aligned}&C_1^{(N)}\det(M) \\ &\quad \times \det\left(I + M^{-1}\left(\int [f\psi_j\varepsilon\psi_k - f\psi_k\varepsilon\psi_j - f\psi_k\varepsilon(f\psi_j)]\,\mathrm{d}x\right)_{j,k=0}^{N-1}\right)_{j,k=0}^{N-1}\end{aligned} \tag{7.146}$$

where the dot denotes matrix multiplication of M^{-1} and the matrix with the integral as its (j,k)-entry. define $\eta_j = \sum_j \mu_{jk}\psi_k$ and use it to simplify the result of carrying out the matrix multiplication. From (7.143) it follows that $\det(M)$ depends only on N we lump it into $C_1^{(N)}$. Thus $G_{1,N}^2(t,\lambda)$ equals

$$C_1^{(N)} \det\left(I + \left(\int [f\eta_j \varepsilon\psi_k - f\psi_k \varepsilon\eta_j - f\psi_k \varepsilon(f\eta_j)]\,\mathrm{d}x\right)_{j,k=0}^{N-1}\right)_{j,k=0}^{N-1} . \tag{7.147}$$

Recall our remark at the very beginning of the section that if $f \equiv 0$ then the integral we started with evaluates to 1 so that

$$C_1^{(N)} \det(I) = C_1^{(N)} , \tag{7.148}$$

which implies that $C_1^{(N)} = 1$. Now $G_{1,N}^2(t,\lambda)$ is of the form $\det(I+AB)$ where $A\colon L^2(J) \times L^2(J) \to \mathbb{C}^N$ is a $N \times 2$ matrix

$$\begin{pmatrix} A_1 \\ A_2 \\ \vdots \\ A_N \end{pmatrix} ,$$

whose jth row is given by

$$A_j = A_j(x) = \begin{pmatrix} -f\varepsilon\eta_j - f\varepsilon(f\eta_j) & f\eta_j \end{pmatrix} .$$

Therefore, if

$$g = \begin{pmatrix} g_1 \\ g_2 \end{pmatrix} \in L^2(J) \times L^2(J) ,$$

then Ag is a column vector whose jth row is $(A_j, g)_{L^2 \times L^2}$

$$(Ag)_j = \int [-f\varepsilon\eta_j - f\varepsilon(f\eta_j)] g_1 \,\mathrm{d}x + \int f\eta_j g_2 \,\mathrm{d}x .$$

Similarly, $B\colon \mathbb{C}^N \to L^2(J) \times L^2(J)$ is a $2 \times N$ matrix

$$B = \begin{pmatrix} B_1 & B_2 & \cdots & B_N \end{pmatrix} ,$$

whose jth column is given by

$$B_j = B_j(x) = \begin{pmatrix} \psi_j \\ \varepsilon\psi_j \end{pmatrix} .$$

Thus if

$$h = \begin{pmatrix} h_1 \\ \vdots \\ h_N \end{pmatrix} \in \mathbb{C}^N ,$$

then Bh is the column vector of $L^2(J) \times L^2(J)$ given by

$$Bh = \begin{pmatrix} \sum_j h_i \psi_j \\ \sum_j h_i \varepsilon \psi_j \end{pmatrix} .$$

Clearly $AB\colon \mathbb{C}^N \to \mathbb{C}^N$ and $BA\colon L^2(J) \times L^2(J) \to L^2(J) \times L^2(J)$ with kernel

$$\begin{pmatrix} -\sum_j \psi_j \otimes f\varepsilon\eta_j - \sum_j \psi_j \otimes f\varepsilon(f\eta_j) & \sum_j \psi_j \otimes f\eta_j \\ -\sum_j \varepsilon\psi_j \otimes f\varepsilon\eta_j - \sum_j \varepsilon\psi_j \otimes f\varepsilon(f\eta_j) & \sum_j \varepsilon\psi_j \otimes f\eta_j \end{pmatrix} .$$

Hence $I + BA$ has kernel

$$\begin{pmatrix} I - \sum_j \psi_j \otimes f\varepsilon\eta_j - \sum_j \psi_j \otimes f\varepsilon(f\eta_j) & \sum_j \psi_j \otimes f\eta_j \\ -\sum_j \varepsilon\psi_j \otimes f\varepsilon\eta_j - \sum_j \varepsilon\psi_j \otimes f\varepsilon(f\eta_j) & I + \sum_j \varepsilon\psi_j \otimes f\eta_j \end{pmatrix} ,$$

which can be written as

$$\begin{pmatrix} I - \sum_j \psi_j \otimes f\varepsilon\eta_j & \sum_j \psi_j \otimes f\eta_j \\ -\sum_j \varepsilon\psi_j \otimes f\varepsilon\eta_j - \varepsilon f & I + \sum_j \varepsilon\psi_j \otimes f\eta_j \end{pmatrix} \begin{pmatrix} I & 0 \\ \varepsilon f & I \end{pmatrix} .$$

Since we are taking the determinant of this operator expression, and the determinant of the second term is just 1, we can drop it. Therefore

$$\begin{aligned} G^2_{1,N}(t,\lambda) &= \det \begin{pmatrix} I - \sum_j \psi_j \otimes f\varepsilon\eta_j & \sum_j \psi_j \otimes f\eta_j \\ -\sum_j \varepsilon\psi_j \otimes f\varepsilon\eta_j - \varepsilon f & I + \sum_j \varepsilon\psi_j \otimes f\eta_j \end{pmatrix} \\ &= \det(I + K_{1,N} f) , \end{aligned}$$

where

$$\begin{aligned} K_{1,N} &= \begin{pmatrix} -\sum_j \psi_j \otimes \varepsilon\eta_j & \sum_j \psi_j \otimes \eta_j \\ -\sum_j \varepsilon\psi_j \otimes \varepsilon\eta_j - \varepsilon & \sum_j \varepsilon\psi_j \otimes \eta_j \end{pmatrix} \\ &= \begin{pmatrix} -\sum_{j,k} \psi_j \otimes \mu_{jk}\varepsilon\psi_k & \sum_{j,k} \psi_j \otimes \mu_{jk}\psi_k \\ -\sum_{j,k} \varepsilon\psi_j \otimes \mu_{jk}\varepsilon\psi_k - \varepsilon & \sum_{j,k} \varepsilon\psi_j \otimes \mu_{jk}\psi_k \end{pmatrix} \end{aligned}$$

and $K_{1,N}$ has matrix kernel

$$K_{1,N}(x,y) = \begin{pmatrix} -\sum_{j,k} \psi_j(x)\mu_{jk}\varepsilon\psi_k(y) & \sum_{j,k} \psi_j(x)\mu_{jk}\psi_k(y) \\ -\sum_{j,k} \varepsilon\psi_j(x)\mu_{jk}\varepsilon\psi_k(y) - \varepsilon(x-y) & \sum_{j,k} \varepsilon\psi_j(x)\mu_{jk}\psi_k(y) \end{pmatrix} .$$

We define

$$S_N(x,y) = -\sum_{j,k=0}^{N-1} \psi_j(x)\mu_{jk}\varepsilon\psi_k(y) .$$

Since M is antisymmetric,

$$\begin{aligned} S_N(y,x) &= -\sum_{j,k=0}^{N-1} \psi_j(y)\mu_{jk}\varepsilon\psi_k(x) \\ &= \sum_{j,k=0}^{N-1} \psi_j(y)\mu_{kj}\varepsilon\psi_k(x) = \sum_{j,k=0}^{N-1} \varepsilon\psi_j(x)\mu_{jk}\psi_k(y) . \end{aligned}$$

Note that

$$\varepsilon S_N(x,y) = \sum_{j,k=0}^{N-1} \varepsilon\psi_j(x)\mu_{jk}\varepsilon\psi_k(y) ,$$

whereas

$$-\frac{\mathrm{d}}{\mathrm{d}y} S_N(x,y) = \sum_{j,k=0}^{N-1} \psi_j(x)\mu_{jk}\psi_k(y) .$$

So we can now write succinctly

$$K_{1,N}(x,y) = \begin{pmatrix} S_N(x,y) & -\mathrm{d}S_N(x,y)/\mathrm{d}y \\ \varepsilon S_N(x,y) - \varepsilon & S_N(y,x) \end{pmatrix} . \tag{7.149}$$

So we have shown that

$$G_{1,N}(t,\lambda) = \sqrt{D_{1,N}(t,\lambda)} \tag{7.150}$$

where

$$D_{1,N}(t,\lambda) = \det(I + K_{1,N} f)$$

where $K_{1,N}$ is the integral operator with matrix kernel $K_{1,N}(x,y)$ given in (7.149).

7.7.2 Gaussian Specialization

We specialize the results above to the case of a Gaussian weight function

$$w(x) = \exp(-x^2/2) \tag{7.151}$$

and indicator function

$$f(x) = -\lambda\chi_J , \quad J = (t,\infty) .$$

Note that this does not agree with the weight function in (7.17). However it is a necessary choice if we want the technical convenience of working with exactly the same orthogonal polynomials (the Hermite functions) as in the $\beta = 2,4$ cases. In turn the Painlevé function in the limiting distribution will be unchanged. The discrepancy is resolved by the choice of standard deviation. Namely here the standard deviation on the diagonal matrix elements is taken to be 1, corresponding to the weight function (7.151). In the $\beta = 2,4$ cases the standard deviation on the diagonal matrix elements is $1/\sqrt{2}$, giving the weight function (7.88). Now we again want the matrix

$$M = \left(\iint \varepsilon(x-y)\psi_j(x)\psi_k(y)\,\mathrm{d}x\,\mathrm{d}y \right)_{j,k=0}^{N-1} = \left(\int \psi_j(x)\varepsilon\psi_k(x)\,\mathrm{d}x \right)_{j,k=0}^{N-1}$$

to be the direct sum of $N/2$ copies of

$$Z = \begin{pmatrix} 0 & 1 \\ -1 & 0 \end{pmatrix}$$

so that the formulas are the simplest possible, since then μ_{jk} can only be 0 or ± 1. In that case M would be skew-symmetric so that $M^{-1} = -M$. In terms of the integrals defining the entries of M this means that we would like to have

$$\int \psi_{2m}(x)\varepsilon\psi_{2n+1}(x)\,\mathrm{d}x = \delta_{m,n}\ , \qquad \int \psi_{2m+1}(x)\varepsilon\psi_{2n}(x)\,\mathrm{d}x = -\delta_{m,n}\ ,$$

and otherwise

$$\int \psi_j(x)\frac{\mathrm{d}}{\mathrm{d}x}\psi_k(x)\,\mathrm{d}x = 0\ .$$

It is easier to treat this last case if we replace it with three nonexclusive conditions

$$\int \psi_{2m}(x)\varepsilon\psi_{2n}(x)\,\mathrm{d}x = 0\ , \qquad \int \psi_{2m+1}(x)\varepsilon\psi_{2n+1}(x)\,\mathrm{d}x = 0$$

(so when the parity is the same for j, k, which takes care of diagonal entries, among others), and

$$\int \psi_j(x)\varepsilon\psi_k(x)\,\mathrm{d}x = 0\ .$$

whenever $|j-k|>1$, which targets entries outside of the tridiagonal. Define

$$\varphi_n(x) = \frac{1}{c_n}H_n(x)\exp(-x^2/2) \quad \text{for } c_n = \sqrt{2^n n!\sqrt{\pi}}$$

where the H_n are the usual Hermite polynomials defined by the orthogonality condition

$$\int H_j(x)H_k(x)e^{-x^2}\,\mathrm{d}x = c_j^2\delta_{j,k}\ .$$

It follows that

$$\int \varphi_j(x)\varphi_k(x)\,\mathrm{d}x = \delta_{j,k}\ .$$

Now let

$$\psi_{2n+1}(x) = \frac{\mathrm{d}}{\mathrm{d}x}\varphi_{2n}(x)\ , \quad \psi_{2n}(x) = \varphi_{2n}(x)\ . \tag{7.152}$$

This definition satisfies our earlier requirement that $\psi_j = p_j w$ for

$$w(x) = \exp(-x^2/2)\ .$$

In this case for example

$$p_{2n}(x) = \frac{1}{c_n}H_{2n}(x)\ .$$

With ε defined as in (7.84), and recalling that, if D denote the operator that acts by differentiation with respect to x, then $D\varepsilon = \varepsilon D = I$, it follows that

$$\begin{aligned}\int \psi_{2m}(x)\varepsilon\psi_{2n+1}(x)\,\mathrm{d}x &= \int \varphi_{2m}(x)\varepsilon\frac{\mathrm{d}}{\mathrm{d}x}\varphi_{2n+1}(x)\,\mathrm{d}x \\ &= \int \varphi_{2m}(x)\varphi_{2n+1}(x)\,\mathrm{d}x \\ &= \int \varphi_{2m}(x)\varphi_{2n+1}(x)\,\mathrm{d}(x) = \delta_{m,n},\end{aligned}$$

as desired. Similarly, integration by parts gives

$$\begin{aligned}\int \psi_{2m+1}(x)\varepsilon\psi_{2n}(x)\,dx &= \int \frac{\mathrm{d}}{\mathrm{d}x}\varphi_{2m}(x)\varepsilon\varphi_{2n}(x)\,\mathrm{d}x \\ &= -\int \varphi_{2m}(x)\varphi_{2n}(x)\,\mathrm{d}x \\ &= -\int \varphi_{2m}(x)\varphi_{2n+1}(x)\,\mathrm{d}(x) = -\delta_{m,n}\,.\end{aligned}$$

Also ψ_{2n} is even since H_{2n} and φ_{2n} are. Similarly, ψ_{2n+1} is odd. It follows that $\varepsilon\psi_{2n}$, and $\varepsilon\psi_{2n+1}$, are respectively odd and even functions. From these observations, we obtain

$$\int \psi_{2n}(x)\varepsilon\psi_{2m}(x)\,\mathrm{d}x = 0\,,$$

since the integrand is a product of an odd and an even function. Similarly

$$\int \psi_{2n+1}(x)\varepsilon\psi_{2m+1}(x)\,\mathrm{d}x = 0\,.$$

Finally it is easy to see that if $|j-k|>1$, then

$$\int \psi_j(x)\varepsilon\psi_k(x)\,\mathrm{d}x = 0\,.$$

Indeed both differentiation and the action of ε can only "shift" the indices by 1. Thus by orthogonality of the φ_j, this integral will always be 0. Thus by our choice in (7.152), we force the matrix

$$M = \left(\int \psi_j(x)\varepsilon\psi_k(x)\,\mathrm{d}x\right)_{j,k=0}^{N-1}$$

to be the direct sum of $N/2$ copies of

$$Z = \begin{pmatrix} 0 & 1 \\ -1 & 0 \end{pmatrix}\,.$$

This means $M^{-1} = -M$ where $M^{-1} = \{\mu_{j,k}\}$. Moreover, $\mu_{j,k} = 0$ if j,k have the same parity or $|j-k| > 1$, and $\mu_{2j,2k+1} = \delta_{jk} = -\mu_{2k+1,2j}$ for $j,k = 0,\ldots,N/2-1$. Therefore

$$\begin{aligned} S_N(x,y) &= -\sum_{j,k=0}^{N-1} \psi_j(x)\mu_{jk}\varepsilon\psi_k(y) \\ &= -\sum_{j=0}^{N/2-1} \psi_{2j}(x)\varepsilon\psi_{2j+1}(y) + \sum_{j=0}^{N/2-1} \psi_{2j+1}(x)\varepsilon\psi_{2j}(y) \\ &= \left[\sum_{j=0}^{N/2-1} \varphi_{2j}\left(\frac{x}{\sigma}\right)\varphi_{2j}\left(\frac{y}{\sigma}\right) - \sum_{j=0}^{N/2-1} \frac{\mathrm{d}}{\mathrm{d}x}\varphi_{2j}\left(\frac{x}{\sigma}\right)\varepsilon\varphi_{2j}\left(\frac{y}{\sigma}\right)\right]. \end{aligned}$$

Manipulations similar to those in the $\beta = 4$ case (see (7.90) through (7.95)) yield

$$S_N(x,y) = \left[\sum_{j=0}^{N-1} \varphi_j(x)\varphi_j(y) - \sqrt{\frac{N}{2}}\varphi_{N-1}(x)(\varepsilon\varphi_N)(y)\right].$$

We redefine

$$S_N(x,y) = \sum_{j=0}^{N-1} \varphi_j(x)\varphi_j(y) = S_N(y,x)\,,$$

so that the top left entry of $K_{1,N}(x,y)$ is

$$S_N(x,y) + \sqrt{\frac{N}{2}}\varphi_{N-1}(x)(\varepsilon\varphi_N)(y)\,.$$

If S_N is the operator with kernel $S_N(x,y)$ then integration by parts gives

$$S_N Df = \int S(x,y)\frac{\mathrm{d}}{\mathrm{d}y}f(y)\,\mathrm{d}y = \int \left(-\frac{\mathrm{d}}{\mathrm{d}y}S_N(x,y)\right)f(y)\,\mathrm{d}y\,,$$

so that $-\mathrm{d}/\mathrm{d}y S_N(x,y)$ is in fact the kernel of $S_N D$. Therefore (7.150) now holds with $K_{1,N}$ being the integral operator with matrix kernel $K_{1,N}(x,y)$ whose (i,j)-entry $K_{1,N}^{(i,j)}(x,y)$ is given by

$$\begin{aligned} K_{1,N}^{(1,1)}(x,y) &= \left[S_N(x,y) + \sqrt{\frac{N}{2}}\varphi_{N-1}(x)(\varepsilon\varphi_N)(y)\right], \\ K_{1,N}^{(1,2)}(x,y) &= \left[SD_N(x,y) - \frac{\mathrm{d}}{\mathrm{d}y}\Big(\sqrt{\frac{N}{2}}\varphi_{N-1}(x)\big(\varepsilon\varphi_N)(y)\big)\Big)\right], \\ K_{1,N}^{(2,1)}(x,y) &= \varepsilon\left[S_N(x,y) + \sqrt{\frac{N}{2}}\varphi_{N-1}(x)(\varepsilon\varphi_N)(y) - 1\right], \\ K_{1,N}^{(2,2)}(x,y) &= \left[S_N(x,y) + \sqrt{\frac{N}{2}}(\varepsilon\varphi_N)(x)\varphi_{N-1}(y)\right]. \end{aligned}$$

Define

$$\varphi(x) = \left(\frac{N}{2}\right)^{1/4} \varphi_N(x)\,, \quad \psi(x) = \left(\frac{N}{2}\right)^{1/4} \varphi_{N-1}(x)\,,$$

so that

$$\begin{aligned}
K_{1,N}^{(1,1)}(x,y) &= \chi(x)[S_N(x,y) + \psi(x)\varepsilon\varphi(y)]\chi(y)\,,\\
K_{1,N}^{(1,2)}(x,y) &= \chi(x)[SD_N(x,y) - \psi(x)\varphi(y)]\chi(y)\,,\\
K_{1,N}^{(2,1)}(x,y) &= \chi(x)[\varepsilon S_N(x,y) + \varepsilon\psi(x)\varepsilon\varphi(y) - \varepsilon(x-y)]\chi(y)\,,\\
K_{1,N}^{(2,2)}(x,y) &= \chi(x)[S_N(x,y) + \varepsilon\varphi(x)\psi(y)]\chi(y)\,.
\end{aligned}$$

Note that

$$\begin{aligned}
\chi(S + \psi\otimes\varepsilon\varphi)\chi &\doteq K_{1,N}^{(1,1)}(x,y)\,,\\
\chi(SD - \psi\otimes\varphi)\chi &\doteq K_{1,N}^{(1,2)}(x,y)\,,\\
\chi(\varepsilon S + \varepsilon\psi\otimes\varepsilon\varphi - \varepsilon)\chi &\doteq K_{1,N}^{(2,1)}(x,y)\,,\\
\chi(S + \varepsilon\varphi\otimes\varepsilon\psi)\chi &\doteq K_{1,N}^{(2,2)}(x,y)\,.
\end{aligned}$$

Hence

$$K_{1,N} = \chi \begin{pmatrix} S + \psi\otimes\varepsilon\varphi & SD - \psi\otimes\varphi \\ \varepsilon S + \varepsilon\psi\otimes\varepsilon\varphi - \varepsilon & S + \varepsilon\varphi\otimes\varepsilon\psi \end{pmatrix} \chi\,.$$

Note that this is identical to the corresponding operator for $\beta = 1$ obtained by Tracy and Widom in [33], the only difference being that φ, ψ, and hence also S, are redefined to depend on λ.

7.7.3 Edge Scaling

7.7.3.1 Reduction of the Determinant

The above determinant is that of an operator on $L^2(J)\oplus L^2(J)$. Our first task will be to rewrite these determinants as those of operators on $L^2(J)$. This part follows exactly the proof in [33]. To begin, note that

$$[S, D] = \varphi\otimes\psi + \psi\otimes\varphi \tag{7.153}$$

so that (using the fact that $D\varepsilon = \varepsilon D = I$)

$$\begin{aligned}
[\varepsilon, S] &= \varepsilon S - S\varepsilon = \varepsilon SD\varepsilon - \varepsilon DS\varepsilon\\
&= \varepsilon[S, D]\varepsilon = \varepsilon\varphi\otimes\psi\varepsilon + \varepsilon\psi\otimes\varphi\varepsilon\\
&= \varepsilon\varphi\otimes\varepsilon^t\psi + \varepsilon\psi\otimes\varepsilon^t\varphi = -\varepsilon\varphi\otimes\varepsilon\psi - \varepsilon\psi\otimes\varepsilon\varphi\,,
\end{aligned} \tag{7.154}$$

where the last equality follows from the fact that $\varepsilon^t = -\varepsilon$. We thus have

$$D(\varepsilon S + \varepsilon\psi \otimes \varepsilon\varphi) = S + \psi \otimes \varepsilon\varphi\,,$$
$$D(\varepsilon SD - \varepsilon\psi \otimes \varphi) = SD - \psi \otimes \varphi\,.$$

The expressions on the right side are the top entries of $K_{1,N}$. Thus the first row of $K_{1,N}$ is, as a vector,

$$D(\varepsilon S + \varepsilon\psi \otimes \varepsilon\varphi, \varepsilon SD - \varepsilon\psi \otimes \varphi)\,.$$

Now (7.154) implies that

$$\varepsilon S + \varepsilon\psi \otimes \varepsilon\varphi = S\varepsilon - \varepsilon\varphi \otimes \varepsilon\psi\,.$$

Similarly (7.153) gives

$$\varepsilon[S, D] = \varepsilon\varphi \otimes \psi + \varepsilon\psi \otimes \varphi\,,$$

so that

$$\varepsilon SD - \varepsilon\psi \otimes \varphi = \varepsilon DS + \varepsilon\varphi \otimes \psi = S + \varepsilon\varphi \otimes \psi\,.$$

Using these expressions we can rewrite the first row of $K_{1,N}$ as

$$D(S\varepsilon - \varepsilon\varphi \otimes \varepsilon\psi, S + \varepsilon\varphi \otimes \psi)\,.$$

Applying ε to this expression shows the second row of $K_{1,N}$ is given by

$$(\varepsilon S - \varepsilon + \varepsilon\psi \otimes \varepsilon\varphi, S + \varepsilon\varphi \otimes \psi)\,.$$

Now use (7.154) to show the second row of $K_{1,N}$ is

$$(S\varepsilon - \varepsilon + \varepsilon\varphi \otimes \varepsilon\psi, S + \varepsilon\varphi \otimes \psi)\,.$$

Therefore,

$$\begin{aligned} K_{1,N} &= \chi \begin{pmatrix} D(S\varepsilon - \varepsilon\varphi \otimes \varepsilon\psi) & D(S + \varepsilon\varphi \otimes \psi) \\ S\varepsilon - \varepsilon + \varepsilon\varphi \otimes \varepsilon\psi & S + \varepsilon\varphi \otimes \psi \end{pmatrix} \chi \\ &= \begin{pmatrix} \chi D & 0 \\ 0 & \chi \end{pmatrix} \begin{pmatrix} (S\varepsilon - \varepsilon\varphi \otimes \varepsilon\psi)\chi & (S + \varepsilon\varphi \otimes \psi)\chi \\ (S\varepsilon - \varepsilon + \varepsilon\varphi \otimes \varepsilon\psi)\chi & (S + \varepsilon\varphi \otimes \psi)\chi \end{pmatrix}\,. \end{aligned}$$

Since $K_{1,N}$ is of the form AB, we can use the fact that $\det(I-AB) = \det(I-BA)$ and deduce that $D_{1,N}(s,\lambda)$ is unchanged if instead we take $K_{1,N}$ to be

$$\begin{aligned} K_{1,N} &= \begin{pmatrix} (S\varepsilon - \varepsilon\varphi \otimes \varepsilon\psi)\chi & (S + \varepsilon\varphi \otimes \psi)\chi \\ (S\varepsilon - \varepsilon + \varepsilon\varphi \otimes \varepsilon\psi)\chi & (S + \varepsilon\varphi \otimes \psi)\chi \end{pmatrix} \begin{pmatrix} \chi D & 0 \\ 0 & \chi \end{pmatrix} \\ &= \begin{pmatrix} (S\varepsilon - \varepsilon\varphi \otimes \varepsilon\psi)\chi D & (S + \varepsilon\varphi \otimes \psi)\chi \\ (S\varepsilon - \varepsilon + \varepsilon\varphi \otimes \varepsilon\psi)\chi D & (S + \varepsilon\varphi \otimes \psi)\chi \end{pmatrix}\,. \end{aligned}$$

Therefore

$$D_{1,N}(s,\lambda) = \det \begin{pmatrix} I - (S\varepsilon - \varepsilon\varphi \otimes \varepsilon\psi)\lambda\chi D & -(S + \varepsilon\varphi \otimes \psi)\lambda\chi \\ -(S\varepsilon - \varepsilon + \varepsilon\varphi \otimes \varepsilon\psi)\lambda\chi D & I - (S + \varepsilon\varphi \otimes \psi)\lambda\chi \end{pmatrix}\,.$$

Now we perform row and column operations on the matrix to simplify it, which do not change the Fredholm determinant. Justification of these operations is given in [33]. We start by subtracting row 1 from row 2 to get

$$\begin{pmatrix} I - (S\varepsilon - \varepsilon\varphi \otimes \varepsilon\psi)\lambda\chi D & -(S + \varepsilon\varphi \otimes \psi)\lambda\chi \\ -I + \varepsilon\lambda\chi D & I \end{pmatrix} .$$

Next, adding column 2 to column 1 yields

$$\begin{pmatrix} I - (S\varepsilon - \varepsilon\varphi \otimes \varepsilon\psi)\lambda\chi D - (S + \varepsilon\varphi \otimes \psi)\lambda\chi & -(S + \varepsilon\varphi \otimes \psi)\lambda\chi \\ \varepsilon\lambda\chi D & I \end{pmatrix} .$$

Then right-multiply column 2 by $-\varepsilon\lambda\chi D$ and add it to column 1, and multiply row 2 by $S + \varepsilon\varphi \otimes \psi$ and add it to row 1 to arrive at

$$\det \begin{pmatrix} I - (S\varepsilon - \varepsilon\varphi \otimes \varepsilon\psi)\lambda\chi D + (S + \varepsilon\varphi \otimes \psi)\lambda\chi(\varepsilon\lambda\chi D - I) & 0 \\ 0 & I \end{pmatrix} .$$

Thus the determinant we want equals the determinant of

$$I - (S\varepsilon - \varepsilon\varphi \otimes \varepsilon\psi)\lambda\chi D + (S + \varepsilon\varphi \otimes \psi)\lambda\chi(\varepsilon\lambda\chi D - I) . \tag{7.155}$$

So we have reduced the problem from the computation of the Fredholm determinant of an operator on $L^2(J) \times L^2(J)$, to that of an operator on $L^2(J)$.

7.7.3.2 Differential Equations

Next we want to write the operator in (7.155) in the form

$$(I - K_{2,N})\Big(I - \sum_{i=1}^{L} \alpha_i \otimes \beta_i\Big) , \tag{7.156}$$

where the α_i and β_i are functions in $L^2(J)$. In other words, we want to rewrite the determinant for the GOE case as a finite dimensional perturbation of the corresponding GUE determinant. The Fredholm determinant of the product is then the product of the determinants. The limiting form for the GUE part is already known, and we can just focus on finding a limiting form for the determinant of the finite dimensional piece. It is here that the proof must be modified from that in [33]. A little simplification of (7.155) yields

$$I - \lambda S\chi - \lambda S(1 - \lambda\chi)\varepsilon\chi D - \lambda(\varepsilon\varphi \otimes \chi\psi) - \lambda(\varepsilon\varphi \otimes \psi)(1 - \lambda\chi)\varepsilon\chi D .$$

Writing $\varepsilon[\chi, D] + \chi$ for $\varepsilon\chi D$ and simplifying $(1 - \lambda\chi)\chi$ to $(1 - \lambda)\chi$ gives

$$\begin{aligned} I &- \lambda S\chi - \lambda(1 - \lambda)S\chi - \lambda(\varepsilon\varphi \otimes \chi\psi) - \lambda(1 - \lambda)(\varepsilon\varphi \otimes \chi\psi) \\ &\qquad - \lambda S(1 - \lambda\chi)\varepsilon[\chi, D] - \lambda(\varepsilon\varphi \otimes \psi)(1 - \lambda\chi)\varepsilon[\chi, D] \\ &= I - (2\lambda - \lambda^2)S\chi - (2\lambda - \lambda^2)(\varepsilon\varphi \otimes \chi\psi) - \lambda S(1 - \lambda\chi)\varepsilon[\chi, D] \\ &\qquad - \lambda(\varepsilon\varphi \otimes \psi)(1 - \lambda\chi)\varepsilon[\chi, D] . \end{aligned}$$

Define $\tilde{\lambda} = 2\lambda - \lambda^2$ and let $\sqrt{\tilde{\lambda}}\varphi \to \varphi$, and $\sqrt{\tilde{\lambda}}\psi \to \psi$ so that $\tilde{\lambda}S \to S$ and (7.155) goes to

$$I - S\chi - (\varepsilon\varphi \otimes \chi\psi) - \frac{\lambda}{\tilde{\lambda}}S(1-\lambda\chi)\varepsilon[\chi, D] - \frac{\lambda}{\tilde{\lambda}}(\varepsilon\varphi \otimes \psi)(1-\lambda\chi)\varepsilon[\chi, D] \,.$$

Now we define $R := (I - S\chi)^{-1}S\chi = (I - S\chi)^{-1} - I$ (the resolvent operator of $S\chi$), whose kernel we denote by $R(x,y)$, and $Q_\varepsilon := (I - S\chi)^{-1}\varepsilon\varphi$. Then (7.155) factors into

$$A = (I - S\chi)B \,.$$

where B is

$$I - (Q_\varepsilon \otimes \chi\psi) - \frac{\lambda}{\tilde{\lambda}}(I + R)S(1-\lambda\chi)\varepsilon[\chi, D] - \frac{\lambda}{\tilde{\lambda}}(Q_\varepsilon \otimes \psi)(1-\lambda\chi)\varepsilon[\chi, D] \,, \qquad \lambda \neq 1 \,.$$

Hence

$$D_{1,N}(s, \lambda) = D_{2,N}(s, \tilde{\lambda}) \det(B) \,.$$

Note that because of the change of variable $\tilde{\lambda}S \to S$, we are in effect factoring $I - (2\lambda - \lambda^2)S$, rather that $I - \lambda S$ as we did in the $\beta = 4$ case. The fact that we factored $I - (2\lambda - \lambda^2)S\chi$ as opposed to $I - \lambda S\chi$ is crucial here for it is what makes B finite rank. If we had factored $I - \lambda S\chi$ instead, B would have been

$$\begin{aligned} B = I - \lambda\sum_{k=1}^{2}(-1)^k(S + RS)(I - \lambda\chi)\varepsilon_k \otimes \delta_k - \lambda(I + R)\varepsilon\varphi \otimes \chi\psi \\ - \lambda\sum_{k=1}^{2}(-1)^k\big(\psi, (I - \lambda\chi)\varepsilon_k\big)\big((I + R)\varepsilon\varphi\big) \otimes \delta_k \\ - \lambda(1-\lambda)(S + RS)\chi - \lambda(1-\lambda)\big((I + R)\varepsilon\varphi\big) \otimes \chi\psi \,. \end{aligned}$$

The first term on the last line is not finite rank, and the methods we have used previously in the $\beta = 4$ case would not work here. It is also interesting to note that these complications disappear when we are dealing with the case of the largest eigenvalue; then is no differentiation with respect to λ, and we just set $\lambda = 1$ in all these formulae. All the new troublesome terms vanish! In order to find $\det(B)$ we use the identity

$$\varepsilon[\chi, D] = \sum_{k=1}^{2m}(-1)^k\varepsilon_k \otimes \delta_k \,, \qquad (7.157)$$

where ε_k and δ_k are the functions $\varepsilon(x - a_k)$ and $\delta(x - a_k)$ respectively, and the a_k are the endpoints of the (disjoint) intervals considered, $J = \cup_{k=1}^{m}(a_{2k-1}, a_{2k})$. We also make use of the fact that

$$a \otimes b \times c \otimes d = (b, c) \times a \otimes d \tag{7.158}$$

where $(\cdot, \cdot)$ is the usual L^2-inner product. Therefore

$$\begin{aligned}(Q_\varepsilon \otimes \psi)(1 - \lambda\chi)\varepsilon[\chi, D] &= \sum_{k=1}^{2m} (-1)^k Q_\varepsilon \otimes \psi \times (1 - \lambda\chi)\varepsilon_k \otimes \delta_k \\ &= \sum_{k=1}^{2m} (-1)^k (\psi, (1 - \lambda\chi)\varepsilon_k) Q_\varepsilon \otimes \delta_k \,.\end{aligned}$$

It follows that

$$\frac{D_{1,N}(s, \lambda)}{D_{2,N}(s, \tilde{\lambda})}$$

equals the determinant of

$$I - Q_\varepsilon \otimes \chi\psi - \frac{\lambda}{\tilde{\lambda}} \sum_{k=1}^{2m} (-1)^k [(S + RS)(1 - \lambda\chi)\varepsilon_k + (\psi, (1 - \lambda\chi)\varepsilon_k), Q_\varepsilon] \otimes \delta_k \,. \tag{7.159}$$

We now specialize to the case of one interval $J = (t, \infty)$, so $m = 1$, $a_1 = t$ and $a_2 = \infty$. We write $\varepsilon_t = \varepsilon_1$, and $\varepsilon_\infty = \varepsilon_2$, and similarly for δ_k. Writing the terms in the summation and using the facts that

$$\varepsilon_\infty = -\tfrac{1}{2} \,, \tag{7.160}$$

and

$$(1 - \lambda\chi)\varepsilon_t = -\tfrac{1}{2}(1 - \lambda\chi) + (1 - \lambda\chi)\chi \,, \tag{7.161}$$

then yields

$$\begin{aligned}I - Q_\varepsilon \otimes \chi\psi &- \frac{\lambda}{2\tilde{\lambda}} [(S + RS)(1 - \lambda\chi) + \big(\psi, (1 - \lambda\chi)\big) Q_\varepsilon] \otimes (\delta_t - \delta_\infty) \\ &+ \frac{\lambda}{\tilde{\lambda}} [(S + RS)(1 - \lambda\chi)\chi + (\psi, (1 - \lambda\chi)\chi) Q_\varepsilon] \otimes \delta_t\end{aligned}$$

which, to simplify notation, we write as

$$\begin{aligned}I - Q_\varepsilon \otimes \chi\psi &- \frac{\lambda}{2\tilde{\lambda}} [(S + RS)(1 - \lambda\chi) + a_{1,\lambda} Q_\varepsilon] \otimes (\delta_t - \delta_\infty) \\ &+ \frac{\lambda}{\tilde{\lambda}} [(S + RS)(1 - \lambda\chi)\chi + \tilde{a}_{1,\lambda} Q_\varepsilon] \otimes \delta_t \,,\end{aligned}$$

where

$$a_{1,\lambda} = \big(\psi, (1 - \lambda\chi)\big), \quad \tilde{a}_{1,\lambda} = (\psi, (1 - \lambda\chi)\chi) \,. \tag{7.162}$$

Now we can use the formula:

$$\det\left(I - \sum_{i=1}^{L} \alpha_i \otimes \beta_i\right) = \det\big(\delta_{jk} - (\alpha_j, \beta_k)\big)_{j,k=1}^{L} \,. \tag{7.163}$$

In this case, $L=3$, and

$$\begin{aligned}\alpha_1 = Q_\varepsilon, \quad \alpha_2 &= \frac{\lambda}{\tilde{\lambda}}[(S+RS)(1-\lambda\chi)+a_{1,\lambda}Q_\varepsilon \,, \\ \alpha_3 &= -\frac{\lambda}{\tilde{\lambda}}[(S+RS)(1-\lambda\chi)\chi+\tilde{a}_{1,\lambda}Q_\varepsilon] \,, \\ \beta_1 = \chi\psi \,, \quad \beta_2 &= \delta_t-\delta_\infty \,, \quad \beta_3=\delta_t \,.\end{aligned} \tag{7.164}$$

In order to simplify the notation, define

$$\begin{aligned} Q(x,\lambda,t) &:= (I-S\chi)^{-1}\varphi \,, & P(x,\lambda,t) &:= (I-S\chi)^{-1}\psi \,, \\ Q_\varepsilon(x,\lambda,t) &:= (I-S\chi)^{-1}\varepsilon\varphi \,, & P_\varepsilon(x,\lambda,t) &:= (I-S\chi)^{-1}\varepsilon\psi \,, \end{aligned} \tag{7.165}$$

$$\begin{aligned} q_N &:= Q(t,\lambda,t) \,, & p_N &:= P(t,\lambda,t) \,, \\ q_\varepsilon &:= Q_\varepsilon(t,\lambda,t) \,, & p_\varepsilon &:= P_\varepsilon(t,\lambda,t) \,, \\ u_\varepsilon &:= (Q,\chi\varepsilon\varphi)=(Q_\varepsilon,\chi\varphi) \,, & v_\varepsilon &:= (Q,\chi\varepsilon\psi)=(P_\varepsilon,\chi\psi) \,, \\ \tilde{v}_\varepsilon &:= (P,\chi\varepsilon\varphi)=(Q_\varepsilon,\chi\varphi) \,, & w_\varepsilon &:= (P,\chi\varepsilon\psi)=(P_\varepsilon,\chi\psi) \,, \end{aligned} \tag{7.166}$$

$$\begin{aligned} \mathcal{P}_{1,\lambda} &:= \int(1-\lambda\chi)P\,\mathrm{d}x \,, & \widetilde{\mathcal{P}}_{1,\lambda} &:= \int(1-\lambda\chi)\chi P\,\mathrm{d}x \,, \\ \mathcal{Q}_{1,\lambda} &:= \int(1-\lambda\chi)Q\,\mathrm{d}x \,, & \widetilde{\mathcal{Q}}_{1,\lambda} &:= \int(1-\lambda\chi)\chi Q\,\mathrm{d}x \,, \\ \mathcal{R}_{1,\lambda} &:= \int(1-\lambda\chi)R(x,t)\,\mathrm{d}x \,, & \widetilde{\mathcal{R}}_{1,\lambda} &:= \int(1-\lambda\chi)\chi R(x,t)\,\mathrm{d}x \,. \end{aligned} \tag{7.167}$$

Note that all quantities in (7.166) and (7.165) are functions of t alone. Furthermore, let

$$c_\varphi = \varepsilon\varphi(\infty) = \frac{1}{2}\int_{-\infty}^{\infty}\varphi(x)\,\mathrm{d}x \,, \quad c_\psi = \varepsilon\psi(\infty) = \frac{1}{2}\int_{-\infty}^{\infty}\psi(x)\,\mathrm{d}x \,. \tag{7.168}$$

Recall from the previous section that when $\beta=1$ we take N to be even. It follows that φ and ψ are even and odd functions respectively. Thus $c_\psi=0$ for $\beta=1$, and computation gives

$$c_\varphi = (\pi N)^{1/4}2^{-3/4-N/2}\frac{(N!)^{1/2}}{(N/2)!}\sqrt{\lambda} \,. \tag{7.169}$$

Hence computation yields

$$\lim_{N\to\infty} c_\varphi = \sqrt{\frac{\lambda}{2}} \,, \tag{7.170}$$

and at $t=\infty$ we have

$$\begin{gathered} u_\varepsilon(\infty)=0 \,, \quad q_\varepsilon(\infty)=c_\varphi \,, \\ \mathcal{P}_{1,\lambda}(\infty)=2c_\psi \,, \quad \mathcal{Q}_{1,\lambda}(\infty)=2c_\varphi \,, \quad \mathcal{R}_{1,\lambda}(\infty)=0 \,, \\ \widetilde{\mathcal{P}}_{1,\lambda}(\infty)=\widetilde{\mathcal{Q}}_{1,\lambda}(\infty)=\widetilde{\mathcal{R}}_{1,\lambda}(\infty)=0 \,. \end{gathered}$$

Hence

$$(\alpha_1,\beta_1) = \tilde{v}_\varepsilon\,, \quad (\alpha_1,\beta_2) = q_\varepsilon - c_\varphi\,, \quad (\alpha_1,\beta_3) = q_\varepsilon\,, \tag{7.171}$$

$$(\alpha_2,\beta_1) = \frac{\lambda}{2\tilde{\lambda}}[\mathcal{P}_{1,\lambda} - a_{1,\lambda}(1-\tilde{v}_\varepsilon)]\,, \tag{7.172}$$

$$(\alpha_2,\beta_2) = \frac{\lambda}{2\tilde{\lambda}}[\mathcal{R}_{1,\lambda} + a_{1,\lambda}(q_\varepsilon - c_\varphi)]\,, \tag{7.173}$$

$$(\alpha_2,\beta_3) = \frac{\lambda}{2\tilde{\lambda}}[\mathcal{R}_{1,\lambda} + a_{1,\lambda}q_\varepsilon]\,, \tag{7.174}$$

$$(\alpha_3,\beta_1) = -\frac{\lambda}{\tilde{\lambda}}[\widetilde{\mathcal{P}}_{1,\lambda} - \tilde{a}_{1,\lambda}(1-\tilde{v}_\varepsilon)]\,, \tag{7.175}$$

$$(\alpha_3,\beta_2) = -\frac{\lambda}{\tilde{\lambda}}[\widetilde{\mathcal{R}}_{1,\lambda} + \tilde{a}_{1,\lambda}(q_\varepsilon - c_\varphi)]\,, \tag{7.176}$$

$$(\alpha_3,\beta_3) = -\frac{\lambda}{\tilde{\lambda}}[\widetilde{\mathcal{R}}_{1,\lambda} + \tilde{a}_{1,\lambda}q_\varepsilon]\,. \tag{7.177}$$

As an illustration, let us do the computation that led to (7.173) in detail. As in [33], we use the facts that $S^t = S$, and $(S+SR^t)\chi = R$ which can be easily seen by writing $R = \sum_{k=1}^{\infty}(S\chi)^k$. Furthermore we write $R(x,a_k)$ to mean

$$\lim_{\substack{y\to a_k\\ y\in J}} R(x,y)\,.$$

In general, since all evaluations are done by taking the limits from within J, we can use the identity $\chi\delta_k = \delta_k$ inside the inner products. Thus

$$\begin{aligned}
(\alpha_2,\beta_2) &= \frac{\lambda}{\tilde{\lambda}}\big[((S+RS)(1-\lambda\chi),\delta_t-\delta_\infty) + a_{1,\lambda}(Q_\varepsilon,\delta_t-\delta_\infty)\big] \\
&= \frac{\lambda}{\tilde{\lambda}}\big[((1-\lambda\chi),(S+R^tS)(\delta_t-\delta_\infty)) + a_{1,\lambda}\big(Q_\varepsilon(t)-Q_\varepsilon(\infty)\big)\big] \\
&= \frac{\lambda}{\tilde{\lambda}}\big[((1-\lambda\chi),(S+R^tS)\chi(\delta_t-\delta_\infty)) + a_{1,\lambda}(q_\varepsilon-c_\varphi)\big] \\
&= \frac{\lambda}{\tilde{\lambda}}\big[((1-\lambda\chi),R(x,t)-R(x,\infty)) + a_{1,\lambda}(q_\varepsilon-c_\varphi)\big] \\
&= \frac{\lambda}{\tilde{\lambda}}[\mathcal{R}_{1,\lambda}(t)-\mathcal{R}_{1,\lambda}(\infty) + a_{1,\lambda}(q_\varepsilon-c_\varphi)] \\
&= \frac{\lambda}{\tilde{\lambda}}[\mathcal{R}_{1,\lambda}(t) + a_{1,\lambda}(q_\varepsilon-c_\varphi)]\,.
\end{aligned}$$

We want the limit of the determinant

$$\det\big(\delta_{jk} - (\alpha_j,\beta_k)\big)_{j,k=1}^{3}\,, \tag{7.178}$$

as $N\to\infty$. In order to get our hands on the limits of the individual terms involved in the determinant, we will find differential equations for them first as in [33]. Row operation on the matrix show that $a_{1,\lambda}$ and $\tilde{a}_{1,\lambda}$ fall out of the determinant; to see this add $\lambda a_{1,\lambda}/(2\tilde{\lambda})$ times row 1 to row 2 and $\lambda\tilde{a}_{1,\lambda}/\tilde{\lambda}$ times row 1 to row 3. So we will not need to find differential equations for them. Our determinant is

$$\det\begin{pmatrix} 1-\tilde{v}_\varepsilon & -(q_\varepsilon - c_\varphi) & -q_\varepsilon \\ -\lambda\mathcal{P}_{1,\lambda}/(2\tilde{\lambda}) & 1-\lambda\mathcal{R}_{1,\lambda}/(2\tilde{\lambda}) & -\lambda\mathcal{R}_{1,\lambda}3(2\tilde{\lambda}) \\ \lambda\widetilde{\mathcal{P}}_{1,\lambda}/\tilde{\lambda} & \lambda\widetilde{\mathcal{R}}_{1,\lambda}/\tilde{\lambda} & 1+\lambda\widetilde{\mathcal{R}}_{1,\lambda}/\tilde{\lambda}\end{pmatrix}. \tag{7.179}$$

Proceeding as in [33] we find the following differential equations

$$\frac{\mathrm{d}}{\mathrm{d}t}u_\varepsilon = q_N q_\varepsilon\,, \qquad \frac{\mathrm{d}}{\mathrm{d}t}q_\varepsilon = q_N - q_N\tilde{v}_\varepsilon - p_N u_\varepsilon\,, \tag{7.180}$$

$$\frac{\mathrm{d}}{\mathrm{d}t}\mathcal{Q}_{1,\lambda} = q_N(\lambda - \mathcal{R}_{1,\lambda})\,, \qquad \frac{\mathrm{d}}{\mathrm{d}t}\mathcal{P}_{1,\lambda} = p_N(\lambda - \mathcal{R}_{1,\lambda})\,, \tag{7.181}$$

$$\frac{\mathrm{d}}{\mathrm{d}t}\mathcal{R}_{1,\lambda} = -p_N\mathcal{Q}_{1,\lambda} - q_N\mathcal{P}_{1,\lambda}\,, \qquad \frac{\mathrm{d}}{\mathrm{d}t}\widetilde{\mathcal{R}}_{1,\lambda} = -p_N\widetilde{\mathcal{Q}}_{1,\lambda} - q_N\widetilde{\mathcal{P}}_{1,\lambda}\,, \tag{7.182}$$

$$\frac{\mathrm{d}}{\mathrm{d}t}\widetilde{\mathcal{Q}}_{1,\lambda} = q_N(\lambda - 1 - \widetilde{\mathcal{R}}_{1,\lambda})\,, \qquad \frac{\mathrm{d}}{\mathrm{d}t}\widetilde{\mathcal{P}}_{1,\lambda} = p_N(\lambda - 1 - \widetilde{\mathcal{R}}_{1,\lambda})\,. \tag{7.183}$$

Let us derive the first equation in (7.181) for example. From [31, (2.17)], we have

$$\frac{\partial Q}{\partial t} = -R(x,t)q_N\,.$$

Therefore

$$\begin{aligned}\frac{\partial\mathcal{Q}_{1,\lambda}}{\partial t} &= \frac{\mathrm{d}}{\mathrm{d}t}\left[\int_{-\infty}^{t} Q(x,t)\,\mathrm{d}x - (1-\lambda)\int_{\infty}^{t} Q(x,t)\,\mathrm{d}x\right] \\ &= q_N + \int_{-\infty}^{t}\frac{\partial Q}{\partial t}\,\mathrm{d}x - (1-\lambda)\left[q_N + \int_{\infty}^{t}\frac{\partial Q}{\partial t}\,\mathrm{d}x\right] \\ &= q_N - q_N\int_{-\infty}^{t} R(x,t)\,\mathrm{d}x - (1-\lambda)q_N + (1-\lambda)q_N\int_{\infty}^{t} R(x,t)\,\mathrm{d}x \\ &= \lambda q_N - q_N\int_{-\infty}^{\infty}(1-\lambda)R(x,t)\,\mathrm{d}x = \lambda q_N - q_N\mathcal{R}_{1,\lambda} = q_N(\lambda - \mathcal{R}_{1,\lambda})\,.\end{aligned}$$

Now we change variable from t to s where $t = \tau(s) = 2\sigma\sqrt{N} + (\sigma s)/(N^{1/6})$. Then we take the limit $N\to\infty$, denoting the limits of q_ε, $\mathcal{P}_{1,\lambda}$, $\mathcal{Q}_{1,\lambda}, \mathcal{R}_{1,\lambda}$, $\widetilde{\mathcal{P}}_{1,\lambda}$, $\widetilde{\mathcal{Q}}_{1,\lambda}$, $\widetilde{\mathcal{R}}_{1,\lambda}$ and the common limit of u_ε and $\tilde{v}_\varepsilon$ respectively by $\bar{q}$, $\overline{\mathcal{P}}_{1,\lambda}$, $\overline{\mathcal{Q}}_{1,\lambda}$, $\overline{\mathcal{R}}_{1,\lambda}$, $\overline{\overline{\mathcal{P}}}_{1,\lambda}$, $\overline{\overline{\mathcal{Q}}}_{1,\lambda}$, $\overline{\overline{\mathcal{R}}}_{1,\lambda}$ and $\bar{u}$. We eliminate $\overline{\mathcal{Q}}_{1,\lambda}$ and $\overline{\overline{\mathcal{Q}}}_{1,\lambda}$ by using the facts that $\overline{\mathcal{Q}}_{1,\lambda} = \overline{\mathcal{P}}_{1,\lambda} + \lambda\sqrt{2}$ and $\overline{\overline{\mathcal{Q}}}_{1,\lambda} = \overline{\overline{\mathcal{P}}}_{1,\lambda}$. These limits hold uniformly for bounded s so we can interchange lim and d/ds. Also $\lim_{N\to\infty} N^{-1/6}q_N = \lim_{N\to\infty} N^{-1/6}p_N = q$, where q is as in (7.26). We obtain the systems

$$\frac{\mathrm{d}}{\mathrm{d}s}\bar{u} = -\frac{1}{\sqrt{2}}q\bar{q}\,, \qquad \frac{\mathrm{d}}{\mathrm{d}s}\bar{q} = \frac{1}{\sqrt{2}}q(1-2\bar{u})\,, \tag{7.184}$$

$$\frac{\mathrm{d}}{\mathrm{d}s}\overline{\mathcal{P}}_{1,\lambda} = -\frac{1}{\sqrt{2}}q(\overline{\mathcal{R}}_{1,\lambda}-\lambda)\,, \qquad \frac{\mathrm{d}}{\mathrm{d}s}\overline{\mathcal{R}}_{1,\lambda} = -\frac{1}{\sqrt{2}}q(2\overline{\mathcal{P}}_{1,\lambda}+\sqrt{2\tilde{\lambda}})\,, \tag{7.185}$$

$$\frac{\mathrm{d}}{\mathrm{d}s}\overline{\overline{\mathcal{P}}}_{1,\lambda} = \frac{1}{\sqrt{2}}q(1-\lambda-\overline{\overline{\mathcal{R}}}_{1,\lambda})\,, \qquad \frac{\mathrm{d}}{\mathrm{d}s}\overline{\overline{\mathcal{R}}}_{1,\lambda} = -q\sqrt{2}\overline{\overline{\mathcal{P}}}_{1,\lambda}\,. \tag{7.186}$$

The change of variables $s \to \mu = \int_s^\infty q(x,\lambda)\,\mathrm{d}x$ transforms these systems into constant coefficient ordinary differential equations

$$\frac{\mathrm{d}}{\mathrm{d}\mu}\bar{u} = \frac{1}{\sqrt{2}}\bar{q}\,, \qquad \frac{\mathrm{d}}{\mathrm{d}\mu}\bar{q} = -\frac{1}{\sqrt{2}}(1-2\bar{u})\,, \tag{7.187}$$

$$\frac{\mathrm{d}}{\mathrm{d}\mu}\overline{\mathcal{P}}_{1,\lambda} = \frac{1}{\sqrt{2}}(\overline{\mathcal{R}}_{1,\lambda}-\lambda\,), \qquad \frac{\mathrm{d}}{\mathrm{d}\mu}\overline{\mathcal{R}}_{1,\lambda} = \frac{1}{\sqrt{2}}(2\overline{\mathcal{P}}_{1,\lambda}+\sqrt{2\tilde{\lambda}})\,, \tag{7.188}$$

$$\frac{\mathrm{d}}{\mathrm{d}\mu}\overline{\overline{\mathcal{P}}}_{1,\lambda} = -\frac{1}{\sqrt{2}}(1-\lambda-\overline{\overline{\mathcal{R}}}_{1,\lambda})\,, \qquad \frac{\mathrm{d}}{\mathrm{d}\mu}\overline{\overline{\mathcal{R}}}_{1,\lambda} = \sqrt{2}\overline{\overline{\mathcal{P}}}_{1,\lambda}\,. \tag{7.189}$$

Since $\lim_{s\to\infty}\mu = 0$, corresponding to the boundary values at $t=\infty$ which we found earlier for $\mathcal{P}_{1,\lambda}$, $\mathcal{R}_{1,\lambda}$, $\widetilde{\mathcal{P}}_{1,\lambda}$, $\widetilde{\mathcal{R}}_{1,\lambda}$, we now have initial values at $\mu = 0$. Therefore

$$\overline{\mathcal{P}}_{1,\lambda}(0) = \overline{\mathcal{R}}_{1,\lambda}(0) = \overline{\overline{\mathcal{P}}}_{1,\lambda}(0) = \overline{\overline{\mathcal{R}}}_{1,\lambda}(0) = 0\,. \tag{7.190}$$

We use this to solve the systems and get

$$\bar{q} = \frac{\sqrt{\tilde{\lambda}}-1}{2\sqrt{2}}\mathrm{e}^{\mu} + \frac{\sqrt{\tilde{\lambda}}+1}{2\sqrt{2}}\mathrm{e}^{-\mu}\,, \tag{7.191}$$

$$\bar{u} = \frac{\sqrt{\tilde{\lambda}}-1}{4}\mathrm{e}^{\mu} - \frac{\sqrt{\tilde{\lambda}}+1}{4}\mathrm{e}^{-\mu} + \frac{1}{2}\,, \tag{7.192}$$

$$\overline{\mathcal{P}}_{1,\lambda} = \frac{\sqrt{\tilde{\lambda}}-\lambda}{2\sqrt{2}}\mathrm{e}^{\mu} + \frac{\sqrt{\tilde{\lambda}}+\lambda}{2\sqrt{2}}\mathrm{e}^{-\mu} - \sqrt{\frac{\tilde{\lambda}}{2}}\,, \tag{7.193}$$

$$\overline{\mathcal{R}}_{1,\lambda} = \frac{\sqrt{\tilde{\lambda}}-\lambda}{2}\mathrm{e}^{\mu} - \frac{\sqrt{\tilde{\lambda}}+\lambda}{2}\mathrm{e}^{-\mu} + \lambda\,, \tag{7.194}$$

$$\overline{\overline{\mathcal{P}}}_{1,\lambda} = \frac{1-\lambda}{2\sqrt{2}}(\mathrm{e}^{\mu}-\mathrm{e}^{-\mu})\,, \quad \overline{\overline{\mathcal{R}}}_{1,\lambda} = \frac{1-\lambda}{2}(\mathrm{e}^{\mu}+\mathrm{e}^{-\mu}-2)\,. \tag{7.195}$$

Substituting these expressions into the determinant gives (7.28), namely

$$D_1(s,\lambda) = D_2(s,\tilde{\lambda})\frac{\lambda-1-\cosh\mu(s,\tilde{\lambda})+\sqrt{\tilde{\lambda}}\sinh\mu(s,\tilde{\lambda})}{\lambda-2}\,, \tag{7.196}$$

where $D_\beta = \lim_{N\to\infty} D_{\beta,N}$. As mentioned in Sect. 7.2.1, the functional form of the $\beta = 1$ limiting determinant is very different from what one would expect, unlike in the $\beta = 4$ case. Also noteworthy is the dependence on $\tilde{\lambda} = 2\lambda - \lambda^2$ instead of just λ. However one should also note that when λ is set equal to

1, then $\tilde{\lambda} = \lambda = 1$. Hence in the largest eigenvalue case, where there is no prior differentiation with respect to λ, and λ is just set to 1, a great deal of simplification occurs. The above formula then nicely reduces to the $\beta = 1$ Tracy–Widom distribution.

7.8 An Interlacing Property

The following series of lemmas establish Cor. 7.3.1:

Lemma 7.8.1. *Define*

$$a_j = \frac{\mathrm{d}^j}{\mathrm{d}\lambda^j}\sqrt{\frac{\lambda}{2-\lambda}}\Bigg|_{\lambda=1} . \tag{7.197}$$

Then a_j satisfies the following recursion

$$a_j = \begin{cases} 1 & \textit{if } j = 0, \\ (j-1)a_{j-1} & \textit{for } j \geq 1,\ j \textit{ even}, \\ j a_{j-1} & \textit{for } j \geq 1,\ j \textit{ odd}. \end{cases} \tag{7.198}$$

Proof. Consider the expansion of the generating function $f(\lambda) = \sqrt{\lambda/2-\lambda}$ around $\lambda = 1$

$$f(\lambda) = \sum_{j\geq 0} \frac{a_j}{j!}(\lambda-1)^j = \sum_{j\geq 0} b_j(\lambda-1)^j .$$

Since $a_j = j!b_j$, the statement of the lemma reduces to proving the following recurrence for the b_j

$$b_j = \begin{cases} 1 & \text{if } j = 0, \\ \dfrac{j-1}{j}b_{j-1} & \text{for } j \geq 1,\ j \text{ even}, \\ b_{j-1} & \text{for } j \geq 1,\ j \text{ odd}. \end{cases} \tag{7.199}$$

Let

$$f^{\text{even}}(\lambda) = \frac{1}{2}\left(\sqrt{\frac{\lambda}{2-\lambda}} + \sqrt{\frac{2-\lambda}{\lambda}}\right), \quad f^{\text{odd}}(\lambda) = \frac{1}{2}\left(\sqrt{\frac{\lambda}{2-\lambda}} - \sqrt{\frac{2-\lambda}{\lambda}}\right).$$

These are the even and odd parts of f relative to the reflection $\lambda-1 \to -(\lambda-1)$ or $\lambda \to 2-\lambda$. Recurrence (7.199) is equivalent to

$$\frac{\mathrm{d}}{\mathrm{d}\lambda}f^{\text{even}}(\lambda) = (\lambda-1)\frac{\mathrm{d}}{\mathrm{d}\lambda}f^{\text{odd}}(\lambda) ,$$

which is easily shown to be true. □

Lemma 7.8.2. *Define*

$$f(s,\lambda) = 1 - \sqrt{\frac{\lambda}{2-\lambda}} \tanh \frac{\mu(s,\tilde{\lambda})}{2}\,, \tag{7.200}$$

for $\tilde{\lambda} = 2\lambda - \lambda^2$. *Then*

$$\left.\frac{\partial^{2n}}{\partial\lambda^{2n}} f(s,\lambda)\right|_{\lambda=1} - \frac{1}{2n+1}\left.\frac{\partial^{2n+1}}{\partial\lambda^{2n+1}} f(s,\lambda)\right|_{\lambda=1} = \begin{cases} 1 & \text{if } n=0, \\ 0 & \text{if } n \geq 1. \end{cases} \tag{7.201}$$

Proof. The case $n = 0$ is readily checked. The main ingredient for the general case is Faá di Bruno's formula

$$\frac{\mathrm{d}^n}{\mathrm{d}t^n} g(h(t)) = \sum \frac{n!}{k_1! \cdots k_n!} \left(\frac{\mathrm{d}^k g}{\mathrm{d}h^k}(h(t))\right) \left(\frac{1}{1!}\frac{\mathrm{d}h}{\mathrm{d}dt}\right)^{k_1} \cdots \left(\frac{1}{n!}\frac{\mathrm{d}^n h}{\mathrm{d}t^n}\right)^{k_n}, \tag{7.202}$$

where $k = \sum_{i=1}^n k_i$ and the above sum is over all partitions of n, that is all values of $k_1, \ldots, k_n$ such that $\sum_{i=1}^n i k_i = n$. We apply Faá di Bruno's formula to derivatives of the function $\tanh\bigl(\mu(s,\tilde{\lambda})\bigr)/2$, which we treat as some function $g\bigl(\tilde{\lambda}(\lambda)\bigr)$. Notice that for $j \geq 1$, $(d^j\tilde{\lambda})/(d\lambda^j)\big|_{\lambda=1}$ is nonzero only when $j = 2$, in which case it equals -2. Hence, in (7.202), the only term that survives is the one corresponding to the partition all of whose parts equal 2. Thus we have

$$\left.\frac{\partial^{2n-k}}{\partial\lambda^{2n-k}} \tanh \frac{\mu(s,\tilde{\lambda})}{2}\right|_{\lambda=1} = \begin{cases} 0 & \text{if } k = 2j+1,\ j \geq 0, \\ \dfrac{(-1)^{n-j}(2n-k)!}{(n-j)!} \left.\dfrac{\partial^{n-j}}{\partial\tilde{\lambda}^{n-j}} \tanh \dfrac{\mu(s,\tilde{\lambda})}{2}\right|_{\tilde{\lambda}=1} & \text{for } k = 2j,\ j \geq 0 \end{cases}$$

$$\left.\frac{\partial^{2n-k+1}}{\partial\lambda^{2n+1-k}} \tanh \frac{\mu(s,\tilde{\lambda})}{2}\right|_{\lambda=1} = \begin{cases} 0 & \text{if } k = 2j,\ j \geq 0, \\ \dfrac{(-1)^{n-j}(2n+1-k)!}{(n-j)!} \left.\dfrac{\partial^{n-j}}{\partial\tilde{\lambda}^{n-j}} \tanh \dfrac{\mu(s,\tilde{\lambda})}{2}\right|_{\tilde{\lambda}=1} & \text{for } k = 2j+1,\ j \geq 0. \end{cases}$$

Therefore, recalling the definition of a_j in (7.197) and setting $k = 2j$, we obtain

$$\begin{aligned} \left.\frac{\partial^{2n}}{\partial\lambda^{2n}} f(s,\lambda)\right|_{\lambda=1} &= \sum_{k=0}^{2n} \binom{2n}{k} \frac{\partial^k}{\partial\lambda^k} \sqrt{\frac{\lambda}{2-\lambda}} \left.\frac{\partial^{2n-k}}{\partial\lambda^{2n-k}} \tanh \frac{\mu(s,\tilde{\lambda})}{2}\right|_{\lambda=1} \\ &= \sum_{j=0}^{n} \frac{(2n)!(-1)^{n-j}}{(2j)!(n-j)!} a_{2j} \left.\frac{\partial^{n-j}}{\partial\tilde{\lambda}^{n-j}} \tanh \frac{\mu(s,\tilde{\lambda})}{2}\right|_{\tilde{\lambda}=1}. \end{aligned}$$

Similarly, using $k = 2j+1$ instead yields

$$\begin{aligned}\frac{\partial^{2n+1}}{\partial\lambda^{2n+1}} f(s,\lambda)\bigg|_{\lambda=1} &= \sum_{k=0}^{2n+1} \binom{2n+1}{k} \frac{\partial^k}{\partial\lambda^k} \sqrt{\frac{\lambda}{2-\lambda}} \frac{\partial^{2n+1-k}}{\partial\lambda^{2n+1-k}} \tanh \frac{\mu(s,\tilde{\lambda})}{2}\bigg|_{\lambda=1} \\ &= (2n+1) \sum_{j=0}^{n} \frac{(2n)!(-1)^{n-j}}{(2j)!(n-j)!} \frac{a_{2j+1}}{2j+1} \frac{\partial^{n-j}}{\partial\tilde{\lambda}^{n-j}} \tanh \frac{\mu(s,\tilde{\lambda})}{2}\bigg|_{\tilde{\lambda}=1} \\ &= (2n+1) \frac{\partial^{2n}}{\partial\lambda^{2n}} f(s,\lambda)\bigg|_{\lambda=1},\end{aligned}$$

since $a_{2j+1}/(2j+1) = a_{2j}$. Rearranging this last equality leads to (7.201).

Lemma 7.8.3. *Let $D_1(s,\lambda)$ and $D_4(s,\tilde{\lambda})$ be as in (7.28) and (7.29). Then*

$$D_1(s,\lambda) = D_4(s,\tilde{\lambda}) \left(1 - \sqrt{\frac{\lambda}{2-\lambda}} \tanh \frac{\mu(s,\tilde{\lambda})}{2}\right)^2. \tag{7.203}$$

Proof. Using the facts that $-1 - \cosh x = -2\cosh^2 x/2$, $1 = \cosh^2 x - \sinh^2 x$ and $\sinh x = 2 \sinh x/2 \cosh x/2$ we get

$$\begin{aligned}&D_1(s,\lambda) \\ &= \frac{-1-\cosh\mu(s,\tilde{\lambda})}{\lambda-2} D_2(s,\tilde{\lambda}) + D_2(s,\tilde{\lambda}) \frac{\lambda + \sqrt{\tilde{\lambda}} \sinh\mu(s,\tilde{\lambda})}{\lambda-2} \\ &= \frac{-2}{\lambda-2} D_4(s,\tilde{\lambda}) \\ &\quad + D_2(s,\tilde{\lambda}) \frac{\lambda\left(\cosh^2(\mu(s,\tilde{\lambda})/2) - \sinh^2(\mu(s,\tilde{\lambda})/2)\right) + \sqrt{\tilde{\lambda}} \sinh\mu(s,\tilde{\lambda})}{\lambda-2} \\ &= D_4(s,\tilde{\lambda}) + \frac{D_4(s,\tilde{\lambda})}{\cosh^2(\mu(s,\lambda)/2)} \frac{\lambda \sinh^2(\mu(s,\tilde{\lambda}))/2 - \sqrt{\tilde{\lambda}} \sinh\mu(s,\tilde{\lambda})}{2-\lambda} \\ &= D_4(s,\tilde{\lambda}) \left(1 + \frac{\lambda \sinh^2(\mu(s,\tilde{\lambda})/2) - 2\sqrt{\tilde{\lambda}} \sinh(\mu(s,\lambda)/2) \cosh(\mu(s,\lambda)/2)}{(2-\lambda)\cosh^2(\mu(s,\lambda)/2)}\right) \\ &= D_4(s,\tilde{\lambda}) \left(1 - 2\sqrt{\frac{\lambda}{2-\lambda}} \tanh \frac{\mu(s,\tilde{\lambda})}{2} + \frac{\lambda}{2-\lambda} \tanh^2 \frac{\mu(s,\tilde{\lambda})}{2}\right) \\ &= D_4(s,\tilde{\lambda}) \left(1 - \sqrt{\frac{\lambda}{2-\lambda}} \tanh \frac{\mu(s,\tilde{\lambda})}{2}\right)^2. \qquad \square\end{aligned}$$

For notational convenience, define $d_1(s,\lambda) = D_1^{1/2}(s,\lambda)$, and $d_4(s,\lambda) = D_4^{1/2}(s,\lambda)$. Then

Lemma 7.8.4. *For $n \geq 0$,*

$$\left[-\frac{1}{(2n+1)!} \frac{\partial^{2n+1}}{\partial\lambda^{2n+1}} + \frac{1}{(2n)!} \frac{\partial^{2n}}{\partial\lambda^{2n}}\right] d_1(s,\lambda)\bigg|_{\lambda=1} = \frac{(-1)^n}{n!} \frac{\partial^n}{\partial\lambda^n} d_4(s,\lambda)\bigg|_{\lambda=1}.$$

Proof. Let
$$f(s,\lambda) = 1 - \sqrt{\frac{\lambda}{2-\lambda}}\tanh\frac{\mu(s,\tilde{\lambda})}{2}$$
by the previous lemma, we need to show that
$$\left[-\frac{1}{(2n+1)!}\frac{\partial^{2n+1}}{\partial\lambda^{2n+1}} + \frac{1}{(2n)!}\frac{\partial^{2n}}{\partial\lambda^{2n}}\right] d_4(s,\tilde{\lambda}) f(s,\lambda)\bigg|_{\lambda=1} \tag{7.204}$$
equals
$$\frac{(-1)^n}{n!}\frac{\partial^n}{\partial\tilde{\lambda}^n} d_4(s,\tilde{\lambda})\bigg|_{\lambda=1}.$$
Now formula (7.202) applied to $d_4(s,\tilde{\lambda})$ gives
$$\frac{\partial^k}{\partial\lambda^k} d_4(s,\tilde{\lambda})\bigg|_{\lambda=1} = \begin{cases} 0 & \text{if } k = 2j+1,\ j \ge 0, \\ \dfrac{(-1)^j k!}{j!}\dfrac{\partial^j}{\partial\tilde{\lambda}^j} d_4(s,\tilde{\lambda}) & \text{if } k = 2j,\ j \ge 0. \end{cases}$$
Therefore
$$\begin{aligned} -\frac{1}{(2n+1)!}\frac{\partial^{2n+1}}{\partial\lambda^{2n+1}} & d_4(s,\tilde{\lambda}) f(s,\lambda)\bigg|_{\lambda=1} \\ &= -\frac{1}{(2n+1)!}\sum_{k=0}^{2n+1}\binom{2n+1}{k}\frac{\partial^k}{\partial\lambda^k} d_4 \frac{\partial^{2n+1-k}}{\partial\lambda^{2n+1-k}} f\bigg|_{\lambda=1} \\ &= -\sum_{j=0}^{n}\frac{(-1)^j}{(2n-2j+1)!j!}\frac{\partial^j}{\partial\tilde{\lambda}^j} d_4 \frac{\partial^{2n-2j+1}}{\partial\lambda^{2n-2j+1}} f\bigg|_{\lambda=1}. \end{aligned}$$
Similarly,
$$\begin{aligned} \frac{1}{(2n)!}\frac{\partial^{2n}}{\partial\lambda^{2n}} d_4(s,\tilde{\lambda}) f(s,\lambda)\bigg|_{\lambda=1} &= \frac{1}{(2n)!}\sum_{k=0}^{2n}\binom{2n}{k}\frac{\partial^k}{\partial\lambda^k} d_4 \frac{\partial^{2n-k}}{\partial\lambda^{2n-k}} f\bigg|_{\lambda=1} \\ &= \sum_{j=0}^{n}\frac{(-1)^j}{(2n-2j)!j!}\frac{\partial^j}{\partial\tilde{\lambda}^j} d_4 \frac{\partial^{2n-2j}}{\partial\lambda^{2n-2j}} f\bigg|_{\lambda=1}. \end{aligned}$$
Therefore the expression in (7.204) equals
$$\sum_{j=0}^{n}\frac{(-1)^j}{(2n-2j)!j!}\frac{\partial^j}{\partial\tilde{\lambda}^j} d_4(s,\tilde{\lambda})\left[\frac{\partial^{2n-2j}}{\partial\lambda^{2n-2j}} f - \frac{1}{2n-2j+1}\frac{\partial^{2n-2j+1}}{\partial\lambda^{2n-2j+1}} f\right]\bigg|_{\lambda=1}.$$
Now Lemma 7.8.2 shows that the square bracket inside the summation is zero unless $j = n$, in which case it is 1. The result follows. □

In an inductive proof of Corollary 7.3.1, the base case $F_4(s,2) = F_1(s,1)$ is easily checked by direct calculation. Lemma 7.8.4 establishes the inductive step in the proof since, with the assumption $F_4(s,n) = F_1(s,2n)$, it is equivalent to the statement
$$F_4(s,n+1) = F_1(s,2n+2)\,.$$

7.9 Numerics

7.9.1 Partial Derivatives of $q(x, \lambda)$

Let

$$q_n(x) = \frac{\partial^n}{\partial \lambda^n} q(x, \lambda)\bigg|_{\lambda=1}, \tag{7.205}$$

so that q_0 equals q from (7.24). In order to compute $F_\beta(s, m)$ it is crucial to know q_n with $0 \le n \le m$ accurately. Asymptotic expansions for q_n at $-\infty$ are given in [31]. In particular, we know that, as $t \to +\infty$, $q_0(-t/2)$ is given by

$$\frac{1}{2}\sqrt{t}\left(1 - \frac{1}{t^3} - \frac{73}{2t^6} - \frac{10657}{2t^9} - \frac{13912277}{8t^{12}} + \mathrm{O}\left(\frac{1}{t^{15}}\right)\right),$$

whereas $q_1(-t/2)$ can be expanded as

$$\frac{\exp\left(\frac{1}{3}t^{3/2}\right)}{2\sqrt{2\pi}t^{1/4}}\left(1 + \frac{17}{24t^{3/2}} + \frac{1513}{2^7 3^2 t^3} + \frac{850193}{2^{10}3^4 t^{9/2}} - \frac{407117521}{2^{15}3^5 t^6} + \mathrm{O}\left(\frac{1}{t^{15/2}}\right)\right).$$

These expansions are used in the algorithms below.

7.9.2 Algorithms

Quantities needed to compute $F_\beta(s, m)$, $m = 1, 2$, are not only q_0 and q_1 but also integrals involving q_0, such as

$$I_0 = \int_s^\infty (x - s) q_0^2(x)\,\mathrm{d}x\,, \quad J_0 = \int_s^\infty q_0(x)\,\mathrm{d}x\,. \tag{7.206}$$

Instead of computing these integrals afterward, it is better to include them as variables in a system together with q_0, as suggested in [28]. Therefore all quantities needed are computed in one step, greatly reducing errors, and taking full advantage of the powerful numerical tools in MATLAB®. Since

$$I_0' = -\int_s^\infty q_0^2(x)\,\mathrm{d}x, \quad I_0'' = q_0^2\,, \quad J_0' = -q_0, \tag{7.207}$$

the system closes, and can be concisely written

$$\frac{\mathrm{d}}{\mathrm{d}s}\begin{pmatrix} q_0 \\ q_0' \\ I_0 \\ I_0' \\ J_0 \end{pmatrix} = \begin{pmatrix} q_0' \\ sq_0 + 2q_0^3 \\ I_0' \\ q_0^2 \\ -q_0 \end{pmatrix}. \tag{7.208}$$

We first use the MATLAB® built-in Runge–Kutta–based ODE solver `ode45` to obtain a first approximation to the solution of (7.208) between $x = 6$, and

$x = -8$, with an initial values obtained using the Airy function on the right hand side. Note that it is not possible to extend the range to the left due to the high instability of the solution a little after -8. (This is where the transition region between the three different regimes in the so-called "connection problem" lies. We circumvent this limitation by patching up our solution with the asymptotic expansion to the left of $x = -8$.) The approximation obtained is then used as a trial solution in the MATLAB® boundary value problem solver `bvp4c`, resulting in an accurate solution vector between $x = 6$ and $x = -10$. Similarly, if we define

$$I_1 = \int_s^\infty (x-s) q_0(x) q_1(x)\, \mathrm{d}x\ , \quad J_1 = \int_s^\infty q_0(x) q_1(x)\, \mathrm{d}x\ , \tag{7.209}$$

then we have the first-order system

$$\frac{\mathrm{d}}{\mathrm{d}s} \begin{pmatrix} q_1 \\ q_1' \\ I_1 \\ I_1' \\ J_1 \end{pmatrix} = \begin{pmatrix} q_1' \\ s q_1 + 6 q_0^2 q_1 \\ I_1' \\ q_0 q_1 \\ -q_0 q_1 \end{pmatrix}, \tag{7.210}$$

which can be implemented using `bvp4c` together with a "seed" solution obtained in the same way as for q_0.

The MATLAB® code is freely available, and may be obtained by contacting the first author.

7.9.3 Tables

Table 7.1. Mean, standard deviation, skewness and kurtosis data for first two edge-scaled eigenvalues in the $\beta = 2$ Gaussian ensemble. Compare to Table 1 in [35].

Eigenvalue	Statistic			
	μ	σ	γ_1	γ_2
λ_1	−1.771087	0.901773	0.224084	0.093448
λ_2	−3.675440	0.735214	0.125000	0.021650

Table 7.2. Mean, standard deviation, skewness and kurtosis data for first four edge-scaled eigenvalues in the $\beta = 1$ Gaussian ensemble. Corollary 7.3.1 implies that rows 2 and 4 give corresponding data for the two largest eigenvalues in the $\beta = 4$ Gaussian ensemble. Compare to Table 1 in [35], keeping in mind that the discrepancy in the $\beta = 4$ data is caused by the different normalization in the definition of $F_4(s, 1)$.

Eigenvalue	**Statistic**			
	μ	σ	γ_1	γ_2
λ_1	−1.206548	1.267941	0.293115	0.163186
λ_2	−3.262424	1.017574	0.165531	0.049262
λ_3	−4.821636	0.906849	0.117557	0.019506
λ_4	−6.162036	0.838537	0.092305	0.007802

Table 7.3. Percentile comparison of F_1 vs. empirical distributions for 100×100 and 100×400 Wishart matrices with identity covariance.

F_1**-Percentile**	**100 × 100**			**100 × 400**		
	λ_1	λ_2	λ_3	λ_1	λ_2	λ_3
0.01	0.008	0.005	0.004	0.008	0.006	0.004
0.05	0.042	0.033	0.025	0.042	0.037	0.032
0.10	0.090	0.073	0.059	0.088	0.081	0.066
0.30	0.294	0.268	0.235	0.283	0.267	0.254
0.50	0.497	0.477	0.440	0.485	0.471	0.455
0.70	0.699	0.690	0.659	0.685	0.679	0.669
0.90	**0.902**	**0.891**	**0.901**	**0.898**	**0.894**	**0.884**
0.95	**0.951**	**0.948**	**0.950**	**0.947**	**0.950**	**0.941**
0.99	**0.992**	**0.991**	**0.991**	**0.989**	**0.991**	**0.989**

Acknowledgement

The authors wish to thank Harold Widom without whom none of this would have been possible. We thank John Harnad for the invitation to participate in the Program on *Random Matrices, Random Processes and Integrable Systems* at the Centre de recherches mathématiques on the campus of the Université de Montréal. This work was supported by the National Science Foundation under grant DMS–0304414.

References

1. T.W. Anderson: *An Introduction to Multivariate Statistical Analysis*, 3rd edn (Wiley-Interscience, Hoboken NJ 2003)

2. G.E. Andrews, R. Askey, R. Ranjan: Special functions. In: *Encyclopedia of Mathematics and its Applications*, vol 71 (Cambridge University Press, Cambridge 1999)
3. J. Baik, E.M. Rains: Duke Math. J. **109** (1), 1 (2001)
4. M. Bilodeau, D. Brenner: *Theory of Multivariate Statistics* (Springer, New York 1999)
5. C. Chevalley: *Theory of Lie Groups* (Princeton University Press, Princeton NJ 1946)
6. P.A. Clarkson J.B. McLeod: Arch. Rational Mech. Anal. **103** (2), 97 (1988)
7. N.G. de Bruijn: J. Indian Math. Soc. **19**, 133 (1955)
8. P. Deift, D. Gioev: `arXiv:math-ph/0507023`
9. P. Deift, X. Zhou: Commun. Pure Appl. Math. **48**, 277 (1995)
10. M. Dieng: Inter. Math. Res. Notices **37** 2263 (2005)
11. M. Dieng: Distribution functions for edge eigenvalues in orthogonal and symplectic ensembles: Painlevé representations. Ph.D. Thesis, UC Davis (2005)
12. I. Dumitriu, A. Edelman: J. Math. Phys. **43** (11), 5830 (2002)
13. A. Edelman, N.R. Rao: Acta Numerica **14**, 233 (2005)
14. N. El Karoui: `arXiv:math.ST/0309355`
15. P.L. Ferrari: Commun. Math. Phys. **252**, 77 (2004)
16. P.J. Forrester, T. Nagao, G. Honner: Nucl. Phys. B **553**, 601 (1999)
17. P.J. Forrester, E.M. Rains: Interrelationships between orthogonal, unitary and symplectic matrix ensembles. In: *Random Matrix Models and their Applications*, ed by P. Bleher, A. Its, S. Levy (Cambridge University Press, Cambridge 2001) pp 171–207
18. I. Gohberg, S. Goldberg, M.A. Kaashoek: *Classes of Linear Operators*, vol **I**, (Birkhäuser Verlag, Basel 1990)
19. S.P. Hastings, J.B. McLeod: *Arch. Rational Mech. Anal.* **73** (1), 31 (1980)
20. A.T. James: *Ann. Math. Statist.* **31**, 40 (1954)
21. A.T. James: *Ann. Math. Statist.* **35**, 475 (1964)
22. I.M. Johnstone: *Ann. Stats.* **29** (2), 295 (2001)
23. I.G. Macdonald: *Symmetric Functions and Hall Polynomials*, 2nd edn (Oxford University Press, New York 1995)
24. M.L. Mehta: *Random Matrices, Revised and Enlarged*, 2nd edn (Academic Press, New York London 1991)
25. M.L. Mehta, F.J. Dyson: J. Math. Phys. **4**, 713 (1963)
26. R.J. Muirhead: *Aspects of Multivariate Statistical Theory* (John Wiley & Sons Inc., New York 1982)
27. K. Pearson: Phil. Mag. **50**, 157 (1900)
28. P. Persson: Numerical methods for random matrices. Course project for MIT 18.337, MIT (2002)
29. A. Soshnikov: J. Stat. Phys. **108** (5-6), 1033 (2002)
30. R. Stanley: *Enumerative Combinatorics*, vol 2 (Cambridge University Press, Cambridge 1999)
31. C.A. Tracy, H. Widom: Commun. Math. Phys. **159**, 151 (1994)
32. C.A. Tracy, H. Widom: Commun. Math. Phys. **163**, 33 (1994)
33. C.A. Tracy, H. Widom: Commun. Math. Phys. **177** 727 (1996)
34. C.A. Tracy, H. Widom: J. Stat. Phys. **92** (5-6), 809 (1998)
35. C.A. Tracy, H. Widom: Airy kernel and Painlevé. In: *Isomonodromic Deformations and Applications in Physics* (Amer. Math. Soc., Providence RI 2002) pp 85–98

36. C.A. Tracy, H. Widom: Ann. Inst. Fourier, Grenoble, **55**, 2197 (2005)
37. H. Widom: J. Stat. Phys. **94**, 347 (1999)
38. J. Wishart: Biometrika A **20**, 32 (1928)
39. P. Zinn-Justin, J.-B. Zuber: J. Math. Phys. A **36** 3173 (2003)

Index

H

S

T

U

V